NUCLEIC ACIDS IN IMMUNOLOGY

NUCLEIC ACIDS IN IMMUNOLOGY

Proceedings of a Symposium
Held at the Institute of Microbiology
of Rutgers, The State University

Edited by

Otto J. Plescia *and* Werner Braun

Institute of Microbiology
Rutgers, The State University

SPRINGER SCIENCE+BUSINESS MEDIA, LLC

Originally published by Springer-Verlag NewYork Inc. in 1968
Library of Congress Catalog Card Number 68-24013
Softcover reprint of the hardcover 1st edition 1968
Title Number 1507

ISBN 978-3-642-87670-7 ISBN 978-3-642-87668-4 (eBook)
DOI 10.1007/978-3-642-87668-4

LIST OF AUTHORS, COAUTHORS AND DISCUSSANTS

Abramson, C., Pennsylvania College of Podiatry, Philadelphia, Pa.

Adler, F., Public Health Research Institute, New York, N. Y.

Alexander, P., Department of Radiobiology, Chester Beatty Research Institute, Royal Cancer Hospital, London, England

Arana, R., Service de Chimie Microbienne, Institut Pasteur, Paris, France

Atchley, W., University of California Medical School, San Francisco, Calif.

Avrameas, S., Institut de Recherches Scientifiques sur le Cancer, Villejuif, Val de Marne, France

Barnett, E. V., School of Medicine, UCLA Rehabilitation Center, Los Angeles, Calif.

Bautz, E. K., Institute of Microbiology, Rutgers—The State University, New Brunswick, N. J.

Beers, R. F., School of Hygiene and Public Health, The Johns Hopkins University, Baltimore, Md.

Beiser, S. M., Department of Microbiology, College of Physicians & Surgeons, Columbia University, New York, N. Y.

Benacerraf, B., Department of Pathology, New York University School of Medicine, New York, N. Y.

Benson, R. C., Shanklin Laboratory of Biology, Wesleyan University, Middletown, Conn.

Bigley, N. J., Department of Microbiology, Ohio State University, Columbus, Ohio

Braun, W., Institute of Microbiology, Rutgers—The State University, New Brunswick, N. J.

Block-Cora, E., Institute of Microbiology, Rutgers—The State University, New Brunswick, N. J.

Capra, J. D., Rockefeller University, New York, N. Y.

Cinader, B., Biology Division, The Ontario Cancer Institute, Toronto, Ontario, Canada

Cohen, E., Medical School, Rutgers—The State University, New Brunswick, N. J.

COHN, M., The Salk Institute for Biological Studies, San Diego, Calif.
COLE, L. J., Experimental Pathology Branch, U. S. Naval Radiol. Def. Lab., San Francisco, Calif.
CURTIS, S., Institute of Microbiology, Rutgers–The State University, New Brunswick, N. J.
DELORME, E. J., Department of Radiobiology, Chester Beatty Research Institute, Royal Cancer Hospital, London, England
DODD, M. C., Department of Bacteriology, Ohio State University, Columbus, Ohio
DREYER, W., Division of Biology, California Institute of Technology, Pasadena, Calif.
DUBISKI, S., Toronto Western Hospital, Toronto, Ontario, Canada
EDELMAN, G. M., The Rockefeller University, New York, N. Y.
ERICKSON, R. J., Institute of Microbiology, Rutgers–The State University, New Brunswick, N. J.
ERLANGER, B. F., Department of Microbiology, College of Physicians & Surgeons, Columbia University, New York, N. Y.
ESTRADA-PARRA, S., Instituto Politecnico Nacional, Escuela Nacional De Ciencias Biologicas, Mexico 17, D.F., Mexico
FIRSHEIN, W., Shanklin Laboratory of Biology, Wesleyan University, Middletown, Conn.
FISHMAN, M., Public Health Research Institute, New York, N. Y.
FRANZL, E., Rockefeller University, New York, N. Y.
FRIEDMAN, H., Department of Microbiology, Albert Einstein Medical Center, Philadelphia, Pa.
FRIOU, J., Department of Medicine, University of Southern California, Los Angeles, Calif.
GARRO, A. J., Department of Microbiology, College of Physicians & Surgeons, Columbia University, New York, N. Y.
GARVEY, J., California Institute of Technology, Pasadena, Calif.
GILL, T. J., Laboratory of Chemical Pathology, Harvard Medical School, Boston, Mass.
GLICK, J. L., Cell Labs., Roswell Park Memorial Institute, Buffalo, N. Y.
GOTTLIEB, A. A., Department of Chemistry, Harvard University, Cambridge, Mass.
GRABAR, P., Institut de Recherches Scientifiques sur le Cancer, Villejuif, Val de Marne, France
GRAY, W. R., Division of Biology, California Institute of Technology, Pasadena, Calif.
GREEN, I., Laboratory of Immunology, NIAID, Bethesda, Md.
HALL, J. G., Department of Radiobiology, Chester Beatty Research Institute, Royal Cancer Hospital, London, England
HALLORAN, M. J., Department of Internal Medicine, Washington University School of Medicine, St. Louis, Mo.

HAN, I., Department of Bacteriology, University of Michigan, Ann Arbor, Mich.

HAMILTON, L. D., Department of Radiobiology, Chester Beatty Research Institute, Royal Cancer Hospital, London, England

HAUROWITZ, F., Department of Chemistry, Indiana University, Bloomington, Ind.

HEIDELBERGER, M., Department of Pathology, New York University Medical School, New York, N. Y.

HOLUB, M., Public Health Research Institute, New York, N. Y.

IMPERATO, S., Institute of Microbiology, Rutgers–The State University, New Brunswick, N. J.

JAROSKOVA, L., Institute of Microbiology, Rutgers–The State University, New Brunswick, N. J.

JAROSLOW, B. N., Argonne National Laboratory, Argonne, Ill.

JIMENEZ, L., Institute of Microbiology, Rutgers–The State University, New Brunswick, N. J.

JOHNSON, A. G., Bacteriology Department, University of Michigan, Ann Arbor, Mich.

JOHNSON, H. G., Department of Microbiology, The Upjohn Co., Kalamazoo, Mich.

JOHNSON, V., Institute of Microbiology, Rutgers–The State University, New Brunswick, N. J.

KANAZIR, T., Laboratory of Radiobiology, Boris Kidrich Institute of Nuclear Sciences, Belgrade, Yugoslavia

KAROL, M. H., Department of Microbiology, College of Physicians & Surgeons, Columbia University, New York, N. Y.

KLEINSCHMIDT, A. K., Department of Biochemistry, New York University School of Medicine, New York, N. Y.

KUNZ, H. W., Laboratory of Chemical Pathology, Harvard Medical School, Boston, Mass.

KWIATKOWSKI, Z. A., Institute of Microbiology, Rutgers–The State University, New Brunswick, N. J.

LACOUR, F., Institut Gustave-Roussy, Villejuif (Seine) France

LANDY, M., Laboratory of Immunology, NIAID, NIH, Bethesda, Md.

LESKOWITZ, S., Immunology Laboratory, Massachusetts General Hospital, Boston, Mass.

LEVINE, B. B., Department of Medicine, New York University School of Medicine, New York, N. Y.

LEVINE, L., Graduate Department of Biochemistry, Brandeis University, Waltham, Mass.

MAALØE, O., University Institute of Microbiology, Copenhagen K, Denmark

MANNICK, J. A., Department of Surgery, University Hospital, Boston, Mass.

MARFEY, P. S., Laboratory of Chemical Pathology, Harvard Medical School, Boston, Mass.

MAURER, P. H., Department of Biochemistry, Jefferson Medical College, Philadelphia, Pa.

MCNAMARA, P., Institute of Microbiology, Rutgers–The State University, New Brunswick, N. J.

MERRITT, K., Department of Bacteriology, University of Michigan, Ann Arbor, Mich.

MICHAELSON, A. M., Institut Gustave-Roussy, Villejuif (Seine), France

MOROZ, C., Irvington House Institute and Department of Medicine, New York University School of Medicine, New York, N. Y.

NAHON, E., Institut Gustave-Roussy, Villejuif (Seine), France

NAKANO, M., Institute of Microbiology, Rutgers–The State University, New Brunswick, N. J.

NOLTENIUS, H., Institute of Microbiology, Rutgers–The State University, New Brunswick, N. J.

NOVELLI, G. D., Biology Division, Oak Ridge National Laboratories, Oak Ridge, Tenn.

OLGUIN-PALACIOS, E., Instituto Politecnico Nacional, Escuela Nacional De Ciencias Biologicas, Mexico 17, D.F., Mexico

ORTIZ-ORTIZ, L., Bureau of Biological Research, Rutgers–The State University, New Brunswick, N. J.

PANIJEL, J., Institut Pasteur, Paris XV[e], France

PAPERMASTER, D. S., Laboratory of Chemical Pathology, Harvard Medical School, Boston, Mass.

PARKER, C. W., Department of Internal Medicine, Washington University School of Medicine, St. Louis, Mo.

PAUL, W. E., Department of Pathology, New York University School of Medicine, New York, N. Y.

PIH, K. D., Institute of Microbiology, Rutgers–The State University, New Brunswick, N. J.

PINCHUCK, P., Department of Biochemistry, Jefferson Medical College, Philadelphia, Pa.

PLESCIA, O. J., Institute of Microbiology, Rutgers–The State University, New Brunswick, N. J.

PRESSMAN, D., Biochemistry Department, Roswell Park Memorial Institute, Buffalo, N. Y.

RASKA, K., Medical School, Rutgers–The State University, New Brunswick, N. J.

RAO, C. V., Institute of Microbiology, Rutgers–The State University, New Brunswick, N. J.

RAZAVI, L., Boston City Hospital, Boston, Mass.

REGELSON, W., Medical College of Virginia, Division of Medical Oncology, Department of Medicine, Richmond, Va.

SABBADINI, E., Departments of Experimental Surgery and Chemistry, McGill University, Montreal, Quebec, Canada

SALOMON, J. C., Institut de Recherches Scientifiques sur le Cancer, Villejuif, Val de Marne, France

SCHIMBOR, C., Institute of Microbiology, Rutgers—The State University, New Brunswick, N. J.

SCHMIDTKE, J., Department of Microbiology, University of Michigan Medical School, Ann Arbor, Mich.

SEAMAN, E., Graduate Department of Biochemistry, Brandeis University, Waltham, Mass.

SEHON, A., Department of Chemistry, McGill University, Montreal, Quebec, Canada

SELIGMANN, M., Hôpital Saint-Louis, Paris X^{e}, France

SIGEL, M., University of Miami Medical School, Miami, Fla.

SIMIĆ, M., Laboratory of Radiobiology, Boris Kidrich Institute of Nuclear Sciences, Belgrade, Yugoslavia

STAVITSKY, A., Department of Microbiology, Case Western Reserve University School of Medicine, Cleveland, Ohio

STOLLAR, B. D., Department of Pharmacology, Tufts University Medical School, Medford, Mass.

TANENBAUM, S. W., Department of Microbiology, College of Physicians & Surgeons, Columbia University, New York, N. Y.

TAUDOU, B., Institut de Recherches Scientifiques sur le Cancer, Villejuif, Val de Marne, France

THOR, D., Department of Microbiology, University of Illinois at the Medical Center, Chicago, Ill.

UHR, J. W., New York University Medical Center, New York, N. Y.

VAN VUNAKIS, H., Graduate Department of Biochemistry, Brandeis University, Waltham, Mass.

WILLIAMSON, A., National Institute for Medical Research, Mill Hill, London, England

YAJIMA, Y., Institute of Microbiology, Rutgers—The State University, New Brunswick, N. J.

YANG, W. K., Biology Division, Oak Ridge National Laboratories, Oak Ridge, Tenn.

PREFACE

Two fields have played a leading role in biomedical research in recent years, the biochemistry of nucleic acids and immunology. Yet, with the exception of those aspects which have been concerned with antibody synthesis as an example of protein synthesis, there was until recently a lack of direct association between the two fields. Until quite recently the antigenicity of nucleic acids was still in doubt and indeed represented a controversial subject. Also, the exact role of the various nucleic acids in various stages of antibody synthesis was uncertain. These skepticisms and uncertainties disappeared rapidly in the last few years. New experimental approaches brought the realization that nucleic acids, under appropriate conditions, are indeed immunogenic, and that the resulting antibodies can furnish new tools for the exploration of the molecular structure of the all-important family of nucleic acid molecules. At the same time, the recognition of the antigenicity of nucleic acids brought a new level of understanding to certain auto-immune diseases and provided new material for the exploration of the role of a carrier in immune responses. Side by side with this development was the almost explosive development of new experimental approaches and new ideas pertaining to the problem of antibody formation. Nucleic acids in their various forms were recognized as playing an expected major role in the activation of antibody-forming cells. Perhaps less to be expected was the role they can play as non-specific stimulators of antibody formation.

We felt that the time was ripe to bring these novel and important aspects together in one book, and we felt that it would be useful to provide a forum for assessing accomplishments and for projecting future possibilities. This appeared particularly desirable since so many different disciplines are involved and representatives of these specialized fields do not often have the opportunity to consider as a group problems of mutual interest. Accordingly, we invited representative investigators to New Brunswick. With two exceptions they were all able to come; one of those who was unable to participate (Simić) contributed a manuscript to this volume.

The organization of the symposium followed a tradition of the Institute of Microbiology, Rutgers, The State University, to provide a forum for the discussion and dissemination of information covering timely topics of wide interest. Since the subject matter we had chosen had not been discussed previously at any similar symposium, we anticipated that the resulting book would provide, in survey form, information of considerable interest to a large group of scientists and teachers, including immunologists, microbiologists, molecular biologists, chemists and the medical profession. Our anticipations were surpassed by the fact that much new information, not previously available in print, was presented and discussed at the meeting which took place in New Brunswick on October 16–18, 1967.

There are many people who helped us in making the symposium and this volume possible. The National Science Foundation and the American Medical Association's Education and Research Foundation provided financial support. Dr. J. O. Lampen, Director of the Institute of Microbiology, and Mr. E. Isaacs, Executive Secretary of the Institute, greatly helped us in our planning and provided much moral support. The staff of Springer-Verlag provided diligent and effective aid in preparing the material for publication. To all of them our gratitude.

Finally, we took it upon ourselves, with fair prior warning to the symposium participants, to edit and to rephrase the Discussions. We have done so, and we hope that as a result we have succeeded in making the Discussions more readable without having changed their substance and meaning.

New Brunswick, N. J.
September 1968

Otto J. Plescia
Werner Braun

CONTENTS

Oligo- and Polynucleotides as Haptens (Part II)

Use of Nucleic Acid-Specific Antibodies

Text of Speech Presented at the Symposium Banquet

OLIGO- AND POLYNUCLEOTIDES AS HAPTENS (PART I)

READING THE HELIX BACKWARDS

MICHAEL HEIDELBERGER

Department of Pathology, New York University School of Medicine
New York, New York

Immunochemistry, in its steady expansion, is now accepting the antigenic capacity of the nucleic acids in its stride. Only a few years ago the demonstration that DNA and RNA could be involved in immune reactions was greeted with skepticism; a generation ago, a clear indication that this might be so was considered so preposterous that it was never followed up.

I should like to paraphrase some protocols, dated February 1931, of experiments at the College of Physicians and Surgeons by Forrest E. Kendall and myself. I do this, not with any idea of claiming priority, but as an illustration of the overwhelming force of circumstances in shaping one's attitude toward observations that seemed minor at the time, but which, seen in the light of subsequently developed knowledge, might have changed immunochemical and biochemical history.

We were fractionating the cells of a strain of Group A hemolytic streptococcus. Rebecca Lancefield had already shown that the type-specific and protective antigens of this microorganism were proteins and we accordingly concentrated on a new method for their separation (Heidelberger and Kendall, 1931). One nucleoprotein fraction, obtained by extracting the acidified cells with neutral buffer, was rich in pentose nucleic acid, or RNA. On treatment with dilute alkali much of the RNA was split off, and upon recovery the RNA gave a strong precipitin reaction, actually a coherent disc, with two different rabbit antisera to the same strain. Kendall and I recorded the result with some wonder, but did not mention it in our publication (Heidelberger and Kendall, 1931) as we felt further confirmation was desirable. We never went back to this, for at the time our main effort was devoted to the development of truly quantitative analytical methods in place of the relative ones then current in immunological practice, and even with Elvin A. Kabat in the laboratory we could not exploit every observation. Moreover, eight years later, Henry W. Scherp and I commented: ". . . owing to the high concentration of nucleic acid required for reactivity, the effect was ascribed to an otherwise unrecognized impurity (Heidelberger and Scherp, 1939)." In the meantime, in 1938, Sevag, Lackman, and Smolens, completely ignoring our work on streptococcal nucleoproteins, published a paper on the same subject (Sevag, *et al.,* 1938). Less cautious than we, they appear to have been the first to report the serological activity at higher dilutions of what was most likely denatured streptococcal nucleic acid.

I refer to all this rather ruefully for your entertainment and to illustrate how easily an observation made many years before its time may easily assume a lack of importance. Now, with the hindsight gained through our modern chemical and immunological knowledge of the nucleic acids, it appears that our judgment was very poor. Or wasn't it? After all, we hewed to our main line of work.

References

Heidelberger, M. and F. E. Kendall (1931). Protein fractions of a scarlatinal strain of Streptococcus hemolyticus. J. Exp. Med., *54*:515–531.

Heidelberger, M. and H. W. Scherp (1939). Protein fractions of a strain of Group "A" hemolytic streptococci III. J. Immunol., *37*:563–570.

Sevag, M. G., D. B. Lackman, and J. Smolens (1938). The isolation of the components of streptococcal nucleoproteins in serologically active form. J. Biol. Chem., *124*:425–436.

METHYLATED BOVINE SERUM ALBUMIN AS A CARRIER FOR OLIGO- AND POLY-NUCLEOTIDES *

OTTO J. PLESCIA, WERNER BRAUN, SAVERIO IMPERATO,
ELISA CORA-BLOCK, LYDIE JAROSKOVA,
AND CONNIE SCHIMBOR

Institute of Microbiology, Rutgers, The State University
New Brunswick, New Jersey

I. EARLY STUDIES

A. DNAASE-SENSITIVE ANTIGENS

We initiated an immunochemical study of DNA in 1956, at a time when there was little decisive data dealing with the antigenicity of DNA (Blix *et al.*, 1954; Lackman *et al.*, 1941) and no evidence that chemically pure DNA was capable of eliciting antibody formation in a suitable host. For this reason, we used in our initial experiments DNA-rich preparations extracted from *Brucella abortus* by a phenol procedure (Braun *et al.*, 1957). These preparations contained appreciable protein and carbohydrate material that could not be dissociated easily from the DNA.

The sera of rabbits that had been immunized with such preparations contained precipitating and C′-fixing antibodies to DNA-containing antigens, as indicated by the fact that the reactivity of these antigens was sensitive to the enzymatic action of pancreatic DNAase (Phillips *et al.*, 1958; Plescia *et al.*, 1961; Palczuk *et al.*, 1961; Plescia *et al.*, 1963). Fractionation of DNA-rich preparations by chromatography and by sedimentation in a density gradient of CsCl, in attempts to isolate reactive antigens in a pure state, resulted in a substantial loss of antigenic reactivity. Also, fractions which contained the bulk of the DNA were essen-

* Supported by Grant CA-04786 from the U. S. Public Health Service, National Institutes of Health.

tially unreactive. It was concluded from these data that the reactive antigens were probably complexes of DNA and protein or carbohydrate. The DNA was considered to be an integral part of the reactive antigens, without which the antigens could not precipitate antibody or fix complement.

It was unlikely that any of the antibodies elicited by DNA-rich preparations were specific for purine and pyrimidine bases or their derivatives because none of these was found to inhibit C′-fixation by homologous DNA-rich preparations. Nevertheless, these initial attempts to produce nucleic acid-specific antibodies proved fruitful in that they provided evidence for antigenicity of DNA-complexes and provided impetus for the study of defined complexes of nucleic acids. Further impetus was provided by the findings of several investigators (Ceppellini *et al.*, 1957; Holman and Kunkel, 1957; Seligmann, 1957; Pearson *et al.*, 1958; Deicher *et al.*, 1959; Stollar and Levine, 1961; Rapp, 1962) that an appreciable number of sera from individuals with lupus erythematosus (L.E.) contained antibodies specific for DNA. The experimental production of antibodies to T_4 phage DNA was also reported (Levine *et al.*, 1960). These antibodies, unlike those in L.E. sera, were specific for glucosylated cytosine present in T_4 phage DNA.

These earlier findings, which have been discussed by us in greater detail in a recent review (Plescia and Braun, 1967), made it clear that DNA contains antigenic determinants against which antibodies can be made. It was equally clear that pure DNA was not immunogenic, and a means was therefore sought to use DNA as a hapten.

B. Rationale for Using Methylated Bovine Serum Albumin as a Carrier

The key to the use of DNA as a hapten was to find a suitable method for linking DNA to a carrier protein in a stable manner. The solution proved to be a simple one because DNA is an acidic polymer and therefore interacts readily, presumably by electrostatic forces, with basic proteins to form stable complexes. Since bovine serum albumin (BSA) is used extensively as a carrier for simple haptens, it seemed reasonable to test methylated bovine serum albumin (MBSA), a basic protein, as a carrier for DNA. At the time when we initiated these attempts, there was no evidence that a protein could function as a carrier if it were linked to a hapten non-covalently. It is now a matter of record that MBSA does indeed function as a carrier (Plescia *et al.*, 1964; 1965a, b).

II. PREPARATION OF ANTISERA TO MBSA-POLYNUCLEOTIDE COMPLEXES

A. Preparation of Immunogen (MBSA-Polynucleotide)

1. Variables

In the formation of the complex, the composition of the diluent, the concentration of the reactants (MBSA and polynucleotide) and the ratio of MBSA to polynucleotide need to be considered. A physiological diluent, 0.15M NaCl adjusted to pH 7, is recommended because the stability of complexes formed between MBSA and polynucleotide depends upon the ionic strength and pH of the medium. Dissociation occurs at high and low pH and at high ionic strength. The concentration of the polynucleotide is somewhat arbitrary and may be varied. Primary considerations are its solubility and its antigenicity, i.e., the amount required to elicit an adequate immune response. A concentration of 500 μg/ml has proved effective. In our work, DNA-type polynucleotides have been used in the denatured single-stranded form, the objective being to obtain antibodies specific for constituent bases rather than for the backbone. Native DNA can also be used, however (Lacour *et al.*, this volume). MBSA dissolves more readily in distilled water than in 0.15M NaCl; it is therefore added to the polynucleotide as a 1% aqueous solution. Because of its relatively high concentration, only a small volume needs to be added to the solution of polynucleotide, causing an insignificant change in the final ionic strength of the solution.

To date, the ratio of MBSA to polynucleotide has been chosen arbitrarily as 1:1 on a weight basis, so that competition between antigenic determinants of the two reactants would not be weighted in favor of either. The relative amount of antibody specific for the hapten may well depend on the weight ratio between carrier and hapten; however, this remains to be investigated.

The MBSA used in our studies was prepared with crystalline BSA according to the method cited by Sueoka and Cheng (1962). A commercial preparation of MBSA is now available from Worthington Biochemical Corporation, Freehold, New Jersey.

B. Choice of Host

Experience to date by ourselves and other investigators has been limited to the rabbit. There is, however, evidence (Maurer, this volume; Braun and Plescia, unpublished results) that MBSA functions as a carrier in a given host provided it is immunogenic in that host. Presumably, therefore, complexes of MBSA-polynucleotide should elicit antibodies to polynucleotides in any host capable of responding to MBSA. It may be mentioned that in mice, which are very poor responders to BSA and MBSA, methylated bovine gamma globulin (MBGG) has proved to be an effective carrier for polynucleotides (Braun and Plescia, unpublished data).

C. Immunization Schedule

Immunization procedures are still very much an art, and variations in the amount of antigen and routes of injection are therefore possible. We have found the following schedule to be generally effective: Rabbits are injected weekly for three weeks with freshly prepared antigens containing 0.25 mg polynucleotide/ml. Each week 0.4 ml is given into the foot pad and 1.0 ml intramuscularly so that each rabbit receives a total of about 1 mg polynucleotide in the course of three weeks. The immunogen has to date been given as an emulsion with complete Freund's adjuvant. An absolute requirement for complete Freund's adjuvant has not been established. However, even if it should not be absolutely necessary, its use would be recommended because of its enhancing effect on antibody response.

III. SPECIFICITY OF ANTIBODIES ELICITED BY POLYNUCLEOTIDES AS HAPTENS

A number of polynucleotides have been investigated as haptens complexed with MBSA (Plescia *et al.*, 1964; 1965). These included denatured calf thymus DNA, denatured T_4 bacteriophage DNA, poly-dAT, *E. coli* sRNA, yeast sRNA and homopolymers of ribonucleotides. The antisera from rabbits immunized with these hapten polynucleotides were assayed for antibody by the method of C′-fixation, and the results in terms of antibody specificity are summarized next.

The immune response was typical in that some of the antibodies formed were specific for either MBSA or polynucleotide, but not both, and some reacted only with the MBSA-polynucleotide complex. Of interest to us was the specificity of the antibodies reactive with polynucleotides. Information regarding the specificity of these antibodies was obtained from studies of cross-reactions with a number of nucleic acids of the DNA and RNA type that differed in their composition of purines and pyrimidines and also from results of inhibition analyses using the bases, mononucleosides, mononucleotides, oligonucleotides, D-ribose and 2-deoxyribose as inhibitors.

Generally, the specificity of the antibodies formed depended upon the composition of the nucleic acid in the complex used for immunization. Natural DNA and RNA, which consist principally of four common bases, induced antibodies specific for each of these constituent bases and combinations thereof. Hence, these antibodies cross-reacted extensively with all DNAs and RNAs tested. Single nucleotides or a mixture of mononucleotides failed to give 100% inhibition, an indication that these antibodies were specific for oligonucleotides of undetermined length. Also, owing to the diversity in sequence of nucleotides in DNA and RNA, the antibodies formed were grossly heterogeneous with respect to the composition of oligonucleotides comprising the antigenic determinant. Due to the differences in the sugar moiety, deoxyribonucleotides and ribonucleotides are antigenically different. Nevertheless, many of the antisera reacted with both DNA and RNA polymers, an indication that the antibodies were specific primarily for the bases. Antibodies elicited by homopolymers of nucleotides were specific for the single nucleotide monomer, but these antibodies were nevertheless heterogeneous with respect to the length of the oligonucleotides with which they could react.

In essence, the results just summarized clearly show that antisera produced by rabbits in response to polynucleotides are very heterogeneous and cross-react extensively with nucleic acids that differ markedly in composition and sequence of nucleotides. Exceptions are nucleic acids with unique bases, such as T_4 bacteriophage DNA which has glucosylated 5-hydroxymethyl cytosine instead of cytosine. Antisera to T_4 bacteriophage DNA, obtained after immunization with T_4 DNA-MBSA, reacted only with DNAs containing glucosylated bases.

If antibodies to nucleic acids are to be useful in elucidating structural features of nucleic acids and in relating biological functions of nucleic acids to their structure, it will be necessary to reduce the heterogeneity of antisera and narrow their specificity. Clearly this has not been possible through the use of polynucleotides as haptens. It was for this reason that we directed our attention next to oligonucleotides as haptens.

IV. OLIGONUCLEOTIDES AS HAPTENS

We showed first that oligonucleotides from calf thymus DNA, in the form of a mixture resulting from digestion with DNAase or by depurination, could complex with MBSA and function as haptens (Plescia *et al.,* 1965). The procedure for preparing the immunogen was the same as that described for polynucleotides.

Having thus established the feasibility of using oligonucleotides as haptens complexed with MBSA, we have more recently directed our efforts to the production of antisera against oligonucleotides of defined composition and length, the expectation being that these antisera would be more restricted in specificity. To date, the following triribonucleotides, the generous gift of Dr. S. Mandeles, have been used as haptens: AAC, AGU, AGC, GGU and AAU.

The results of analyses of a rabbit antiserum to AAC are shown in Table 1. The analyses were carried out by the method of C′-fixation, in which a number of different test antigens were employed. The C′-fixing antibodies in this antiserum gave a positive reaction with denatured but not with native DNAs, with yeast sRNA whether heated or not, and with RNAase digests of sRNA. There was no reaction with the SIII polysaccharide which is a linear negatively charged polymer like the polynucleotides. From these results, it is clear that these antibodies are specific for polynucleotides with some single-stranded regions and thus require accessibility of bases for reactivity. This antiserum also gave a strong positive reaction with AAC, the trinucleotide used as hapten, and a significant reaction with certain dinucleotides such as GC. Only those oligonucleotides containing bases capable of interacting to form base pairs tended to show positive complement fixation reactions. A possible and reasonable explanation for these results is that the antibodies in this particular antiserum are specific for the ribose-phosphate backbone and not the bases, leaving the bases free to interact with each other and to form aggregates containing antibody capable of fixing complement.

Other rabbit antisera, prepared against complexes of MBSA and defined trinucleotides, were similarly examined for C′-fixing antibodies. The results are summarized in Table 2. Each of these antisera was found to contain C′-fixing antibodies to denatured, but not to native, calf

Table 1

SPECIFICITY OF C′-FIXING ANTIBODIES IN RABBIT ANTISERUM * TO APAPCP

Test Antigen	C′-Fixation Reaction
ΔCT DNA	+
CT DNA	−
ΔT_4 phage DNA	+
T_4 phage DNA	−
SIII polysaccharide	−
yeast sRNA	+
Δ yeast sRNA	+
yeast sRNA digest	+
Dialysed yeast sRNA digest	+
Dialysed CT DNA digest	−
(Mononucleosides conjugated to rabbit serum albumin)	−
GpCp	+
ApApCp	+

* Antiserum prepared against MBSA-ApApCp.

Table 2

SPECIFICITY OF C′-FIXING ANTIBODIES IN RABBIT ANTISERA * TO TRINUCLEOTIDES

	Test Antigens †										
Antiserum to	ΔCT DNA	CT DNA	Yeast s RNA	RNA Digest	A-RSA	G-RSA	C-RSA	U-RSA	AGC-RNA	AGU-RSA	AUG-RSA
ApApUp	+	−	−	±	−	NT	NT	−	NT	NT	NT
ApGpUp	+	−	−	±	−	−	NT	−	NT	NT	NT
ApGpCp	+	−	−	±	−	−	−	NT	+	±	−
GpGpUp	+	−	−	±	NT	−	NT	−	NT	NT	NT
ApUpGp	+	−	−	−	−	−	NT	−	NT	NT	NT

* Antisera prepared against MBSA-trinucleotide.

† A-RSA, G-RSA, C-RSA and U-RSA are chemical conjugates of mononucleosides with rabbit serum albumin; AGC-RSA, AGU-RSA and AUG-RSA are conjugates of trinucleotides.

thymus DNA. Unlike the anti-AAC serum, these sera were unreactive with yeast sRNA and gave little or no reaction with RNAase digests of sRNA. They were also tested for antibodies to individual monoribonucleosides conjugated chemically to rabbit serum albumin (RSA) as an inert matrix. The reactions were uniformly negative. In contrast, positive reactions were observed between these monoribonucleoside-conjugates and rabbit antisera prepared against monoribonucleosides conjugated to bovine serum albumin (BSA) as a carrier. One of the antisera was also tested for C′-fixing antibodies specific for trinucleotides conjugated to RSA. The homologous trinucleotide, AGC, against which the antiserum was prepared, gave a significant though weak reaction; AGU gave less of a reaction, and AUG no reaction at all. Conjugates of the dinucleotides AG and GC with RSA also gave negative reactions. It is apparent from these results that these antisera contain antibodies with specificity for oligonucleotides consisting of more than two nucleotide residues. The degree of specificity of these antisera with respect to composition and sequence of nucleotides in reactive oligonucleotides remains to be established.

In contrast to the anti-AAC serum, none of the antisera against the other trinucleotides fixed complement on addition of di- and tri-nucleotides. These oligonucleotides were, therefore, tested as possible inhibitors

Table 3

ENHANCEMENT BY TRINUCLEOTIDES OF THE C′-FIXATION REACTION BETWEEN RABBIT ANTISERUM * TO ApGpCp AND ΔCT DNA

Added to Reaction †	Test Antigen ‡ ΔCT DNA	CT DNA
ApGpCp (20 γ)	+	–
ApUpGp (20 γ)	+	–
ApGp (20 γ)	–	–
GpCp (20 γ)	–	–
Monoribonucleosides	–	–
Monodeoxyribonucleosides	–	–
Bases	–	–
Ribose	–	–
Deoxyribose	–	–
Ribose-5′ PO_4	–	–

* Produced against MBSA-ApGpCp.

† Added to 0.1 ml antiserum (1/100), let stand 24 hrs. at 0°C before test antigen and C′ added.

‡ No fixation of C′ with CT DNA.

of the C′-fixation reaction between one of the trinucleotide sera and denatured calf thymus DNA. Included in this analysis were monoribonucleosides, monodeoxyribonucleosides, purine and pyrimidine bases, ribose, ribose-5′-phosphate and deoxyribose. Of all these substances, only the trinucleotides had any effect on the fixation of complement by antiserum and denatured calf thymus DNA. However, in contrast to our initial expectations, these trinucleotides did not inhibit the reaction, but instead they showed an enhancing effect, causing the degree of C′-fixation to increase.

This analysis was repeated, using native as well as denatured calf thymus DNA as test antigen. The results shown in Table 3 confirm that only the trinucleotides have an enhancing effect and only when denatured calf thymus DNA is the test antigen.

Having established the reproducibility of this phenomenon, we next examined the enhancing effect by trinucleotides as a function of their concentration. The results, shown in Figure 1, clearly indicate the dependence of enhancement on the concentration and composition of the trinucleotide. Of the three trinucleotides tested, the homologous trinucleotide gave the greatest degree of enhancement per unit weight. The finding that trinucleotides with base compositions different from that of the trinucleotide used to prepare the antiserum also gave an appreciable enhancement excluded the possibility that the antibodies are strictly base-specific.

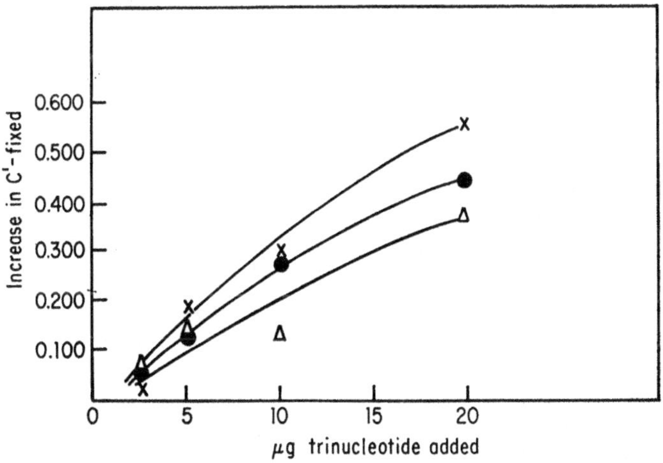

Fig. 1. Enchancement of the C′-fixation reaction between a rabbit anti-ApGpCp serum (0.1 ml, 1/100) and denatured calf thymus DNA (4 μg) by the following trinucleotides: ApGpCp, ×; UpUpCp, ●; ApUpGp, △. The C′ fixed is given in O.D. units at 541 mμ, a measure of the hemoglobin liberated from lysed sensitized sheep erythrocytes.

How then do we explain this phenomenon of enhancement? Let us examine the facts in search of clues. We know that DNA must be in a single stranded form for enhancement of C′-fixation to occur. This suggests that the bases of the DNA participate directly in interactions between DNA, trinucleotide and antibody. If antibody were to react directly with the bases of denatured DNA, there could be no increase in C′-fixation after addition of a trinucleotide. It is likely, therefore, that antibody reacts primarily with trinucleotides. We also know that enhancement occurs only with oligonucleotides containing a minimum of three residues, and that enhancement is not strictly dependent on the composition of these oligonucleotides. It would seem, therefore, that if antibody can react with such oligonucleotides, as suggested above, the specificity of such antibody cannot be directed to the bases directly. Finally, we know that antibody can fix C′ provided antibody can be made to aggregate. This is essentially the function of antigen in C′-fixation. What we need, therefore, is to think of ways in which antibody, denatured DNA, and trinucleotides might form aggregates capable of fixing complement.

One possibility is that the anti-trinucleotide serum has antibodies that can react primarily with the ribose-phosphate backbone of trinucleotides, thus leaving the bases free to hybridize with homologous triplets on individual strands of denatured DNA. We already know that antiribosomal sera contain such antibodies (Panijel, this volume). Such antibodies could thus form cross-links between strands of DNA, as suggested in Figure 2, and cause the formation of antibody-antigen aggregates capable of fixing complement. The stability of such aggregates would

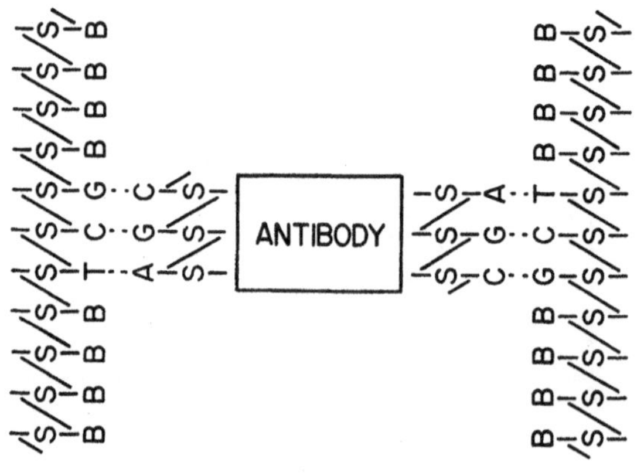

Fig. 2. A possible model showing a cross-linking of single strands of DNA by the trinucleotide ApGpCp and antibody to ApGpCp; S denotes sugar residues linked by 3′,5′ phosphodiester bonds; B denotes a purine or pyrimidine base; and A, G, C are adenine, guanine, and cytosine respectively. Such cross-linking could result in the formation of aggregates able to fix C′.

depend on the affinity between antibody and trinucleotide and also on the number and accessibility of complementary triplets on the DNA strands with which the trinucleotide can hybridize.

If the above explanation for the enhancement phenomena is indeed correct, we would predict that the extent of enhancement of C′-fixation by a given trinucleotide would be a function of the composition and sequence of bases of the polynucleotide used as test antigen. Also, we would predict that for a given polynucleotide as test antigen the extent of enhancement would vary with the composition of the trinucleotide because the configuration of a trinucleotide is subject to change as a result of intramolecular forces between pairs of bases. This change in configuration may influence the affinity between antibody and trinucleotide. Also, the extent of enhancement could vary as a function of differences in the ability of different trinucleotides to hybridize with a given polynucleotide.

These predictions are subject to experimental verification, and if they should prove correct, we would have a relatively simple, sensitive and specific method for ascertaining the presence and frequency of defined trinucleotides in a given DNA, and also a method for the characterization and possible quantitation of trinucleotides.

V. CONCLUDING REMARKS

Only a few years ago the single point in question was whether or not nucleic acid-specific antibodies could be induced in suitable hosts by any means. Once it was recognized that purified nucleic acids are reactive antigens, although they cannot function as immunogens, it was inevitable that sooner or later means would be found to link oligo- and poly-nucleotides to carriers and their hapten nature discovered. We now have an abundance of ways in which oligo- and poly-nucleotides can be coupled to a carrier. Each gives essentially similar results, the chief difficulty being that antisera produced in response to oligo- and poly-nucleotides as haptens are heterogeneous and cross-react extensively. One of the major questions to be resolved now is whether or not homogeneous populations of antibodies with defined and desired specificity can be obtained by any means and to what extent such antibodies can be utilized in studies on structure and function of nucleic acids.

References

Blix, U., C. N. Iland, and M. Stacey (1954). The serological activity of deoxypentosenucleic acids. Brit. J. Exptl. Pathol., *35*:241–251.

Braun, W., J. Burrows, and J. Phillips (1957). A phenol-extracted bacterial deoxyribonucleic acid. Nature, *180*:1356–1357.

Ceppellini, R., E. Polli, and F. Celada (1957). A DNA-reacting factor in serum of a patient with lupus erythematosus diffusus. Proc. Soc. Exptl. Biol. Med., *96*:572–574.

Deicher, H. R. G., H. R. Holman, and H. G. Kunkel (1959). The precipitin reaction between DNA and a serum factor in systemic lupus erythematosus. J. Exptl. Med., *109*:97–114.

Holman, H. R. and H. Kunkel (1957). Affinity between the lupus erythematosus serum factor and cell nuclei and nucleoprotein. Science, *126*:162–163.

Lackman, D., S. Mudd, M. G. Sevag, J. Smolens, and M. Wiener (1941). The serological reactivity of nucleic acid. J. Immunol., *40*:1–20.

Lacour, F., A. M. Michelson, and E. Nahon (1968). Specific antibodies to polynucleotide complexes and their reaction with nucleic acids: Importance of the secondary structure of the antigen. This volume, p. 32.

Levine, L., W. T. Murakami, H. Van Vunakis, and L. Grossman (1960). Specific antibodies to thermally denatured deoxyribonucleic acid of phage T_4. Proc. Natl. Acad. Sci. U. S., *46*:1038–1043.

Maurer, P. H. and P. Pinchuck (1968). Importance of immunogenicity of the carrier in inducing a response against carrier-synthetic polymer aggregates. This volume, p. 301.

Palczuk, N. C., O. J. Plescia, and W. Braun (1961). Studies on DNAase-sensitive antigens of *Brucella abortus* by complement-fixation. Proc. Soc. Exptl. Biol. Med., *107*:982–987.

Panijel, J. (1968). Use of RNA-specific antibodies in the study of the structure of polynucleotides and ribosomal particles. This volume, p. 174.

Pearson, C. M., C. G. Craddock, and N. S. Simmons (1958). Complement fixation reactions with DNA and leukocyte material in systemic lupus erythematosus. J. Lab. Clin. Med., *52*:580–587.

Phillips, J., W. Braun, and O. J. Plescia (1958). Immunochemical study of a bacterial DNA. J. Am. Chem. Soc., *80*:2710–2714.

Plescia, O. J., J. Noval, N. C. Palczuk, and W. Braun (1961). Fractionation and immunological properties of a DNA-rich preparation from *Brucella abortus*. Proc. Soc. Exptl. Biol. Med., *106*:748–752.

Plescia, O. J., W. Braun, and N. C. Palczuk (1963). DNA-associated antigens. In "Conceptual Advances in Immunology and Oncology," Harper and Row, pp. 238–256.

Plescia, O. J., W. Braun, and N. C. Palczuk (1964). Production of antibodies to denatured deoxyribonucleic acid (DNA). Proc. Natl. Acad. Sci. U. S. *52*:279–285.

Plescia, O. J., N. C. Palczuk, E. Cora-Figueroa, A. Mukherjee, and W. Braun (1965a). Production of antibodies to soluble-RNA. Proc. Natl. Acad. Sci. U. S., *54*:1281–1285.

Plescia, O. J., N. C. Palczuk, W. Braun, and E. Cora-Figueroa (1965b). Antibodies to DNA and a synthetic polydeoxyribonucleotide produced by oligodeoxyribonucleotides. Science, *148*:1102–1103.

Plescia, O. J. and W. Braun (1967). Nucleic acids as antigens. Adv. Immunol., *6*:231–252.

Rapp, F. (1962). Localization of antinuclear factors from lupus erythematosus sera in tissue culture. J. Immunol., *88*:732–740.

Seligmann, M. (1957). Evidence for the presence in the serum of patients with lupus erythematosus disseminate of a substance determining a precipitation reaction with deoxyribonucleic acid. Compt. Rend. Acad. Sci., *245*:243–245.

Stollar, B. D. and L. Levine (1961). Antibodies to denatured deoxyribonucleic acid in a lupus erythematosus serum. J. Immunol., *87*:477–484.

Sueoka, N. and T. Cheng (1962). Fractionation of nucleic acids with the methylated albumin column. J. Mol. Biol., *4*:161–172.

THE PRODUCTION OF ANTIBODIES TO MONONUCLEOTIDES, OLIGONUCLEOTIDES, AND DNA *

C. W. PARKER † AND M. J. HALLORAN

Division of Immunology, Department of Medicine, Washington University School of Medicine, St. Louis, Missouri

As evidence accumulated indicating that proteins, carbohydrates and lipids could serve as immunogens, attention began to be directed toward nucleic acids as possible immunogens. Early attempts to produce experimentally antibodies specific for nucleic acids involved the use of partially purified nucleoproteins rather than pure DNA, but they were inconclusive for technical reasons. In 1958, highly suggestive evidence for the formation of antibodies to DNA was obtained by Phillips *et al.* (1958). Following immunization with brucella nucleoproteins an antibody response was obtained. The possibility that DNA might be directly implicated in the immune reaction was suggested by the fact that treatment of the antigen with DNAase abolished its *in vitro* reactivity with antibody. Shortly thereafter, Levine *et al.* (1960) obtained convincing evidence that antibodies with nucleic acid specificity were formed in response to immunization with T-even coliphage extracts. Hapten inhibition studies indicated that the bulk of the antibody response was directed toward glycosylated 5-hydroxymethylcytosine (Murakami *et al.*, 1962). During the same period antibodies with nucleotide specificity were demonstrated in the serum of patients with disseminated lupus erythematosus, indicating the spontaneous formation of such antibodies in man (Deicher *et al.*, 1959; Stollar *et al.*, 1962).

* Supported by grants of the U. S. Public Health Service.

† Career Development Awardee of the National Institute of Allergy and Infectious Disease.

I. THE PREPARATION OF DNA- AND NUCLEOTIDE-PROTEIN CONJUGATES

A. Rationale

With available evidence that there was no fundamental barrier to the formation of antibodies to nucleic acids, the question arose as to whether or not such antibodies might be useful for the elucidation of RNA and DNA structure. This application would of course require the availability of antibodies to highly purified nucleic acids. However, quite in contrast to the results with nucleoproteins, highly purified preparations of DNA and polynucleotides, essentially free of protein, appeared to be completely ineffective as immunogens (Yachnin, 1962). In approaching the general problem of how to obtain antibodies to purified DNA and polynucleotides, we considered the feasibility of conjugating these materials with proteins. Protein conjugation is known to convert low molecular weight non-immunogenic simple chemicals (or haptens) into effective immunogens. Since nucleic acids do not undergo covalent bond formation with proteins under ordinary reaction conditions, some form of coupling agent was required. Unfortunately the problem of obtaining polynucleotide-protein conjugates was considerably more formidable than that of preparing base or nucleoside-protein conjugates (Butler *et al.*, 1962; Tanenbaum and Beiser, 1963), because of the variety of side reactions which the many functional groups of the polynucleotide might undergo during conjugation. While it was recognized that some degree of alteration of DNA was probably inevitable, it was felt that this would not necessarily seriously impair the usefulness of the conjugates in the event they proved to be effective in stimulating an antibody response.

B. Carbodiimides as Coupling Agents

Among the possible agents for conjugating RNA and DNA to proteins, the carbodiimides seemed especially attractive. This versatile class of reagents has been useful in the preparation of a wide variety of compounds including peptides, polynucleotides, nucleoside-5′-phosphoramidates, cyclic phosphates, nucleotide coenzymes, and ortho- and pyrophosphoric acids (Smith *et al.*, 1958). In particular, the ability of carbodiimides to promote phosphodiester, phosphoramidate and carboxylate ester bond formation (see below) suggested several possible ways in which they might effect the formation of covalent bonds between proteins and nucleic acids. Moreover, at the time this work was initiated in 1963, studies in this laboratory had already indicated the feasibility of utilizing carbodiimides for conjugating small peptides to proteins under mild conditions.

C. Mononucleotide-Protein Conjugates

The carbodiimides can be classified into two general types, nonpolar and polar. The latter generally have relatively high solubility in water due to the presence of tertiary or quaternary amino groups. Since water solubility seemed to be a desirable property in reactions involving proteins and DNA, initial studies were conducted with water soluble derivatives. As expected, the water soluble carbodiimides were found to promote conjugation between purified proteins and mononucleotides. Thus, when human serum albumin was reacted with thymidylic acid in the presence of 1-ethyl-3-(3-dimethylaminopropyl) carbodiimide, a conjugate which had an average of about 20 thymidylic acid residues/molecule of protein was obtained (Halloran and Parker, 1964; Halloran and Parker, 1966a). Conditions were favorable for the reaction since conjugation occurred readily at neutral pH and at room temperature. Further studies demonstrated that the reaction could take place with other proteins and other nucleotides as well and that a nonpolar carbodiimide, dicyclohexylcarbodiimide (DCC), could also be employed, provided the nucleotide and DCC were incubated together in anhydrous pyridine before their introduction into the aqueous protein solution.

D. Possible Conjugation Mechanisms

Having obtained evidence for a facile conjugation reaction between nucleotide and protein, we carried out further experiments in an attempt to elucidate the mechanism of conjugation. Four possible mechanisms were considered: (1) Phosphodiester bond formation between aliphatic hydroxyl groups on the protein and the phosphate moiety of the nucleotide (Fig. 1, reaction 1); (2) phosphoramidate bond formation between primary amino groups on the protein and the phosphate moiety of the nucleotide (Fig. 1, reaction 2); (3) carboxylate ester formation between the 3′OH group of the nucleotide and carboxylate groups of the protein (Fig. 1, reaction 3); (4) a reaction between nucleotide bases and functional groups on the protein (not shown).

In order to distinguish between these possibilities, four major kinds of evidence were considered (Halloran and Parker, 1966a):

(1) The results obtained on digestion of conjugates with phosphodiesterase and phosphomonoesterase.

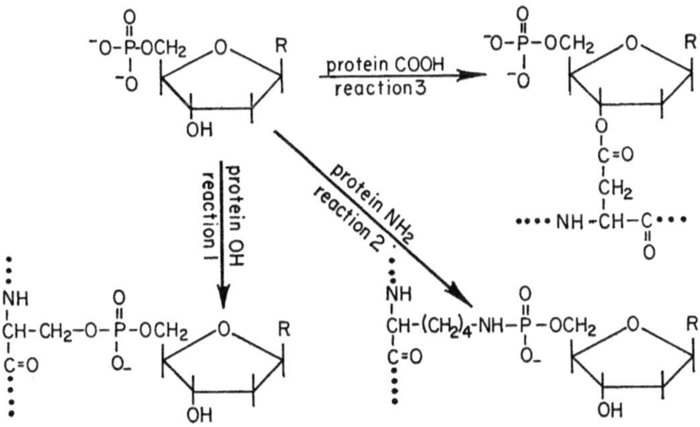

Fig. 1. Possible reactions of nucleotides with proteins in the presence of carbodiimides. Very unstable products such as acyl phosphates are not shown. Taken from Halloran and Parker (1966a).

(2) The relative degree of coupling of nucleotides with polylysine, hydroxyethylated polylysine, polyglutamic acid and proteins in which the primary amino groups were blocked.

(3) The relative extent of coupling to protein of nucleotides, nucleotides blocked at the 3′OH position, and nucleosides.

(4) The relative stability of nucleotide-protein bonds to chemical hydrolysis.

On the basis of the information obtained, it seemed clear that phosphodiester and carboxylate ester bonds were not involved to any significant extent in conjugation. Considering the evidence as a whole, and in particular the finding that extensive conjugation of mononucleotides occurred with primary amino groups of polylysine, the predominant bond appeared to be of the phosphoramidate type (pathway 2). However, rigorous chemical proof for this linkage was not obtained. Bonds of the N–P–O type are known to occur naturally in α-casein (Perlmann, 1954). From results of enzymatic digestion it was inferred that 40% of the total phosphorus of α-casein was bound as N–P–O, 40% as a phosphomonoester, and 20% as pyrophosphate.

The remaining possible linkage between nucleotide bases and protein cannot be excluded since nucleosides do conjugate with protein to a limited extent in the presence of carbodiimides. The alternative possibility is that the 5′OH group of the nucleoside rather than the base is involved in this reaction. Unfortunately, problems with solubility make it difficult to compare nucleotide and base conjugation to proteins under identical reaction conditions.

E. Oligonucleotide and DNA-Protein Conjugates

Although mononucleotides could be coupled to proteins by means of carbodiimides, the applicability of the method to oligonucleotides and DNA still had to be established. Not unexpectedly, subsequent studies with synthetic tetrathymidylic acid and several dinucleotides revealed that coupling occurred readily. High molecular weight single-stranded DNAs, and other polynucleotides also formed stable complexes with protein in the presence of coupling agents (Halloran and Parker, 1964; Halloran and Parker, 1966a). Since efforts to dissociate the complexes at high concentrations of urea and salt were unsuccessful, it was assumed that covalent bond formation had occurred. Because of the complex structures of DNA and protein, however, the types of bonds involved here could not be readily established. In view of the limited number of free phosphomonoester groups on DNA and the large number of purine and pyrimidine bases, conjugation of DNA to protein would not necessarily occur via the same pathways involved in mono- and oligo-nucleotide conjugation. Thus, a comparatively slow protein-base interaction might assume major importance in a conjugation reaction involving high molecular weight DNA.

II. RESULTS OF IMMUNIZATION

Immunization of rabbits and guinea pigs with conjugates of foreign proteins with mononucleotides, oligonucleotides, and DNA (Halloran and Parker, 1964; Halloran and Parker, 1966b), in complete adjuvant produced an intense nucleotide-specific immunological response in each instance. Native protein conjugates were less effective as immunogens, indicating that the immunogenicity of the protein carrier is one of the determining factors in the amount of the nucleotide-directed antibody produced. Antisera could be analyzed by quantitative precipitin analysis and complement-fixation. Nucleotide-containing conjugates with heterologous proteins and denatured DNA were used as the test antigens.

A. Mononucleotide- and Polynucleotide-protein Conjugates

By means of hapten inhibition, it was demonstrated that antibodies obtained by immunization with homopolymers or mononucleotides reacted much more extensively with homologous than with heterologous nucleotides (cf. Figs. 2 and 3). Comparative inhibition studies utilizing mononucleotides, nucleosides and mixtures of the base, sugar, and phosphate established that the antibody response was to the entire nucleotide (cf. Figs. 2 and 3). Antibodies with mononucleotide specificity were

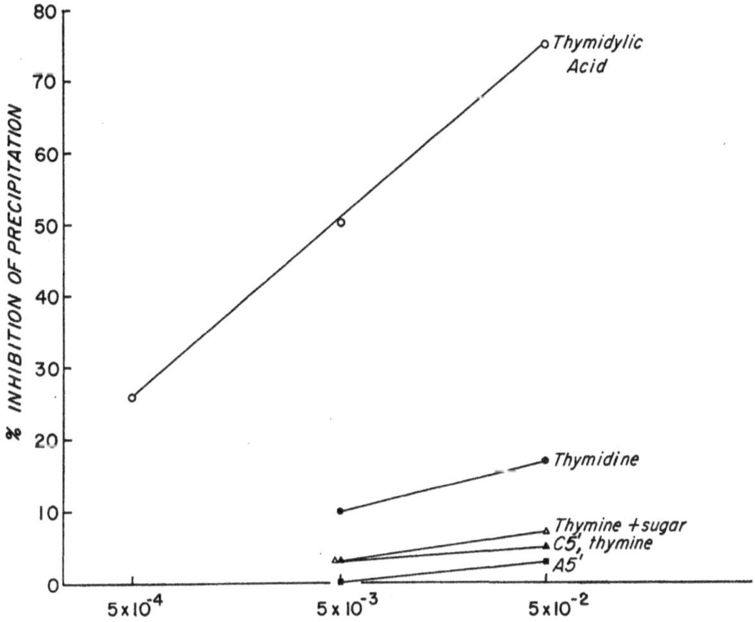

Fig. 2. Hapten inhibition of precipitation of a globulin fraction of rabbit antiserum to a thymidylate-bovine γ-globulin conjugate using thymidylate-human serum albumin as the precipitating antigen (at equivalence, 0.1 mg antigen protein). Each point represents the average of duplicate determinations. Control precipitates (no hapten) contained 0.2 mg antibody protein. The figures on the abscissa indicate the final hapten concentration in mmoles/ml. Analysis of precipitates was carried out by dissolution in aqueous detergent solution and measurement of ultraviolet absorbancy at 280 mμ. ○—○ = thymidylic acid; ●—● = thymidine; △—△ = thymine + deoxyribose (each at the concentration indicated on the abscissa); ▲—▲ = deoxycytidylic acid or thymine; ■—■ = deoxyadenylic acid. Thymidylic acid did not inhibit precipitation at 0.06 *M* in two unrelated antigen antibody systems. Taken from Halloran and Parker (1966b).

able to combine with single-stranded DNA also as demonstrated by complement-fixation.

B. DNA-protein Conjugates

1. Specificity of Antisera

As expected, antibodies obtained by immunization with denatured DNA-protein conjugates were specific for a mixture of nucleotides, rather than single nucleotides. This was demonstrated by hapten inhibition of complement-fixation using homologous DNA as the antigen. The antibody response in rabbits to calf thymus DNA exhibited an inhibition pattern which varied to some extent with the individual antiserum. With most antisera, thymidylic acid, deoxyadenylic acid, and deoxycytidylic acid were equivalent as inhibitors of complement-fixation, whereas

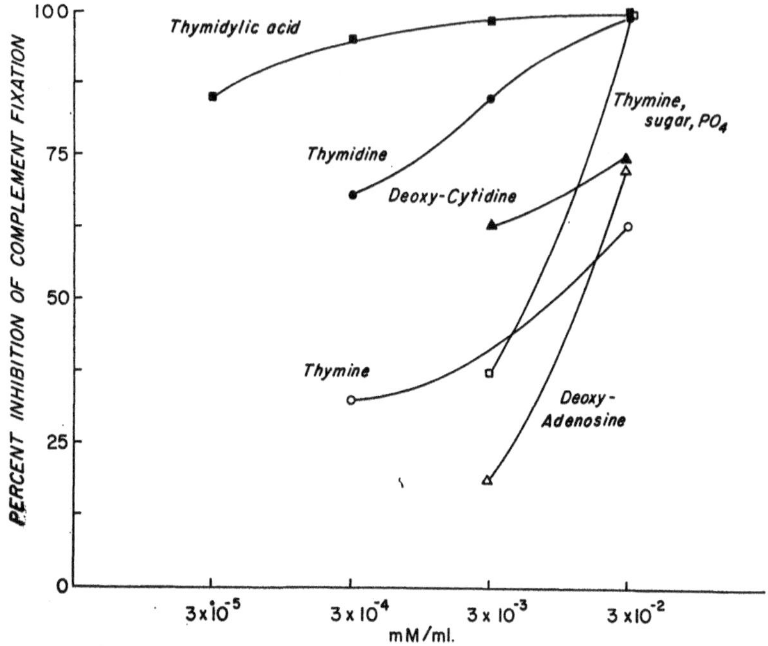

Fig. 3. Hapten inhibition of complement-fixation. A 1:100 dilution of a globulin fraction of rabbit antiserum to a thymidylic acid-human serum albumin conjugate was incubated with 1 μgm of a thymidylate-bovine γ-globulin conjugate. Hapten concentrations are expressed in mmoles/ml (abscissa). ■—■ = thymidylic acid; ●—● = thymidine; ▲—▲ = deoxycytidine; ○—○ = thymine; △—△ = deoxyadenosine; □—□ = a mixture of thymine,2-deoxy-d-ribose, and inorganic phosphate, each at the concentrations indicated on the abscissa. Taken from Halloran and Parker (1966b).

deoxyguanylic acid was somewhat less effective. This average pattern differed somewhat from the one obtained by Plescia *et al.* (1964), following immunization of rabbits with calf thymus DNA electrostatically complexed with methylated bovine serum albumin. They found that deoxyadenylic acid was the poorest inhibitor of the 4 deoxynucleotides by a considerable margin. The difference in inhibition pattern observed in the two studies may reflect variations in the degree of accessibility of individual nucleotides in the two kinds of immunizing complexes. However, diversification in the pattern of immunologic responsiveness between individual rabbits also must be considered.

Sera specific for salmon sperm or calf thymus DNA reacted indistinguishably with the homologous and heterologous DNAs in complement-fixation (see further discussion below; Halloran and Parker, 1966b). Efforts to produce antibodies specific for double-stranded DNA using "native" DNA-protein conjugates were unsuccessful. Although some antibody was formed, its specificity was toward denatured DNA rather than native DNA (cf. Fig. 4). It is not known whether denaturation had oc-

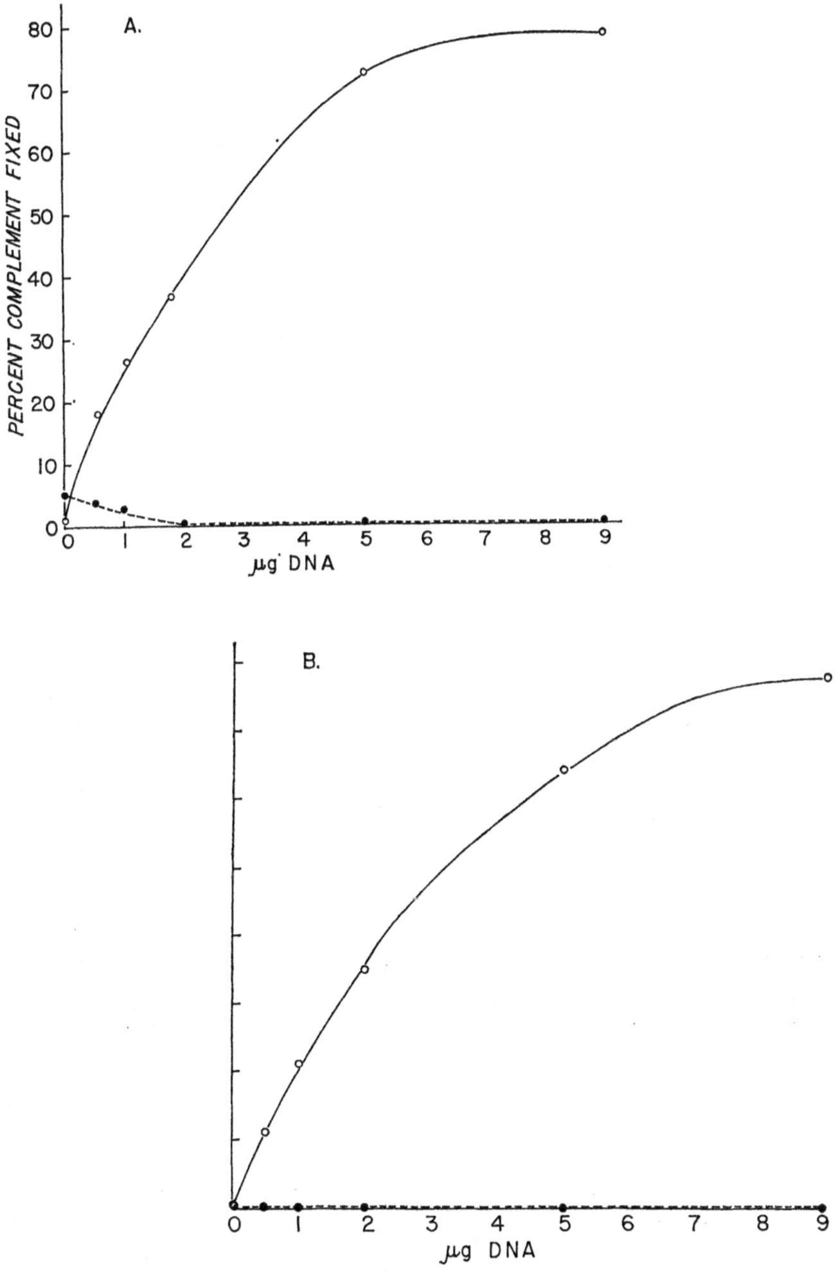

Fig. 4. Complement-fixation by a 1:100 dilution of a globulin fraction of rabbit antiserum to native (curve B) and denatured (curve A) calf thymus DNA (conjugated with bovine γ-globulin). The indicated amounts of native (●—●) and denatured (○—○) calf thymus DNA were used as the complement-fixing antigen. Taken from Halloran and Parker (1966b).

curred *in vivo* or *in vitro,* during conjugation or incorporation of the conjugate into adjuvant.

2. The Basis for Immunization

While the stable association of DNA with protein would be expected to account for the formation of anti-DNA antibody, at least one alternative explanation must be considered. The water soluble carbodiimides, which are currently available commercially and have been used in our conjugation studies, contain tertiary amino groups. Carbodiimides with quaternary amino groups close to the carbodiimide radical are known to be capable of reacting directly with nucleotide bases altering DNA and RNA structure (Gilham, 1962; Augusti-Tocco and Brown, 1965; Girshovich *et al.,* 1966). The reaction occurs with uracil, guanine, and thymine bases which are not protected by hydrogen bonding. It has been suggested that this special class of quaternary diimides may be useful in the selective inactivation of the acceptor activity of certain sRNAs (Girshovich *et al.,* 1966). To what extent carbodiimides with tertiary amino groups would undergo the same reaction is not known at this time. If a reaction of this kind does take place, alteration of DNA structure during conjugation might impart antigenicity to the DNA without any protein being involved, assuming a natural tolerance to DNA. Loss of immunological tolerance to DNA in this event would be analogous to disruption of immunological tolerance to native proteins after exposure to the protein in a chemically modified form. However, this possibility seems unlikely because preliminary data suggest that animals immunized with carbodiimide-treated DNA (modified chemically in the absence of protein) do not produce significant amounts of DNA-specific antibody. The fact that an excellent antibody response can be obtained with DNA-protein conjugates in which the coupling agent is DCC, a carbodiimide which does not contain an amino group and should have little if any tendency to undergo a base substitution reaction, is further evidence that chemical modification of DNA is not a major factor in antibody formation.

C. Other Methods for Producing Anti-DNA Antibody

At about the same time as our initial communication on the usefulness of the carbodiimide method for producing antibody to DNA (Halloran and Parker, 1964), Plescia *et al.* (1964) reported the formation of anti-DNA antibody by another method in which insoluble complexes of DNA with oppositely charged proteins were used for immunization. The relative merits of the two methods have not yet been fully evaluated. The carbodiimide technique is applicable to units of any size including mononucleotides. The minimal nucleotide size for antigenicity in oligonucleotide-methylated serum albumin complexes is in the range of 4–8 units (Plescia *et al.,* 1964), and may be larger in certain instances. Another possible advantage of the carbodiimide technique is that the conjugates obtained are usually completely soluble. Large sections of the DNA molecule in methylated serum albumin-DNA insoluble complexes may be buried and unable to act as an antigenic stimulus. On the other hand, the methylated serum albumin procedure does not require chemical manipulation of the DNA and thus precludes modification of nucleotide bases, a distinct possibility in the presence of the carbodiimides. Although base substitution by carbodiimide should be largely controlled by the use of nonpolar carbodiimides like DCC, one must consider the very real possibility that side reactions of this and other types may never be entirely eliminated.

III. POSSIBLE APPLICATIONS OF ANTI-DNA AND ANTI-RNA ANTIBODIES IN STUDIES OF THE STRUCTURE OF NUCLEIC ACIDS

One of the most interesting applications of anti-nucleic acid antibodies would be in the purification and structural evaluation of RNA and DNA. This requires that antibodies to various kinds of DNA or RNA be able to distinguish homologous from heterologous nucleic acids. The limited number of individual nucleotides and the structural similarity between purines and pyrimidines greatly limits the potential antigenic variability of nucleic acids, compared with proteins. In the absence of unusual bases like those found in T-even phage DNA, the ability of an antiserum to distinguish one DNA from another would depend in large measure on the extent to which the serum antibody response was adapted to large oligonucleotide units. Thus, antibodies with specificity for trinucleotides or smaller units would be expected to react readily with many different DNAs. The presence of antibodies to units of this size presumably explains the extensive cross-reactivity exhibited by different DNAs with SLE sera and anti-DNA sera produced experimentally. On the other hand, if some of the antibody was directed toward oligonucleotides of five or more residues, differentiation of one DNA from another might be possible. Certain SLE sera have been shown to have antibodies specific for units of this size (Stollar *et al.,* 1962). Two sera were found with antibodies to a tetramer or pentamer of thymidylic acid, since these oligonucleotides inhibited complement-fixation with denatured DNA much more effectively than other oligo- or mononucleotides, including thymidylic acid.

Even if we accept the possibility of having antibodies specific for relatively large oligonucleotide units in antisera to DNA, the relatively higher concentrations of antibodies with specificity for smaller units might prevent their usefulness. This may explain the failure of antisera to calf thymus DNA and salmon sperm DNA to distinguish between homologous and heterologous DNAs by complement-fixation, as noted above. It should be possible to circumvent this difficulty, at least in part, by absorption of anti-DNA sera with cross-reacting DNAs. However, to date, this has not proved successful; some degree of cross-reactivity persists after absorption of anti-calf thymus DNA serum with salmon sperm DNA.

In view of the complexity of DNA as an antigen, further attempts were made to demonstrate the experimental induction of antibodies to relatively large oligonucleotide units, using rabbit antisera to tetrathymidylic acid (Halloran and Parker, 1966b). As a control, rabbit antiserum to the monomer, thymidylic acid, was used. Prior to absorption, patterns of inhibition of precipitation by haptens were similar for the two kinds of antisera. Tetrathymidylic acid was about twice as effective an inhibitor as was thymidylic acid on a molar basis, suggesting that it was able to act as a multi-determinant hapten. However, the tetramer was unable to mediate complement-fixation or form insoluble complexes with antibody. After absorption of antisera with the monomer coupled to protein or aminoethyl cellulose, some change in the inhibition pattern was evident. With several of the anti-tetrathymidylic acid sera, the relative inhibitory capacity of tetrathymidylic acid increased significantly. With one antiserum, tetrathymidylic acid became 50–100 times as effective an inhibitor as thymidylic acid on a molar basis. Interestingly enough, a definite but much more limited increase in the relative inhibitory capacity of tetrathymidylic acid also occurred with an anti-thymidylic acid antiserum. Taking into consideration the hapten inhibition data as a whole, it seems clear that certain of the absorbed anti-tetrathymidylic acid sera do have some degree of specificity for the entire tetranucleotide. However, the degree of adaptation here is much less striking than has been observed with certain of the SLE sera, for which the thymidylic acid pentamer may be 10,000 to 20,000 times as effective an inhibitor of complement-fixation as the mononucleotide (Stollar *et al.*, 1962). Possibly, part of the reason for this difference may be that different kinds of assays were employed. In inhibition of precipitation by hapten most or all of the antibody population is evaluated. Complement-fixation, on the other hand, is carried out with very dilute antibody solutions and may therefore focus attention on that portion of the antibody population with a relatively high affinity for antigen. Although further absorption studies are needed, there is as yet no clear indication that a suitably absorbed antiserum would be able to distinguish subtle differences in fine structure of two DNAs with a similar base composition. This apparent lack of individuality in antibody response to DNA does not, of course, apply to polynucleotides containing unusual bases (Murakami *et al.,* 1962; Levine *et al.,* 1966).

IV. ANIMAL MODELS FOR HUMAN DISEASE

In addition to the possible application of anti-nucleic acid antibodies in structural studies, the experimental production of antibodies specific for DNA is of interest because of the occurrence of DNA-specific antibodies in certain pathological human sera. Antibodies to various nuclear components occur in a number of different diseases; however, antibodies with specificity for purified single-stranded DNA are largely confined to individuals with SLE. The role of anti-DNA antibodies in the pathogenesis of SLE is not clear at this time. None of the animals we have immunized with DNA have developed lupus-like manifestations. Nor are antisera from these animals able to promote the formation of the LE inclusion body. This result is not unexpected because the LE phenomenon appears to depend on antibodies with specificity for nucleoprotein rather than pure DNA (Deicher *et al.,* 1959). Since patients with SLE have antibodies to many different tissue components in addition to DNA, any experimental model based on immunization with DNA alone will have distinct limitations. There is the additional problem that many of the pathological changes associated with SLE simulate those of serum sickness and could be caused by complexes of foreign serum proteins and antibody. Thus, histological criteria for a lupus-like disease would have to be rigidly defined.

If attention is focused too rigidly on the serum antibody response to nucleic acids, interesting possibilities of a role of cellular immunity in SLE and similar diseases may be neglected. For example, many patients with SLE develop slowly evolving cutaneous reactions to various nuclear antigens, including calf thymus DNA (Friedman *et al.,* 1960; Azoury and Hess, 1967). Patients with other diseases rarely react in this way to purified nuclear antigens. Although the significance of these cutaneous responses is not yet clear, their manner of evolution suggests that they may be manifestations of delayed hypersensitivity. If this is so, then the extensive immunological cross-reactivity exhibited by DNA from many sources with anti-DNA antibody apparently extends to the delayed hypersensitivity response as well. In regard to the evidence suggesting an unusual form of delayed hypersensitivity in patients with SLE, a recent report by Trayanova *et al.* (1966) is of some interest. They found that peripheral blood leukocytes from patients with SLE had a cytopathic effect on human dermal fibroblasts in tissue culture. These two observations taken together raise the possibility that delayed hypersensitivity may be important in the pathogenesis of SLE and other connective tissue diseases. In view of this possibility, attempts to develop a delayed hypersensitivity response to nuclear antigens in experimental animals seem warranted. Whether or not DNA-protein conjugates can elicit a DNA-specific delayed hypersensitivity response remains to be established, however. There is some doubt that they can on the basis of the experience of Green *et al.* (1966) with guinea pigs genetically unable to respond to 2,4 dinitrophenyl-(DNP)-polylysine. They found that although electrostatic complexes of the DNP-polylysine hapten with foreign albumins were able to produce a DNP-specific serum antibody response in these animals, delayed hypersensitivity to the DNP-polylysine moiety could not be demonstrated. Experiments designed to test whether or not there will be the same restriction to the development of delayed hypersensitivity to nucleic acid antigens combined with foreign proteins are currently in progress in this laboratory.

References

Augusti-Tocco, G., and G. L. Brown (1965). Reaction of N-cyclo-hexyl, N′-β(4-methylmorpholinium) ethyl carbodiimide iodide with nucleic acids and polynucleotides. Nature, *206*:683–685.

Azoury, F. J., and J. W. Hess (1967). Cutaneous reactivity in rheumatic and nonrheumatic disease. J.A.M.A., *201*:357–361.

Butler, V. P., S. M. Beiser, B. F. Erlanger, S. W. Tanenbaum, S. Cohen, and A. Bendich (1962). Purine-specific antibodies which react with deoxyribonucleic acid (DNA). Proc. Nat. Acad. Sci., *48*:1597–1602.

Deicher, H. R. G., H. R. Holman, and H. G. Kunkel (1959). The precipitin reaction between DNA and a serum factor in lupus erythematosus. J. Exp. Med., *109*:97–114.

Friedman, E. A., W. A. Bardawil, J. P. Merrill, and C. H. Chim (1960). "Delayed" cutaneous hypersensitivity to leukocytes in disseminated lupus erythematosus. New Eng. J. Med., *262*:486–491.

Gilham, P. T. (1962). An addition reaction specific for uridine and guanosine nucleotides and its application to the modification of ribonuclease action. J. Am. Chem. Soc., *84*:687–688.

Girshovich, A. S., D. G. Knorre, O. D. Nelidova, and M. N. Ovander (1966). Inactivation of acceptor activity of sRNA by means of N-cyclo-hexyl, N′-β(4-methylmorpholinium) ethyl carbodiimide p-toluene-sulphonate. Biochim. Biophys. Acta., *119*:216–218.

Green, I., W. E. Paul, and B. Benacerraf (1966). The behavior of hapten-poly-L-lysine conjugates as complete antigens in genetic responder and as haptens in nonresponder guinea pigs. J. Exp. Med., *123*:859–879.

Halloran, M. J., and C. W. Parker (1964). Production of antibodies to nucleotides, polynucleotides and DNA. J. Lab. Clin. Med., *64*:865.

——— (1966a). The preparation of nucleotide-protein conjugates: Carbodiimides as coupling agents. The J. Immunol., *96*:373–378.

——— (1966b). The production of antibodies to mononucleotides, oligonucleotides and DNA. The J. Immunol., *96*:379–385.

Levine, L., W. T. Murakami, H. Van Vunakis, and L. Grossman (1960). Specific antibodies to thermally denatured deoxyribonucleic acid of phage T_4. Proc. Nat. Acad. Sci., *46*:1038–1043.

Levine, L., E. Seaman, E. Hammerschlag, and H. Van Vunakis (1966). Antibodies to photoproducts of deoxyribonucleic acids irradiated with ultraviolet light. Science, *153*:1666–1667.

Murakami, W. T., H. Van Vunakis, H. I. Lehrer, and L. Levine (1962). Immunochemical studies on bacteriophage deoxyribonucleic acid. The J. Immunol., *89*:116–123.

Perlmann, G. E. (1954). Phosphorus linkages in α-casein. Nature, *174*:273–274.

Phillips, J. H., W. Braun, and O. J. Plescia (1958). Immunochemical study of a bacterial DNA. J. Am. Chem. Soc., *80*:2710–2714.

Plescia, O. J., W. Braun, and N. C. Palczuk (1964). Production of antibodies to denatured deoxyribonucleic acid (DNA). Proc. Nat. Acad. Sci., *52*:279–285.

Smith, M., J. G. Moffatt, and H. G. Khorana (1958). Carbodiimides. VIII. Observations on the reactions of carbodiimides with acids and some new applications in the synthesis of phosphoric acid esters. J. Am. Chem. Soc., *80*:6204–6213.

Stollar, D., L. Levine, H. I. Lehrer, and H. Van Vunakis (1962). The antigenic determinants of denatured DNA reactive with lupus erythematosus serum. Proc. Nat. Acad. Sci., *48*:874–880.

Tanenbaum, S. W., and S. M. Beiser (1963). Pyrimidine-specific antibodies which react with deoxyribonucleic acid (DNA). Proc. Nat. Acad. Sci., *49*:662–668.

Trayanova, T. G., V. V. Sura, and G. J. Svet-Moldavsky (1966). Destruction of human cells in tissue culture by lymphocytes from patients with systemic lupus erythematosus. Lancet, *1*:452–454.

Yachnin, S. (1962). Non-antigenicity of synthetic polyribonucleotides and apurinic acid. Nature, *195*:1319.

SPECIFIC ANTIBODIES TO POLYNUCLEOTIDE COMPLEXES AND THEIR REACTION WITH NUCLEIC ACIDS: IMPORTANCE OF THE SECONDARY STRUCTURE OF THE ANTIGEN *

F. Lacour, A. M. Michelson, and E. Nahon

Institut Gustave Roussy, Villejuif, and Institut de Biologie Physicochimique

It has been shown by various authors that immunization with purines, pyrimidines, nucleosides, mono-, di-, and oligo-nucleotides, or single-stranded polynucleotides (conjugated or complexed to proteins or polypeptides) gives antibodies which are specific for the bases (Butler *et al.*, 1962; Tanenbaum and Beiser, 1963; Karol and Tanenbaum, 1967; Erlanger and Beiser, 1964; Sela *et al.*, 1964; Plescia *et al.*, 1964; Lacour and Harel, 1965; Halloran and Parker, 1966; Seaman *et al.*, 1965). These antibodies react principally with denatured DNA, but not with native DNA, an indication that the immune reaction involves nucleotides that are freely accessible and not involved in a hydrogen-bonded helical secondary structure. Some of these antibodies could also react with RNA.

These results and the differences in the immunochemical reactivity between single-stranded synthetic polynucleotides and double-stranded polynucleotide complexes with anti-ribosome antibodies drew our attention to the importance of secondary structure for the immunological reactivity of RNA. Ribosomal RNA, transfer RNA, and messenger RNA behaved differently in their reaction with anti-ribosome antibodies (Lacour *et al.*, 1962; Nahon *et al.*, 1965). We were thus prompted to undertake a study of the role of macromolecular conformation and helical secondary structure in immunological reactions of nucleic acids. Such a study was possible owing to the availability of double- and triple-stranded complexes containing synthetic homopolynucleotides (Table 1).

* This work was supported in part by CNRS and INSERM.

Table 1

POLYNUCLEOTIDE COMPLEXES USED FOR IMMUNIZATION

Complex *	Molar Ratio (A/U or I/C)	Rabbit Number
Poly A•poly U	1	372, 373, 375
Poly A + poly A•poly U	1.3	310, 311, 313
Poly A + poly A•poly U	2	343, 344, 345
Poly I•poly C	1	327, 328, 331
Poly A•2 poly U	0.5	346, 347, 348, 349

* Three types of polynucleotide complexes containing methylated bovine serum albumin were used for immunization:

(1) Poly A•poly U double-stranded complexes with or without an excess of poly A.

(2) Poly I•poly C double-stranded complex.

(3) Poly A•2 poly U triple-stranded complex.

I. REACTIONS OF ANTISERA TO DOUBLE- AND TRIPLE-STRANDED COMPLEXES WITH POLYNUCLEOTIDES AND POLYNUCLEOTIDE COMPLEXES

Immunization of rabbits with the double-stranded complex poly A• * poly U-MBSA (in an equimolar ratio of A and U containing an excess of poly A (cf. Table 1) produced antibodies capable of reacting specifically with poly A•poly U (Nahon *et al.,* 1967). A reaction was also observed with poly A but no precipitation occurred with poly U, poly C or poly G. One out of nine antisera cross-reacted with poly I. All these sera reacted not only with the homologous poly A•poly U but also with triple-stranded poly A•2 poly U (Fig. 1).

Also, cross-reactions were observed with the heterologous double-stranded complex poly I•poly C, whereas no reaction was noted with poly C or poly I (with the exception of antiserum #343). However, these antibodies were unreactive with the double- or triple-stranded complexes poly G•poly C or 2 poly G•poly C (Table 2). In contrast, anti-poly A-MBSA sera (#304 and 305) reacted only with poly A and failed to precipitate any of the double- or triple-stranded polynucleotide complexes (Table 3).

Antibodies produced following immunization with the double-stranded complex poly I•poly C precipitated the homologous complex poly I•poly C (Fig. 2). They also reacted with poly I, but did not precipitate poly A,

* Abbreviations used: polyriboadenylic acid, poly A; polyribouridylic acid, poly U; polyinosinic acid, poly I; polyribocytidylic acid, poly C; polyriboguanylic acid, poly G. Complex formation is denoted by a •.

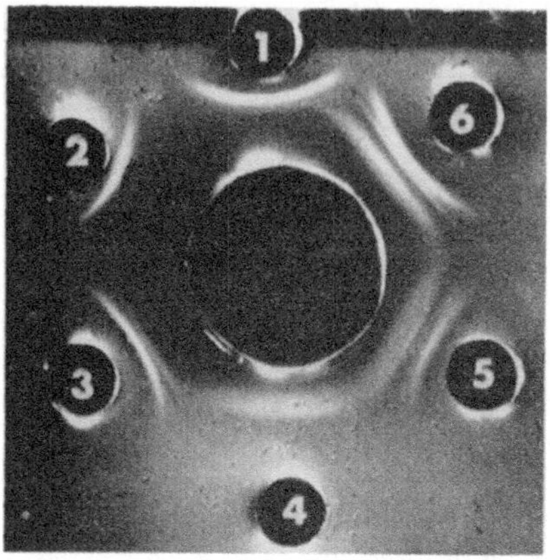

Fig. 1. Double diffusion micro test in agar. Reaction of anti-poly A●poly U-MBSA antiserum number 313 (central reservoir) with: poly A●poly U in peripheral wells 1 and 3; poly A●2 poly U (well no. 2); poly A (well no. 4); poly A + poly A●poly U molar ration A/U = 2 well no. 5 and poly A●poly U molar ratio A/U = 1.3 well no. 6.

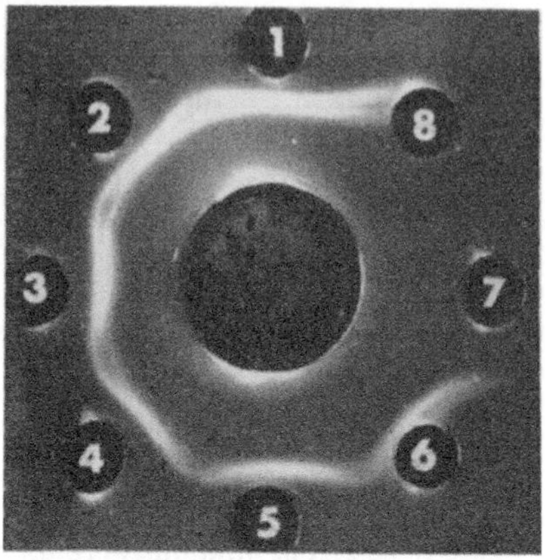

Fig. 2. Double diffusion micro test in agar. Reaction of anti-poly I●poly C-MBSA antiserum number 328 (central reservoir) with increasing dilutions of poly I●poly C in peripheral wells 1–6. Wells 7–8 contain NaCl, 0.14 *M*.

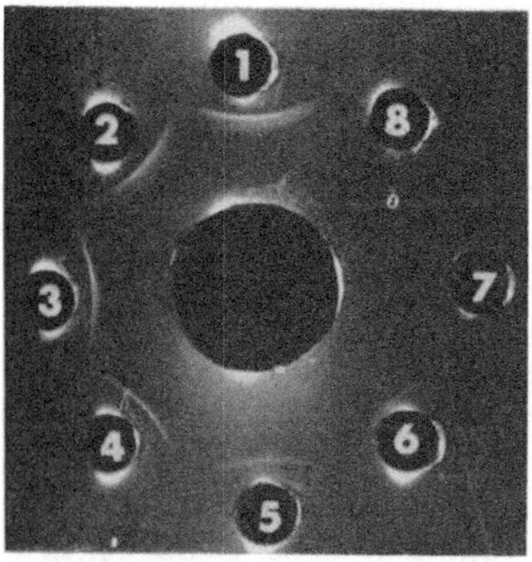

Fig. 3. Double diffusion micro test in agar. Reaction of anti-poly A●2 poly U-MBSA antiserum number 347 (central reservoir) with increasing dilutions of poly A●2 poly U in peripheral wells 1–6. Wells 7–8 contain NaCl, 0.14 *M*.

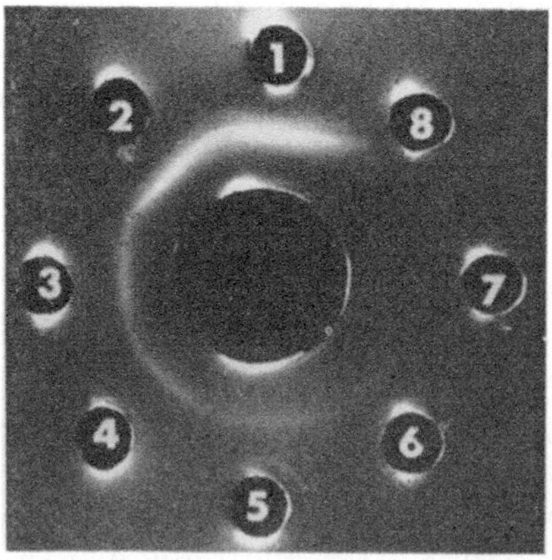

Fig. 4. Double diffusion micro test in agar. Reaction of anti-poly I●poly C-MBSA antiserum number 327 (central reservoir) with RNA from mouse ascites cells. Peripheral wells 1–6 contain increasing dilutions of RNA. Wells 7–8 contain NaCl, 0.14 *M*.

Table 2

REACTIONS OF ANTI-POLY A•POLY U-MBSA SERA WITH POLYNUCLEOTIDES AND POLYNUCLEOTIDE COMPLEXES

	Immunizing Antigen								
	Poly A•Poly U (A/U = 1)			Poly A•Poly U (A/U = 1.3)			Poly A•Poly U (A/U = 2)		
	Rabbit Number								
Antigen Tested	372	373	375	310	311	313	343	344	345
Poly A	+	+	+	+	+	+	+	+	+
Poly U	–	–	–	–	–	–	–	–	–
Poly I	–	–	–	–	–	–	+	–	–
Poly C		–	–	–	–	–	–	–	–
Poly A•poly U	+	+	+	+	+	+	+	+	+
Poly I•poly C	+	+	+	+	+	+	+	+	+
Poly G•poly C	–		–	–	–	–	–	–	
Poly A•2 poly U	+	+	+	+	+	+	+	+	+
2 Poly G•poly C	–		–	–	–	–	–	–	

Immunization was carried out according to the technique of Plescia *et al.* (1964) but the rabbits were given 2 or 3 booster injections without Freund's adjuvant intravenously.

The gel double diffusion method previously described (Lacour *et al.*, 1962) was used to follow the reaction between the antibodies and the homopolymers, the double- and triple-stranded polynucleotide complexes and various nucleic acids.

Poly A and poly U were purchased from Miles Chemical Company.

poly C or poly G, as shown in Table 4. Cross-reactions were obtained with the heterologous double- or triple-stranded helical complexes which are formed between poly A and poly U, although no reaction was obtained with poly A or poly U. No cross-reactions with poly G•poly C or 2 poly G•poly C could be demonstrated with anti-poly I•poly C sera (cf. Table 4).

Anti-poly I-MBSA sera (#781, 784, 550, and 551) precipitated poly I but did not react with poly A, poly C, or poly U (cf. Table 3). One serum (#781) cross-reacted with methylated poly G but no reaction occurred between anti-poly I-MBSA sera and double-stranded poly I•poly C.

Immunization with the triple-stranded complex poly A•2 poly U-MBSA gave antibodies which reacted specifically with the homologous antigen poly A•2 poly U (Fig. 3). No precipitation was noted with the constituent homopolymers poly A or poly U, in contrast with the results obtained

Table 3

REACTIONS OF ANTI-POLY A-MBSA AND ANTI-POLY I-MBSA SERA

	Immunizing Antigen					
	Poly A-MBSA		Poly I-MBSA			
	Rabbit Number					
Antigen Tested	304	305	781	784	550	551
Poly A	+	+	–	–	–	–
Poly I	–	–	+	+	+	+
Poly A•poly U	–	–	–	–	–	–
Poly A•2 poly U	–	–	–	–	–	–
Poly I•poly C	–	–	–	–	–	–
Poly U	–	–	–	–	–	–
Poly C	–	–	–	–	–	–
Methylated poly G			+	–	–	–

Methylated poly G was prepared by methylation of poly G with dimethyl sulfate and tri-n-butylamine in aqueous solution.

Table 4

REACTIONS OF ANTI-POLY I•POLY C-MBSA SERA WITH SINGLE-STRANDED POLYNUCLEOTIDES AND POLYNUCLEOTIDE COMPLEXES

	Rabbit Number		
Antigens	327	328	331
Poly I	+	+	+
Poly A	–	–	–
Poly C	–	–	–
Poly G	–	–	–
Poly I•poly C	+	+	+
Poly A•poly U	+	+	+
Poly G•poly C	–	–	–
Poly A•2 poly U	+	+	+
2 Poly G•poly C	–	–	–

Poly I, poly G and poly C were synthesized with polynucleotide phosphorylase from *Azotobacter vinlandii.*

Table 5

REACTIONS OF ANTI-POLY A•2 POLY U-MBSA SERA WITH POLYNUCLEOTIDES AND POLYNUCLEOTIDE COMPLEXES

Antigens	Rabbit Number			
	346	347	348	349
Poly A	–	–	–	–
Poly U	–	–	–	–
Poly I	–	–	–	–
Poly C	–	–	–	–
Poly A•poly U	–	–	–	–
Poly I•poly C	–	–	–	–
Poly G•poly C	–	–	–	–
Poly A•2 poly U	+	+	+	+
2 Poly G•poly C	–	–	–	–

with both groups of anti-double-stranded sera. No reaction was observed with poly I or poly C (Table 5). In addition, anti-poly A•2 poly U antibodies did not react with any of the double-stranded complexes. Such antisera are thus specifically directed towards triple-stranded helical complexes and show no reaction with single- or double-stranded structures.

II. REACTIONS OF IMMUNE SERA ANTI-POLY A•POLY U-MBSA AND ANTI-POLY A-MBSA WITH ADENYLATE OLIGOMERS OF VARYING CHAIN LENGTH

An attempt was made to determine the size of the antigenic determinant for antibodies to poly A, using oligomers of riboadenylic acid of varying chain length: A5, A7, A8, A12, A30, and A45.*

Anti-poly A•poly U sera (#372 and 375), anti-poly A + poly A•poly U sera (#310, 313, 343, and 344) and two anti-poly A sera (#304 and 305) were tested. The results showed a great variability in reactivity of the antisera. None of the sera showed any precipitation line with the pentamer, but the immune sera #375, 343, and 344 reacted with A7, #372 with A12, and #310 and 313 with A30. The anti-poly A serum #304 reacted with duodeca-adenylate, whereas in order to obtain a positive reaction with serum #305, chain lengths of some 30 adenylic residues were necessary.

* The oligo-adenylates are represented as A5, A7 . . . A*N* where *N* represents the number of adenine residues in the chain.

III. REACTIONS OF ANTISERA TO DOUBLE-STRANDED COMPLEXES WITH RNA

Both groups of anti-double-stranded sera reacted with RNA. This cross-reaction was observed with seven of the nine anti-poly A•poly U sera and two of the three anti-poly I•poly C sera. Total RNA, ribosomal, and transfer RNAs all cross-reacted with the immune sera. Table 6 summarizes experiments where total RNA (before and after thermal treatment), ribosomal and transfer RNA from mouse ascites cells, yeast RNA and methylated yeast RNA, and *E. coli* tRNA were tested.

One or two precipitating bands developed (Fig. 4). They could be distinguished from the lines obtained with polynucleotide complexes (Fig. 5) or with single-stranded polynucleotides both by their intensity and their rate of migration. Thus there was only a partial antigenic identity between RNA and poly A and between RNA and poly A•poly U, as illustrated in Figure 6. In contrast, no reaction was observed between any of the poly A•2 poly U antisera and RNA, regardless of the source or nature of the ribonucleic acid (total RNA, methylated RNA, rRNA, or

Table 6

REACTIONS BETWEEN RNA AND ANTI-POLY A•POLY U-MBSA, ANTI-POLY I•POLY C-MBSA AND ANTI-POLY A•2 POLY U-MBSA SERA

Immune Sera Anti-	Rabbit Number	RNA	RNA (heated) *	rRNA †	tRNA †
Poly A•poly U	372	+	+	+	+
(A/U = 1)	373	+	+	±	+
	375	+	+	+	+
Poly A + poly A•poly U	310	+	+	+	+
(A/U = 1.3)	311	–	–	–	–
	313	+	+	+	+
Poly A + poly A•poly U	343	+	+	+	+
(A/U = 2)	344	+	+	+	±
	345	–	–	–	–
Poly I•poly C	327	+	+	+	+
	328	+	+	+	+
	331	–	–	–	–
Poly A•2 poly U	346	–	–	–	–
	347	–	–	–	–
	348	–	–	–	–
	349	–	–	–	–

* Heated at 100°C and quickly cooled.

† rRNA and tRNA from mouse ascites cells were isolated by sucrose gradient centrifugation.

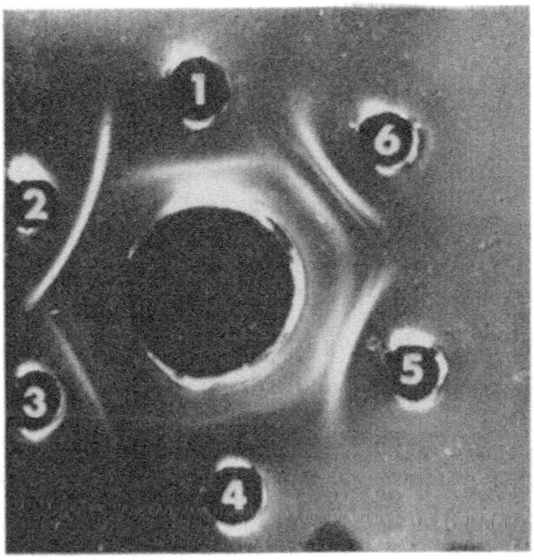

Fig. 5. Double diffusion micro test in agar. Reaction of anti-poly A●poly U antiserum number 313 (central reservoir) with: RNA from mouse ascites cells in peripheral well no. 1; the same RNA after heating and quick cooling in well no. 4; poly A●2 poly U in well no. 2; poly I●poly C in well no. 3; poly A + poly A●poly U molar ration A/U = 1.3 in well no. 5; poly A + poly A●poly U molar ratio A/U = 2 in well no. 6.

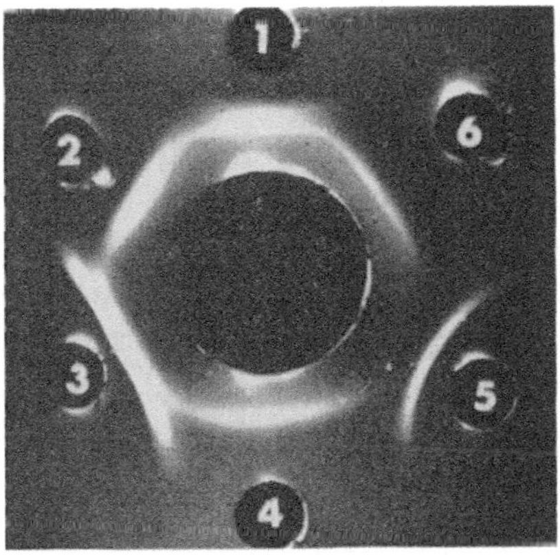

Fig. 6. Double diffusion micro test in agar. Reaction of anti-poly A●poly U antiserum no. 343 (central reservoir) with: total RNA (well no. 1), rRNA (well no. 2) and tRNA (well no. 6) from mouse ascites cells; poly A in well no. 3, *Micrococcus lysodeikticus* native DNA in well no. 4, poly A●poly U in well no. 5.

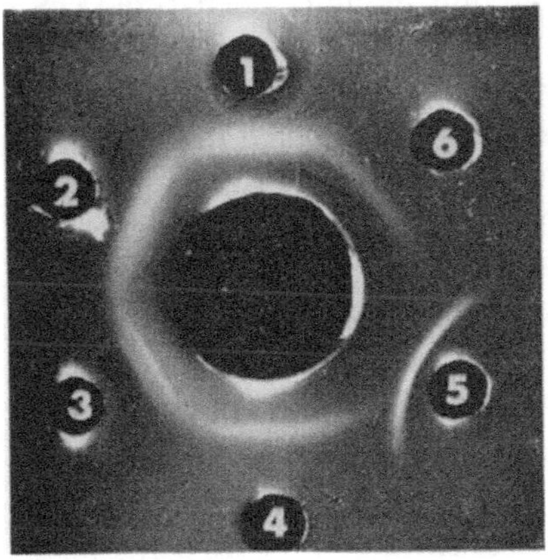

Fig. 7. Double diffusion micro test in agar. Reaction of anti-poly A●poly U antiserum no. 343 (central reservoir) with: calf thymus denatured DNA in peripheral well no. 1; *Clostridium perfringens* denatured DNA in well no. 2; RNA from mouse ascites cells in well no. 3; *Micrococcus lysodeikticus* native DNA in well no. 4; denatured DNA in well no. 6; and poly A●poly U in well no. 5. Note the antigenic identity between the different DNAs (1, 2, 4, 6) and the RNA (3).

tRNA). Similarly, the presence of RNA-precipitating antibodies could not be demonstrated in either the anti-poly A sera or the anti-poly I sera.

IV. REACTIONS OF ANTI-POLY A•POLY U-MBSA SERA WITH DNA

Both groups of anti-double-stranded complex antibodies precipitated RNA, but only the anti-poly A•poly U group of sera reacted with native or denatured DNA. Anti-poly I•poly C sera were inactive.

Samples of DNA from various sources and of different G-C content were tested. Table 7 illustrates the diversity in reactivity of antisera in terms of the DNA tested. Seven anti-poly A•poly U sera precipitated native DNA from bacteriophage 2C and five sera precipitated native DNA from *Micrococcus lysodeikticus.* Crab poly dA-T, which was only tested with four immune sera, reacted with three of them. On the whole, anti-poly A•poly U sera were unreactive with native DNAs from calf thymus or *Clostridium perfringens.* This cross-reaction with native DNA does not seem to depend on the base composition since the G-C content of the DNAs reacting with anti-poly A•poly U antibodies varied from 3 to 72%.

Thermal denaturation of the various DNAs (DNA samples were heated at 100°C for 10 minutes and rapidly chilled) modified their reactivity in diverse ways. It may be noted (cf. Table 7) that two antisera (#310 and 313) were specific for native DNA, and that they did not react with any of the heat-denatured DNA samples. The other five antisera reacted with both native and denatured DNA from *Micrococcus lysodeikticus* and phage 2C.

Depending on the sera tested, the minimum amount of denatured DNA required to give a visible precipitation band was equal to or greater than the quantity of native DNA necessary for the same purpose. However, one exception was observed; antiserum #372 precipitated denatured but not native *Micrococcus lysodeikticus* DNA. Thermal denaturation of calf thymus and *Clostridium perfringens* DNAs rendered the antigenic sites accessible to four immune sera: 372, 375, 343, and 344.

Depending on the antisera used, one to three precipitation lines were obtained, but the number of bands observed is rather difficult to define as in most experiments they rapidly became confluent. As was the case in reactions with RNA, the formation of a spur indicated only partial antigenic identity between the DNA and the poly A•poly U double-stranded complex (Fig. 6). However, the continuity of at least one of the precipitation bands observed when the antibodies react with nucleic acids suggests an antigenic identity between DNA and RNA (Fig. 7).

Anti-poly A•2 poly U sera were also tested with native and denatured DNA. No reaction was obtained with any of the nucleic acids tested so

Table 7

REACTIONS BETWEEN DNA AND ANTI-POLY A•POLY U-MBSA SERA

DNA Antigen		Immunizing Antigen											
		Poly A•Poly U (A/U = 1)			Poly A + Poly A•Poly U (A/U = 1.3)			Poly A + Poly A•Poly U (A/U = 2)			Poly I•Poly C		
		Rabbit Number											
		372	373	375	310	311	313	343	344	345	327	328	331
Micrococcus	native *	−		+	+	−	+	+	+		−	−	−
lysodeikticus	denatured †	+	−	+	−	−	−	+	+				
Phage 2C	native	+	+	+	+	−	+	+	+		−	−	
	denatured †	±	±	+	−	−	−	+	±		−	−	
Calf thymus	native	−	−	−	−	−	−	−	±	−	−	−	−
	denatured †	+	−	+	−	−	−	+	+	−	−	−	−
Clostridium	native	±	−	−	−	−	−	−	−		−	−	−
perfringens	denatured †	+	−	+	−	−	−	+	+		−	−	−
Poly d A-T				+			+	+	±		−	−	

* DNA samples were obtained by standard procedures; absence of denatured material was based on temperature-ultraviolet absorption profiles.

† Heated at 100°C for 10 min. and quickly cooled.

far. In contrast, among the anti-homopolymer sera, anti-poly A serum #304 was able to react with denatured DNAs. However, the precipitation reaction was difficult to demonstrate and could only be obtained with very concentrated solutions of DNA. Nevertheless, the same antibodies gave very strong complement-fixation reactions with heat-denatured DNAs from calf thymus, phage 2C, and *Bacillus subtilis*.

Reaction between anti-poly I-MBSA sera and denatured DNA could not be demonstrated using either the agar diffusion procedure or the complement-fixation technique.

V. DISCUSSION

In immunological reactions involving polynucleotides several possible levels of specificity may be expected to occur. Among these are a dependence on the sugar of the sugar phosphate chain (and hence specificity for ribose or 2-deoxyribose), specificity for a particular base or a certain group of atoms in a purine or pyrimidine, and the macromolecular structural character, e.g., single-stranded polynucleotide, double- or triple-stranded helices, left- or right-handed helices. At present, some characteristics are undoubtedly modified owing to the binding of polynucleotide with MBSA to prepare the immunogen. This binding is presumably a simple polyelectrolyte interaction between the basic protein and the phosphate groups of the polynucleotide.

The present examination of the reactions of antisera prepared against double- and triple-stranded polynucleotide complexes has shown that the specificity of these antisera is determined by the macromolecular conformation of the antigen which stimulated the immunogenic process. Thus, immunization with double-stranded poly A•poly U or poly I•poly C complexed with MBSA produced antibody that was specific for double-stranded structures. Such antibody precipitated homologous and heterologous double-stranded polynucleotides, and was reactive against the homologous but not the heterologous constituent single-stranded polynucleotides when tested individually. Thus, the poly A•poly U antisera reacted with heterologous poly I•poly C but not with poly I or poly C.

Although there was a cross-reaction between the poly A•poly U and poly I•poly C antisera and their respective immunizing antigens, none of the nine anti-poly A•poly U and three anti-poly I•poly C sera reacted against poly G•poly C. This suggests that the polyribose phosphate chain per se is not the exclusive antigenic determinant in these reactions. To account for the absence of poly G•poly C reactivity, it is necessary to postulate differences in stereochemical structure, such as helix pitch, interplanar spacing of the bases, and the inclination of bases to the helix axis; all of these factors will, of course, affect the external geometry of the sugar phosphate chains. Antigenic similarity between poly A•poly U and poly I•poly C is probably reflected in molecular similarity at this level since the common chemical groupings in the bases are situated at N7 and C8 of the purine (adenine or hypoxanthine), and are also present in poly G•poly C. Alternatively, antigenic similarity is related to C2 and N3, both of which are accessible, though poorly so, in poly A•poly U and poly I•poly C but not in poly G•poly C (Fig. 8). In addition, rabbits immunized with double-stranded polynucleotide complexes also produced antibodies against one of the polynucleotide constituents of the complex. Thus, anti-poly A•poly U sera reacted with poly A, and poly I was precipitated by anti-poly I•poly C sera. In this case, since C2 and N3 of the purine bases are common, it is necessary to postulate a molecular specificity in the region of N7 and the 6-amino or 6-keto group.

When a triple-stranded complex such as poly A•2 poly U bound to MBSA was used as immunogen, the rabbit sera were specific only for the homologous triple-stranded antigen, and did not react with double- or single-stranded polynucleotides even though these were components of the original complex. Since the bases are conceivably more accessible for the reaction in the double-stranded rather than in the triple-stranded complexes, one cannot exclude the possibility that the bases play an immunogenic role. Although the identity of the antigenic determinants cannot be resolved by the techniques used in the present study, a more sensitive approach, such as inhibition analysis, might aid in ascertaining whether antibodies are produced against the polyribosephosphate chain only and to what extent the bases contribute to antigenic reactivity.

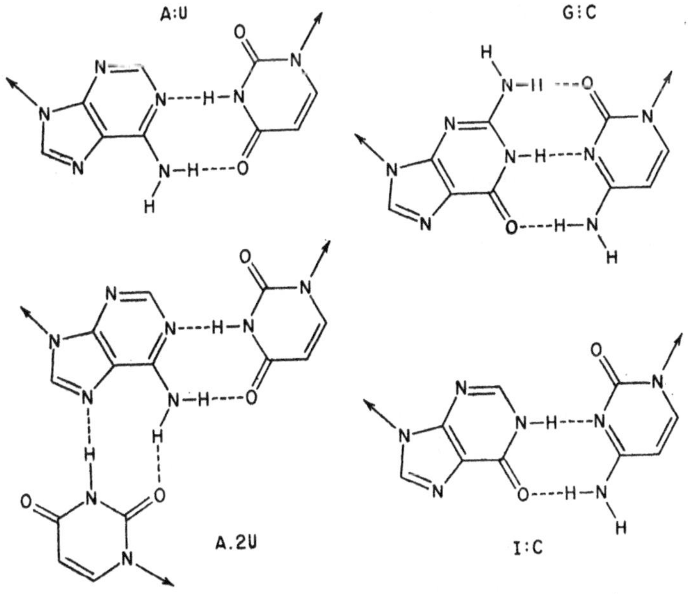

Fig. 8. Configurations resulting from base-pairing.

Nevertheless, it may be concluded that the macromolecular structure and conformation of the antigen provoking the immunogenic response does determine the specificity of the antibody.

Antibodies produced by immunization with single-stranded polynucleotides are specific only for the homologous immunizing antigen. Such antibodies are incapable of recognizing the homologous antigen when it is incorporated into a more organized complex structure, with loss of availability of hydrogen-bonded groupings (6-CO, 6-NH_2, N1, etc.).

Antisera against double-stranded poly A•poly U have been shown to react with several native and denatured DNA preparations, and can distinguish between different DNAs when the native secondary structure was intact. Thus many of these sera could react with DNAs from *M. lysodeikticus,* phage 2C, and with crab poly dAT, but none gave any positive reaction with native DNAs from calf thymus or *Cl. perfringens.* After denaturation, all the DNAs reacted with the antisera. However, antibodies reacting with denatured DNAs have been obtained also by immunization with poly A-MBSA. Since immunization with the complex poly A•poly U produced antibody reactive against both the homologous double-stranded poly A•poly U and single-stranded poly A, it is necessary to determine the relative roles of both of these groups of antibodies in the reaction with DNA. Inhibition absorption tests, utilizing the Bjorklund technique, showed a total inhibition by poly A•poly U of the reactivity of serum #343 with native phage 2C and denatured calf thymus DNA. With the same serum, poly A inhibited only the denatured calf thymus DNA reaction. It appears, therefore, that antibodies to the double-stranded polynucleotide are necessary to recognize native DNA, while anti-poly A antibody interferes more in reactions with denatured DNA. However, the single-stranded antisera did show some basic variations, since some of the sera that were reactive against homologous poly A failed to react with the denatured DNAs.

In studying the minimal size of the antigenic determinants reactive with the poly A antisera, it was found that this varied from 7 to 30 adenine residues, depending on the particular serum tested. Because of its insensitivity, the direct precipitation method used in this study probably overestimates the minimal size of the reactive oligomer sequences. Inhibition techniques would probably show positive results with shorter chain fragments. This individual variability in immunologic response to the same antigen with respect to the length of a given oligonucleotide is of some interest. Thus anti-poly A serum 304, which reacted with duodeca-adenylate, also reacted with all the denatured DNAs, whereas anti-poly A serum 305, which reacted only with chains containing 30 or more adenylate residues, did not react with any of the denatured DNAs tested. Individual antisera apparently measure different lengths of adenine residues in sequence.

In spite of the extensive cross-reactions of the poly A•poly U and poly A antisera, the anti-poly I•poly C and poly I sera failed to react with any of the native or denatured DNA preparations tested. However, antibodies reacting with RNA were found in both the poly A•poly U and poly I•poly C antisera (9 of 12 sera were active), but not in the anti-single-strand sera (2 anti-poly A and 4 anti-poly I) or anti-triple-strand sera (poly A•2 poly U). The latter two groups of sera failed to react with any of the RNAs tested, including total, ribosomal, or transfer RNAs. Furthermore, poly A•poly U completely inhibited the reaction of RNA with anti-poly A•poly U sera, whereas poly A had no inhibitory effect. These results suggest that recognition of RNA involves antibodies specific for a double-stranded helical structure.

The observations described above lead us to conclude that the macromolecular conformation of the immunogen plays an important role in the specificity of the interaction between antibody and double- or triple-stranded polynucleotide complexes. Certain groups of atoms in the purine or pyrimidine bases also appear to determine the immunochemical specificity of the antibodies. This would explain the importance of the bases in the immune reaction, whether the polynucleotide involved occurs as a single-strand or as a structured hydrogen-bonded double helix.

The antisera directed against the double-stranded complexes poly A• poly U and poly I•poly C reacted in a similar fashion with different RNAs. This is not the case for the reaction of anti-poly A•poly U sera and DNA, where variation of antibody specificity was observed. Such variations are due to the interaction of antibodies with different native or denatured DNAs as well as the size of the antigenic site. Similar observations have been reported by Stollar *et al.* (1962) in the case of natural anti-DNA antibodies found in certain cases of lupus erythematosus.

References

Butler, V. P., S. M. Beiser, B. F. Erlanger, S. W. Tanenbaum, S. Cohen, and A. Bendich (1962). Purine-specific antibodies which react with deoxyribonucleic acid (DNA). Proc. Natl. Acad. Sci. U. S., *48*:1597–1602.

Erlanger, B. F., and S. M. Beiser (1964). Antibodies specific for ribonucleosides and ribonucleotides and their reaction with DNA. Proc. Natl. Acad. Sci. U. S., *52*:68–74.

Halloran, M. J., and C. W. Parker (1966). The production of antibodies to mononucleotides, oligonucleotides and DNA. J. Immunol., *96*:379–385.

Karol, M. H., and S. W. Tanenbaum (1967). Antibodies to hapten-conjugated proteins which cross-react with RNA. Proc. Natl. Acad. Sci. U. S., *57*:713–720.

Lacour, F., J. Harel, L. Harel, and E. Nahon (1962). Etude immunologique de l'acide ribonucleique (RNA) par la microméthode de double diffusion dans l'agar. Compt. Rend., *255*:2322–2324.

Lacour, F., and L. Harel (1965). Réaction des anticorps, anti conjugé adenosine monophosphate-serum albumine bovine, avec les polynucléotides. Biochim. Biophys. Acta, *103*:358–360.

Nahon, E., F. Lacour, and J. Harel (1965). Identification immunologique de diverses fractions de RNA cellulaire et de polynucléotides de synthèse. Compt. Rend., *260*:357–360.

Nahon, E., F. Lacour, and M. Michelson (1967). Effet de la structure secondaire des polynucléotides sur leur réaction immunologique. Compt. Rend., *264*: 515–518.

Plescia, O. J., W. Braun, and N. C. Palczuk (1964). Production of antibodies to denatured deoxyribonucleic acid (DNA). Proc. Natl. Acad. Sci. U. S., *52*: 279–285.

Seaman, E., H. Van Vunakis, and L. Levine (1965). Antigenicity of polyribonucleotides. Biochemistry, *4*:1312–1318.

Sela, M., H. Ungar-Waron, and Y. Schechter (1964). Uridine-specific antibodies obtained with synthetic antigens. Proc. Natl. Acad. Sci. U. S., *52*:285–292.

Stollar, D., L. Levine, and J. Marmur (1962). Antibodies to denatured deoxyribonucleic acid in lupus erythematosus serum. II. Characterization of antibodies in several sera. Biochim. Biophys. Acta, *61*:7–18.

Tanenbaum, S. W., and S. M. Beiser (1963). Pyrimidine-specific antibodies which react with deoxyribonucleic acid (DNA). Proc. Natl. Acad. Sci. U. S., *49*: 662–668.

SPECIFICITY IN THE REACTION BETWEEN ANTI-PYRIMIDINE NUCLEOSIDE ANTIBODIES AND DNA *

A. J. GARRO,† B. F. ERLANGER, AND S. M. BEISER
College of Physicians and Surgeons, Columbia University
New York

I. INTRODUCTION ‡

The chemical coupling of nucleosides and nucleotides to proteins to form antigenic conjugates has been reported (Erlanger and Beiser, 1964). The antibodies elicited by such conjugates have been shown to be specific for the coupled moiety, and can react with denatured, but not native, DNA. One of the potential applications of such antibodies is their use as probes for the detection of localized single-stranded regions (in which the paired bases are not hydrogen-bonded) within native DNA molecules. Such regions might be of biological importance in replication or transcription, or might provide flexibility in a double-stranded helical molecule. Before such studies were undertaken, it was considered desirable to investigate the reaction of anti-nucleoside antibodies, specifically anti-cytidine and anti-thymidine, with native DNA molecules which had been modified chemically so as to contain exposed or nonhydrogen-bonded bases. These studies would serve as a model for future investigations.

Methylene blue-sensitized photooxidation specifically alters guanine residues in DNA (Simon and Van Vunakis, 1962). In T-even phage DNA, this alteration has been shown to expose the opposed glucosylated hy-

* This work was supported by a grant (AI-06860) from the U. S. Public Health Service and contracts NOnr 266(40) and NOnr 4259(11) between the Office of Naval Research and Columbia University.

† Predoctoral trainee, grant 5-F1-GM-31,689-02 from the National Institutes of Health. This report will form part of a dissertation to be submitted in partial fulfillment of the requirements for the degree of Doctor of Philosophy, Columbia University.

‡ Abbreviations used: DNA, deoxyribonucleic acid; A, adenosine; C, cytidine; T, thymidine; G, guanosine; BSA, bovine serum albumin; C′, complement.

droxymethylcytosine, which can then be detected by reaction with specific antibody to T-even phage DNA (Simon and Van Vunakis, 1962).

We have extended the study of partially denatured DNA produced by methylene blue photooxidation by investigating the development of reactivity with specific anti-cytidine and anti-thymidine antibodies following DNA photooxidation. In this way, "local" denaturation could be distinguished from what may be termed "generalized" denaturation. Alteration of guanine with exposure of cytosine residues should result in reactivity only with anti-cytidine serum if strand separation were strictly limited. On the other hand, more widespread strand separation should expose both cytosine and thymine to reaction with antibody. This report deals with development of reactivity between pyrimidine-specific antisera and photooxidized methylene blue-sensitized DNA as measured by quantitative precipitin or complement-fixation reactions. The appearance of serologic reactivity was correlated with the extent of guanine destruction in the photooxidized DNA.

II. MATERIALS AND METHODS

A. DNA Preparations

(1) Calf thymus DNA was obtained from Nutritional Biochemicals Corporation and used without further purification.

(2) Methylene blue was obtained from the Allied Chemical Corporation.

(3) DNA was photooxidized according to the procedure of Simon and Van Vunakis (1962). We used 450–500 $\mu g/cm^3$ of DNA and 20 $\mu g/cm^3$ of methylene blue in 0.1 N Tris buffer pH 8.5. The temperature of the reaction mixture was maintained at 15°C.

(4) DNA was denatured either by heating a 5–6 $\mu g/cm^3$ solution in 0.15 M saline at 100°C for ten minutes followed by quenching in an ice bath or by heating an 800–1000 $\mu g/cm^3$ solution in 1% HCHO-saline at 100°C for ten minutes followed by quenching in an ice bath.

(5) Base ratios were determined by chromatographic analyses of formic acid hydrolyzates of DNA (Wyatt, 1951).

B. Antisera and Serologic Procedures

(1) The preparation of nucleoside conjugates, the protocol of immunization, and the schedule and manner of bleeding have been described by Erlanger and Beiser (1964). Rabbits were immunized with BSA conjugates of cytidine (C) and 1-*β*-*D*-ribofuranosyl thymine (T).

(2) Individual antisera were examined for specificity by the gel diffusion method and those showing strict specificity for the immunizing

antigen were pooled. Antisera showing cross reactions with nucleoside conjugates other than the immunogen were pooled separately and absorbed with the appropriate antigens before use. Globulin fractions were prepared either by salt fractionation (Strauss *et al.*, 1960), or by DEAE-sephadex chromatography using conditions described by Dedmon *et al.* (1965).

(3) Precipitin and Kjeldahl analyses (Kabat, 1961) were carried out and specific precipitates analyzed using a modified Lowry-Folin method described by Mage and Dray (1965).

III. RESULTS

A. Antisera Specificity

Quantitative C′-fixation and precipitin reactions were used to characterize the antisera according to their specificity of reaction with nucleoside conjugates of the four major bases of DNA (Figs. 1–4). The globulin fractions used in the C′-fixation reactions, anti-cytidine-BSA-Pool I (globulin) and anti-thymidine-BSA-Pool I (globulin), were prepared from rabbit antisera which showed strict specificity for the immunizing conjugates in gel diffusion studies and were used without prior absorption. The globulin fractions used in the precipitin reactions, anti-cytidine-BSA-Pool II (globulin) and anti-thymidine-BSA-Pool II (globulin), were prepared from rabbit antisera which reacted with conjugates other than the immunogens and were absorbed with the cross-reacting antigens before use in this study.

As can be seen in Figures 1 and 2, both the anti-cytidine preparations

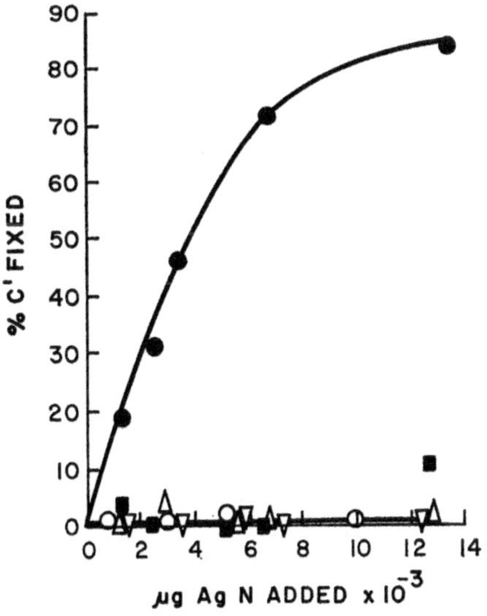

Fig. 1. C′-fixation reactions between nucleoside-BSA conjugates and anti-cytidine-BSA-Pool I (globulin) 1:840. ● C-BSA; ■ T-BSA; △ A-BSA; ○ G-BSA; ▽ BSA.

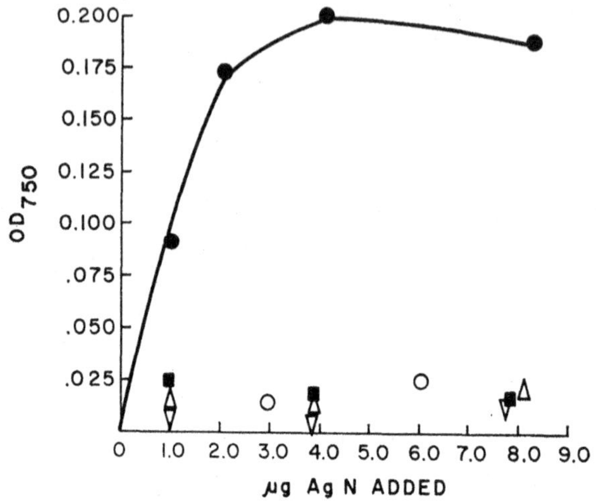

Fig. 2. Precipitin reactions between nucleoside-BSA conjugates and 0.5 cm^3 of a 1:10 dilution of anti-cytidine-BSA-Pool II (globulin). ● C-BSA; ■ T-BSA; △ A-BSA; ○ G-BSA; ▽ BSA.

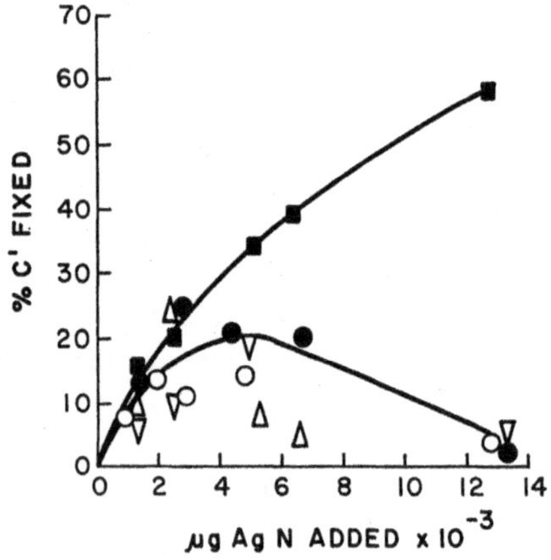

Fig. 3. C′-fixation reactions between nucleoside-BSA conjugates and anti-thymidine-BSA-Pool I (globulin) 1:1750. ● C-BSA; ■ T-BSA; △ A-BSA; ○ G-BSA; ▽ BSA.

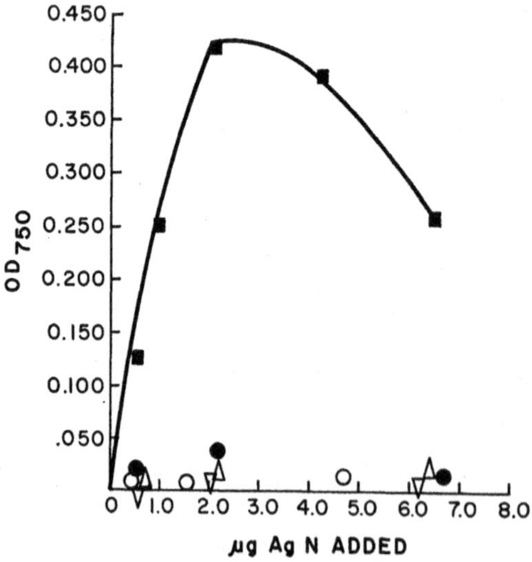

Fig. 4. Precipitin reactions between nucleoside-BSA conjugates and 0.5 cm^3 of a 1:20 dilution of anti-thymidine-BSA-Pool II (globulin). ● C-BSA; ■ T-BSA; △ A-BSA; ○ G-BSA; ▽ BSA.

tested reacted only with C-BSA. The anti-thymidine-BSA-Pool I (globulin), when tested by quantitative C′-fixation (Fig. 3), reacted best with T-BSA but exhibited some cross-reaction with the other nucleoside conjugates. However, since approximately the same amount of reaction is observed with the BSA carrier, the reaction with the other nucleoside conjugates may be due to their BSA backbone. The anti-thymidine-BSA-Pool II (globulin) which had been absorbed with cross-reacting antigens reacted only with T-BSA (Fig. 4).

B. Reaction with Denatured DNA

Before attempting to detect areas of "local" denaturation in photooxidized DNA, the concentrations of antisera were adjusted to levels at which the anti-thymidine antisera reacted with completely denatured DNA as well or better than the anti-cytidine antisera (Figs. 5 and 6). At these concentrations one would expect to detect thymine as well as cytosine residues exposed as a result of the photooxidative process. It should be noted that the calf thymus DNA, used in the precipitin reaction shown in Figure 5, had been denatured in the presence of 1% formaldehyde. Since formaldehyde reacts with amino and imino groups which become available during DNA denaturation some masking or alteration of the antigenic configuration of the purine and pyrimidine residues may take place. However, based on the precipitin reactions observed between these antisera and the nucleoside conjugates (Figs. 2 and 4), we feel that the curves depicted in Figure 6 reflect the ability of these antisera to precipitate with denatured DNA. The nucleoside-conjugate precipitin curves show that the anti-thymidine serum pool at

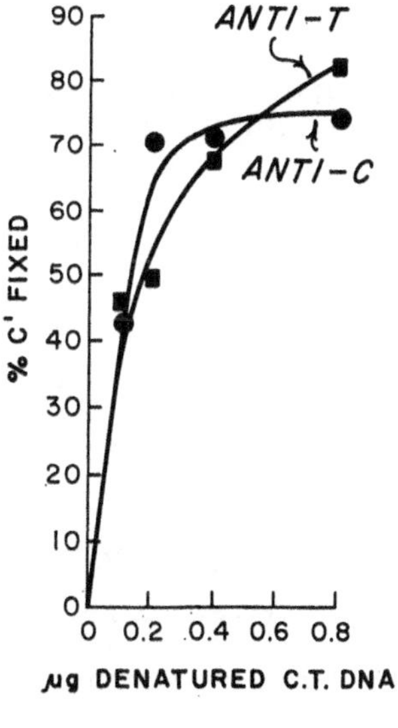

Fig. 5. C′-fixation reactions between heat denatured calf thymus DNA and anti-cytidine-BSA-Pool I (globulin) 1:300 and anti-thymidine-BSA-Pool I (globulin) 1:1000. ● Anti-C; ■ Anti-T.

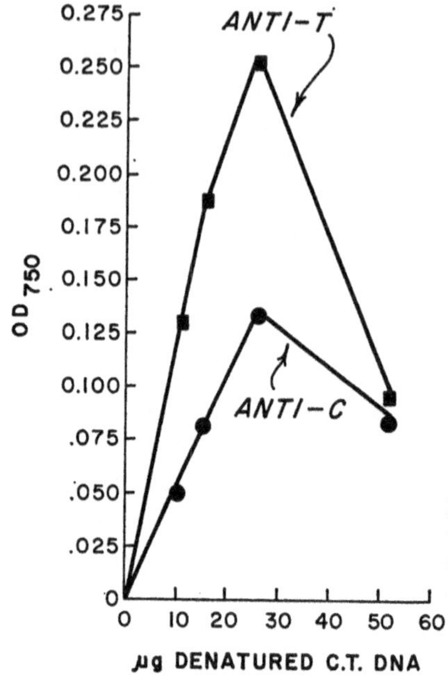

Fig. 6. Precipitin reactions between calf thymus DNA thermally denatured in the presence of 1% formaldehyde and 0.25 cm^3 of 1:4 dilutions of anti-cytidine-BSA-Pool II (globulin), and anti-thymidine-BSA-Pool II (globulin). ● Anti-C; ■ Anti-T.

a 1:20 dilution precipitates twice as much protein as a 1:10 dilution of the anti-cytidine serum pool. Thus, the use of anti-thymidine and anti-cytidine antisera at the same dilution (1:4) for precipitation of the DNA preparations should actually favor reactions with thymine residues (cf. Fig. 6).

C. Reactions with Photooxidized DNA

Photooxidized DNA was tested by the quantitative precipitin method at a concentration approximately three times that required for maximal precipitation with "completely" denatured DNA in order to increase the possibility of detecting small locally denatured regions. At this con-

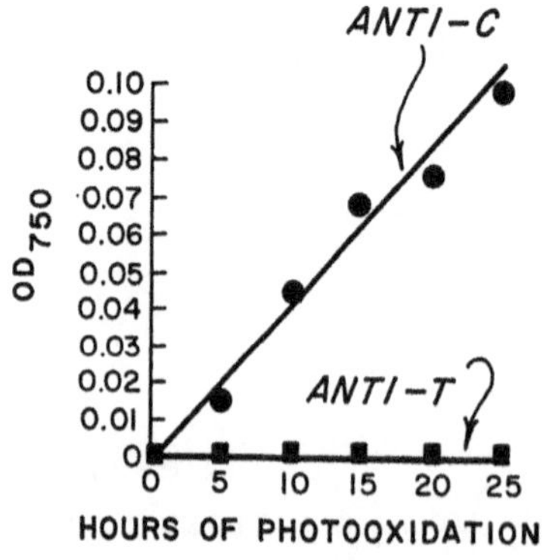

Fig. 7. Precipitin reactions between photooxidized calf thymus DNA, 70 μg, and 0.25 cm^3 of 1:4 dilutions of anti-cytidine-BSA-Pool II (globulin) and anti-thymidine-BSA-Pool II (globulin). ● Anti-C; ■ Anti-T.

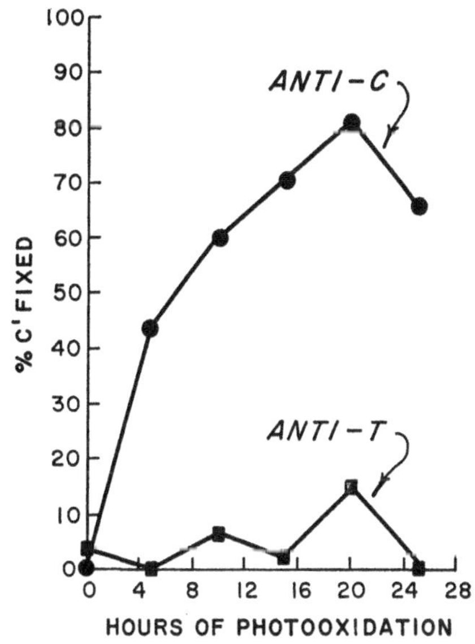

Fig. 8. C′-fixation reactions between photooxidized calf thymus DNA, 0.1 μg, and anti-cytidine-BSA-Pool I (globulin) 1:100 and anti-thymidine-BSA-Pool I (globulin) 1:300. ○ Anti-C; ■ Anti-T.

centration the reaction between the photooxidized DNA and the anti-cytidine antisera increased with increasing periods of photooxidation, while the anti-thymidine antisera did not give a detectable reaction (Fig. 7).

Similar results were obtained by quantitative C′-fixation analysis (Fig. 8), in which the antisera concentrations used were three times greater than those used for reaction with completely denatured DNA. The reaction with the anti-cytidine antisera increased up to the 20th

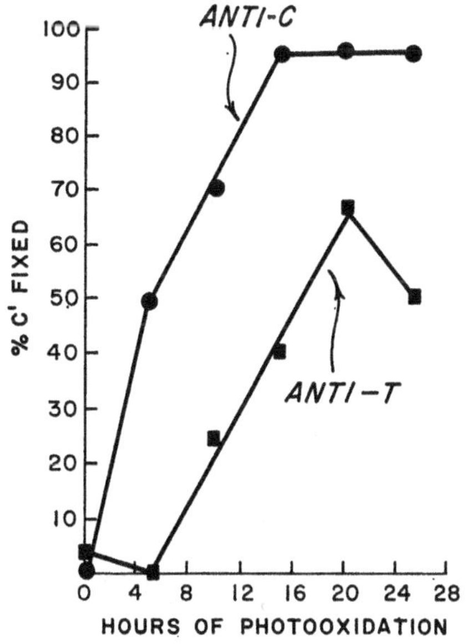

Fig. 9. C′-fixation reactions between photooxidized calf thymus DNA, 0.2 μg, and anti-cytidine-BSA-Pool I (globulin) 1:100; and anti-thymidine-BSA-Pool I (globulin) 1:300. ● Anti-C; ■ Anti-T.

hour of photooxidation and then decreased. This decrease may be the result of DNA fragmentation resulting from the photooxidation. Photooxidative fragmentation of denatured DNA has been reported by Simon and Van Vunakis (1962) and Seaman *et al.* (1966), while Waskell *et al.* (1965) have reported the photooxidative destruction of the ribose moiety of guanosine during methylene blue photooxidation and postulated its possible role in polynucleotide chain scission.

When the photooxidized DNA was tested at a higher concentration, the anti-thymidine antisera did fix C′ (Fig. 9). At this concentration, however, a 17-hour period of photooxidation was required to reach a 50% level of C′ fixation with anti-thymidine sera, but only 5 hours were needed for the anti-cytidine antisera. The extent of reaction was also less for the anti-thymidine antisera relative to the anti-cytidine antisera.

D. Base Ratio Analyses

The molar ratios of the bases recovered from untreated and photooxidized DNA, relative to cytosine, are shown in Figure 10. After 5 hours of photooxidation approximately 25% of the guanine has been lost. At this time and at all concentrations of DNA tested by both quantitative precipitation and C′-fixation, only cytosine residues are detectable (Figs. 7, 8, and 9). After 10 hours of photooxidation and 56% guanine destruction, thymine residues are detectable by C′-fixation but

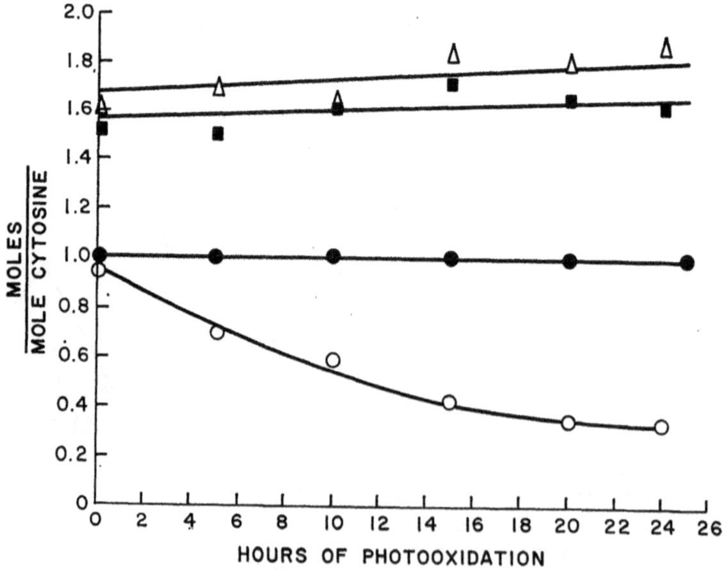

Fig. 10. Base ratio analyses of photooxidized calf thymus DNA. ● cytosine; △ adenine; ○ guanine; ■ thymine.

not by precipitin analysis. As the loss of guanine becomes asymptotic after 20 hours of photooxidation, no further increase in reactivity is detected by C′-fixation; however, the extent of precipitation continues to increase past this time, reflecting possibly the occurrence of further changes within the DNA.

IV. DISCUSSION

Evidence has been presented for the ability of antinucleoside antibodies to detect local areas on a DNA molecule in which base pairing has been disrupted. This conclusion derives from the finding that anti-cytosine, but not anti-thymine antibodies react with photooxidized DNA under conditions in which unpaired thymine residues would also have been detected.

The antisera used in this study were obtained by immunization with chemically defined antigens and were shown to be specific for either cytosine or thymine, and to react with denatured but not with native DNA. Their specificity for the denatured conformation of DNA is understandable in terms of our current knowledge of DNA structure. In the native molecule the hydrophobic bases, to which the antibodies are directed, are hydrogen-bonded A=T and G≡C, within the interior of the molecule. Denaturation disrupts long range helical order and exposes the bases to reaction with antibody. The actual extent of bihelical disruption may vary with the denaturation procedure and the initial state of the DNA (Kohn *et al.*, 1966).

Serologic methods have been used previously to follow the transition from native to denatured states produced by thermal and/or chemical treatment of DNA. The results obtained closely paralleled the observations made by physico-chemical methods with respect to observed melting temperatures and to the effectiveness of the denaturing agents. However, these studies, which were summarized in a review by Levine and Van Vunakis (1966), were not designed to distinguish between what we have defined as "local" and "generalized" denaturation.

Although we were able to detect only cytosine residues in photooxidized DNA after 25% destruction of guanine, the lower limits of detection are not known. By working within the 5-hour time interval, during which 25% of the guanine has been destroyed, it may be possible to set such lower limits.

No attempt has been made to correlate the actual size(s) of the lesions produced with the extent of guanine destruction. Electron microscopic techniques which have been used to visualize partially denatured phage DNA (Inman, 1966) could prove useful in making this correlation.

Although no precipitin reaction was observed with the anti-thymine antisera during the entire period of photooxidation, it was possible to demonstrate exposed thymine residues by C′-fixation after 10 hours of photooxidation. These results probably reflect differences in sensitivity between these methods. The continued increase in reactivity of the photooxidized DNA with the anti-cytidine sera observed in the precipitin reaction beyond the point where guanine destruction reaches a maximum may reflect further changes occurring within the DNA bihelix at that time. Studies of the kinetics and end products of the methylene blue-sensitized photooxidation of guanosine by Waskell *et al.* (1966) indicate that there are several steps involved in the reaction which continues after the disruption of the glycosidic bond between guanine and ribose. If the same situation prevails in the photooxidative destruction of the deoxyguanosine residues of DNA, it is conceivable that further changes in the bihelix may occur after guanine destruction has reached a maximum.

The type of serologic investigation of DNA cited in this report also should be suitable for the study of DNA which has been partially denatured under conditions which would preferentially disrupt A=T hydrogen bonding. Heating DNA at temperatures below its melting temperature should provide such conditions. Marmur and Lane (1960) have shown that, upon slow cooling, denatured pneumococcal DNA renatures and a significant portion of its original transforming activity can be restored. If selective immunologic precipitation of partially denatured DNA occurs, similar procedures for renaturing fractionated partially denatured DNA should also be possible and might be useful in the separation of various transforming activities.

References

Dedmon, R. E., A. W. Holmes, and F. Deinhardt (1965). Preparation of fluorescein isothiocyanate-labeled γ-globulin by dialysis, gel filtration and ion exchange chromatography in combination. J. Bacteriol., *89*:734–747.

Erlanger, B. F., and S. M. Beiser (1964). Antibodies specific for ribonucleosides and ribonucleotides and their reaction with DNA. Proc. Natl. Acad. Sci. U. S., *52*:68–74.

Inman, R. B. (1966). A denaturation map of the phage DNA molecule determined by electron microscopy. J. Mol. Biol., *18*:464–476.

Kabat, E. A. (1961). Kabat and Mayer's Experimental Immunochemistry, 2nd ed. Charles C Thomas Co., Springfield, Illinois.

Kohn, K. W., C. L. Spears, and P. Doty (1966). Interstrand crosslinking of DNA by nitrogen mustard. Appendix. Terminology of configurational changes of DNA in solution. J. Mol. Biol., *19*:266–288.

Levine, L., and H. Van Vunakis (1966). Serologic Activities of the nucleic acids. Proc. Fed. European Biochem. Soc., *1*:25–60.

Mage, R., and S. Dray (1965). Persistent altered phenotypic expression of allelic

G-immunoglobulin allotypes in heterozygous rabbits exposed to isoantibodies in fetal and neonatal life. J. Immunol., *95*:525–535.

Marmur, J., and D. Lane (1960). Strand separation and specific recombination in deoxyribonucleic acids: Biological studies. Proc. Natl. Acad. Sci. U. S., *46*:453–461.

Seaman, E., L. Levine, and H. Van Vunakis (1966). Antibodies to the methylene blue sensitized photooxidation product in deoxyribonucleic acid. Biochemistry, *5*:1216–1223.

Simon, M. I., and H. Van Vunakis (1962). The photodynamic reaction of methylene blue with deoxyribonucleic acid. J. Mol. Biol., *4*:488–499.

Strauss, A. J. L., B. C. Seegal, K. C. Hsu, P. M. Burkholder, W. L. Nastuk, and K. E. Osserman (1960). Immunofluorescence demonstration of a muscle-binding, complement-fixing serum globulin fraction in myasthenia gravis. Proc. Soc. Exp. Biol. Med., *105*:184–191.

Waskell, L. A., K. S. Sastry, and M. P. Gordon (1966). Studies on the photosensitized breakdown of guanosine by methylene blue. Biochim. Biophys. Acta, *129*:49–53.

Wyatt, G. R. (1951). The purine and pyrimidine composition of deoxypentose nucleic acids. Biochem. J., *48*:584–590.

ANTIBODIES TO THE TERMINAL MOIETIES OF GUANINE MONONUCLEOTIDES AND GUANINE-CONTAINING DINUCLEOTIDES *

HELEN VAN VUNAKIS, EDNA SEAMAN,
AND LAWRENCE LEVINE

Graduate Department of Biochemistry, Brandeis University
Waltham, Massachusetts

Antibodies specific for nucleic acids have been obtained by immunizing rabbits with conjugates composed of bases, nucleosides or nucleotides covalently linked to proteins or polyamino acids. The chemical methods used to prepare such immunogens have been reviewed (Levine and Stollar, 1967) and the characteristics of some of these antibodies are considered in this volume.

In most of the syntheses, the nucleic acid component was linked to the carrier via the sugar or phosphate residue. In general, the antibodies produced against these immunogens exhibited specificity toward the purine or pyrimidine base. The classic studies of Landsteiner (1936) in which haptens were rendered immunogenic by coupling to large macromolecules provide the precedent for such investigations.

In this paper, a novel synthesis is described by which guanine-containing mono- and dinucleotides are coupled to the amino groups of polyamino acids rich in lysine through a photointermediate produced during the photooxidation of guanine. The antibodies which result from immunizing rabbits with the mononucleotide conjugate show specificity towards the sugar and sugar-phosphate linkage, whereas antibodies produced in response to immunization with dinucleotide conjugates in which guanosine is linked to a second nucleotide via its 3′-OH group show a specificity towards the nucleoside-5′-phosphate.

* Aided by grants from the NIH (AI-01940 and AI-02792) and the Charles Simon Memorial Grant for Cancer Research from the American Cancer Society (E.222). L. L. is an American Cancer Society Professor of Biochemistry and H. V. V. is a recipient of a Public Health Service Research Career Award K6-AI-2372. Publication No. 564 from the Graduate Department of Biochemistry, Brandeis University.

When guanine derivatives are photooxidized in the presence of visible light, methylene blue and compounds containing primary amino groups, the base is destroyed to yield a photointermediate of unknown structure which can condense with available amino groups (Van Vunakis *et al.*, 1966). Therefore, if GMP-5′ is photooxidized in the presence of a random copolymer composed of glutamic acid and lysine residues in a mole ratio of 60:40 (Poly Glu,Lys), the addition product (Poly Glu,Lys-[G]MP5′) shown in Figure 1 may be formed. The photooxidation of a guanine-containing dinucleotide such as guanylyl-(3′→5′)-adenosine (GpA) in the presence of Poly Glu,Lys, results in a product, Poly Glu,Lys-[G]pA, illustrated in Figure 2.

I. ANTI-POLY GLU,LYS-[G]MP5′

Complement-fixation curves of the reaction between anti-Poly Glu,Lys-[G]MP5′ serum and the carrier, the homologous antigen and a deoxyribo-analog are shown in Figure 3. The serum reacts well with Poly Glu,Lys-[G]MP5′ and gives no reaction with the carrier at the dilution of antiserum used. The antibodies produced are therefore specific for the nucleotide portion of the immunogen. Photooxidized Poly Glu,Lys-

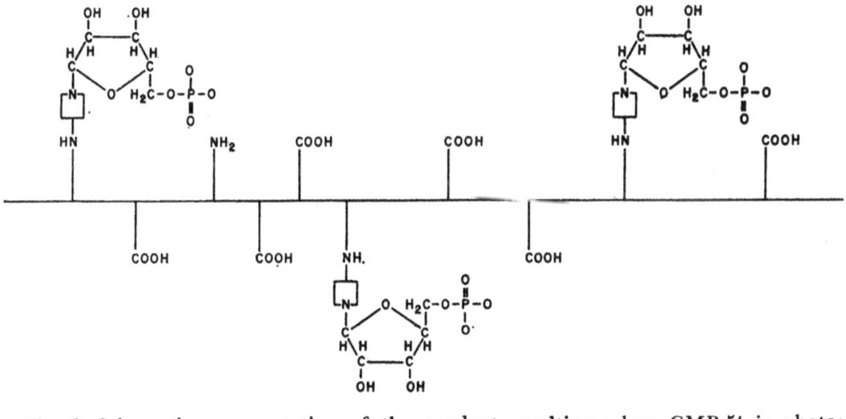

Fig. 1. Schematic representation of the product resulting when GMP-5′ is photooxidized in the presence of Poly $Glu_{60}Lys_{40}$ (M.W. = 120,000). [G] represents the unknown photo-intermediate formed during the photooxidation of guanine which condenses with the ε-amino groups of lysine. The photooxidation was carried out as described by Simon and Van Vunakis (1962). 5 mgs/ml of GMP-5′ were photooxidized in the presence of 5 mgs/ml of the Poly Glu,Lys at pH 8.5, using 20 μgs/ml of methylene blue as the photosensitizing dye. The reaction was terminated when there was no longer any change in the optical density at 250 mμ. After dialysis, aliquots of the reaction mixture were assayed for phosphate content (Chen *et al.*, 1956). A ratio of one phosphate to ten lysine residues was found. Rabbits were immunized with a complex of the adduct and methylated bovine serum albumin (Plescia *et al.*, 1964) and the sera thus obtained were characterized.

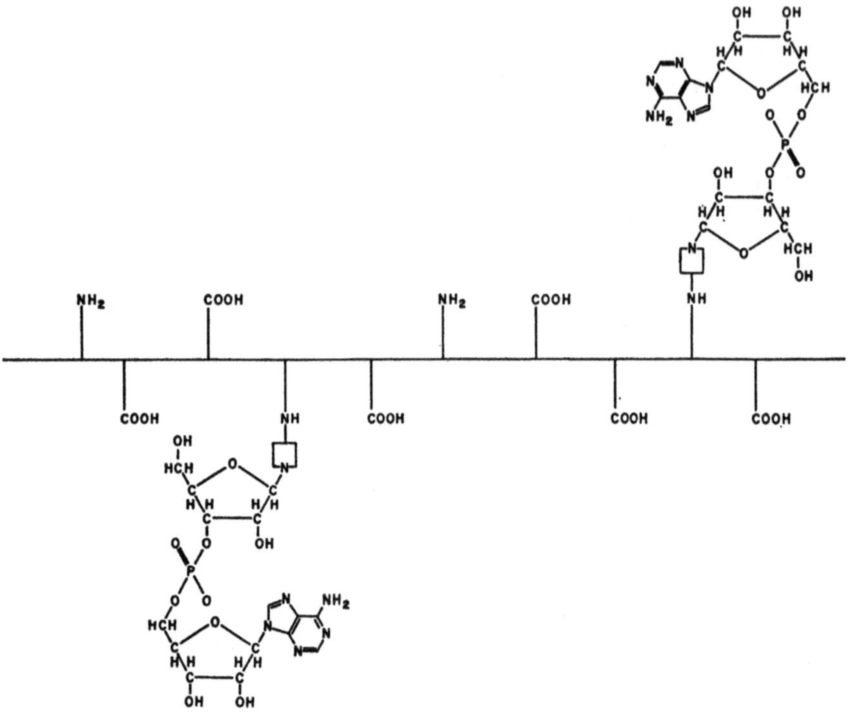

Fig. 2. Schematic representation of the product resulting when GpA is photooxidized in the presence of Poly Glu,Lys. The dinucleotides were photooxidized as described (see legend to Fig. 1). The linking of the dinucleotide to macromolecule occurred at a ratio of one [G]pA to 13 lysine residues.

d[G]MP5′ is less reactive than the homologous antigen Poly Glu,Lys-[G]MP5′ indicating that the sugar moiety of the nucleotide is part of the antigenic determinant.

Inhibition studies of the homologous immune system were undertaken in order to characterize further its specificities (Table 1). All of the ribonucleoside-5′-phosphates(GMP5′, AMP5′, UMP5′ and CMP5′) were equally effective as inhibitors. The deoxynucleoside-5′-phosphates reacted to a lesser extent than the ribonucleotides (compare CMP5′ and dCMP5′). [G]MP5′ which had been photooxidized in the presence of different buffers inhibited the immune system to the same extent as unirradiated GMP5′. Even GMP5′ photooxidized in the presence of ε-aminocaproic acid to form an adduct which would resemble the ε-NH Lys-[G]MP5′ linkage in the immunogen was no more effective as an inhibitor than GMP5′. The specificity, therefore, does not extend to the base, nor does a purine or pyrimidine structure have to be present in order to inhibit the immune reaction effectively.

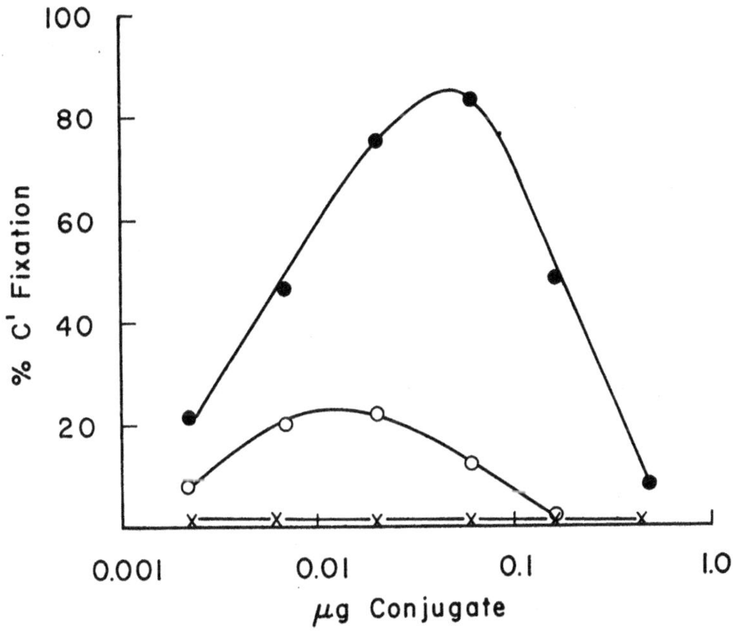

Fig. 3. C′-fixation of anti-Poly Glu,Lys-[G]MP5′ and Poly Glu,Lys-[G]MP5′ (●), Poly Glu,Lys d[G]MP5′ (○) and Poly Glu,Lys (×). The complement-fixation reaction was carried out according to the procedure of Wasserman and Levine (1960).

Table 1

INHIBITION OF THE POLY GLU,LYS-[G]MP5′ IMMUNE SYSTEM

Inhibitor	mμMoles Required for 50% Inhibition
GMP5′	13
AMP5′	8
UMP5′	15
CMP5′	16
dCMP5′	90
Guanosine	>370 *
Ribose-5′-phosphate	40
Ribose-1′-phosphate	>470 *
GpU	>105 *
AMP3′	>100 *
AMP2′	>100 *
AMP cyclic 3′5′	>100 *

* No inhibition at these levels of hapten.

Guanosine does not inhibit the immune system indicating that the phosphate group is an essential part of the antigenic determinant. The question arises, however, whether the phosphate is part of the antigenic site or whether it simply facilitates the interaction of the hapten with the antibody through a charge effect. To answer this question, ribose-1′-phosphate and ribose-5′-phosphate were used as inhibitors. While ribose-5′-phosphate is a potent inhibitor, ribose-1′-phosphate has no effect on the system. The stringent requirement for a phosphate group on C_5 of ribose is further illustrated by the complete lack of inhibition by AMP2′ and AMP3′. Furthermore, the antibody can distinguish between a mono- and diesterified phosphate. AMP cyclic 3′5′ and the dinucleotide GpU, both of which contain the required ribose-5′-phosphate moiety in a diesterified form fail to inhibit the reaction.

II. ANTI-POLY GLU,LYS-[G]pA

Antibodies in the Poly Glu,Lys[G]pA antiserum appear to be directed toward the adenosine-5′-phosphate moiety of the [G]pA. This serum reacts only with the homologous antigen (Fig. 4) and in contrast to the Poly Glu,Lys-[G]MP5′ immune system, it is inhibited best by AMP5′

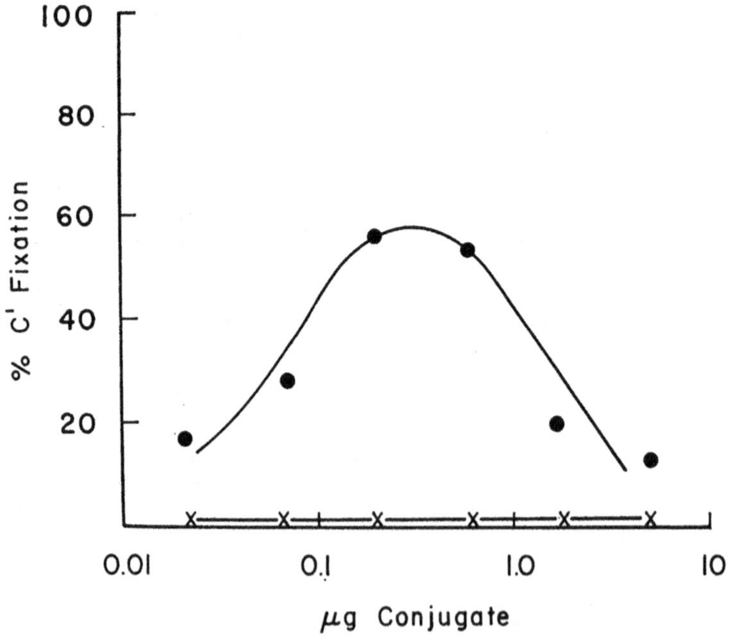

Fig. 4. C′-fixation of anti-Poly Glu,Lys-[G]pA with Poly Glu,Lys-[G]pA (●); Poly Glu,Lys; Poly Glu,Lys-Ap[G]; Poly Glu,Lys-Cp[G]; Poly Glu,Lys-[G]pC or Poly Glu,-Lys-[G]pU (×).

Table 2

INHIBITION OF THE POLY GLU,LYS-[G]PA IMMUNE SYSTEM

Inhibitor	mμMoles Required for 50% Inhibition
AMP5′	24
GMP5′	580 *
CMP5′	>250 †
UMP5′	>225 †
dAMP	>300 †
Adenosine	140
AMP2′	>300 †
AMP3′	>300 †
AMP cyclic 3′5′	>288 †

* Extrapolated value.
† 10% inhibition or less at these levels of hapten.

(Table 2). GMP5′ is inhibitory at a concentration 20 times higher than AMP5′, but the pyrimidine nucleoside-5′-phosphates are ineffective as inhibitors. The sugar component of the nucleotide is involved in the specificity, since the deoxynucleotide (dAMP) is ineffective even at ten-fold greater concentration than the corresponding ribotide. With respect to the sugar, the specificity parallels that observed with the Poly Glu,Lys-[G]MP5′ immune system (Table 1). Adenosine is also capable of inhibiting the immune system though not as efficiently as AMP5′ (Table 2). The phosphate moiety is a part of the antigenic determinant but its presence is not essential for the antibody-hapten interaction. AMP2′, AMP3′, and AMP cyclic 3′5′ are extremely poor inhibitors of this system. This observation leads us to believe that the presence of the phosphate group on carbons 2 or 3 changes the conformation of the nucleotide sufficiently to render it unrecognizable by the antibody. There is no cross-reaction with an antigen in which the sequence of the bases in the dinucleotide is reversed so that the unaltered nucleotide is connected to the photooxidized nucleoside via its 3′OH group, i.e., Poly Glu,Lys-Ap[G].

III. ANTI-POLY GLU,LYS-[G]pU

The serologic specificity of anti-Poly Glu,Lys-[G]pU serum is analogous to anti-Poly Glu,Lys-[G]pA serum. The C′ fixation curves in Figure 5 show the reaction of this serum with the homologous antigen and the

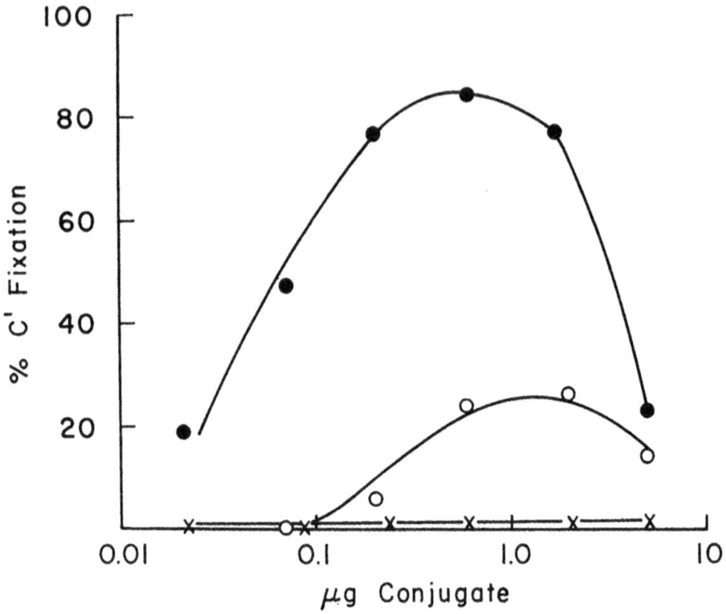

Fig. 5. C′-fixation of anti-Poly Glu,Lys-[G]pU with Poly Glu,Lys-[G]pU (●) or Poly Glu,Lys-[G]pC (○); or Poly Glu,Lys; Poly Glu,Lys-[G]pA; Poly Glu,Lys-Ap[G] or Poly Glu,Lys-Cp[G] (×).

weak cross-reaction with the heterologous antigen Poly Glu,Lys-[G]pC in which cytosine has replaced uracil. There is no reaction with Poly Glu,Lys-[G]pA in which the pyrimidine has been replaced by a purine, adenine.

The pyrimidine nucleotides CMP5′ and rTMP5′, though less effective than UMP5′, can inhibit the Poly Glu,Lys-[G]pU reaction with the homologous antibody (Table 3), whereas the purine nucleotides are not inhibitory at the concentrations tested. The importance of the sugar moiety is again noted by comparing the effectiveness of dUMP5′ to UMP5′ as inhibitors.

The phosphate group is part of the antigenic determinant in both dinucleotide immune systems. Approximately, twenty-five times more uridine than UMP5′ is required for equivalent inhibition. This system is not inhibited significantly by UMP3′(2′). The inhibitory capacity of ribonucleoside mono-, di- and triphosphate is similar indicating that the addition of phosphate groups does not hinder the antibody-hapten interaction.

The fact that UMP5′, GpU and UpU inhibit the anti-Poly Glu,Lys-[G]pU system to approximately the same extent indicates that the 5′ phosphate can be either mono- or diesterified and that the combining site of the antibody does not extend beyond the mononucleotide.

Table 3

INHIBITION OF POLY GLU,LYS-[G]PU IMMUNE SYSTEM

Inhibitor	mμMoles Required for 50% Inhibition
UMP5′	24
CMP5′	330
AMP5′	>288 †
GMP5′	>276 †
rTMP5′	560
dUMP5′	>325 *
Uridine	640
UMP3′(2′)	>310 †
UDP	65
UTP	90
GpU	30
UpU	23

* 25% inhibition at this level of hapten.

† 10% inhibition or less at this level of hapten.

IV. DISCUSSION

The antigens used in this study were prepared by photooxidizing guanine-containing nucleotides or dinucleotides in the presence of a macromolecule rich in available amino groups. A condensation reaction occurs between a guanine photo-intermediate and the available amino groups to form covalent bonds between the hapten and the carrier. The sugar, phosphate and other non-photooxidizable bases of the hapten are not altered nor are they in any way involved in forming bonds with the carrier. They are, therefore, exposed in the antigen.

The antibodies formed against Poly Glu,Lys-[G]MP5′ are directed primarily towards the ribose-5′ monoesterified phosphate moiety of the nucleotide; the presence and the position of the phosphate on the ribose being of prime importance. The antibodies formed against Poly Glu,Lys-[G]pA or Poly Glu,Lys-[G]pU are directed toward the purine or the pyrimidine nucleotide attached to the photooxidized guanosine residues.

There is thus a significant difference between these antibodies and those of Erlanger and Beiser (1964) and Halloran and Parker (1966). Sela *et al.* (1964) did find that one of their five nucleoside antisera recognized the sugar moieties, which have their specificities directed primarily toward the bases. The antibodies to Poly Glu,Lys-[G]pA and Poly Glu,Lys[G]pU also differ from those produced by Panijel and his co-workers (1966) which are specific for the polyphosphate backbones rather than the bases.

The antibodies described here are specific for terminal phosphate, sugar, nucleoside and nucleotide residues. Thus far, the limiting factor in obtaining antibodies to different terminal residues by this procedure appears to be the availability of the suitable guanine derivatives. These antibodies may prove useful in detecting and/or blocking specific terminal groups in nucleic acids. It would be interesting to test the effect of such antibodies on nucleic acids whose biological activities are dependent upon a specific grouping on their terminal ends, e.g., the effect of anti-Poly Glu,Lys[G]pA on the ability of soluble RNA (terminating in pA) to accept amino acids.

References

Chen, Jr., P. S., T. Y. Toribara, and H. Warner (1956). Microdetermination of Phosphorus. Analytical Chemistry, *28*:1756–1758.

Erlanger, B. F., and S. M. Beiser (1964). Antibodies specific for ribonucleosides and ribonucleotides and their reaction with DNA. Proc. Natl. Acad. Sci. U. S., *52*:68–74.

Halloran, M. J., and C. W. Parker (1966). The production of antibodies to mononucleotides, oligonucleotides and DNA. J. Immunol., *96*:379–385.

Landsteiner, K. (1936). The specificity of serological reactions (Harvard University Press, Cambridge, Mass.).

Levine, L., and D. Stollar (1967). Nucleic acid immune systems in "Progress in Allergy." Vol. 12 in press, (S. Karger, Basel, Switzerland).

Panijel, J., C. Souleil, and P. Cayeux (1966). Immunochemical characterization of polyribonucleotides. Science, *132*:773–775.

Plescia, O. J., W. Braun, and N. C. Palczuk (1964). Production of antibodies to denatured deoxyribonucleic acid (DNA). Proc. Natl. Acad. Sci. U. S., *52*: 279–285.

Sela, M., H. Ungar-Waron, and Y. Shechter (1964). Uridine specific antibodies obtained with synthetic antigens. Proc. Natl. Acad. Sci. U. S., *52*:285–292.

Simon, M. I., and H. Van Vunakis (1962). The photodynamic reaction of methylene blue (MB) with deoxyribonucleic acid (DNA). J. Mol. Biol., *4*:488–499.

Van Vunakis, H., E. Seaman, L. Kahan, J. W. Kappler, and L. Levine (1966). Formation of an adduct with tris (hydroxymethyl) aminomethane during the photooxidation of deoxyribonucleic acid and guanine derivatives. Biochemistry, *5*:3986–3991.

Wasserman, E., and L. Levine (1960). Quantitative micro complement fixation and its use in the study of antigenic structure by specific antigen-antibody inhibition. J. Immunol., *87*:290–295.

DISCUSSION

Chairman: M. HEIDELBERGER

PLESCIA: I think I will start the discussion by making a general observation. It is clear from what we have heard this morning that the production of antibodies specific for nucleotides is no longer a problem. Rather the problem is how to produce antibodies of defined specificity. Therefore, I would like to pose the question of how best to prepare such antibodies. How good do you think are the prospects of accomplishing this task? How likely is it that we may produce antibodies specific for a defined sequence of nucleotide residues so that we can use them as meaningful tools?

L. LEVINE: A number of years ago I thought we could do this, but now I don't think it is possible or I think it is very unlikely that we shall be able to get an antibody that will be unique for any particular DNA. Dr. Kabat has provided data on the size of the antigenic determinant in polysaccharides, and David Stollar and I have found the size of antigenic determinants of polynucleotides to be essentially the same, namely a pentanucleotide. This suggests that an antibody may recognize maybe two codons. The probability for the occurrence of any such two codons in sequence is sufficiently high to preclude the use of such antibodies for distinguishing one DNA from another. Of course with DNAs that have unique bases, such as glucosylated 5-hydroxymethyl cytosine, antibodies specific for such DNAs can be produced.

HEIDELBERGER: Earlier results on the specificity of small segments of polysaccharides might make a small contribution to this problem. Dr. Levine brought out that his sera will recognize ribose or ribose with phosphate linked in a definite position. My suggestion is that in view of the heterogeneity of antibodies, a comparison of two different antisera in which the heterogeneity is different, might specify the location of some of the nucleotides. The point that I am thinking of particularly is the case of two different antipneumococcal type VI antisera; one of them was very good at recognizing non-reducing end groups of galactose although the type VI determinant does not have such end groups. It has galactose with the 2 position occupied by phosphate and most of the rest of the molecule sticks out and behaves immunologically more or less like an end group. We had another type VI anti-pneumococcal serum which was very much better in recognizing the multiples of 1,3-linked L-rhamnose. From differential results with two antisera of this sort it might be possible to specify something about the arrangement and number of nucleotides in DNA.

GARRO: Concerning Dr. Plescia's comment, we have attempted to use specific anti-pyrimidine antibodies to differentiate between DNAs on the basis of their base composition. At first we were very encouraged by the results. When we took two DNAs which varied greatly in their base composition, for example *Micrococcus lysodeikticus* whose DNA has a GC ratio in the 70% range, and calf thymus DNA which has about 50% GC, the types of curves we obtained showed that the anti-C serum would react best with the high GC DNA, and less with the low GC DNA. However, lately we have been having problems. When we tried to fit more than two or three DNAs into this pattern, they did not all fit. However, all of these studies have been done by complement-fixation, and we are not sure whether the complement-fixing system may not be sensitive to differences in DNA other than base composition. We intend now to look at DNAs of different base composition, using specific anti-pyrimidine antibodies and other methods, for example the precipitin method.

PARKER: With respect to the first of Dr. Plescia's questions regarding the best method of immunization, I think that this is entirely empirical at the present time. I think that the method of immunization will have to depend on the application that you are interested in. We don't know which areas of DNA are accessible to the immunological apparatus in terms of antibody formation. I think we simply have to try different immunogens to see which seems to work best for the particular application that we have in mind. I would share Dr. Levine's pessimism as to the possibility of recognizing different DNAs by antisera, even by absorbed antisera.

BARNETT: We had the opportunity to look at rabbit antibodies induced by single-stranded calf thymus DNA complexed with methylated rabbit serum albumin, methylated bovine serum albumin or linked by carbodiimide to rabbit albumin. One such group of antibodies did not react in complement-fixation with single-stranded T_4 phage DNA although it reacted very well with single-stranded calf thymus DNA and did not react in radioimmunoelectrophoresis with C^{14}-labelled single-stranded T_4 phage DNA. Using such rabbit antibodies we have been able to detect a precipitable antigen in certain human sera; this antigen was not detectable by antibodies in sera from human patients with LE. We have evidence that these results are not just quantitative, because the human sera that failed to detect the precipitable antigen did contain antibodies that would precipitate single-stranded calf thymus DNA. So we now have about 15 sera from patients with LE that appear to have an antigen precipitable by rabbit anti-DNA antibody. This provides some optimism regarding the use of antisera to detect unique species of DNA.

STOLLAR: I would like to make just a couple of comments that might temper the latest swing in optimism. I think there is an as yet unmentioned complicating factor in the use of antibodies specific for a given base in the study of specific sequences. For example, denatured DNA is usually material that has been boiled and rapidly cooled and is used at a temperature at which it exists as a randomly coiled structure. It would seem that in this state factors other than just the base composition will determine the overall configuration. I think this is clear from Dr. Garro's comments and also from an examination of an anti-adenosine and an anti-guanosine serum in our laboratory, and from studies with an anti-thymidine serum in Dr. Levine's laboratory. The reactivity of DNA samples of widely differing base composition went in the same direction for all these three sera. This suggests that something as yet unknown, perhaps the degree of collapse in solution at the temperatures at which the DNA samples are used, determines their configuration and the degree of reactivity. A second comment in regard to examining sera by inhibition: I also have noticed a phenomenon similar to the enhancement described by Dr. Plescia and I must admit that I did not pursue this at all and remain completely puzzled by it. I wonder about the specificity, however, since this was an example of *E. coli* RNA enhancing the complement-fixation by calf thymus DNA.

One question I would like to ask Dr. Lacour concerns the antiserum which reacted with native DNA from two species and did not react with native DNA from calf thymus. I am wondering what she feels might be the basis for this specificity and whether there is any possibility that in fact there was a small amount of denatured DNA in the native samples from the species that did react. We have found, for example, that some sera which react with denatured DNA only, will react with samples of presumably native DNA which have as little as 1% denatured material in them, and one can tell this from the quantitative relationship.

LACOUR: I don't think there was any denatured DNA in our native DNA preparation. This was carefully checked. In fact much less of the native DNA is required than of the denatured DNA to get this reaction. An explanation is very difficult because we thought at first that base composition might explain it, but since we got positive reaction with crab DNA that has only 7% GC and with *M. lysodeikticus* DNA that has

72% GC, it seems that the overall base composition cannot explain the non-reactivity of a DNA. We are obviously dealing with a mixed population of antibodies against double-stranded and against single-stranded DNA molecules, and we shall have to conduct some quantitative investigations to determine the exact content of antibodies that can react with double-stranded and single-stranded DNA, respectively.

MAURER: I would like to ask whether consideration has been given to tolerance studies in order to increase the specificity of the antisera. This question is particularly relevant to Dr. Lacour's work. Could injections of single-stranded DNA and subsequent immunization with double-stranded or triple-stranded materials eliminate the possibility that the antisera contained antibodies to denatured single-stranded DNA?

LACOUR: I have never tried this. However, we have some observations that suggest that some rabbits are tolerant towards certain polynucleotides. We have, for instance, immunized rabbits with poly-A and found that a large percentage of these rabbits do not produce antibodies against poly-A.

GARRO: Let me point out that our experience shows that different rabbits respond differently. Some rabbits respond with a great deal of cross-reaction between all the nucleoside conjugates, some respond with very strict specificity for the immunogen. This is biological variability. I would like to suggest one thing in regard to testing antibodies to polyribose phosphate as the backbone of these polynucleotides. It was shown a long time ago that the type-specific substance of *Hemophilus influenzae* type B is a highly cross-linked polyribose phosphate, and this is the type-specific substance for the capsule. It is highly cross-linked but it is susceptible to RNAase digestion and it may be possible to get out small polymers of polyribose phosphate by treating the capsular material of Hemophilus which may then be useful for inhibition studies or for conjugation with methylated bovine serum albumin for the production of antibodies.

GLICK: I would like to ask Dr. Plescia with respect to his mechanism of complementarity whether he has tried to react antibodies against the various trinucleotides with much larger artificial polymers of nucleotides instead of just with DNA or RNA. Also, it should be possible to test your theory by using an RNA of known nucleotide sequence, e.g., one of the soluble RNAs such as the one analyzed by Holley.

PLESCIA: So far we have not had an opportunity to examine the two points you raised.

SEHON: I would like to ask one general question concerning oligonucleotides and mononucleotides as haptens. What I would like to ask here is what is the role of the carrier? The second question is directed to Dr. Plescia and is a comment on his work. He observed no inhibition with the mononucleotides conjugated to rabbit serum albumin. This was a chemical conjugation, in all probability a conjugation that differed from that resulting when methylated bovine serum albumin is used. Therefore, a different portion, perhaps a different size and configuration of the nucleotide may have faced the antibody. This could explain the lack of reaction between the antiserum to a trinucleotide and mononucleotide conjugated to rabbit serum albumin.

PLESCIA: The mononucleosides, the dinucleotides and the trinucleotides were conjugated to rabbit serum albumin precisely in the same fashion as was the mononucleotide conjugated to the bovine serum albumin as the immunogen. Therefore, I do not think that this is the reason why there is lack of reactivity. In regard to the first question, I do not think that the role of the carrier in producing antibodies to the mono-, di-, and other oligonucleotides is any different than it is in producing antibodies to the many haptens investigated in the past, and therefore it is a subject better left to the session that deals specifically with the carrier (p. 277).

ESTRADA-PARRA: I want to comment on Dr. Levine's remarks regarding inhibition studies with ATP and DNP. We were interested in having antibodies that react with co-enzymes. Since we know that adenosine is present in some of these co-enzymes, we conjugated 5′-adenylic acid to proteins and obtained antibodies against adenylic acid. These react also with adenosine. To our surprise, when we tried inhibition with ATP and DPN, we obtained only 30% inhibition compared with 75% or more inhibition with ADP or adenylic acid. On looking in the literature we found that Szent-Georgy had postulated that some of these compounds, especially ATP, are folded in such a way that the last phosphate approaches the base and this is probably the reason for our having gotten less inhibition than we expected on a molar basis. In other words, there are interactions in the co-enzyme structure between pyrimidines and other bases. This might explain why there is less inhibition in the case of ATP or DPN compared to adenylic acid on a molar basis. ADP and AMP inhibit the same but ATP inhibits much less.

GOTTLIEB: I would like to suggest that there is a rather direct way of testing Dr. Plescia's intriguing model of the interaction of specific antibody with DNA and the trinucleotides. My suggestion would be to use equilibrium centrifugation in a density gradient of cesium chloride. It is known that native DNA has a distinctly lighter density than denatured DNA in cesium chloride gradients. Therefore, Dr. Plescia's model predicts that there should be a shift from the density of denatured DNA to either the density of native DNA or, by virtue of the presence of globulin in the complex, to densities lighter than native DNA. It also may turn out that the extent of the shift is related to the amount of protein in the complex.

LANDY: In these discussions we have paid a great deal of attention to the nature of the structures in the immunogen and the heterogeneity of the antibodies obtained. Aside from the comments by Dr. Garro on the extreme variability of the rabbit responsiveness, we have ignored a major factor and that is the use of random-bred rabbit populations. I assume that such rabbits were used throughout, and it is perhaps inevitable that one would see a great diversity of responses to a given product, let alone the range of the products with which you are working as an immunizing agent. We have examples of this problem in the work with inbred lines of guinea pigs by Benacerraf, Maurer, Ben-Ephraim, and many others. Their studies have shown that responses can vary all the way from high responsiveness to no responsiveness, the magnitude being correlated with the genetic constitution of the host. Perhaps this problem can be solved in the future through the availability of inbred rabbits. Programs to provide such animals are currently underway in Bethesda, Bar Harbor, and other locations.

TANENBAUM: Let me come back to the problem of using complement-fixation for testing interactions between specific immunoglobulins and informational macromolecules. I should like to point out that in studying anti-hapten antibodies which cross-react with RNA, Dr. Karol and I noticed that complement-fixation occurred between RNA cross-reactive immunoglobulins and ribonucleotides and ribonucleosides as well as oligonucleotides and oligonucleosides. Therefore, if one uses such materials in order to study inhibition of complement-fixation by haptens, one might actually observe, as Dr. Plescia pointed out, a certain amount of enhancement of complement-fixation. This enhancement may be due to the combination of the presumed inhibitor with the antibody-combining region, leading to fixation of complement, and this, of course, is going to complicate the picture inordinately and make the interpretation of experimental results rather difficult.

HEIDELBERGER: Doesn't combination with the antibody site usually result in inhibition?

TANENBAUM: If you omit the antigen and merely mix the nucleotide or the inhibitor with the antibody you will, in certain cases, observe fixation of complement.

HEIDELBERGER: Well that might be a serious objection to all inhibition experiments.

TANENBAUM: I wouldn't say that this is necessarily so, because the reaction of antibody with hapten does not always lead to complement-fixation, but it certainly is something that one must keep in mind when one is dealing with ribonucleotides and ribonucleosides. I am not sure that this phenomenon would be observed, for example, with antibody to 2,4-dinitrophenol hapten or something like that.

L. LEVINE: Obviously, in any complement-fixation inhibition you run an inhibitor control, that is, inhibitor plus antibody and complement. There is no fixation of complement under such circumstances; we have ascertained this with tetranucleotides, pentanucleotides, and hexanucleotides. This is true not only for oligonucleotides, it is also true for oligomers of glutamic acid and lysine. If you see a reaction between the inhibitor and the antibody, such as we have seen with polylysine, it is just a charge effect and such data cannot be used.

TANENBAUM: I wasn't trying to negate any of the experimental results that have been presented nor did I intend to imply that adequate controls have not been used. I just wanted to point out that under certain circumstances, when many reactants capable of interactions are mixed together in the presence of complement, results may be obtained that are difficult to interpret.

PLESCIA: My only comment at this point would be that we must remember that polynucleotides and oligonucleotides are not just ordinary substances. They are capable of unique interactions. For example, in our tests with oligonucleotides, denatured DNA, and antibody, it made little difference whether the three were added simultaneously or whether one added the third some time after the other two. This suggests that one is dealing with a three-component system, any two of which by themselves do nothing in terms of complement-fixation. But you put that third party in there and then fireworks begin, and you begin to see complement-fixation.

HAUROWITZ: We heard so much about the conformation of the antigen as being important in antibody formation. The conformation is of course that of the polyribosephosphate, and I wonder whether polyribosephosphate bound, for instance, to different carriers like methylated bovine serum albumin and other protein carriers would be antigenic. Would the antibody formed be identical under these conditions? If the response to the different complexes would be different, we would suspect that it may be as a result of differences in the conformation of the polyribosephosphate. I don't know whether bases, which as such have no affinity for the protein, can be responsible for the antigenic activity or whether the whole thing is not just a reflection of the conformation of the polyribosephosphate.

PLESCIA: I think that the conformation of the polyribosephosphate depends very much upon whether or not it exists merely as a polyribosephosphate or whether, attached to this backbone, there are bases capable of interacting with each other. Therefore, it may not be sufficient to produce an antiserum against the polyribosephosphate backbone, if this is indeed possible. One should compare the specificity of an antiserum prepared against backbone with those prepared against oligonucleotides.

SELIGMANN: I do agree with Dr. Lacour who emphasized the importance of the molecular conformation of the antigenic DNA. She is probably dealing in her antisera with a mixture of different antibodies of different specificities. I would like to ask her: Is the reaction with native DNA inhibited by single-stranded DNA and have you any quantitative data regarding this? And if it is inhibited by denatured DNA, would it also be inhibited by samples of DNA that do not give positive reactions in the native state?

LACOUR: I am sorry but I have no data regarding this. I have tested only inhibitions by single-stranded polynucleotides, and these do not inhibit reactions with native DNA.

BAUTZ: I had the impression from Dr. Plescia that the antibody recognizes the overall conformation of the backbone of the trinucleotide. Now, if one accepts this model, I am concerned about the possible difficulty in having the antibodies still recognize the trinucleotide when it is hydrogen-bonded to DNA. In order to form the hydrogen bond, the trinucleotide would probably have to undergo a conformational change and would therefore have a different overall conformation. I was wondering, Dr. Plescia, whether you have given any thought to this possibility.

PLESCIA: Yes, indeed we have. You remember that I stated earlier that it makes no difference whether the three components are added simultaneously or not. I think that perhaps the antibodies begin to react only when the trinucleotide has interacted with the DNA, and perhaps the antibodies are recognizing this particular conformation.

BAUTZ: But this would imply that when the antibody is formed the trinucleotide is processed in a hydrogen-bonded form.

PLESCIA: Yes, this would imply this.

PRESSMAN: In connection with Dr. Plescia's increased complement-fixation by the trinucleotides, I wonder if the possible participation of other proteins in the system has been considered. I ask this because when we used whole sera rather than purified globulins in precipitation reactions with antibodies to the para-azobenzoate, we observed that the presence of aliphatic carboxylic acid in the serum resulted in increased precipi-tation. This increased precipitation did not occur when purified globulins were used. I wonder if you have considered the possibility that other serum proteins may be responsible for the increased complement-fixation that you observed.

PLESCIA: Yes, we have considered this. This does not appear to be the explanation, since normal rabbit serum gives no reactions and the extent of enhancement differs for different antisera.

DREYER: The antibodies that have been discussed here should be biological dynamite. They are directed against co-factors and other vital materials. What happens to the rabbits? Do they show any signs of deficiencies, diseases, or anything of that sort?

LACOUR: Nothing seems to happen to the rabbits, except when we use antigens that are not sterile.

KLEINSCHMIDT: After injection of DNA or RNA, we found high levels of DNAase and RNAase in sera. I would, therefore, like to ask: What happens to native DNA or native RNA during such enhanced production of DNAase or RNAase?

PLESCIA: You are asking what the likelihood might be that the injected DNA becomes depolymerized in the course of immunization. Well, one of the virtues of the methylated bovine serum albumin method is that once native or denatured DNA is complexed with it, it becomes resistant to the action of DNAase, and this may well be one of the reasons why the methylated bovine serum albumin functions as it does.

CINADER: It is interesting to hear that the effect of methylated albumin may possibly be due to an effect on resistance to nucleases, but probably more so to an effect on the configuration. Thus, if the substance is non-antigenic and it undergoes a configurational change, it may become immunogenic. Now there's more and more evidence that antibody can also superimpose configurational changes, as exemplified by activating antibodies to enzymes. Therefore, I would like to suggest that one attempt to use, as potential carrier, antibodies that have been obtained against nucleic acid.

GARRO: One final word regarding nuclease activity. Dr. Beiser, in studying antisera to dinucleotides conjugated to proteins, observed what might be an effect of nucleases in the serum. If you have a purine-pyrimidine dinucleotide conjugated to protein, the resulting antisera often have higher specificity for the internal pyrimidine rather than for the external base. What may have occurred is that the serum nucleases may have split the dinucleotide, leaving only the pyrimidine attached to the protein, and this could account for the observed specificity.

OLIGO- AND POLYNUCLEOTIDES AS HAPTENS (PART II)

FORMATION AND ISOLATION OF ANTIBODIES SPECIFIC FOR NUCLEOTIDES

P. GRABAR, S. AVRAMEAS, B. TAUDOU, AND J. C. SALOMON

Institut de Recherches Scientifiques sur le Cancer
(94) Villejuif, France

The first reports on the formation of antibodies reacting with nucleic acids were published in 1957 (Seligmann, 1957; Ceppellini *et al.*, 1957; Robbins *et al.*, 1957), and ever since many publications concerned with this problem have appeared. Some of the important activities of such antibodies have been discussed previously by one of us (Grabar, 1964) and by others. It seemed to us particularly interesting to investigate activities of antibodies that are strictly specific for given nucleosides, and in the last few years we have made attempts to prepare and isolate such antibodies.

I. MATERIAL AND METHODS

A. PREPARATION OF NUCLEOSIDE-PROTEIN CONJUGATES

P-aminobenzoate conjugates of thymidine and uridine were prepared according to previously published techniques (Coat *et al.*, 1965). The 3′p-aminobenzoate conjugate of thymidine (I) was prepared by treating tritylated O–5′ thymidine with p-nitrobenzoyl-chloride. After detritylation and subsequent reduction the 3′p-aminobenzoate of thymidine was obtained.

2,6 dimethyl-4 amino-benzoyl O–5′ uridine (II) was prepared in a similar manner by treating isopropylidene-uridine with 2,6 dimethyl-4 nitrobenzoyl-chloride.

1 meq of the nucleosides derivatives was dissolved in a minimal quantity of water to which 3 meq of HCl were added. The solution was cooled at 4°C and 1.1 meq of $NaNO_2$ was added. After half an hour, the diazo-

tized derivatives were added to 10 ml of a cold solution of 2 M carbonate buffer pH 9,5 containing 0.02 meq of bovine serum albumin. The coupling reaction was allowed to proceed at this pH for 30 minutes; the solution was then adjusted to pH 8 and left at 4°C overnight. After precipitation with ammonium sulfate and dialysis, the nucleoside-protein conjugate was lyophilized. Pyrimidine is released when such azoproteins are hydrolyzed with 0.5 N perchloric acid for forty-five minutes at 70°C. By use of this procedure it was found that 33 molecules of thymidine and 25 molecules of uridine were coupled per molecule of bovine serum albumin.

Nucleosides also were coupled to other proteins, such as human IgG, horse radish peroxidase, ovalbumin, or glucose oxidase; this was done by the technique mentioned above but different proportions of nucleoside to protein were used.

B. Immunization of Rabbits

Rabbits (1 year old Giant Holland rabbits) were injected intradermally in the hind foot pad with 20 mg of the thymidine-bovine serum albumin complex in 0.7 ml of 0.15 M NaCl and 0.3 ml of complete Freund's adjuvant. Immunization of the animals by intramuscular and intravenous routes was continued for two months, each animal receiving a total of 60 mg of antigen. The animals were bled 5 days after the last intravenous injection.

C. Immunological Methods

(1) Passive cutaneous anaphylaxis was carried out according to the technique of Ovary (1964).

(2) Passive haemagglutination techniques were performed as described elsewhere (Kabat, 1961). Sensitization of red blood cells with thymidine was accomplished either by diazotization of 3′p-aminobenzoyl thymidine and direct coupling to red cells or by coupling ovalbumin thymidine to red cells via tetraazotized o-dianisidine. Heat-denatured DNA was coupled to red cells using the same compound (tetraazotized ortho-dianisidine).

(3) Double diffusion in agarose gel was performed according to the technique of Ouchterlony (1948).

(4) Immunoelectrophoretic analysis in 0.8% agarose gel was performed according to Grabar and Williams (1953).

(5) Identification of antibodies: antibodies present in whole serum or isolated by immunoadsorbents were identified, after immunoelectrophoresis, by a method that employs labeling of antigens with enzymes.

In this case thymidine was coupled to trypsin (Avrameas and Uriel, 1966).

(6) Quantitative precipitation tests were carried out according to Heidelberger and Kendall (1935). Hapten inhibition of the precipitin reaction was done in the zone of equivalence.

(7) Preparation and use of the immunoadsorbents: the immunoadsorbent used for the purification of the anti-thymidine antibodies was prepared by coupling the diazotized 3'p-aminobenzoyl thymidine to a water-insoluble polymer of rabbit serum albumin. The water-insoluble protein polymer was obtained by polymerization of the rabbit serum albumin at pH 4,8, using ethyl chloroformate as a polymerizing agent (Avrameas and Ternynck, 1967a). For the coupling of thymidine, one gram of the protein polymer was suspended in 10 ml of cold carbonate buffer (2 M, pH 9.5) and 1 meq of the diazotized 3'p-aminobenzoyl thymidine, prepared as described above, was added to the suspension with stirring. The reaction was allowed to proceed at pH 9.5 for 30 minutes. The solution then was adjusted to pH 8 and stirred gently at 4°C overnight. The thymidine-protein polymer immunoadsorbent was washed thoroughly with successive solutions of saline and 2 M sodium iodide (Avrameas and Ternynck, 1967b) until the eluates gave zero optical density at 280 mμ. The succeeding steps, for the absorption and elution of antibodies, were performed as described in a prior publication (Avrameas and Ternynck, 1967a).

The immunoadsorbent used for the purification of the anti-DNA antibodies was prepared by coupling the heat-denaturated DNA to the water insoluble rabbit serum albumin via a tetraazotized o-dianisidine. For the coupling, 200 mg of tetraazotized o-dianisidine were added to 5 ml of 0.5 M bicarbonate solution containing 200 mg of protein polymer and 100 mg of heat denaturated DNA. The reaction mixture was allowed to stand at room temperature overnight and then was treated and used as mentioned above for thymidine immunoadsorbents.

II. RESULTS AND DISCUSSION

A. Reaction of the Anti-thymidine Antibodies with Thymidine

(1) Sera of rabbits immunized with thymidine-BSA gave positive reactions when tested by passive cutaneous anaphylaxis using thymidine-ovalbumin as antigen.

(2) In passive haemagglutination tests, using erythrocytes coupled with thymidine, positive reactions were obtained with the anti-thymidine-BSA sera (Table 1).

(3) In double diffusion tests these sera gave one, and sometimes two, precipitin lines with thymidine conjugated to different proteins (Fig. 1A).

Table 1

TITERS OF ANTI-THYMIDINE ANTIBODIES IN TESTS WITH HEAT-DENATURED DNA OR RED CELLS SENSITIZED WITH THYMIDINE-OVALBUMIN

Antiserum Used	Titers in Tests with Ovalbumin-Thymidine	DNA
Whole anti-thymidine-rabbit serum albumin	1:64	1:32
Purified antibodies *	1:6144	1:640

* 20 mg of protein in 5 ml physiological saline; isolated from 50 ml of whole anti-thymidine-albumin rabbit serum.

The precipitation was at least partially, and sometimes completely, inhibited by an incorporation of thymidine into the gel. Similar inhibition was also obtained with uridine, but only if much larger quantities were used; other nucleosides were inactive.

(4) Quantitative estimations of antibodies in the immune sera using a thymidine-ovalbumin complex (containing 10 haptenic groups per mole-

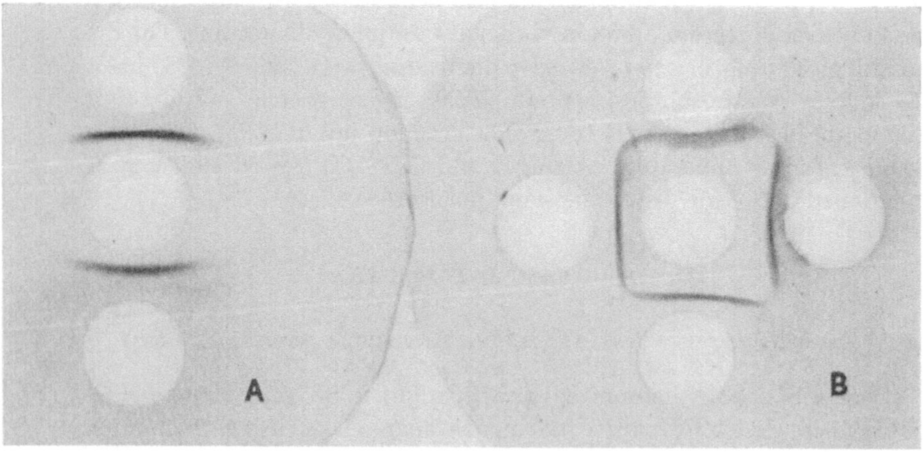

Fig. 1. A: Reaction of anti-thymidine antibodies with thymidine coupled to ovalbumin (center well, 2 mg/ml; top well, anti-thymidine-rabbit serum albumin complex; lower well, purified anti-thymidine antibodies). **B:** Reaction of anti-thymidine antibodies with heat-denatured DNA (center well, 1 mg/ml; upper and lower wells, whole anti-thymidine-rabbit serum albumin; left and right wells, purified anti-thymidine antibodies).

cule) showed the presence of 0.2–0.3 mg of anti-thymidine antibodies per ml.

Inhibition of the precipitation reaction, by different nucleosides at the zone of equivalence, using, whenever possible, a 0.1 M concentration or saturated solutions of nucleosides, was negative with cytidine, inosine, adenosine, guanosine, deoxyuridine and xanthosine. Similar negative results were obtained with deoxyribose and with ribose. Uridine inhibited, but only at a concentration of 0.2 M, whereas thymidine produced 50% inhibition even at a concentration of 0.001 M.

From these results we conclude that the antibodies obtained were specific for the entire molecule of thymidine. Analogous results have recently been published by Ungar-Waron *et al.* (1967).

A series of experiments was undertaken to determine to which family of immunoglobulins these antibodies belonged. The insoluble rabbit serum albumin-thymidine immunoadsorbent was used for the absorption of anti-thymidine antibodies, and after washing, antibodies were eluted either (1) with solutions of thymidine at various concentrations, or (2) with potassium iodide at 2 M concentration (Avrameas and Ternynck, 1967b). Ten to 15 mgms of protein have been isolated from the immunoadsorbent from 50 ml of the whole rabbit anti-thymidine serum. No protein can be extracted from the regenerated immunoadsorbent when it is treated a second time with the supernatant of the first adsorption. Immuno-electrophoretic analysis of these eluates has shown that the antibodies belong to IgG, IgA and IgM-families (Fig. 2). By the use of thymidine coupled with trypsin it was shown that all precipitation lines possess antibody activity. In some cases 3 or 4 precipitin arcs were observed in the IgG zone and sometimes a precipitation line was detected even in the α_2-zone (Fig. 2).

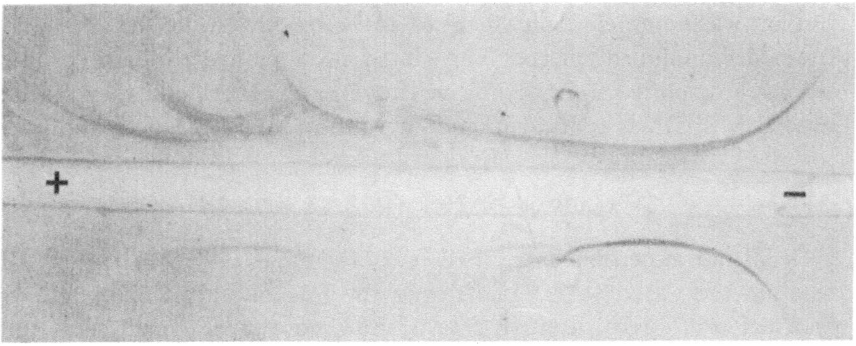

Fig. 2. Immunoelectrophoretic analysis of whole anti-thymidine-rabbit serum albumin (top) and purified antibodies (bottom). Preparations were developed by a horse anti-rabbit serum.

B. Reactions of Anti-thymidine Antibodies with DNA

Anti-thymidine sera were tested using various preparations of DNA (calf-thymus, salmon sperm, *Esch. coli,* etc.). Positive haemagglutination tests were obtained with erythrocytes sensitized with denatured DNA. Quantitative precipitin curves obtained with denatured DNA have shown that the precipitates were not soluble in excess antigen; this has also been observed by other authors (Stollar and Levine, 1963; Buttler *et al.,* 1965). One or two precipitin lines were formed in gel-diffusion with heat-denatured DNA (Fig. 1B), whereas native DNA had to be used in very high concentrations (5–10 mgrs par ml) in order to obtain even a slight precipitation. Treatment of denatured DNA by DNAase diminished the intensity of the precipitation, but even a prolonged treatment (24 hours) did not abolish the precipitation completely. Similar findings were obtained when thymidine was previously incorporated into the gel. A complete inhibition of precipitation was obtained with thymidylic acid. The precipitin lines could be solubilized when the gel was incubated at room temperature with 0.5 M phosphate buffer at pH 7.

These results indicate that in the reaction of anti-thymidine antibodies with DNA the charges and also the structure of the hapten play an important role. These observations are analogous to those obtained with antibodies that had been formed in animals immunized with ribosomes (Barbu *et al.,* 1963).

Our immune sera, when absorbed on protein-polymer-thymidine immunoadsorbent, do not react with DNA, which proves that they contain no antibodies for other groups on the DNA. However, antibodies eluted from the immunoadsorbent precipitate DNA. Antibodies reacting with DNA were also isolated using an immunoadsorbent formed by an insoluble polymer of rabbit serum albumin and DNA. Ten to 15 mgms of protein was obtained from 50 ml of antiserum by elution of antibodies from the immunoadsorbent. The eluted proteins had the same properties as those eluted from thymidine-protein complexes, and they reacted as well with DNA as with thymidine-ovalbumin (Table 1).

C. Reaction of Normal Sera with DNA

In all our experiments we used sera of normal rabbits, or pre-immunization sera, as control sera. During the first period of our work, all "normal sera" gave negative results. But more recently, all sera from so-called "normal" rabbits have reacted with DNA in gel precipitation and in passive haemagglutination tests. With one exception, sera from different rabbits still reacted with DNA even after adsorption on thymidine-protein-polymer immunoadsorbent. It seems then that most of the antibodies that react with DNA and are present in "normal" sera differ from those formed in animals after their immunization with nucleoside derivatives.

It is possible that the observed reactions of "normal" sera may be due to some prior infections of the experimental animal. In order to verify this assumption, experiments have been performed on axenic (germfree) mice. It could be observed that most conventionally grown mice had anti-DNA antibodies, whereas axenic mice did not. Four groups of axenic mice of the same age have been utilized. One was immunized with the thymidine-BSA complex, the second with BSA alone, the third with killed Gram-negative bacteria, and the fourth with a complex of methylated BSA and DNA (Plescia *et al.*, 1964). These substances were injected with an incomplete Freund's adjuvant (without bacteria) by intraperitoneal and sub-cutaneous routes for 45 days. A group of untreated axenic mice served as a control. The serum of the latter animals and the sera of mice injected with BSA alone did not react with denatured DNA. In contrast, sera of mice injected with killed bacteria, or with the methylated BSA-DNA complex, or with the BSA-thymidine complex reacted with denatured DNA in double-diffusion and in passive haemagglutination. Particularly strong reactions were observed with mice immunized with killed bacteria.

Thus we may conclude that the reactions observed with sera from so-called "normal" animals are in fact due to their prior contact with DNA-containing agents, and it seems justified to recommend that appropriate controls be included in experiments concerned with the production of antibodies to nucleosides. Such controls may avoid erroneous interpretations of reactions due to the presence of antibodies formed in the experimental animals in response to DNA of infectious origin.

III. SUMMARY

(1) Antibodies specific for thymidine have been obtained. Their reactions are inhibited by small quantities of thymidine and by much higher concentrations of uridine, but not by other nucleosides.

(2) Antibodies to thymidine isolated from sera of hyperimmunized rabbits belong to IgG, IgA and IgM families.

(3) The antisera react well with denatured DNA; absorption of the sera by a complex of thymidine and polymerized proteins eliminates this reactivity.

(4) Sera of so-called "normal" animals frequently react with DNA. Experiments with axenic (germfree) mice, whose sera do not react with DNA, have shown that these mice form antibodies to DNA when immu-

nized with DNA-containing complexes or with killed bacteria. Thus, such DNA-reactive antibodies when present in "normal" sera may be the consequence of prior apparent or inapparent infections.

References

Avrameas, S., and J. Uriel (1966). Méthode de marquage d'antigène et d'anticorps avec des enzymes et son application en immunodiffusion. C. R. Acad. Sci. Paris, *262*:2543–2545.

Avrameas, S., and T. Ternynck (1967a). Biologically Active Water-insoluble Protein Polymers. I. Their use for isolation of antigens and antibodies. J. Biol. Chem., *242*:1651–1659.

Avrameas, S., and T. Ternynck (1967b). Use of Iodide Salts in the Isolation of Antibodies and the Dissolution of Specific Immune Precipitates. Biochem. J., *102*:37c–39c.

Barbu, E., G. Quash, and J. P. Dandeu (1963). Spécificité des Anticorps anti-RNA. Inst. Pasteur, *105*:849–865.

Braun, W., O. J. Plescia, and N. Palczuk (1959).: in Proc. 5th Intern. Meet. Biol. Stand., Jérusalem: p. 367.

Butler, Jr., V. P., M. D. Stuart, W. Tanenbaum, and S. M. Beiser (1965). A Study of the Cross-Reactivity of Antipurin-6-Oyl Serum with Deoxyribonucleic Acid (DNA). J. Exp. Med., *121*:19–38.

Ceppellini, R., E. Polli, and F. Celada (1957). A DNA Reacting Factor in Serum of a Patient with Lupus Erythematosus Diffusus. Proc. Soc. Exptl. Biol. Med., *96*:572–574.

Christian, C. L., A. R. Desimone, and J. L. Arbuzzo (1965). Anti-DNA Antibodies in Hyperimmunized Rabbits. J. Exp. Med., *121*:309–321.

Coat, J. P., S. David, and J. C. Fischer (1965). Préparation de combinaisons protéines-nucléosides en vue d'études immunochimiques. Bull. Soc. Chim. France, 2489–2493.

Grabar, P., and C. A. Williams (1953). Méthode permettant l'étude conjuguée des propriétés électrophorétiques et immunochimiques d'un mélange de protéines. Application au sérum sanguin. Biochim. Biophys. Acta, *10*:193.

Grabar, P. (1964). Quelques Aspects du Problème des Anticorps Anti-Acides Nucléiques. Nouv. Rev. Fr. d'Hémat., *4*:729–734.

Heidelberger, M., and F. E. Kendall (1935). A quantitative theory of the precipitin reaction. III. The reaction between crystalline egg albumin and its homologous antibody. J. Exp. Med., *62*:697.

Kabat, E. (1961), in Kabat, E., and M. Mayer. Experimental immunochemistry. Charles C Thomas, Springfield, Illinois.

Ouchterlony, O. (1949). Antigen-Antibody Reaction in Gel. Arkiv. Kemi., *26B*:14.

Ovary, Z. (1964). Passive Cutaneous Anaphylaxis in Immunological Methods (J. F. Ackroyd, ed.), Blackwell Scient. Public., Oxford, p. 259.

Plescia, O. J., W. Braun, and N. C. Palczuk (1964). Production of Antibodies to Denatured Deoxyribonucleic Acid (DNA). Proc. Nat. Acad. Sci. N. Y., *52*: 279–285.

Poverenny, A. M., A. S. Saenko, N. V. Stefani, and N. T. Saenko (1967). The

Immunochemical Methods of Determination of Polymeric DNA. Vopr. Med. Chim. (URSS), *13*:203–7.

Robbins, W. C., R. H. Holman, H. R. G. Deicher, and H. G. Kunkel (1957). Complement Fixation with Cell Nuclei and DNA in Lupus Erythematosus. Proc. Soc. Exptl. Biol. Med., *96*:575–579.

Seligmann, M. (1957). Mise en évidence dans le sérum de Malades atteints de Lupus Erythémateux Disséminé d'une Substance déterminant une Réaction de précipitation avec l'Acide Désoxyribonucléique. Compt. Rend., *245*: 243–245.

Stollar, D., and L. Levine (1963). Antibodies to Denatured Deoxyribonucleic Acid in Lupus Erythematosus Serum. IV. Evidence for purine determinants in DNA. Arch. Biochem. Biophys., *101*:417.

Timakov, V. D., A. G. Skavronskaya, N. B. Borisova, and L. A. Zamchok (1963). Antigenic Properties of Desoxyribonucleic Acid of S. Typhimurium. Zh. Mikrobiol. Epidemiol. i. Immunobiol., *40*:5.

Ungar-Waron, H., E. Hurwitz, J. C. Jaton, and M. Sela (1967). Antibodies elicited with conjugates of nucleosides with synthetic polypeptides. Bioch. Biophys. Acta, *138*:513–531.

THE REACTION OF ANTI-DEOXYRIBONUCLEIC ACID SERA WITH NUCLEOSIDE AND NUCLEOTIDE CONJUGATES *

SERGIO ESTRADA-PARRA AND EUGENIA OLGUIN-PALACIOS

Departamento de Microbiologia, Escuela Nacional de Ciencias Biologicas, Instituto Politecnico Nacional, Mexico, D.F., Mexico

As is well known, the specificity and sensitivity of immunochemical reactions have been of great value in the study of polymers such as polysaccharides and proteins (Heidelberger, 1960, 1967). Also well known is the role of nucleic acids as carrier of genetic information and as a participant in protein synthesis and other biological events (Vogel *et al.*, 1963). There is, therefore, considerable interest and value in having available antibodies capable of reacting specifically with nucleic acids. Recently, the production of such antibodies has been described; it is based on the use of nucleic acid-protein complexes as antigens or the use of conjugation of nucleic acid derivatives and macromolecules.

Antibodies against nucleic acids have been elicited in rabbits by heat-denatured DNA or by RNA that had been complexed with methylated bovine serum albumin (Plescia *et al.*, 1964; Plescia *et al.*, 1965a). Antibodies that react with nucleic acids have also been obtained by coupling purinoyl derivatives to proteins (Butler *et al.*, 1962), by coupling periodate-oxidized nucleosides or nucleotides to proteins (Erlanger and Beiser, 1964; Estrada-Parra *et al.*, 1965), by binding uridine-5-carboxylic acid to a synthetic polypeptide (Sela *et al.*, 1964), or by the use of oligonucleotides complexed with methylated bovine serum albumin (Plescia *et al.*, 1965b). All the above mentioned antibodies are potentially useful in studies of processes in which nucleic acids are involved.

Now that it is possible to prepare conjugates of nucleosides, nucleotides and oligonucleotides, it is possible to explore, with these conjugates, the

* This work was carried out, in part, with the aid of a grant from the Brown-Hazen fund of the Research Corporation. We also wish to thank graduate students Oscar Rojas and Carlos Arroyave for their help.

specificity of anti-DNA antibodies, especially of antibodies in sera of patients with diseases such as systemic lupus erythematosus (SLE), rheumatoid arthritis (Seligmann, 1963), and tuberculosis (Burns and Rheins, 1966). Recently it was found that anti-nuclear antibodies appear in the serum of patients when extensive necrosis occurs and that some of these sera show specificity for DNA (Dr. Carlos Biro, Instituto de Cardiologia, personal communication).

In the present paper, the reaction of anti-DNA sera from patients with nucleoside, nucleotide and dinucleotide conjugates is reported.

I. RESULTS AND DISCUSSION

Sera from patients with SLE and tuberculosis, and from an individual who had a pulmonary infarct, were obtained through the courtesy of Drs. C. Biro, D. Alarcon, H. Alvarez, F. Martinez, and E. Escarzaga.

The nucleoside, nucleotide and dinucleotide conjugates were prepared by the method described by Erlanger and Beiser (1964) or by the one described by us (Estrada-Parra *et al.*, 1965). Human serum albumin, HSA, was used as carrier protein and to this periodate-oxidized nucleosides, nucleotides or dinucleotides were coupled (see Fig. 1). The ultraviolet

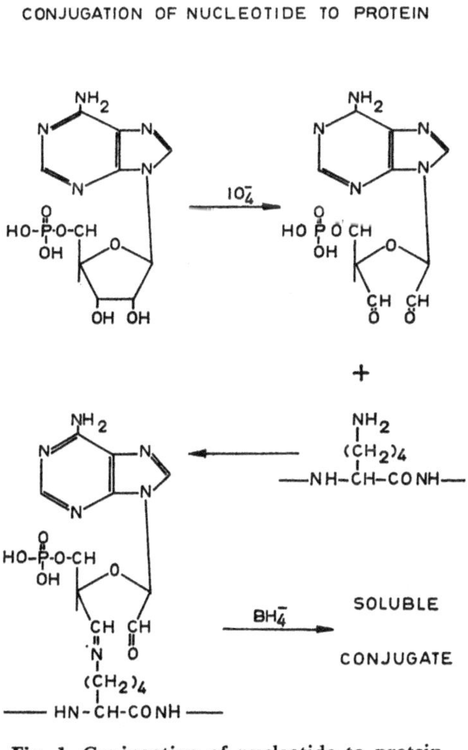

Fig. 1. Conjugation of nucleotide to protein.

absorbance spectra of the conjugates are given in Figures 2, 3, 4 and 5. The reactivity of these conjugates was determined by the capillary precipitin test. This was done by mixing a test serum with an equal volume of conjugate in saline, at a concentration of 250 or 1000 μg/ml, and incubating the mixture in a refrigerator up to one week. Daily observations were made.

In a preliminary study (Estrada-Parra and Olguin-Palacios, 1966) we found that the serum from a patient with pulmonary infarct had antibodies against DNA and that this serum also precipitated with nucleoside conjugates. This serum showed specificity for the bases present in DNA since it precipitated with the adenosine, cytosine, guanosine and thymidine conjugates but not with the uridine conjugate (Fig. 6). Sera from 10 healthy people failed to precipitate with the conjugates or with DNA. After this finding it was decided to investigate the reaction of nucleoside conjugates with sera from patients with SLE. The results of

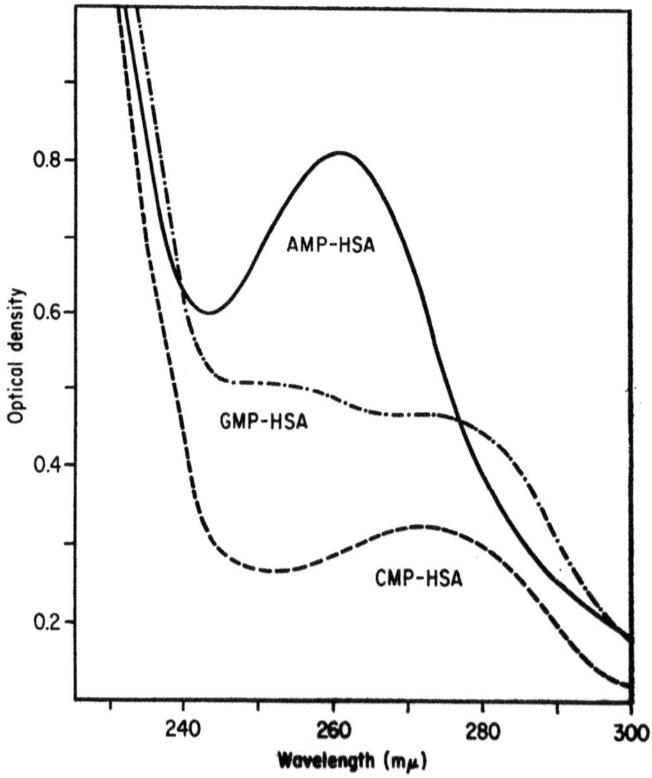

Fig. 2. Ultraviolet absorbance spectra of various nucleotidde conjugates.
——— Adenosine monophosphate–HSA
—·—·— Guanosine monophosphate–HSA
– – – – Cytidine monophosphate–HSA

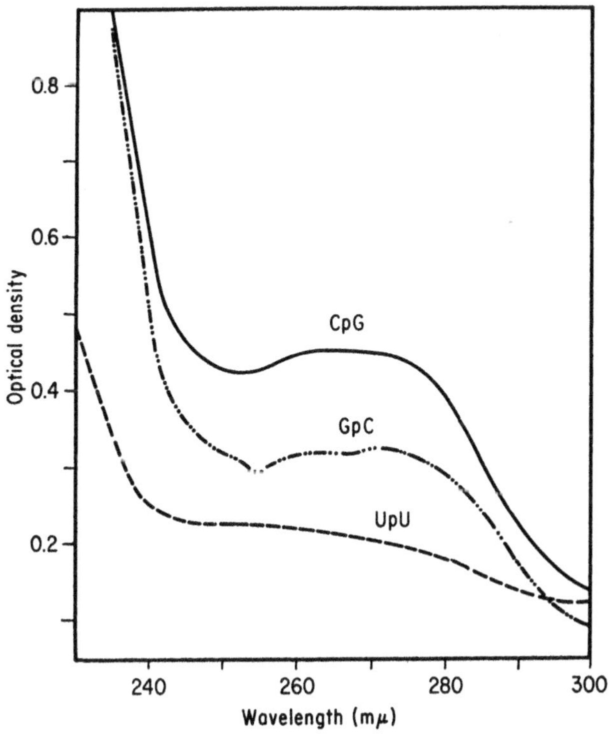

Fig. 3. Ultraviolet absorbance spectra of various dinucleotide conjugates.
———— Cytidylyl-guanosine—HSA
—··—··— Guanylyl-cytidine—HSA
— — — — Uridylyl-uridine—HSA

this study are summarized in Table 1. In view of the results obtained with mononucleosides, and in view of the fact that several dinucleotides became available commercially recently, it was of interest to investigate the reaction of anti-DNA sera with conjugates of nucleotides and dinucleotides as well. The results of these reactions are given in Table 2. Only those sera which precipitated with DNA are shown in Table 2 since it is well known that sera from patients with SLE do not all precipitate with DNA.

Of all the sera tested so far, the serum from the patient with a pulmonary infarct gave the strongest precipitation reaction by the capillary tube method. This serum also precipitated with DNA and with the thymidine conjugate by the immuno-diffusion method of Ouchterlony. Unfortunately, this serum was not tested with nucleotide or dinucleotide conjugates because at the time we did not have these reagents available. The anti-nuclear antibodies in this serum may be similar in origin to those found by Dr. Biro (personal communication) in patients with ex-

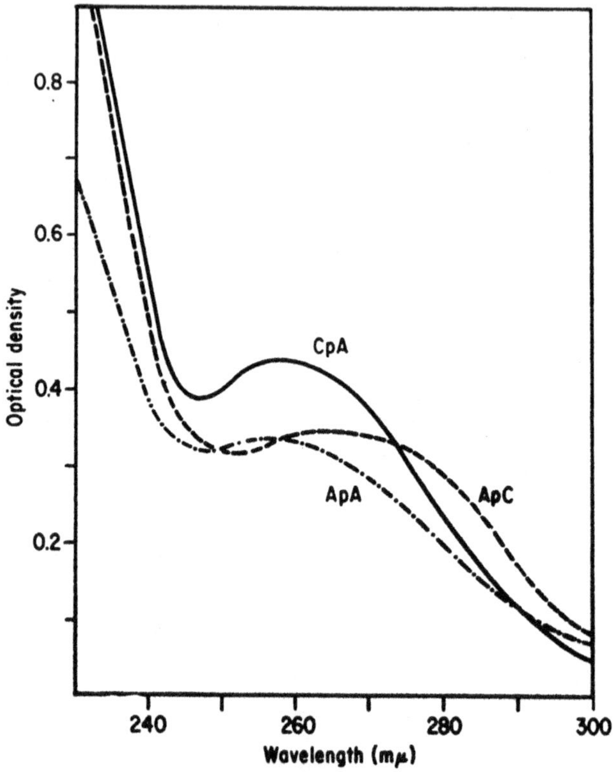

Fig. 4. Ultraviolet absorbance spectra of various dinucleotide conjugates.
——— Cytidylyl-adenosine–HSA
—·—·— Adenylyl-adenosine–HSA
– – – – Adenylyl-cytosine–HSA

tensive necrosis. The level of such antibodies in the serum decreases soon after the patients have recovered. The antibody is of the IgM class, resulting presumably from a primary response, and this probably accounts for the decline in antibody.

As the results in Table 1 show, the reactivity of SLE sera with the nucleoside conjugates is variable. This finding was as expected. Serum 1 gave a good precipitate with native and denatured DNA and with the cytidine, guanosine and thymidine conjugates. It precipitated also, although with less intensity, with the uridine conjugate, but it failed to precipitate with the adenosine conjugate. Sera 2 and 7 behaved very similarly, both precipitating with the conjugates of adenosine, cytidine and uridine. They also reacted with denatured DNA but failed to precipitate with native DNA. These two sera may have antibodies with specificity for the bases only, and probably this is why they do not precipitate with native DNA. Serum 3 behaved very similarly to serum 8, which was obtained from a patient with a pulmonary infarct and was discussed above. Serum 6 did not precipitate either with DNA or with the conjugates. Serum 4 precipitated with both native and denatured DNA but failed to precipitate with any of the conjugates. This could be due to the presence of antibodies which have specificity for the sugar-phosphate backbone of DNA.

All the above results strongly suggest that sera from patients with SLE and the serum of at least one patient with a pulmonary infarct contain antibodies specific for the constituent bases of nucleic acids.

From the results in Table 2 it can be seen that the reactivity of anti-DNA sera with nucleoside, nucleotide and dinucleotide conjugates is also, as expected, quite variable. Serum 9 precipitated with all conjugates, whereas sera 10 and 11 did not precipitate with any conjugate. Serum 11 precipitated better with native DNA than with denatured

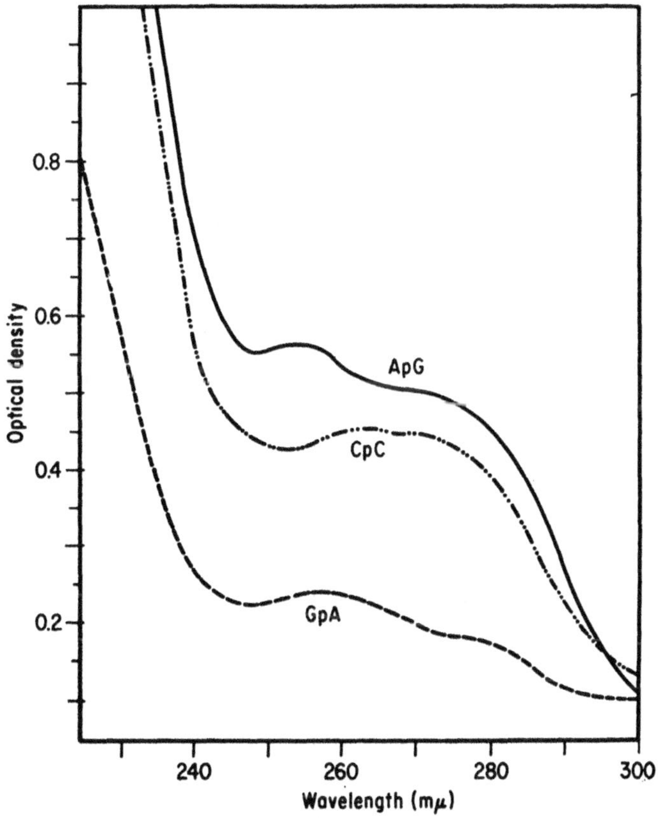

Fig. 5. Ultraviolet absorbance spectra of various dinucleotide conjugates.
———— Adenylyl-guanosine—HSA
—··—··— Cytidylyl-cytidine—HSA
– – – – – Guanylyl-adenosine—HSA

Fig. 6. Reaction of a serum from an individual with a pulmonary infarct with nucleoside conjugates and DNA. From the left: (1) Serum blank, (2) Adenosine, (3) DNA, (4) Cytosine, (5) Guanosine, (6) Thymidine, (7) Uridine. With permission, Estrada-Parra and Olguin-Palacios. Rev. Lat. Amer. Microbiol. Parasitol., *8,* 215 (1966).

Table 1

CAPILLARY PRECIPITIN REACTION OF ANTI-DNA SERA, FROM PATIENTS WITH SLE, WITH DNA AND NUCLEOSIDE CONJUGATES

Conjugate or DNA	Sera							
	1	2	3	4	5	6	7	8 *
A-HSA	(−)	+	+	(−)	±	(−)	+	+
C-HSA	++	(+)	+	(−)	(−)	(−)	+	++
G-HSA	++	(−)	+	(−)	(−)	(−)	+	+++
Denatured DNA	++	(+)	+	+	(−)	(−)	+	+++
Native DNA	++	(−)	+	+	(−)	(−)	(−)	++++
T-HSA	++	(−)	+	(−)	(−)	(−)	(−)	+++
U-HSA	+	+	(−)	(−)	(−)	(−)	+	(−)

HSA, human serum albumin; A, adenosine; C, cytidine; G, guanosine; T, thymidine; U, uridine; SLE, systemic lupus erythematosus.

* Serum from a patient with pulmonary infarct.

Table 2

Capillary Precipitin Reaction of Anti-DNA Sera, from Patients with SLE and Tuberculosis, with DNA and Nucleoside, Nucleotide and Dinucleotide-conjugates

Conjugate or DNA	Sera 9	10	11	12	13	14	15	16	17	18 *	19 *
Native DNA	±	±	++	+	++	++	+	++	++	+	+
Denatured DNA	+	+	±	+	+	+	+	±	++	+	++
A-HSA	+	(−)	(−)	(−)	(−)	(−)	(−)	(−)	±	±	(−)
C-HSA	+	(−)	(−)	(−)	±	+	+	+	+	+	+
G-HSA	+	(−)	(−)	(−)	(−)	+	±	±	(−)	(−)	+
T-HSA	+	(−)	(−)	(−)	(−)	±	(−)	(−)	(−)	(−)	(−)
U-HSA	+	(−)	(−)	(−)	(−)	+	(−)	+	+	(−)	+
AMP-HSA	+	(−)	(−)	(−)	(−)	±	(−)	(−)	+	+	+
CMP-HSA	+	(−)	(−)	(−)	(−)	(−)	±	(−)	+	±	+
GMP-HSA	+	(−)	(−)	(−)	(−)	±	(−)	(−)	±	+	±
UMP-HSA	+	(−)	(−)	(−)	(−)	±	(−)	(−)	±	±	±
ApA-HSA	+	(−)	(−)	++	++	(−)	(−)	(−)	+	+	+
ApC-HSA	+	(−)	(−)	+	++	(−)	±	(−)	+	+	+
ApG-HSA	+	(−)	(−)	(−)	++	(−)	±	(−)	+	(−)	(−)
CpA-HSA	+	(−)	(−)	+	(−)	(−)	±	(−)	+	(−)	(−)
CpC-HSA	+	(−)	(−)	±	(−)	(−)	+	(−)	±	(−)	(−)
CpG-HSA	+	(−)	(−)	++	++	(−)	+	(−)	+	±	+
GpA-HSA	+	(−)	(−)	++	++	(−)	+	(−)	+	(−)	+
GpC-HSA	±	(−)	(−)	+	±	±	(−)	(−)	+	(−)	(−)
UpU-HSA	+	(−)	(−)	+	+	+	(−)	(−)	+	(−)	+

HSA, human serum albumin; A, adenosine; C, cytidine; G, guanosine; T, thymidine; U, uridine; AMP, adenosine monophosphate; CMP, cytidine monophosphate; GMP, guanosine monophosphate; UMP, uridine monophosphate; ApA, adenylyl-adenosine; ApC, adenylyl-cytidine; ApG, adenylyl-guanosine; CpA, cytidylyl-adenosine; CpC, cytidylyl-cytidine; CpG, cytidylyl-guanosine; GpA, guanylyl-adenosine; GpC, guanylyl-cytidine; UpU, uridylyl-uridine; SLE, systemic lupus erythematosus.

* Sera from patients with tuberculosis.

DNA. This serum probably had antibodies directed against the ribose phosphate backbone of the DNA. Sera 12 and 13 are interesting ones since they failed to precipitate with nucleoside or nucleotide conjugates but precipitated with the dinucleotide conjugates, an indication that they contain antibodies with specificity for more than one base. It should be noted also that serum 13 gave a good precipitate with the adenylyl-cytidine conjugate but did not precipitate with the cytidylyl-adenosine conjugate. This serum also precipitated better with the cytidylyl-guanosine conjugate than with guanylyl-cytidine conjugate. Moreover, serum 12 failed to precipitate with the adenylyl-guanosine conjugate, but it precipitated well with the guanylyl-adenosine conjugate. These results suggest the possible importance of sequence on specificity.

Serum 14 precipitated, though not too strongly, with conjugates of nucleosides and nucleotides but in general it did not precipitate with the dinucleoside conjugates. Serum 15 behaved, although less clearly, like sera 12 and 13; serum 16 like 10 and 11, and serum 17 like serum 9.

Of the sera tested, the two sera from patients with tuberculosis were the only ones which precipitated with DNA. Both of these sera precipitated with some of the nucleoside conjugates, with all of the nucleotide conjugates and with some of the dinucleotide conjugates.

The fact that some of the sera precipitated with uridine, uridine monophosphate and with uridylyl-uridine conjugates may be due to a cross-reaction with antibodies against thymine or due to the presence of antibodies elaborated against RNA. The latter is likely because, as a matter of fact, some of these sera precipitated with RNA.

We feel that more tests, especially quantitative tests, should be carried out with more sera in order to distinguish clearly the differences in the reactivity of the different conjugates. However, the above described results together with the results of inhibition studies reported by Stollar *et al.* (1962, 1963) clearly show the variable specificity of antibodies in SLE sera against the constituent bases. Some are specific for one or more bases of nucleic acids and some of them are capable of reacting with more than one base in certain sequences. It is entirely possible that among the many SLE sera available some will be found to have a limited, desired specificity. It would be worthwhile to search these out because they represent invaluable tools for the study of structure and function of nucleic acids.

References

Burns, R. M., and M. S. Rheims (1966). Hemagglutinins for DNA in tuberculosis and histoplasmosis. Proc. Soc. Exptl. Biol. Med., *122:*714–717.

Butler, V. P., S. M. Beiser, B. F. Erlanger, S. W. Tanenbaum, S. Cohen, and A. Bendich (1962). Purine-specific antibodies which react with deoxyribonucleic acid (DNA). Proc. Natl. Acad. Sci. U. S., *48:*1597–1602.

Erlanger, B. F., and S. M. Beiser (1964). Antibodies specific for ribonucleosides and ribonucleotides and their reaction with DNA. Proc. Natl. Acad. Sci. U. S., *52:*68–74.

Estrada-Parra, S., A. Schmill, and R. M. Martínez (1965). Reaction of adenine-specific antibodies with denatured deoxyribonucleic acid. Nature, *208:*1010–1012.

Estrada-Parra, S., and E. Olguin-Palacios (1966). The reaction of anti-deoxyribonucleic acid sera with nucleoside-conjugates. Rev. Lat-Amer. Microbiol., *8:*215.

Heidelberger, M. (1960). Structure and immunological specificity of polysaccharides. Fortsch. der Chem. Org. Naturstoffe, *18:*503–536.

Heidelberger, M. (1967). Some contributions of immunochemistry to biochemistry and biology. Annual Review Bioch., *36*:1–12.

Plescia, O. J., W. Braun, and N. C. Palczuk (1964). Production of antibodies to denatured deoxyribonucleic acid (DNA). Proc. Natl. Acad. Sci. U. S., *52*: 279–285.

Plescia, O. J., N. C. Palczuk, E. Cora-Figueroa, A. Mukherjee, and W. Braun (1965a). Productions of antibodies to soluble RNA. Proc. Natl. Acad. Sci., *54*:1281–1285.

Plescia, O. J., N. C. Palczuk, W. Braun, and E. Cora-Figueroa (1965b). Antibodies to DNA and a synthetic polydeoxyribonucleotide produced by oligodeoxyribonucleotides. Science, *148*:1102–1105.

Sela, M., H. Ungar-Waron, and Y. Shechter (1964). Uridine-specific antibodies obtained with synthetic antigens. Proc. Natl. Acad. Sci. U. S., *52*:285–292.

Seligman, M. (1963). Immunologic aspects of rheumatoid arthritis and systemic lupus erythematosus: DNA antibodies. Arthritis Rheum., *6*:542–546.

Stollar, D., L. Levine, H. I. Lehrer, and H. Van Vunakis (1962). The antigenic determinants of denatured DNA reactive with lupus erythematosus serum. Proc. Natl. Acad. Sci. U. S., *48*:874–880.

Stollar, D., and L. Levine (1963). Antibodies to denatured deoxyribonucleic acid (DNA) in lupus erythematosus serum IV. Evidence for purine determinants in DNA. Arch. Biochem. Biophys., *104*:417–422.

Vogel, H. J., V. Bryson, and J. O. Lampen (1963). Informational Macromolecules. Academic Press, New York.

THE VARIOUS TYPES OF DNA ANTIBODIES IN LUPUS SERA *

M. SELIGMANN AND R. ARANA †

Research Institute for Blood Diseases of the University of Paris Hôpital Saint-Louis, Paris, France

The presence of antibodies reacting with deoxyribonucleic acid (DNA) in the sera of some patients with systemic lupus erythematosus (SLE) was demonstrated ten years ago (Seligmann, 1957; Robbins *et al.,* 1957; Ceppellini *et al.,* 1957; Seligmann and Milgrom, 1957). The early studies on these antibodies (Seligmann, 1958; Deicher *et al.,* 1959) were performed with native DNA. Subsequently Barbu *et al.* (1960) showed that some lupus antibodies also react with DNA previously denatured or degraded by various procedures. Stollar and Levine (1961) and Stollar *et al.* (1962a) reported that all but one of their DNA-reactive SLE sera reacted exclusively or more effectively with thermally denatured DNA. Most of the experimentally induced antibodies capable of reacting with DNA were shown to react only with denatured DNA. Therefore, in recent years, most reports on DNA antibodies have been devoted to immunological reactions with denatured DNA.

The study of the relative reactivities with native and denatured DNA of a series of SLE sera seemed warranted in order to determine (1) to what extent the previously observed reactions with native DNA were due to a denatured contaminant, (2) whether some lupus antibodies reacted preferentially or exclusively with native DNA, and (3) whether different anti-DNA antibodies with various specificities could be demonstrated in a single lupus serum.

I. METHODS

Highly polymerized calf thymus DNA (Worthington Biochemical Corp.) was used for all experiments. Samples of *E. coli,* T7 phage and

* We thank the Editors of the Journal of Clinical Investigation for allowing us to reproduce some of the figures of an article to be published in the November 1967 issue.

† Fellow of the National Council for Scientific Investigations (Argentina) and of the World Health Organization. Present address: C.E.M.I.C. Bustamante 2560, Buenos Aires.

C2 phage DNA were used for some experiments. Preparations of purified native DNA were obtained by chromatography on Kieselguhr impregnated with methylated albumin, according to the method of Mandel and Hershey (1960).

Routine complement-fixation tests were performed using $3C'H_{50}$ units and two different concentrations of native and denatured DNA. The reactions were considered as positive when $2C'H_{50}$ units had been fixed. The quantitative microcomplement-fixation technique was performed according to Wasserman and Levine (1961) in a total volume of 3.5 ml. Inhibition experiments were performed under standard conditions, utilizing concentrations of antiserum and antigen giving maximal complement fixation. Because of the anticomplementary activity of our denatured DNA preparations, the highest concentrations used were 0.6 to 0.8 μg/ml. All comparative experiments were performed simultaneously.

Double diffusion tests were performed in 0.6% agarose. Purified DNA antibodies were obtained by digestion of specific precipitates by deoxyribonuclease, as described previously (Seligmann, 1958). Rabbit antisera specific for γ, μ, α, κ and λ chains of immunoglobulins were used for the determination of the class and light chain types of the purified antibodies.

II. RELATIVE REACTIVITIES OF SLE SERA WITH NATIVE AND DENATURED DNA

A. Qualitative Complement-fixation Tests

As shown in Table 1, the incidence of positive reactions with denatured DNA was higher than that of positive reactions with native DNA. Forty-five per cent of the positive sera reacted in routine tests with both native and denatured DNA. Amongst the sera reacting only with denatured *or* native DNA, the former were more frequent than the latter.

Table 1

Results of Routine Complement-fixation Tests of a Series of SLE Sera with Native (N) and Denatured (D) Calf Thymus DNA

Number of Sera	Total Number of Positive Reactions with: Native DNA	Denatured DNA	N+D+	N+D−	N−D+	N−D−
215	46	63	34	12	29	140

Two concentrations of DNA were used: 1 and 5 μg per ml of serum for native DNA; 0.1 and 1 μg per ml of serum for denatured DNA.

Comparative studies using the routine technique and the more sensitive micromethod were carried out on eighty sera. The results are shown in Table 2. As expected, the percentage of positive sera and the incidence of positive reactions with both native and denatured DNA was higher with the microtechnique. However, some sera were positive only with native or with denatured DNA.

B. Quantitative Complement-fixation Studies

The 35 sera tested by quantitative complement-fixation showed great variations in their relative reactivities with native and denatured DNA. This is reflected by exclusive, preferential or identical reactions with native or/and denatured DNA, as judged by the amount of complement fixed and the antigen concentration at peak fixation. With some sera, peak complement-fixation was obtained with a concentration of native DNA many-fold greater than that required for denatured DNA (Fig. 1). This type of pattern strongly suggested that this reaction was due to a small amount of denatured DNA contaminating the native preparation. The use of purified native DNA, free of denatured contaminant, enabled us to show that this was indeed the case for some sera (Fig. 1A). However, in the case of other such sera (Fig. 1B), the possibility of contamination with a small amount of single-stranded denatured DNA could be excluded even though a concentration of native DNA 100 times greater than that of denatured DNA was required for reactivity: the complement-fixation curve with purified native DNA was

Table 2

Comparative Results of a Study of 80 SLE Sera Tested with Native (N) and Denatured (D) DNA by Routine and Microcomplement-fixation Techniques

Type of Serum	Routine Technique	Microtechnique
N+D+	14	27
N+D−	5	2
N−D+	7	7
N−D−	54	44

Two concentrations of DNA were used in the microtechnique: 0.5 and 2 μg per ml of serum for native DNA; 0.0125 and 0.1 μg per ml of serum for denatured DNA.

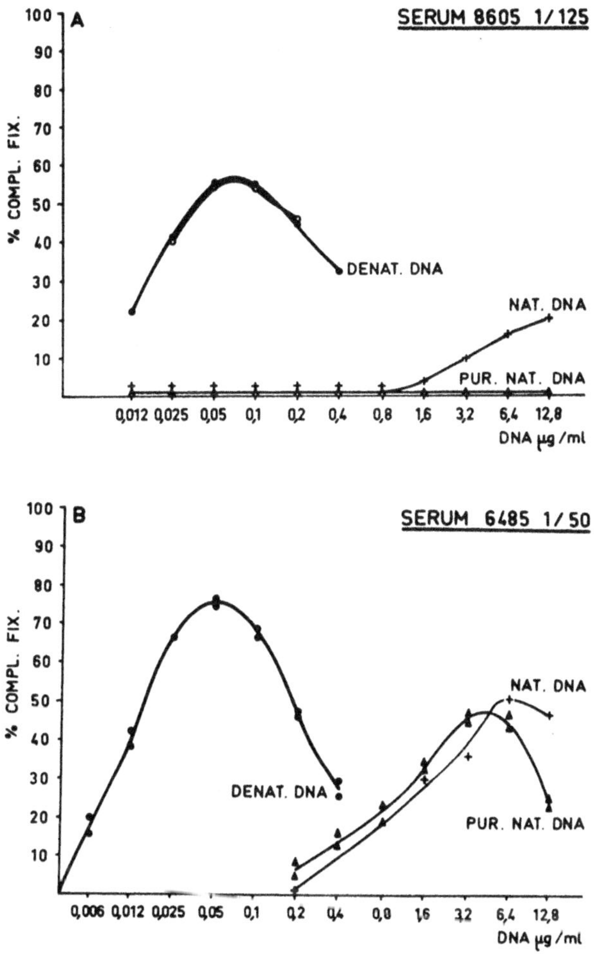

Fig. 1. Complement-fixation curves. **A:** Serum reacting with a denatured contaminant of the native preparation. o—o Complement-fixation with denatured DNA in the presence of 20 μg/ml of purified native DNA. **B:** Serum reacting with native and purified native DNA at concentrations 100 times higher than that of denatured DNA. (From the Journal of Clinical Investigation, 1967.)

similar to that of native DNA, except in the zone of antigen excess. Altogether 30% of the sera containing antibodies to DNA were demonstrated to react only with denatured DNA. In quantitative inhibition experiments, up to 20 μg/ml of purified native DNA did not interfere with the complement-fixation reaction of sera that reacted with denatured DNA. The complement-fixation curve of one of these sera is shown in Figure 2A. Another type of pattern is illustrated in Figure 2B: these sera reacted with a 20 to 30 times higher concentration of purified native DNA than denatured DNA, giving the same maximal comple-

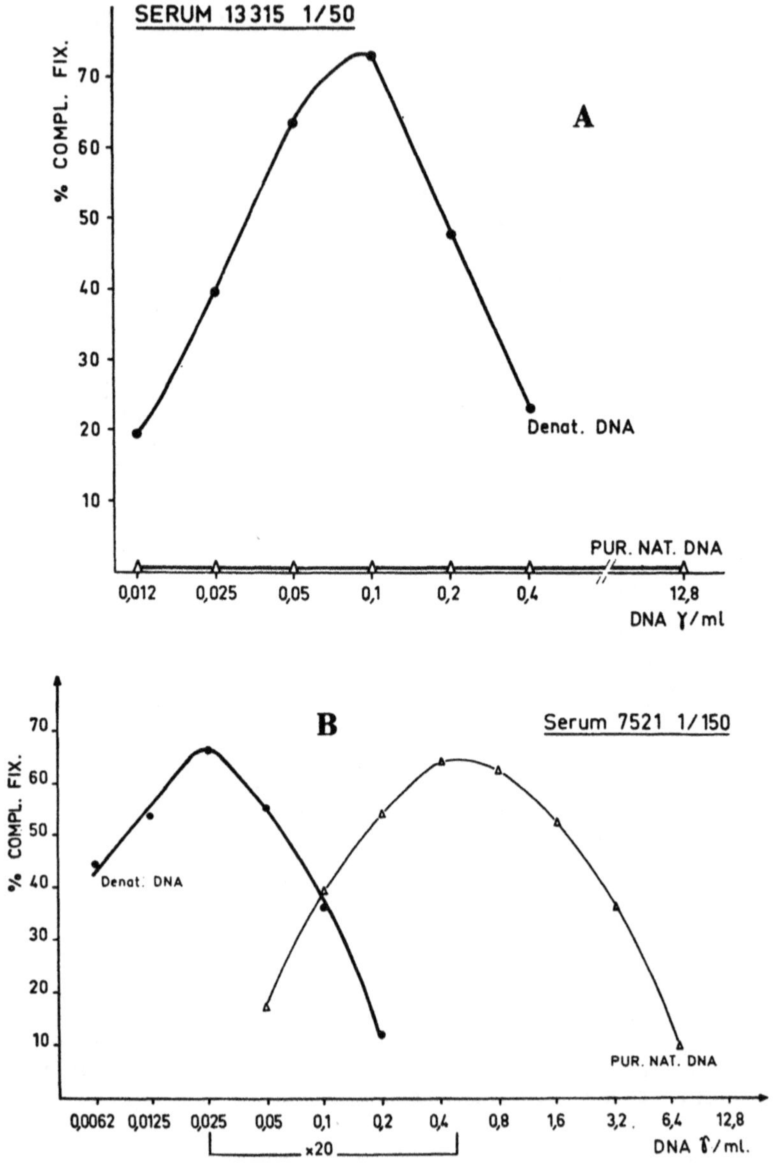

Fig. 2. Complement-fixation curves of 6 different SLE sera with native and thermally denatured calf thymus DNA. These patterns show the great variability in relative reactivities with denatured (Denat.) and purified native (Pur. Nat.) DNA.

ment-fixation. In other cases the serum showed a peak reaction for similar concentrations of native and denatured DNA, but denatured DNA fixed more complement (Fig. 2C). A fourth type of pattern, illustrated in Figure 2D, is defined by similar complement-fixation curves for purified native and denatured DNAs. In all these instances purified

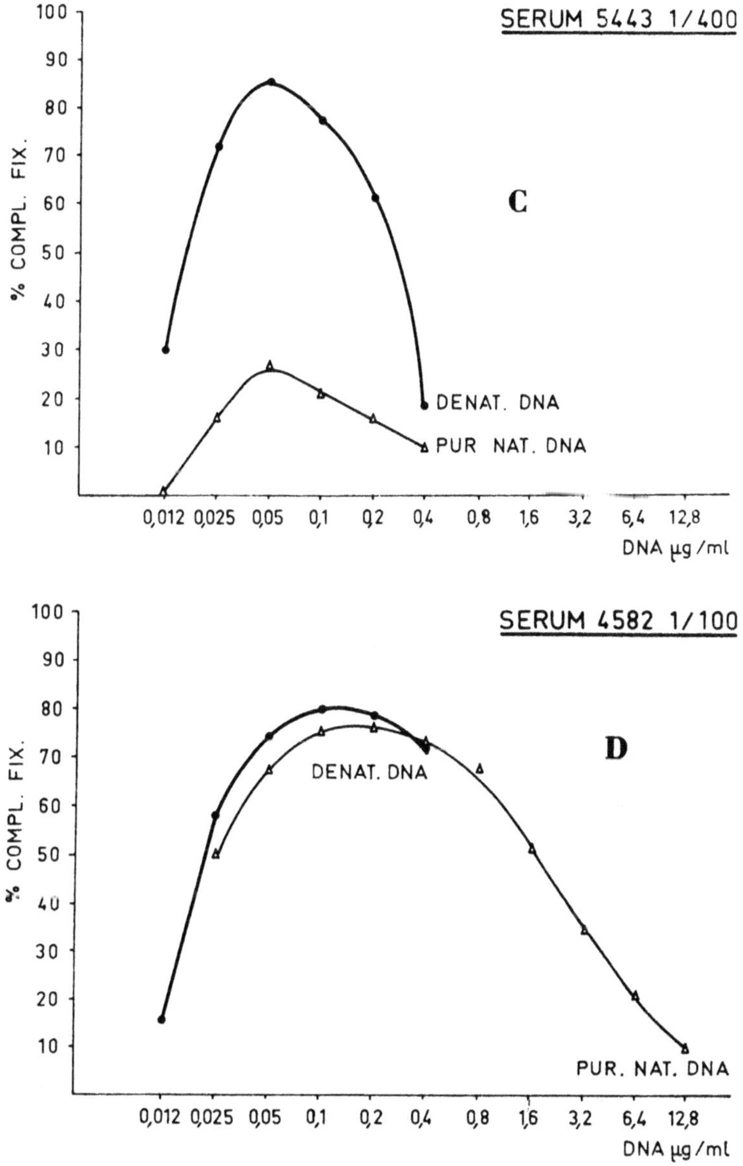

Fig. 2. (continued)

native DNA in extreme antigen excess was an effective inhibitor of the reaction with denatured DNA. Figure 2E gives an example of another type of pattern where the serum reacts more effectively with native than with denatured DNA; for this particular serum, the difference in reactivity was more striking with T7 phage DNA. Finally, serum 10,466 (Fig. 2F) reacts strongly with native DNA, although denatured DNA shows little if any reaction at the concentrations tested.

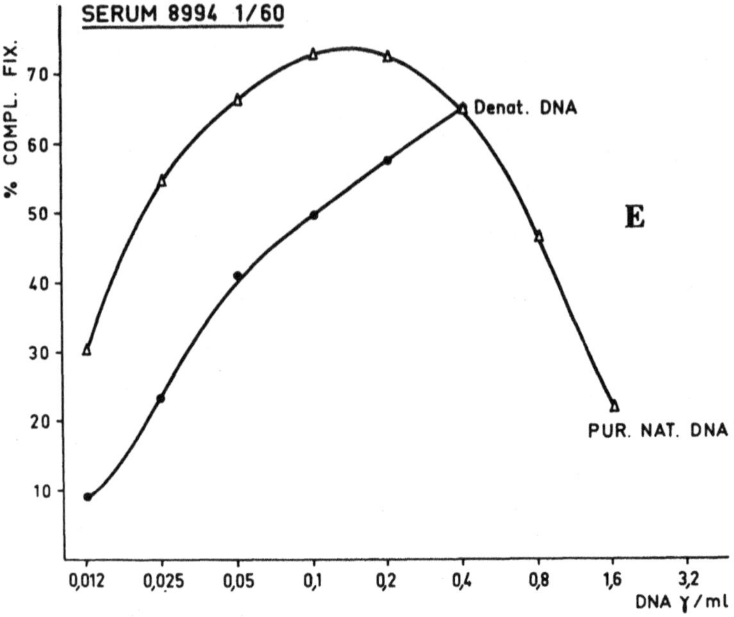

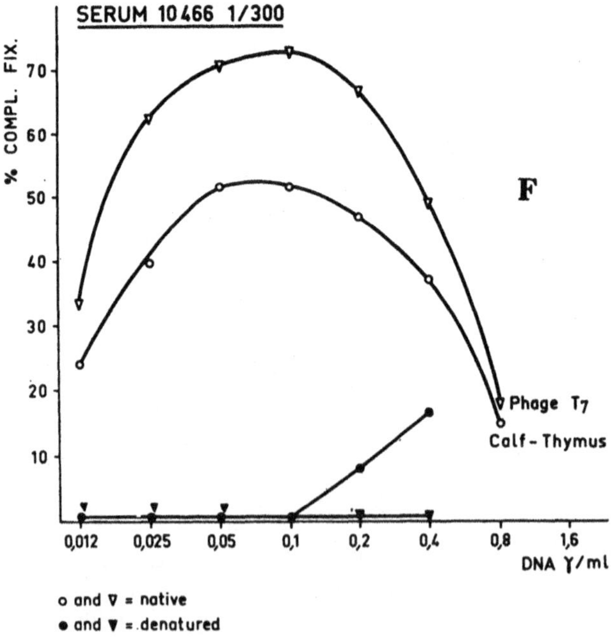

Fig. 2. (continued)

C. Precipitin Patterns in Double Diffusion Tests

As shown in Figure 3, similar variability in the relative reactivities of SLE sera with native and denatured DNA can be demonstrated by double diffusion tests with various concentrations of native or denatured sonicated DNA. With some sera, precipitin lines were only seen with denatured (Fig. 3A) or native (Fig. 3C) DNA, whereas other sera gave identity (Fig. 3B) or partial identity (Fig. 3D) reactions with single- and double-stranded DNAs. A single serum in this series showed precipitin lines with native and denatured DNA that crossed each other with a decrease in intensity (Fig. 3E).

III. THE THREE MAIN VARIETIES OF DNA-REACTIVE ANTIBODIES IN SLE

The SLE sera contained three main types of DNA-reactive antibodies: those reacting only with denatured DNA, those reacting to the same extent with both forms of DNA, and those reacting preferentially with native DNA.

A. Antibodies Reacting Only with Denatured DNA

These antibodies react with determinants that are inaccessible on the native molecule, as confirmed by the absence of inhibition by purified native DNA. Stollar *et al.* (1962b) and Stollar and Levine (1963) have studied this type of antibodies extensively. They have shown that they differ in specificity and, for some of them, they identified the nature of the antigenic determinant. As observed by these authors, the denatured end products of digestion by DNAase are effective inhibitors of these antibodies. In our test series, denatured yeast RNA did not inhibit the reaction of these antibodies with denatured DNA.

B. Antibodies Reacting to the Same Extent with Native and Denatured DNA

These antibodies react with antigenic determinants common to both forms of DNA. This is reflected by identical complement-fixation curves or by reaction of identity in precipitin tests. Similar patterns have been described for some lupus sera by Stollar *et al.* (1962a) and by Tan *et al.* (1966). The results of cross absorption experiments prove that the same antibodies react with both forms of DNA, presumably with "back-bone" determinants. Since native and denatured RNA do not inhibit the re-

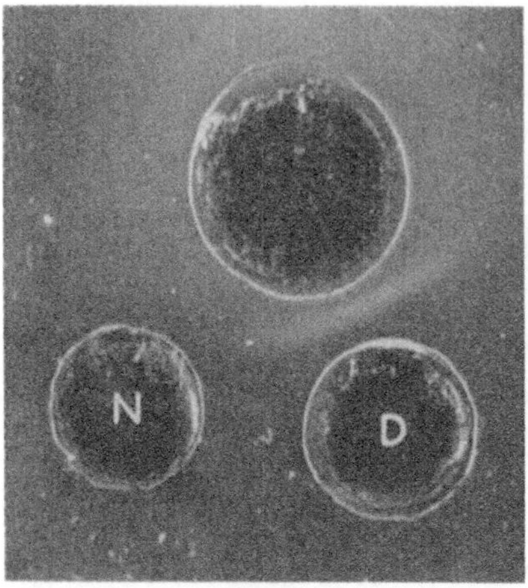

(A)

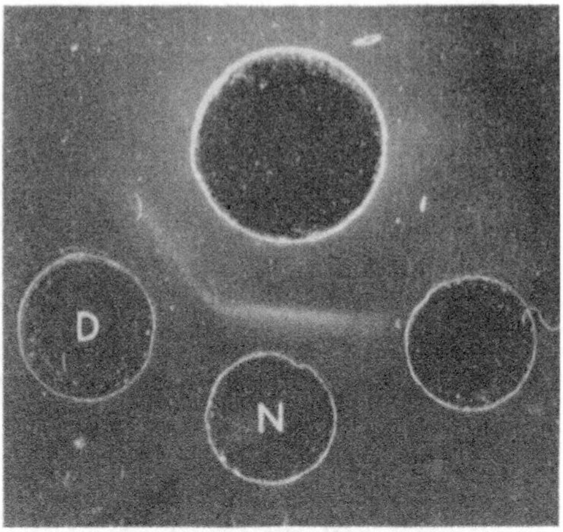

(B)

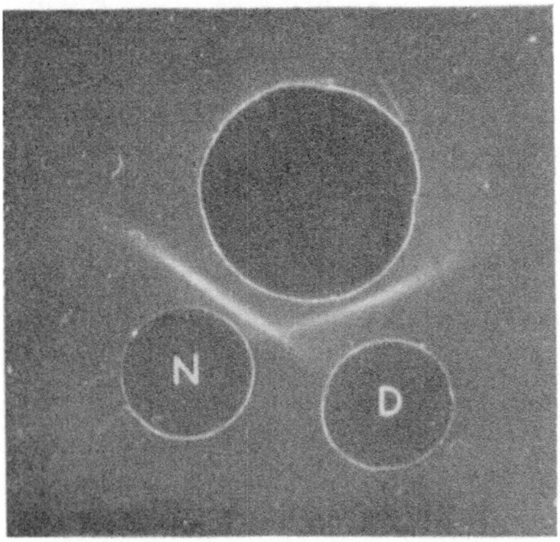

(C)

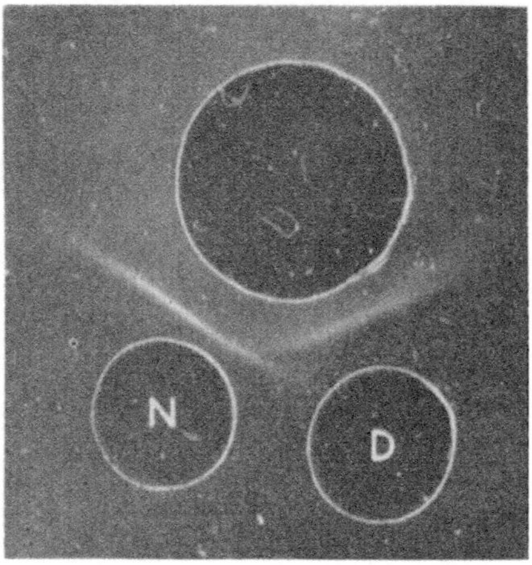

(D)

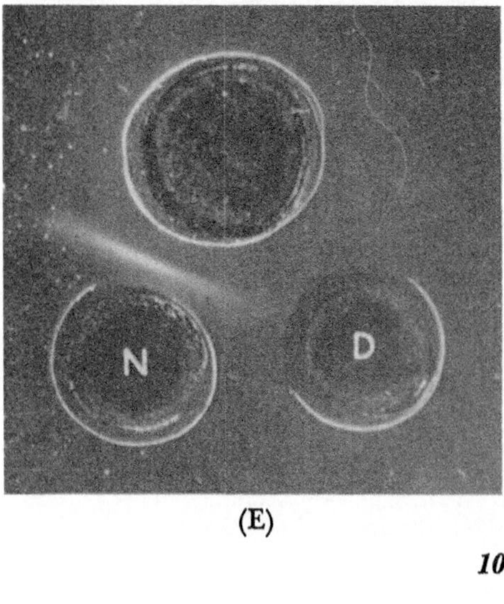

(E)

Fig. 3. Precipitin patterns produced 5 different SLE sera when tested by double diffusion in gel with native (N) and denatured (D) DNA.

action, deoxyribose might be involved in the antigenic site(s). The lack of inhibition by DNAase digests could reflect a requirement for relatively large oligonucleotide fractions.

C. Antibodies Reacting Preferentially with Native DNA

Several sera in this series reacted more strongly with native than with denatured DNA in complement-fixation. Similarly, in double diffusion tests a definite spur of native over denatured DNA has been observed in several instances. Analogous patterns have been described recently for two lupus sera by Tan *et al.* (1966) and Stollar and Sandberg (1966). One of our sera reacted exclusively with native DNA in complement-fixation and precipitin tests. Another serum gave similar precipitin reactions in liquid and in gel but complement-fixation data could not be obtained because of high anticomplementary activity of this serum. Such findings have not been reported previously; they point to the importance of conformation in the antigenic structure of DNA. However, it should be emphasized that denatured DNA was a potent inhibitor of antibodies to native DNA. It thus appears that the antigenic sites may be present on each chain of the DNA molecule, despite the inability to detect them by precipitin and complement-fixation techniques. The firm binding of these antibodies may require the rigid double helix structure of native DNA. Variations in the degree of reactivity of this type of antibodies with native DNA samples from different sources have been observed.

D. Renaturation Experiments

In order to show the different specificities of antibodies, the kinetics of renaturation of T7 phage DNA have been followed with selected sera. These experiments, illustrated in Figure 4, readily demonstrate the three main types of DNA antibodies described above. Serum 10,466, reacting only with native DNA, shows maximal reactivity with the sample prior to denaturation. No reaction occurs with the denatured sample at the beginning of incubation (zero time). Ten minutes later, 40% of the antigen activity is recovered, and 100% activity is recovered after 20 minutes of incubation. In contrast, serum 13,315, reacting only with denatured DNA, shows maximal activity at zero time. Thirty-five per cent of the antigenic activity is lost after 10 minutes of incubation, then there is a slow decrease in activity and, with the 3 hour sample, no reactivity is detectable. Serum 4,582, reacting to the same extent with native and denatured DNA, does not show any change in the antigenic activity upon renaturation.

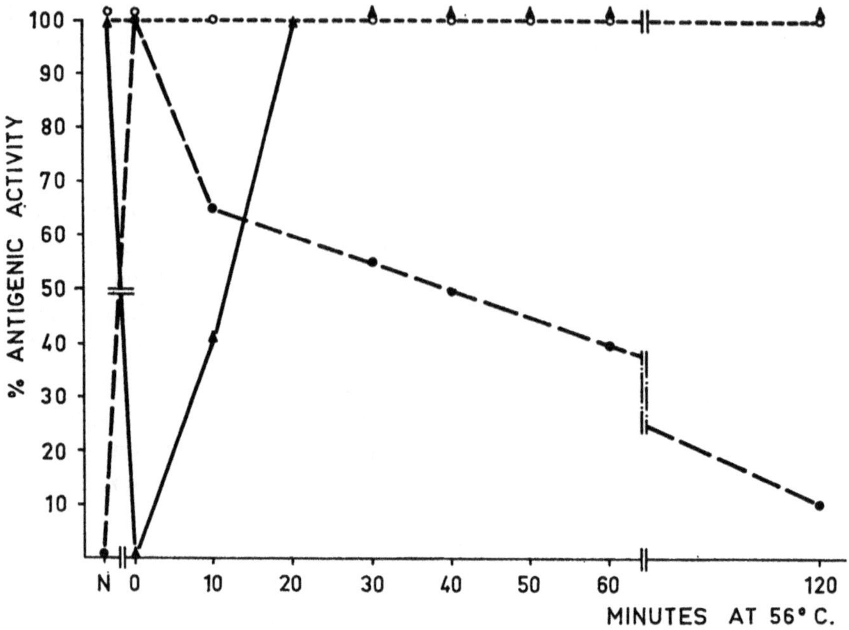

Fig. 4. T7 phage DNA renaturation, as determined by complement-fixation with 3 different sera containing antibodies to: ▲—▲ native DNA (10,466), ·——· denatured DNA (13,315) and o--o common antigenic determinants to native and denatured DNA (4,582). 100% antigenic activity was determined, with the sample prior to denaturation (N) for serum 10,466, with the zero sample (0) for serum 13,315 and with both for serum 4,582. (From the Journal of Clinical Investigation, 1967.)

IV. THE SIMULTANEOUS OCCURRENCE OF DIFFERENT VARIETIES OF ANTI-DNA ANTIBODIES IN LUPUS SERA

The occurrence in a single serum of several DNA-reactive antibodies with different specificities has been demonstrated in many instances by precipitin patterns in double diffusion tests, cross-absorption experiments, analysis of purified antibodies and inhibition of complement-fixation. It should be emphasized that no definite conclusions can be drawn in this regard solely on the basis of complement-fixation curves. Thus, the interpretation of patterns similar to those of Figures 5 and 6 is impossible in the absence of additional tests. For serum 3,258, no complement-fixation with denatured DNA occurred in the presence of 6.4 μg/ml of purified native DNA (Fig. 5). Although serum 11,752 gave an analogous reaction pattern, 30 μg/ml of purified native DNA were unable to inhibit more than 28% of the reaction with denatured DNA (Fig. 6). This weak inhibition suggested that we were dealing with different kinds of antibodies. This hypothesis was confirmed by the following experiment: the addition of 0.4 μg/ml of purified native DNA (in

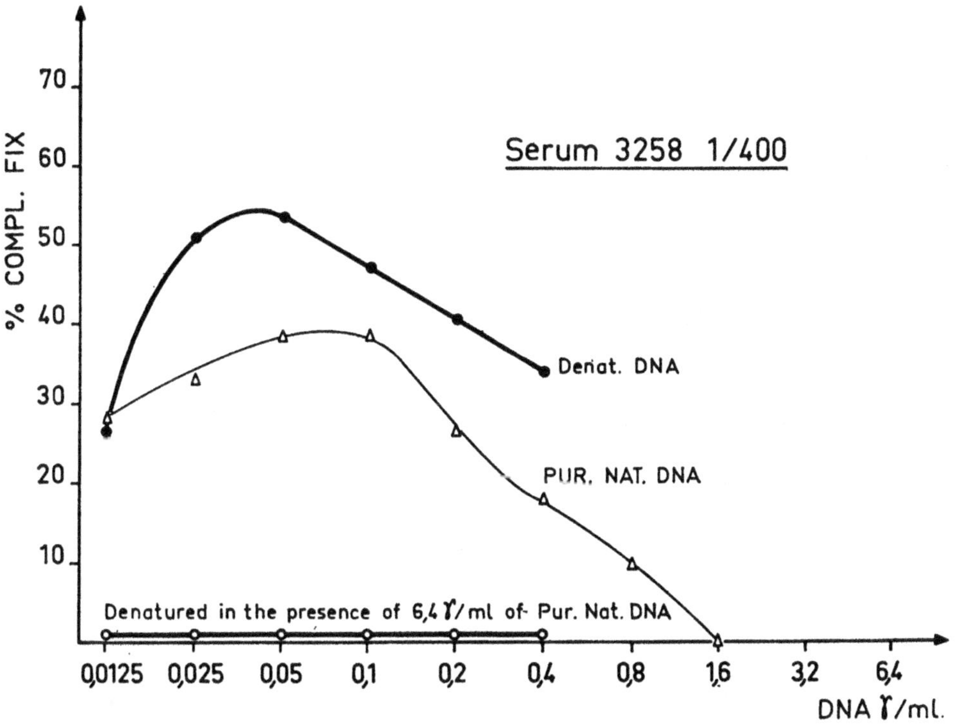

Fig. 5. Complement-fixation curves of serum no. 3,258 with purified native (Pur. Nat.) DNA, with denatured (Denat.) DNA and with denatured DNA in the presence of an excess of purified native DNA.

slight antigen excess) and 0.025 μg/ml of denatured DNA (in large antibody excess) to a fixed amount of antibody, resulted in a definite increase of complement-fixation. Moreover, the patterns obtained from complement-fixation curves do not always correlate with those obtained from gel diffusion studies. For example, the serum illustrated in Figure 3D gave identical complement-fixation curves with native and denatured DNA, although double diffusion studies showed a definite spur of native over denatured DNA and a persistence of a precipitin line with native DNA after absorption with denatured DNA. It should be remembered that the microcomplement-fixation technique uses diluted antiserum and that the results reflect only the reaction(s) given by the most concentrated population(s) of antibodies.

Although serum no. 5,443 gave the complement-fixation curves illustrated in Figure 2C, the simultaneous occurrence of at least three kinds of DNA antibodies in this serum was suggested by its precipitin pattern (Fig. 3E). This interpretation was confirmed by a single plate depicted in Figure 7. Moreover, the purified antibodies obtained from a precipitate formed at equivalence with native DNA reacted with both native

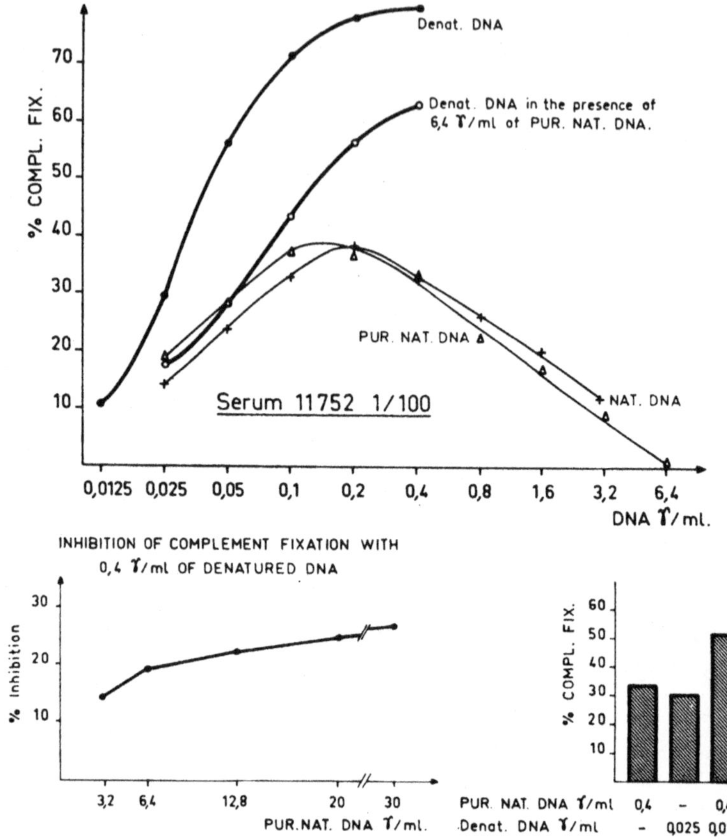

Fig 6. Complement-fixation and inhibition reactions of serum no. 11,752. For the additional experiment illustrated at the right bottom corner of the figure, see text. (From the Journal of Clinical Investigation, 1967.)

Table 3

IMMUNOGLOBULIN CLASSES AND LIGHT CHAIN TYPES OF SOME PURIFIED DNA ANTIBODIES

Serum Number	Precipitated by	Light Chain Type	Ig. Class	Reacting with
8,400	N	K+L	G	N
12,271	D	K+L	G	D
6,969	N	K+L	M	N+D
	D	K+L	M	N+D
5,443	N	K+L	G	N
			M	N+D

N = Native DNA.
D = Denatured DNA.

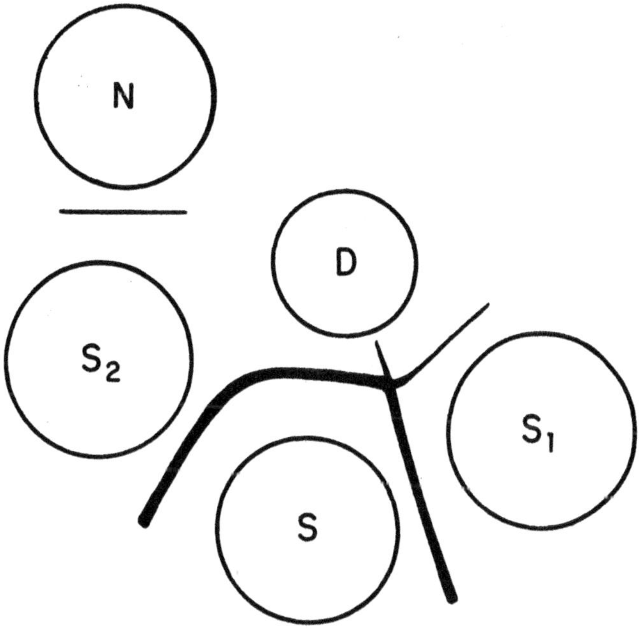

Fig. 7. Demonstration of the simultaneous occurrence of three different kinds of antibodies in serum no. 5,443. S: untreated serum; S_1: supernatant 1; S_2: supernatant 2; D: DNA denatured in the presence of formaldehyde; N: native DNA. For this experiment, one sample of serum was absorbed with native DNA at equivalence and a small excess of native DNA was then added (supernatant 1); another serum sample was absorbed with denatured DNA at equivalence and a small excess of denatured DNA was then added (supernatant 2). Both supernatants contain free antibodies and antigen excess. Supernatant 1 reacts at the same time with denatured DNA and with the antibodies to native DNA of the untreated serum. Supernatant 2 reacts at the same time with antibodies to denatured DNA of the untreated serum and with native DNA. Furthermore, the line of native DNA (between serum and supernatant 1) clearly decreases in intensity after crossing the denatured DNA line. This line with denatured DNA is weaker with supernatant 1 than with whole serum, because of the previous absorption of antibodies reacting with determinants common to both forms of DNA. (From the Journal of Clinical Investigation, 1967.)

and denatured DNA, the line of native DNA spurring over denatured DNA; the supernatant reacted exclusively with denatured DNA.

V. IMMUNOGLOBULIN CLASSES OF PURIFIED DNA ANTIBODIES

Although a limited number of purified antibodies was tested, it was shown that DNA-reactive antibodies in human sera can belong to the IgM globulin class as well as to IgG (Table 3). This finding is in contrast with the recent data of Stollar and Sandberg (1966) and with earlier observations from this (Seligmann, 1958) and other (Deicher *et al.*, 1959) laboratories. In all these instances, anti-DNA antibodies were shown to be IgG globulins. As shown in Table 3, the study of one serum, containing both IgG and IgM anti-DNA antibodies, demonstrated that different specificities could be assigned to each immunoglobulin class. The finding that purified antibodies contain molecules with both types of light chains is not unexpected since we are dealing with a heterogeneous system.

VI. CONCLUSIONS

Anti-DNA antibodies from different SLE patients, as well as from a single patient, vary greatly with respect to the specificity of their receptor sites. This broad heterogeneity probably explains the discrepancies between some of our early results and those of others (Seligmann, 1963).

Antibodies to denatured DNA have been applied successfully to biochemical and physico-chemical studies, and a promising list of potential applications to biological problems has been recently drawn up (Beiser and Erlanger, 1966). Similarly, antibodies in SLE sera that react preferentially or only with native DNA might prove valuable in many respects and especially for conformational studies.

It should be noted that no antibodies reacting equally well with native and denatured DNA have been detected in the serum of animals immunized with denatured DNA or polynucleotides complexed with methylated bovine serum albumin (Plescia *et al.,* 1964, 1965), with T-even bacteriophage lysates (Levine *et al.,* 1960) or with Gram-negative bacteria (Christian *et al.,* 1965). Although native calf thymus DNA is immunogenic in rabbits previously immunized with DNAase, these antibodies also react only with denatured DNA (Beiser and Erlanger, 1966). Thus, at the present time no experimental model is available for the broad spectrum of anti-DNA antibodies present in sera of SLE patients.

References

Barbu, E., M. Seligmann, and M. Joly (1960). Réactions entre des acides désoxyribonucléiques diversement dénaturés ou dégradés et les anticorps anti-acide désoxyribonucléique du sérum de malades atteints de lupus érythémateux disséminé. Ann. Inst. Pasteur, *99:*695.

Beiser, S. M., and B. F. Erlanger (1966). Antibodies which react with nucleic acids. Cancer Research, *26:*2012.

Ceppellini, R., E. Polli, and F. Celada (1957). A DNA reacting factor in serum of a patient with lupus erythematosus diffusus. Proc. Soc. Exp. Biol. Med., *96:*572.

Christian, C. L., A. R. Desimone, and J. L. Arbuzzo (1965). Anti-DNA antibodies in hyperimmunized rabbits. J. Exp. Med., *121:*309.

Deicher, H. R. G., H. R. Holman, and H. G. Kunkel (1959). The precipitin reaction between DNA and a serum factor in systemic lupus erythematosus. J. Exp. Med., *109:*97.

Levine, L., W. T. Murakami, H. Van Vunakis, and L. Grossmann (1960). Specific antibodies to thermally denatured DNA of phage T4. Proc. Natl. Ac. Sci., *46*:1038.

Mandel, J. D., and A. D. Hershey (1960). A fractionating column for analysis of nucleic acids. Anal. Biochem., *1*:66.

Plescia, O. J., W. Braun, and N. C. Palczuk (1964). Production of antibodies to denatured deoxyribonucleic acid (DNA). Proc. Natl. Ac. Sci., *52*:279.

Plescia, O. J., N. C. Palczuk, W. Braun, and M. E. Cora-Figueroa (1965). Antibodies to DNA and a synthetic polydeoxyribonucleotide produced by oligodeoxyribonucleotides. Science, *148*:1102.

Robbins, W. C., H. R. Holman, H. Deicher, and H. G. Kunkel (1957). Complement fixation with cell nuclei and DNA in lupus erythematosus. Proc. Soc. Exp. Biol. Med. *96*:575.

Seligmann, M. (1957). Mise en évidence dans le sérum de malades atteints de lupus érythémateux disséminé d'une réaction de précipitation avec l'acide désoxyribonucléique. C. R. Acad. Sci., *245*:243.

Seligmann, M. (1958). Etudes immunologiques sur le lupus érythémateux disséminé. Rev. Fr. Et. Clin. Biol., *3*:558.

Seligmann, M. (1963). DNA antibodies. Arthritis and Rheumatism, *6*:542.

Seligmann, M., and F. Milgrom (1957). Mise en évidence par la fixation du complement de la réaction entre acide désoxyribonucléique et le serum de malades atteints de lupus érythémateux disséminé. C. R. Acad. Sci., *245*:1472.

Stollar, D., and L. Levine (1961). Antibodies to denatured DNA in a lupus erythematosus serum. J. Immun., *87*:477.

Stollar, D., L. Levine, and J. Marmur (1962a). Antibodies to denatured DNA in lupus erythematosus serum. II. Characterization of antibodies in several sera. Biochem. Biophys. Acta, *61*:7.

Stollar, D., L. Levine, H. I. Lehrer, and H. Van Vunakis (1962b). The antigenic determinants of denatured DNA reactive with lupus erythematosus serum. Proc. Natl. Ac. Sci., *48*:874.

Stollar, D., and L. Levine (1963). Antibodies to denatured deoxyribonucleic acid in lupus erythematosus serum. IV. Evidence for purine determinants in DNA. Arch. Biochem. Biophys., *101*:417.

Stollar, B. D., and A. L. Sandberg (1966). Comparison of antibodies reacting with DNA. I. Systemic lupus erythematosus sera and rabbit antibodies induced by DNA-methylated bovine serum albumin complexes. J. Immun., *96*:755.

Tan, E. M., P. H. Schur, R. I. Carr, and H. G. Kunkel (1966). Deoxyribonucleic acid (DNA) and antibodies to DNA in the serum of patients with systemic lupus erythematosus. J. Clin. Invest., *45*:1732.

Wasserman, E., and L. Levine (1961). Quantitative micro-complement fixation and its use in the study of antigenic structure by specific antigen-antibody inhibition. J. Immun., *87*:290.

DNA AND NUCLEOPROTEIN DETERMINANTS FOR SYSTEMIC LUPUS ERYTHEMATOSUS SERA *

B. DAVID STOLLAR

Department of Biochemistry, Tufts University School of Medicine Boston, Massachusetts

Studies of deoxyribonucleic acid (DNA) and nucleoprotein determinants for systemic lupus erythematosus (SLE) sera have had a two-fold objective. One goal has been to learn what structures act as immunizing agents in this disease, and the second, to characterize sera for use as immunochemical reagents in studies of DNA and nucleoprotein structure. The determinants have been explored mainly by: (1) comparisons of direct complement (C′)-fixation reactivity of native and denatured DNA, of nucleoprotein, and of artificial DNA-histone complexes; and (2) inhibition of C′-fixation by fragments of the antigens.

I. SUMMARY OF STUDIES OF DNA DETERMINANTS FOR SLE SERA

It has become clear that SLE sera may contain a variety of anti-DNA antibodies, as judged by their reactivity with native or denatured DNA or both (Stollar *et al.*, 1962a; Arana and Seligmann, 1966; Tan *et al.*, 1966). The C′-fixation patterns we have obtained with whole sera are summarized in Figure 1, the most common sera being those which reacted more strongly with denatured DNA than with native DNA.

These variations, and variations in susceptibility to inhibition by oligonucleotides (Stollar *et al.*, 1962b) reflect a wide variety of possible determinant structures. For some sera, these structures have been at least partially identified by hapten inhibition experiments, and the bases themselves shown to determine specificity (Stollar *et al.*, 1962a; Stollar

* Research described in this article was supported in part by grants #GB2770 and GB5606 from the National Science Foundation.

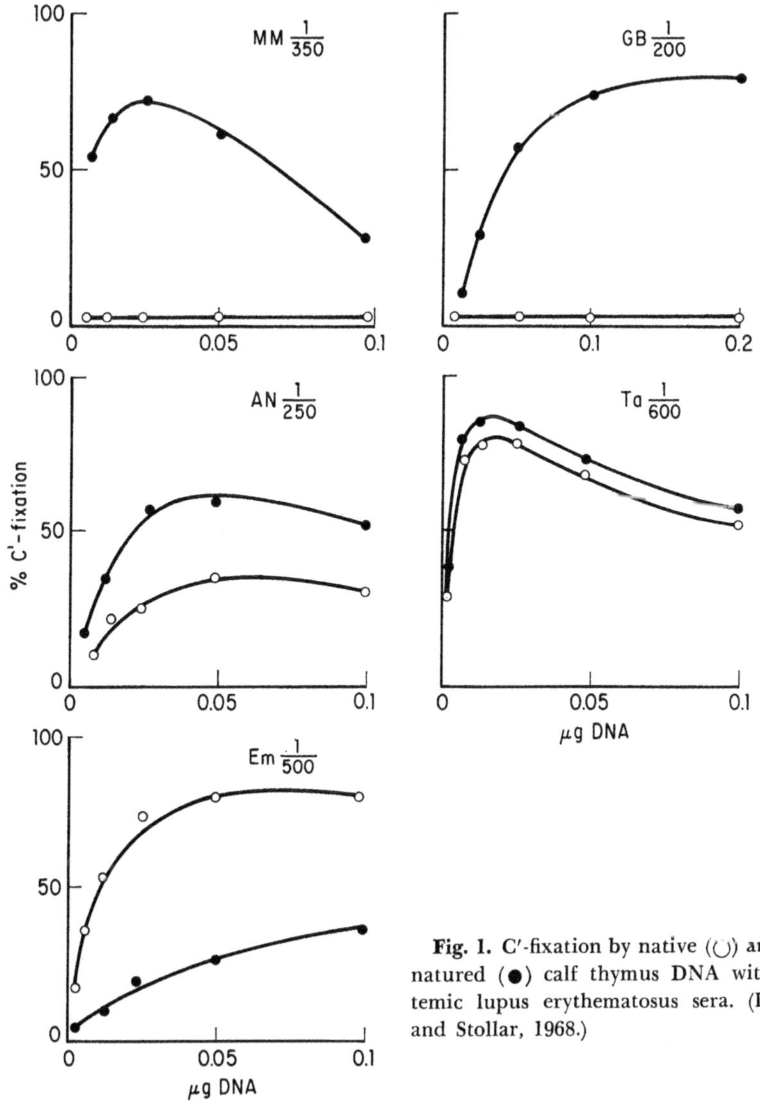

Fig. 1. C′-fixation by native (○) and denatured (●) calf thymus DNA with systemic lupus erythematosus sera. (Levine and Stollar, 1968.)

and Levine, 1963). In relatively simple instances single purine bases or oligonucleotides containing either thymidylic acid alone or cytidylic acid residues alone caused the most effective inhibition. These instances all involved sera which reacted with denatured but not with native DNA. With other sera, mixed purine-pyrimidine sequences were probably involved as determinants since partially digested DNA, containing all bases, inhibited more effectively than apurinic acid, containing only pyrimidines. The variations in inhibition by these two preparations are shown for several sera in Table 1.

Table 1

INHIBITION OF SLE SERA BY DNA DIGEST AND APURINIC ACID
(Levine and Stollar, 1968)

SLE Serum	μg of Inhibitor Required for 50% Inhibition by	
	DNA Digest	Apurinic Acid
M.M.	0.007	0.005
M.T.	0.07	0.25
N.A.	0.04	0.07 [a]
Ab.	0.03	[b]
J.W.	0.05	[c]
S.R.	0.05	[d]

[a] Maximum inhibition 60%.
[b] Maximum inhibition 20%, at 0.15 μg.
[c] Maximum inhibition 30%, at 0.2 μg.
[d] Maximum inhibition 20%, at 0.25 μg.

It has not been possible to prepare well-defined oligonucleotide fragments which could cause potent inhibition of the reaction of some SLE sera with native DNA. This is to be expected, if the determinants for these sera involve the helical configuration of the DNA, as small DNA fragments probably lack such structure. The reactive configurations have therefore not been characterized.

II. STUDIES OF NUCLEOPROTEIN DETERMINANTS

SLE serum antibodies to nucleoprotein have been demonstrated in many laboratories (Holman and Kunkel, 1957; Friou, 1958; Weir and Holborow, 1962; Tan, 1967). It has been shown that both histone and DNA are required for maximal reactivity with these antibodies, and partial reactivity has been observed with artificial histone-DNA complexes having a protein/DNA ratio similar to that of isolated nucleoprotein (Holman and Deicher, 1959; Tan, 1967).

There are several ways in which the complexed histone and DNA could build a reactive nucleoprotein structure. Either the DNA or the histone could provide the reacting group, with the other component serving either to form a suitable aggregate or to influence suitably the configuration of the reacting group. Alternatively, a true junctional region could involve both the nucleic acid and the protein in one determinant configuration. On the basis of quantitative C′-fixation experiments, it appears that SLE sera may contain a variety of antibodies reactive with these different possible antigenic forms.

A. Sera Which React with Free Native DNA

The C′-fixation reactions of serum Em. with native DNA and with calf thymus nucleoprotein (prepared by the method of Zubay and Doty, 1959) are shown in Figure 2. The maxima of the C′-fixation curves were similar, but the reaction with nucleoprotein required much higher antigen concentrations, in terms of DNA content. This serum did not react with free histone. The reaction of the nucleoprotein was markedly inhibited by the presence of excess free native DNA, indicating that the anti-DNA antibody was indeed reacting with DNA in the nucleoprotein.

From measurements of the degree of shift to higher antigen concentration for the C′-fixation reaction with nucleoprotein, it was estimated that 1–2% of the DNA within the nucleoprotein was available for reaction with the antibody. In the presence of added histone, the C′-fixation curve was shifted further toward higher DNA concentrations, suggesting further masking of the DNA (Fig. 2). Unlike excess DNA, the histone did not reduce the amount of C′ fixed. Upon exposure of nucleoprotein to 3 *M* NaCl under appropriate conditions followed by dialysis against

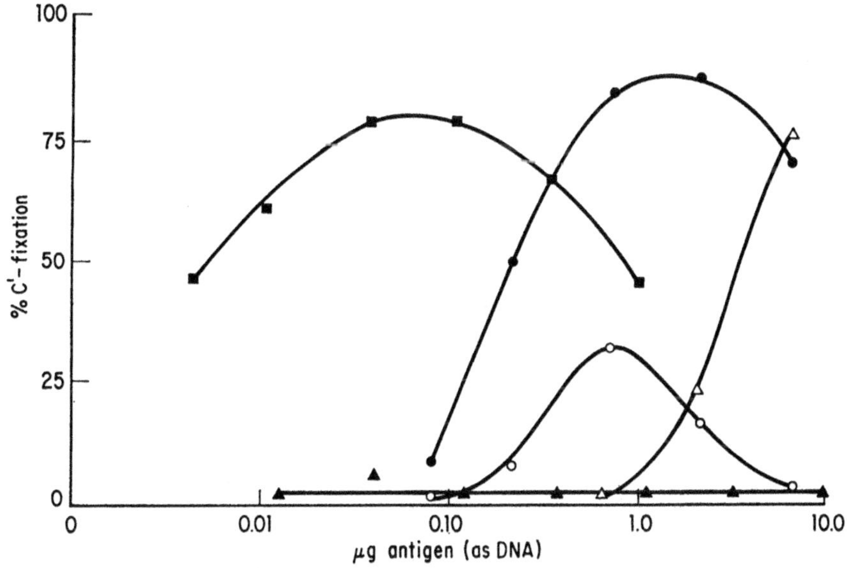

Fig. 2. C′-fixation by SLE serum Em. and: histone (▲), native calf thymus DNA (■), calf thymus nucleoprotein, alone (●), and in the presence of excess DNA (○), and in the presence of excess histone (△).

an isotonic saline-Tris buffer (.01 *M* Tris, 0.14 *M* NaCl, pH 7.4, with 5×10^{-4} *M* Mg^{++} and 1.5×10^{-4} Ca^{++}), the nucleoprotein curve shifted toward lower antigen concentration. The extent of this shift was variable, but up to 20% of the DNA content could remain exposed or available to the antibody after this dissociation and reassociation. In a reverse procedure, formation of artificial histone-DNA complexes caused a shift of the C′-fixation curve with DNA to higher concentrations, causing it to become similar to the nucleoprotein curve. In the presence of excess free native DNA, the C′-fixation by this histone-DNA complex was markedly inhibited just as with nucleoprotein. Thus, for this serum the DNA was a major reacting moiety of the nucleoprotein.

B. Reactivity of Serum Absorbed with DNA

Another serum (Ja.), which reacted with both native DNA and nucleoprotein, was absorbed with native DNA to allow examination of the residual reactivity with nucleoprotein. This absorbed serum fixed C′ with free histone, indicating that the protein could be a major contributor to the nucleoprotein determinant. The reaction of the absorbed serum with artificial histone-DNA complexes was greater than that with either the free histone or the nucleoprotein (Fig. 3). Thus, although DNA was not a direct reactant, its presence did enhance the reaction of the histone, possibly by inducing a more favorable size or configuration of the histone

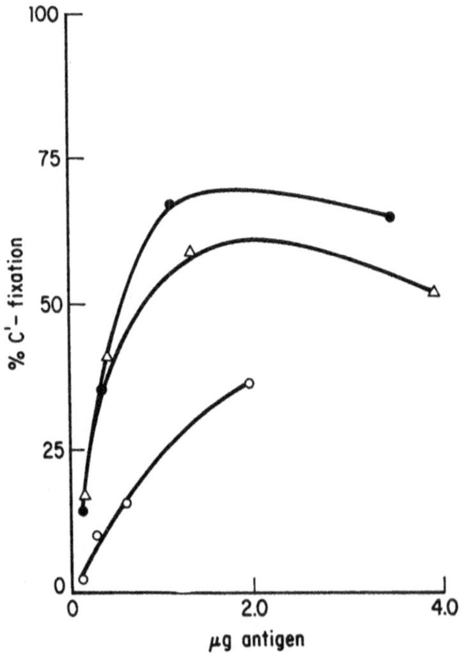

Fig. 3. C′-fixation by SLE serum Ja., which had been absorbed with native DNA, and calf thymus histone (○), nucleoprotein (△), and histone-DNA complexes consisting of 15 μg/ml and 2.5 μg/ml of histone and DNA respectively (●).

or causing a new type of determinant to be formed at a junctional region in the complex.

C. Serum Which Did Not React with Free DNA or Free Histone

Serum Pc. did not fix C′ with either free native or denatured DNA or with histone, but it did react with nucleoprotein and, even more strongly, with artificial histone-DNA complexes (Fig. 4) (Stollar, 1967). In further experiments, complexes were prepared with varying fractions of histones according to the method of Johns *et al.* (1964a, 1964b) and characterized by their arginine and lysine contents, as shown in Table 2.

Whole histone formed more reactive complexes than fractions of histone. When native DNA was used to form complexes, the lysine-rich

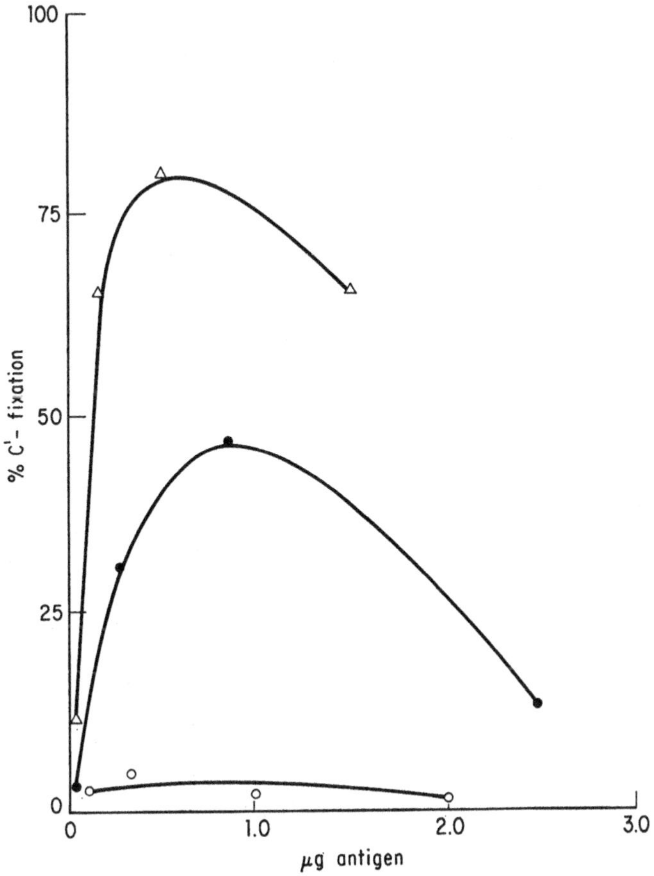

Fig. 4. C′-fixation by SLE serum Pc. and: native calf thymus DNA or calf thymus histone (○); calf thymus nucleoprotein (●); and a histone-DNA complex containing 5.0 μg/ml and 2.5 μg/ml of histone and native DNA respectively (△). (Stollar, 1967.)

Table 2

LYSINE AND ARGININE CONTENTS OF HISTONE FRACTIONS

	Whole Histone	F1(A)	F2a	F2b	F3
Lysine [a]	16.2	26.7	9.8	12.1	11.0
Arginine [a]	6.6	2.2	10.9	7.9	10.3
Lysine/arginine	2.5	14.8	0.9	1.5	1.1

[a] Expressed as moles/100 moles of amino acids.

histone fraction, F1(A), was much more effective than any of the others in forming serologically reactive structures (Fig. 5). The reaction of such histone-native DNA complexes was abolished by an excess of free native DNA but not by free histone. Reactivity with these histone-native DNA complexes was lost when the serum was absorbed with native DNA. Thus, it appeared that DNA was a major contributor to this determinant even though it did not react directly by itself in C′-fixation.

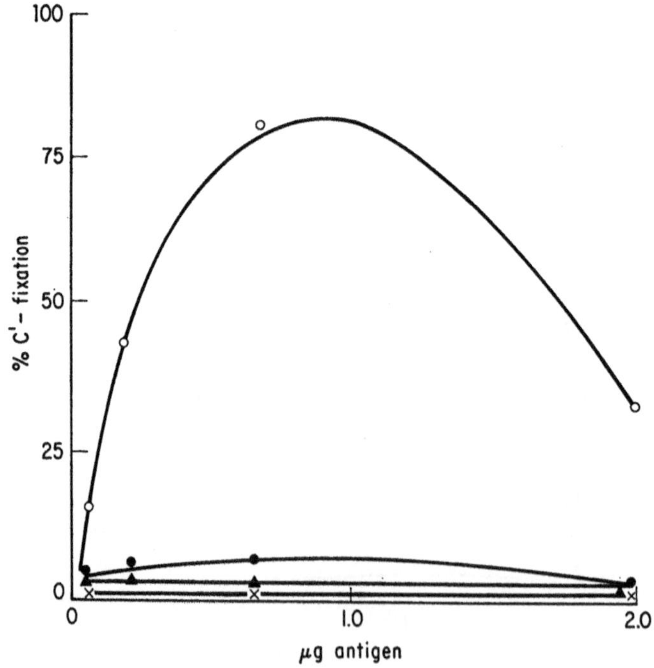

Fig. 5. C′-fixation of serum Pc (1/100) with complexes containing 2.5 μg/ml of native DNA and 7.5 μg/ml of histone fractions F1(A) (○); F2a (●); F2b (▲); and F3 (×). (Stollar, 1967.)

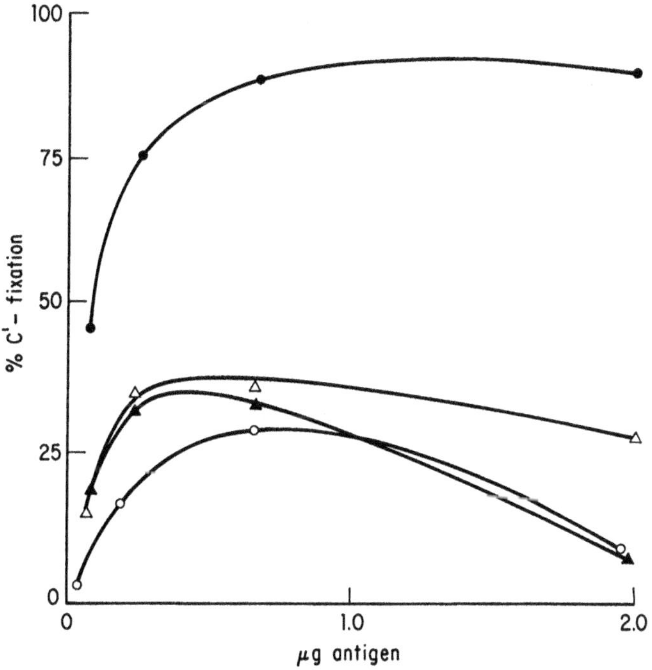

Fig. 6. C′-fixation of serum Pc (1/100) with complexes containing 2.5 μg/ml of denatured DNA and 7.5 μg/ml of histone fractions F1(A) (○); F2a (●); F2b (△); and F3 (▲). (Stollar, 1967.)

A second pattern of reactivity for the complexes was observed when denatured DNA or RNA was used in place of native DNA. Whole histone was still more effective than the histone fractions. However, the arginine-rich F2a histone fraction was much more effective than the lysine-rich F1(A) fraction in producing serologically reactive complexes (Fig. 6), in contrast to the results obtained with native DNA. Further, excess F2a histone completely abolished the reaction of F2a-denatured DNA complexes, while an excess of free native DNA or of RNA had no inhibitory effect. When the serum was absorbed with native DNA, it still reacted with F2a histone-denatured DNA complexes even though it no longer reacted with histone-native DNA complexes. Absorption with denatured DNA did not remove either reactivity. It was evident therefore that a second determinant for this serum consisted mainly of a histone structure.

III. STUDIES OF DETERMINANTS OF ANTI-NUCLEAR ANTIBODIES BY IMMUNOFLUORESCENT REACTIONS

Sera from virtually all SLE patients contain anti-nuclear antibodies detectable by immunofluorescence (Hamard *et al.*, 1964; Gonzalez and Rothfield, 1966). Four major morphological patterns of fluorescent staining have been observed and are described as being either diffuse (homogeneous), speckled, nucleolar, or peripheral (shaggy). There is evidence that the diffuse staining pattern correlates with anti-nucleoprotein activity (Friou, 1958; Beck, 1961), and that the speckled pattern correlates with antibodies to a nuclear glycoprotein (Lachman and Kunkel, 1961). The peripheral pattern has been related to anti-DNA antibodies (Casals *et al.,* 1958). Hamard *et al.* (1964) also found that sera with anti-DNA antibodies gave a peripheral staining pattern, but he also found that some sera gave this pattern although they did not contain anti-DNA antibodies detectable by C′-fixation. Such apparent discrepancies require more study since a variety of determinants may actually be involved. Also, all classes of immunoglobulin may take part in immunofluorescence reactions (Barnett *et al.,* 1964; Gonzalez and Rothfield, 1966), whereas IgA antibodies do not fix C′ (Cohen and Porter, 1964) and macroglobulins do not fix C′ with all antigens (Cunniff and Stollar, 1968).

In a recent further study of these relationships (Rothfield and Stollar, 1967), it has been found that all 16 sera which reacted by C′-fixation with both DNA and nucleoprotein uniformly caused a peripheral nuclear staining pattern detectable with fluorescent sera specific for IgG. All 31 sera which did not fix C′ with either DNA or nucleoprotein were negative in tests for IgG peripheral staining activity. On the other hand, three sera contained IgG antibody which gave peripheral fluorescence and fixed C′ with nucleoprotein but not with free DNA. One of these sera was studied in more detail, and is the example described in section II.C of this article. As indicated, native DNA did contribute strongly to a nucleoprotein determinant for C′-fixation even though it did not react directly itself. Both the peripheral staining and the C′-fixing reactivity with this DNA determinant were removed by absorption of the serum with native DNA. Thus it appeared that native DNA within the nucleoprotein structure was indeed a determinant for the peripheral staining pattern.

It was further found in this study that the peripheral pattern was not produced by either rabbit or SLE sera which fixed C′ with only denatured DNA and not with native DNA or nucleoprotein.

SLE serum Ja., described in section II.B of this article, was also examined by immunofluorescence (Rothfield and Stollar, 1967). Whole serum, which fixed C′ with both native DNA and nucleoprotein, produced peripheral nuclear staining due to IgG antibody. Absorption of the serum with native DNA removed this peripheral staining activity. The remaining anti-nucleoprotein antibody, which was at least partly reactive with histone determinants by C′-fixation (but possibly with junctional structures as well, as described above), was accompanied by a diffuse antinuclear staining activity. Many more detailed studies will be required to dissect such sera further and to match more precisely the various possible nucleoprotein determinants with specific staining morphology.

IV. SUMMARY

It has been shown that a wide variety of DNA determinants exist for SLE sera, though these determinants have been characterized in only a few cases. The bases themselves are clearly involved in some instances, in which reactivity with denatured DNA is predominant. Determinants for sera reactive with native DNA have not been clarified.

Nucleoprotein also contains several types of determinants. Evidence has been obtained that both DNA alone and histone alone may contribute the major parts of some of these determinant structures. The other component may be necessary to cause suitable aggregation or configuration for maximal reactivity. There is no clear evidence for or against the existence of truly junctional determinants consisting of parts of both protein and nucleic acid. As the nature of possible determinants becomes clarified, it is hoped to gain a better understanding of the immunological aspects of SLE and at the same time to provide useful reagents for the study of the organization of nucleoproteins and other nuclear material.

References

Arana, R., and M. Seligmann (1966). Présence, dans certains sérums de lupus, d'anticorps anti-acide deoxyribonucléique natif. Compt. Rend. Acad. Sci., *262*:2665–2668.

Barnett, E. V., J. J. Condemi, J. P. Leddy, and J. H. Vaughan (1964). $Gamma_2$, $Gamma_{1A}$ and $Gamma_{1M}$ antinuclear factors in human sera. J. Clin. Invest., *43*:1104–1115.

Beck, J. S. (1961). Variations in the morphological patterns of autoimmune nuclear fluorescence. Lancet, (*i*):1203–1205.

Casals, S. P., G. J. Friou, and P. O. Teague (1963). Specific nuclear reaction pattern of antibody to DNA in lupus erythematosus sera. J. Lab. Clin. Med., *62*:625–631.

Cohen, S., and R. R. Porter (1964). Structural and biological activity of immunoglobulins. Advan. Immunol., *4*:287–349.

Cunniff, R., and B. D. Stollar (1968). Properties of 19S antibodies in complement fixation. I. Temperature dependence and role of antigen structure. J. Immun., *100*:7–14.

Friou, G. J. (1958). Identification of the nuclear component of the interaction of lupus erythematosus globulin and nuclei. J. Immun., *80*:476–481.

Gonzalez, E., and N. Rothfield (1966). Immunoglobulin class and pattern of nuclear fluorescence in systemic lupus erythematosus. New Eng. J. Med., *274*:1333–1338.

Hamard, M., A. Cannat, and M. Seligmann (1964). La recherche d'anticorps anti-nucléaires par immunofluorescence. (Étude de 1430 serums.) Rev. franç. d'études clin et biol., *9*:716–728.

Holman, H. R., and H. R. Deicher (1959). The reaction of lupus erythematosus (LE) cell factor with deoxyribonucleoprotein of the cell nucleus. J. Clin. Invest., *38*:2059–2072.

Holman, H. R., and H. G. Kunkel (1957). Affinity between the lupus erythematosus serum factor and cell nuclei and nucleoprotein. Science, *126*:162–163.

Johns, E. W. (1964a). Studies on histone. 7. Preparative methods for histone fractions from calf thymus. Biochem. J., *92*:55–59.

Johns, E. W. (1964b). Studies on lysine-rich histones, in The Nucleohistones, J. Bonner and P. T'so, eds., Holden-Day, San Francisco, pp. 52–58.

Lachman, P., and H. G. Kunkel (1961). Correlation of antinuclear antibodies to nuclear staining patterns. Lancet, (*ii*):436–437.

Levine, L., and B. D. Stollar (1968). Nucleic acid immune systems. Progress in Allergy, in press.

Rothfield, N., and B. D. Stollar (1967). The relation of immunoglobulin class, pattern of antinuclear antibody and complement fixing antibodies to DNA in sera from patients with systemic lupus erythematosus. J. Clin. Invest., *46*:1785–1794.

Stollar, B. D. (1967). Studies on nucleoprotein determinants for systemic lupus erythematosus serum. J. Immun., *99*:959–965.

Stollar, D., and L. Levine (1963). Antibodies to denatured deoxyribonucleic acid in lupus erythematosus serum. IV. Evidence for purine determinants in DNA. Arch. Biochem. Biophys., *101*:417–422.

Stollar, D., L. Levine, H. I. Lehrer, and H. Van Vunakis (1962a). The antigenic determinants of denatured DNA reactive with lupus erythematosus serum. Proc. Nat. Acad. Sci. (U. S.), *48*:874–880.

Stollar, D., L. Levine, and J. Marmur (1962b). Antibodies to denatured deoxyribonucleic acid in lupus erythematosus serum. II. Characterization of antibodies in several sera. Biochim. Biophys. Acta, *61*:7–18.

Tan, E. (1967). An immunologic precipitin system between soluble nucleoprotein and serum antibody in systemic lupus erythematosus. J. Clin. Invest., *46*:735–745.

Tan, E., P. Schur, R. I. Carr, and H. G. Kunkel (1966). Deoxyribonucleic acid (DNA) and antibodies to DNA in the serum of patients with systemic lupus erythematosus. J. Clin. Invest., *45*:1732–1740.

Weir, D. M., and E. J. Holborow (1962). Serum anti-nuclear factor. Ann. Rheum. Dis., *21*:40–44.

Zubay, G., and P. Doty (1959). The isolation and properties of deoxyribonucleoprotein particles containing single nucleic acid molecules. J. Mol. Biol., *1*: 1–20.

CELLULAR PHASES OF THE IMMUNE RESPONSE FOLLOWING IMMUNIZATION WITH MAMMALIAN AND BACTERIAL RIBOSOMES

M. C. DODD AND N. J. BIGLEY
Ohio State University, Columbus, Ohio

I. INTRODUCTION

With the avalanche of information concerning nucleic acids currently accumulating from many scientific disciplines, a more complete knowledge of the biological role of nucleic acids is evolving simultaneously. As immunologists, it is necessary to consider that any given animal host is actually a complete and functioning ecological unit of specialized cells, all of which were derived from one fertilized ovum. As yet, we are still uncertain as to what particular cells function as immunocompetent cells in maintaining the integrity of the animal in a continually changing environment. All somatic cells of a given individual in interphase supposedly contain the same genetic material (DNA) of which only a small portion is involved or "turned on" in the transcription of messenger RNA (Frenster, 1965). It is tempting to speculate that the "turning-on" and "turning-off" of genes might well be one physiological function of the host's immune system. With this perspective, we wish to present a portion of our research experience in the immunization of experimental animals with mammalian and bacterial ribosomes.

We have demonstrated that the injection of ribosomes results in anti-RNA reactivities, as described by others (Plescia and Braun, 1967), and can produce concurrent autoimmune disease. Other phenomena associated with ribosome injections also have been observed, namely, the appearance of antibodies to other antigens and cellular immunity. These observations will be discussed in this communication.

II. REVIEW OF PREVIOUS WORK

The specificities of antibodies to nucleic acids have been studied in our laboratory, for the most part, by passive hemagglutination and hemagglutination-inhibition techniques (Dodd *et al.*, 1962; Bigley *et al.*, 1963). Similar to the findings of others (Plescia *et al.*, 1965a, b, c; Plescia and Braun, 1967; Sela and Ungar-Waron, 1965; Sela *et al.*, 1964; Sela, 1966; Barbu and Panijel, 1960; Barbu *et al.*, 1963; Beiser and Erlanger, 1966; Tannenbaum and Beiser, 1963) the hemagglutinins measured in antisera to ribosomes and sRNA are specific for nucleotides, nucleosides, purine and pyrimidine bases, as well as for the ribose component. Even the presence or absence of a phosphate group produces a serologic difference in hemagglutination-inhibition studies. Hemagglutinins with similar specificities for nucleotides, nucleosides, and bases were found in antisera to nuclei. Antisera to ribosomes, nuclei and ribosomal protein also contain hemagglutinins for ribosomal protein indicating the possible antigenic similarity of the protein associated with ribo- and deoxyribonucleic acids.

The hapten nature of nucleic acid antigens has been used in our laboratory to explain the development of agglutinins for normal human erythrocytes in rabbits immunized with multiple injections of either homologous liver ribosomes or with normal and malignant human nuclei, mitochondria and ribosomes. Since nucleotides are present in erythrocyte stromata (Everett, 1946; Behrendt, 1957; Prankhard, 1961; Pennell, 1964) treatment of the human red cell with either RNAase or DNAase rendered them inagglutinable by these sera. Recently, Troendly (1967) and Kurent (1967), using methylated BSA-nucleotide and MBSA-nucleoside triphosphates as adsorbents, showed that the common antigenic reactivity of human red cells, colon ribosomes and mitochondria is due to the existence of adenylic and uridylic acid moieties in all of these materials, while Gaffar (1967a, b) was able to adsorb anti-human erythrocyte reactivity from anti-human nuclear sera with nuclear sap fractions containing protein, RNA and DNA.

As noted in the review by Plescia and Braun (1967), rabbits immunized with nuclei, MBSA-RNA, or nucleosides and nucleotides coupled to protein or polymerized amino acid carriers, produce antibodies directed against the immunizing nucleic acids and their structural components, but no detectable disease manifestation ensues. However, in our investigations it is apparent that rabbits immunized with either homologous liver or rat liver ribosomes incorporated in complete Freund's adjuvant develop not only antibodies directed against the structural components of the immunizing nucleic acids but also an autoimmune hemolytic anemia 70–77 days after immunization. This autoimmune disease was passively transferable to normal, non-immunized recipient rabbits with immune serum or its crude 19S globulin component but not with the 7S globulin or albumin fractions. Adsorption of such sera with either sRNA or ribosomes removed the nucleic acid hemagglutinins as well as the capacity to produce autoimmune disease in normal recipients. Intramuscular injection of rabbits with homologous reticulocyte ribosomes prepared by Dr. Paul H. Aldenderfer (now of Chas. P. Pfizer Co., Maywood, N. J.) in complete Freund's adjuvant did not produce hemagglutinating antibodies but did stimulate complement-fixing antibodies, in marked contrast to results obtained with animals immunized with liver ribosomes in which both hemagglutinins and complement-fixing antibodies were produced. This curious finding is, as yet, not understood. However, animals actively immunized with reticulocyte ribosomes also developed a hemolytic anemia characterized by a marked decrease in hematocrit, hemoglobin and red cell values as well as anti-globulin positive erythrocytes. This hemolytic anemia was also transferable with sera or its 19S globulin component.

Although hemagglutinins for the homologous antigen were not demonstrable in antisera to reticulocyte ribosomes, they were detected in antisera to liver ribosomes. In some sera, they were equally reactive, while in others a quantitative difference was shown, indicated by a 4–6-fold difference in titer. With both types of antisera, the reticulocyte ribosomes were the more effective complement-fixing antigen.

III. APPEARANCE OF ANTIBODIES TO OTHER ANTIGENS AFTER INJECTION OF ANTI-RIBOSOME SERUM

Antibodies to a variety of tissue substances are detectable in sera of humans suffering from so-called autoimmune collagen-type diseases (Holman, 1965). In a comparison of DNA and RNA reactivities in the sera of patients with rheumatoid arthritis with those in the sera of supposedly normal individuals by bentonite flocculation, Scheetz (1965) and Larkin (1966) of our laboratory were able to demonstrate antibodies for both DNA and RNA in rheumatoid sera which were specific for nucleotide components of the nucleic acids.

In the sera of normal humans, Larkin was able to demonstrate a complement-dependent reactivity for the RNA-bentonite complex. Although, as Holman (1965) has stated, no evidence exists to indicate that autoantibodies are of any biological or clinical significance, it is tempting to suggest that such antibodies might serve a regulatory role in the efficient removal of dead or effete cells or of cellular debris. Fidalgo and Najjar (1967) have discussed the presence of an erythrophilic gamma globulin in canines which protects the red cell membrane from shearing forces and from potassium loss and described the existence in canines of a separate leucophilic gamma globulin which appears to function in the phagocytic mechanism.

The major portion of this report describes the appearance of antibodies for which there was no known parenteral stimulus in the sera of ribosome-injected animals as well as in the sera of normal rabbits injected with anti-ribosome sera. Furthermore, an abolition of specifically acquired immune tolerance to bovine serum albumin (BSA), to human gamma globulin (HGG) and to rabbit iso-antigen A was noted in rabbits after the injection of anti-ribosome rabbit serum containing the auto-antibodies described earlier by Dodd *et al.* (1962) and by Dodd (1963). Currently, Miller of our laboratory is studying such effects after transfer of rabbit anti-human ribosome sera, anti-sRNA (bovine liver) sera and anti-methylated bovine serum (MBSA)-RNA sera to immunized recipient rabbits. All of these antisera were prepared with antigens incorporated in complete Freund's adjuvant.

The anti-ribosome sera used in this investigation were the same sera described previously by Dodd *et al.* (1962) and Bigley *et al.* (1963). They were prepared by injecting rabbits with rat or rabbit liver ribosomes or with rabbit reticulocyte ribosomes. From a total of approximately 60 such animals, the sera of 20 have been used in this investigation. Fifty normal, healthy rabbits were injected with these anti-ribosome sera collected at various intervals after injection of ribosomes and the findings with sera of twenty of these animals are included in this report. Sixteen animals were used in the tolerance experiments of which 3 were immunized and 13 made tolerant to either BSA, HGG or rabbit erythrocyte iso-antigen A. The rabbits tolerant to rabbit iso-antigen were provided by Dr. Carl Cohen, Western Reserve University, Cleveland, Ohio, and the procedure has been described by him (Cohen, 1958, 1959).

Rabbits were injected intraperitoneally with 110 mg of BSA on each of the first 5 days of life. No antibody to BSA was detected 100 days later after 5 intravenous injections on alternate days totaling 550 mg BSA. Two normal adult animals were immunized at the same time with 5 injections (110 mg BSA each) of BSA. Rabbits were injected with 100 mg HGG intraperitoneally on the second day of life and challenged at 45 and 73 days of age with 50 mg doses of HGG (Garvey *et al.,* 1960; Hanan and Oyama, 1954; Terres and Hughes, 1959; Sercarz and Coons, 1963). Unless otherwise stated, the anti-ribosome sera used were obtained from rabbits 69 days after intramuscular injection of 77 mg of ribosomes in Freund's complete adjuvant (Dodd *et al.,* 1962).

Forty-three days after the BSA injections, tolerant rabbits were challenged; they were injected intravenously with 5.0 ml of anti-ribosome rabbit serum, which was repeated 4 days later. The technique of passive hemagglutination, using bis-diazotized benzidine (BDB) as coupling agent and normal human group O Rh+ red cells as carrier, was used in the examination of these sera (Bigley *et al.*, 1963).

Antibody for BSA was demonstrable at dilutions of 1:64 and 1:128 in the sera of animals 19 hr after injection of the anti-ribosome serum (Table 1). Most of this antibody disappeared rapidly from the serum of the tolerant animals, the titer being 1:2 on the fourth day after injection. Similar results were observed after the second injection of anti-ribosome sera, but in this case the titer again increased to 1:16 between 7 and 13 days and dropped again to 1:4 during the next week. At this time the marked hematologic changes produced in normal rabbits by anti-ribosome serum (Dodd *et al.*, 1962; Dodd, 1963) were evident.

Hemagglutination of BSA-red cells by a 1:4 dilution of serum obtained from an immunized animal 19 hours after injection of anti-ribosome serum was inhibited completely with 17.2 μg of BSA and partially by 4.3 μg. Hemagglutination by the same dilution of serum taken at the same time from a tolerant animal was inhibited completely by a concentration of 68.8 μg of BSA and partially by 17.2 μg.

The antibodies in the sera of both animals were not good precipitins. When BSA in concentrations ranging from 1100 μg/0.1 ml to 2.1 μg/0.1 ml were added to 1:4 dilutions of the serum of the immunized animal, some precipitate was observed with 550 μg of BSA after 3 hr at room

Table 1

BSA Antibodies in Sera from BSA-tolerant and Immune Rabbits After Injection of Anti-ribosome Serum

Serum	Time After Injection	Reciprocal of Serum Dilution							
		2	4	8	16	32	64	128	C
Tolerant	0	–	–	–	–	–	–	–	–
	19 hr	4+	4+	4+	3+	2+	+	–	–
	4 days	2+	–	–	–	–	–	–	–
	19 hr after 2nd serum injection	S	4+	3+	2+	+	+	–	–
	2 days	2+	–	–	–	–	–	–	–
	5 days	+	–	–	–	–	–	–	–
	7 days	+	–	–	–	–	–	–	–
	13 days	4+	3+	2+	+	–	–	–	–
	21 days	3+	+	–	–	–	–	–	–
Immune	0 hr	2+	+	–	–	–	–	–	–
	19 hr	S	4+	4+	3+	2+	+	+	–

temperature. The same dilution of serum from the tolerant animal showed at this time only a trace of precipitate with 8.6 μg of BSA. After 6 days in the refrigerator, serum from the immunized animal had formed only slight amounts of precipitate with BSA concentrations of 550, 275, and 138 μg, and only traces of precipitate were observed in the serum of the tolerant animal with 8.6 and 4.3 μg. Precipitates in both sera were dissolved by addition of excess antigen.

These experiments were repeated with 8 other rabbits tolerant to BSA and with 4 rabbits tolerant to HGG. Twenty-four days after the second challenge dose, the animals made tolerant to HGG were injected intravenously with 5.0 ml anti-ribosome serum and eight days later with a second injection of the same. The data presented in Table 2 show the resulting appearance of anti-HGG after two separate injections of anti-ribosome serum in the sera of two animals tolerant to HGG. Antibodies for HGG were detected in the sera of both animals 17 hours after injec-

Table 2

HGG Antibodies in Sera from HGG-Tolerant Rabbits After Injection of Anti-ribosome Serum

Time After	Reciprocal of Serum Dilution										
Injection	2	4	8	16	32	64	128	256	512	1024	C
0	−	−	−	−	−	−	−	−	−	−	−
After 1st											
17 hr	S	4+	−	−	−	−	−	−	−	−	−
2 day	4+	3+	+	−	−	−	−	−	−	−	−
3 day	2+	−	−	−	−	−	−	−	−	−	−
After 2nd											
3 hr	3+	2+	+	+	+	+	±	±	−	−	−
24 hr	3+	3+	−	−	−	−	−	−	−	−	−
2 day	3+	+	−	−	−	−	−	−	−	−	−
3 day	4+	3+	2+	+	−	−	−	−	−	−	−
0	−	−	−	−	−	−	−	−	−	−	−
After 1st											
17 hr	2+	+	−	−	−	−	−	−	−	−	−
2 day	−	−	−	−	−	−	−	−	−	−	−
4 day	4+	4+	3+	2+	+	+	−	−	−	−	−
8 day	4+	3+	+	+	−	−	−	−	−	−	−
After 2nd											
3 hr	4+	4+	4+	3+	2+	+	+	−	−	−	−
1 day	S	S	4+	2+	+	+	+	+	−	−	−
2 day	3+	+	+	−	−	−	−	−	−	−	−

tion of the anti-ribosome serum in titers of 1:4. Antibody was absent from one serum by the second day and minimally present in the other. Even higher titers were obtained 4 to 8 days later with the one serum tested. Seventeen hours after injection of anti-ribosome serum, the sera of the recipients demonstrated titers of 1:4 for the HGG-coupled red cells and in 2 days, a titer of 1:8 was detected in one serum, while no antibody was present in the serum of the other. At 3 days, the serum of one animal contained only a reactivity for the HGG-cells in a serum dilution of 1:2 while at 4 days, the other animal's serum exhibited reactivity for the HGG-cells in a dilution of 1:64 and by 8 days, this titer had diminished to a dilution of 1:16. At this time (8 days) all animals received a second injection of anti-ribosome serum. Antibody was now detectable three hours later at titers of 1:64 to 1:128. These titers had decreased somewhat by the second and third days. Of the results shown in Table 2, the one serum exhibited reactivity in a dilution as high as 1:64 and the other animal's serum reactivity had risen from a dilution of 1:16 to 1:128 for the HGG-coupled red cells by three hours after injection of anti-ribosome sera. By 24 hours, titers of 1:4 and 1:256 were evident and by 48 hours, the recipient sera contained reactivity for HGG-cells in dilutions of 1:4 and 1:8.

Similar results were obtained in two rabbits tolerant to iso-antigen A (Table 3). Two hours after the injection of anti-ribosome serum, anti-A hemagglutinins were detected in the sera of both animals, but these hemagglutinins almost completely disappeared within 5 to 11 days. By the 21st day there seemed to be a slight increase in titer. A second injection of serum into one of these animals on the 24th day did not cause an increase in the titer of iso-hemagglutinins but the agglutination was stronger. It was necessary to sacrifice this animal 4 days after the second injection because of the severity of the auto-immune disease present. Comparing these data with the results for the similar experiments with BSA or HGG in Tables 1 and 2, it can be noted that reactions with the protein antigens appeared in higher titers, usually disappeared more rapidly and the secondary rise in titer was more pronounced than with the iso-hemagglutinin.

Further investigations on the effects of injecting anti-ribosome serum into recipient animals showed that: (1) the sera of the donor animals, the ribosome-injected animals, contained antibodies for both BSA and HGG; (2) passive transfer of such sera into normal animals was followed by the occurrence of antibodies to BSA and HGG in much greater titers than would be expected by the transfer of 5.0 ml and subsequent dilution of this volume in the body of the recipient; and (3) after injection of anti-ribosome serum the sera of the recipients were found on periodic re-examination to contain reactivities for BSA or HGG.

Table 3

SPECIFIC ISO-ANTIBODIES IN SERA OF RABBITS TOLERANT TO RABBIT ISO-ANTIGEN AFTER INJECTION OF ANTI-RIBOSOME SERUM

Serum	Time After Injection	Reciprocal of Serum Dilution 0	2	4	8	16	32	C
Tolerant	0	–	–	–	–	–	–	–
(1)	2 hr	4+	4+	2+	–	–	–	–
	22 hr	4+	2+	+	+	–	–	–
	3 day	3+	3+	+	–	–	–	–
	5 day		+	–	–	–	–	–
	11 day		2+	–	–	–	–	–
	21 day		2+	–	–	–	–	–
	24 day		2+	+	–	–	–	–
	25 day		2+	+	–	–	–	–
	27 day		2+	+	+	–	–	–
	32 day		2+	–	–	–	–	–
Tolerant	0	–	–	–	–	–	–	–
(2)	2 hr	2+	2+	+	–	–	–	–
	22 hr	3+	3+	–	–	–	–	–
	3 day	2+	2+	–	–	–	–	–
	5 day		2+	–	–	–	–	–
	11 day		–	–	–	–	–	–
	21 day		2+	+	+	–	–	–
	24 day *		2+	+	+	+	–	–
	1 day		3+	2+	+	+	–	–
	3 day		4+	3+	2+	+	–	–
	4 day		4+	3+	+	–	–	–

* Second injection of anti-ribosome serum given on 24th day.

The results shown in Table 4 indicate that HGG and BSA antibodies were detected in the serum of an animal that had received 230 mg of rabbit liver ribosomes in 3 injections over a 108-day period. The reactivity in the serum for BSA-coupled cells persisted to a dilution of 1:64. Absorption with BSA-coupled cells removed this agglutinin from the serum, leaving the agglutinin for HGG-cells in the serum. Conversely, absorption with HGG-cells removed the HGG-agglutinins but not those for BSA-coupled cells. Absorption with cells that had been treated with the coupling agent (BDB) only, did not remove much agglutinating activity for either BSA- or HGG-coupled erythrocytes. Likewise, the serum contained agglutinins for HGG-cells in a dilution of 1:32 which were reduced to a titer of 1:2 after one absorption with HGG-cells. Absorption with BSA-coupled cells or BDB-treated normal cells did not appreciably

Table 4

ANTI-BSA AND ANTI-HGG PRESENT IN SERUM OF RIBOSOME-INJECTED ANIMAL

Serum	BSA-Coupled Cells. Reciprocal of Serum Dilutions							HGG-Coupled Cells. Reciprocal of Serum Dilutions						
	C	2	4	8	16	32	64	C	2	4	8	16	32	64
Unabsorbed	–	4+	4+	3+	2+	2+	+	–	4+	3+	3+	3+	+	–
Abs. 1x with BSA-NHO$^+$	–	–	–	–	–	–	–	–	4+	2+	+	+	–	–
Abs. 1x with HGG-NHO$^+$	–	4+	4+	3+	2+	+	–	–	+	–	–	–	–	–
Abs. 1x with BDB-NHO$^+$	–	4+	4+	2+	+	–	–	–	4+	3+	2+	2+	±	–

diminish reactivity for HGG-coupled erythrocytes. From these data it is apparent that the HGG and BSA antibodies which developed in the ribosome-injected animals possessed marked specificity for their respective antigen.

Further evidence for the specificity of the agglutinins for BSA and HGG is observable from the data in Table 5. Nineteen hours after injection of 5.0 ml of anti-ribosome serum, the serum of recipient tolerant animals contained agglutinins in dilutions as high as 1:64 for BSA and HGG. Serum dilutions of 1:2 were selected for hemagglutination-inhibition studies since all reactions were maximal (4+) at this dilution. One thousand μg quantities of HGG completely inhibited HGG-cell aggluti-

Table 5

INHIBITION OF HEMAGGLUTININS IN SERA OF TOLERANT RABBITS WITH SPECIFIC ANTIGENS

Serum	Test Cells	Inhibitors: None	HGG	BSA
19 Hours After Anti-ribosome Serum				
Tolerant to BSA	HGG	4+	–	4+
	BSA	4+	4+	+
Tolerant to HGG	HGG	4+	–	2+
	BSA	4+	3+	–

nation in the sera of both BSA and HGG-"tolerant" animals while they did not interfere with the BSA agglutination. The same quantity of BSA decreased from a 4+ agglutination to a 1+ titer in the serum from the BSA-"tolerant" rabbit and did not interfere with the HGG-cell agglutination. However, in the serum from the HGG-"tolerant" animal, 1000 μg amounts of BSA caused complete inhibition of the 4+ agglutination of BSA-coupled cells and at the same time diminished the 4+ agglutination of HGG-cells from 4+ to 2+. Nevertheless, a striking degree of specificity for the respective antigens was again demonstrated.

In an attempt to examine further the mechanism by which passive transfer apparently produced antibodies in the sera of the recipient animals for antigens to which the animals were never intentionally immunized, two rabbits were injected intravenously with 3.0 ml of the anti-ribosome serum described in Table 4. One sample had been adsorbed with BSA-coupled cells and the other, with HGG-coupled erythrocytes prior to injection into recipients. In the animals injected with anti-ribosome serum, absorbed with BSA-coupled cells, no antibody to BSA was detected at 1-day and only a minimal amount by the second day (1:2), while reactivity for HGG-cells was present both days in low serum dilutions (1:4) (Table 6). However, by the 7th day, the serum titer was 1:64 for BSA-cells and greater than 1:64 for HGG-cells. By 34 days, long after any direct effects of transferred antibody might be detectable, titers of 1:256 for both BSA- and HGG-coupled cells were observed. At this time the serum of the recipient of BSA-absorbed anti-ribosome serum also was found to contain agglutinins (at titers of 1:64 and 1:128) for normal human red cells, both for cells treated with BDB as well as for untreated cells, and also for hemocyanin-treated red cells. The animal injected with anti-ribosome serum, absorbed with HGG-cells, exhibited serum reactivity for both BSA- and HGG-coupled cells prior to injection. One day later a difference was noted in that there was no increase in titer for the absorbing cell (HGG-cell), but agglutination with BSA-coupled cells occurred in a dilution of serum as high as 1:32. At 7 and 34 days, the titers of the recipient animal's serum for either HGG- and BSA-coupled cells were the same (at least 1:64 and 1:1024). Also, at 34 days, titers of 1:512 were obtained with human red cells, both with BDB-treated and untreated cells; agglutination also occurred in a dilution with hemocyanin-treated cells of 1:64.

When unabsorbed anti-ribosome serum was injected into rabbits, reactivities for both BSA and HGG were detected by 24 hours after injection. Normal rabbit serum injected into recipient animals produced, at no time after injection, any altered reactivities for BSA, HGG, hemocyanin or human red cells.

The results shown in Table 6 indicated that: (1) absorption of anti-ribosome serum with BSA- or HGG-coupled erythrocytes delayed the

Table 6

THE EFFECTS OF ANTIGEN ABSORPTIONS ON THE ANTI-RIBOSOME SERA USED IN PASSIVE TRANSFER

Serum Transferred	Time after Transfer (Days)	Antigens-Coupled to Erythrocytes				
		BSA	HGG	NHO	NHO-BDB	KLH
Normal rabbit	0	—	—			
	1	—	—			
	2	—	—			
	7	—	—			
	34	—	—			
Anti-ribosome absorbed with BSA-coupled cells	0	—	—			
	1	—	4			
	2	2	4			
	7	64	>64			
	34	256	256	128	64	>64
Anti-ribosome absorbed with HGG-coupled cells	0	4	4			
	1	32	4			
	2	8	4			
	7	>64	64			
	34	1024	1024	512	512	>64

BSA, bovine serum albumin; HGG, human gamma globulin; NHO, normal human O, Rh positive red cells; NHO-BDB, Bis-diazotized benzidine treated normal, human O, Rh positive red cells; KLH, keyhole limpet hemocyanin.

appearance of agglutinins for the absorbing antigen in the serum of the recipients for as long as 24 to 48 hours, and (2) the reactions produced by these anti-ribosome sera in recipient animals included, in addition to the auto-immune disease, a development of agglutinins for a variety of antigens. The process of absorption of anti-ribosome serum with coupled erythrocytes carries with it the possibility of a release of red cell material and of absorbed proteins which may be left in the serum even after removal of the cells by centrifugation. However, unabsorbed samples of anti-ribosome sera also produced in recipient animals agglutinins for BSA and HGG. When the 34-day serum samples shown in Table 6 were absorbed with normal human, O, Rh-positive red cells, agglutinins remained for the HGG- and BSA-coupled red cells; they were greatly diminished but still present in dilutions of 1:8. Titers for the hemocyanin-treated cells were at least 1:32. An occasional reactivity for human erythrocytes in animals injected with ribosomes has been described elsewhere (Gaffar, 1967; Kurent, 1967; Troendly, 1967) and has been attributed to antibodies to nucleotides present in erythrocyte stroma. However, since other reactivities developed in these animals for antigens to which they have not been immunized, the development of a reactivity for erythrocytes is not surprising.

By use of the bentonite flocculation test, James C. Darner of our laboratory has shown that reactions with BSA or horse serum albumin (HrA) are also demonstrable in the sera of animals injected with liver or reticulocyte ribosomes and in animals injected with such sera. Some of these results are illustrated in Table 7. The sera of actively immu-

Table 7

DEMONSTRATION BY BENTONITE FLOCCULATION OF ANTIBODIES FOR OTHER ANTIGENS IN THE SERA OF ACTIVELY AND PASSIVELY IMMUNIZED RABBITS

Serum Dilution	Reticulocyte Ribosome Animal		Liver Ribosome Animal	
	BSA	HrA	BSA	HrA
1:2	1+	2+	3+	2+
1:4	2+	2+	2+	2+
1:8	2+	1+	2+	1+
1:16	2+	3+	3+	2+
1:32	3+	4+	3+	3+
1:64	3+	2+	2+	1+
1:128	1+	1+	1+	+
1:256	+	1+	1+	+
1:512	–	1+	1+	–
1:1024	–	1+	1+	–
Control	–			

Sera of Rabbits Passively Immunized

Serum Dilution	Anti-Reticulocyte Ribosome		19S Glob. from Same	
	BSA	HrA	BSA	HrA
1:2	2+	2+	2+	2+
1:4	2+	2+	2+	2+
1:8	1+	2+	2+	2+
1:16	1+	1+	2+	1+
1:32	1+	–	X	1+
1:64	X	+	1+	1+
1:128	1+	–	1+	1+
1:256	+	–	1+	1+
1:512	1+	–	1+	1+
1:1024	+	–	1+	1+
Control	–			

X = Tubes were dropped, antigen-antibody flocculate was lost.

nized rabbits, when assayed by the bentonite flocculation test, gave titers as high as 1:1024 for both BSA and HrA. It should also be noted that these sera gave maximal flocculation (4+, 3+) in a dilution of 1:32. These titers are noticeably higher than those detectable by hemagglutination with coupled red cells. Antibody-like reactivities could also be detected in the sera of rabbits that had received anti-ribosome sera. However, these sera were less potent (maximum reaction of 2+) than the sera of actively immunized rabbits, and no inhibition of flocculation was observed in the lower serum dilutions as was noted in the actively immunized animals. Table 7 shows the reactivities resulting from the passive transfer of anti-reticulocyte serum and of the 19S globulins of this serum. The 19S fraction which contains the auto-immune disease factor as well as a complement-fixing reactivity for reticulocyte ribosomes also produced a greater effect than did the whole serum in terms of releasing BSA and HrA activities in normal, recipient animals. The sera and the fractions were obtained from the passively transferred animals 14 and 7 days, respectively, after intravenous injections of whole serum and 19S globulin. Prior to the injection of either ribosomes or sera, the sera of these animals contained no flocculating capabilities for the bentonite antigens.

It is interesting to note that Aldenderfer *et al.* (1963) reported the existence of erythrophagocytosis and leucophagocytosis by histiocytic cells in the peripheral circulation of rabbits that had been immunized either actively with ribosomes or passively with anti-ribosome sera.

IV. INDUCTION OF MACROPHAGE IMMUNITY BY BACTERIAL RIBOSOMES

Many of the intricacies involved in the process of phagocytosis and in the immunologic phenomena attributable to the phagocytic cell, i.e., antibody production (Fishman and Adler, 1963) and cellular immunity (Lurie, 1932, 1938), are still poorly understood. The classical studies of Metchnikoff (1893, 1905) led him to believe that the process of phagocytosis and, subsequently, the fate of the ingested particle within the phagocytic cell was the primary "immune" mechanism of both the vertebrate and invertebrate host. For many years, however, there existed much controversy over whether or not phagocytosis, and in particular the phagocytic cells of the reticuloendothelial system, were responsible for any self-contained immunity void of classical antibody mediation. It was not until ingenious experiments by Lurie in 1933 demonstrated that the mononuclear phagocyte was involved in *in vivo* immunity to tuberculosis, and was responsible for reducing the bacterial population within the tuberculous lesion, that the potential role of the phagocyte as an immune entity was realized. Subsequent experimentation by Lurie (1942), Suter (1953), and Mackaness (1954) further demonstrated that this "cell-associated immune phenomenon" also could be obtained under *in vitro* conditions although only in the presence of immune sera from animals immunized with a living or attenuated strain of *Mycobacterium tuberculosis.*

Recent evidence indicates that the phenomenon of acquired cellular immunity is induced in monocytic phagocytes (macrophages and histiocytes) and possibly in lymphocytes. It is induced principally by living parasites that can exist chronically within host cells, such as *Mycobacterium tuberculosis, Listeria monocytogenes, Francisella tularensis, Salmonella typhi,* certain strains of *Salmonella enteritidis, Salmonella typhimurium,* as well as by certain fungi and protozoa. The response is not as specific as in the case of humoral antibody, e.g., cells "immune" to tubercle bacilli also inhibit the intracellular growth of *Brucella,* etc., on primary exposure, yet they display a more rapid inhibition of the specific parasite on reexposure (Mackaness, 1964). This cellular immunity is associated with the ribosomes of the host cell or with extractable RNA; it is transferable *in vitro* in cell culture or *in vivo* by immune macrophages or their ribosomes. Thus Fong and associates (1961, 1963) showed that ribosomes or phenol-extracted RNA, from peritoneal cells of rabbits immunized 13 days previously with BCG, could be used to transfer cellular immunity serially. Saito and Mitsuhashi (1965a, b) also demonstrated a transfer of immunity in experimental salmonellosis of mice using ribosomes from immune macrophages or immune macrophages themselves. Both groups of investigators have shown that this ability of ribosomes to transfer immunity is sensitive to RNAase, while Jenkin and Rowley (1964) have reported that the transfer of cellular immunity by whole macrophages is diminished by trypsin. Youmans and Youmans (1965) have shown that ribosomes of tubercle bacilli plus adjuvant, or microbial microsomes alone, could produce protective immunity against tuberculosis in mice. Lurie and Davenport (1965), working with inbred rabbits, have shown that animals that are more resistant to tuberculosis are endowed with a greater cellular immune potential as demonstrated by increased macrophage activity and greater delayed hypersensitivity.

Venneman (of this laboratory) has shown that F1 hybrid (C3H × DBA) mice are more susceptible to lethal infection with *S. typhimurium* than are either of the parental strains. Macrophages from the F1 hybrids immunized with ribosomes from virulent *S. typhimurium* (incorporated in incomplete Freund's adjuvant) not only inhibit multiplication of the virulent bacilli but apparently are more resistant to damage by this bacterium than are macrophages from uninfected or from adjuvant-injected F1 hybrid mice. Furthermore, this increased resistance to *S. typhimurium* was passively transferred to other F1 hybrids with macrophages from mice immunized with bacterial ribosomes in incomplete Freund's adjuvant. For this effect the time of macrophage transfer after initial immunization was critical. Macrophages removed from mice 14 days after immunization protected 80% of recipient mice for 10 days against a 10,000 LD_{50} dose of virulent bacilli, while macrophages removed from mice 30 days after immunization demonstrated no passive protection in recipients receiving the same massive challenge dose. However, mice initially immunized with 5 mg of bacterial ribosomes in incomplete Freund's adjuvant survived this massive dose (10,000 LD_{50}) of virulent bacilli indefinitely.

As noted by Saito and Mitsuhashi (1965a, b) the most striking feature of this macrophage "immunity" is its occurrence *in vitro* in the absence of immune sera. Furthermore, spheroid bodies similar to those found in intracellular infections with brucellae (Elberg, 1960) were observable 12 days post-infection in cultures of immune macrophages indicating a possible intracellular L-form of the viable bacilli. This response may be an expression of the immune response by the macrophages. Recently it has been shown that brucella-infected macrophages can contain brucellar L-forms which after ten days were no longer reversible to vegetative cells and eventually resembled an intracellular *Mycoplasma* residing near the nuclear membrane of the macrophage (Hatten and Sulkin, 1966a, b).

There is no reason not to believe that this group of facultative intracellular parasites might function in infected macrophages in a manner similar to bacteriophages, i.e., the bacterial RNA or DNA may affect host cell nucleic acids and/or protein synthesis.

V. DISCUSSION

The significance of the effects of anti-ribosome sera on the intact host appear to be of a greater scope than thought initially. Recently, Zatti and Revoltella (1966) were able to demonstrate immunity to Yoshida ascites tumor in rats following injection of homologous ribosomes. From our experience regarding the effects of mammalian and bacterial ribosomes on the intact animal, it is apparent that: (1) ribosomes in complete Freund's adjuvant are adequate immunogens in rabbits; (2) the antibodies that result display specificities for the immunogen and its components, a situation that is no different than that described by Plescia and Braun (1967) and Sela *et al.* (1964, 1965, 1966) for carrier-linked polynucleotides, nucleosides, and nucleotides; (3) autoantibodies and concurrent auto-immune hemolytic anemia result from immunization of rabbits with ribosomes extracted from rat or rabbit liver or from reticulocytes; (4) in addition to the auto-immune phenomenon, antibodies are found in the sera of immunized rabbits for antigens that have not been injected; (5) both of the latter two are passively transferable with whole serum or its macroglobulin component to non-immunized recipients; and (6) mice and their peritoneal macrophages can be rendered immune to *Salmonella typhimurium* following immunization with *Salmonella* ribosomes in incomplete Freund's adjuvant.

Can all of these observations be attributed to properties of the immunizing RNA, since others who have immunized animals with RNAs coupled to carriers have not observed autoimmune phenomena or the occurrence of antibodies to antigens other than those that were administered?

For the purpose of discussing the significance of this research, the role of gene-controlled protein synthesis of mammalian cells must be considered. Frenster (1965) recently discussed a mechanism of specific derepression within interphase chromatin. Since only one strand of DNA encodes a specific messenger RNA (mRNA), the question was raised as to what happens to the other strand of DNA. Frenster presents data supporting the fact that specific derepressor RNA (dRNA) can hybridize with the complementary nucleotide sequences on the other separated DNA strand, tending to stabilize or even extend the loop of DNA strand separation, thereby freeing the remaining strand of DNA to serve as template for continuous synthesis of specific mRNA. Most pertinent to this discussion are the effects of derepression on immunogenesis. "How might the RNA of a phagocytic cell containing antigen stimulate antibody synthesis and immunologic memory in lymphoid cells contiguous to that phagocyte?" Frenster suggests that some of the "immunologic" RNA transferred could be retained in an immune clone by hybridization with single strands of the lymphoid or plasma cell's DNA, functioning there as a stable derepressor through successive cell divisions. Moreover, Trakatellis *et al.* (1964) reported the occurrence of ribosomal antibodies in a female mouse rendered tolerant to male skin isografts by the neonatal injection of splenic microsomal RNA from isogenic males.

Investigators such as Campbell and Garvey (1963), Saha *et al.* (1964), Askonas and Rhodes (1965a, b), Friedman *et al.* (1965), Fishman and Adler (1963), Fishman (1961, 1963, 1965, 1966a, b), Stavitsky and Gusdon (1966), Braun (1966, 1967), and Thor (1967) have contributed greatly to understanding the genesis of the immune response and their results, though pertinent here, are discussed elsewhere in this symposium.

Compatible with the findings presented by us is the possibility that anti-ribosome sera are capable of penetrating, or affecting from without, precommitted antibody-producing cells influencing dRNA at the gene level and thereby releasing mRNA coding for the production of specific IgM globulins. The possibility that homologous anti-ribosome sera may derepress inherent, but quiescent, cellular capabilities offers an exciting tool for use in the study of the mechanism of macrophage immunity induced by facultative intracellular parasites.

For the phase of our investigation concerned with the mechanism of acquired cellular immunity in experimental salmonellosis, a possible relationship between the processing of antigen by macrophages resulting in the release of an antigen-RNA complex and the process of establishment of non-antibody immunity in macrophages is exciting.

Are antibodies to nucleic acid capable of penetrating, or affecting from without, immunocompetent cells, thereby altering the immunologic behavior of such cells? Rosenkranz *et al.* (1964) have noted that purine- and pyrimidine-specific antibodies penetrate sea urchin embryos inhibiting development at various stages. As yet no satisfactory explanation exists to explain the phenomenon of maternal suppression in rabbits (Dray, 1962; Mage and Dray, 1965) in which the paternal genotype is phenotypically suppressed in offspring after birth for varying periods, in some cases for as long as 2–3 years. However, Gell and Sell (1965) and Sell and Gell (1965a, b) have shown that blast transformation and DNA synthesis result from an *in vitro* exposure of rabbit lymphoid cells to immune sera against the appropriate allotypic specificities. If antibodies to nucleic acids are capable of modifying the immunologic reactivity of competent cells, are antibodies directed against certain base sequences more effective than antibodies to single nucleotides? In relation to this inquiry, Miller (1967) of our laboratory has recently observed that the transfer of rabbit anti-methylated BSA-sRNA (from bovine liver) to non-immunized animals stimulates the elaboration of antibodies to bovine serum albumin (BSA) within 24 hours. This reactivity in the recipient was shown not to be due to the anti-BSA of the donor because the passive transfer of the same quantity of rabbit anti-BSA elicited no such response in the recipients.

VI. SUMMARY

Both bacterial and mammalian ribosomes have been shown to be effective immunogens with the major portion of the resulting antibodies being specific for the nucleic acid components. This report describes several additional facets of the influence of ribosomes on the immunologic mechanisms of the recipient. In addition to an autoimmune hemolytic anemia which develops in a certain percentage of rabbits immunized with ribosomes from rat or rabbit liver or from reticulocytes, antibodies are present in the sera for antigens that were never knowingly introduced into these rabbits. Both of these phenomena are transferable to non-immunized recipients with either whole serum or with its 19S globulin component. In fact, states of specific immunologic tolerance to bovine serum albumin, human gamma globulin and rabbit isoantigen A can be abolished, with the appearance of detectable serum antibody, in tolerant rabbits within 5–19 hours after an injection of antiribosome serum. Furthermore, upon immunization of mammalian hosts with bacterial ribosomes from certain of the facultative intracellular microbial parasites, not only the intact animal but also the monocytic cells (macrophages) display increased resistance to microbial parasitism. The experimental system described herein involves the strikingly increased resistance to inbred mice and their peritoneal monocytes to salmonellosis after immunization with ribosomes extracted from virulent *Salmonella typhimurium.*

References

Ada, G. L., G. J. V. Nossal, J. Pye, and A. Abbot (1963). Behaviour of active bacterial antigens during the induction of the immune response. (1) Properties of flagellar antigens from salmonella. Nature, *199*:1257–1259.

Aldenderfer, P. H., M. C. Dodd, F. W. McCoy, and L. Liss (1963). Histologic changes in rabbits injected with homologous liver ribosomes and antiribosome serum. Federation Proc., *22*:338.

Askonas, B. A., and J. M. Rhodes (1965a). In "Molecular and Cellular Basis of Antibody Formation" (J. Sterzl *et al.*, eds.), p. 503. Publishing House Czech. Acad. Sci., Prague (distrib., Academic Press, New York).

——— (1956). Immunogenicity of antigen-containing ribonucleic acid preparations from macrophages. Nature, *205*:470–474.

Barbu, E., and J. Panijel (1960). Presence d'anticorps antiacid ribonucleique dans les immunserums antiribosomes. Compt. Rend. Acad. Sci., *250*:1382–1384.

Barbu, E., J. Panijel, and G. Quash (1961). Immunochemical characterization of ribosomes. Ann. Inst. Pasteur, *100*:725–746.

Barbu, E., G. Quash, and J. P. Dandeu (1963). Spécificité des anticorps anti-RNA. Ann. Inst. Pasteur, *105*:849–865.

Baron, S., and H. B. Levy (1966). Interferon. Ann. Rev. Microbiol., *20*:291–318.

Behrendt, H. (1957). Chemistry of erythrocytes, p. 69, Charles C. Thomas, Springfield, Ill.

Beiser, S. M., and B. F. Erlanger (1966). Antibodies which react with nucleic acids. Cancer Res., *26*:2012–2017.

Bigley, N. J., M. C. Dodd, and V. B. Geyer (1963). The immunologic specificity of antibodies to liver ribosomes and nuclei. J. Immun., *90*:416–423.

Braun, W. (1966). Microbiological approaches to problems of antibody formation, Ohio Branch American Soc. for Microbiology, April 23.

Braun, W., and L. J. Lasky (1967). Antibody formation in newborn mice initiated through adult macrophages. Federation Proc., *26*:2162.

Campbell, D. H., and J. S. Garvey (1963). Nature of retained antigen and its role in immune mechanisms. Adv. Immun., *3*:261–313.

Cohen, C. (1958). Factors influencing isoimmunization in rabbits. Effect of early postnatal immunization of specific isoantibody response. Proc. Soc. Exptl. Biol. and Med., *99*:607–610.

——— (1959). Anamnestic reactions in rabbits showing unresponsiveness to erythrocyte isoantigens. Transplant. Bull., *6*:426–428.

Darner, J. C. (1965). The use of bentonite for the detection of antibodies in tolerant and immune rabbits. M.Sc. Thesis, Ohio State University.

Dodd, M. C., N. J. Bigley, V. B. Geyer, F. W. McCoy, and H. E. Wilson (1962). Autoimmune response in rabbits injected with rat and rabbit liver ribosomes. Science, *137*:688–689.

Dodd, M. C. (1963). Conference on the immunologic aspects of rheumatoid arthritis and systemic lupus erythematosus, pp. 501–511, published by The Arthritis and Rheumatism Foundation in cooperation with The Nat. Inst. of Arthritis and Metabolic Diseases, PHS. Grune and Stratton, Inc., New York, N. Y.

Dray, S. (1962). Effect of maternal isoantibodies on the quantitative expression of two allelic genes controlling γ-globulin allotypic specificities. Nature, *195*:677–680.

Elberg, S. S. (1960). Cellular immunity. Bacteriol. Rev., *24*:67–95.

Everett, M. R. (1946). Medical Biochemistry, p. 517. Paul B. Hoeber, Inc., New York, N. Y.

Fidalgo, B. V., and V. A. Najjar (1967). The physiological role of the lymphoid system. III. Leucophilic γ-globulin and the phagocytic activity of the polymorphonuclear leucocyte. Proc. Nat. Acad. Sci. U. S., *57*:957–964.

Fishman, M. (1961). Antibody formation *in vitro*. J. Exptl. Med., *114*:837–856.

Fishman, M., and F. L. Adler (1963). Antibody formation initiated *in vitro*. II. Antibody synthesis in x-irradiated recipients of diffusion chambers containing nucleic acid derived from macrophages incubated with antigen. J. Exptl. Med., *117*:595–602.

Fishman, M., R. A. Hammerstrom, and V. P. Bond (1963). *In vitro* transfer of macrophage RNA to lymph node cells. Nature, *198*:549–551.

Fishman, M., J. J. van Rood, and F. L. Adler (1965). In "Molecular and Cellular Basis of Antibody Formation" (J. Sterzl *et al.*, eds.), p. 491. Publishing House of Czech. Acad. Sci., Prague (distrib., Academic Press, New York).

Fishman, M. (1966). 1st Internat. Symposium–Immunity, Cancer and Chemotherapy. Buffalo, N. Y., Sept. 20–22.

Fong, J., P. Schneider, and S. S. Elberg (1956). Studies of tubercle bacillus-monocyte relationship. I. Quantitative analysis of effect of serum of animals vaccinated with BCG upon bacterium-monocyte system. J. Exptl. Med., *104*:455–465.

Fong, J., D. Chin, and S. S. Elberg (1961). Studies on tubercle bacillus-histiocyte relationship. V. Passive transfer of cellular resistance. J. Exptl. Med., *115*: 475–489.

——— (1963). Studies of tubercle bacillus-histiocyte relationships. VI. Induction of cellular resistance by ribosomes and ribosomal RNA. J. Exptl. Med., *118*:371–386.

Frenster, J. H. (1965). Mechanisms of repression and de-repression within interphase chromatin. In Vitro, *1*:78–101.

Friedman, H. P., A. B. Stavitsky, and J. M. Solomon (1965). Induction *in vitro* of antibodies to phage T2: Antigens in the RNA extract employed. Science, *149*:1106–1107.

Gaffar, A., N. J. Bigley, J. P. Minton, and M. C. Dodd (1967a). Serologic analysis of nuclear antigens from normal and malignant human tissues. Federation Proc., *26*:753.

——— (1967b). Immunochemical properties of nucleoprotein antigens from normal and malignant human tissues. Ph.D. Dissertation, Ohio State University.

Garvey, J. S., D. V. Eitzman, and R. T. Smith (1960). The distribution of S^{35}-labeled bovine serum albumin in newborn and immunologically tolerant adult rabbits. J. Exptl. Med., *112:*533–549.

Gell, P. G. H., and S. Sell (1965). Studies of rabbit lymphocytes *in vitro.* II. Induction of blast transformation with antisera to six IgG allotypes and summation with mixtures of antisera to different allotypes. J. Exptl. Med., *122:*813–821.

Hanan, R., and Oyama (1954). Inhibition of antibody formation in mature rabbits by contact with antigen at an early age. J. Immun., *73:*49–53.

Hatten, B. A., and S. E. Sulkin (1966a). Intracellular production of brucella L forms. I. Recovery of L forms from tissue culture cells infected with *Brucella abortus.* J. Bacteriol., *91:*285–296.

——— (1966b). Intracellular production of brucella L forms. II. Induction and survival of *Brucella abortus* L forms in tissue culture. J. Bacteriol., *91:*14–20.

Holman, H. R. (1965). Systemic lupus erythematosus in "Immunologic Diseases." (Eds.—M. Samter, H. L. Alexander, D. W. Talmage, B. Rose, W. B. Sherman, and J. H. Vaughan.) Little, Brown and Co., Boston.

Jenkin, C. R., and D. Rowley (1964). The basis for immunity to mouse typhoid. II. Cell-bound antibody. Austral. J. Exptl. Biol. and Med. Sci., *42:*237–248.

Kurent, J. E. (1967). The nucleotide nature of a common antigenicity existing in human erythrocytes and subcellular organelles. M.Sc. Thesis, Ohio State University.

Larkin, G. F. (1966). Antibodies to RNA in the sera of individuals with rheumatoid arthritis. M.Sc. Thesis, Ohio State University.

Lurie, M. B. (1932). The correlation between the histological changes and the fate of living tubercle bacilli in the organs of tuberculous rabbits. J. Exptl. Med., *55:*31–54.

——— (1938). Nature of inherited resistance to tuberculosis. Proc. Soc. Exptl. Biol. Med., *39:*181–187.

——— (1942). Studies on the mechanism of immunity in tuberculosis. The fate of tubercle bacilli ingested by mononuclear phagocytes derived from normal and immunized animals. J. Exptl. Med., *75:*247–268.

Lurie, M. B., and A. M. Davenport, Jr. (1965). Macrophage function in infectious disease with inbred rabbits. Bacteriol. Rev., *29:*466–476.

Mackaness, G. B. (1954). The growth of tubercle bacilli in monocytes from normal and vaccinated animals. Amer. Rev. Tuberc., *69:*495–510.

——— (1962). Cellular resistance to infection. J. Exptl. Med., *116:*381–406.

——— (1964a). The immunologic basis of acquired cellular resistance. J. Exptl. Med., *120:*105–109.

——— (1964b). The behaviour of microbial parasites in relation to phagocytic cells *in vitro* and *in vivo.* Fourteenth Symp. of The Soc. for Gen. Microbiol., 213–240, Cambridge University Press, Cambridge, England.

Mage, R., and S. Dray (1965). Persistent altered phenotypic expression of altered γG-immunoglobulin allotypes in heterozygous rabbits exposed to isoantibodies in fetal and neonatal life. J. Immun., *95:*525–535.

Metchnikoff, E. (1893). Lectures on the Comparative Pathology of Inflammation. London.

—— (1905). Immunity in Infective Disease. Cambridge Univ. Press, Cambridge, England.

Miller, S. B. (1967). The effects of passive transfer of anti-RNA in unimmunized recipients. M.Sc. Thesis, Ohio State University.

Pennell, R. B. (1964). In "Red Blood Cell"—A comprehensive treatise. Ed. Bishop, C., Academic Press, New York.

Plescia, O. J., N. C. Palczuk, W. Braun, and E. Cora-Figueroa (1965a). Antibodies to DNA and a synthetic polydeoxyribonucleotide produced by oligodeoxyribonucleotides. Science, *148*:1102–1103.

Plescia, O. J., N. C. Palczuk, E. Cora-Figueroa, A. Mukherjee, and W. Braun (1965b). Production of antibodies to soluble RNA (sRNA). Proc. Natl. Acad. Sci. U. S., *54*:1281–1285.

Plescia, O. J., N. C. Palczuk, A. Mukherjee, and E. Cora-Figueroa (1965c). Production of antibodies to t-RNA. Federation Proc., *24*:185.

Plescia, O. J., and W. Braun (1967). Nucleic acids as antigens. Adv. Immun., *6*:231–252.

Prankard, T. A. J. (1961). The red cell, p. 70. Charles C. Thomas, Springfield, Ill.

Rosenkranz, H. S., B. F. Erlanger, S. W. Tannenbaum, and S. Beiser (1964). Purine- and pyrmidine-specific antibodies: Effect on the fertilized sea urchin egg. Science, *145*:282–283.

Saha, A., J. S. Garvey, and D. H. Campbell (1964). The physico-chemical characterization of the ribonucleic acid-antigen complex persisting in the liver of immunized rabbits. Arch. Biochem. Biophys., *105*:179–192.

Saito, K., and S. Mitsuhashi (1965a). Experimental salmonellosis. VI. *In vitro* transfer of cellular immunity of mouse mononuclear phagocytes. J. Bacteriol., *90*:629–634.

——— (1965b). Experimental salmonellosis. VII. *In vitro* transfer of cellular immunity by a ribosomal fraction of mouse mononuclear phagocytes. J. Bacteriol., *90*:1194–1199.

Scheetz, II, M. E. (1965). Flocculation of sensitized particles by various human sera. M.Sc. Thesis, Ohio State University.

Sela, M., H. Ungar-Waron, and X. Schechter (1964). Uridine-specific antibodies obtained with synthetic antigens. Proc. Natl. Acad. Sci. U. S., *52*:285–292.

Sela, M., and H. Ungar-Waron (1965). Nucleoside-specific antibodies elicited by synthetic antigens. Federation Proc., *24*:1438.

Sela, M. (1966). Immunological studies with synthetic polypeptides. Adv. Immun., *5*:29–129.

Sell, S., and P. G. H. Gell (1965a). Studies on rabbit lymphocytes *in vivo*. I. Stimulation of blast transformation with antiallotypic serum. J. Exptl. Med., *122*:423–440.

——— (1965b). Studies on rabbit lymphocytes *in vitro*. IV. Blast transformation of the lymphocytes from the newborn rabbit induced by antiallotype serum to a paternal IgG allotype not present in the lymphocytes donors. J. Exptl. Med., *122*:923–942.

Sercarz, E. E., and A. H. Coons (1963). The absence of antibody-producing cells during unresponsiveness to BSA in the mouse. J. Immun., *90:*478–491.

Stavitsky, A. B., and J. P. Gusdon, Jr. (1966). Symposium on *in vitro* studies of the immune response. IV. Role of nucleic acids in the anamnestic antibody response. Bacteriol. Rev., *30:*418–426.

Suter, E. (1953). Multiplication of tubercle bacilli within mononuclear phagocytes in tissue cultures derived from normal animals and animals vaccinated with BCG. J. Exptl. Med., *97:*235–255.

Tannenbaum, S., and S. Beiser (1963). Pyrimidine-specific antibodies which react with deoxyribonucleic acid (DNA). Proc. Natl. Acad. Sci. U. S., *49:* 662–668.

Terres, G., and W. L. Hughes (1959). Acquired immune tolerance in mice to crystalline bovine serum albumin. J. Immun., *83:*459–467.

Thor, D. E. (1967). Delayed hypersensitivity in man: A correlate *in vitro* and transfer by an RNA extract. Science, *157:*1567–1569.

Trakatellis, A. C., A. E. Axelrod, M. Montjar, and F. Lamy (1964). Induction of immune tolerance with ribosomes and ribonucleic acid extracts in newborn mice. Nature, *202:*154–157.

Troendly, B. C. (1967). A serologic comparison of the nucleic acid specificities of antisera to normal and malignant human ribosomes. M.Sc. Thesis, Ohio State University.

Venneman, M. R. (1967). The nature of immunity in experimental salmonellosis. M.Sc. Thesis, Ohio State University.

Youmans, A. S., and G. P. Youmans (1965a). Immunogenic activity of a ribosonal fraction obtained from *Mycobacterium tuberculosis.* J. Bacteriol., *89:*1291–1298.

——— (1965b). Nonspecific factors in resistance of mice to experimental tuberculosis. J. Bacteriol., *90:*1675–1681.

Youmans, A. S., and G. P. Youmans (1966a). Preparation of highly immunogenic ribosomal fractions of *Mycobacterium tuberculosis* by use of sodium dodecyl sulfate. J. Bacteriol., *91:*2139–2145.

——— (1966b). Effect of trypsin and ribonuclease on the immunogenic activity of ribosomes and ribonucleic acid isolated from *Mycobacterium tuberculosis.* J. Bacteriol., *91:*2146–2154.

Zatti, M., and R. Revoltella (1966). Tumor immunity in rats following injection of homologous ribosomes. Experentia, *15:*467.

DISCUSSION

Chairman: B. CINADER

CINADER: There is one problem that I would like to ask about and it is the following: There were very marked discrepancies in different sera between the antigen quantity which brought about maximum precipitation in the presence of denatured or renatured DNA. Is this due to two different classes of antibodies?

SELIGMANN: The results obtained both by Dr. Stollar and ourselves clearly point out that people working with rabbit antisera are dealing with a mixture of antibodies with different specificities. This heterogeneity can only be resolved by quantitative analyses, including inhibition studies, such as those performed some years ago by Drs. Stollar and Levine. This is particularly relevant to situations where there are reactions both with native and denatured DNA.

In some cases we seem to be dealing with a single kind of antibody, as indicated by cross absorption experiments. In some instances, however, we seem to be dealing with two different kinds of antibodies with two different specificities. We showed that in some instances we had lupus sera reacting with purified native DNA, but we needed a hundred times more native than denatured DNA to block the reaction, but this does not happen with all lupus sera.

STOLLAR: We have, just as Dr. Seligmann has reported, found sera that do have distinct antibodies reacting with the native and with the denatured forms of DNA. In some instances the reaction with denatured DNA is still present in the presence of an excess of native DNA. In other instances, the excess of native DNA completely abolishes the reaction of denatured DNA.

GRABAR: We have found antibodies of different classes—IgG, IgA and IgM in rabbits. Since many investigators test their antibodies only by complement-fixation, it is possible that some of these antibodies would not react.

STOLLAR: I think this is a very important point. Barnett and others have correlated immunofluorescence with other methods of measuring anti-DNA and anti-nucleoprotein antibodies. Their studies clearly demonstrated that three classes of immunoglobulins can be involved in antinuclear staining by LE sera. We have found a particularly close correlation between the presence of serum antibodies fixing complement with DNA (with native DNA or DNA determinants within a nucleoprotein structure) and the presence of peripheral staining of the nucleus due to IgG antibodies. However, this correlation is absent when immunofluorescence is due to IgM or IgA. In terms of complement-fixation, I think it has been clearly shown that IgA antibodies do not fix complement, and in some extensive studies with various types of antigens in our laboratories we have found that not all antigens will fix complement with IgG and IgA, and when complement-fixation does occur with such globulins it is usually much more efficient when carried out at 37° than at 4°. For example, antibodies against denatured DNA reacted with macroglobulins in complement-fixation and did so better at 37° than at 4°. But in some situations where we knew that macroglobulin antibody was present with a specificity towards adenosine, such antibody would not react with an adenosine-human serum albumin conjugate in complement-fixation but would react with denatured DNA. The reaction was inhibited by adenosine (so this is how we knew that macroglobulin antibodies were present) but the reaction occurred only when the adenosine determinant was present in the proper form, namely, in denatured DNA. Thus, I think that, in addition to the class of antibodies, the form of the antigen also is very important in determining whether or not complement-fixation reactions may occur.

SELIGMANN: Since even with the very sensitive technique of micro-complement fixation it is possible that antibodies present in low concentrations may be hard to detect, especially if there is some anti-complementary activity, it is advisable to use simultaneously other methods such as precipitation.

I was interested to hear that Dr. Estrada-Parra has studied sera of patients with T.B. and found DNA-reactive antibodies. I wanted to ask him if these patients had been treated with drugs for a long time since we demonstrated that anti-nuclear factors are rather common in T.B. patients and this phenomenon seems to be related to treatment with isoniazid. This drug is able to produce serologically and clinically lupus-like syndromes, and very recently we have been able to demonstrate the LE-inducing activity of both hydralazine and isoniazid in some inbred strains of mice. So I wonder if Dr. Estrada-Parra thinks that these DNA antibodies are related to the disease or to treatment since both in humans and mice we have been able to find anti-nuclear factors only by immuno-fluorescence but never true DNA antibodies.

ESTRADA-PARRA: All of our sera were furnished by the hospital and the donors may have been under treatment; however, I am not certain about this.

SELIGMANN: I would like to ask Dr. Stollar regarding the correlation between the LE phenomenon and the nature of anti-nuclear antibodies. There has been some speculation in the past that there are different kinds of anti-nuclear antibodies which are able to induce the LE phenomenon including possibly some DNA antibodies. Purified DNA-specific antibodies have been shown in our lab to give a positive LE reaction although there remained a lot of activity in the supernate after absorption of DNA-specific antibodies. Pullman has thought the antibodies giving the LE phenomenon are actually of one kind and react with determinants at the junction of DNA and histone.

STOLLAR: We have examined the correlation between complement-fixation data, immunofluorescence and the presence or absence of positive LE cells. We did not find any significant correlation in this regard. It turned out that if the serum had anti-DNA antibodies, it was much more likely to produce a positive LE cell phenomenon, but at the same time it would also react with nucleoprotein. Thus most of the antisera that we have found capable of producing LE cells would react both with native DNA and with nucleoprotein. We found many LE-positive sera, however, which did not have a high titer of DNA-specific antibody, and we also found some that would not react directly in complement-fixation with nucleoprotein. I might say also that human or rabbit sera which reacted only with denatured DNA and not with native DNA or nucleoprotein did not cause peripheral staining, and I think we can conclude that LE cells are not produced by antibodies to denatured DNA.

DODD: Some time ago we looked for RNA- and DNA-specific antibodies in rheumatoid arthritis sera and we used in this particular type of study both bentonite flocculation and hemagglutination. When we tested the sera by bentonite flocculation they were positive with DNA, but so were a very large group of sera from normal humans and normal rabbits. Now if we heated all these sera then only about 35% still had antibody to DNA and these sera were from individuals with rheumatoid arthritis. There was no activity with normal sera. However, if we froze and immediately thawed these sera, then the antibody reappeared. If you then heated the sera to 63°, again there was no antibody except in the sera of rheumatoid patients; the heat-labile antibody never reappeared whether the sera were frozen and thawed many times.

SIGEL: Dr. Seligmann, how consistent are the variations in different individual sera? In other words, if you take repeated bleedings from the same patient, would you get the same variation with regard to reactivity with native and denatured DNA?

SELIGMANN: We have taken repeated bleedings from individual patients and almost always found the same pattern of specificity. We also have serial samples from a patient under treatment. Under the influence of the treatment the antibodies to native DNA disappeared first and antibodies to denatured DNA disappeared much later. In other words, all kinds of antibodies to DNA seem to be correlated with the disease but this seems particularly true for antibodies to native DNA.

GLICK: A question to Dr. Stollar. Have you tried to determine whether your sera react to a different extent with nucleoproteins from different tissues and whether or not nucleoproteins from active chromatin react differently than nucleoproteins from repressed chromatin? And just one more point: do you know whether or not your nucleoprotein fractions contain any phospholipids?

STOLLAR: In answer to your first question, we have found that when nucleoprotein was prepared from calf thymus or from mouse thymus or from dog fish erythrocyte nuclei, the reaction was very similar when tested with the serum that reacts with native DNA.

We have not examined possible differences between repressed and free nucleoproteins. We have not assayed our system for phospholipids. However, we had tried to rule out a participation of factors other than nucleoprotein itself by testing artificial histone-DNA complexes in which case phospholipids do not play a role.

FRIOU: I have two comments. One is in relation to this morning's discussion about the possibility of using inbred animals. We have shown that certain strains of mice, particularly the A/J strain, spontaneously develop antibodies to nucleoprotein with age. But in carrying out this study we also found that the DBA strain never develops such antibodies. We have here a clear-cut strain difference in the spontaneous appearance of anti-nuclear factors. We have immunized these animals in two different ways, namely with denatured DNA-MBSA in adjuvant and by hyperimmunization with *E. coli*. Although the studies with these animals are still incomplete, thus far none of the strains developed anti-DNA antibodies.

The other comment is to say that I have found the present discussion especially interesting because it demonstrates how premature the generalizations based on rather crude, earlier studies of anti-nuclear antibodies actually were. The discussion here casts some light on differences in results that were reported previously from different laboratories.

CINADER: How did A/J × DBA hybrids behave?

FRIOU: The A/J × DBA crosses, the F_1 hybrids, also never spontaneously developed anti-nucleoprotein factors even up to a couple of years of age.

PANIJEL: I have two comments for Dr. Bigley. The first concerns the assertion that the majority of the antibodies in the anti-ribosome sera are specific for nucleic acids. I do not believe that this is true. In our experience the major fraction of the anti-ribosome sera are anti-ribosomal protein antibodies. Perhaps this point can explain some of the variation in your results with tolerance. The second point concerns the problem of the increased resistance to salmonellosis. If you used crude ribosomes, you have in your preparations a great amount of endotoxin and perhaps the increased resistance is due to the injection of your animals with endotoxin.

BIGLEY: I can answer the question about endotoxin. We had a great deal of trouble getting our experiments off the ground because we always killed our mice when we followed the technique used by the Youmans with tubercle bacilli to achieve resistance with bacterial microsomes. There is some question whether the endotoxin of Gram-negative bacteria is associated with the cell wall or the cell membrane. We have come to the conclusion that it has to be the membrane because no matter what we did we killed mice with it. So we removed the microsomal membrane with Duponol, washed the ribosome preparations well, took electron micrographs of it and we had polysome aggregates which we could disperse into single ribosome units as shown by electron microscopy. These then were filtered, freed of any contamination and these preparations were used for the induction of immunity in mice. Another factor of this immunity was that the immunized macrophages displayed in cell culture a resistance to endotoxin which the normal cell did not display.

BRAUN: I would like to ask Drs. Dodd and Bigley whether, in view of their demonstration of unexpected antibody proteins in animals exposed to anti-ribosome sera, they have considered the possibility that their phenomenon might be related to the mistranslation phenomenon that occurs in bacteria, and apparently also in mammalian cells, in the presence of agents that "distort" the ribosomes. As you know, Gorini and Davis and others have shown that in the presence of streptomycin, of ethyl alcohol, of DMSO and other such agents, a false translation may occur at the ribosome level. If one made the assumption that the anti-ribosome antibodies can indeed reach the ribosome, then perhaps the unexpected antibody protein might be related to false translation of a specific mRNA in antibody-forming cells. Alternately, one might also speculate that you might be dealing with a phenomenon in which we have become interested in recent months, namely the ability to turn on wrong antibody-forming cells with a specific antigen if at the time of antigen exposure the animal is also exposed to an agent that alters the permeability of lymphocytes. I wonder whether you have considered the possibility that the strange effects that you have observed in your anti-ribosome-treated animals might be related to either mistranslation or permeability alteration.

BIGLEY: We have considered both of these possibilities. We have no data to support either one at this point, but we would like to examine these possibilities.

NOVELLI: I have a question for Dr. Seligmann. In one of the sera that reacted with native DNA, denaturation of the DNA resulted in a loss of most of its activity. On renaturation full activity was restored within ten minutes. With sera that reacted with denatured DNA, when that DNA was renatured, it took 120 minutes for a loss of full reactivity. What may account for this difference?

SELIGMANN: We did indeed regain, on renaturation, 100% of antigenic reactivity after 40 minutes in the case of antisera reacting only with native DNA; in order to have no reactivity on renaturation with denatured DNA, in the case of antisera reacting only with denatured DNA, we had to wait for three hours or even longer. I do not have any explanation for this. However, I might say that we have seen in these renaturation experiments that the antigenic activity of renatured DNA is not exactly the same as native DNA and that it is independent of any denatured contaminant included in the renatured sample. It is possible that the differences are related to some difference in conformation of the renatured DNA molecule as compared to the native molecule.

L. LEVINE: And I might add that using antibodies to denatured DNA, if one is measuring renaturation of T-even phage DNA, one never gets complete renaturation to native double strands. Using various nucleases with different specificities one can demonstrate that the renatured product has some single-stranded 3′-ends and also some single-stranded 5′-ends. Thus, one never gets a complete match of the two strands, there is always some single-stranded region left over.

SELIGMANN: All I can say is that in view of such observations it is surprising that after renaturation, we have no reactions at all with antibody to denatured DNA. That is, at least at the concentrations that we tested.

I would like to discuss briefly the question of whether DNA antibodies are harmful for the patient with lupus. Here are some facts:

First, there is surely some good correlation between the activity of the disease and the level of antibodies to DNA, especially to native DNA. Secondly, I think there has been no good direct demonstration that antibodies to either native or denatured DNA are harmful to a living cell. And thirdly and lastly, it seems established that soluble complexes of anti-nuclear antibodies and nuclear constituents are harmful for the kidney and are responsible for many of the lesions in the glomeruli.

STOLLAR: In reviewing the literature, and summarizing the data of several laboratories, I found that among several hundred patients with systemic lupus, whose sera have been examined for antibodies to native DNA, the incidence of such antibodies is approximately 35–40%. When similar examinations were made on three thousand and some odd number of samples of patients with other diseases, including many tissue-destructive diseases, only four or five examples of antibodies to native DNA were clearly demonstrated. This suggests a rather unique association between anti-native DNA antibodies and LE. It is true that a very much larger number of positive sera were found when techniques such as passive hemagglutination were used with denatured DNA as the test antigen, but at least in some of these cases it included up to 75% of normal sera. I think we must keep in mind the possibility of basic serum protein reacting nonspecifically in such systems. Still another way in which the presence of nuclear antibodies has been related to the disease is the study of complement levels. It was found to be low in the sera of LE patients. Finally, in some elegant studies on the timing of appearance of anti-DNA antibodies, both in Friou's lab and in Dr. Kunkel's laboratory, the time of occurrence of anti-DNA antibodies has been correlated very significantly with the presence of acute clinical episodes.

ATCHLEY: I would just like to make the suggestion that if the serological abnormalities of lupus are important, the ultimate molecular pathology may be due to the carrier, either the occurrence of a new carrier or an alteration of some endogenous carrier.

CAPRA: In our experience, DNA-reactive antibodies seem to have great specificity for systemic lupus. In most instances certainly well over 98% of anti-nuclear antibodies present in other diseases, such as rheumatoid arthritis, can be absorbed completely with a nuclear extract (at least in tests with immunofluorescence). This extract is devoid of DNA activity. We have not been able to show the presence of anti-DNA antibodies in sera from many of the so-called autoimmune diseases. The incidence of anti-DNA antibody in systemic lupus determined in our laboratory was somewhere in the range of 50–60%. I would point out that there is a very marked change during the course of the disease for a number of reasons, one of which is steroid therapy. Also, the appearance of DNA antigen in the serum may significantly depress the appearance of antibody. One of the most significant studies in the lupus field was the demonstration of a rather good correlation between the development of nephritis and the appearance of DNA antigen in the circulation without any concomitant demonstration of antibody. To reiterate, in our experience anti-DNA is quite specific for systemic lupus, and we also believe that it has importance in the pathogenesis through the indirect mediation of immuno-complexes.

USE OF NUCLEIC ACID-SPECIFIC ANTIBODIES

IMMUNOLOGICAL EVIDENCE FOR THE EXISTENCE OF THYMINE DIMERS IN ULTRAVIOLET IRRADIATED DNA *

EDNA SEAMAN, LAWRENCE LEVINE
AND HELEN VAN VUNAKIS

Graduate Department of Biochemistry, Brandeis University, Waltham, Massachusetts

Rabbits immunized with complexes of methylated bovine serum albumin (MBSA) and denatured DNA produce antibodies directed toward DNA (Plescia *et al.*, 1964). These antibodies, like those described for the glucosylated DNA of the T-even phages (Levine *et al.*, 1960) react with the denatured antigen; a conformation in which the bases are not involved in hydrogen bonding and are therefore available for interaction with the antibody. The antibodies to the T-even phage DNA's are highly specific for the glucosylated cytosine moieties (Murakami *et al.*, 1961; Townsend *et al.*, 1965) and can be used to determine the extent of phage DNA synthesis in T_4-infected *E. coli* cells even though host cell DNA is present in large excess (Levine *et al.*, 1960).

Despite many attempts in our laboratory, antibodies formed to DNAs containing the *normal bases* were not able to distinguish among DNAs of different base composition. If, however, rabbits are immunized with complexes of MBSA and DNA which have been irradiated or chemically treated, antibodies directed toward the modified base can be obtained (Levine *et al.*, 1966; Seaman *et al.*, 1966; and Van Vunakis *et al.*, 1966). These antibodies are able to distinguish the lesion in the presence of unaltered DNA and, therefore, become powerful tools for detecting a specific structure. An antibody specific for one of these base modifica-

* Aided by grants from the NIH (AI-01940 and AI-02792) and the Charles Simon Memorial Grant for Cancer Research from the American Cancer Society (E.222). L. L. is an American Cancer Society Professor of Biochemistry and H. V. V. is a recipient of a Public Health Service Research Career Award K6-AI-2372. Publication No. 565 from the Graduate Department of Biochemistry, Brandeis University.

tions, ultraviolet induced damage of pyrimidine bases, will be considered here.

For the production of antisera, DNA was subjected to 5×10^5 ergs/mm^2 of radiation at 260 mμ, complexed to MBSA and injected into rabbits (Levine *et al.,* 1966). When native or denatured DNA from *P. vulgaris* was exposed to various doses of UV light and assayed by complement fixation, the serologic response was observed at dilutions of antisera that did not react with unirradiated DNA (Fig. 1). The denatured *P. vulgaris* DNA showed a somewhat faster increase in C′-fixing activity following varying doses of UV irradiation. The photoproducts, therefore, are formed in *double-* and *single-*stranded DNA although the *rate* of their formation, as measured serologically, appears to be influenced by the conformation of the DNA.

Several photoproducts can be produced in DNA irradiated with ultraviolet light. Among them are pyrimidine dimers (Wacker *et al.,* 1960; Setlow *et al.,* 1965; Freeman *et al.,* 1965), hydrates (Shugar and Wierzchowski, 1957) and peroxides (Wang and Alcantara, 1965). There is some controversy concerning the exact chemical nature of the photoproduct(s) responsible for the biological effects of UV-irradiation. Much evidence exists implicating thymine and cytosine as sites of damage in the UV-irradiated DNA (Setlow and Setlow, 1962; Johns *et al.,* 1962). Since thymine dimers are formed more readily than cytosine dimers, and are more stable under physiological conditions than are cytosine hydrates, it was anticipated that antibodies would be directed primarily to thymine photoproducts. Thus the serologic activity of DNA exposed to UV-irradiation should reflect the thymine content of the DNA, much as the UV sensi-

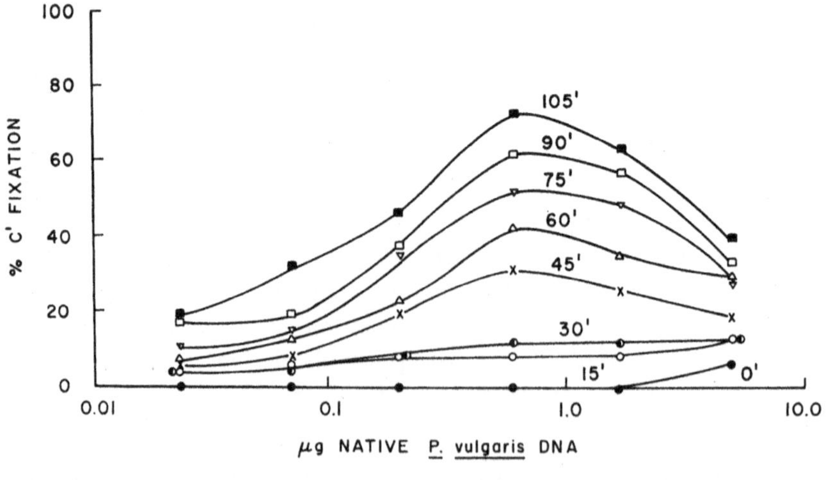

Fig. 1. Fixation of C′ by increments of native *P. vulgaris* DNA exposed to varying doses (seconds) of ultraviolet light from a germicidal lamp at a distance of 5 cm. Antiserum diluted 1/2000.

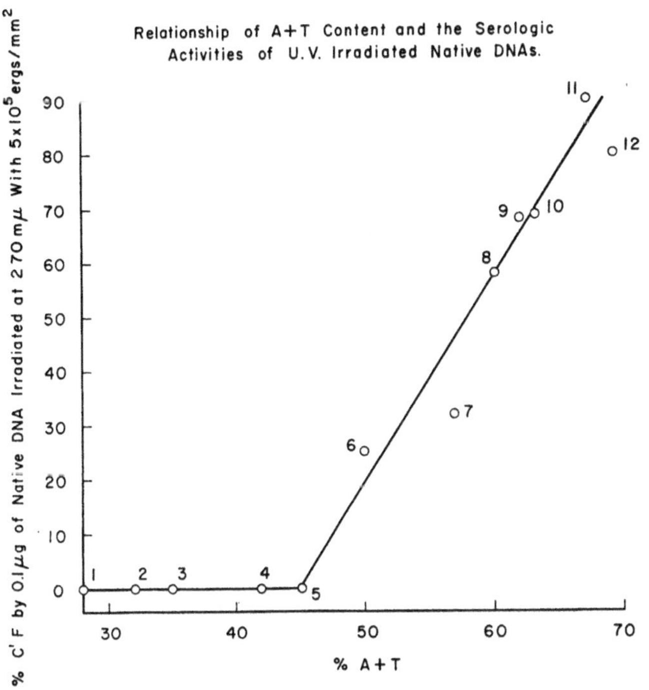

Fig. 2. Complement fixation reactions with DNA from various bacterial sources. All DNA samples were irradiated with 1×10^5 ergs/mm2 at 270 mμ. The adenine plus thymine content is indicated in parentheses. 1. *M. lysodeikticus* (28%); 2. *Ps. aeurginosa* (32%); 3. *Vibrio alkaligenes* (35%); 4. *S. marcescens* (42%); 5. *Alkaligenes faecalis* (45%); 6. *E. coli* (50%); 7. *B. subtilis* (57%); 8. *B. pantothenicus* (60%); 9. *P. mirabilis* (62%); 10. *P. vulgaris* (63%); 11. *B. cereus* (67%); 12. *Cl. perfringens* (69%).

tivity of bacteria reflects the thymine content of the DNA (Haynes, 1963). When DNAs varying in A+T content from 69 to 28% were irradiated, the resulting serologic activity was highest with *B. cereus* DNA (67% A+T) and *Cl. perfringens* DNA (69% A+T) and lowest with *M. lysodeikticus* DNA (28% A+T). DNAs with intermediate A+T content were intermediate with respect to serologic activities (Fig. 2).

The antibodies directed toward the UV-irradiated DNA were capable of being inhibited with UV-irradiated thymine oligonucleotides. In a test system consisting of 0.1 μg of *P. vulgaris* DNA irradiated with 5×10^5 ergs/mm² at 270 mμ and a 1/2000 dilution of rabbit antiserum to irradiated DNA from calf thymus, 0.11 mμ moles of $(Tp)_2$, 0.0035 mμ moles of $(Tp)_3$ and 0.0045 mμ moles of $(Tp)_4$ were required for 50% inhibition. Irradiated tri- and tetra-cytosine oligonucleotides were 100 to 1000 times less effective than the thymine photoproducts of equivalent chain length, while a mixed pyrimidine oligonucleotide C_3Tp5 was eight times more effective than the tetra nucleotide, C_4p_5. The fact that irra-

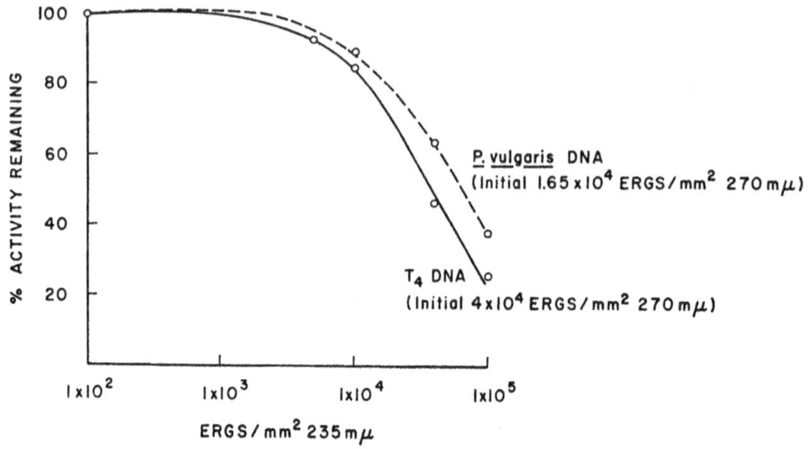

Fig. 3. Low wavelength reactivation of *P. vulgaris* DNA and T_4 DNA which have been irradiated at 270 mμ.

diated $(Tp)_3$ and $(Tp)_4$ are more effective inhibitors than $(Tp)_2$ indicated that the antigenic site is a trinucleotide consisting of two thymine residues and a third as yet unspecified base. It is more likely that the third base is a pyrimidine rather than a purine since if it were a purine, $(Tp)_3$ would not necessarily be much more effective than $(Tp)_2$ due to lack of structural similarity between the purines and thymine. The unequivocal determination of the nature of the antigenic site awaits the availability of all the possible thymine containing trinucleotides.

The dimerization of pyrimidines by ultraviolet irradiation is a reversible reaction and the equilibrium between the dimer formation and dimer splitting depends, in part, on the wavelength of irradiation; monomerization is favored at low wavelengths, while dimerization is favored at higher wavelengths (Setlow, 1961). As previously shown (Levine *et al.*, 1966), irradiation of DNA at 270 mμ leads to increased serologic activity, while exposure of the serologically active DNA to re-irradiation at 235 mμ leads to a loss of activity (Fig. 3). Irradiation damage in DNA can also be reversed by a photoreactivating enzyme isolated from yeast (Rupert, 1960) which splits the thymine dimer to yield the original monomers. In Figure 4, we see the anticipated decrease in activity following incubation of the irradiated DNA with the photoreactivating enzyme (kindly supplied by Dr. Jane K. Setlow) in the presence of visible light.

Two different lines of evidence implicate the thymine photoproducts as part of the antigenic determinants of UV-irradiated DNA: (1) the direct relationship of the serological response to the thymine content of the DNA, and (2) the more potent inhibition by thymine oligonucleotides compared to the inhibitory effects of cytosine oligonucleotides.

Recently Lamola and Yamane (1967) have reported that if DNA in aqueous acetophenone solutions was irradiated with visible light, thymine dimers were formed. After acid hydrolysis of the treated DNA, the hydrolysate was chromatographed and a single peak corresponding in position to the thymine dimers of UV-treated DNA was obtained. Using the antibody against UV-irradiated DNA, it was shown that irradiation of DNA in acetophenone solutions with visible light gave a positive serologic response (Fig. 5). This experiment provides information about the *lesion* in the *macromolecule* and indicates that the photoproducts formed by the two procedures are similar.

As shown in Figure 1, the antisera prepared against UV-irradiated DNA can react with native as well as with denatured DNA. The formation of the UV lesion in native DNA disrupts the hydrogen bonding primarily between the A-T pairs of the double-stranded helix and forms local areas of denaturation which might be envisioned as "bubbles" in the double helix. Thus, while the anti-UV DNA can react with the photoproduct, a serum specific for the unaltered exposed base on the opposite strand (in this case adenine) should also be capable of reacting with irradiated DNA. Using anti-A prepared by the method of Erlanger and Beiser (1964), the curves shown in Figure 6 were obtained.

These two sera were also used to determine if an endonuclease isolated from the livers of the smooth dogfish (*Mustelus canis*) which is

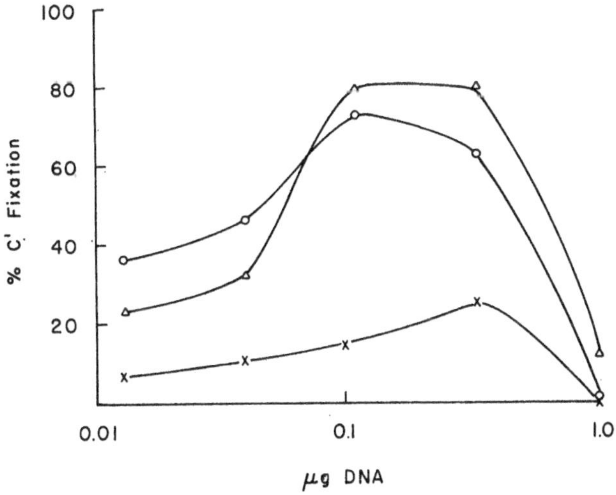

Fig. 4. Photoreactivation of *P. vulgaris* DNA. *P. vulgaris* DNA was irradiated with 5×10^5 ergs/mm^2 (△) and then treated with photoreactivating enzyme from yeast in the dark (○) and in the presence of visible light (×) from a 15 watt Sylvania "cool white" bulb.

capable of degrading denatured DNA (Ashe *et al.,* 1965) might be able to function as an excision enzyme, i.e., remove the damaged area from native DNA. Native DNA which had been irradiated and shown to give a good reaction with anti-UV DNA was, therefore, incubated with the endonuclease and reassayed with the antibody. The serological reaction progressively decreased as might be expected if the photoproduct was eliminated by excision. The model proposed for *in vivo* irradiated systems which undergo repair is that the excision enzyme removes the damaged area because it either recognizes the damage or, more likely, distortions of the helix in the damaged area (cf. Lehman, 1967). Subsequently, nucleotides complementary to those of the intact opposite strand are presumably inserted into the gap by a DNA polymerase, thereby leading to a restoration of the double helix. If the dogfish endonuclease resembled this model of an excision enzyme, the serologic reaction between the enzyme-treated irradiated DNA and anti-A should remain constant. This particular enzyme, however, also excises the non-damaged area on the opposite strand as shown by the loss of serological reaction with anti-A. These events, i.e., digestion of both strands of irradiated DNA, have been verified by ultracentrifugal analysis. Thus, the dogfish endonuclease distinguishes only single-stranded areas whether or not they contain damaged bases.

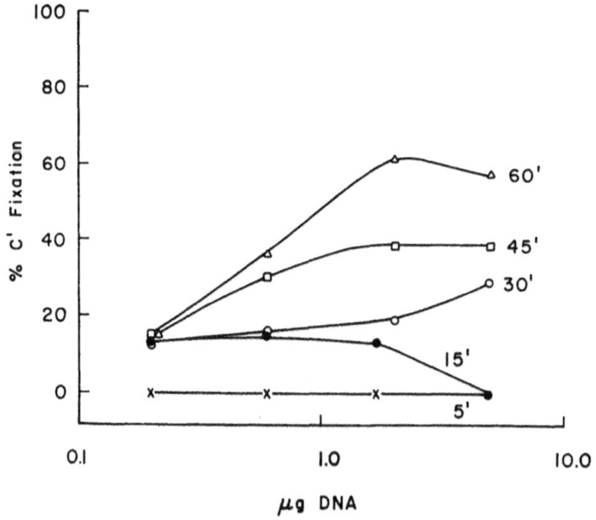

Fig. 5. Complement fixation reaction of *P. vulgaris* DNA irradiated for various periods of time in the presence of acetophenone. The acetophenone was present at a concentration of 0.02 *M*. A 65 watt GE medium pressure mercury arc was used as the light source. Light of wavelengths below 300 mμ was filtered out by the use of Pyrex filters. DNA irradiated in the absence of acetophenone gave a negative complement fixation reaction.

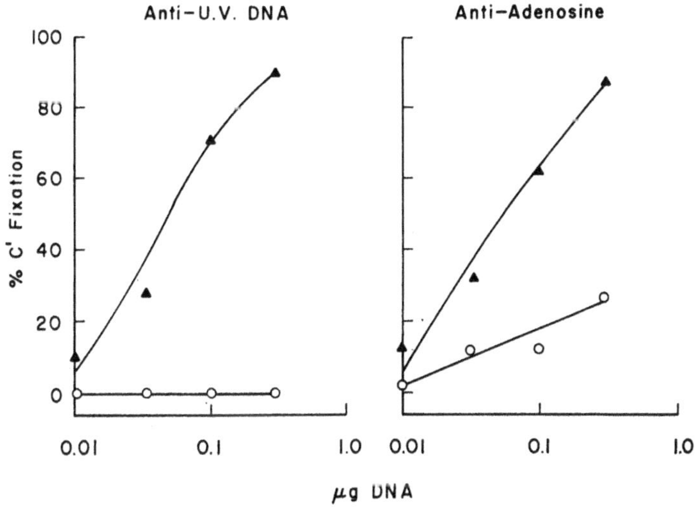

Fig. 6. The serological reaction of native *P. vulgaris* DNA before (○) and after (▲) irradiation at 270 mμ. The samples were assayed with an anti-UV serum (left) and anti-adenosine serum (right).

It is clear that the immunological approach outlined above can prove useful in the detection of modifications in DNA. It eliminates the necessity of subjecting the nucleic acid to harsh hydrolytic procedures and thus reduces greatly the possibility of artificial changes. It also provides a sensitive assay system in which attention can be focused on one of the two DNA strands and on the fate of that strand during repair processes.

References

Ashe, H., E. Seaman, H. Van Vunakis, and L. Levine (1965). Characterization of a deoxyribonuclease of *Mustelus canis* liver. Biochim. Biophys. Acta, *99*:298–306.

Erlanger, B. F., and S. M. Beiser (1964). Antibodies specific for ribonucleosides and ribonucleotides and their reaction with DNA. Proc. Natl. Acad. Sci. U. S., *52*:68–74.

Freeman, K. B., P. V. Hariharan, and H. E. Johns (1965). The ultraviolet photochemistry of cytidylyl (3′–5′) cytidine. J. Mol. Biol., *13*:833–848.

Haynes, Robert H. (1963). Molecular localization of radiation damage relevant to bacterial inactivation. In "Physical Processes in Radiation Biology," L. Augenstein, R. Mason, B. Rosenberg, eds. (Academic Press, New York), pp. 51–67.

Johns, H. E., S. A. Rapaport, and M. Delbrück (1962). Photochemistry of thymine dimers. J. Mol. Biol., *4*:104–114.

Lamola, A. A., and T. Yamane (1967). Sensitized photodimerization of thymine in DNA. Proc. Natl. Acad. Sci. U. S., *58*:443–446.

Lehman, I. R. (1967). Deoxyribonucleases: Their relationship to deoxyribonucleic acid synthesis. Ann. Rev. of Biochem., *36*:645–668.

Levine, L., W. T. Murakami, H. Van Vunakis, and L. Grossman (1960). Specific antibodies to thermally denatured deoxyribonucleic acid of phage T_4. Proc. Natl. Acad. Sci. U. S., *46*:1038–1043.

Levine, L., E. Seaman, E. Hammerschlag, and H. Van Vunakis (1966). Antibodies to photoproducts of deoxyribonucleic acids irradiated with ultraviolet light. Science, *153*:1666–1667.

Murakami, W. T., H. Van Vunakis, L. Grossman, and L. Levine (1961). Immunochemical studies of bacteriophage deoxyribonucleic acid. II. Characterization of the active antigen. Virology, *14*:190–197.

Plescia, O. J., W. Braun, and N. C. Palczuk (1964). Production of antibodies to denatured deoxyribonucleic acid (DNA). Proc. Natl. Acad. Sci. U. S., *52*:279–285.

Rupert, C. S. (1960). Photoreactivation of transforming DNA by an enzyme from baker's yeast. J. Gen. Physiol., *43*:573–595.

Seaman, E., L. Levine, and H. Van Vunakis (1966). Antibodies to the methylene blue sensitized photooxidation product in deoxyribonucleic acid. Biochemistry, *5*:1216–1223.

Setlow, R. B. (1961). The action spectrum for the reversal of the dimerization of thymine induced by ultraviolet light. Biochim. Biophys. Acta, *49*:237–238.

Setlow, R. B., W. L. Carrier, and F. J. Bollum (1965). Pyrimidine dimers in UV-irradiated poly dI:dC. Proc. Natl. Acad. Sci. U. S., *53*:1111–1118.

Setlow, R. B., and J. K. Setlow (1962). Evidence that ultraviolet-induced thymine dimers in DNA cause biological damage. Proc. Natl. Acad. Sci. U. S., *48*:1250–1257.

Shugar, D., and K. L. Wierzchowski (1957). Reversible photolysis of pyrimidine derivatives including trials with nucleic acids. Biochim. Biophys. Acta, *23*:657–658.

Townsend, E. E., H. Van Vunakis, and L. Levine (1965). Immunochemical studies on bacteriophage deoxyribonucleic acid. IV. Hapten inhibition. Biochemistry, *4*:943–948.

Van Vunakis, H., E. Seaman, L. Kahan, J. W. Kappler, and L. Levine (1966). Formation of an adduct with Tris (hydroxymethyl) aminomethane during the photooxidation of deoxyribonucleic acid and guanine derivatives. Biochemistry, *5*:3986–3991.

Wacker, A. H., H. Dellweg, and D. Weinblum (1960). Strahlenchemische Veranderung der Bakterien-deoxyribonucleinsaure *in vivo*. Naturwissenschaften, *47*:477.

Wang, S. Y., and R. Alcantara (1965). The possible formation of dithymine peroxide in irradiated DNA. Photochemistry and Photobiology, *4*:477–481.

IMMUNOLOGICAL EVIDENCE FOR THE IDENTITY OF A PHOTOPRODUCT FORMED DURING PHOTOOXIDATION OF DNA WITH METHYLENE BLUE, ROSE BENGAL, THIONIN, AND ACRIDINE ORANGE *

LAWRENCE LEVINE, EDNA SEAMAN AND HELEN VAN VUNAKIS

Graduate Department of Biochemistry, Brandeis University, Waltham, Massachusetts

Irradiation of DNA with visible light, in the presence of oxygen and a photosensitizing dye such as methylene blue, is known to destroy specifically the guanine residues in nucleic acids (Simon and Van Vunakis, 1962). The chemistry of the photooxidation reaction is being studied actively in several laboratories (cf. Bellin and Grossman, 1965) but it is not yet understood. It is known that the photoproducts may vary depending upon the composition of the reaction mixture. Immunological experiments provided the first evidence that, during the photooxidation of DNA, a reactive photointermediate condensed with the available amino groups of the buffer, Tris(hydroxymethyl)amino methane (Tris) to yield an adduct (Van Vunakis *et al.,* 1966). The evidence that established the existence of this adduct will be reviewed only briefly. DNA was photooxidized in Tris buffer at pH 8.5 using methylene blue as the photosensitizing dye. DNA thus treated ($DNA_{Tris\text{-}MB}$) was complexed with methylated bovine serum albumin (Plescia *et al.,* 1964) and injected into rabbits. The serologic activities of native and denatured DNAs with the anti-$DNA_{Tris\text{-}MB}$ became apparent as the reaction mixture containing DNA, Tris buffer and methylene blue was irradiated with visible light.

* Aided by grants from the NIH (AI-01940 and AI-02792) and the Charles Simon Memorial Grant for Cancer Research from the American Cancer Society (E.222). L. L. is an American Cancer Society Professor of Biochemistry and H. V. V. is a recipient of a Public Health Service Research Career Award K6-AI-2372. Publication No. 566 from the Graduate Department of Biochemistry, Brandeis Universitv.

Irradiation of DNA in bicarbonate buffer did not give a serological reaction with the anti-$DNA_{Tris\text{-}MB}$ although base analysis revealed that guanine had been destroyed preferentially. DNA, irradiated in the presence of C^{14} Tris for various times, showed an increased uptake of Tris as irradiation proceeded. The light control (i.e., the reaction mixture, irradiated in the absence of methylene blue) and the dark control (i.e., the complete reaction mixture which was never irradiated) showed no significant uptake of C^{14} Tris. DNA photooxidized in $NaHC^{14}O_3$ buffer showed no incorporation of the isotope.

The uptake of Tris during the photooxidation of DNA is dependent upon the relative concentrations of the reactants and under certain conditions reaches a value of 1 mole Tris incorporated for each mole of guanine destroyed. If the four nucleotides are irradiated in Tris buffer in the presence of methylene blue, only the photooxidized deoxyguanylic acid ($d[G]MP_{Tris\text{-}MB}$) inhibits the serologic activity of the immune system. The guanine ribonucleotide is as effective an inhibitor as the deoxyribonucleotide, indicating that the specificity of the antibodies does not extend to the sugar moieties. Guanosine 5′-phosphate (GMP) or deoxyguanosine 5′-phosphate (dGMP) photooxidized in borate, phosphate or bicarbonate buffers do not yield inhibitory photoproducts although the nucleotides are destroyed.

While methylene blue has served as the photosensitizing dye in most of our studies, several other dyes belonging to different chemical classes are known to be capable of photooxidizing guanine derivatives (Simon and Van Vunakis, 1964) and other susceptible substrates (Bellin and Grossman, 1965). Although the structure of the Tris adduct has not yet been identified, the antibodies permit us to detect its formation during the photooxidation of DNA and guanine mononucleotides. We can then ask the following question: Does the same photoproduct form during photooxidation of DNA or guanine nucleotides in the presence of *different* photosensitive dyes?

I. PHOTOOXIDATION OF DENATURED DNA USING DIFFERENT DYES AS SENSITIZERS

Denatured T4 DNA was photooxidized in Tris buffer in the presence of different dyes and assayed for serological activity using anti-$DNA_{Tris\text{-}MB}$. The rate of appearance of the serologic response varied with the dye used; methylene blue and rose bengal were very efficient and thionin was markedly less so (Fig. 1). Rose bengal, methylene blue and thionin are almost equally effective in photooxidizing guanosine but acridine orange is essentially inert (Simon and Van Vunakis, 1964).

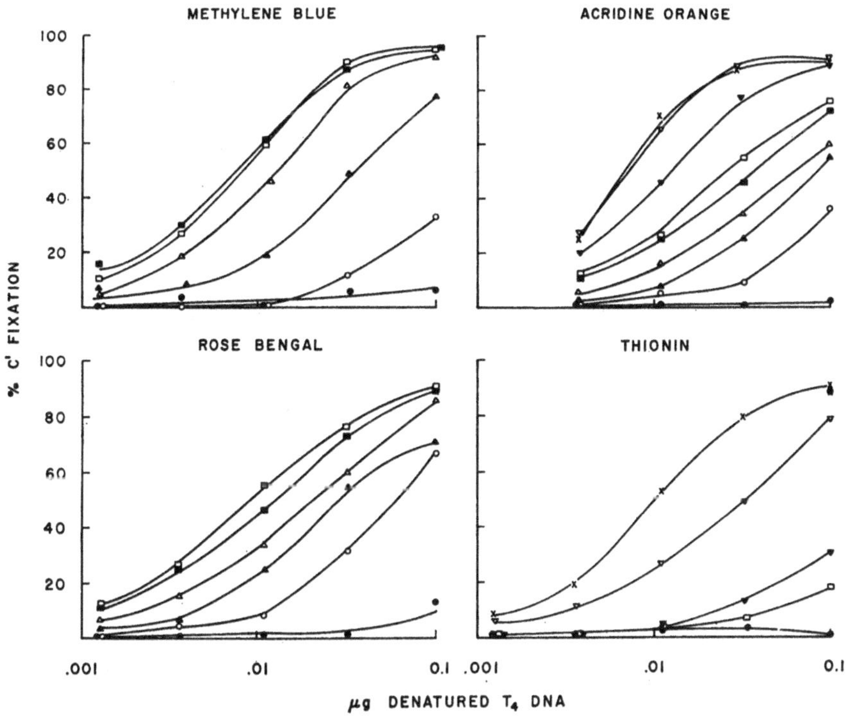

Fig. 1. C'-fixation of photooxidized denatured T_4 DNA and anti-$DNA_{Tris\text{-}MB}$ diluted 1:100. The preparation of the antibody and the immunization procedures are described in Seaman *et al.*, 1966. 100 μg/ml of denatured T_4 DNA were photooxidized in the presence of 20 μg/ml of methylene blue, rose bengal, thionin and 62 μg/ml of acridine orange in 0.1 *M* Tris buffer, pH 8.5. O_2 was bubbled through the reaction mixtures and the irradiation was carried out as described by Simon and Van Vunakis, 1962. Samples were withdrawn at various times and assayed for serologic activities. The times of irradiation are (●) unirradiated sample, (○) 10 min., (▲) 20 min., (△) 40 min., (■) 50 min., (□) 60 min., (▼) 120 min., (▽) 240 min., and (×) 360 min.

Although three times as much acridine orange as methylene blue was used during the photooxidation of DNA, the rate of appearance of the lesion in $DNA_{Tris\text{-}AO}$ was surprisingly rapid when compared to the ability of the two dyes to photooxidize guanosine. Shastry and Gordon (1966) have noted that acridine orange was only 5–6 times less effective than methylene blue in inactivating the ribonucleic acid of tobacco mosaic virus and that it also was appreciably less efficient in destroying guanosine.

Antibodies were also obtained by immunizing rabbits with denatured DNA which had been photooxidized in Tris buffer using acridine orange as the sensitizer (anti-$DNA_{Tris\text{-}AO}$). Assay of identical samples of DNA photooxidized with different dyes showed the same relative order of serological response whether anti-$DNA_{Tris\text{-}MB}$ or anti-$DNA_{Tris\text{-}AO}$ was used

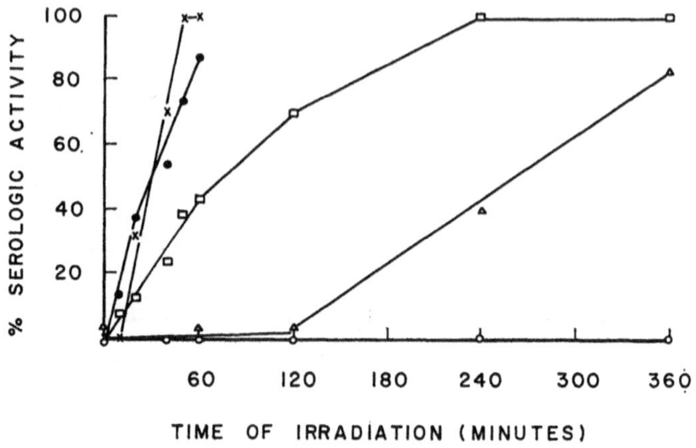

Fig. 2. The rate of appearance of serologic activity of denatured T_4 DNA irradiated in the presence of different dyes, with anti-DNA$_{\text{Tris-MB}}$ diluted 1:100. The photosensitizing dyes were: methylene blue (×), rose bengal (●), acridine orange (□), thionin (△) and methyl orange, alizarin red (○). Conditions of irradiation as in Fig. 1.

(Figs. 2 and 3). The dark controls (i.e., the complete reaction mixtures deprived of light) and the light controls (i.e., the reaction mixtures lacking the dye) gave no serological response. In addition, methyl orange and alizarin red, dyes that fail to photooxidize guanosine as measured by spectral changes and O_2 uptake (Simon and Van Vunakis, 1964), and fail to inactivate transforming DNA (Bellin and Grossman, 1965), also are inert in this system.

In the experiments shown in Figures 1–3, the antibodies were used at

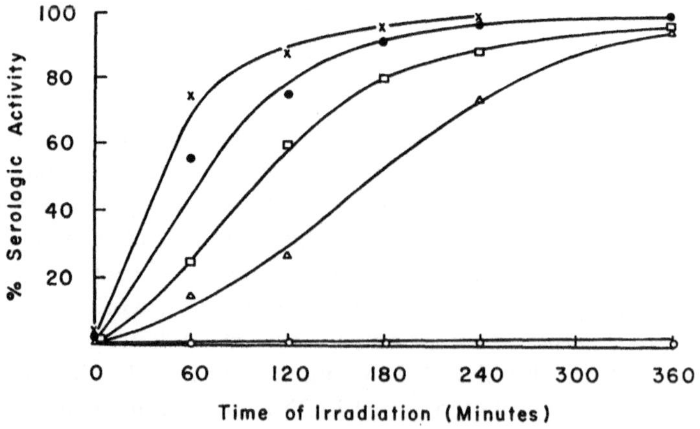

Fig. 3. The rate of appearance of serologic activity of denatured T_4 DNA, irradiated in the presence of different dyes, with anti-DNA$_{\text{Tris-AO}}$. DNA samples and photosensitizing dyes are the same as Fig. 2.

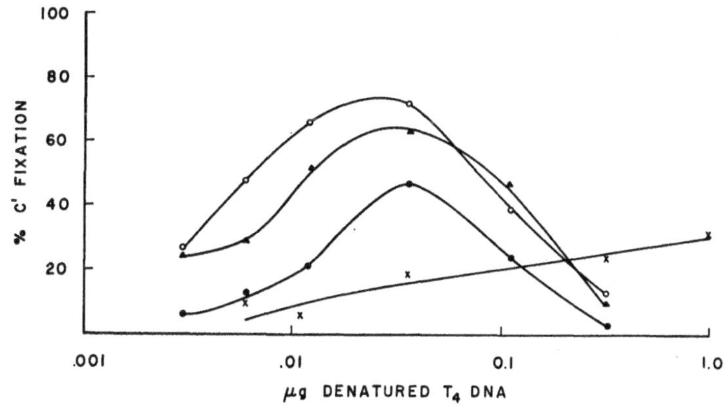

Fig. 4. Extent of maximum C′-fixation by denatured T_4 DNA photooxidized in the presence of methylene blue (○), rose bengal (▲), acridine orange (●) and thionin (×) with anti DNA$_{\text{Tris-MB}}$ diluted 1:650.

a high concentration in order to detect the early appearance of the photoproduct, but as a result, measurements of the extent of maximum reaction (i.e., analyses with this antiserum at dilutions with which a complete C′ fixation curve can be obtained) could not be made. The two antibodies were therefore used at a higher dilution (Figs. 4 and 5) and the order of maximum C′ fixation was found to be similar to the rate of appearance of the lesions.

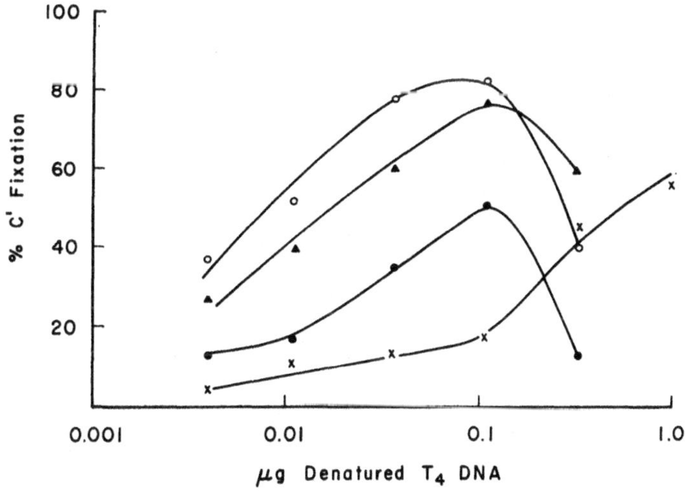

Fig. 5. Extent of maximum C′-fixation by denatured T_4 DNA photooxidized in the presence of different dyes (same samples as in Fig. 4) with anti-DNA$_{\text{Tris-AO}}$ diluted 1:4000.

II. PHOTOOXIDATION OF NATIVE DNA USING DIFFERENT DYES AS SENSITIZERS

As shown previously, the rate of formation of the methylene blue-sensitized photoproduct in denatured DNA is more rapid than in native DNA (Seaman *et al.*, 1966). The same situation was found to hold with rose bengal. With acridine orange or thionin, however, the formation of the photoproduct was not detected after irradiation of the native DNA for 24 hours. The capacity of the dye to sensitize photooxidation may be influenced by its ability to bind to the susceptible substrate (cf. Bellin and Grossman, 1965).

Native DNA undergoes partial and eventually complete denaturation as photooxidation proceeds (Simon and Van Vunakis, 1962; Bellin and Grossman, 1965; and Seaman *et al.*, 1966). An immunologic technique has been employed to demonstrate the opening of the A-T hydrogen bonds in ultraviolet irradiated DNA (Seaman *et al.*, 1967). The formation of the thymine lesion is detected by antibodies specific for the photoproduct while the adenine on the opposite strand is now available for reaction with anti-A serum. The destruction of G residues during photooxidation should render cytosine residues on the opposite strand available for reaction with anti-C serum and lead to the detection of small areas of denaturation. The results of an experiment showing the concomitant appearance of a positive serological reaction with anti-$DNA_{Tris\text{-}MB}$ and anti-C are shown in Figure 6.

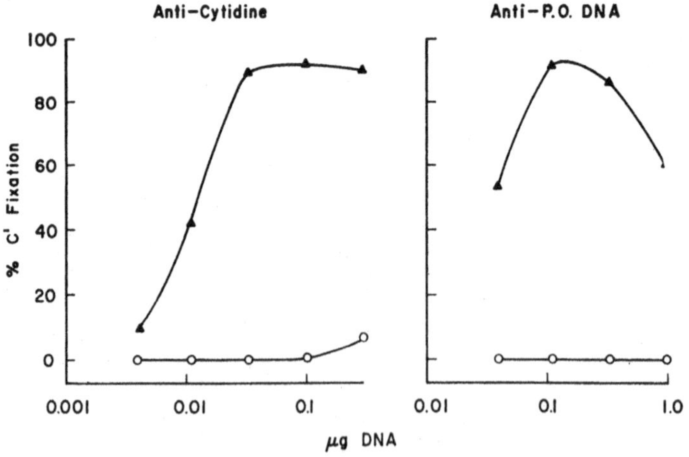

Fig. 6. The C′-fixation reaction of native *Ps. aeruginosa* DNA, which has been photooxidized in the presence of methylene blue, with anti-cytidine (left) and anti-$DNA_{Tris\text{-}MB}$ (right). (○) represents DNA sample before irradiation, (▲) represents DNA sample after irradiation. The concentration of methylene blue in this experiment was 5 µg/ml and the time of irradiation two hours. All other conditions as in Fig. 1.

III. PHOTOOXIDATION OF dGMP USING DIFFERENT DYES AS SENSITIZERS

As already pointed out, the ability of a particular dye to sensitize the photooxidation of guanine nucleosides and nucleotides does not necessarily correspond to its ability to photooxidize guanine residues in nucleic acids. In addition, while the photosensitized destruction of guanine nucleosides and nucleotides (as measured by loss of spectra, O_2 uptake, etc.) can be compared with the destruction of guanine residues in the macromolecule (as determined by base analyses), the comparisons apply only to the overall reaction. The antibodies to the specific photoproduct offer the possibility of determining its formation not only in the macromolecule (Figs. 1–5) but also on the nucleotide level. The latter assay depends on the ability of the photooxidized dGMP to inhibit the specific immune system. With the two antibodies available (anti-$DNA_{Tris\text{-}MB}$ and anti-$DNA_{Tris\text{-}AO}$), and the four $DNA_{Tris\text{-}Dye}$ antigens, eight antigen-antibody reactions were analyzed by inhibition of dGMP photooxidized with the four different dyes (Table 1). If the antigenically reactive determinants of the photoproducts are identical regardless of the nature of the photosensitizing reagent, the pattern of inhibition of both antisera with these photooxidized reaction mixtures would be similar.

Table 1

INHIBITION OF ANTI-$DNA_{Tris\text{-}MB}$ AND ANTI-$DNA_{Tris\text{-}AO}$ BY D[G]MP PHOTOOXIDIZED WITH SEVERAL PHOTOSENSITIZING DYES

dGMP Photo-oxidized with	μg of d[G]MP Required for 50% Inhibition of the Anti-$DNA_{Tris\text{-}MB}$ and DNA Photooxidized with				μg of d[G]MP Required for 50% Inhibition of the Anti-$DNA_{Tris\text{-}AO}$ and DNA Photooxidized with			
	MB	Thio	AO	RB	MB	Thio	AO	RB
Methylene blue	4.0	7.4	9.0	1.5	0.95	1.4	1.6	0.6
Thionin	3.5	7.0	9.0	2.0	0.8	1.4	1.4	0.6
Acridine orange	15.0	20.0	35.0	9.0	4.0	8.0	7.0	1.2
Rose bengal	3.0	5.7	6.0	1.6	0.8	1.9	1.9	—

5 mg/ml of dGMP were photooxidized in the presence of methylene blue, rose bengal, and thionin (20 μg/ml) and acridine orange (62 μg/ml) in 0.1 M Tris buffer, pH 8.5. All reaction mixtures except that containing acridine orange were irradiated until the absorbance at 250 mμ was 40% of the initial value. Even after 21 hours of irradiation, the destruction of dGMP in the acridine orange sensitized reaction mixture was only 15% (i.e., the absorbance was 85% of the initial value).

It can be seen (Table 1) that the inhibition patterns with the two antisera are indeed similar. With both anti-$DNA_{Tris\text{-}AO}$, and anti-$DNA_{Tris\text{-}MB}$, the inhibitory effectiveness of the photoproducts is $d[G]MP_{Tris\text{-}RB} > d[G]MP_{Tris\text{-}MB} > d[G]MP_{Tris\text{-}Thio} = d[G]MP_{Tris\text{-}AO}$. Within the experimental error of the analyses, this pattern is repeated regardless of the photosensitizing dye used to prepare the immunogen. In general, the anti-$DNA_{Tris\text{-}AO}$ serum is more susceptible to inhibition with d[G]MP than the anti-$DNA_{Tris\text{-}MB}$ serum. This could be due to the fact that the DNA used as immunogen in the acridine orange system was photooxidized to a lesser extent than that used in the methylene blue system. Thus, statistically there should be a higher proportion of single photoproduct residues in the $DNA_{Tris\text{-}AO}$ as compared to $DNA_{Tris\text{-}MB}$. The latter would be expected to contain a greater proportion of photoproduct residues in sequence rendering the methylene blue system less susceptible to inhibition by the photooxidized monomer.

It is known that photooxidation of GMP in the presence of methylene blue results in the formation of more than one photoproduct (Van Vunakis *et al.,* 1966). It is likely that the other photosensitizing dyes also lead to the production of several products but the methods used in this work allowed us to focus only on the immunologically active species. Using the immunochemical approach, the nature of this photoproduct has been shown to be identical regardless of the photosensitizing agent used.

References

Bellin, J. S., and L. I. Grossman (1965). Photodynamic degradation of nucleic acids. Photochemistry and Photobiology, *4*:45–53.

Plescia, O. J., W. Braun, and N. C. Palczuk (1964). Production of antibodies to denatured deoxyribonucleic acid (DNA). Proc. Natl. Acad. Sci. U. S., *52:* 279–285.

Sastry, K. S., and M. P. Gordon (1966). The photodynamic inactivation of Tobacco Mosaic Virus and its ribonucleic acid by acridine orange. Biochim. et Biophys. Acta, *129*:32–48.

Seaman, E., L. Levine, and H. Van Vunakis (1966). Antibodies to the methylene blue sensitized photooxidation product in deoxyribonucleic acid. Biochemistry, *5*:1216–1223.

Seaman, E., L. Levine, and H. Van Vunakis (1968). Immunological evidence for the existence of thymine dimers in ultraviolet irradiated DNA. In *Nucleic Acids in Immunology*. Springer-Verlag, N. Y.

Simon, M. I., and H. Van Vunakis (1962). The photodynamic reaction of methylene blue (MB) with deoxyribonucleic acid (DNA). J. Mol. Biol., *4:* 488–499.

Simon, M. I., and H. Van Vunakis (1964). The dye sensitized photooxidation of purine and pyrimidine derivatives. Arch. Biochem. Biophys., *105*:197–206.

Van Vunakis, H., E. Seaman, L. Kahan, J. W. Kappler, and L. Levine (1966). Formation of an adduct with tris (hydroxymethyl) aminomethane during the photooxidation of deoxyribonucleic acid and guanine derivatives. Biochemistry, *5*:3986–3991.

USE OF ANTI-RNA ANTIBODIES IN THE STUDY OF THE STRUCTURE OF POLYNUCLEOTIDES AND RIBOSOMAL PARTICLES *

JACQUES PANIJEL
Institut Pasteur, Paris, France

I. INTRODUCTION

Immunization of an animal with bacterial ribosomes generally elicits the formation of antibodies capable of precipitating not only the homologous ribosomes used for immunization but also ribosomes of different origin (Barbu *et al.*, 1961; Panijel and Barbu, 1961). However, for a given antiserum, the qualitative aspects of the immune reaction vary with the species of bacteria being tested. In addition, as shown in Figure 1 (A, B, C), the amount of antibody precipitable by a given bacterial ribosome varies with the antiserum used. It is evident that ribosomes even from distant species possess common antigenic determinants, thus explaining the cross-reactions observed. With a given antiserum, such as the *E. coli* K12 ribosome antiserum, one can distinguish the following: (a) homologous ribosomes, e.g., those from various strains of *E. coli* or even other enteric bacteria; (b) close heterologous ribosomes, e.g., ribosomes of *Proteus vulgaris;* and (c) distant heterologous ribosomes, e.g., ribosomes of *Clostridia.* Such a "classification" of ribosomes relative to a given antiserum seems to be genetically significant. Indeed, McCarthy and Bolton (1963), who used the hybridization technique to study the relationship between messenger RNAs extracted from *E. coli* and DNAs of other origin, subsequently arrived at a classification in agreement with ours.

* Some of the data presented here were obtained in collaboration with different coworkers, particularly Dr. C. Souleil and Miss M. C. Delaunay. Acknowledgment is also due to the skilled technical assistance of Miss M. Lenepveu and Mr. J. Le Goff. This work was supported by grants from the C.N.R.S. (Centre National de la Recherche Scientifique) and the D.G.R.S.T. (Grant 6600449 Biologie Moléculaire).

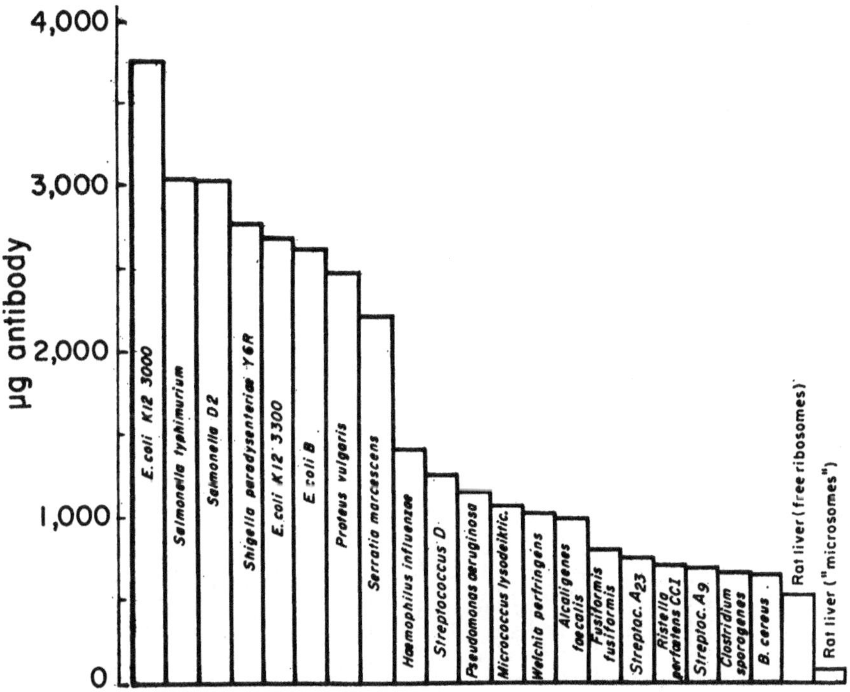

Fig. 1A. Results obtained with an antiserum to ribosomes of *E. coli K12*. Amounts of antibody precipitated by ribosomes of various origins from antisera obtained by immunization of rabbits with bacterial ribosomes of three different sources. In each experiment, 830 μg of ribosomes were added to 1 ml of antiserum. Estimation of the amount of antibody precipitated was done according to methods previously described (Panijel and Barbu, 1961).

Finally, it was found that even ribosomes of animal origin, e.g., ribosomes from rat liver, precipitate to some extent antibody from anti-bacterial ribosome sera (cf. Fig. 1). However, if the animal ribosomes are tested in the form of microsomes, no precipitation occurs.

Thus the "basal" reactivity of *free* animal ribosomes with a bacterial ribosome antiserum must be due to universal antigen present in all ribosomes regardless of origin. It was reasonable to assume that the universal ribosomal antigen mentioned above could be RNA (Barbu and Panijel, 1960a and b; Panijel *et al.,* 1963).

This prediction was verified experimentally, and we were able to obtain subsequently, from horses immunized with bacterial ribosomes, purified anti-RNA antibodies to which we shall refer as NG I antibodies (Panijel, 1963a, b, d; 1966a, b).

The availability of these antibodies allowed us to study: (1) the structural immunochemistry of RNA and synthetic polyribonucleotides (Panijel *et al.,* 1966a, b; Souleil *et al.,* 1965, 1966); (2) the possible specific

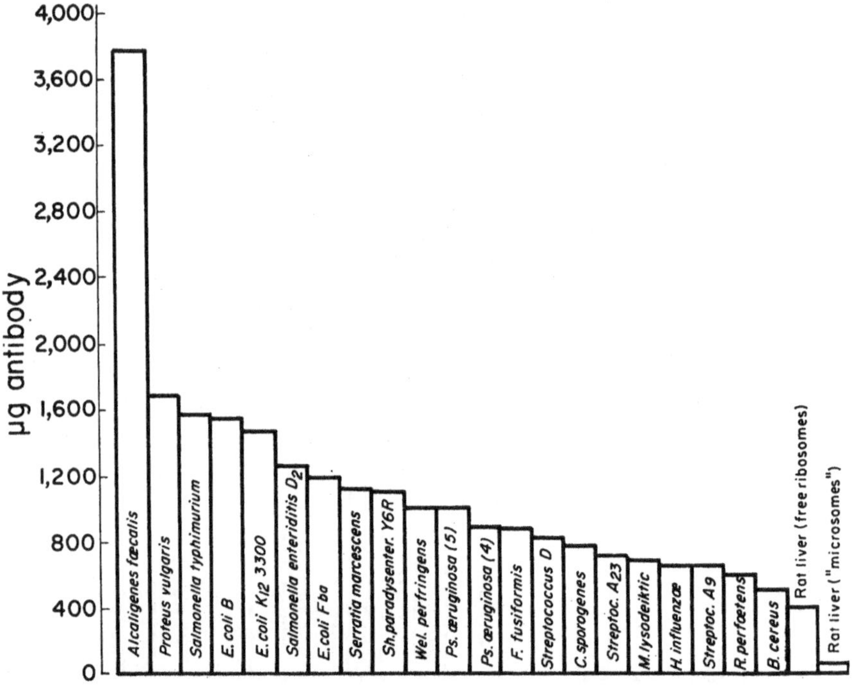

Fig. 1B. Results obtained with an antiserum to ribosomes of *Alcaligenes faecalis.*

interactions between antibodies and both cellular or subcellular constituents (Cayeux *et al.,* 1965; Panijel and Delaunay, 1967); and (3) the biosynthesis of these antibodies and their possible relationship to the problems of auto-antibody formation (Panijel, 1963c; Panijel *et al.,* 1963; Panijel *et al.,* 1966c).

In this report, we shall examine some results dealing with the first two areas of investigation, with particular emphasis on the methodological aspects of this research.

II. IMMUNOCHEMISTRY OF POLYRIBONUCLEOTIDES

A. Statement of the Problem

The overall structure of various RNAs differs because it is determined by the sequence of nucleotides and the level of their organization.

The NG I antibodies precipitate completely all RNAs whatever their origin (viral, bacterial, vegetal or animal) and whether they be soluble or ribosomal RNA. In addition, they precipitate all synthetic polyribonucleotides. This fact indicates that these antibodies have no specificity for the bases. Nevertheless it is possible that the nature of the base influ-

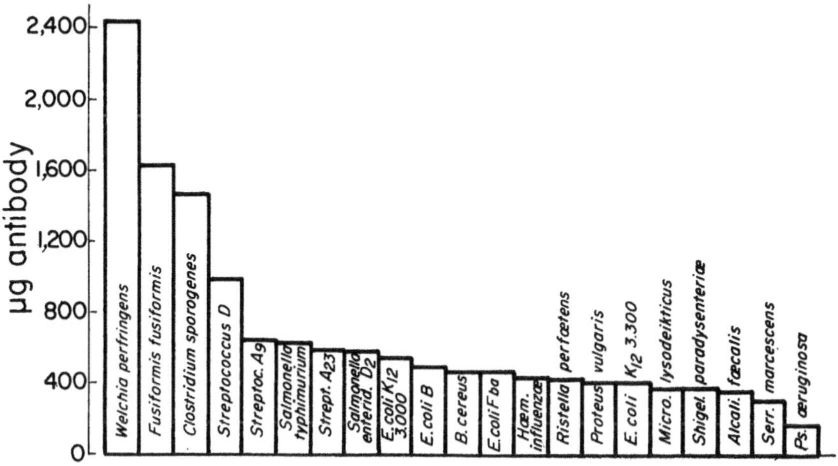

Fig. 1C. Results obtained with an antiserum to ribosomes of *Welchia perfringens.*

ences the quantitative aspects of the reaction. If this is so, the precipitation curve will vary with the polynucleotide used.

To ascertain the influence of the base composition of polynucleotides, it is necessary to determine: (1) the influence of the structure and the frequency of antigenic sites on the quantity of precipitated antibody, the frequency of antigenic sites depending perhaps on the distribution of ordered regions; (2) the quantity of precipitated polynucleotide which is indicative of what we call "the molecular carrier" of the antigenic sites; (3) the ratio of "precipitated antibody to the precipitated antigen" (Ab/Ag) which permits an overall appraisal of the precipitation reaction. In this report we shall describe the results obtained using poly U * and poly A as models of single-stranded polynucleotides, poly (A + U) and poly (A + 2U) as models of complexes of different homopolymers, and poly I and poly G as multi-stranded homopolymers.

B. Single-stranded Homopolymers

1. *Poly U as an Immunochemical Reference* (Panijel *et al.*, 1966a, b)

At low temperature poly U possesses a significant amount of single-stranded helical structure arising from interplanar interactions among successive stacked or partially stacked uracil residues (Michelson and Monny, 1966). But this amount decreases rapidly as the temperature increases and, at 20°C, there is no longer evidence of any organized structure.

* Abbreviations used: Poly U: polyribouridylic acid; poly A: polyriboadenylic acid; poly C: polyribocytidylic acid; poly G: polyriboguanylic acid; poly I: polyriboinosinic acid; Ab: antibody; Ag: antigen; P.F.U.: plaque-forming units; i.c.: infection centers.

The immunochemical behavior of poly U under the latter conditions presents the following features:

(a) The addition of 5×10^{-4} *M* Mg^{++} is reflected by an increase in the maximum amount of antibody precipitated (Fig. 2A).

(b) In NaCl, the maximum values of precipitable antibody and antigen are equal to those obtained in Mg^{++} and stay unchanged when the salt concentration is increased from 0.05 to 0.14 *M* (Fig. 2A and B).

(c) The maximum amount of antibody precipitated at 40°C is the same as that at 20°C.

(d) The immunochemical reactions are identical at pH 7.4 and 6.5.

(e) The values of Ab/Ag are very high in the presence of excess antibody, i.e., for the smallest amounts of poly U added.

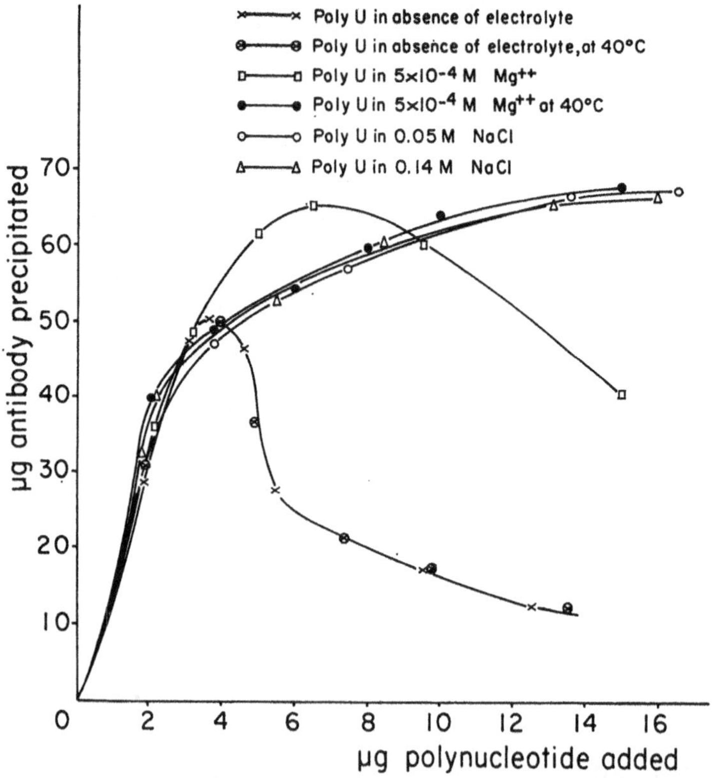

Fig. 2A. Amounts of antibody precipitated. Precipitin curves of poly U. The polyribonucleotide was added in increasing amounts on a mononucleotidic basis, the amount of the antibody being kept constant (100 μg). The specific precipitates were collected, washed three times in ten volumes of 0.1 *M* NaCl, 0.0033 *M* Mg^{++} acetate, 0.0017 *M* Tris-HCl buffer, pH 7.4, and then dissolved in 0.3 ml of 0.1 *N* NaOH containing 1% Na_2CO_3. One portion of the specific precipitate was used for estimating the proteins according to the technique of Lowry *et al.* (1951). The other served for the estimation of nucleotides: optical density measurements were made at 260 mμ after hydrolysis by 1 *N* perchloric acid.

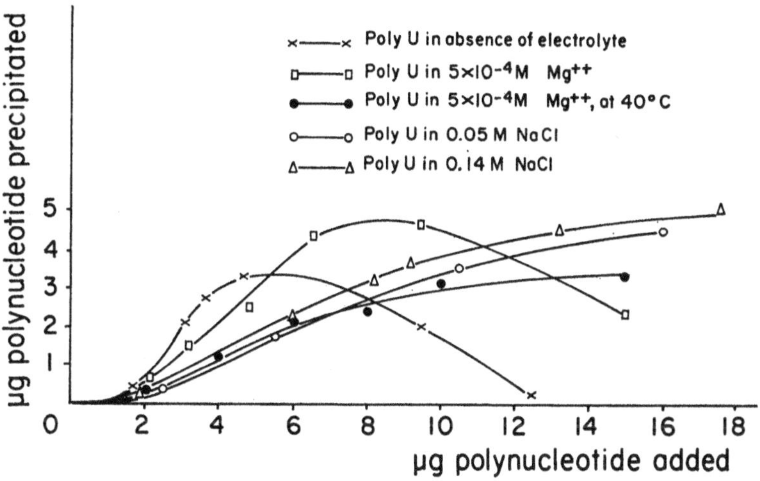

Fig. 2B. Amounts of polynucleotide precipitated.

As will be shown later, when the conditions of medium or temperature give rise to instability of organized regions, all the polyribonucleotides tested give precipitin curves that are closely similar to those of poly U. Thus, it is possible to assert that the nature of the base is neither qualitatively nor quantitatively a determining factor in the precipitation reaction.

The immunochemical behavior of poly U has to be related to its random coiled conformation. Consequently, the fundamental principle of the method we shall use will be necessarily a comparative one. For each medium, the results of precipitation of poly U must be taken as the norms for a system devoid of order, and all modification from these norms would imply a certain level of organization.

2. *The Case of Poly A at pH 7.4* (Panijel *et al.,* 1965; 1966a)

Poly A possesses a helical double-stranded structure at acid pH which apparently persists in solution (Fresco, 1959; Lipsett, 1960). However, in neutral solutions the conformation is that of a single-stranded helical structure with base stacking (Witz and Luzzati, 1965; Van Holde *et al.,* 1965).

Most of the experiments that showed such a structure were performed at neutral pH in 0.14 M NaCl. An immunochemical study of poly A was carried out under these same conditions. The results reflected the structure based on physical chemical data. Thus, the maximum amount of antibody precipitated in 0.14 M NaCl is much less than that obtained with poly U. This decrease indicates that some of the antigenic sites are no longer accessible to the antibodies; we use the term "masking of antigenic sites" to designate this phenomenon (Fig. 3A). The amount of

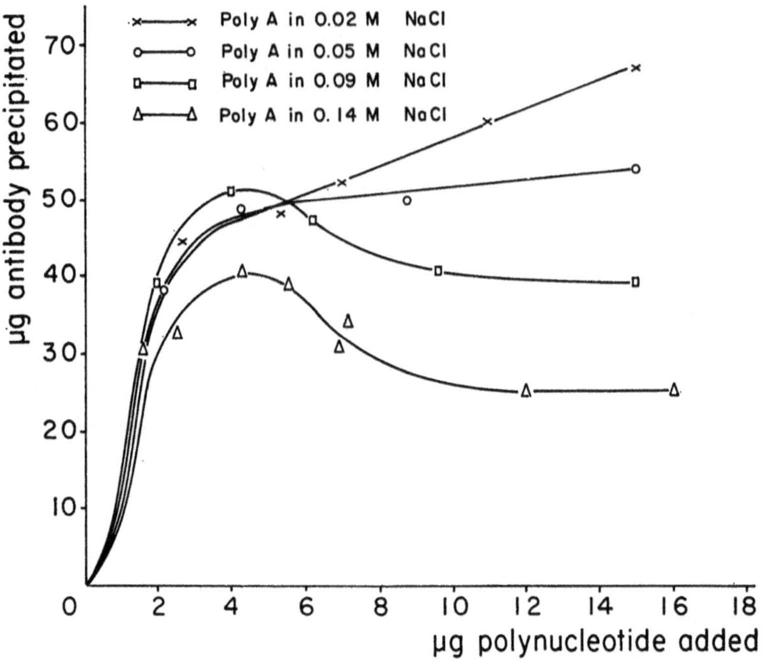

Fig. 3A. Amounts of antibody precipitated. Precipitin curves of poly A in increasing NaCl concentration. Experimental conditions as in Fig. 2. Optical density measurements for the estimation of nucleotides in the precipitate were made at 257 mμ after hydrolysis by 1 *N* perchloric acid.

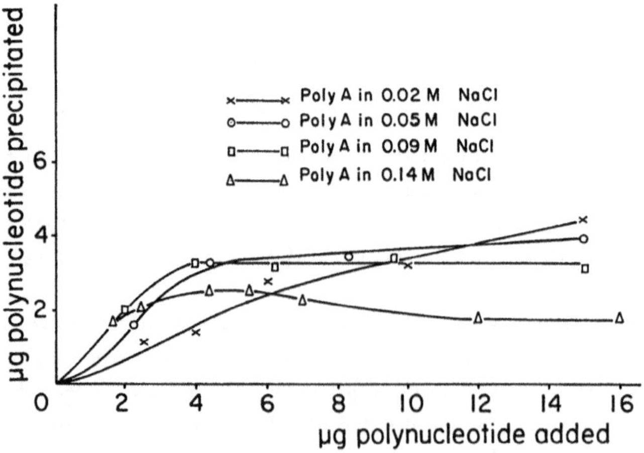

Fig. 3B. Amounts of polynucleotide precipitated.

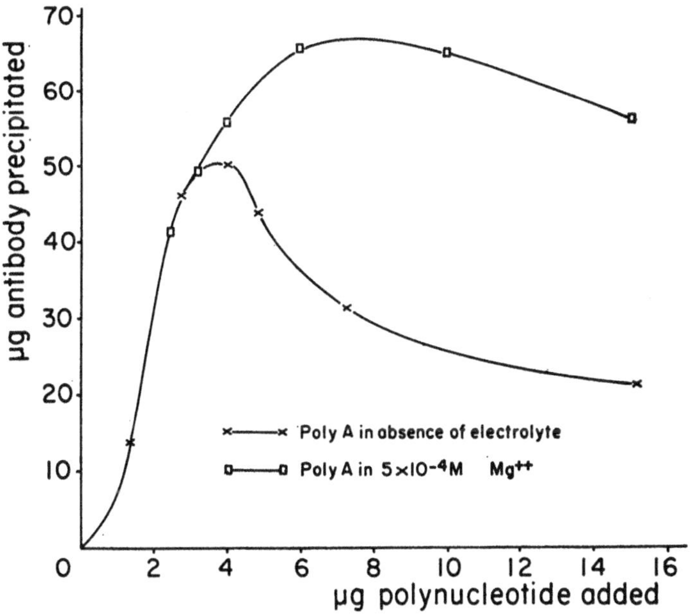

Fig. 4A. Amounts of antibody precipitated. Precipitin curves of poly A in absence of electrolytes or in 5×10^{-4} *M* Mg++. Experimental conditions as in Fig. 3.

precipitated antigen is also less compared with the precipitation of poly U (Fig. 3B). However, when the NaCl concentration is decreased, the immunological behavior of the poly A resembled more and more that of poly U; the amount of antibody and antigen precipitated increases regularly and at a 0.02 M concentration the precipitin curve is identical to that obtained for poly U, just as in the absence of electrolytes or in the presence of 5×10^{-4} M Mg++ (Fig. 4A and B).

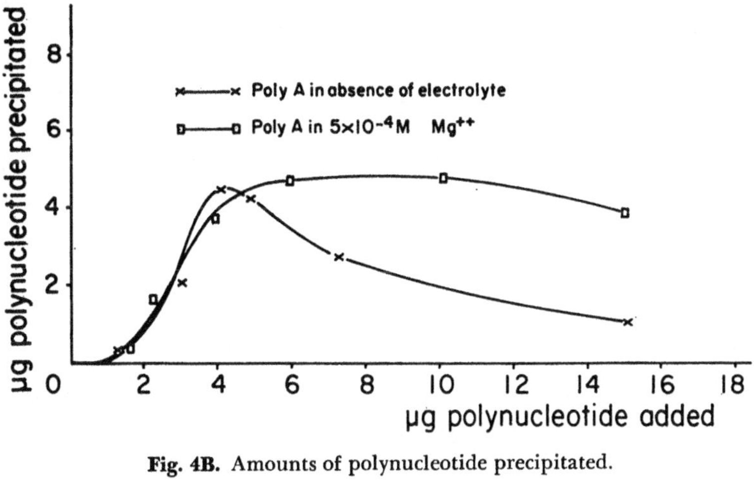

Fig. 4B. Amounts of polynucleotide precipitated.

The physico-chemical properties of poly A indicate that the amount of helical single-stranded structure at neutral pH decreases when the ionic strength of the solvent decreases. Hence, the increase in the quantity of antibody precipitated when the molarity of NaCl is decreased from 0.14 to 0.002 must be attributed to a reduction in the helical content of the molecule, which favors the unmasking of new antigenic sites. Since, under the same conditions, the precipitin curve of an unorganized polynucleotide remains unchanged, the variations observed with poly A must be considered as highly significant. We have obtained the same kind of results using poly C in solution at pH 7.4.

C. Complexes of Different Homopolymers (Panijel *et al.,* 1966a; Souleil *et al.,* 1966)

1. *Methodological Aspects of the Analysis*

It is known that complexes result from the pairing of strands of different homopolymers in the presence of electrolytes, notably, low concentrations of divalent cations. The availability of such complexes makes it possible to study the immunochemical behavior of highly ordered systems consisting of 2 or 3 strands of polynucleotide depending on the molar ratio of the constituent homopolymers. Such a study involves a comparison between these complexes and mixtures of homopolymers that do not complex. This comparison presents two principal problems:

(1) The simultaneous presence of different homopolymers, though they might not form complexes, can result in a reciprocal inhibition of precipitation. We encountered this phenomenon in reactions performed with anti-RNA antibodies other than the NG I antibodies (Panijel, 1963c). One must therefore ascertain by a study of precipitation in the absence of electrolytes that, in the case of NG I antibodies, the reaction with a mixture of different homopolymers is quantitatively equal to the average of the reactions with each homopolymer separately.

(2) The presence of salt, especially Mg^{++}, influences the extent of precipitation of each of the homopolymers used. Since precipitation reactions cannot be carried out in the absence of salt, it is necessary to establish theoretical curves of precipitation on the basis of the average of the values obtained for the precipitation in 5×10^{-4} *M* Mg^{++} of each of the homopolymers in the mixture in amounts equal to those in the complex itself. Identity of theoretical and experimental curves indicates the absence of formed complexes.

2. *The Poly (A + U) Complex in* 5×10^{-4} *M* Mg^{++}

In the absence of electrolytes, the experimentally obtained precipitin curves of a 1:1 mixture of poly A and poly U coincide exactly with theoretical curves calculated from the values obtained separately for each polynucleotide, not only for the precipitated antibodies (Fig. 5A) but also for the polynucleotide precipitated (Fig. 5B). This result answers the question previously raised regarding reciprocal inhibition; clearly, there is no reciprocal inhibition of precipitation between two different polynucleotides mixed in a given medium.

We have previously seen that in the presence of Mg++ the amount of

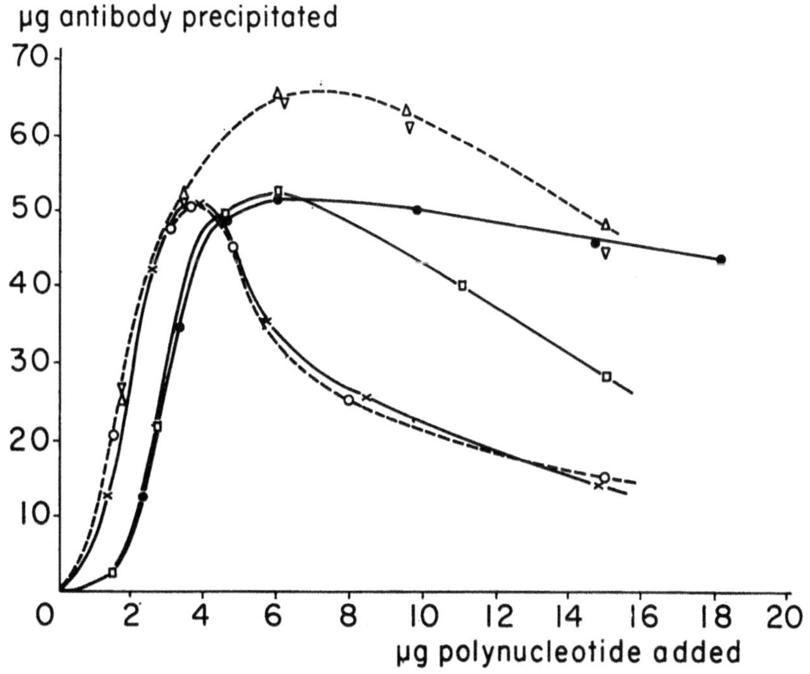

Fig. 5A. Amounts of antibody precipitated. ×—× = observed values for the poly A + poly U mixture in absence of electrolytes. ○--○ = calculated values of a theoretical mixture of poly A + poly U in absence of electrolytes. □—□ = poly (A + U) complex in 5 × 10−4 *M* Mg++. △--△ = calculated values of a theoretical mixture of poly A + poly U in 5 × 10−4 *M* Mg++. ●—● = poly (A + 2U) complex in 5 × 10−4 *M* Mg++. ▽--▽ = calculated values of a theoretical mixture of poly A + 2 poly U in 5 × 10−4 *M* Mg++. Precipitin curves obtained in the case of mixture of poly A and poly U in absence of electrolytes and in 5 × 10−4 *M* Mg++. The theoretical curves were constructed from the average between the individual precipitation of each constituent polynucleotide in the corresponding medium. Poly A and poly U were mixed at pH 7.4 in 5 × 10−4 *M* Mg++ in the following concentration: poly A to poly U equal to 1:1 for the formation of the poly (A + U) complex, and poly A to poly U to 1:2 for the formation of the poly (A + 2U) complex. These concentration ratios were those used for the calculation of the theoretical curves. The complexes in Mg++ were used after a formation time of 90 minutes in the case of the di-stranded complex (25% final hypochromia at 260 mµ) and 120 minutes for the triple-stranded complex (34% final hypochromia at 260 mµ). The theoretical curves in Mg++ represent the expected precipitin curves of non-associating mixtures of poly A and poly U and, immunochemically, indicate the level of precipitation of a non-ordered system.

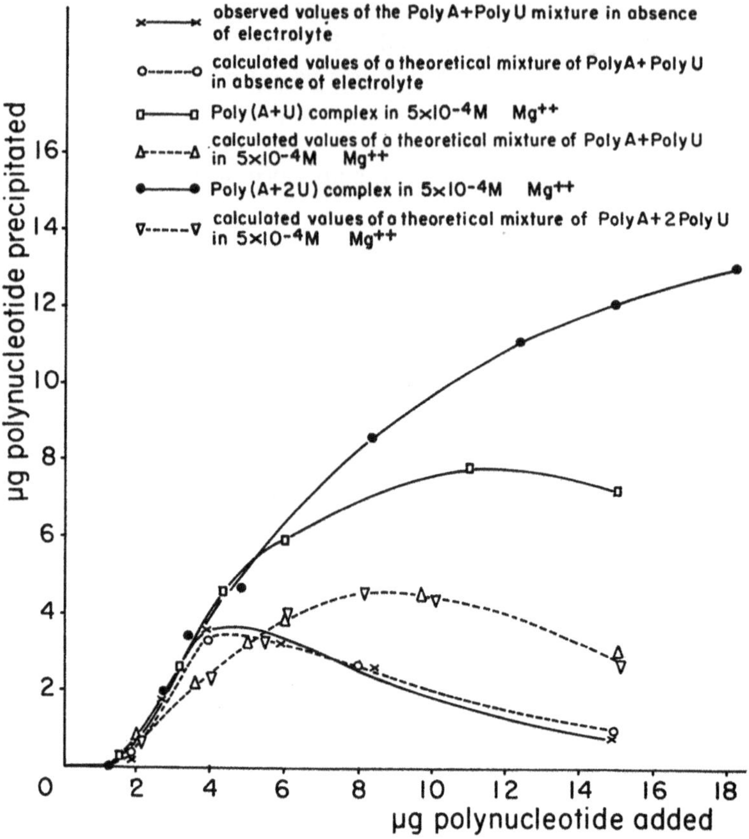

Fig. 5B. Amounts of polynucleotidic material precipitated.

antibody precipitated by poly A was the same as that precipitated by poly U. The theoretical curves therefore indicate the level of precipitation, in the case of the non-associating mixture, of a polyribonucleotide system devoid of order. The experimental curves of the highly ordered complex were clearly different. Indeed, it was found that: (a) The amount of antibody precipitated is always less (about 20% at the point of equivalence) than the amount expected from the theoretical curve. The formation of a helical di-stranded structure results in a masking of antigenic sites, compared with polyribonucleotide systems devoid of order (Fig. 5A). (b) The maximal amount of precipitated antigen is about twice that expected on the basis of the theoretical curve for a single-stranded system devoid of order (Fig. 5B).

3. *The Three-stranded Complex of Poly (A + 2U)*

In the absence of electrolytes, the theoretical and experimental curves coincide exactly (Fig. 5A and B). In this case, as in that of a 1:1 mixture, the polynucleotides used produce no reciprocal inhibition of precipitation. In the presence of 5×10^{-4} M Mg^{++}, the complex poly (A + 2U) shows the following immunochemical characteristics:

(1) The amount of antibody precipitated (Fig. 5A) is less than the theoretical value, but is equal to that precipitated by the di-stranded complex up to the zone of equivalence; in excess of antigen, the inhibition is less marked in the case of poly (A + 2U). Thus, the addition of a supplementary strand to the poly (A + U) complex does not increase the masking of antigenic sites. This is consistent with the hypothesis that the new strand of poly U is inserted into the wide helical groove of the poly (A + U) complex, without structural reorganization of the di-stranded molecule (Rich and Davies, 1956; Felsenfeld *et al.*, 1957).

(2) The maximal amount of antigen precipitated is three times the theoretical value for a single-stranded polynucleotide (Fig. 5B).

D. Multi-stranded Homopolymers (Panijel *et al.*, 1965; Souleil *et al.*, 1966; and Panijel, 1967)

It is an established fact that the conformation of poly I depends on salt concentration. In 1 *M* NaCl and at low temperature, several methods, e.g., circular dichroism (Brahms and Sadron, 1966), optical rotatory dispersion (Sarkar and Yang, 1965a) and infrared analysis (Miles, 1964a and b), indicate the presence of a highly ordered structure which, according to X-ray diffraction data (Rich, 1958), is consistent with a tri- or tetra-stranded model. At lower salt concentrations (0.1 *M* NaCl) poly I is thought to have a helical single-stranded conformation (Sarkar and Yang, 1965a; Brahms and Sadron, 1966). Although poly G has been studied less, it is believed to have a structure comparable to that of poly I. From the immunochemical point of view, it can be expected that the conformation and the appearance in solution of multi-stranded structures of one polynucleotide will exhibit the same characteristics of reactivity shown previously by the multi-stranded complexes of several polynucleotides.

Thus, if poly U is chosen as the immunological standard for a single-stranded molecule, it should be possible to describe the precipitation of poly I or poly G by determining, for each of several salt concentrations, the maximal amounts of poly I or poly G precipitable relative for the amount of poly U that is precipitable. The results shown in Figure 6 were obtained in media of increasing concentrations of $MgCl_2$ or NaCl. For each medium, the amounts of precipitated poly U, poly I and poly G were determined at the zone of equivalence and the ratios of these maximal amounts plotted as a log function of the salt concentration. With no electrolytes added, or up to a concentration of 10^{-5} *M* Mg^{++} and 10^{-3} *M* NaCl, the immunological behavior of poly I and poly G is indistinguishable from that of the single-stranded poly U. At higher salt concen-

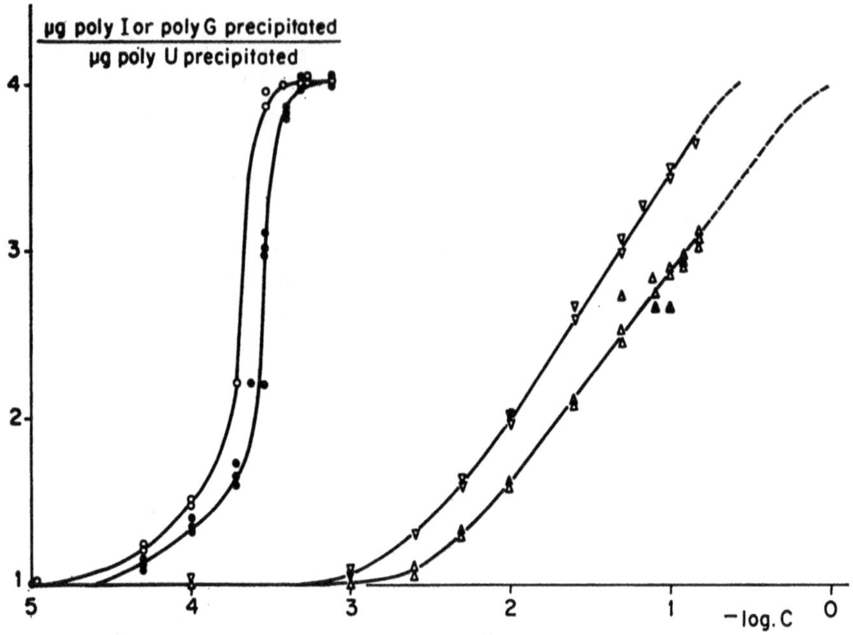

Fig. 6. Ratios of the maximal amount of poly I, or poly G, precipitated to the maximal amount of poly U precipitated under the same conditions. The experiments were carried out at 20°C in 0.005 *M* Tris, pH 7.1, in increasing concentrations of either $MgCl_2$ or NaCl plus 5×10^{-4} *M* EDTA. The experiments were repeated in water at pH 6.5 with the same results. Immunological reactions and estimations of the precipitates were carried out as in the previous figures. UV readings were performed at 248 mμ for poly I and 254 mμ for poly G. ●—● = poly I in $MgCl_2$. ○—○ = poly G in $MgCl_2$. △—△ = poly I in NaCl. ▽—▽ = poly G in NaCl.

trations, however, the maximal amount of poly I and poly G precipitated exceeds that of poly U so that the ratio follows a sigmoid curve and reaches a value of 4 for both poly G and poly I at 5×10^{-4} *M* Mg^{++}. It remains constant, up to a concentration of 8×10^{-4} *M* Mg^{++}, an indication that the structure precipitated is stable under these conditions. The effect of concentration of NaCl was studied only up to 0.15 *M* because the immunological reaction is inhibited at high ionic strength. Nevertheless, extrapolation of the curves clearly indicates that the concentration of NaCl needed to obtain a value of 4 differs for poly G and poly I, 1 *M* being required for poly I and 0.25 *M* for poly G.

These results establish that poly I and poly G are able to assume a four-stranded structure in solution. Other immunochemical data also indicate that:

(1) The formation of multi-stranded structures is accompanied by the masking of antigenic sites (up to 20%), which we have shown to be a general character of ordered molecules.

(2) The methylation of poly I, which is known to affect markedly its conformation (Van Holde *et al.*, 1965; Michelson and Pochon, 1966), results in a spectacular modification of the precipitin curves that clearly reflects the inability of methylated poly I to form defined ordered structures.

(3) The multi-stranded structure of poly I is unstable at relatively low temperatures (Sarkar and Yang, 1965a); indeed, at 45°C in both 0.1 *M* NaCl and 5×10^{-4} *M* Mg^{++}, the maximal amount of poly I precipitable is equal to that of poly U, reflecting the dissociation on heating of multi-stranded molecules into single-stranded molecules of poly I.

E. Significance of the Ab/Ag Curves

The results discussed above provide evidence for the importance of the conformation in determining the immunological reactivity of polynucleotides with the NG I antibodies. In the absence of electrolytes, the curves obtained for the precipitation of both the antibodies and the antigens are identical up to the zone of equivalence. At higher ionic strengths, the curves of poly A, poly I and poly G differ from those of poly U; the amount of polynucleotide precipitated exceeded that of poly U, whereas the amount of antibodies precipitated decreased compared with the amount precipitated by poly U. This was also the case for the complexes poly (A + U) or poly (A + 2U). However, these differences disappear when a loss of secondary structure occurs, either through methylation (as for poly I) or by "helix → random coil" transition brought about by heating (as for poly A, poly I or poly G). Thus, it appears that the Ab/Ag ratio is lower for an ordered system compared with a random-coiled polymer. This feature is particularly true for multi-stranded polynucleotides, but it is also discernible in the case of single-stranded molecules like poly A or poly C which possess a certain level of organization as a result of base-stacking.

At the zone of equivalence, the value of Ab/Ag[$(Ab/Ag)_{eq}$] represents an estimation of the maximum number of antigenic sites accessible to the antibodies in a given medium. In our experiments, the calculation of the (Ab/Ag) values do not take into account the molecular weight of the antigen tested, instead they are based on the amount of nucleotide material obtained after hydrolysis. Obviously, the reference to the random-coiled poly U may be invalid in the case of synthetic co-polymers or RNAs. However, the $(AB/Ag)_{eq}$ value in a given medium conserves the same significance and represents a value proportional to the maximum number of accessible sites. If the molecular weights of both the antibodies and the antigens are known, it will be possible to interpret the variations in the number of accessible sites.

For a given polynucleotide or RNA variations may result from changes in the conditions of the reaction (salt concentration of medium, temperature, etc.) but they may also be brought about by modifying, in a given reaction medium, the configuration of the ribonucleotide component and consequently the accessibility of the specific sites to antibody. This approach can be applied to viral or ribosomal particles whose immunochemical behavior can then be compared to that of the RNA they contain. The following section will discuss such experiments.

III. IMMUNOCHEMICAL APPROACH TO THE ANALYSIS OF THE STRUCTURE OF VIRAL AND RIBOSOMAL PARTICLES

A. Methodological Considerations

As already shown, the "basal" precipitation of very distant heterologous ribosomes by an antiserum prepared against ribosomes correlates with the anti-RNA antibodies present in this serum. However, when extracts of animal tissue were reacted with an antiserum against bacterial ribosomes, reactivity was obtained with ribosomes but not with microsomes. This difference in reactivity must be due to the presence of fragments of the ergastoplasmic lamellae which cover the ribosomes, causing the antigens to be inaccessible to the antibodies. Thus, it appears that a positive or negative reactivity of a biological RNA-containing structure provides information on the topology of this structure, a positive reaction indicating an external superficial arrangement of some part of the RNA molecule.

B. Study of Viral Particles

When a lysate of an RNA phage is treated with NG I antibodies, a precipitate which contains RNA is produced. However, the number of plaque-forming units in the supernatant remains unchanged as compared to the control (a lysate which received, instead of antibodies, the same volume of salt solution). Thus, treatment with NG I antibodies results in some "purification" of the infectious particles by precipitating other RNA compounds present in the lysate and decreasing the characteristic ratio of the amount of RNA/number of plaque-forming units (Cayeux *et al.*, 1965). Table 1 shows the successive steps in the "purification" of a f2 phage lysate with RNA-specific antibodies. The results clearly show that the viral particles cannot be precipitated or inactivated by the antibodies since their RNA, completely enclosed in a protein shell, is not accessible to the anti-RNA antibodies.

A different result was obtained with certain defective particles produced by amber mutants, for example by mutants of fr, as described by Heisenberg (1966). These virus-like particles are not infectious but have the same buoyant density and the same RNA-content as fr phages. However, a con-

Table 1

IMMUNOCHEMICAL PURIFICATION OF THE PHAGE f2

Steps in the Purification of Infectious Units	$(NA/P.F.U.) \times 10^{11}$
1. Suspension containing: 6×10^{11} P.F.U /ml; 210 μg prot./ml; 149 μg NA/ml	24.8
Addition of DNAase 45 min, 30°C	
2. Suspension containing: 6×10^{11} P.F.U./ml; 235 μg prot./ml; 96 μg NA/ml	16
Addition of 900 μg NG I (0.2 ml)	
3. a) Precipitate containing: 210 μg protein; 41 μg nucleic acid	
+	
b) Supernatant containing: 4.7×10^{11} P.F.U /ml; 33 μg NA/ml	7.08
Addition to the supernatant 3 of 675 μg NG I (0.15 ml)	
4. a) Precipitate containing: 163 μg protein; 19 μg nucleic acid	
+	
b) Supernatant containing: 3.9×10^{11} P.F.U./ml; 3.2 μg NA/ml	0.82

stant fraction of their RNA, about 70%, is hydrolyzed by RNAase. Also, they show no difference in their protein coat. As shown in Table 2, 96% of the defective particles was precipitated, whereas only a small per cent of fr particles was precipitated. After RNAase treatment, precipitation of the defective particles does not take place. From this result, which shows that the part of the RNA that is sensitive to RNAase is also accessible to the antibodies, the conclusion can be drawn that the defect in these particles is due to a lack of encapsulation of the RNA by a protein shell.

C. STUDY OF RIBOSOMAL PARTICLES (Panijel and Delaunay, 1967)

1. *Immunological Behavior of 70, 50 and 30 S Particles of E. Coli*

Ribosomal particles are particularly suitable for an immunochemical analysis by the NG I antibodies since these antibodies, as previously de-

Table 2

PRECIPITATION OF DEFECTIVE BACTERIOPHAGE PARTICLES PRODUCED BY fr AMBER MUTANTS WITH THE NG I ANTI-RNA ANTIBODIES

(Data from Heisenberg, 1967)

	Per Cent Precipitation of the Labelled Material
1. Defective particles purified by sucrose gradient centrifugation	96
2. fr phages purified by sucrose gradient centrifugation	8
3. Defective particles, no antibodies	4
4. fr phages, no antibodies	2
5. Defective particles after treatment with RNAase at 37°C for 10 min	3

The reaction mixture, containing labelled fr phages or labelled defective particles, was allowed to precipitate for 30 min at room temperature. The precipitate was collected on Whatman glass filters, washed and the radioactivity determined.

scribed, interact with the 70 S and 80 S ribosomes of bacterial and animal origin.

The experiments reported here were carried out with particles of *E. coli* RNAase I^-_{10}. Crude ribosomes, prepared by conventional methods (containing 63% RNA and 37% protein), and washed ribosomes treated with 0.5 M NH_4Cl (Shin and Moldave, 1966) (containing 67% RNA and 33% protein), were used. The precipitin reactions were performed in potassium buffer (0.1 M KCl, 5 mM Tris-HCl, pH 7.4) containing 1 *mM* Mg^{++} (K_2 buffer) or 4 *mM* Mg^{++} (K_3 buffer). These buffers assured stability of the particles and reactions between antigens and antibodies. Crude or washed particles, suspended either in K_2 or in K_3 buffers, were precipitated at 95–100% in excess antibody (Fig. 7). This result not only confirms that ribosomes contain a certain fraction of their RNA, referred to as "reactive RNA," in an accessible form but also eliminates the possibility that differences in the immunochemical reactivity of different particles might be due to the presence of antibodies with distinct specificities directed against one of the particles and not the others. The curves for the crude particles (Fig. 8) and for the washed particles (Figs. 9 and 10) show clearly that differences in reactivity do in fact exist.

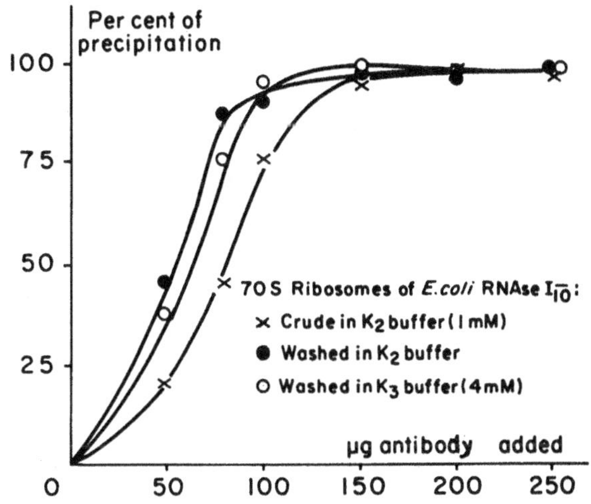

Fig. 7. Percentage of precipitation of crude or washed 70S ribosomes from *E. coli* RNAase I$-_{10}$ by NG I antibodies. In this experiment, the amount of ribosomes either in K_2 (1 mM Mg++) or in K_3 (4 mM Mg++) buffers was kept constant (120 µg of ribosomes containing about 80 µg of RNA). The antibodies were added in increasing amounts. The specific precipitates were collected after 30 min at 4°C, washed three times in ten volumes of 0.1 *M* NaCl, 0.0033 *M* Mg++ acetate, 0.0017 *M* Tris buffer, pH 7.4 and then dissolved in 0.3 ml of 0.1 *N* NaOH containing 1% Na_2CO_3. One portion of the specific precipitate was used for estimating the proteins (Lowry *et al.*, 1951); the other served for the estimation of RNA after hydrolysis by 1 *N* perchloric acid. From the latter, the amount of precipitated ribosomes was deduced.

Thus, it can be seen that: (1) the results are identical in 1 *mM* Mg++ (K_2 buffer) or in 4 *mM* Mg++ (K_3 buffer); (2) the amount of 70 S or 50 S particles precipitated by a given quantity of antibodies is greater than that of 30 S particles. Consequently, the $(Ab/Ag)_{eq}$ values are 2 to 2.5 times greater for the latter, as indicated in Table 3. Pertinent to this point also is the fact that the Ab/Ag values for the washed particles are greater by 15–20% than those of the corresponding crude particles. Since the treatment with 0.5 *M* NH_4Cl splits non-structural proteins, this result shows that these detachable proteins, associated with the "reactive" RNA, hinder the access of the antibodies to a certain number of specific antigenic sites; (3) the $(Ab/Ag)_{eq}$ values for 70 S ribosomes are less than the theoretical values calculated from the values obtained with the sub-units. However, this calculated value is in fact the one obtained when a mixture of equal parts of 30 and 50 S particles is used for the reaction under conditions preventing any reassociation (Fig. 11).

2. *Structural Interpretation*

A structural interpretation of these results is rendered possible by a comparison of the immunochemical behavior of corresponding RNAs.

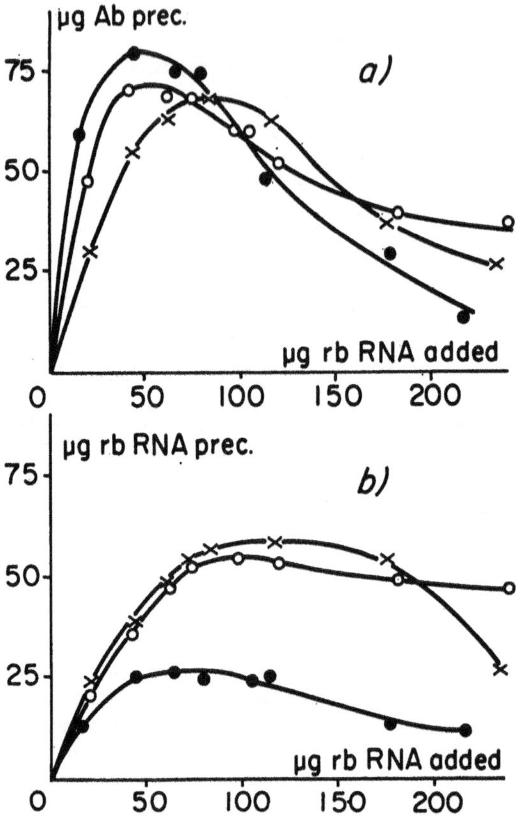

Fig. 8. Amounts of antibody (a) and antigen (b) precipitated by crude ribosomal particles in K_2 medium. The 30 and 50S sub-units were prepared from the 70S ribosomes of *E. coli* RNAase I^{-}_{10} by dissociation and separation through a sucrose gradient according to Pestka and Nirenberg (1966). The different particles estimated in terms of their RNA content (rb RNA) were added in increasing amounts, the amount of antibody being kept constant (110 µg). The specific precipitates were collected and treated as in the previous figure. From the RNA content of the precipitates, the amounts of ribosomal proteins were calculated and the amounts of antibody precipitated could be obtained by subtracting the ribosomal proteins from protein content of the precipitates. ●—● = 30S sub-units. ○—○ = 50S sub-units. ×—× = 70S ribosomes.

These RNAs were prepared by the phenol-SDS method, from either the bacterial homogenate or from the particles themselves, taking all the precautions recommended by Midgley (1965a, b, c) for obtaining the correct separation between the 16 and 23 S RNAs in sucrose gradients. Whatever the source used, the immunochemical behavior of corresponding ribosomal RNAs was similar with the values of $(Ab/Ag)_{eq}$ close to one another: 3 for the 23 S fraction, 3.5 for the 16 S fraction (Table 3). Thus, the difference between the 50 S particles and the 23 S RNA isolated from it was very striking, whereas the values of $(Ab/Ag)_{eq}$ were identical in the case of the 30 S particles and the 16 S RNA isolated from it. In other words, all the 16 S RNA remains accessible to the antibodies when it forms part of the 30 S sub-units; in contrast, the inclusion of the 23 S RNA in the 50 S structures results in a loss of 55–60% of the antigenic sites.

We have tried to use these results for a calculation of the number of antigenic sites per molecule. From the average $S_{w,20}$ value of 9.5 for NG I antibodies, a molecular weight of $2.5 - 3 \times 10^5$ would be indicated by comparison with other immunoglobulins. Taking the molecular weight of ribosomal RNAs of *E. coli* to be that established by Kurland (1960)

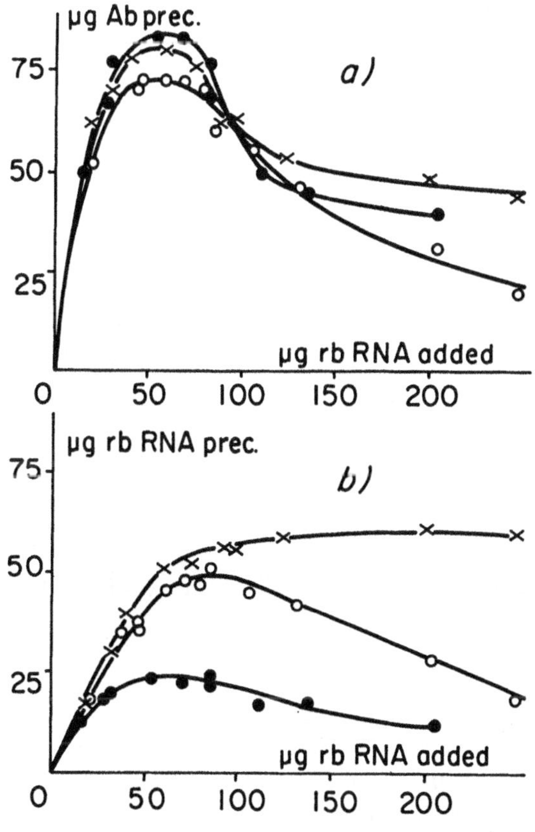

Fig. 9. Amounts of antibody (a) and antigen (b) precipitated by washed particles in the K_2 medium. The washed 70S ribosomes were obtained from the crude 70S ribosomes by treating three times with 0.5 *M* NH_4Cl as described by Shin and Moldave (1966). The washed sub-units were obtained from the washed ribosomes as indicated in the previous figure. Methods and symbols as in Fig. 8.

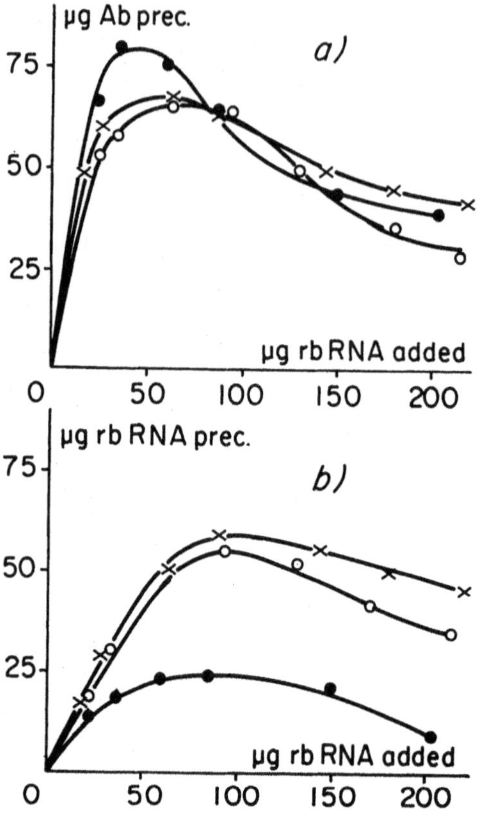

Fig. 10. Amounts of antibody (a) and antigen (b) precipitated in the K_3 medium. Methods and symbols as in Fig. 8.

and taking the precipitating antibodies to be divalent, the number, N, of specific sites per RNA molecule would be:

$$2 \times (\text{mol. weight RNA/mol. weight NG I}) \times (\text{Ab/Ag})_{eq}$$

This gives 25–31 accessible sites for the 23 S RNA and only 10–12 for the 50 S washed sub-units, whereas the 14–19 accessible sites of the 16 S RNA all remain equally accessible in the washed 30 S sub-units.

A similar calculation shows that, in the case of the crude particles, the non-structural proteins are implicated in masking 2–3 antigenic sites on each fraction of "reactive" RNA. In addition, the fact that the association of a 50 S sub-unit and a 30 S sub-unit to a 70 S ribosome gives rise to a value for $(\text{Ab/Ag})_{eq}$ lower than that of a non-associating mixture means that the association is accompanied by the loss of a certain number of antigenic sites. It is reasonable to attribute the latter to the pairing between the "reactive" fractions of the respective RNAs of each sub-unit.

Table 3

RATIOS OF PRECIPITATED ANTIBODY TO THE PRECIPITATED ANTIGEN (ESTIMATED IN TERMS OF RNA PRECIPITATED) AT THE EQUIVALENCE ZONE ($(Ab/Ag)_{eq}$) FOR THE DIFFERENT RIBOSOMAL PARTICLES AND RIBOSOMAL RNAS, IN K_2 (1 mM Mg^{++}) OR K_3 (4 mM Mg^{++}) BUFFERS

Antigen	30 S	50 S	70 S	30 S + 50 S (non-associated)	16 S	23 S
Crude particles	3.07	1.2	1.17	1.9		
Washed particles						
in K_2	3.5	1.4	1.4	2		
in K_3	3.3	1.2	1.2	1.9		
RNAs						
in K_2					3.5	3
in K_3					3.4	3

Thus, we find here again the phenomenon of "masking" already described for the complexes of polynucleotides.

The implications of these results, in so far as they concern ribosomal structure, are of some interest; indeed, it is known that when the dissociation of 70 S ribosomes into sub-units takes place, about 60% of the RNA no longer binds Mg^{++} (Goldberg, 1966; Choi and Carr, 1967). This percentage is precisely that which would result from the pairing of the two ribosomal RNAs, since the pairing necessarily takes place between nucleotidic sequences of equal length. It should therefore correspond, at a maximum, to the length of the "reactive" fraction of the 23 S RNA and should form a double-stranded segment corresponding to a molecular weight of about 1×10^6, i.e., 60% of the sum of the molecular weights of the two ribosomal RNAs. Under these conditions, the single-stranded sequences which remain accessible in the ribosomes should correspond to a small molecular weight, no more than $5\text{–}6 \times 10^4$, because nearly all of the "reactive" RNA of the 70 S ribosomes is in di-stranded form. This situation could explain, for example, the apparent paradox that messenger RNAs associate specifically with 30 S sub-units but interact both with the 16 S or 23 S RNAs (Watson, 1964); this may be so because on the 70 S ribosome the single-stranded "reactive" RNA is exclusively a part of the 16 S RNA.

Other studies of this same type could be performed using certain particles like those obtained from chloramphenicol-treated bacteria (CM particles) or the so-called "core-particles" resulting from treatment of the sub-units with CsCl (cf. Nomura and Traub, 1966). Such studies could give some information on the relationship between the different kinds of ribosomal proteins and the corresponding ribosomal RNAs.

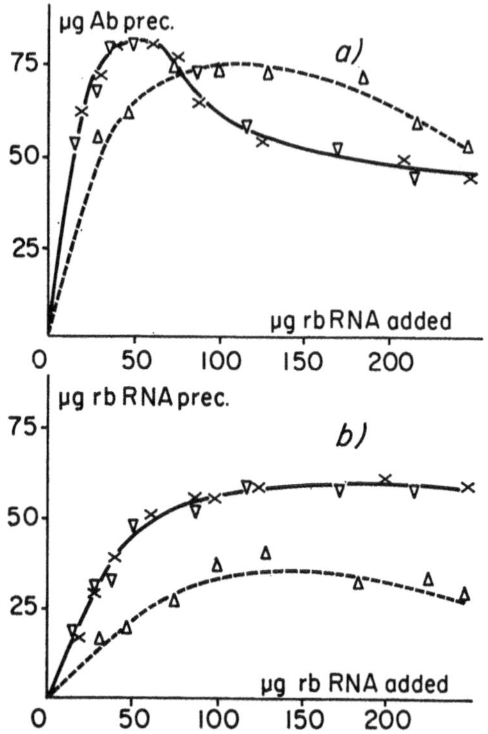

Fig. 11. Comparison between the amounts of antibody (a) and antigen (b) precipitated in K_2 medium by 70S ribosomes (native or reassociated) and a mixture in a 1:1 ratio of 30 and 50S non-associated sub-units. The theoretical curves were calculated from the individual precipitation of each sub-unit in the K_2 medium. The washed sub-units showed no sign of reassociation in such a buffer. ×—× = native washed 70S ribosomes. ▽—▽ = reassociated ribosomes. △—△ = 30S sub-units + 50S sub-units mixed in a 1/1 ratio. -- = theoretical curve for the mixture 30S sub-units + 50S sub-units calculated from the individual precipitation of each sub-unit under the same conditions.

3. *Biochemical Approach*

On the basis of our results, one can ask whether the specific association of NG I antibodies with the accessible "reactive" RNA of ribosomes would impair the latter in playing their role in protein biosynthesis. We are carrying out experiments on this subject and here report some of our preliminary observations. The experiments are performed using two different types of Ab–Ag complexes: the precipitate itself, obtained at the zone of equivalence, carefully washed and dispersed, and a soluble complex resulting from reactions at large antigen excess. The two complexes are added, instead of ribosomes themselves, to the usual medium used for protein biosynthesis, and the incorporation of C^{14}-phenylalanine directed by poly U is measured at regular intervals. However, in these preliminary experiments, the ionic conditions were not the best possible. Indeed, for preventing some aggregation of antibodies, we used 0.025 M KCl. Nevertheless, even under these conditions it appears that for the same amount of ribosomal RNA reacted (in either complex or ribosome form): (1) the precipitate ribosome-NG I is able to perform a certain level of biosynthesis (about 25–30% of the control), and (2) the soluble complex seems more active than the ribosomes themselves (from about 30% upwards). These observations mean that the determining sites of the ribosomal RNA are distinct from the sites of translation. In addition, it could be assumed that the binding of antibodies under the conditions of a soluble complex allows some limited aggregation of ribosomes more appropriate to protein biosynthesis than the ribosomes in form of monomers. Obviously, further experiments are necessary to support this hypothesis.

IV. CONCLUSIONS

In the experiments reported in this paper, the NG I antibodies have been used as a tool for the exploration of biochemical or biological polyribonucleotidic structures. Such studies have proved to be of analytical value because the NG I antibodies represent a homogenous population of antibodies of defined specificity against the polyphosphate backbone of synthetic polynucleotides and RNAs. Of course, many problems other than those developed here will require investigation: for example, the problem of the nature of the NG I themselves with respect to the different types of immunoglobulins, and the question of the structure of the NG I, whose fragments, produced by enzymatic or chemical treatments, could present particular immunological properties. Of interest also would be kinetic studies of the reaction between the various antigens and the NG I; a point, directly related to the latter, concerns the problem of finding a model of association between RNA and protein, which would play in some way, a pilot role analogous to that of the model histone-DNA. However, in the present framework we would like to examine, in conclusion, whether the immunochemical method described here presents some special utility when compared with established physicochemical methods.

The utility of immunochemical methods to reveal structural features of proteins and polysaccharides has been fully proved and constitutes a great success for immunology. In the case of nucleic acids, the principal difficulty emphasized by Plescia and Braun (1967) lies with the heterogeneity of the population of antibodies with respect to their ability to react with defined structures. It seems that, as concerns the NG I antibodies, this difficulty is overcome, since it has been demonstrated that the bases do not play any role in the reaction, either qualitatively or quantitatively. In fact, this property allows first of all a quantitative comparison of the precipitin curves of the different synthetic polyribonucleotides performed in the same medium and from this comparison a direct correlation between the weight of polyribonucleotide precipitable by a given weight of antibody and the number of strands per molecule of tested antigen can be derived. This property also gives significance to the ratio Ab/Ag at the zone of equivalence and allows a comparison between the results obtained with a given RNA reacted in the same medium but in different states.

Thus, it is our feeling that the NG I antibodies should be considered of special interest for at least three uses: (1) to determine the number of strands of multi-stranded polyribonucleotides in solution; (2) to study the relationships between the RNA constituent and the different kinds of coat or shell associated with it, whether they be proteins (as for the phages) or lipoproteins (as in the case of the ergastoplasmic lamellae or the membrane-associated structures); (3) to intervene in biochemical systems at the level of both isolated RNA and RNA-containing structures by controlling the association between the polyribosephosphate backbone and a specifically bound protein.

Work is now in progress to explore these different aspects in more detail.

References

Barbu, E., and J. Panijel (1960). Présence d'anticorps anti-acide ribonucléique dans les immunsérums anti-ribosomes. C.R. Acad. Sc., *250:*1382–1384.

Barbu, E., and J. Panijel (1961). Concentration et propriétés des anticorps anti-acide ribonucléique. C.R. Acad. Sc., *252:*3157–3159.

Barbu, E., J. Panijel, and G. Quash (1961). Charactérisation immunochimique des ribosomes. Ann. Inst. Pasteur, *100:*725–746.

Brahms, J. (1963). Optical activity and the conformation of polynucleotides. Jl. Am. Chem. Soc., *85:*3298–3300.

Brahms, J., and C. Sadron (1966). Ionic and hydrogen bonds contributing to the conformational stability of polynucleotides. Nature, *212:*1309–1312.

Cayeux, P., J. Panijel, and C. Souleil (1965). Action des anticorps anti-RNA sur les ribosomes et les phages. Bioch. Bioph. Acta, *108:*605–618.

Choi, Y. S., and C. W. Carr (1967). Ion-binding studies of ribonucleic acid and *E. coli* ribosomes. Jl. Mol. Biol., *25:*331–345.

Felsenfeld, G., D. Davies, and A. Rich (1957). Formation of a three-stranded polynucleotide molecule. Jl. Am. Chem. Soc., *79:*2023–2024.

Fresco, J. (1959). Polynucleotides II. The X-ray diffraction patterns of solutions of the random-coiled and helical forms of polyriboadenylic acid. Jl. Mol. Biol., *1:*106–110.

Goldberg, A. (1966). Magnesium binding by *E. coli* ribosomes. Jl. Mol. Biol., *15:*663–673.

Heisenberg, M. (1966). Formation of defective bacteriophage particles by fr *Amber* mutants. Jl. Mol. Biol., *17:*136–144.

——— (1967). Formation of defective bacteriophage particles by fr *Amber* mutants (II). Bioch. Bioph. Res. Comm., *27:*131–137.

Kurland, C. G. (1960). Molecular characterization of ribonucleic acid from *E. coli* ribosomes. Jl. Mol. Biol., *2*:83–91.

Lipsett, M. N. (1960). Evidence for helical structure in polyuridylic acid. Proc. Nat. Acad. Sc., *46*:445–446.

Lowry, O., N. J. Rosenbrough, A. L. Farr, and R. J. Randall (1951). Protein measurement with the Folin-phenol reagent. Jl. Biol. Chem., *193*:265–275.

McCarthy, B. J., and E. T. Bolton (1963). An approach to the measurement of genetic relatedness among organisms. Proc. Nat. Acad. Sc., *50*:156–164.

Michelson, A. M., and C. Monny (1966). Polynucleotides VIII. Base stacking in polyuridylic acid. Proc. Nat. Acad. Sc., *56*:1528–1534.

Michelson, A. M., and F. Pochon (1966). Polynucleotides analogs VII. Methylation of polynucleotides. Bioch. Bioph. Acta, *114*:469–480.

Midgley, J. E. M. (1965a). Effects of different extraction procedures on the molecular characteristics of bacterial ribosomal ribonucleic acid. Bioch. Bioph. Acta, *95*:232–243.

——— (1965b). The estimation of polynucleotide chain length by a chemical method. Bioch. Bioph. Acta, *108*:340–347.

——— (1965c). Studies on the structure of 23-S ribosomal ribonucleic acid from *E. coli*. Bioch. Bioph. Acta, *108*:348–354.

Miles, H. T. (1964). The structure of the three-stranded helix, poly (A+2U). Proc. Nat. Acad. Sc., *51*:1104–1109.

Miles, H. T., and J. Frazier (1964). Infra-red spectra in H_2O_2 solution of guanylic acid helices and poly G. Bioch. Bioph. Acta, *79*:216–220.

Nomura, M., and P. Traub (1966). Structure and function of ribosomes and subribosomal particles. Proceedings Symp. Rutgers, in "Organizational Biosynthesis," Academic Press, N. Y., p. 459.

Panijel, J. (1963a). Diversité immunologique des RNA et anticorps anti-RNA. C.R. Acad. Sc., *256*:4130–4133.

——— (1963b). Précipitation immunologique des polynucléotides synthétiques et multiplicité des anticorps anti-RNA. C.R. Acad. Sc., *256*:4540–4543.

——— (1963c). Anticorps anti-poly I spécifiques et anticorps anti-RNA dans les sérums des sujets immunisés ou non contre les ribosomes. C.R. Acad. Sc., *257*:798–801.

——— (1963d). Précipitation des ribopolynucléotides homopolymères par les divers anticorps du serum anti-RNA. C.R. Acad. Sc., *257*:3064–3067.

Panijel, J., and E. Barbu (1961). Étude immunologique des ribosomes. Pathol. Biol., *9*:2243–2255.

Panijel, J., E. Barbu, and G. Quash (1963). Tolérance immunitaire et constituants autologues, in "La tolérance acquise et la tolérance naturelle à l'égard de substances antigéniques définies" Colloques Intern. du C.N.R.S., Editions du C.N.R.S., 308–334.

Panijel, J., C. Souleil, and P. Cayeux (1966a). Immunochemical characterization of polyribonucleotides. Science, *152*:773–775.

——— (1966b). Immunochimie des polyribonucléotides I. Homopolymères à une ou plusieurs chaines. Bioch. Bioph. Acta, *123*:231–234.

Panijel, J., P. Cayeux, E. Sacquet, and H. Charlier (1966c). Anti-ribonucleic acid factors in the sera of germ-free animals. Nature, *211*:94–96.

Panijel, J., and M. C. Delaunay (1967). An immunochemical approach of the structure of the ribosomal particles. Bioch. Bioph. Res. Com., *129*:158–165.

Pestka, S., and M. Nirenberg (1966). Regulatory mechanisms and protein synthesis X. Codon recognition on 30 S ribosomes. Jl. Mol. Biol., *21*:145–171.

Plescia, O. J., and W. Braun (1967). Nucleic acids as antigens. Adv. in Immun., *6*:231–252. Acad. Press.

Rich, A. (1958). The molecular structure of polyinosinic acid. Bioch. Bioph. Acta, *29*:502–509.

Rich, A., and D. Davies (1956). A new two-stranded helical structure: polyadenylic acid and polyuridylic acid, *78*:3548–3549.

Sarkar, P. K., and J. T. Yang (1965a). Optical activity and the conformation of polyinosinic acid and several other polynucleotide complexes. Biochemistry, *4*:1238–1244.

——— (1965b). Rotatory dispersion and conformation of polyadenylic acid and polyuridylic acid. Jl. Biol. Chem., *240*:2088–2100.

Shin, D. H., and K. Moldave (1966). Effects of ribosomes on the biosynthesis of RNA *in vitro*. Jl. Mol. Biol., *21*:231–245.

Souleil, C., P. Cayeux, and J. Panijel (1966). Immunochimie des polynucleotides II. Complexes d'homopolymères différents, *123*:235–245.

——— (1968). Immunochemistry of polyribonucleotides. Study of poly I and poly G. Biochemistry, *17*:7–13.

Van Holde, K. E., J. Brahms, and A. M. Michelson (1965). Base interactions of polynucleotide polymers in solution. Jl. Mol. Biol., *12*:726–739.

Watson, J. D. (1964). The synthesis of proteins upon ribosomes. Bull. Soc. Chim. Biol., *46*:1399–1425.

Witz, J., and V. Luzzati (1965). La structure des acides polyadénylique et polyuridylique en solution. Etude par diffusion centrale des rayons X. Jl. Mol. Biol., *11*:620–630.

INHIBITION OF BACTERIAL TRANSFORMATION BY NUCLEIC ACID-SPECIFIC ANTIBODIES *

ROBERT ERICKSON, WERNER BRAUN, OTTO J. PLESCIA, AND ZBIGNIEW KWIATKOWSKI

Institute of Microbiology, Rutgers, The State University
New Brunswick, New Jersey

I. INTRODUCTION

Shortly after the successful production of antisera that react with DNA and synthetic polynucleotides (Plescia *et al.,* 1964, 1965; Plescia and Braun, 1967) it was decided to determine the influence of such sera on a system in which the biological activity of DNA can be assayed with relative ease. Transformation in bacteria represents such a system in which it is possible to determine quantitatively the ability of naked DNA to transfer information from one cell population to related, but genotypically slightly different, recipient cells (cf. Schaeffer, 1964; Braun, 1965; Spizizen *et al.,* 1967). Accordingly, studies were initiated with *Diplococcus pneumoniae* and *B. subtilis* (Braun *et al.,* 1965), two species in which methods for assaying transformation have been well standardized. The early tests revealed that antisera against DNA and against synthetic polynucleotides can indeed inhibit transformation in pneumococci and *B. subtilis.* The early tests also showed that an antiserum reacting only with two of the four bases in DNA (anti-poly dAT) inhibited different non-linked markers to a different extent. This implied that the inhibition may depend on the relative frequency of a given nucleotide in the region with which the information of the marker under scrutiny is associated. The early results also indicated that the inhibition was principally dependent on the time of exposure of the recipient cells to antiserum and was relatively independent of the time of exposure of the double-stranded transforming DNA to antiserum. These findings suggested that additional studies with transforming systems and with anti-

* Supported by NIH grant CA-04786.

sera to oligo- and poly-nucleotides may be rewarding in two respects: (1) they might result in a method for estimating the relative frequency of a given nucleotide in a genetically defined DNA region, i.e., in a cistron, and (2) they might furnish additional information on the process of transformation itself. *Bacillus subtilis* was chosen as the principal test system since this bacterium can be grown in a chemically defined medium and since numerous mutant strains, providing suitable markers, are available (Spizizen, 1958; Anagnostopoulos and Spizizen, 1961; Young and Spizizen, 1961).

II. THE *BACILLUS SUBTILIS* TRANSFORMATION SYSTEM

Since this volume deals primarily with immunological aspects, it may be helpful to summarize briefly some general aspects of the *B. subtilis* transformation system.

Braun (1965) has defined transformation as "a process of intercellular transfer of information in which a fraction of a donor cell's total DNA, obtained by chemical extraction or natural cell lysis, can penetrate into a related bacterial cell and replace there, through a process of recombination, a specific nucleotide sequence of the recipient's genome." This definition contains a number of points of particular importance to the present study. First, the donor DNA must be extracted from the donor cells. This extraction process generally generates enough shearing forces to fragment the DNA into smaller molecules; a cell's DNA may thus break into 50–200 pieces (Schaeffer, 1964). This fragmentation necessitates a redefinition of the concept of genetic linkage, which in bacterial transformation is expressed as the probability that distinct genetic markers will end up on the same DNA fragment. Thus, the smaller the distance between cistrons controlling given markers the higher is their linkage value.

In actual transformation, a population of isolated DNA molecules is added to a competent recipient culture. "Competence" is a function of the ability of the recipient cells to incorporate extracellular DNA (Lerman and Tolmach, 1957; Young and Spizizen, 1963). The actual mechanism of uptake of DNA by competent cells is still controversial (see Spizizen *et al.*, 1966 for a recent review) but undoubtedly involves some alteration of the cell surface (McCarty *et al.*, 1946). In the process of associating with the recipient cell, the transforming DNA becomes DNAase-insensitive, presumably after penetration through the cell wall, and this insensitivity develops sequentially for linked markers, suggesting a lengthwise uptake, the actual orientation of the entering molecule being randomly determined (Strauss, 1965). There is evidence that in the course of the "entry process" the incoming DNA is reduced to a single-stranded state in pneumococcal transformation (Lacks, 1962); however, there has been no definite demonstration of such events in *B. subtilis.* In any case, single-stranded DNA appears to be involved in the integration process (Venema *et al.,* 1965; Bodmer, 1966) and since either strand can transform with equal efficiency when denatured and separated (Chilton, 1967), it appears likely that a denaturation step is involved at some point in transformation. Recent observations have suggested that transforming DNA, once inside the cell wall, may associate with a site on the bacterial membrane (Kammen *et al.,* 1966) at or near a postulated replication point of the recipient cell's DNA (Ganesan and Lederberg, 1965). Autoradiographs (Wolstenholme *et al.,* 1966; Young, 1966) using ^{3}H-labeled DNA have indicated that the activated grains are localized in the space between cell wall and cell membrane.

The final step in transformation is the synapsis of the transforming DNA piece with the homologous segment of the recipient's DNA, and subsequent integration of part of the donor's DNA through an exchange process (Bodmer and Ganesan, 1964; Pené and Romig, 1964). Recent data (Bodmer, 1965, 1966) have suggested that integration may occur only at time of replication of the homologous region of the recipient's DNA. However, this conclusion has been challenged (Ganesan, 1967).

III. STUDIES BY OTHERS ON ALTERATIONS IN THE FREQUENCY OF TRANSFORMATION

DNA-mediated transformation of bacteria can be inhibited either by a change in the state of competence of the recipient cells or by physicochemical changes in the DNA molecules to which the recipient cells are exposed. Inhibitions that are the consequence of an alteration in the state of competence (i.e., of the cells' ability to take up DNA) inhibit all markers equally, whereas alterations in the transforming DNA molecules can alter the transforming capacities of different markers to a different extent. The latter type of differential alteration of transforming activity has been studied by others with a variety of methods; these include inactivation by X- and U.V.-irradiation (Lerman and Tolmach, 1959; Louarn and Sicard, 1965; Marmur *et al.,* 1961; Zamenhof *et al.,* 1957), by various mutagens (Lerman and Tolmach, 1959; Zamenhof *et al.,* 1957; Ephrati-Elizur and Zamenhof, 1959; Canner and Spizizen, 1965), by heating (Lerman and Tolmach, 1959; Canner and Spizizen, 1965; Saito and Masamune, 1964). The frequency of transformation for non-linked markers has been altered by fractionation on methylated albumin columns (Saito and Masamune, 1964), and by using donor DNA extracted from synchronized cultures (Sueoka and Yoshikawa, 1963). The following generalizations may be made on the basis of such studies: A differential alteration of transforming activity of non-linked markers can be produced: (1) by variations in the size of the DNA pieces carrying the markers; (2) by differences in the base content and base sequences in such DNA fragments; and (3) by the location of the cistron in relation to points of induced lesions on the DNA fragment (Marmur *et al.,* 1961). These studies also showed that linked markers are usually inactivated to the same extent.

An interesting way for altering competence has been achieved through the use of antisera prepared against surface antigens of competent and noncompetent pneumococci (Tomasz and Beiser, 1965). Only antisera produced against competent cells (anti-Co+) inhibited transformation. The inhibitory anti-Co+ did not react with transforming DNA but decreased the binding of the DNA by competent cells. Anti-Co+ inhibited transformation of a closely related α-hemolytic streptococcus but had no effect on transformation in *B. subtilis* or *H. influenzae.*

IV. EFFECTS OF ANTISERA TO OLIGO- AND POLY-NUCLEOTIDES ON TRANSFORMATION

A. General Procedures

1. *Extraction of DNA*

Transforming DNA was extracted from *B. subtilis* 168+ by the method of Anagnostopoulos and Spizizen (1961). The saturating concentration of DNA was determined by titration to be approximately 5 μg/ml under the conditions employed by us. This concentration was employed routinely in the experiments to be described.

2. *Preparation of Competent Cells*

Recipient cells were brought to the competent state by either of two methods. In the first of these methods (Anagnostopoulos and Spizizen, 1961), vegetative cells were grown for 18 hours in minimal medium supplemented with 0.1% yeast extract. They were then diluted $1/10$, transferred to a similar medium and after a four-hour growth period were again diluted $1/10$ and transferred into a minimal medium supplemented with 0.01% yeast extract. Competence was maximal 60–90 minutes after transfer to the "step down" medium. DNA was then added to the recipient culture and after the culture was incubated for an additional 30 minutes at 37°C it was assayed for transformants by plating on appropriate growth media.

In the second procedure (Ephrati-Elizur, 1965), spores were used as the initial inoculum and the cultures were maintained in minimal medium supplemented with 0.1% yeast extract for 18 hours at 37°C. The cells were then diluted $1/10$ into a fresh medium of identical composition; competence appeared in this medium 90–150 minutes later. Transfer of the 18 hour cells to a "step down" medium destroyed competence (the 18 hour culture being slightly competent).

3. *Addition of Antiserum*

When antiserum was used, it was added at various times before or after the time of addition of DNA, or it was added simultaneously. In the experiments to be described the time of addition of the DNA will be designated as zero time.

B. Early Studies on Antiserum Effects

The early studies on the effect of antisera on transformation (Braun *et al.*, 1965) utilized antisera prepared in rabbits against calf thymus DNA or against the synthetic co-polymer poly-dAT. Anti-DNA sera were found to inhibit the transformation of three non-linked markers (thr$^-$, leu$^-$, met$^-$) almost equally while the anti-poly-dAT serum showed a very definite and reproducible differential inhibition of these markers (Table 1). The extent of inhibition in a decreasing order was thr+ → met+ → leu+.

Since anti-DNA serum contains antibodies against all four types of nucleotides, the equal inhibition of transformation of the three markers was to be expected. However, the frequency with which sequences of dAT appear in the different DNA fragments carrying the information for the three markers may vary to a considerable extent and thus could explain the differential inactivation by antibodies directed against these bases. This explanation was supported by the finding that shearing of

Table 1

The Effects of Antisera Produced Against Calf-thymus DNA and Poly-dAT on Transformation of *B. subtilis* 168 (thr$^-$, met$^-$, leu$^-$)

Antisera	Time of Antisera Addition in Relation to DNA Addition	Per Cent Inhibition for *		
		thr+	met+	leu+
anti-DNA	−60′	100	100	100
anti-DNA	0	92	90	84
anti-DNA	+15′	27	30	0
anti-poly-dAT	−60′	58	49	23
anti-poly-dAT	0	23	16	9

* Percentage of inhibition in comparison to normal serum controls. Final dilution of all sera was 1:200.

Table 2

EFFECTS OF SHEARING OF DNA ON THE INHIBITION OF *B. subtilis* TRANSFORMATION BY ANTISERUM TO POLY-dAT *

Number of Times That DNA Was Forced Through a Syringe Needle to Shear It	Per Cent Inhibition of Transformation		
	thr+	met+	leu+
0	44	54	24
3	24	38	59
10	15	34	11

* Recipient cells (Mu5u8u5) were exposed to antiserum 60 minutes before DNA addition. Final serum concentration was 1:200.

the transforming DNA altered the pattern of differential inhibition by anti-poly-dAT serum (Table 2) presumably by removing reactive nucleotides from areas of the transforming DNA pieces that are beyond the cistron under scrutiny. Shearing was accomplished by forcing the DNA a number of times through a 27-gauge needle.

The early data also demonstrated that the extent of inactivation by antisera increases with the length of time that the recipient cells are exposed to the antiserum and that the inhibitory effects are independent of the length of exposure of the double-stranded transforming DNA to the antibodies. This finding suggested that the inhibitory effects may be the result of interactions between single-stranded DNA and antibody occurring at some stage in the association of transforming DNA with the recipient cell, and that until this interaction occurs the antiserum-exposed cell may "store" the antibodies (between cell wall and membrane?). As we shall see, this assumption was supported by a number of later findings (see p. 214).

The early experiments also indicated an apparent escape of the leu+ marker from the action of the antiserum when antiserum was added late (Table 1).

C. REMOVAL OF NON-SPECIFIC SERUM FACTORS

It was recognized in the course of the studies summarized above that non-specific effects of rabbit serum on frequencies of transformation can hinder a precise quantitative analysis of inhibitory effects. In some experiments normal serum was inhibitory, and in others stimulatory. Figure 1 shows that the non-specific modifying effects of normal serum (added 15′ prior to DNA) are dependent both on the concentration of the serum and on the batch of serum used.

To remove the non-specific serum factors from antiserum, it was fractionated on DEAE-cellulose. The elution was started with 0.03 M NaCl, in 0.01 M Tris buffer (pH 7.4) and the concentration of the NaCl was increased stepwise to 0.15 M. Fractions were collected, and their optical density at 280 mμ was determined. Fractions corresponding to individual peaks were pooled, heated to 65°C for 60 min and distributed into sterile test tubes by filtration through a millipore filter (HA 0.45 μ).

Complement-fixation tests were carried out with these fractions using denatured *B. subtilis* DNA and denatured calf thymus DNA to determine which fraction contained C′-fixing antibodies. Figure 2 shows that the first fraction eluted from the column fixed complement in tests with both calf thymus and *B. subtilis* DNA. The apparent fixation occurring with Fraction II when denatured *B. subtilis* DNA was used may be non-specific since *B. subtilis* DNA tends to be anti-complementary at a concentration at which calf thymus DNA is not anti-complementary.

Figure 2 also indicates that the first fraction, containing γ-globulins, was consistently inhibitory in tests on transformation. Fraction III, presumably containing serum albumin, also produced a high degree of inhibition in agreement with Spizizen's (1958) earlier report that serum

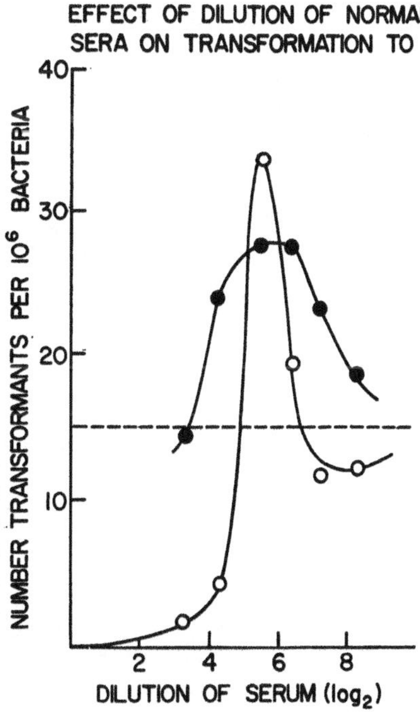

Fig. 1. The effect of different concentrations of two normal rabbit sera on the frequency of transformation. (—●—) normal serum #29; (—○—○—) normal serum #26. Recipient cells exposed to serum 15 minutes before DNA addition. ---- = saline control.

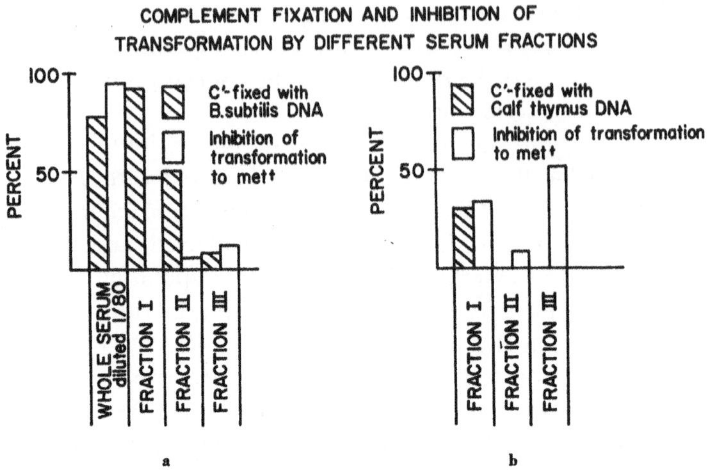

Fig. 2. A: Complement-fixation and inhibition of transformation by different fractions of an anti-calf thymus DNA serum. **B:** Complement-fixation and inhibition of transformation by different antiserum fractions of an anti-poly-dAT serum. Fractionation was carried out on DEAE-cellulose (0.01*M* Tris buffer, pH 7.4). Elution was with: (I) 0.03*M* NaCl; (II) 0.10*M* NaCl; (III) 0.15*M* NaCl. 0.1 ml of a fraction (O.D. 280 mμ = 0.40) used in all cases. DNA was heat-denatured and used at a concentration of 2 μg/0.1 ml for C′-fixation tests. Recipient cells exposed to serum fractions 15 minutes before DNA addition.

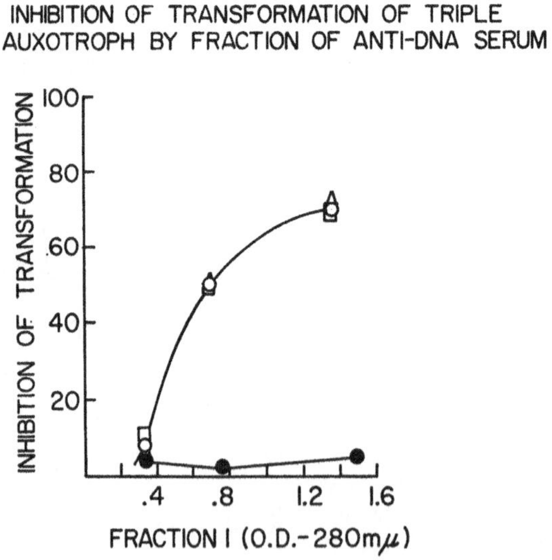

Fig. 3. The effect of antiserum fraction I and normal serum fraction I on transformation. (—○—) transformation to leu+; (—□—) transformation to met+; (—△—) transformation to thr+; (—●—) average response of all markers. Open symbols represent inhibition by fraction I of anti-calf thymus DNA and closed symbols represent inhibition by fraction I of normal serum.

albumin is a potent inhibitor of transformation. We found that in the presence of the albumin fraction transformation of three non-linked markers is equally inhibited.

D. Inhibitory Effects of Serum Fractions

Serum fractions were obtained from normal serum and antisera to DNA and poly-dAT, as described above, and their effects on transformation were studied as a function of serum concentration (Fig. 3 and Table 3). It can be seen that the non-specific effect of normal serum was reduced by fractionation to a minor (<10%) but consistent level of inhibition. Fraction I of anti-DNA produced an equal inhibition of transformation of all three non-linked markers (Fig. 3), whereas Fraction I of anti-poly-dAT caused a differential inhibition of the three non-linked markers (Table 3), as did the unfractionated sera. However, a different pattern of inhibition was observed with fractionated anti-poly-dAT serum compared to the unfractionated antisera (see Table 1).

When antisera, produced by immunization of rabbits with the oligonucleotides ApApCp, TpTpTp, TpTpTpTpTp, or poly-C, were fractionated and the fractions tested for their effects on the three non-linked markers leu$^-$, met$^-$, and thr$^-$, different patterns of inhibition, depending on the antiserum employed, were obtained. Figure 4 shows the results for two types of experiments: (1) when the inhibitory antiserum fraction is added to the suspension of competent recipient cells 15 minutes prior to the addition of transforming DNA, or (2) at the time of DNA addition. The different effects of different antisera therefore should be compared

Table 3

Inhibition of Transformation of Triple Auxotroph (thr$^-$, leu$^-$, met$^-$) by Fractionated Anti-poly-dAT Serum

Exp.* #	Time of Preincubation of Recipient Cells with Antiserum	Fraction Concentr.†	Per Cent of Inhibition of Transformation to thr+	leu+	met+
1	15 minutes	0.5	38	48	58
	0 minutes	0.5	14	17	29
2	15 minutes	0.5	17	19	28
	15 minutes	0.7	28	37	47
	15 minutes	0.9	39	50	55

* Competence level in experiment 1 was relatively high (0.3), in experiment 2 it was much lower (0.01).

† In terms of O.D. at 280 mμ.

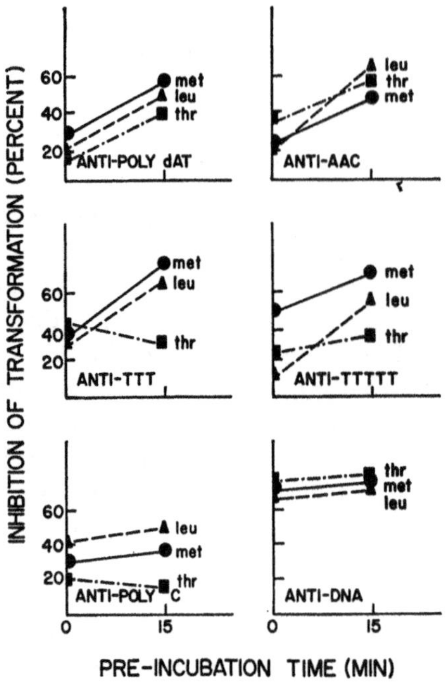

Fig. 4. Inhibition of transformation of *B. subtilis* 168 MU5U8U5 (Sueoka) by various antisera (fractions I used throughout). Two sets of data have been plotted. The points nearest the ordinate represent results obtained when recipient cells were exposed to the antiserum fraction at time of addition of DNA; the points to the right of the ordinate represent results obtained when recipient cells were exposed to antiserum fraction 15 minutes prior to the addition of DNA.

by noting the differences at the 15 minute point (e.g., order of inhibition for anti-TTTTT: met → leu → thr; order for anti-poly C: leu → met → thr). Changes in the slopes in Figure 4 reflect differences in the rate of escape of different markers from inhibition by different antisera; we shall later return to a further consideration of this phenomenon.

Repeated tests have shown that for any given antiserum identical patterns of differential inhibition of non-linked markers can be obtained when experiments are run with the same antiserum several months apart.

As indicated earlier, differential inhibitions of non-linked markers by a given antiserum may reflect the frequency of occurrence of the reactive nucleotide in the transforming DNA piece containing the cistron for the marker that is being scored. If this interpretation is correct, it would be anticipated that transformation of linked markers, involving cistrons likely to end up on the same transforming DNA piece, would be inactivated equally by a given antiserum. Table 4 illustrates that this is indeed the case. All the linked markers studied were inhibited almost equally and the co-transfer index (Nester and Lederberg, 1961), did not change significantly.

If the overall nucleotide content of a piece of transforming DNA determines the ability of this DNA to react with a given antiserum, one could anticipate little influence of the actual site of a mutation on the inhibition resulting from the exposure of DNA to antiserum. To test this assumption, a variety of independently isolated leucine-requiring mutants were collected from several laboratories and used as recipients

Table 4

THE INHIBITION OF TRANSFORMATION OF LINKED MARKERS (tyr_1, his_2) IN *B. subtilis* 168 SB 202 *

Antisera	Per Cent Inhibition of Transformation † his_2	tyr_1	his_2 tyr_1	Change in Co-transfer Index ‡
anti-poly-dAT	64	67	63	.01
anti-AAC	17	20	20	.02
anti-AAC	18	19	20	.00
anti-5T	29	28	27	.01

* *B. subtilis* 168 SB 202 was obtained from Dr. E. Nester.

† Recipient cells were exposed to the antiserum fraction (Fraction I) 15 min. before DNA addition.

‡ See Nester and Lederberg (1961).

for leu+-carrying DNA in tests in which anti-poly-dAT served as the inhibitory antiserum. Mutations affecting the biosynthesis of leucine were used since in mapping studies to date only one genetic locus has been shown to be concerned with the synthesis of this amino acid (Dubnau *et al.*, 1967). Figure 5 shows that in all mutants studied transformation to methionine independence is always inhibited more than

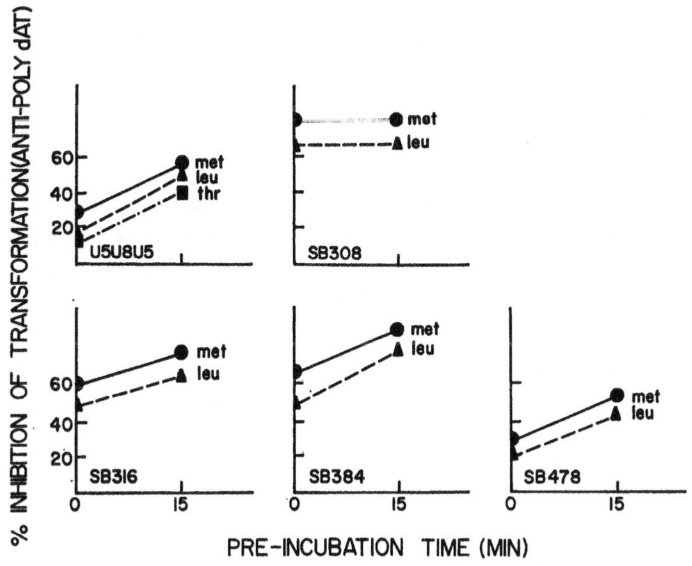

Fig. 5. Inhibition of transformation by anti-poly-dAT serum fraction I (O.D. 280 mμ = 0.40) using various, independently isolated *B. subtilis* auxotrophs. See legend to Fig. 4 for an explanation of the points plotted. Origin of auxotrophs: MU5U8U5 (N. Sueoka); SB 308, SB 316, SB 384 (E. Nester); SB 478 (E. Lederberg).

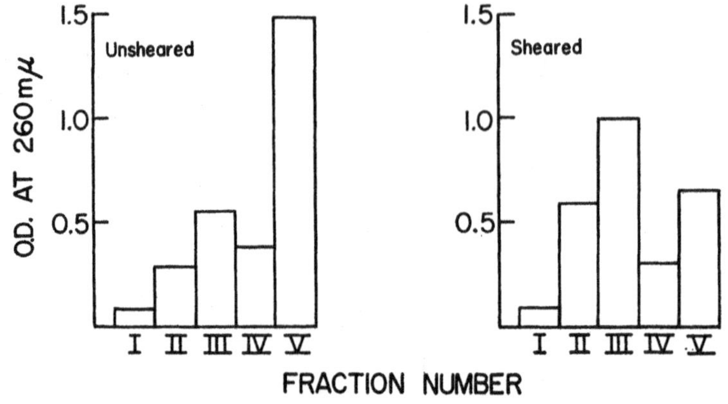

Fig. 6. Distribution of DNA in various chromatographic fractions obtained from unsheared and sheared DNA preparations. Shearing of the transforming DNA was accomplished by forcing DNA through a 27 gauge needle. Fractionation was carried out according to the method of Davila *et al.* (1965). Eluants: fraction I, 0.01*M* PO_4 buffer, pH 7.0; fraction II, same + 0.5*M* NaCl; fraction III, same as I + 1.0*M* NaCl; fraction IV, 2*M* NaCl + .2*M* NH_3; fraction V, 1*M* NaOH.

transformation to leucine independence. Of particular interest are mutants SB308, SB316 and SB384, all of which were derived from the same methionine-requiring mutant. Although we do not have direct proof that the sites of mutation in these different leu^- strains are different, it is very likely that at least some of the mutants employed are changed at non-identical sites of a cistron. We may therefore conclude from the results obtained that the relative extent of inhibition by antisera is dependent on the overall nucleotide content of the DNA piece carrying the information for the marker being scored and that the actual site of mutation does not affect the result.

Additional data on sheared DNA indicate that the inhibition of transformation by antisera is dependent on the nucleotide content of transforming DNA pieces, and apparently is not a phenomenon that can be attributed to differences in the size of DNA pieces with which information for different non-linked markers is associated. A sample of transforming DNA was sheared, as described under B of Section IV above, and fractionated on DEAE-cellulose by the method of Davila *et al.* (1965). The distribution of DNA pieces of different size in the unsheared and sheared DNA preparations is shown in Figure 6. Transforming activity was associated only with fractions III, IV and V of both types of DNA, and since fraction III contains the smallest of the active pieces and, in terms of the starting DNA, a more uniform population of transforming molecules, it was chosen for studies with antiserum. (The dependence of transformation on a minimum size of the DNA molecule is a well-established phenomenon.) Table 5 indicates that with the use of Fraction

Table 5

INHIBITION OF TRANSFORMATION BY ANTI-TTTTT SERUM FRACTION * USING FRACTIONATED AND UNFRACTIONATED TRANSFORMING DNA (see Fig. 6 for fractions)

Transforming DNA	Per Cent Inhibition for		
	thr+	met+	leu+
Fraction III of sheared DNA †	33	46	0
Unsheared DNA	37	69	58

* Added to recipient culture 15 min. before addition of DNA.
† Eluted by method of Davila, *et al.* (1965) with 0.01M phosphate buffer, pH 7.0 and 1.0M NaCl. Mu5u8u5 (N. Sueoka) used as recipient.

III, with recipient cells exposed to TTTTT-antiserum, differential inhibition of transformation of non-linked markers again occurred. The occurrence of differential inhibition with a population of DNA pieces more homogeneous in size than the original population indicates that differential inhibitions by the antisera are not due to reactions controlled solely or principally by the differing sizes of transforming DNA pieces in non-sheared DNA.

Now let us return to the "escape phenomenon" mentioned on p. 210 and shown in Table 1 and Figure 4. This phenomenon involves the ability of some markers, particularly leu+, to escape from antibody inhibition when the addition of the antiserum fraction is delayed until the time of DNA addition or shifted to an even later period. The escape from inhibition implies that the DNA strand carrying the information for the non-inhibited marker has reached a site or a state where it can no longer react with the inhibitory antibodies. Comparative tests have indicated that, whatever this site or state might be, it differs from the one accessible to DNAase. Pancreatic DNAase also can inactivate transforming DNA (Levine and Strauss, 1965), but we have found that at a time after DNA addition, when a given marker has become insensitive to DNAase inhibition, it may still be inhibited by antibody.

E. MODE OF ACTION OF INHIBITORY ANTISERA

The phenomenon just described, namely the ability of antibody to inhibit DNA at a time when it has become resistant to DNAase effects, already suggests that antibodies may react with transforming DNA at a site internal to the cell wall. Possible storage of antibodies in such a

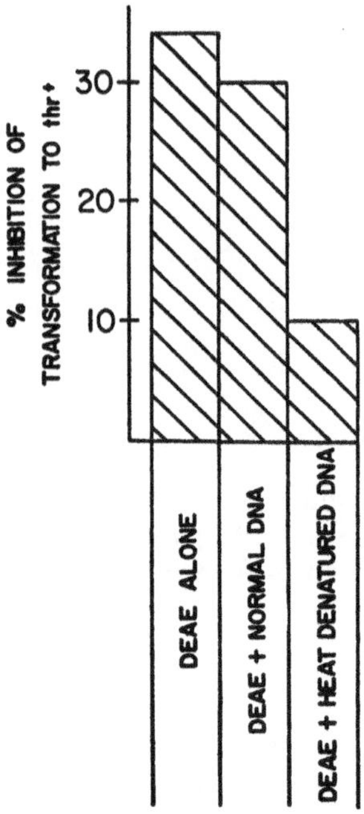

Fig. 7. Influence of passage of anti-poly-dAT serum (fraction I) through a DEAE-cellulose column, saturated with native or heat-denatured DNA, on its inhibitory activity. DEAE-cellulose (in 0.01*M* Tris buffer) columns were equilibrated with 0.15*M* NaCl and then saturated with *B. subtilis* DNA and washed thoroughly. Fraction I of anti-poly-dAT serum was then passed through the column, collected, and added to recipient cells 15 minutes before the addition of transforming DNA.

reaction site is suggested by the finding that the length of pre-incubation of antiserum and recipient cells significantly increases the extent of inhibition of transformation (Table 1, Figs. 4 and 5). Pre-incubation of transforming DNA with antiserum does not cause a significant alteration of the extent of inhibition. Also, pre-exposure of the antiserum to normal DNA or sheared DNA does not significantly affect its ability to inhibit transformation.

It is known that the antibodies employed in these studies react principally with single-stranded nucleic acids (Plescia *et al.,* 1964). That this is also the case in the present series of experiments was demonstrated in the following manner: DEAE-cellulose columns were saturated with either normal transforming DNA or with heat-denatured DNA. Anti-poly-dAT serum was layered on top of the two columns and elution started with 0.15 M NaCl (in 0.01 M Tris, pH 7.4). The fractions collected were then tested in the transformation system. The results showed that the majority of the inhibitory activity was removed after passage of the antiserum through the column saturated with denatured DNA, whereas little activity was removed after its passage through the column containing native double-stranded DNA (Fig. 7).

A possible causal relationship between sites or factors responsible for uptake of DNA (competence factors) and those responsible for participation in the antiserum effect is indicated by the data shown in Figure 8. Cultures just developing competence or at the peak of competence show

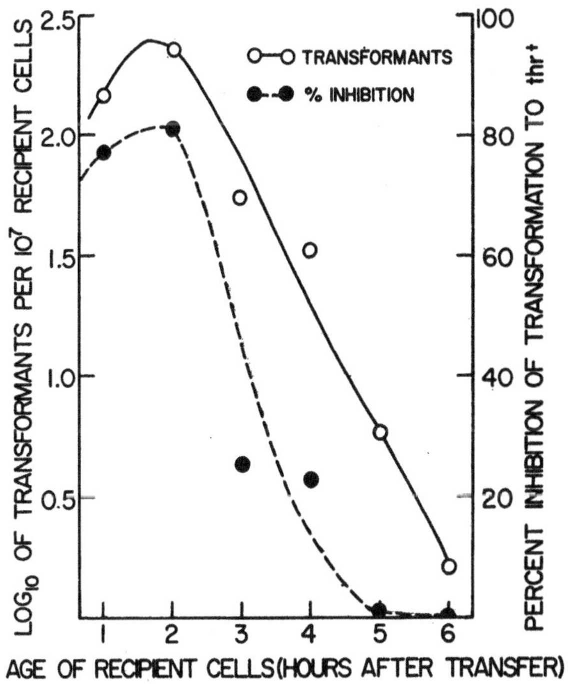

Fig. 8. The influence of the stage of competence of the recipient culture on the inhibition of transformation by anti-poly-dAT serum (fraction I). Recipient cells were exposed to the antiserum fraction 15 minutes before addition of DNA. Age of recipient culture represents hours after transfer of an 18-hour-old culture, initiated with spores, of *B. subtilis* 168 Mu5u8u5 (Sueoka), to a fresh culture medium.

a much greater degree of inhibition by antiserum than cultures in which the ability to be transformed is decreasing. This may be due to a parallel capacity of the recipient cells to incorporate antibody and transforming DNA.

F. New Information on Events in the Transformation Process

In most of the studies described so far, Spizizen's (1961) procedure for the development of competence, described under IV A above, was employed. In an effort to reduce continued uncontrollable variability in this procedure, we decided to shift to a modification of a method first popularized by Ephrati-Elizur (1965), namely to initiate competent cultures from spore suspensions. Spores were grown for 18 hours in a minimal medium supplemented with 0.1% yeast extract and were then transferred to a similar medium. This transfer from exhausted to fresh medium resulted in the development of competence within 90–150 minutes after transfer. When antiserum was added to such cultures at different periods during the competence state, unexpected changes in relative frequencies of transformation and inhibition of different non-linked markers were encountered. This finding prompted us to explore this phenomenon further in the absence of antiserum. As shown in Figure 9, the relative frequency of transformation of non-linked markers was found to fluctuate in a cyclic fashion as a function of time of addition of transforming DNA to the recipient culture. These results can best be explained on the basis of Bodmer's (1965) hypothesis that integration of DNA will occur at the point of DNA replication. If one assumes that the transforming DNA associates with a fixed point at which DNA replication proceeds, presumably at the cell membrane, as indicated by Ganesan's and Lederberg's data (1965), and if one takes into consideration a decay of the cell-associated transforming DNA with time, then one can explain the results shown in Figure 9 by the scheme illustrated

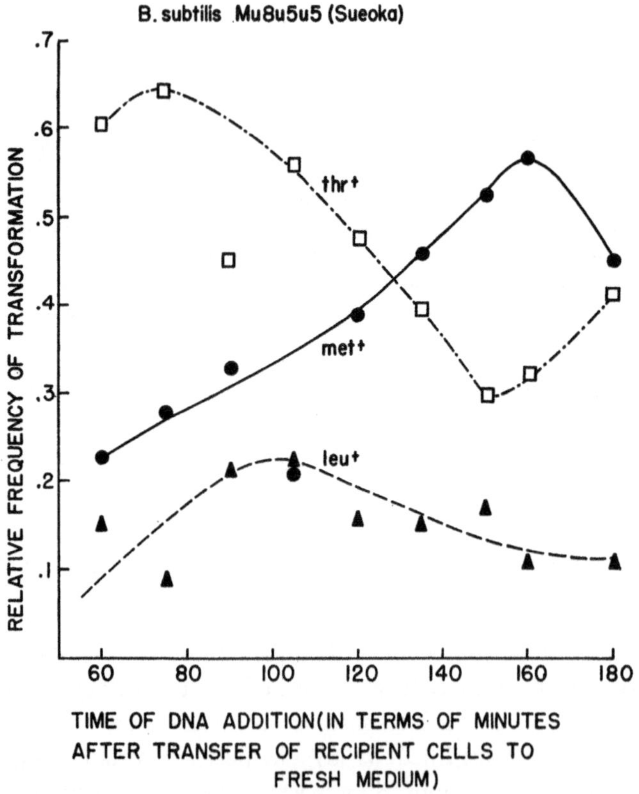

Fig. 9. Fluctuation of the relative frequency of transformation of three non-linked markers as a function of time of addition of DNA to recipient cells. Recipient culture was initiated with spores and time of DNA addition represents minutes after transfer of an 18-hour-old culture to fresh culture medium.

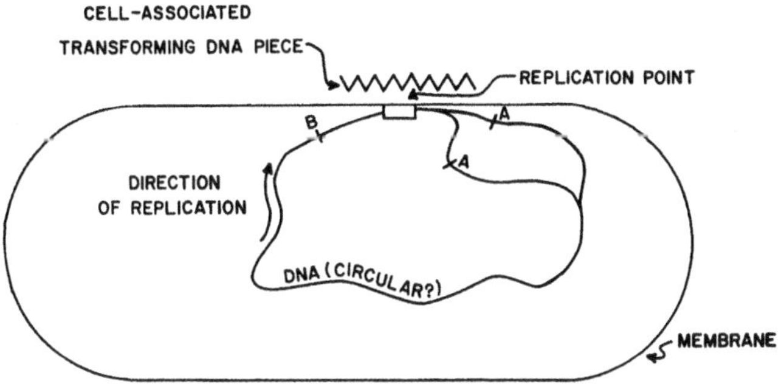

Fig. 10. Model of some possible initial stages in transformation.

in Figure 10. The competent recipient cells, grown from spores, may be fairly well synchronized in respect to DNA replication and possibly, but not necessarily, may be at an identical stage of temporarily suspended chromosomal replication (Bodmer, 1965). Thus, taking a recipient population at a time when the DNA region controlling marker B happens to be near the replication point will assure a high relative frequency of transformation for this marker, in contrast to a distant marker A which will not reach the replication point until some time later. If the recipient population is exposed to DNA at a later time when region A has become proximal to the replication point, and B has become distal, the conditions would be reversed and the relative frequency of transformation for A would be higher than that for B. This is exactly what the experimental data showed, and in fact by comparing relative frequencies of transformation for non-linked markers when DNA was added at different times to the recipient culture, it was possible to estimate the relative position of these markers on a circular DNA. The relative positions did indeed correspond to those previously ascertained, for example by Sueoka and Yoshikawa (1963), with the aid of entirely different techniques.

G. Effects of Antisera on Bacteriophage Transformation

One of the aims of the use of antisera to oligo- and poly-nucleotides in transformation studies is to obtain information on relative frequencies of particular nucleotides and nucleotide sequences in genetically defined regions of biologically active DNA. The experience with the *B. subtilis* transformation system has indicated that it may indeed be possible to obtain, with the aid of antisera, some clues to nucleotide content of DNA pieces of which the cistron under observation represents but a part. It should be remembered that in bacterial transformation, DNA pieces of a minimum size of approximately 7 cistrons are required. Thus, unless this requirement can be circumvented by some tricks, such as the hybridization of a labelled cistron-sized DNA strand to a larger "carrier" strand, the goal of assaying nucleotide content of cistrons appears remote. However, recent studies by Bautz and associates (Bautz and Reilly, 1966; Mazaitis and Bautz, 1967) with T_4 phage DNA have suggested a feasible approach to our goal. In the studies by Bautz and associates, messenger RNA specific for a gene was isolated with the aid of T_4 bacteriophage DNA from a mutant having a cistron-sized deletion. The gene-specific mRNA was then employed for the isolation of a corresponding gene-sized piece of DNA and this DNA was used successfully in bacteriophage transformation (Bautz, 1966; Mazaitis and Bautz, 1967). A procedure and material thus became available for testing the influence of antibodies to oligo- and poly-nucleotides on a genetically defined piece of DNA. In preliminary trials, we have employed unfractionated T_4 phage DNA and transformation involving the rII region of the bacteriophage genome. Such transformation can be obtained with the aid of *E. coli* spheroplasts and urea-treated helper phage and can be scored with relative ease by determining whether the phage can lyse both *E. coli* B and *E. coli* K-12; the untransformed rII mutant can only lyse *E. coli* B. As illustrated in Table 6, the usual frequency of transformation was significantly inhibited when the *E. coli* cells were exposed to

Table 6

EFFECT OF ANTISERA ON TRANSFORMATION IN A BACTERIOPHAGE SYSTEM

Treatment	Average Number of Plaques Per Plate on Indicator *E. coli* Strains: K 12(λ)	BA	K/B
no serum	50	163	0.30
normal rabbit serum	65	200	0.32
anti-TTTTT serum	18	220	0.08
anti-poly-dAT serum	25	174	0.14

E. coli B spheroplasts were mixed with 0.1 ml serum and incubated at 37°C for 10 min. Denaturated r+ DNA (2 μg/ml) was added and the mixture was incubated at 37°C for 10 min. Urea-treated r 1519 helper phage carrying a deletion in rII region was then added. The incubation time of the transformation mixture was usually 2 hours. Samples were lysed with chloroform and 0.1 ml of the undiluted mixture was plated on indicator *E. coli* K 12 and 0.1 ml of 10^{-6} dilution on indicator *E. coli* B. The K/B values give the ratio of the r+ titer on K strain to the rII titer on strain B.

anti-poly-dAT sera or anti-TTTTT sera 5–10 minutes prior to infection of the spheroplasts with denatured transforming phage DNA. We hope to utilize this system now for exploring differential effects of antisera to different oligo- and poly nucleotides, employing DNA pieces of gene size.

V. SUMMARY

Studies with fractionated antisera to oligo- and poly-nucleotides have confirmed that linked markers can be inhibited equally by a given antiserum, whereas non-linked markers are inhibited to an unequal extent. The actual degree of inhibition is dependent on the type of antiserum used and may reflect quantitative differences in the base content of DNA pieces with which non-linked markers are associated. The alteration of inhibitory effects with the use of recipient cells from cultures of different age has suggested a possible dependence of transforming events on the stage of DNA replication of recipient cells. We are now attempting to apply similar techniques to bacteriophage transformation where it will be possible to utilize DNA pieces representing no more than one cistron. Work with such material may, hopefully, enable us to determine, with the aid of inhibitory antisera, the relative frequency of given nucleotides, perhaps even of particular nucleotide sequences, in a defined region of the genome.

References

Anagnostopoulos, C., and J. Spizizen (1961). Requirements for transformation in *Bacillus subtilis*. J. Bacteriol., *81*:741–746.

Bautz, E. K. F., and E. Reilly (1966). Gene-specific messenger RNA: Isolation by the deletion method. Science, *151*:328–330.

Bautz, F. A. (1966). Expression of the rII function by native transforming DNA of bacteriophage T_4. Genetics, *53*:913–921.

Bodmer, W. F. (1965). Recombination and integration in *Bacillus subtilis* transformation: Involvement of DNA synthesis. J. Mol. Biol., *14*:534–557.

Bodmer, W. F. (1966). Integration of deoxyribonuclease-treated DNA in *Bacillus subtilis* transformation. J. Gen. Physiol., *49*:233–258.

Bodmer, W. F., and A. T. Ganesan (1964). The molecular basis for recombination in *Bacillus subtilis* transformation. Genetics, *50*:717–738.

Braun, W. (1965). Bacterial Genetics, 2nd ed. Philadelphia: W. B. Saunders.

Braun, W., O. J. Plescia, M. Kohoutová, and J. Grellner (1966). Inhibition of transforming activities by antisera against DNA and poly-dAT. In "The physiology of gene and mutation expression." Prague: Academia, 181–185.

Canner, M. K., and J. Spizizen (1965). Inactivation of *Bacillus subtilis* DNA by mutagenic agents. J. Bacteriol., *89*:915–916.

Chilton, M.-D. (1967). Transforming activity in both complementary strands of *Bacillus subtilis* DNA. Science, *157*:817–819.

Davila, C., P. Charles, and L. Ledoux (1965). The Chromatography of nucleic acid preparations on DEAE-cellulose paper. I. Fractionation of Deoxyribonucleic acid on paper strips or on centrifuged paper pulp. J. Chromatog., *19*:382–395.

Dubnau, D., C. Goldwaite, I. Smith, and J. Marmur (1967). Genetic mapping in *Bacillus subtilis*. J. Mol. Biol., *27*:163–185.

Ephrati-Elizur, E. (1965). Development of competence for transformation experiments in an overnight culture of germinating spores of *Bacillus subtilis*. J. Bacteriol., *90*:550–551.

Ephrati-Elizur, E., and S. Zamenhof (1959). Incorporation of 5-bromouracil into transforming principle of *Bacillus subtilis* and its biological effects. Nature, *184*:472–473.

Ganesan, A. T. (1967). Particulate fractions in macromolecular synthesis and genetic transformation. In "Organizational Biosynthesis" (H. J. Vogel, J. O. Lampen and V. Bryson, eds.). New York: Academic Press.

Ganesan, A. T., and J. Lederberg (1965). A cell-membrane bound fraction of bacterial DNA. Biochem. Biophys. Res. Comm., *18*:824–835.

Kammen, H. O., R. J. Wojnar, and E. S. Canellakis (1966). Transformation in *Bacillus subtilis*. II. The development and maintenance of the competent state. Biochim. Biophys. Acta, *123*:56–65.

Lacks, S. (1962). Molecular fate of DNA in genetic transformation of pneumococcus. J. Mol. Biol., *5*:119–131.

Lerman, L. S., and L. J. Tolmach (1957). Genetic transformation. I. Cellular incorporation of DNA accompanying transformation in *Pneumococcus*. Biochim. Biophys. Acta, *26*:68–82.

Lerman, L. S., and L. J. Tolmach (1959). Genetic Transformation. II. The significance of damage to the DNA molecule. Biochim. Biophys. Acta, *33*:371–387.

Levine, J. S., and N. Strauss (1965). Lag period characterizing the entry of transforming DNA into *Bacillus subtilis*. J. Bacteriol., *89*:281–287.

Louarn, J.-M., and A. M. Sicard (1965). Sur la sensibilité différentielle des marqueurs génétiques aux rayonnements ultraviolets, et les mecanismes de réactivation au cours de la transformation chez *Diplococcus pneumoniae*. C. R. Acad. Sc. Paris, *261*:852–855.

McCarty, M., H. E. Taylor, and O. T. Avery (1946). Biochemical studies of environmental factors essential in transformation of pneumococcal types. Cold Springs Harbor Symp. Quant. Biol., *11*:177–183.

Marmur, J., W. F. Anderson, L. Matthews, K. Berns, E. Gajewska, D. Lane, and P. Doty (1961). The effects of ultraviolet light on the biological and physical chemical properties of deoxyribonucleic acid. J. Cell. Comp. Physiol., *58* (Suppl. 1):33–55.

Mazaitis, A. J., and E. K. F. Bautz (1967). Partial isolation of an RIIB segment of T_4 DNA by hybridization with homologous RNA. Proc. Nat. Acad. Sci., *57*:1633–1637.

Nester, E. W., and J. Lederberg (1961). Linkage of genetic units of *Bacillus subtilis* in DNA transformation. Proc. Nat. Acad. Sci., *47*:52–55.

Pene, J. J., and W. R. Romig (1964). On the mechanism of genetic recombination in transforming *B. subtilis*. J. Mol. Biol., *9*:236–245.

Plescia, O. J., and W. Braun (1967). Nucleic acids as antigens. Adv. in Immunol., *6:*231–252.

Plescia, O. J., W. Braun, and N. C. Palczuk (1964). Production of antibodies to denatured DNA. Proc. Nat. Acad. Sci., *52:*279–285.

Plescia, O. J., N. C. Palczuk, W. Braun, and E. Cora-Figueroa (1965). Antibodies to DNA and a synthetic polydeoxyribonucleotide produced by oligodeoxyribonucleotides. Science, *148:*1102–1103.

Saito, H., and Y. Masamune (1964). Fractionation of transforming DNA of *Bacillus subtilis* with the methylated albumin column. Biochim. Biophys. Acta, *91:*344–347.

Schaeffer, P. (1964). Transformation, in "The Bacteria" vol. 5 (I. C. Gunsalus and R. Y. Stanier, eds.). New York: Academic Press, pp. 87–153.

Spizizen, J. (1958). Transformation of biochemically deficient strains of *Bacillus subtilis* by deoxyribonucleate. Proc. Nat. Acad. Sci., *44:*1072–1078.

Spizizen, J., B. E. Reilly and E. H. Evans (1966). Microbial transformation and transfection. Ann. Rev. Microbiol., *20:*371–400.

Strauss, N. (1965). Configuration of transforming DNA during entry into *Bacillus subtilis.* J. Bacteriol., *89:*288–293.

Sueoka, N., and H. Yoshikawa (1963). Regulation of chromosome replication in *Bacillus subtilis.* Cold Spring Harbor Sym. Quant. Biol., *28:*47–54.

Tomasz, A., and S. M. Beiser (1965). Relationship between the competence antigen and the competence-activator substance in *Pneumococci.* J. Bacteriol., *90:*1226–1232.

Venema, G., R. H. Pritchard, and T. Venema-Schroder (1965). Fate of transforming deoxyribonucleic acid in *Bacillus subtilis.* J. Bacteriol., *89:*1250–1255.

Wolstenholme, D. R., C. A. Vermeulen, and G. Venema (1966). Evidence for the involvement of membranous bodies in the processes leading to genetic transformation in *B. subtilis.* J. Bacteriol., *92:*1111–1121.

Young, F. E. (1967). Competence in *Bacillus subtilis* transformation system. Nature, *213:*773–775.

Young, F. E., and J. Spizizen (1961). Physiological and genetic factors affecting transformation of *Bacillus subtilis.* J. Bacteriol., *81:*823–829.

——— (1963). Incorporation of DNA in the *Bacillus subtilis* transformation system. J. Bacteriol., *86:*392–400.

Zamenhof, S., G. Leidy, S. Greer, and E. Hahn (1957). Differential stabilities of individual hereditary determinants in transforming principle. J. Bacteriol., *74:*194–199.

EFFECTS OF ANTI-HAPTEN ANTIBODIES ON THE BIOLOGICAL ACTIVITIES OF NUCLEIC ACIDS *

S. W. TANENBAUM AND MERYL H. KAROL †

Department of Microbiology, Columbia University College of Physicians and Surgeons, New York, New York

I. INTRODUCTION

The serological activities of DNA, RNA, and of various free and complexed polynucleotides has been comprehensively reviewed by Levine and Van Vunakis (1966) and by Plescia and Braun (1967).

One of the alternative initial methods that was developed for obtaining nucleic acid-reactive antibodies involved the use of hapten-conjugated proteins. Thus, Butler *et al.* (1962) reacted 6-trichloromethylpurine with BSA to obtain a protein carrier substituted with approximately 25 haptenic groups per mole. This conjugated antigen induced antibodies specific for the purin-6-oyl determinant. These antibodies were shown to cross-react with various DNAs by precipitation (Beiser *et al.,* 1964), by complement-fixation (Butler *et al.,* 1962; 1965) by radio-immunoelectrophoresis, and by passive cutaneous anaphylaxis (Butler *et al.,* 1962; 1965). In view of the double-helical structure of native DNA and the configurational changes caused by heating (cf. Watson, 1965), it was not surprising that these antibodies reacted preferentially with thermally denatured DNA. However, they failed to exhibit cross-reaction with RNA or denatured RNA preparations.

Pyrimidine-specific antibodies were obtained by Tanenbaum and Beiser (1963) by immunization of rabbits with 5-acetyluracil-BSA. These anti-

* Studies herein described which originated in the author's laboratory were supported by Grants 5-F1-GM-22807 from the U. S. Public Health Service, and GB-5316x from the National Science Foundation.

† Present address: Department of Biological Sciences, State University of New York, Stoney Brook.

sera, although reactive with both native and denatured DNAs, again did not appear to react with RNA. Ribonucleosides and ribonucleotides were subsequently caused to react with proteins for use as immunogens (Erlanger and Beiser, 1964). The reaction sequence included splitting of the ribose moiety by periodate oxidation, followed by condensation with proteins and reduction of the resultant intermediary eneamines. The antibodies elicited by these immunogens reacted with denatured DNA from a number of sources. More striking was the observation that these latter specific antibodies did not appear to react directly with RNA. More recently, however, evidence has been adduced for their complexing with RNA by virtue of its ability to inhibit antibody interactions with DNA (Beiser and Erlanger, 1966), or with the homologous AMP-BSA antigen (Lacour and Harel, 1965).

The experiments reported here were therefore undertaken to develop additional synthetic protein conjugates, which would contain the intact nucleosides, and which might induce the production of RNA-reactive antibodies in rabbits. For presumptive RNA specificity, a (poly)uridine-conjugate was the obvious first choice. However, since pseudouridine (5-ribosyluracil, ψ) and 5,6-dihydrouridine are found predominantly in the transfer-RNA species (Holley *et al.*, 1965), it was anticipated that antibodies specific for these pyrimidine nucleosides might also prove useful in anti-RNA specific reactions. Pseudouridine, and uridine were each oxidized to their 5'carboxyl derivative (Moss *et al.*, 1963). Subse-

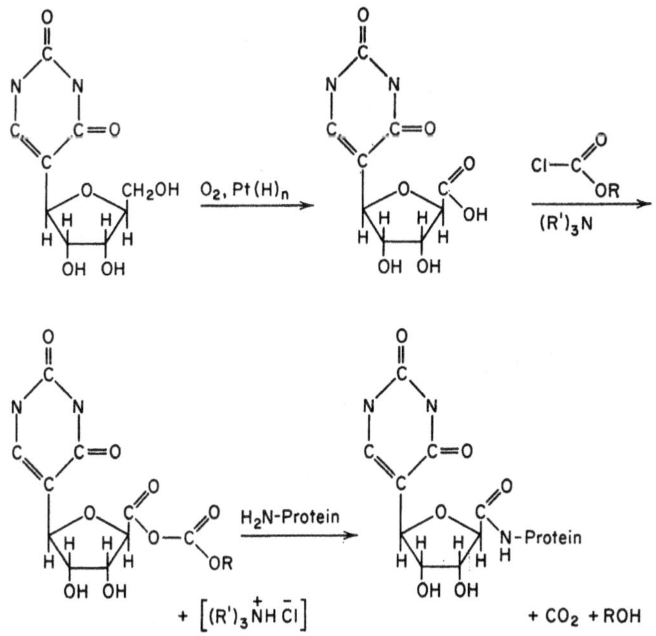

Fig. 1. Synthesis of uridine- and pseudouridine-albumins.

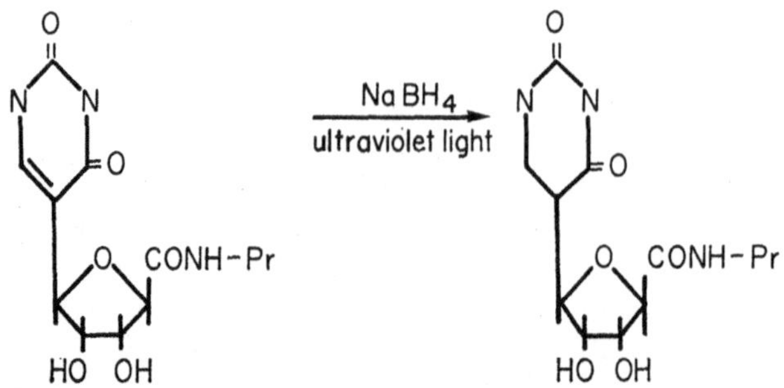

Fig. 2. Photochemical reduction of uridine-conjugates to "dihydrouridine"-proteins (red.U; for clarity, several hydrogen atoms are omitted from formulae).

quent conjugation of each sugar acid to serum albumins, using Boissonnas' (1951) mixed anhydride method provided (poly)uridine-5′-CO-albumins and (poly)pseudouridine-5′-CO-albumins (Fig. 1). Dihydrouridine was not conjugated in the same way since this pyrimidine ring probably would have ruptured (Batt *et al.,* 1954) under the alkaline conditions used for protein coupling. This particular haptenic grouping was therefore provided by photochemical reduction (Cerutti *et al.,* 1965) of the uridine antigen (Fig. 2). Another approach devised for the purpose of obtaining RNA-reactive antibodies involved synthesis of the hitherto unreported 6-amino-quinazoline-2,4(1H, 3H)-dione, and its subsequent coupling to serum albumins (Fig. 3). This antigen possessed known potent immunogenic characteristics: aromatic azo linkages (Landsteiner, 1946), and rigidity (Sela and Arnon, 1960). It was anticipated that the similarity of the heterocyclic ring of this haptenic moiety to that of uracil, as well as resemblance of the entire quinazoline structure to those of purines, might enhance cross-reactions of its complementary antibodies with RNA.

Immunoglobulins from rabbit antisera to each of these four types of conjugate were most reactive with the haptenic group which caused

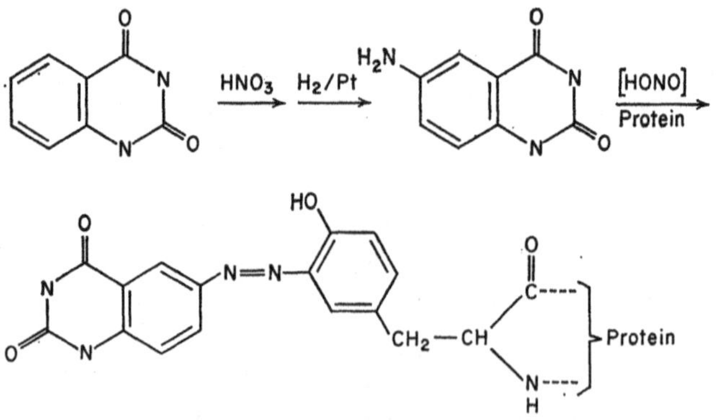

Fig. 3. Synthesis of quinazoline 2,4-dione-6-azoproteins.

their induction. They also cross-reacted with denatured DNA and with several RNAs. Their reaction with appropriate homoribopolymers was noted in gel diffusion tests. An attempt to ascertain their effect upon RNA-programmed *in vitro* protein synthesis will be outlined. Parallel studies by Sela and coworkers (1964, 1965) and by Ungar-Waron *et al.* (1967), have also shown that ribonucleoside-specific antibodies can be obtained by immunization with nucleoside-5′-CO-polypeptides, and that antibodies to these conjugates precipitated with denatured DNA, RNA, and with polyribonucleotides. Halloran and Parker (1966a, 1966b) have prepared deoxynucleotide-protein conjugates, using carbodiimides as protein coupling agents. This method possesses applicability to the synthesis of ribonucleotide-protein antigens as well.

II. SPECIFICITY OF ANTI-HAPTEN IMMUNOGLOBULINS

Quantitative precipitin curves obtained with uridine-specific sera and rabbit albumin-conjugates are exemplified in Figure 4. Both diluted anti-

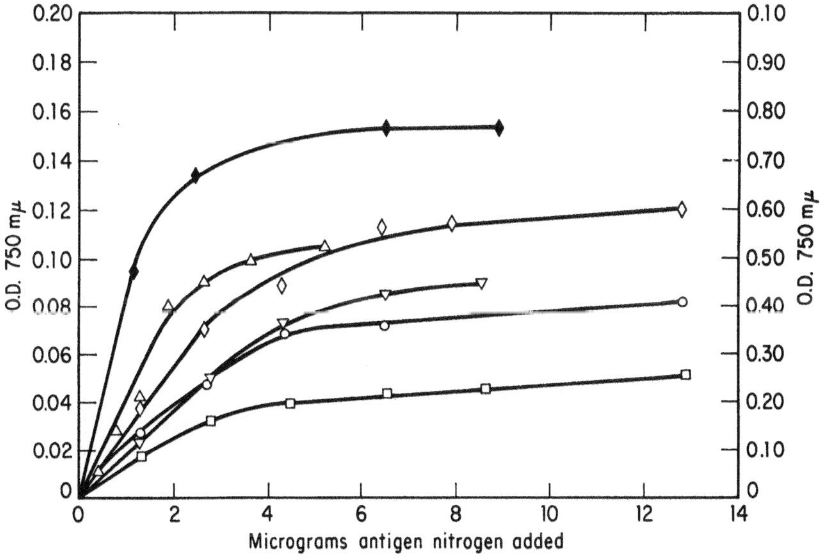

Fig. 4. Precipitin reactions between anti-U solutions and U-RSA. ◆ Anti-U serum (No. 19) diluted 1:2.5 in 0.15 *M* NaCl, right ordinate. ◇ Anti-U globulins from serum No. 19, diluted 1:10 in tris-buffer-saline (TBS), pH 7.5, left ordinate. △ Anti-U serum (No. 16), diluted 1:5 in 0.15 *M* NaCl, left ordinate. ▽ Anti-U serum (No. 24), diluted 1:5 in 0.15 *M* NaCl, left ordinate. ○ Anti-U serum (No. 20), diluted 1:3 in 0.15 *M* NaCl, right ordinate. □ Anti-U serum (No. 22), diluted 1:5 in 0.15 *M* NaCl, right ordinate. U-RSA (concentration determined by the Markham modification of the Kjeldahl procedure) was added to 0.5 ml anti-U solutions. Final reaction volumes were brought to 1.5 ml. The washed specific precipitates were analyzed by the Folin-Ciocalteu method. Negligible precipitation was obtained with these immunoglobulins and RSA.

sera and their sodium sulfate-precipitated globulins reacted with homologous and heterologous antigens to give precipitin curves of the same shape. No indication of inhibition of precipitation in the region of antigen excess was obtained with these immunoglobulins. Anti-ψ-sera and globulins, as well as anti-red.U globulins, precipitated with their corresponding RSA-conjugates (Figs. 5 and 6). Precipitation of an anti-Q globulin with Q-HSA was delineated earlier (Karol and Tanenbaum, 1967). In this instance, the data suggested that antibody precipitation was inhibited in the region of antigen excess. Since negligible reaction occurred between each of the above specific immunoglobulins and the unconjugated carrier protein (RSA or HSA), the observed reactivities of the antibodies indicated specificity for the introduced haptenic groupings.

A more detailed investigation of the specificity of these antibodies was made using the hapten inhibition technique (Table 1A). Amounts of antigen were added to afford maximum antibody precipitation, and results of such studies (Karol and Tanenbaum, 1967) with anti-U preparations showed in all cases that uridine was more effective than the other nucleosides tested. Furthermore, uracil and ribose each possessed little inhibitory ability. Apparently, the entire nucleoside structure is involved

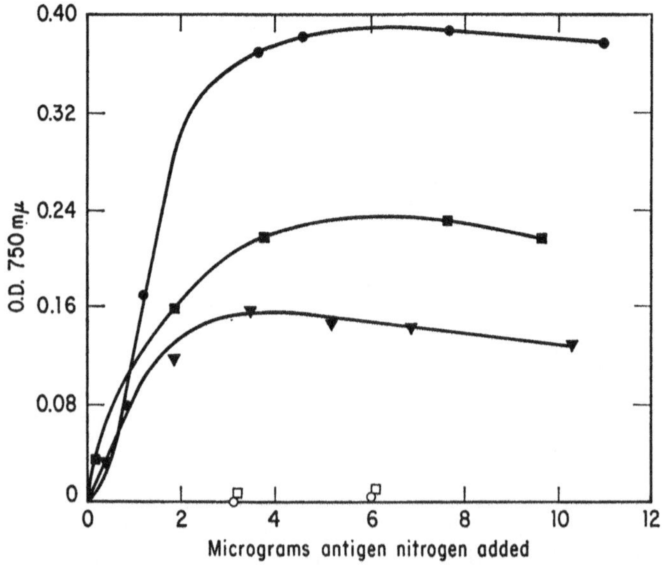

Fig. 5. Precipitin reactions between anti-ψ antibodies and RSA-containing antigens. ● Anti-ψ serum (No. 25), diluted 1:5 in 0.15 *M* NaCl. ■ Anti-ψ globulin from serum No. 25, diluted 1:10 in TBS. ▼ Anti-ψ globulin from serum No. 35, diluted 1:10 in TBS. Increasing amounts of ψ-RSA (concentration determined by the Markham modification of the Kjeldahl procedure) were added to 0.5 ml anti-ψ preparations. Final volume was 1.5 ml. Washed specific precipitates were analyzed by the Folin-Ciocalteu procedure. Corresponding open symbols represent reactions with RSA.

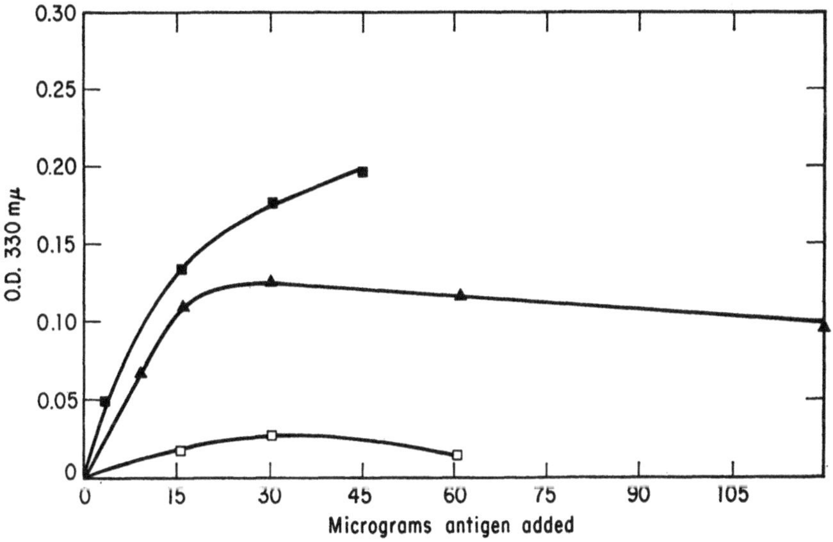

Fig. 6. Precipitin reaction between an anti-red.U globulin fraction (previously absorbed with red.RSA) and red.U-RSA. ▲ Anti-red.U globulins from serum No. 68, 200 μliters of a 1:5 dilution in TBS. ■ 250 μliters of the same globulin dilution. Final reaction volumes were brought to 0.60 ml with TBS. The specific precipitates were analyzed by the microbiuret procedure. Corresponding open symbols represent reactions with red.RSA. The antigen concentrations were determined by direct weighing.

in the U-combining site. The relative importance, as antigenic determinant, of the intact nucleoside as opposed to constituent sugar or base has also been discussed by Ungar-Waron *et al.* (1967). Uridine-5′-COOH was an efficient inhibitor of both the reaction between complementary antisera and U-conjugated proteins and the reaction between BSA and anti-BSA serum. It thus provided no indication of the specificity of anti-U antibodies. A similar finding was reported by Sela *et al.* (1964) with anti-uridine-5′-CO-polypeptide antibodies, and may be related to the findings of Kleinschmidt and Boyer (1952), that heterocyclic carboxylic acids may non-specifically inhibit immune precipitation.

Anti-ψ antibodies were inhibited most efficiently in their precipitation with ψ-RSA by pseudouridine (Table 1B). Since uridine and dihydrouridine were also comparably effective inhibitors in this system, the specificity of anti-ψ antibodies appears to be less restricted than that of anti-U antibodies. Antibodies induced by the red.U-BSA antigen were inhibited almost equally well by dihydrouridine, uridine, and cytidine (Table 1B). Pseudouridine and ribose showed some inhibition in this system; the other haptens tested showed little activity. The unexpected cytidine reactivity, observed with anti-red.U globulins from two different animals, was confirmed in subsequent analyses by C′-fixation and gel

Table 1A

INHIBITION OF QUANTITATIVE PRECIPITIN REACTIONS

	Anti-U					
	Serum No. 24 *		Serum No. 19 *		Globulins from Serum No. 23 (Anti-U_1) †	
Hapten	μmoles	Inhib. (%)	μmoles	Inhib. (%)	μmoles	Inhib. (%)
Uridine	4.9	44	19.3	65	10.0	55
	2.4	38	13.0	51	5.0	62
	1.2	32	9.7	42	1.5	29
			4.8	38		
			3.3	34		
			1.1	13		
Pseudouridine	5.2	25	6.9	0	10.0	16
	2.6	25	3.5	0	5.0	12
	1.3	15	1.7	0	1.2	9
Uridine-5'-PO_4	5.6	29	11.1	26		
	2.8	21	5.6	14		
	1.4	10	2.8	12		
Cytidine	9.5	15	9.5	0	10.0	14
Thymidine	10.7	11	10.7	15	10.0	15
Dihydrouridine					9.9	26
					5.0	22
					1.2	15
Uracil	10.5	10	10.5	0	4.6	9
Ribose	10.2	12	12.2	13	10.0	2

* Total reaction volumes were 1.50 ml containing 0.50 ml antiserum or immunoglobulin dilutions and the amounts of antigen required to give maximum precipitation.

† Total reaction volumes were 0.50 ml containing 0.20 ml immunoglobulin dilutions, and the amounts of antigen required for maximum precipitation.

diffusion with poly C. The inhibition pattern given by anti-Q antibodies, as discussed by Karol and Tanenbaum (1967), indicated that the anti-Q antibodies possess a high degree of specificity for the Q-determinant. Only in 500-fold higher concentration did the pyrimidine nucleosides become effective inhibitors. The finding that Q-tyrosine is the most potent hapten yet tested is in keeping with the presumed presence of monoazotyrosine substitution in Q-HSA. Due to their limited solubilities, larger amounts of purines, or of Q-haptens could not be used in these tests. Increasing the total reaction volumes proved impractical because dilution of the antibody solutions led to decreased antibody precipitation

Table 1B

INHIBITION OF QUANTITATIVE PRECIPITIN REACTIONS

	Anti-ψ				Anti-red.U ‡			
	Globulins from Serum 25 *		Globulins from Serum No. 35 (Anti-ψ_2) †		Globulins from Serum No. 68 (Anti-red.U_1) †		Globulins from Serum No. 64 †	
Hapten	μmoles	Inhib. (%)	μmoles	Inhib. (%)	μmoles	Inhib. (%)	μmoles	Inhib. (%)
Uridine	9.8	18	12.0	24	12.0	16	12.0	30
	4.9	11	5.0	31				
			1.5	16				
Pseudouridine	12.2	26	12.0	44	12.0	12	12.0	20
	6.1	17	5.0	36				
	2.0	4	1.2	23				
Uridine-5'-PO_4	11.6	25						
Cytidine	9.5	3	10.0	15	12.0	25	12.0	33
Thymidine	10.7	0	9.9	1	12.0	2	12.0	8
Dihydrouridine			11.9	20	12.0	20	11.9	34
			5.0	14				
Uracil	10.5	4	5.5	11	5.0	0	5.6	8
Dihydrouracil			5.5	13	5.0	6	5.5	5
Ribose	10.2	0	10.0	0	12.0	9	12.0	15

‡ Globulin preparations were absorbed with red.RSA prior to hapten inhibition reactions.
For other legends, see Table 1A.

with Q-heterologous antigens. However, a control BSA-anti-BSA precipitating system, which contained a comparable amount of antibody, was not inhibited by the Q-haptens.

The specificities of the antibodies discussed above were also studied by the gel diffusion method. Anti-U sera and their globulins precipitated with U-BSA, ψ-BSA, and U-RSA, but not with either RSA, or ψ-RSA (Fig. 7A). The reactivity of these antibodies with U-RSA is therefore a manifestation of their interaction with the uridine determinants and not with the protein backbone. Failure to precipitate with ψ-RSA or with Q-HSA (not depicted) confirms the notion that the entire nucleoside structure is necessary for immunological reactivity of these antibodies. Anti-ψ globulins precipitated with U-RSA, and with ψ-RSA but not with carrier RSA (Fig. 7B). That U-RSA was the more complete antigen of the two conjugates was indicated by the presence of a spur when they were placed in adjacent wells. Anti-Q globulins precipitated

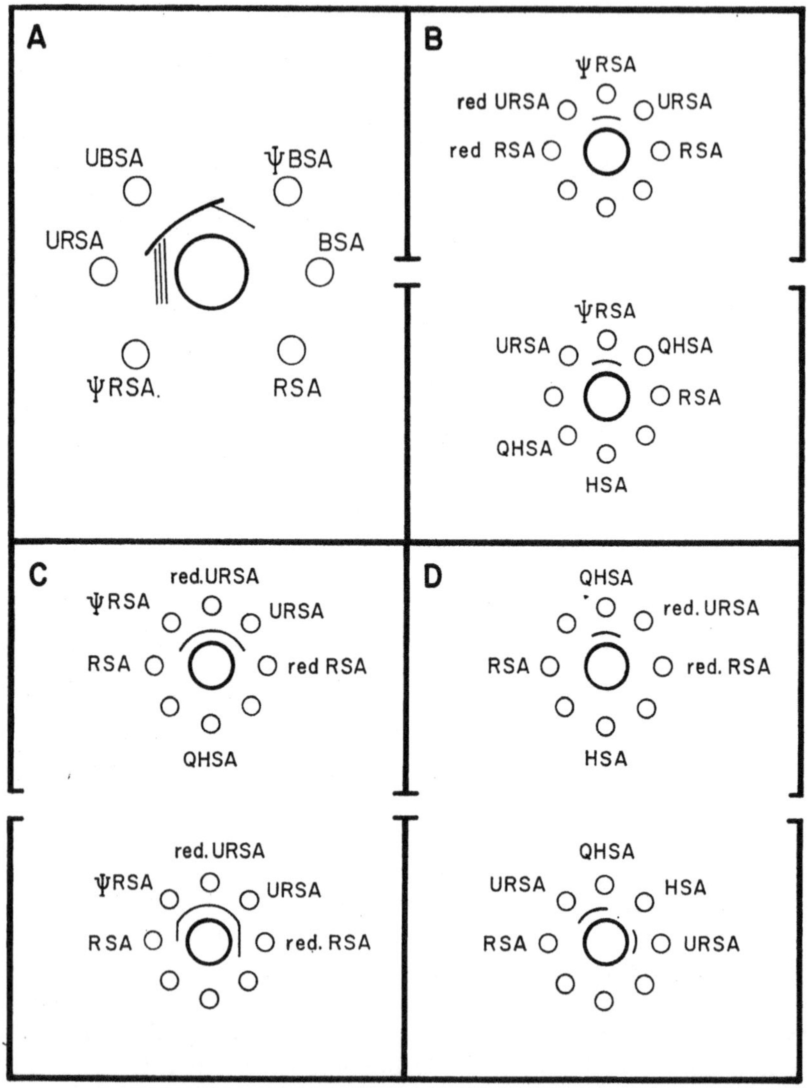

Fig. 7. Gel diffusion analyses of anti-hapten sera, or globulins, with various antigens. **A.** Central well: anti-U serum; peripheral wells: antigens, 1 mg per ml in 0.15 *M* NaCl. **B.** Central wells: anti-ψ globulins; peripheral wells: antigens, 15 mg per ml in TBS. **C.** Central wells: anti-red.U globulins (previously absorbed with red.RSA); peripheral wells: antigens, 15 mg per ml in TBS. Lower pattern depicts the reactions of an incompletely absorbed anti-red.U globulin preparation with the same antigens. **D.** Central wells: anti-Q globulins; peripheral wells: antigens, 15 mg per ml in TBS. All unlabeled wells were filled with TBS.

with Q-HSA but not with HSA. The specificity of anti-Q sera for the uracil moiety was indicated by the reaction of these antibodies with U-RSA (Fig. 7D) and the lack of reaction with ψ-RSA, red.U-RSA, and RSA. The precipitation band formed between these antibodies and Q-HSA appeared to extend beyond that obtained with U-RSA (see lower half of figure). However, little significance is attached to this observation in view of the unusual precipitation behavior of the Q-system and the difficulties encountered in analyses of other anti-hapten systems by the method of gel diffusion (Paul and Benacerraf, 1966a; 1966b).

The photoreduced RSA antigen (red.RSA) precipitated in gel diffusion against anti-BSA and against anti-BSA hapten-conjugated globulins; RSA gave very little or no precipitation under the same conditions. Photochemical reduction of RSA with $NaBH_4$ apparently had caused a change in its antigenic structure. The ultraviolet absorption spectra of irradiated reduced BSA and of RSA indicated peptide bond cleavage (lower absorbance at 230 mμ) and possible exposure of aromatic residues (increased absorbance at 283–295 mμ). Since these altered proteins were less soluble than the native proteins, it is possible that aggregation, breakage and exchange of disulfide bonds (Piras and Vallee, 1966), and formation of amino acid ligands (Smith and Aplin, 1966) resulted during photochemical reduction. For this reason, anti-red.U globulins were absorbed with red.RSA prior to immunochemical analyses. The absorbed anti-red.U immunoglobulins precipitated with U-RSA, ψ-RSA, and red.U-RSA in gel diffusion (Fig. 7C). The band obtained with red.U-RSA extended beyond those obtained with the other RSA conjugates, indicating that here, too, the homologous RSA-conjugate possessed some determinants not present on the other RSA-conjugates.

The results of complement-fixation studies with anti-U and anti-red.U immunoglobulins supported the conclusions regarding antibody specificity based on precipitation data. Thus, curves obtained with an anti-U serum and an anti-red.U globulin preparation are shown in Figure 8. Inhibitions of C′-fixation by haptens revealed furthermore that uridine was an effective inhibitor in the U-RSA-anti-U system, whereas no single hapten effectively inhibited the red.U system. Uridine, dihydrouridine and cytidine each gave approximately 10% inhibition. However, a combination of uridine and dihydrouridine, or of dihydrouridine and cytidine, inhibited the latter up to 30%. These results tend to confirm prior indications that red.U-BSA is an extremely heterogeneous antigen.

III. REACTIONS OF THE ANTIBODIES WITH NUCLEIC ACIDS

A survey of the literature on nucleic acid-antibody interactions (Sela *et al.*, 1964; Panijel *et al.*, 1966; Beiser and Erlanger, 1966) indicated that the experimental demonstration of such reactions might be difficult.

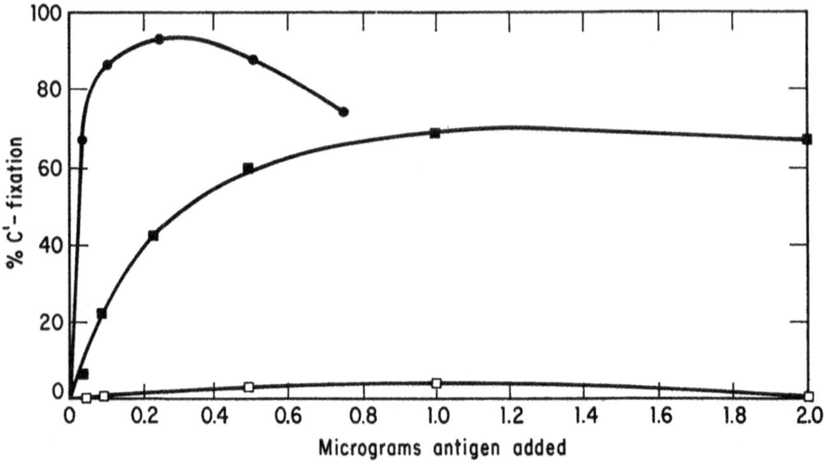

Fig. 8. Complement-fixation of specific antisera, or globulins, with their corresponding heterologous RSA conjugates. ● Anti-U serum (diluted 1:3000) and U-RSA. ■ Anti-red.U globulin fraction (diluted 1:1500, previously absorbed with red.RSA) and red.U-RSA. □ represents reaction of this anti-red.U preparation with red.RSA. At a 1:300 dilution, no C′ was fixed by the anti-U serum when tested with RSA.

For example, precipitation of poly-I with non-immune sera has been reported (Panijel, 1963). Dextran-sulfate and heparin reacted in a similar manner, and these interactions were affected by changes in ionic strength of the sera (Barbu *et al.,* 1964). The ubiquity of nucleases requires that further controls be included since the inhibitory effects of rabbit serum nucleases on precipitation of antibody has been documented (Sela *et al.,* 1964; Ungar-Waron *et al.,* 1967). However, in the present study, removal of nucleases was not obligatory for the manifestation of immunological cross-reactions with informational macromolecules. Moreover, in the preparation of the sodium sulfate-precipitated globulin fraction, up to 90% of the serum nuclease activity was removed.

Commercial preparations of tRNA gave non-specific bands of precipitation in gel diffusion against anti-nucleoside sera or anti-Q immunoglobulins. Therefore, this RNA species was further purified by ion-exchange and molecular-exclusion chromatography to the point where spurious reactions with antibodies no longer occurred. Also, formaldehyde, in a concentration routinely added to the medium during nucleic acid denaturation (Stollar and Grossman, 1962) caused non-specific precipitations with serum globulins. Therefore, all nucleic acid specimens denatured in the presence of this reagent were treated with Sephadex G-25 to effect its removal. This step also served to separate mono- and oligomers whose presence might have caused hapten-inhibition. Using purified materials, the specific globulins reacted with DNA, RNA, and ribohomopolymers as determined by several immunological methods. These reactions were inhibited by nucleosides and by RNA hydrolysates. In particular, chick embryo RNA was specifically precipitated from solution by an anti-Q globulin preparation (Karol and Tanenbaum, 1967).

The precipitin curves obtained with anti-U, anti-ψ, and anti-red.U immunoglobulins and denatured calf thymus DNA are shown in Figure 9. As can be seen, anti-BSA globulins did not react with this DNA sample. With increasing amounts of DNA, the precipitin curve with anti-ψ exhibited a zone of antigen inhibition, while the curves plotted

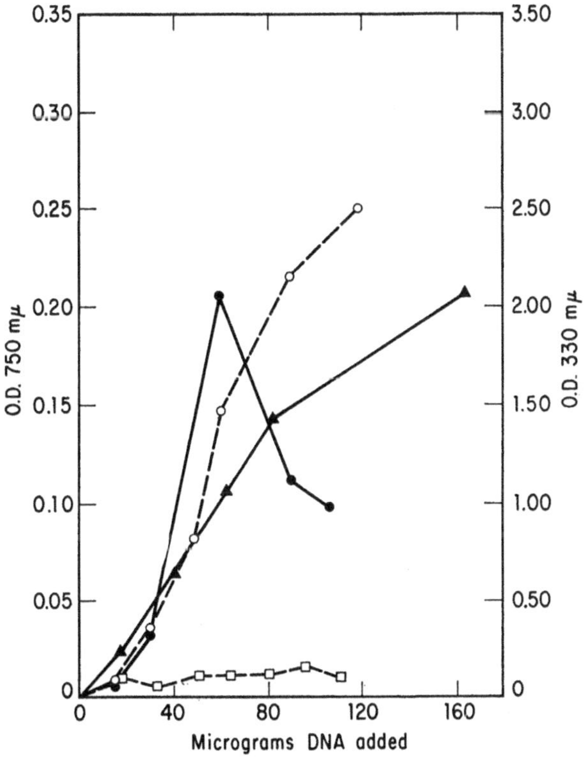

Fig. 9. Precipitin reactions between the globulin fractions of antisera and calf thymus DNA. DNA was thermally denatured in the presence of 1% formaldehyde, then treated with Sephadex G-25. ● Anti-ψ, 125 μliters added per tube. ○ Anti-U, 100 μliters added per tube. □ Anti-BSA, 250 μliters added per tube. ▲ Anti-red.U (previously absorbed with red.RSA), 200 μliters added per tube. To the above specified volumes of globulin preparations was added increasing amounts of denatured DNA. Final reaction volumes were 0.50–0.70 ml. The specific precipitates obtained with anti-U, anti-ψ and anti-BSA globulins were analyzed by the Folin-Ciocalteu method (left ordinate). The precipitates obtained with anti-red.U globulins were analyzed by the microbiuret technique after being quantitatively diluted with 3% NaOH. The right ordinate refers to the corrected values for undiluted anti-red.U globulins.

for the interaction of anti-U or of anti-red.U fail to evince this phenomenon even at relatively high concentrations of antigens. No precipitation was observed in gel diffusion tests between the above globulins and denatured DNA. The reason for this negative result is not apparent.

Anti-U and anti-red.U immunoglobulins fix complement with thermally denatured calf thymus DNA (Fig. 10). Antibodies to the red.U antigen also fixed some complement with unheated DNA. A similar finding was reported for antibodies to 5-acetyluracil-BSA (Tanenbaum and Beiser, 1963) and unheated *E. coli* and *Pneumococcus* DNAs. The preimmunization sera (each diluted 1:700) from the two rabbits immunized with U-BSA and with red.U-BSA, when tested with denatured DNA, fixed a maximum of only 6% of the added complement under identical reaction conditions.

Antibodies with specificity toward U, ψ, Q and red.U showed precipitate reactions in gel (for technics cf. Lacour *et al.*, 1962) with purified formaldehyde-denatured yeast RNA; variable precipitation was obtained with "undenatured" RNA. Typical reactions are shown in Figure 11. In the presence of an alkaline hydrolysate of RNA, inhibition of gel precipitation (Björklund, 1952) of the RNA by anti-U and anti-ψ specific antibodies was observed. With anti-red.U and anti-Q globulins, inhibition was afforded by incorporating an enzymatic RNA core diffusate (Mann)

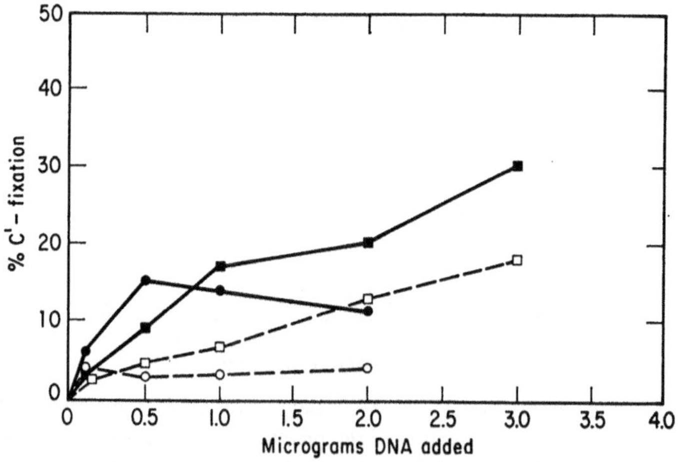

Fig. 10. Complement-fixation reactions of an anti-U serum and of an anti-red.U globulin fraction (previously absorbed with red.RSA) with calf thymus DNA. ● Anti-red.U (1:700) and unheated DNA. ■ Anti-red.U (1:700) and thermally denatured DNA. ○ Anti-U (1:700) and unheated DNA. □ Anti-U (1:700) and thermally denatured DNA. DNA at a concentration of 1 mg per ml was treated with Sephadex G-25. After dilution to 10 μg per ml with TBS, an aliquot was heated at 100° for 10 min., and was immediately cooled in an ice bath.

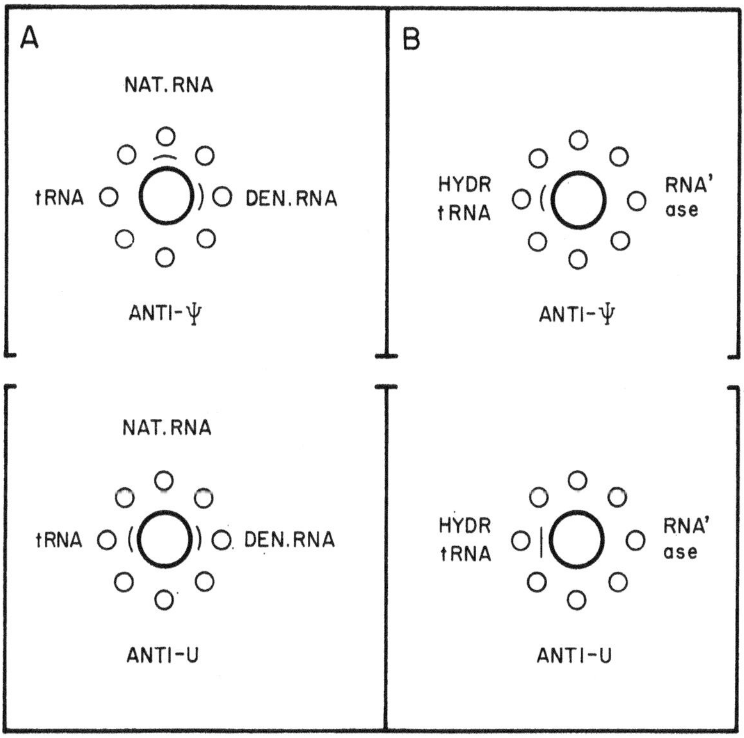

Fig. 11. Gel diffusion analyses of globulins from anti-nucleoside sera and various RNAs. **A.** Top central well: anti-ψ globulins, bottom central well: anti-U globulins. **B.** Top central well: anti-ψ globulins, bottom central well: anti-U globulins. Nat.RNA: yeast RNA, 28 mg per ml TBS, isolated subsequent to Sephadex G-200 chromatography. Den.RNA: formaldehyde-denatured yeast RNA, 20 mg per ml TBS, isolated from Sephadex G-200. tRNA: yeast, purified by DEAE-cellulose and Sephadex G-100 partition chromatography, 24 mg per ml TBS. Hydr. tRNA: yeast tRNA, 24 mg per ml, hydrolyzed with pancreatic RNAase. Unlabeled wells were filled with TBS.

into the gel. The presence of this RNA hydrolysate did not inhibit precipitation in gel of a BSA-anti-BSA system.

Since tRNA contains a higher proportion of "minor" bases than other RNAs, the several specific anti-hapten antibodies were also tested against this macromolecular species. Commercial yeast tRNA, further purified by partition chromatography on DEAE-cellulose (Holley, 1963) and on Sephadex G-100 (Schleich and Goldstein, 1966), precipitated with selected anti-ψ or anti-U immunoglobulins (Fig. 11). However, after treatment with RNAase, the tRNA precipitated with all immunoglobulin preparations as well as with anti-Q and anti-red.U globulins. Under identical experimental conditions anti-BSA globulins and preimmunization globulins did not precipitate in diffusions against this hydrolyzed tRNA. Plescia *et al.* (1965) have previously described C′-fixation between several tRNAs and rabbit antisera to a yeast tRNA-MBSA complex.

Precipitation reactions in gel of each of the specific immunoglobulins with polyribonucleotides can be summarized as follows: all the hapten-specific globulins precipitated with poly X, whereas no precipitation was observed when anti-U and anti-red.U were tested with poly A. Thus, the reaction of these immunoglobulins with purine ribopolymers is a manifestation of their specificity for the diketopyrimidine structure. Although poly U was precipitated in gel with anti-ψ, anti-Q and with anti-red.U globulins, neither of two anti-U preparations could be shown to elicit this cross-reaction. To eliminate the possibility that this failure to react with poly U was due to the presence, in these particular globulins, of RNAase, a highly sensitive radiochemical assay (Spahr, 1964) for the activity of this enzyme was performed. The activities of the two anti-U specific immunoglobulins, as well as immunoglobulin preparations of anti-ψ, anti-Q, and anti-red.U sera which did precipitate with poly U, were analyzed for RNAase. The other immunoglobulins exhibited RNAase activities comparable to those of the anti-U globulins. Thus, failure to observe precipitation of poly-U with anti-U globulins is probably not due to RNAase activity.

Data illustrating the ability of anti-U and of anti-red.U immunoglobulins to fix complement with chromatographically purified yeast and with chick embryo RNAs have been reported (Karol and Tanenbaum, 1967). Anti-red.U antibodies distinguished to a greater degree between thermally denatured and untreated RNAs than did anti-U antibodies. In other experiments, a higher reactivity of anti-U serum with formaldehyde-denatured, than with native, RNA was noted. Thus, antibodies with either of these specificities showed greater reactivity with RNA after it had been denatured. Both anti-U and anti-red.U immunoglobulins fix C′ with a mixture of mono-, di-, tri-, and tetranucleotides (core diffusate, Mann) obtained by the enzymatic hydrolysis of RNA. In addition, these antibodies displayed C′-fixing reactivity with pyrimidine ribonucleotides. For this reason, it was not possible to investigate extensively the specificities of the RNA cross-reactions by hapten inhibition in C′-fixing systems.

IV. EFFECT OF ANTI-HAPTEN ANTIBODIES ON TRANSCRIPTION OF DNA

The availability of anti-purin-6-oyl sera prompted an examination of the possible effects of antibodies with this specificity on biological programming by exposed DNA. It was initially noted (Butler *et al.,* 1962) that anti-purinoyl sera inhibited DNA-coded transformation to streptomycin resistance in the *Pneumococcal* system. Subsequent experiments (Beiser and Galis, unpublished data) have further extended these observations to inhibition of transformation of various markers in competent strains of *Hemophilus* and *B. subtilis* using anti-hapten globulin fractions. Erickson, Braun and Plescia (this Symposium) have also studied the effects of nucleic acid-specific antibodies on bacterial transformation.

An experiment carried out by Bowman and Patnode (1964) demonstrated that the overt DNA of bacteriophage ϕX174, which can mediate the infection of *E. coli* protoplasts, is inhibited in this process by rabbit antibodies to purin-6-oyl-human serum albumin. In one instance, the increase of plaque-formers after incubation of *E. coli* protoplasts with infectious ϕX174 DNA was only 27% of that found in a control which contained normal rabbit serum. Also, the presence of SLE sera diminished the number of plaques to *ca.* 39% of those found in the controls (Bowman, personal communication).

The work of Sinai, Lackman and Cohen (1965), concerning the effect of purin-6-oyl-human serum albumin conjugates on transplanted Ehrlich ascites tumor in mice, suggests that tumor regression which occurs in animals inoculated with this hapten-conjugated protein is due primarily to the production of specific antibodies. This conclusion was further supported by the finding that the antigen itself was not inhibitory to the tumor cells.

Rabbit antisera to purin-6-oyl, 5-acetyluracil, "adenylic acid," and polyuridylic-conjugated albumins, were each shown to inhibit the development of fertilized *Arbacia punctulata* eggs (Rosenkranz *et al.*, 1964). In this work, it was first determined that fluorescein-tagged anti-purinoyl globulin was irreversibly fixed by the sea urchin egg, whereas non-specific fluorescein-labeled rabbit globulin, although taken up by the cell, could be readily washed out. Antisera with the above-mentioned specificities were then shown to stop embryological development in this test system. The stage of arrestment was determined, in part, to be a function of antiserum dilution as well as the specificity of the particular antibody used as inhibitor. As a control, an inhibitory antiserum was absorbed with antigen. The resulting loss of the inhibitory activity minimizes the possibility that inhibition might have been due to the presence of a non-specific toxic factor in rabbit serum.

Since it had been established by *in vitro* experiments that all but one of the anti-hapten antisera used in this work could react with denatured, but not with native, DNA or with RNA, it was hypothesized that the molecular biological site of inhibition of fertilized sea urchin egg development by these antibodies was interference with the functioning of single-stranded DNA. Mention should also be made of a related application of similar anti-hapten antibodies as metabolic probes in cytological investigations. Thus, Klein *et al.* (1967) have presented evidence that fluorescein-labeled "anti-nucleoside" immunoglobulins, which also react selectively with denatured DNA, gave patterns of nuclear fluorescence in mouse L cells. This phenomenon was observed only with populations of cells taken at times when DNA synthesis was occurring maximally. These data also suggested the notion that the active molecular species inside the mouse cell nucleus which bound these antibodies was single-stranded DNA.

Recent work by Wallace *et al.* (1966) indicated that globulin fractions of sheep antisera against "nucleosides" (i.e., thymine-, or guanine-morpholino-albumins), made by the procedure of Erlanger and Beiser (1964), inhibited the incorporation of tritiated thymidine into an enzymatic product primed by denatured DNA. Anti-thymidine antibody was twice as effective as anti-guanosine antibody; control sheep anti-BSA sera were inactive. Since many SLE sera mimic the cross-reactions of anti-hapten immunoglobulins with informational macromolecules, it is of interest that Williams and Bollum (1963) had shown earlier that selected SLE sera inhibit *in vitro* DNA synthesis by DNA polymerase. These inhibitions required preincubation of unfractionated antisera with the thermally denatured calf thymus DNA primer. Control normal sera gave no inhibition.

V. ACTIVITY OF RNA-REACTIVE ANTIBODIES IN AN RNA TRANSLATION PROCESS

Ribosomal and transfer RNAs have now been proven to be intimately involved in the recognition and translation of genetic information (cf. Watson, 1965). If the anti-hapten antibodies, whose specificities have been described above, can selectively recognize their complementary sites on intact RNAs, it was thought that these antibodies might be used to provide an insight into the nature of potentiating regions of these macromolecules which are involved in the transmission of biological information. To test this hypothesis, a biological system with poly A as the "messenger" RNA was considered first. The rationale for this experiment was based on the previous findings that none of the nucleoside-directed antibodies displayed reactivity with poly A and that adenine as a hapten was not inhibitory when tested in heterologous cross-reactions with these antibodies. Any observable effect of specific antibodies in a protein-synthesizing system based on poly-A would therefore be a manifestation of interactions of these antibodies with ribosomal RNA or with tRNA whose biologically active sites comprise nucleotides other than adenosine-phosphate.

In parallel experiments, poly U was also used to direct protein synthesis, since all four types of antibodies had uridine, or uracil specificity.

These experiments served to increase the probability of defining antibody effects on the same biosynthetic system when "messenger" RNA, in addition to ribosomal and transfer RNAs, was amenable to reactivity. Furthermore, reaction of these antibodies with denatured DNA demanded that the enzyme systems which were used as the source of tRNA and of ribosomal RNA contain little DNA. For the above reasons, the *in vitro* protein-synthesizing system of Matthaei and Nirenberg (1961), as modified by Holland and McCarthy (1964, 1965, 1966) for their demonstration of the coding properties of DNA and RNA in the presence of antibiotics, was chosen. Quantitative analyses showed that each ml of enzyme solution contained 9.5 mg RNA, 0.42 mg DNA, and 9.3 mg protein.

Globulin fractions of antisera to U-BSA, ψ-BSA, Q-BSA, and red.U-BSA, after concentration by sodium sulfate precipitation (Strauss *et al.*, 1960), were dialyzed against the Matthaei-Nirenberg buffer (low magnesium). One anti-U serum (No. 19) was first chromatographed on DEAE-cellulose (Rapp, 1964) before its globulins were concentrated by the sodium sulfate technique. After dialysis for 24 hours, the solutions were found to contain between 20–35 mg protein per ml (assuming E_{280} mμ = 1.6×10^3 for rabbit γ-globulin).

The experimental protocol was designed to permit preliminary contact of specific antibodies with RNA. Thus, in amino acid incorporation experiments directed by poly A, the hapten-specific antibodies were preincubated with the S_{30} extract (the source of transfer and ribosomal RNAs). Similarly, in the parallel series, preincubation of antibodies with poly U served to permit observation of antibody interaction with messenger RNA. After a specified time interval, these "antibody-RNA" mixtures were introduced into the protein synthesizing systems which contained salts, buffer, and energy sources. Isotopically labelled lysine (poly A system) or C^{14}-phenylalanine (poly U system) was added last. These complete reaction mixtures were then incubated at 37°. To terminate the syntheses, cold amino acid (0.5 μmoles C^{12}-lysine or phenylalanine), carrier BSA, and trichloroacetic acid (TCA) were added. The resulting precipitates were extracted with hot TCA, were collected by filtration, and were then counted in a liquid scintillation spectrometer.

C^{14}-amino acid incorporation in the presence of poly A or of poly U is shown in Figure 12A. Maximum uptake into poly-amino acid was afforded by 20 μg. of each ribopolymer. Endogenous radioisotope incorporations were approximately one-tenth the maximum values obtained in the presence of the polymers. Using 20 μg. of poly A, C^{14}-lysine uptake was found to vary directly with the amount of S_{30} extract added (Fig. 12B). In the area of high incorporation, the slope of the curve varied sharply with small additions of extract; thus, 50 μliters of S_{30} was selected

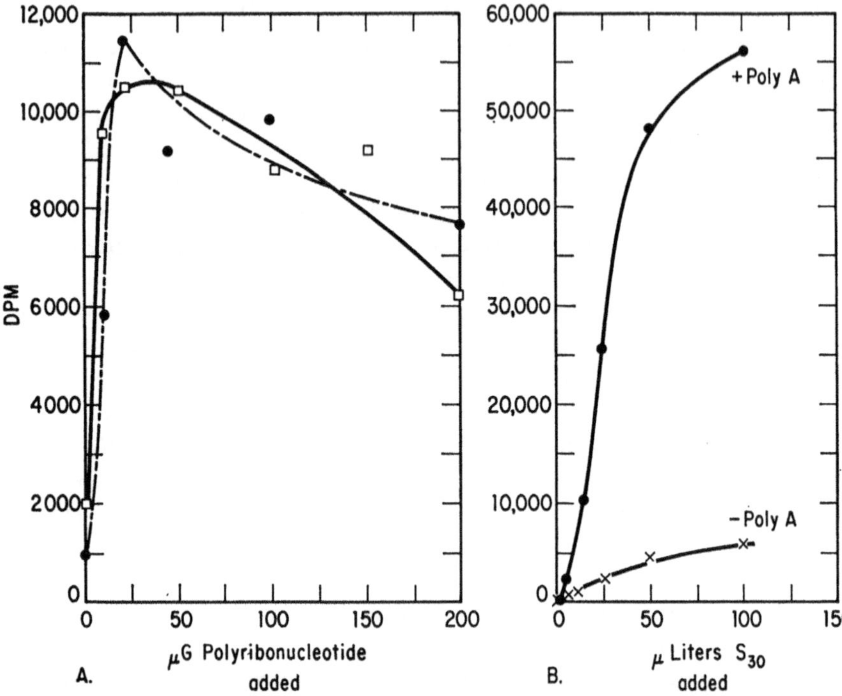

Fig. 12. A. Effect of increasing amounts of poly A, or of poly U, on C^{14}-amino acid incorporation. □ poly A (left ordinate). ● poly U (right ordinate). Each reaction mixture contained 50 μliters S_{30}. **B.** Effect of increasing amounts of S_{30} on C^{14}-lysine uptake. In the experiments which included poly A, 20 μg was present.

for subsequent experiments. The kinetics of C^{14}-lysine incorporation into polypeptide are depicted in Figure 13. In the presence of poly A there is an initial rapid incorporation which levels off and eventually decreases. By 12 minutes, approximately 90% of maximum incorporation has taken place. Synthesis under the direction of endogenous "messengers" displays a slower rate of increase which also levels off after 30 minutes.

The action of anti-U, anti-ψ, anti-red.U, and anti-Q on C^{14}-lysine incorporation is depicted in Figure 14. There appeared to be a causal relationship between increasing amounts of specific immunoglobulin and inhibition of protein synthesis. However, the nature of the antibody specificity was unrelated to the degree of inhibition obtained. Thus, two anti-U preparations of comparable antibody content fell into the bottom and top of the spectrum of inhibitory activity. The effects of anti-U, anti-ψ, anti-Q, and anti-BSA globulins on poly-U-directed C^{14}-phenylalanine polycondensation is depicted in Figure 15. These results resemble those obtained in the poly A-directed system. Among the specific immunoglobulins examined, the anti-U preparations were again at the extremes in

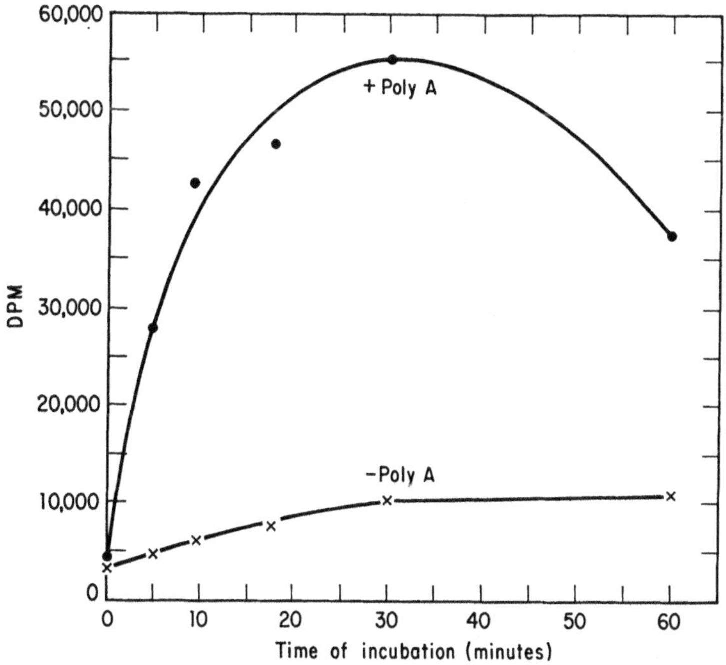

Fig. 13. Rate of incorporation of C^{14}-lysine into polypeptide. In those experiments employing poly A, 20 μg was present.

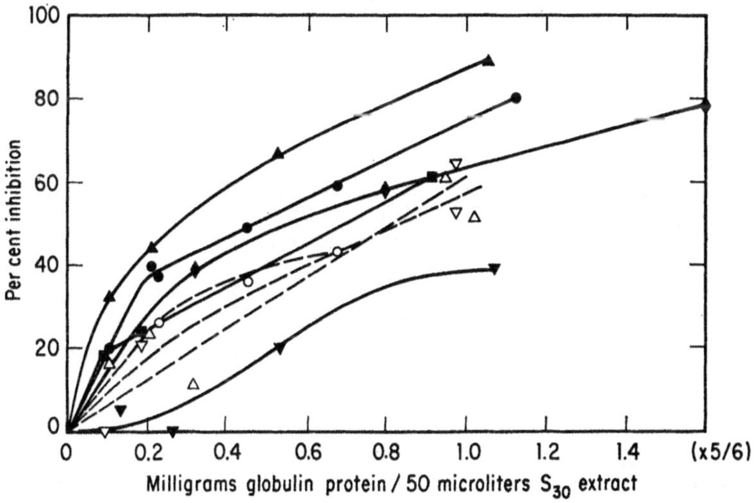

Fig. 14. Effect of immunoglobulin on C^{14}-lysine uptake. Globulins (60 μliters) were preincubated for 25 min. at 0° with an equal volume of the S_{30} extract. A 100 μliter aliquot of this mixture was then added to the protein-generating systems. ▲ Anti-U_1. ▼ Anti-U (No. 19). ◆ Anti-ψ_2. ■ Anti-Q_1. ● Anti-red.U. Open symbols connected by broken lines represent reactions of anti-BSA globulins.

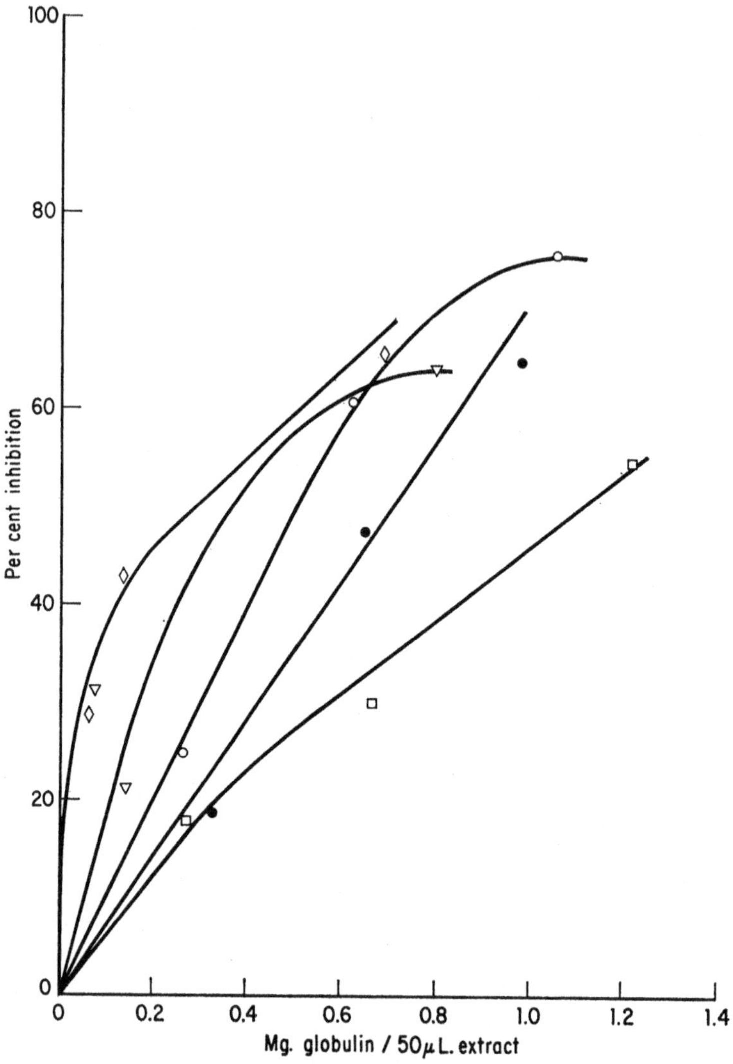

Fig. 15. Effect of immunoglobulin on C^{14}-phenylalanine incorporation. Globulins were preincubated with poly U for 25 min. at 0°C. An aliquot of this mixture, which contained the amount of poly U required for maximum poly-phenylalanine formation, was added to the protein-synthesizing system. ○ Anti-U_1. □ Anti-U (No. 19). ◇ Anti-ψ_2. ▽ Anti-Q_1. ● Anti-BSA.

inhibitory ability, and anti-BSA preparations were equally as effective in inhibiting C^{14}-uptake as were anti-U and anti-Q. Anti-BSA globulins were further tested for their effects on protein synthesis directed by endogenous "messengers." The results of such experiments (Table 2) indicated 38% inhibition of endogenous incorporation, and 36% inhibition of poly A-directed synthesis. Thus, anti-BSA globulins had an effect on

Table 2

EFFECT OF ANTI-BSA GLOBULINS ON *in vitro* PROTEIN BIOSYNTHESIS IN THE PRESENCE AND ABSENCE OF POLY A

Tube	C^{14}-lysine	S_{30} (μliters)	Poly A	Anti-BSA	CPM	Inhibition, Per Cent
1	0.08 μc	50	—	—	2116	
2	0.08 μc	50	20 μg.	—	34900	
3	0.08 μc	50	—	0.67 mg	804	38
4	0.08 μc	50	20 μg.	0.67 mg	12700	36

Preincubation of the S_{30} with anti-BSA globulins, or with buffer, was for 25 min. at 0 deg.

the components of this system roughly equal to that of the specific RNA cross-reactive antibodies.

In precipitin and C′-fixation assays, in which measurable interaction between these same hapten-specific immunoglobulins and RNA were observed, the weight ratio of reactants was at least 10:1. The protocols in the *in vitro* protein-forming system were therefore adjusted to resemble more closely those which evinced optimal immunological manifestation. These changes were accomplished by reducing from 50 μliters to 10 μliters the amount of S_{30} fraction which was preincubated with antibody solutions. Although the immunoglobulins were now found to inhibit radioisotope incorporation some 70–95%, once again anti-BSA was an effective inhibitor as were the anti-hapten preparations. It can be concluded from all these data, contrary to expectations, that little specific effect of the RNA cross-reactive antibodies in this protein-synthesizing system was evident.

VI. CONCLUDING REMARKS

The experiments outlined in the preceding sections have validated the premise that the induction of antibodies against certain relatively simple hapten-conjugated proteins, of the type pioneered by Landsteiner (1946), can evoke a population of immunoglobulins, a portion of which can react with RNA. However, the goal of demonstrating a restricted specificity for RNA remains unrealized. Although all four types of protein-conjugates contained uracil as part of the haptenic moieties, cross-reactions with DNA were also found. Apparently, in an antigenic sense though obviously not in a replicative one, there is little discrimination at the macromolecular level between uridine and thymidine, in terms of either base or of pentose structure. Ungar-Waron and coworkers (1967) reported that the reaction of antibodies elicited by uridine-5′-CO-polypeptides with homologous antigen is inhibited by uridine and ribosylthymine to almost the same extent; whereas thymidine and deoxyuridine inhibit the reaction less. Despite this apparent difference in reactivity between ribose and deoxyribose, these antibodies also precipitated with denatured DNA.

Related inexplicable facets of our studies are the observations that antibodies to pseudouridine, a nucleoside positional isomer predominantly found in tRNA, reacted with denatured DNA, and that antibodies to quinazoline-2,4-dione-6-azo-conjugates, although appearing quantitatively to cross-react better with RNA, also reacted with denatured DNA. Clues to the nature of the complementarity responsible for these phenomena cannot be found in the limited hapten-inhibition patterns (Tables 1A, 1B; Karol and Tanenbaum, 1967) which have been accumulated.

Despite the fact that results thus far obtained indicate clearly the difficulty in producing truly selective anti-RNA antibodies by the use of hapten-conjugates as immunogens, the facile synthesis, e.g., of anti-Q proteins, provides a generally available source of antigens for RNA-reactive, if not specific, immunoglobulins. The initial attempt, herein described, to demonstrate a biochemical application for these antisera has perhaps raised more questions than it has answered. However, inherent risks in the experimental design included: (a) the use of a complex system which involved tRNA, messenger RNA, and ribosomal RNA in the programing process; (b) the known presence of a small amount of DNA in this crude enzyme protein-biosynthesizing system; and (c) the employment of immunoglobulin fractions as opposed to highly purified antibodies. Since these anti-hapten antibodies, whose structures are complementary to well-defined chemical groups on the immunogen, are particularly amenable to purification by hapten-dissolution and chromatographic (Bassett *et al.*, 1961) or specific immunoadsorption (Jaton *et al.*, 1967) technics, such variables as non-specific serum protein-polynucleotide interactions (Panijel, 1963; Panijel *et al.*, 1966) or other adventitious interfering serum factors could be eliminated in this manner. The choice of an RNA informational system, which could be prepared free of DNA, and which optimally, though not necessarily, would depend upon the function of single-stranded RNA, might also be expected to favor the demonstration of a role for these antibodies as a tool at the molecular biological level of organization. Candidate systems for such investigations might include those discovered by Spiegelman and his group (1967), in which the effect of highly purified RNA-reactive antibodies on activity and template specificity of Qβ replicase, or upon efficiency of protoplast infection by RNAs, could be examined.

References

Barbu, E., J. Dandeu, and G. Quash (1964). Immunochimie. Influence de la force ionique sur les réactions des anticorps anti-RNA avec divers polyélectrolytes. C. R. Acad. Sci., *258*:1930–1933.

Bassett, E. W., S. M. Beiser, and S. W. Tanenbaum (1961). Purification of antibody to golactosyl-protein conjugates. Science, *133*:1475–1476.

Beiser, S. M., S. W. Tanenbaum, and B. F. Erlanger (1964). Purine- and pyrimidine-specific antibodies: precipitation with denatured deoxyribonucleic acid. Nature, *203*:1381–1382.

Beiser, S. M., and B. F. Erlanger (1966). Antibodies which react with nucleic acids. Cancer Res., *26*:2012–2017.

Björklund, B. (1952). Specific inhibition of precipitation as an aid in antigen analysis with gel diffusion method. Proc. Soc. Exp. Biol. Med., *79*:319–324.

Boissonnas, R. A. (1951). Une nouvelle méthode de synthèse peptidique. Helv. Chim. Acta, *34*:874–879.

Bowman, B. U., and R. A. Patnode (1964). Further studies of a protoplast-infecting agent from antiserum treated bacteriophage ϕX174. Federation Proc., *23*:295.

Butler, V. P., S. M. Beiser, B. F. Erlanger, S. W. Tanenbaum, S. Cohen, and A. Bendich (1962). Purine-specific antibodies which react with deoxyribonucleic acid. Proc. Nat. Acad. Sci. (U. S.), *48*:1577–1602.

Butler, V. P., S. W. Tanenbaum, and S. M. Beiser (1965). A study of the cross-reactivity of anti-purin-6-oyl serum with deoxyribonucleic acid. J. Exp. Med., *121*:19–38.

Cerutti, P., K. Ikeda, and B. Witkop (1965). The selective photoreduction of uridine in polynucleotides. J. Am. Chem. Soc., *87*:2505–2507.

Erlanger, B. F., and S. M. Beiser (1964). Antibodies specific for ribonucleosides and ribonucleotides and their reaction with DNA. Proc. Nat. Acad. Sci., *52*:68–74.

Halloran, M. J., and C. W. Parker (1966a). The preparation of nucleotide-protein conjugates: carbodiimides as coupling agents. J. Immunol., *96*:373–378.

Halloran, M. J., and C. W. Parker (1966b). The production of antibodies to mononucleotides, oligonucleotides and DNA. J. Immunol., *96*:379–385.

Holland, J. J., and B. J. McCarthy (1964). Stimulation of protein synthesis *in vitro* by denatured DNA. Proc. Nat. Acad. Sci. (U. S.), *52*:1554–1561.

Holland, J. J., C. A. Buck, and B. J. McCarthy (1966). Stimulation of protein synthesis *in vitro* by partially degraded ribosomal ribonucleic acid and transfer ribonucleic acid. Biochemistry, *5*:358–365.

Holley, R. W. (1963). Large scale preparation of yeast "soluble" ribonucleic acid. Biochim. Biophys. Res. Comms., *10*:186–188.

Holley, R. W., J. Apgar, G. A. Everett, J. T. Madison, M. Marquisee, S. H. Merrill, J. R. Penswick, and A. Zamir (1965). Structure of a transfer ribonucleic acid. Science, *147*:1462–1465.

Jaton, J. C., H. Ungar-Waron, and M. Sela (1967). Comparison of IgG and IgM antibodies specific for uridine. European J. Biochem., *2*:106–114.

Karol, M., and S. W. Tanenbaum (1967). Antibodies to conjugated proteins which cross-react with RNA. Proc. Nat. Acad. Sci. (U. S.), *57*:713–720.

Kleinschmidt, W. J., and P. D. Boyer (1952). Interaction of protein antigens and antibodies. J. Immunol., *69*:247–271.

Klein, W. J., S. M. Beiser, and B. F. Erlanger (1967). Nuclear fluorescence employing antinucleoside immunoglobulins. J. Exp. Med., *125*:61–70.

Lacour, F., J. Harel, L. Harel, and E. Nahon (1962). Étude immunologique de l'acide ribonucleique par la micromethode de double diffusion dans l'agar. C. R. Acad. Sci., *255*:2322–2324.

Lacour, F., and L. Harel (1965). Réaction des anticorps, anti conjugué adénosine monophosphate-sérum albumine bovine, avec les polynucléotides. Biochim. Biophys. Acta, *103*:358–360.

Landsteiner, K. (1946). "The specificity of serological reactions," Cambridge, Mass. Harvard University Press.

Levine, L., and H. Van Vunakis (1966). Serologic activities of the nucleic acids. Proc. 2nd Meeting Fed. European Biochem. Soc., Oxford-New York. Pergamon, 25–60.

Matthaei, J. H., and M. W. Nirenberg (1961). Characteristics and stabilization of DNAase-sensitive protein synthesis on *E. coli* extracts. Proc. Nat. Acad. Sci. (U. S.), *47*:1580–1588.

McCarthy, B. J., and J. J. Holland (1965). Denatured DNA as a direct template for *in vitro* protein synthesis. Proc. Nat. Acad. Sci. (U. S.), *54*:880–886.

Moss, G. P., C. B. Reese, K. Schofield, R. Shapiro, and Lord Todd (1963). Nucleotides. Part XLVII. The catalytic oxidation of nucleosides and nucleotides: A projected stepwise degradation of polynucleotides. J. Chem. Soc., 1149–1154.

Panijel, J. (1963). Anticorps anti-poly I spécifiques et anticorps anti-RNA dans les sérums de subjets immunisés ou non contre les ribosomes. C. R. Acad. Sci., *257*:798–804.

Panijel, J. P., Cayeux, E. Sacquet, and H. Charlier (1966). Antiribonucleic acid factors on the sera of germ-free animals. Nature, *211*:94–96.

Paul, W. E., and B. Benacerraf (1966a). Problems encountered in double diffusion analysis in agar of hapten specific immune systems. I. Complement dependent precipitation. J. Immunol., *95*:1067–1073.

Paul, W. E., and B. Benacerraf (1966b). The effect of the relative solubility of the immune complexes formed by cross-reacting systems. J. Immunol., *95*: 1074–1079.

Piras, R., and B. Vallee (1966). Structural changes of carboxypeptidase A on ultraviolet irradiation. Biochemistry, *5*:855–860.

Plescia, O. J., N. Palczuk, E. Cora-Figueroa, A. Mukherjee, and W. Braun (1965). Production of antibodies to soluble RNA. Proc. Nat. Acad. Sci. (U. S.), *54*:1281–1285.

Plescia, O. J., and W. Braun (1967). Nucleic acids as antigens. Adv. in Immunol., *6*:231–252, New York, Academic Press.

Rapp, H. J. (1964). Immunological methods, J. F. Ackroyd, ed. Philadelphia, Davis, 18.

Rosenkranz, H. S., B. F. Erlanger, S. W. Tanenbaum, and S. M. Beiser (1964). Purine- and pyrimidine-specific antibodies: effect on the fertilized sea urchin egg. Science, *145*:282–283.

Schleich, T., and J. Goldstein (1966). Gel filtration heterogeneity of *Escherichia coli* soluble RNA. J. Mol. Biol., *15*:136–146.

Sela, M., and R. Arnon (1960). Studies on the chemical basis of the antigenicity of proteins. Biochem. J., *77*:394–399.

Sela, M., H. Ungar-Waron, and Y. Schechter (1964). Uridine-specific antibodies obtained with synthetic antigens. Proc. Nat. Acad. Sci., *52*:285–292.

Sela, M., and H. Ungar-Waron (1965). Nucleoside-specific antibodies elicited by synthetic antigens. Federation Proc., *24*:1438–1445.

Sinai, Y., C. Lachman, and S. Cohen (1965). Effect of purinoyl human serum albumin on transplanted Ehrlich ascites tumour in mice. Nature, *205*:192–193.

Smith, K. C., and R. T. Aplin (1966). A mixed photoproduct of uracil and cysteine (5-S-cysteine-6-hydrouracil). A possible model for the *in vivo* cross-linking of deoxyribonucleic acid and protein by ultraviolet light. Biochemistry, *5*:2125–2130.

Spiegelman, S. (1967). *In vitro* synthesis of viral RNA. Chem. and Eng. News. *Aug. 7th*:151–156.

Stollar, D., and J. Grossman (1962). The reaction of formaldehyde with denatured DNA: spectrophotometric, immunologic, and enzymatic studies. J. Mol. Biol., *4*:31–38.

Strauss, A. J. L., B. C. Seegal, K. C. Ksu, P. M. Burkholder, W. L. Nastuk, and K. E. Osserman (1960). Immunofluorescence demonstration of a muscle binding, complement-fixing serum globulin fraction on myasthenia gravis. Proc. Soc. Exp. Biol. Med., *105*:184–191.

Tanenbaum, S. W., and S. M. Beiser (1963). Pyrimidine-specific antibodies which react with deoxyribonucleic acid. Proc. Nat. Acad. Sci. (U. S.), *49*:662–668.

H. Ungar-Waron, E. Hurwitz, J. C. Jaton, and M. Sela (1967). Antibodies elicited with conjugates of nucleosides with synthetic polypeptides. Biochim. Biophys. Acta, *138*:513–531.

Wallace, S., B. F. Erlanger, and S. M. Beiser (1966). Purine and pyrimidine nucleoside specific antibodies: Effect on *in vitro* priming ability of DNA. 2d Int. Biophys. Congress (IOPAB), Vienna, 171.

Watson, J. (1965). "Molecular Biology of the Gene." New York: Benjamin.

Williams, A. M., and F. J. Bollum (1963). Inhibition of enzymatic DNA synthesis by certain pathological human sera. Proc. Soc. Exp. Biol. Med., *112*: 701–706.

USE OF FLUORESCEIN-LABELLED ANTI-NUCLEAR ANTIBODIES IN THE STUDY OF CHROMOSOMAL STRUCTURE *

LAWRENCE RAZAVI

Channing Laboratory, Boston City Hospital; and Department of Bacteriology and Immunology, Harvard Medical School, Boston, Mass.

Chromosomes contain four types of macromolecule: deoxyribonucleic acid (DNA); ribonucleic acid (RNA); a basic protein, histone; and acidic so-called "residual protein." It is likely that these macromolecules are distributed in a manner allowing suitable genetic and metabolic activity during the cell cycle, by interaction with divalent cations, lipids and the DNA polymerase which are also found in chromosomes (Swanson *et al.*, 1967). Unfortunately, conventional histochemical and microscopical techniques have not proved specific or sensitive enough to resolve, without controversy, the structure of chromatin in a way that relates to its function (Gall, 1954 and 1963; Ris, 1961; Taylor, 1963). On the other hand, suitably labelled anti-nuclear antibodies offer themselves as an alternative histochemical reagent. Thus, fluorescein-labelled anti-nuclear sera from natural disease have been used by Krooth *et al.* (1961), Beck (1962b), Vaccaro *et al.* (1965), and Rapp *et al.* (1962) to trace, immunohistologically, the pattern of several nuclear components during the cell cycle. The interesting results of these investigators proved the feasibility of using antibodies for cytological analysis. At the same time, there has been sufficient variability in the reproducibility of the techniques and results to warrant cautious interpretations by those who used these techniques.

Krooth *et al.* (1961) used six sera from patients with Lupus erythematosus (LE), three from Sjögren's Syndrome (SS) and two from nephrosis, in a double layer ("sandwich") immunofluorescence technique. All the sera reacted with acetone-fixed mouse liver interphase nuclei, but the SS

* Supported by Grants from U.S.P.H.S. (HD 01288) and the Massachusetts Heart Association, Inc.

sera, one LE and one nephrosis serum failed to react with Carnoy-fixed metaphases from human blood and skin, Hela (S_3 subline) and Chinese hamster cells. The other sera attached to all chromosomes, but in the pictures published by these authors there is considerable variability of uptake from chromosome to chromosome, with a tendency for the attachment to occur at the perimeter of each chromosome and, particularly densely, *between* nearby chromosomes. That is, positive reactions occurred between as well as within chromosomes. The authors concluded that the technique would be more useful for chromosome analysis if the sera used were better defined.

Beck (1961a and b, 1962a), using LE and SS sera on unfixed mouse liver interphase nuclei, described reactions of "homogeneous," "speckled" and "nucleolar" patterns against deoxyribonucleoprotein, a phosphate buffered saline-extractable protein, and an unidentified constituent of nucleoli, respectively. Using mitotic Hela cells (1962b), he obtained evidence that deoxyribonucleoprotein was confined to chromosomes throughout mitosis; buffer-extractable protein was present between chromosomes and in the cytoplasm, and nucleolar substance adhered to early prophase and telophase chromosomes, but passed into the cytoplasm during metaphase. He found that these patterns were only "fairly reproducible," and noted that the cold acetone fixation he used affected nuclear structure in a critical but uncertain fashion.

Rapp (1962) used LE serum on Hela cells. He found that acetone, methanol and ethanol fixation gave poor results; he preferred air-dried or mildly heat-fixed cells, although the degree of heating affected the patterns seen. No reaction occurred if viable cells with intact membranes were used. He described chromosomal and nuclear membranous attachments in mitosis. Frequently both patterns were seen, in addition to cytoplasmic reactions.

No conclusions were made by these authors about chromosomal behavior. Similar experiments made in this laboratory suggest that this omission was warranted because interpretations depend on several uncertain factors. The antisera were only partially specified and, coming from natural diseases, contained an unknown number of different antibodies, including the precipitins (Deicher, 1959; Tan *et al.*, 1966a, b) which are used in immunofluorescence. The resolution of natural antisera is inadequate because the usefulness of interesting antibodies (or removal of unwanted antibodies) is limited if their complementary antigens are unknown. In most cases, the cells used were aneuploid, and since the behavior and composition of their chromatin is neither stable nor predictable, they cannot be used as reliable substrates. Finally, the unsynchronised metabolism and the uncertain effect of fixation from cell to cell made it difficult to rely on the observed patterns as true or representative.

The observations described here were the result of experiments designed (1) to limit nuclear fluorescence to interactions between specific constituents of the nucleus and specific factors in the antiserum, and (2) to permit the attribution of variable fluorescence to natural changes in the site and concentration of nuclear antigens as a function of nuclear metabolism. The results provide evidence for the presence of single-stranded segments of deoxyribonucleic acid (DNA) both in mitotic chromosomes and in interphase nuclei, and suggest that, in lymphocytes, chromosomal genetic activity continues throughout the cell cycle.

I. MATERIALS AND METHODS

A. Cell Preparations

Preliminary experiments used as substrate intact white cells from mixed and unsynchronized short-term primary cultures, washed in globulin-free serum. Smears were made on slides, applying the cells with a glass rod in a spiral fashion. The resulting drying gradient of cells was fixed in cold acetone for five seconds. The cells were then washed in phosphate-buffered saline (pH 7.2, 0.15M–"PBS") for five minutes at room temperature and stained immediately.

Osmotically isolated interphase and metaphase nuclei of lymphocytes were prepared from partially synchronized phytohaemagglutinin-stimulated cultures (Fig. 1) at 45 to 50 hours. The volume of distilled water

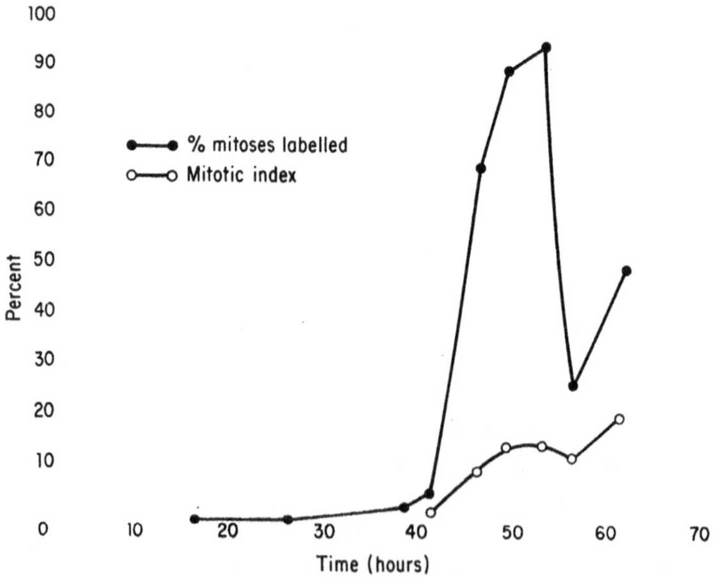

Fig. 1. Rate of uptake of tritiated thymidine and mitotic index of lymphocytes held at room temperature from the eighth to the twentieth hour of primary culture. (Reproduced from Razavi, 1967, by permission of the Editor of Nature.)

required to dilute cultures to 100 milliosmoles was determined in triplicate (separate aliquots) for each batch of medium using an "Advanced Instrument" freezing-point osmometer. After hypotonic swelling for 20 minutes, the cells were centrifuged at 500 rpm for five minutes (International Centrifuge rotor head No. 221) and the supernate was decanted. The cells were then gently agitated in an iced water bath while 2 ml of ice-cold Carnoy's fixative (three parts methanol, one part glacial acetic acid) was added, drop by drop, down the side of the tube. After standing overnight at 4°C, the fixative was changed to two parts methanol and one part glacial acetic acid. Following gentle pipetting, an opalescent suspension was obtained. This suspension was centrifuged and all but 0.2 ml of the fixative was removed. 0.2 ml of 50% acetic acid was added, the cells were resuspended, and placed, in individual drops, onto slides lying on a hot plate at 55°C. The drops were allowed to dry for 30 seconds, and the slides were then washed and stained immediately.

B. Nuclease Treatment

Before staining, some slides were treated with deoxyribonuclease (DNAase, Nutritional Biochemicals Corporation, amorphous), 0.1 mgm/ml, 37°C in PBS containing 0.05 mgm $MgCl_2$/ml. Controls were treated with enzyme-free solutions, and all slides washed in PBS and stained immediately.

C. Antisera

"Reticular" antiserum was provided by Drs. Deborah Doniach and Ivan Roitt, WHO Autoimmune Diseases Reference Laboratory, London; other natural antisera were obtained from patients at the Boston City Hospital. Anti-calf thymus DNA serum was provided by Drs. Werner Braun and Otto Plescia, Institute of Microbiology, Rutgers University (Plescia *et al.,* 1964).

The anti-DNA serum was conjugated at a fluor:protein ratio of 1:100, and the free dye removed by DEAE-Sephadex chromatography using saline (buffered at pH 7.0 by 0.0175M phosphate) as eluant, according to the method of Tokumaru (1962). The eluate was concentrated to its original volume by dialysis against 25% polyvinyl pyrrolidone and stored at 4°C with 1:10,000 merthiolate as preservative.

D. Fluorescent Antibody Staining

Slides, in Petri dishes lined with damp filter paper, were treated with serum for one hour at 37°C. (Lower temperatures were found to produce attachment of normal serum components.) They were then washed in 500 ml PBS for 15 minutes with gentle magnetic stirring. In preliminary experiments, a two-layer technique with unlabelled LE serum and conjugated horse anti-human globulin serum (Sylvana, Millburn, N. J.) was used.

E. Ultraviolet Microscopy and Photomicrography

Coverslips were mounted with 10% PBS in glycerol. The slides were illuminated by a Sylvania L50 mercury vapour lamp attached to a Zeiss GFL microscope with a dark field condenser; they were viewed through apochromatic objectives, containing iris diaphragms, using UG2 (4 mm) exciter and No. 47 barrier filters. The preparations were photographed through BG12 exciter and No. 47 barrier filters, on Agfachrome 35 mm film using three- and six-minute exposures.

II. RESULTS

A. Preliminary Experiments with Natural Anti-Nuclear Sera

Early studies were done with acetone or Carnoy-fixed whole cells and isolated nuclei in interphase and metaphase. The nuclei were stained

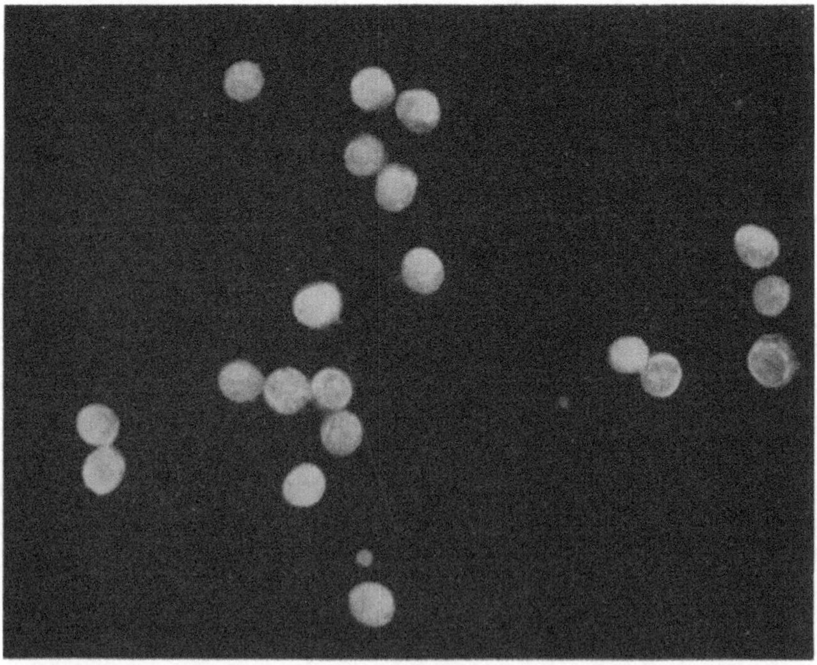

Fig. 2. Typical patterns obtained in interphase and metaphase cells with natural antisera. Agfa-chrome transferred to black and white; six-minute exposure; original magnifications 400× and 1000×. A. "Homogeneous" staining of whole lymphocytes at interphase.

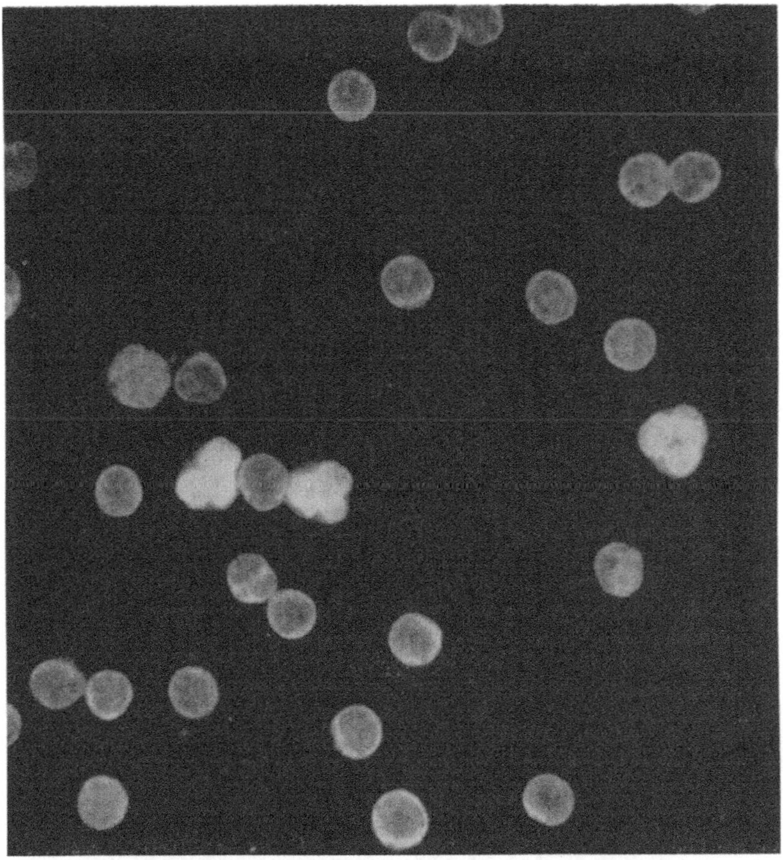

Fig. 2B. "Homogeneous" staining of interphase polymorphonuclear cells, but not lymphocytes, using a different serum from that in 2A.

with natural anti-nuclear sera and labelled anti-globulin conjugates, using the two-layer technique. The results are described to illustrate the problems encountered in interpreting such systems.

The patterns seen with natural anti-nuclear sera in isolated interphase and metaphase nuclei are shown in Figure 2. The reproducibility obtainable with reference sera of high titer is unfortunately poor, and though the patterns they produce are fairly stable, the antigens that are involved can not be fully defined. For these reasons, usefulness of such sera is limited.

When cells in clumps are examined, difficulties of interpretation are emphasized. Figure 3A shows acetone-fixed whole cells stained with an LE serum (which had given a homogeneous pattern in mouse liver nuclei) followed by fluoresceinated horse anti-human globulin. Interference from cellular debris is considerable, and the patterns vary from cell to cell, resulting in different combinations of cytoplasmic, nuclear and nucleolar

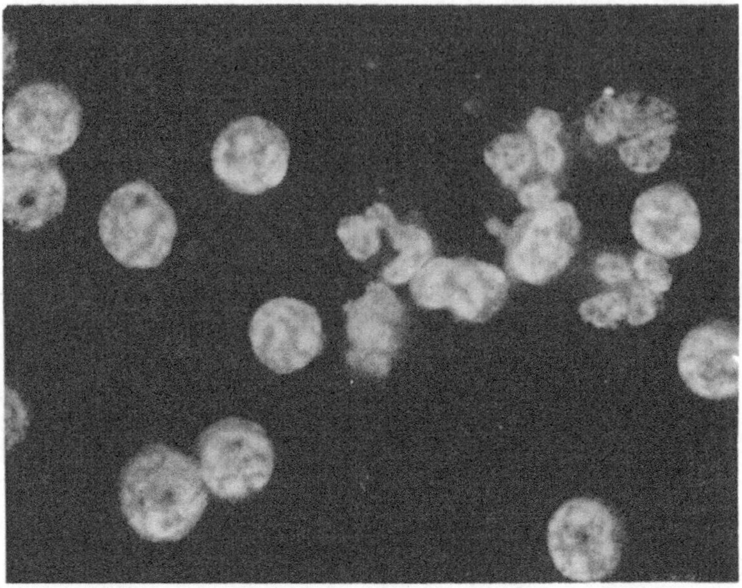

Fig. 2C. "Homogeneous" staining of both lymphocytes and polymorphonuclear cell nuclei, using a different serum from those in 2A and 2B.

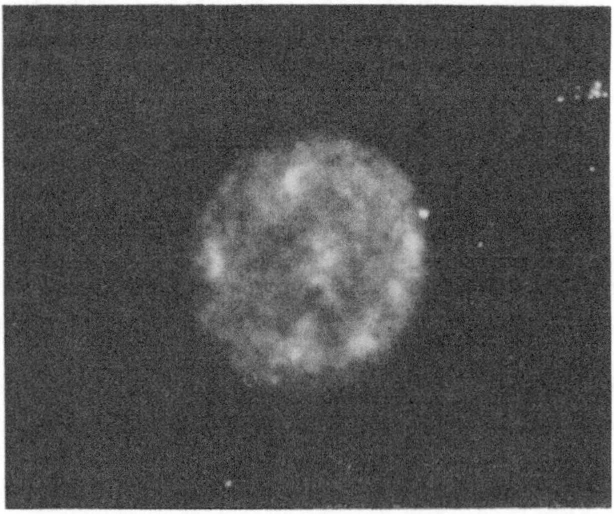

Fig. 2D. "Speckled" staining of isolated nucleus of lymphocyte at interphase.

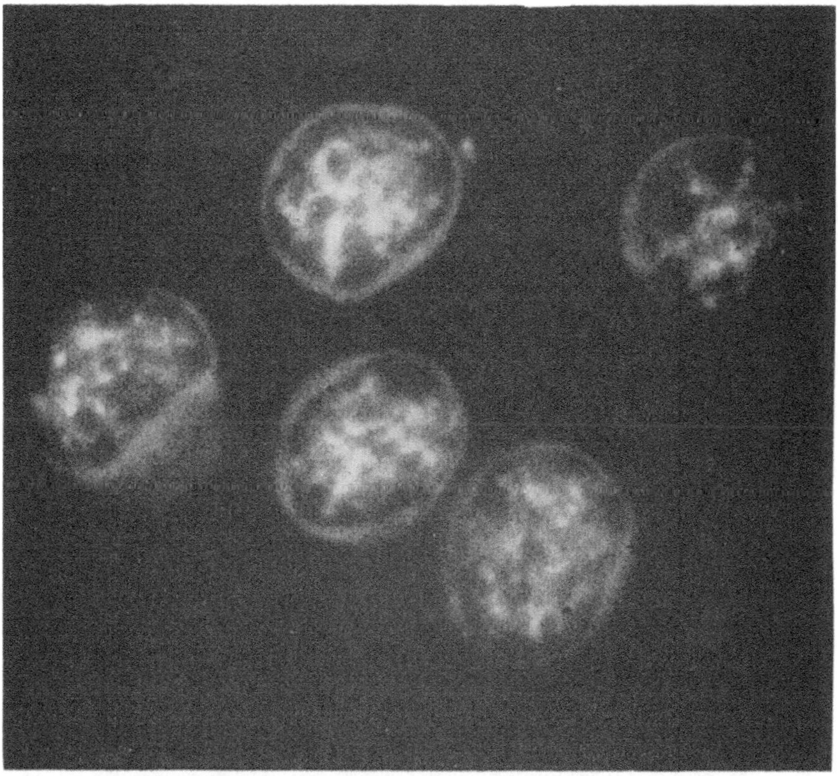

Fig. 2E. "Reticular" staining, distinguished from "speckled" by exclusion of nucleolus from reaction, in whole lymphocytes at interphase.

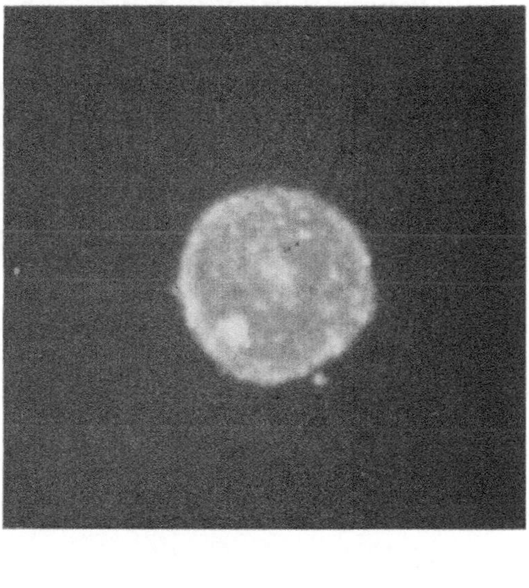

Fig. 2F. "Nucleolar" and "speckled" reactions in a whole lymphocyte at interphase.

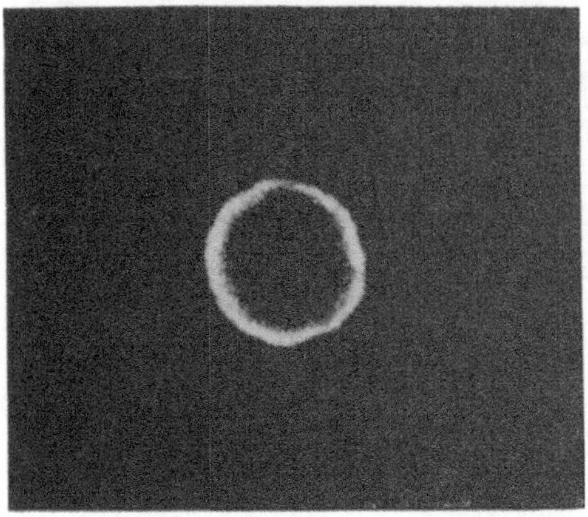

Fig. 2G. "Nuclear membranous" attachment in isolated lymphocyte interphase nucleus.

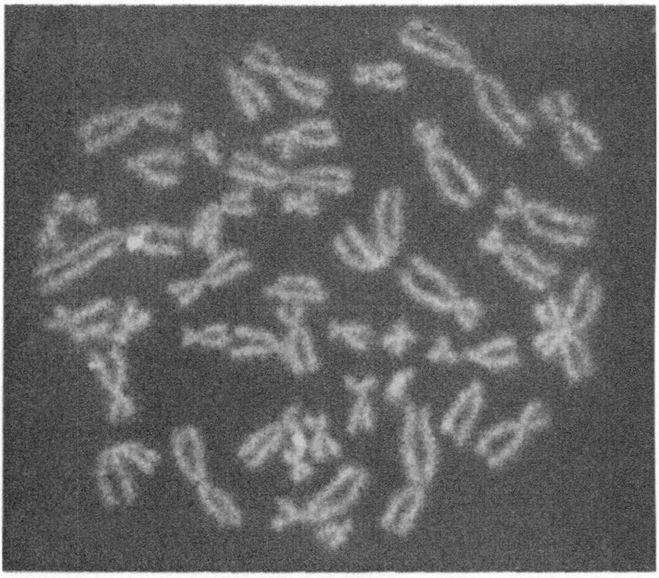

Fig. 2H. "Homogeneous" staining of lymphocyte metaphase spread.

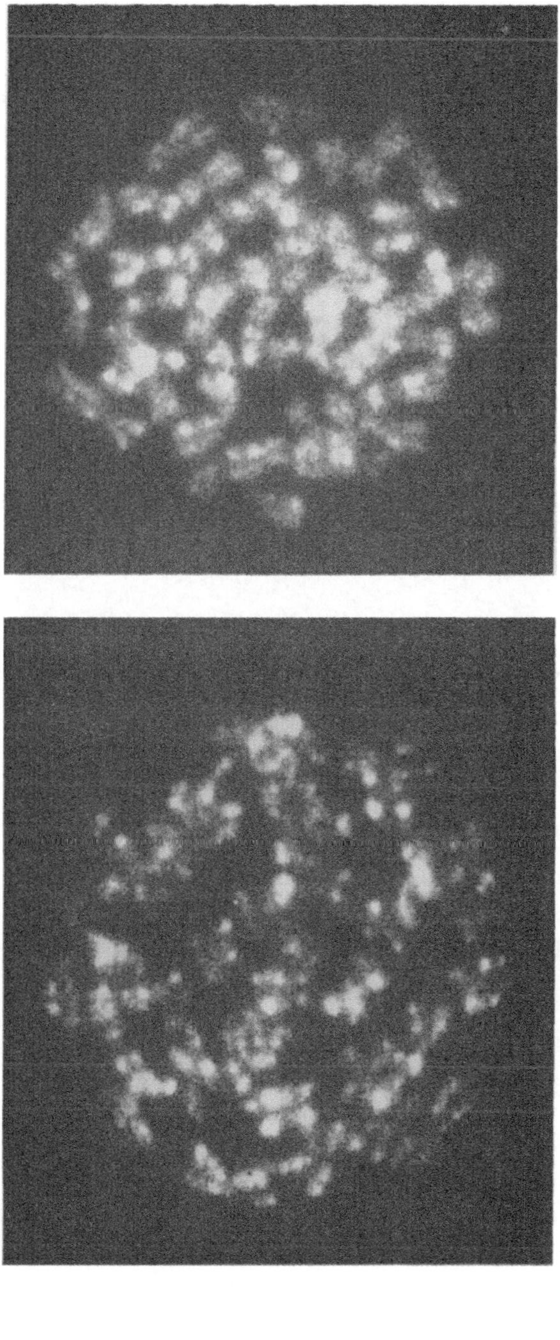

Figs. 2I and J. "Nucleolar" and "speckled" staining of lymphocyte metaphases.

fluorescence. It is difficult to correlate these patterns with the picture seen after conventional staining (Fig. 3B). Figure 4 shows three asynchronous lymphocytes stained with anti-nuclear serum and horse antiglobulin. On the left, a cell with a large nuclear-cytoplasmic ratio shows only two or three small irregular areas of fluorescence. Below, a moderately enlarged cell with increased cytoplasm shows finely stippled nuclear fluorescence and diffuse cytoplasmic uptake. It is not easy to decide whether the nuclear attachment is real or caused by overlying cytoplasm. The fact that the two cells are metabolically separate entities could offer one explanation for their different appearances. The third cell is in mitosis and shows bright cytoplasmic and very bright perinuclear staining. Fluorescence also appears in and among the chromosomes. It is not clear whether this represents limited access of antibody to the nucleus, which otherwise would be uniformly brilliant, or whether the disintegrating mitotic nuclear membrane provides an increased amount of antigen. The probability that the antiserum contains several types of anti-

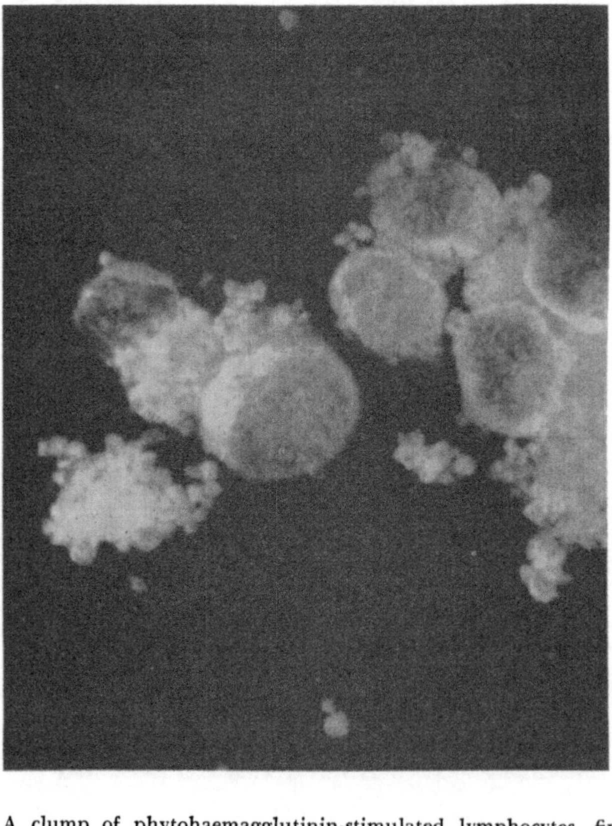

Fig. 3A. A clump of phytohaemagglutinin-stimulated lymphocytes, fixed in cold acetone, and stained with "homogeneous" natural antibody followed by fluorescein-conjugated horse anti-human globulin (fluor:protein ratio 1:40).

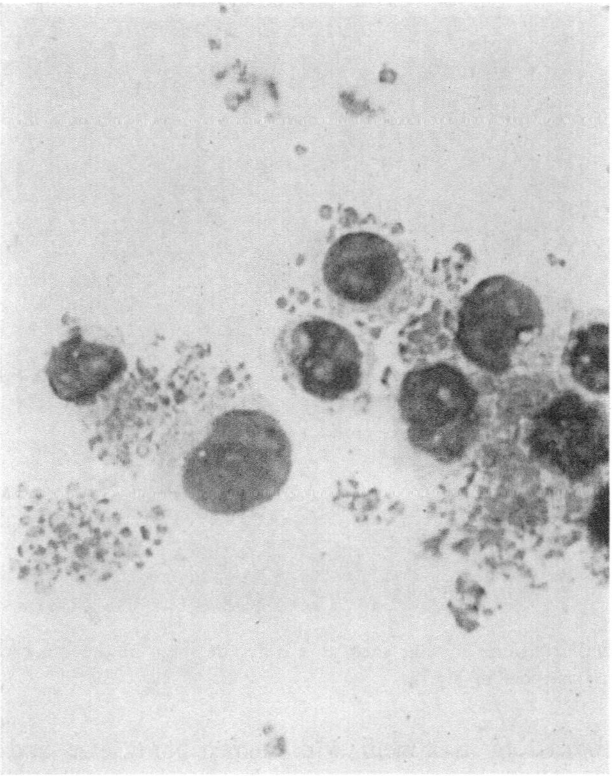

Fig. 3B. The same lymphocytes as in Fig. 3A subsequently stained with Giemsa stain.

body could provide one possible explanation for the uneven fluorescence. At the same time, this consideration vitiates its usefulness for analyzing particular nuclear substances.

B. Experiments with Directly and Minimally Conjugated Anti-DNA

Anti-DNA serum was used in tests with osmotically isolated and Carnoy-fixed nuclei from cultures that had been partially synchronized by holding overnight at room temperature following eight hours of initial incubation at 37°C. In these experiments discrete attachment of antibody occurred in interphase and metaphase nuclei (Fig. 5). In the former, the frequency of attachment concurred with the probable replication of RNA in G_1 cells, and DNA in S cells (Klein *et al.*, 1967; Razavi, 1967). In metaphase chromosomes the most frequent attachment was telomeric or near secondary constrictions, sites that may be assumed to offer suitable opportunities for the unraveling of single-stranded DNA. These patterns were intensified by pre-heating the cells to 80°C for five minutes; they

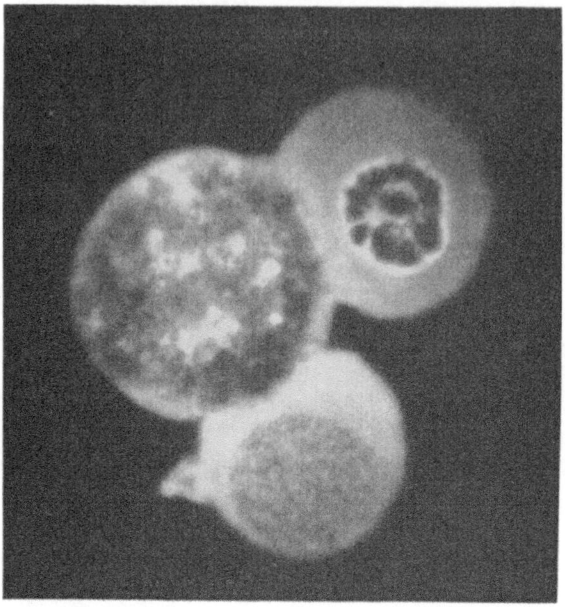

Fig. 4. Smaller clump of cells, each at a different stage in the life cycle, treated in the same way as those in Fig. 2.

were minimized by treatment with deoxyribonuclease and they were abolished by brief preheating in an open flame.

III. DISCUSSION

In the early experiments the problem of identifying antigenic material, whose site and concentration varied with metabolism and genetic activity from cell to cell, was compounded by poor control of fixation, the introduction of unknown ionic forces into the immune reaction, especially in sandwich techniques, and the poorly defined specificity of natural antisera. These difficulties prevented the emergence of a reproducible picture and a chronological analysis of the true underlying processes. In general it has been difficult to refute the surmise (Beck, 1962a) that the antigens concerned are rather labile, and that caution must be used in interpreting results as indicative of chromosomal behavior.

These problems were somewhat alleviated in the later experiments by consideration of the following factors:

(1) *The use of moderately synchronized cells from which slightly swollen nuclei can be isolated free of interfering cytoplasmic debris.* The use of temperature as a synchronizing agent, while not as efficient as chemical synchronization, has avoided aberrations of DNA structure that may attend the use of chemical agents. Slightly swollen nuclei also have

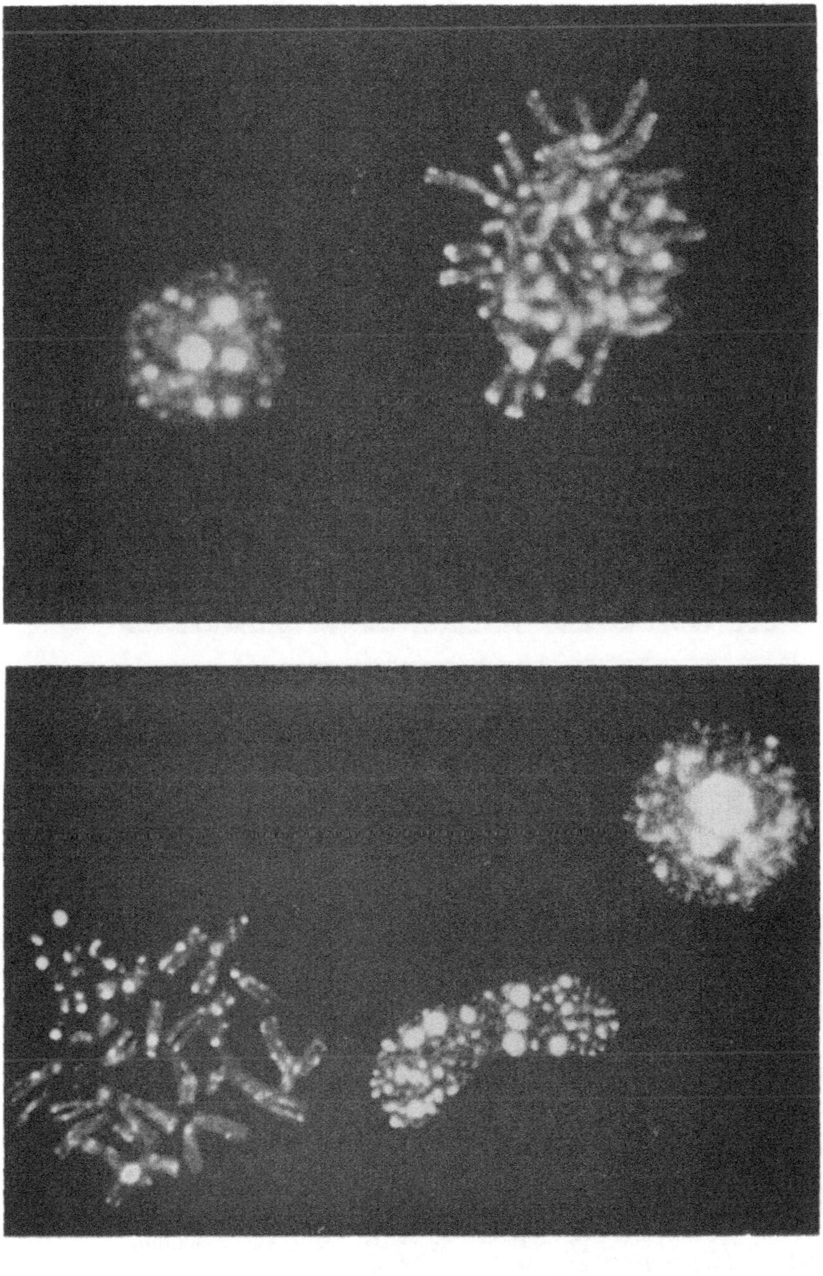

Figs. 5A and B. Directly conjugated (fluor:protein ratio 1:100) anti-DNA serum staining isolated Carnoy-fixed lymphocyte interphase and metaphase nuclei.

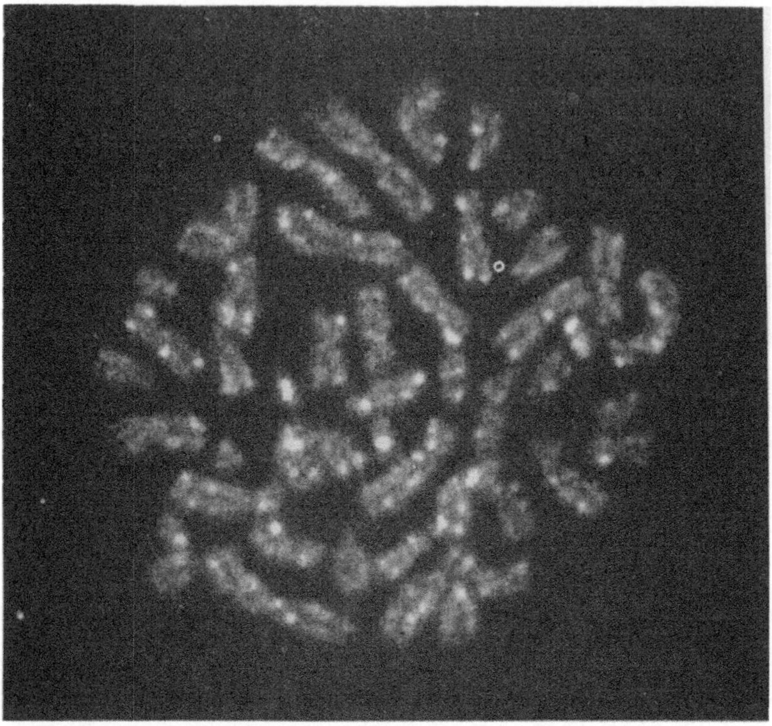

Fig. 5C. Metaphase similarly stained to show asymmetrical distribution of fluorescent areas. (Reproduced from Razavi, 1967, by permission of the Editor of Nature.)

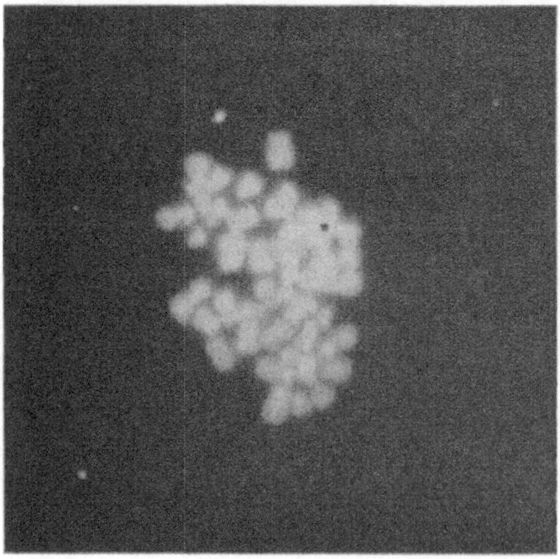

Fig. 5D. Metaphase similarly stained but after being heated to 80°C for five minutes.

been used by Frenster *et al.* (1963) for the study of genetically active DNA in lymphocytes and liver cells.

(2) *The use of an appropriate fixative.* The distribution of antigen among cytoplasm, nuclear membrane and nucleus is difficult to ascertain in the intact cell. The problem is not relieved by the use of a fixative that renders membranes permeable to antibody, because cellular detail is then distorted and difficult to analyze. Carnoy fixation of isolated nuclei, on the other hand, removes all components except nucleoproteins and preserves the latter in a configuration still susceptible to enzymatic digestion. It has been assumed that the steric preservation of groups vulnerable to enzymes also preserves antigenic groups (which may or may not overlap) close to their natural form.

(3) *The use of antiserum directed against single-stranded DNA.* Many natural antisera, particularly those with anti-nucleolar factors, contain antibodies that react similarly, but are always associated with a number of other anti-cellular factors; these anti-cellular factors are detectable by immunofluorescence, but are not definable because the antigens with which they react are not specified. The resolution of natural sera can be increased by dilution, because titers of antibodies reacting with different factors can vary widely—for example, separation of "speckled" and "homogeneous" patterns can be achieved in this way. Also, the resolution can be increased by absorption, permitting a separation of "nucleolar" and "speckled" antibodies. But these methods weaken the sera and are not entirely reliable. The use of a serum prepared against a chemically defined antigen permits both topographic and biochemical determination of sites of attachment and permits more precise inferences as to the functional implications of the observed distribution.

(4) *The use of minimally and directly labelled conjugates.* A particular difficulty with anti-nuclear antibodies is that salt concentration appears to affect the precipitin reactions more strongly than other immunological reactions, perhaps because the acid phosphate groups of the DNA antigen participate in both the specific immune reaction and in the non-specific salt linkage with basic proteins. This can be obviated by maintaining the reaction between pH 7 and 8 where over-all change of DNA and globulins is the same (Deicher *et al.,* 1959). In immunofluorescence work, where the immune reaction essentially involves the formation of a precipitin complex, prozone phenomena may occur because antigen concentration is not controllable. This difficulty is further compounded by changes in the isoelectric point and increases in relative electronegativity (Frommhagen, 1965) as a result of labelling with fluorescein isothiocyanate (FITC). Particularly irregular results are obtained in two-layer techniques with horse anti-globulin conjugates (Krooth

et al., 1961), possibly because the high lysine content of horse globulin encourages reactions with DNA. If such globulin is in addition conjugated with FITC at the usual fluor:protein ratio of 1:40, precipitation from simple ionic interaction will result in false positive attachment of the conjugate alone, without any participation of specific immunoglobulins. This difficulty can be overcome by direct conjugation of the serum at minimal ratios (1:100 or 1:150). Some loss in sensitivity occurs, but similarly labelled *non-specific* conjugates act as truly negative controls, and the retention of specificity and greater ease of control makes the use of such conjugates essential in topographical analysis.

In spite of these precautions, certain difficulties remain unsolved. The variable effect of heat observed here, and also noted by Rapp (1962), is evidence for the persistent influence of mode of preparation on antigen distribution. It may be desirable to substitute squashing for air- and heat-drying in the spreading of metaphase chromosomes, but the amount of cellular debris left after squashing interferes with specific reactions, causing autofluorescence and irregular reactions from cytoplasmic DNA. The debris may be removed by enzymatic digestion, but this merely replaces one variable with another.

A second set of problems arises from the lack of adequate controls for reagents for which the substrate cannot be specified. No histochemical procedures distinguish single-stranded from double-stranded DNA, although methyl green at low pH may distinguish highly- from poorly-polymerized DNA. Similar difficulties attend the use of the enzyme DNAase, which, though it attacks randomly one strand of the Watson-Crick double helix at a time, will nonetheless ultimately digest all DNA, whether single- or double-stranded. Furthermore, any negative results may be as much due to mechanical loss of a DNA-associated antigen as to the degradation of a DNA antigen. The use of drugs such as actinomycin D or chloroquin is vitiated, because of their propensity to cause chromosome aberrations or to influence only polymerization, rather than strandedness; these properties make any results obtained difficult to use in a description of the normal behavior of chromatin.

Finally, an exact interpretation of the results obtained so far with anti-DNA serum is rather difficult. The simplest explanation is an attachment of antibody to the Y junction of replicating strands of DNA, or near the site of formation of mRNA. Antibody may also attach to the exposed ends (telomeres or ends of DNA pieces) of the organized chromoneme or to single-stranded interruptions. However, it is not known if interruptions and pieces occur in eucaryotic chromosomes, and furthermore these and other possible sites may well be shielded by DNA polymerase which needs single-stranded DNA as a primer for its activity.

It may be that this shielding is only partial or involves regions of the nucleotide chain that are distinct from the antigenically active sites.

If, however, it is accepted that the likeliest site of attachment is single-stranded DNA, then the random distribution of stainable sites (-antibody attachment sites) observed through all chromosomes, varying not only between cells but also between isologous chromatids, raises interesting questions about chromosomal gene activity.

With respect to *structure,* the statistical nature of homologue pairing by metrical or autoradiographical methods is readily explainable if chromatids contain constantly varying amounts of uncoiled and single-stranded DNA. With respect to *activity,* it may be that the 20% of lymphocyte chromatin, said to be active by Frenster *et al.* (1963), changes with the time in the cell cycle, different genes being required for different time-related cellular activities. In a similar manner, the observation that, to a greater or lesser extent, *all* chromosomes contain single-stranded DNA suggests that no one chromosome is entirely heterochromatic. This finds support in the detection of RNA attached to, and possibly synthesized by X chromosomes (Pegoraro *et al.,* 1967; Back *et al.,* 1967), the variable replication lag among X chromosome segments (Muckherjee *et al.,* 1967), and variations in the heterochromatization of interphase sex chromatin (Weste *et al.,* 1967). Hence, dosage compensation may depend on changes in rates of DNA uncoiling or on asymmetry between coiled and uncoiled phases in a dynamically oscillating system. The flexibility of hydrogen bonds, which has called into question their role in maintaining the Watson-Crick helix, may however be very suitable for allowing thermodynamically feasible twists between coiled and open states; in the latter fluctuating single nucleotide chains could form transiently and successively, without any disturbance of the overall structure.

Such structural and functional organization of chromosomal single-stranded DNA, as a dynamic rather than a static component, suggests a genetic system in which the DNA molecule is acting as a continually available coding device only. Actual transcription may be modulated through other components of the chromosome or nucleus. Such gene availability is a prerequisite for genetic systems proposed by Stern *et al.* (1963) and others, in which the elements of gene expression are separated from gene structure (see Razavi, 1967, for further references).

Further resolution of these problems can perhaps be obtained by the use of differential labelling—two color fluorescence or isotopic-fluorescent combinations—and by base-specific anti-nucleotide sera. Their use in conjunction with other immunological, isotopic or electron-dense labels appears feasible and may permit studies also on nuclear com-

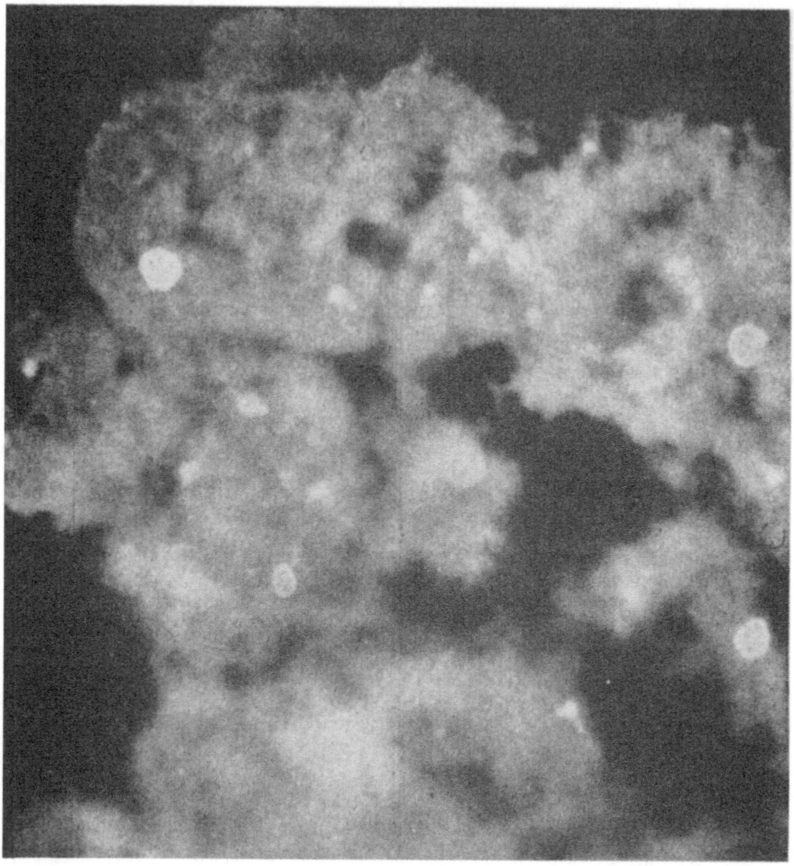

Fig. 6A. Nuclei from cells of the Culicine mosquito gut, squash preparation, stained with "homogeneous" antibody.

ponents such as methylated RNA (hybridizable), DNA polymerase and, once isolated, repressors (Ptashne, 1967). Such techniques may provide information on chromosome organization, including heterochromatin and euchromatin structure, and its concurrence with template or replicative activity in different tissues. These techniques should also permit studies on allotypic hybridization, nuclear membrane-, nucleolar- and virus-associated portions of chromatin, and, finally, of autoantibody-associated portions of aberrant chromosomes (Fialkow, 1964).

Suitable nuclei for such studies readily suggest themselves. The well-marked and mapped polytene chromosomes of *Drosophila melanogaster* are perhaps the most obvious choice, but heterospecificity between natural mammalian anti-nuclear sera and arthropod nuclei is poor (possibly because of different binding proteins), and no anti-Drosophila chromatin sera are yet available. That this problem is not insuperable is shown by the good attachment of natural "homogeneous" antibody to the nuclei of Malpighian tubules of Culicine mosquito (Fig. 6), which have been teased away from their cytoplasm. It has not been possible to obtain similar preparations of polytene chromosomes because of their fragility and poor reactivity. It is possible that the latter may be improved by induction of lampbrush loops with suitable hormones.

Among other useful substrates the following may be suggested: translocations occurring in hamster cell lines, inter-specific heterokaryons, tumor virus-transformed nuclei, anucleate acetabularian protozoa containing episomal factors, and plant cells containing chloroplasts.

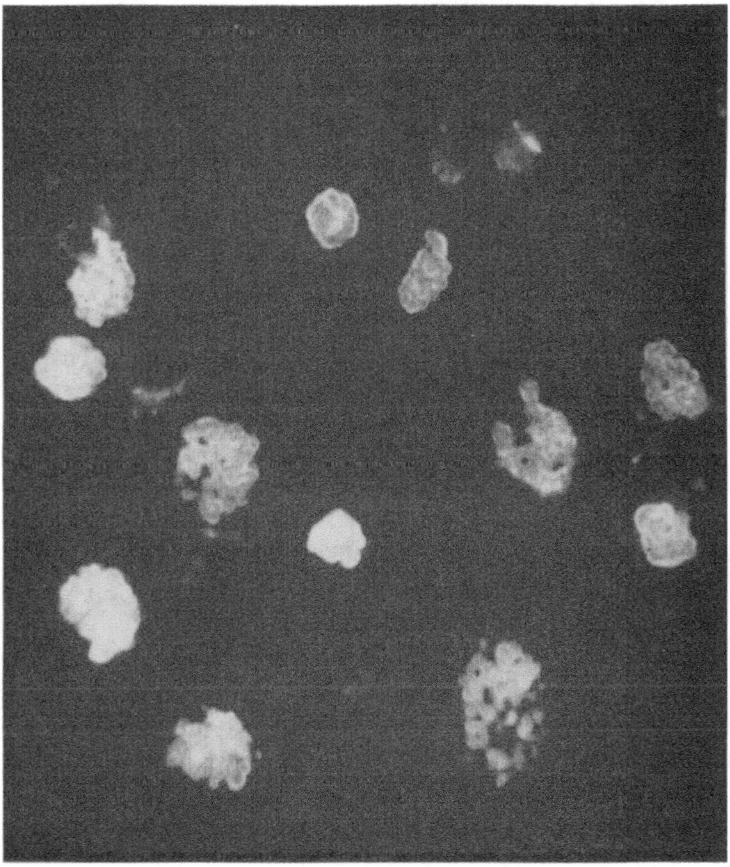

Fig. 6B. The nuclei of Malpighian tubule cells of the Culicine mosquito teased away from cytoplasm and stained with "homogeneous" antibody.

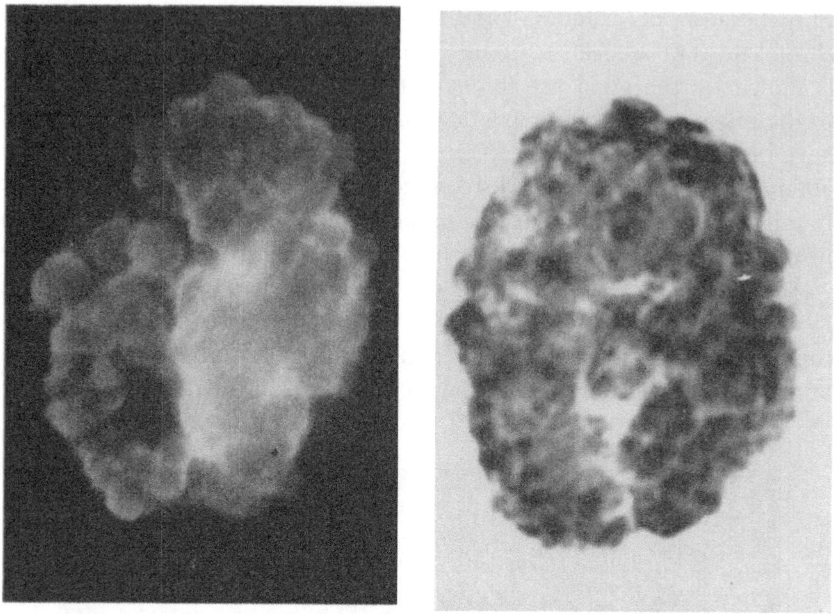

Figs. 6C and D. Enlarged view of nucleus stained with antibody followed by Giemsa.

IV. SUMMARY

Anti-nuclear globulins from patients with collagen diseases attach to several nuclear components in interphase and metaphase. The titer and specificity of the anti-nuclear factors vary from serum to serum, most sera containing factors against two or more cellular constituents. The concentration and distribution of the nuclear substances, to which these serum factors attach, vary with cell type and metabolic activity, and also with mode of preparation, being especially dependent on the type of fixation. For these reasons one sees many different patterns which are difficult to interpret. The analysis of nuclear components is considerably eased by the use of defined antisera and isolated nuclei.

In lymphocyte chromosomes the distribution of single-stranded DNA, as detected by fluorescein-labelled anti-DNA serum, suggests that gene activity is continuous in all chromosomes throughout the cell cycle. This supports the theory that the DNA molecule may act as a coding device only while gene expression is under the control of cytoplasmic and nuclear factors that are regulated by the demands of the environment.

References

Back, F., and P. Dörmer (1967). X-Chromosome activity in lymphocytes. Lancet, *1*:385.

Beck, J. S. (1961). Variations in the morphological patterns of "autoimmune" nuclear fluorescence. Lancet, *1*:1203–1204.

Beck, J. S. (1962a). Partial identification of the "speckled" nuclear antigen. Lancet, *1*:241–243.

Beck, J. S. (1962b). The behaviour of certain nuclear antigens in mitosis. Exp. Cell Res., *28*:406–418.

Deicher, H. R. G., H. R. Holman, and H. G. Kunkel (1959). The precipitin reaction between DNA and a serum factor in systemic lupus erythematosus. J. Exp. Med., *109*:97–114.

Fialkow, P. J. (1964). Autoimmunity: a predisposing factor to chromosomal aberrations? Lancet, *1*:474–475.

Frenster, J. H., V. G. Allfrey, and A. E. Mirsky (1963). Repressed and active chromatin isolated from interphase lymphocytes. Proc. Nat. Acad. Sci., *50:* 1026–1032.

Frommhagen, L. H. (1965). The solubility and other physicochemical properties of human Y-globulin labelled with fluorescein isothiocyanate. J. Immunol., *95*:442–445.

Gall, J. (1954). Lampbrush chromosomes from oöcyte nuclei of the newt. J. Morphol., *94*:283–352.

Gall, J. (1963). Kinetics of deoxyribonuclease action on chromosomes. Nature, *198*:36–38.

Klein, J. M., S. M. Beiser, and B. F. Erlanger (1967). Nuclear fluorescence employing antinucleoside immunoglobulins. J. Exp. Med., *125*:61–70.

Krooth, R. S., J. E. Tobie, J. H. Tjio, and H. C. Goodman (1961). Reaction of human sera with mammalian chromosomes shown by fluorescent antibody technique. Science, *134*:284–285.

Mukherjee, B. B., A. K. Sinha, K. E. Mann, S. K. Ghosal, and W. C. Wright (1967). Replicative behaviour of bovine X-chromosomes during early DNA synthesis. Nature, *214*:710–712.

Pegoraro, L., G. Rovera, and F. Gavosto (1967). X-chromosome activity in lymphocytes. Lancet, *1*:781.

Plescia, O. J., W. Braun, and N. C. Palczuk (1964). Production of antibodies to denatured deoxyribonucleic acid (DNA). Proc. Nat. Acad. Sci., *52*:279–284.

Ptashne, M. (1967). Specific binding of the λ Phage repressor to λ DNA. Nature, *214*:232–234.

Rapp, F. (1962). Localization of antinuclear factors from lupus erythematosus sera in tissue culture. J. Immunol., *88*:732–740.

Razavi, L. M. (1967). Single stranded DNA in lymphocyte chromosomes. Nature, *215*:928–930.

Ris, H. (1961). Ultrastructure and molecular organization of genetics systems. Canad. J. Genet. Cytol., *3*:95–120.

Swanson, C. P., T. Merz, and W. J. Young (1967). Cytogenetics, 1st ed. Englewood Cliffs, New Jersey: Prentice-Hall, Inc.

Stern, H., and Y. Hoffa (1963). "Meristems and Differentiation," Brookhaven Symposia in Biology No. 16, 59–63.

Tan, E. M., and H. G. Kunkel (1966a). Characteristics of a soluble nuclear antigen precipitating with sera of patients with systemic lupus erythematosus. J. Immunol., *96*:464–469.

Tan, E. M., P. H. Schur, R. I. Carr, and H. G. Kunkel (1966b). Deoxyribonucleic acid (DNA) and antibodies to DNA in the serum of patients with systemic lupus erythematosus. J. Clin. Invest., *45*:1732–1740.

Taylor, J. H. (1963). The replication and organization of DNA in chromosomes. Molecular Genetics, Taylor (ed.). New York: Academic Press, 1963, *1*:65–112.

Tokumaru, T. (1962). A kinetic study on the labelling of serum globulin with fluorescein isothiocyanate by means of the gel filtration technique. J. Immunol., *89*:195–203.

Vaccaro, R., and N. Biscati (1965). Immunofluorescenza de cromosomi umani ottenuta medianti impiego di sieri "LE" attivi. Proc. Soc. Ital. Biol. Sper., *41*:597–599.

Weste, S. M., J. S. Barnett, G. M. Garson, and A. G. Baikie (1967). Variations in the frequency and form of sex chromatin in females with burns. Lancet, *2*:745–746.

DISCUSSION

Chairman: P. GRABAR

L. LEVINE: I would like to ask about the specificity of precipitation reactions involving highly charged antigens other than DNA, for example polyglutamic acid or basic proteins.

PANIJEL: Yes, your question is important. One must consider the role of non-specific charge reactions in the case of polyanions such as nucleic acids. In the precipitation reaction which we use in all of our experiments we can wash the precipitates in appropriate buffers and eliminate charge reactions if present. For example, in media with low salt concentration, we observed slight DNA precipitation, but these non-specific precipitates were completely dissociated after three washings with 0.14 M NaCl. In contrast specific precipitates formed with polyribonucleotides are completely undissociated and remain insoluble in 0.14 M NaCl. Thus the way in which we carry out precipitation reactions will eliminate non-specific reactions due to charges. I don't think that this is the case for complement-fixation and perhaps immuno-diffusion analysis. In these cases it is possible to have non-specific reactions because the washing step is absent or insufficient.

To change the subject, I would like to state that I think that everybody who works with nucleic acid-reactive antibodies has hoped that such antibodies may be able to act on cells, but the question is whether these antibodies actually can penetrate into the cells. In the case of our antibodies I feel that they do not penetrate into the cells; I don't know whether other investigators feel the same way.

L. LEVINE: Some of my associates have looked at the effects of nucleic acid-specific antibodies on transformation and found inhibition with such antibody, but when they labelled their transforming DNA with C^{14} tritium they found that they got inhibition because the transforming DNA failed to penetrate the cell. So, neutralization or inhibition of the transforming DNA occurred outside the cell.

GRABAR: Antibody certainly does not usually penetrate but perhaps it may do so under some cellular conditions such as during division.

ERICKSON: We have no direct proof for the site of reaction of DNA with the antibody in *B. subtilis* transformation, but it might be at the DNA receptor site on the bacterial surface. This could explain why an increased period of preincubation of antibodies with the recipient cells can cause an increase in inhibition. In the competent cells there appears to be a partial breakdown of cell wall, and we can assume that such breakdown may facilitate the penetration of the antibody into the space between cell wall and cell membrane.

TANENBAUM: In our demonstration of penetration of antibody into sea urchin cells, we may have a rather specialized system which may not apply to mammalian cells.

GRABAR: Of course, antibodies might get into cells by pinocytosis but in this case enzymes liberated from lysozomes might destroy the antibody.

PLESCIA: Regarding the possible biological activity of nucleic acid-specific antibodies and the problem of penetration, we need go no further than to look at the cells that make the antibodies. In other words, what happens to antibody-forming cells that make an antibody capable of reacting with their own RNA and other nuclear materials? One might well consider the possibility that such cells, once induced to do something that they normally don't do, will find themselves in a rather strange situation and would commit suicide. To test this possibility, Drs. Albright, Makinodan, and I took spleen cells from rabbits immunized with denatured calf thymus DNA conjugated to methylated bovine serum albumin and then transferred these cells, after reexposure to antigen, into X-irradiated recipient mice. What we found was that these cells did indeed behave like any other primed cells. So it appears that cells making antibodies to nucleic acids do not commit suicide. The antibody, once formed, is apparently strictly compartmentalized and unable to react with intracellular components or else antibodies are not reactive until they are excreted from the cell in which they are formed.

PANIJEL: We have tried to inhibit fertilization of sea urchin eggs with antibodies but we did not succeed, presumably because fertilization takes place in high ionic medium and it is impossible to have a good immunochemical reaction in such a medium.

It would be very interesting to achieve a true penetration of antibodies because in the case of RNA-specific antibodies, precipitation of ribosomes by these antibodies in antigen excess yields a supernatant containing a soluble complex of antibodies and ribosomes which is more active in a cell-free Nirenberg system than the ribosomes themselves.

GRABAR: Dr. Tanenbaum, did you try to absorb your active anti-BSA serum with BSA before measuring activity in the cell-free protein-synthesizing system?

TANENBAUM: No, we have not done this as yet.

SELIGMANN: Dr. Tanenbaum, in your previous work you demonstrated that antibodies to 5-acetyl uracil react with native DNA. When you published your paper, you had no satisfactory explanation for this and I wondered if you now have an explanation for this rather paradoxical finding.

TANENBAUM: I have no satisfactory explanation for this particular phenomenon even now.

SIGEL: In view of Dodd's and Bigley's finding of anti-BSA antibody in rabbits injected with ribosomes, and in view of Dr. Tanenbaum's finding that anti-BSA serum interferes with translation, I wonder if anybody has looked for possible common determinants in ribosomal protein and BSA.

BRAUN: Some time ago Thorne and Leonard observed that sera of recently immunized rabbits contain high levels of basic proteins reacting with nucleic acid. I wonder whether this finding has been taken into consideration by those who have encountered unexpected effects of immune sera with nucleic acids. It should be remembered that the basic serum proteins studied by Thorne and Leonard were able to fix complement in the presence of nucleic acid, yielded precipitates and gave all of the typical antigen-antibody reactions yet did not represent specific antibody.

TANENBAUM: I can only state that the preimmunization gamma globulin, prepared by sodium sulfate fractionation, did not show any activity in the cell-free system and the active fractions of anti-BSA immunoglobulins and of anti-hapten immunoglobulins were similarly prepared.

ROLE OF THE CARRIER IN THE PRODUCTION OF HAPTEN-SPECIFIC ANTIBODIES

THE GENETIC CONTROL OF THE IMMUNE RESPONSE TO HAPTEN-POLY-L-LYSINE CONJUGATES IN GUINEA PIGS

B. B. LEVINE,* I. GREEN, AND B. BENACERRAF *

New York University School of Medicine, New York, New York

That the immune response to an immunogen by an individual animal is under genetic control is evident from the results of several studies. For example, Fjord-Scheibel (1943), Carlifanti (1948), and Sang and Sobey (1954) showed a statistically significant relationship between the abilities of parents and their offspring to produce high serum titers of antibody to a given immunogen. Also, Ibsen (1959) showed that inbred strains of mice required different doses of antigen to produce a standardized immune response. Recently, Sobey *et al.* (1966) observed genetically transmissable differences among random bred mice to respond immunologically to bovine serum albumin. Finally, Arquilla and Finn (1965), in studies of the immune response of inbred guinea pigs to bovine insulin, found that strain 2 guinea pigs could make antibodies to an antigenic determinant of insulin to which strain 13 could not make antibodies. However, breeding studies showed complex patterns of inheritance.

In none of the above studies was a simple pattern of inheritance observed, i.e., the immune responses were found to be under the control of several genes at different gene loci. This observed complexity of the genetic mechanisms involved is probably due, in large part, to the structural and antigenic complexity of the immunogens used.

Recently, several groups of investigators have begun to study immune responses to structurally simpler immunogens, synthetic polypeptides, as an approach to the study of the mechanisms involved in immunogenicity (see Maurer, 1964, and Sela, 1966). From the point of view of genetic control of immunogenicity, three systems have been used: the hapten-poly-L-lysine system in guinea pigs (Kantor, Ojeda and Benacerraf, 1963;

* Career Investigator, Health Research Council of the City of New York.

Levine, Ojeda and Benacerraf, 1963a, b), the random linear copolymer L-glutamic acid-L-alanine-L-lysine in Swiss mice (Pinchuck and Maurer, 1965) and the branched multi-chain-L-polypeptide, poly (Tyr-Glu)-poly (DL Ala)-poly-L-lysine in inbred mice (McDevitt and Sela, 1965). The first two systems show Mendelian patterns of inheritance; the last named system is under more complex genetic control.

The purpose of this paper is to summarize the work on the immune responses of guinea pigs to hapten conjugates of poly-L-lysine (PLL), which is the first system in which a strict unigenic control of an immune response was demonstrated.

I. IMMUNOGENICITY OF HAPTEN-PLL

Firstly, it was found that when random bred albino Hartley guinea pigs were immunized with 2,4-dinitrophenyl-poly-L-lysine conjugates (DNP-PLL) in complete Freund's adjuvant, only 25–35% made immune responses (Kantor, Ojeda and Benacerraf, 1963). The responder guinea pigs showed typical delayed hypersensitivity skin reactions to DNP-PLL, produced serum antibodies detectable by PCA and ring precipitation tests, and the immunized animals regularly died in acute anaphylaxis upon intravenous injection of DNP-protein conjugates. When immunized with 100 μg DNP-PLL in complete adjuvant and bled at 18–21 days, responder animals regularly produce 1–2 mg/ml precipitable anti-DNP antibodies (Green, Paul and Benacerraf, 1966). These DNP-specific antibodies, as well as delayed hypersensitivity skin reactions to DNP-PLL, display carrier specificity for the PLL carrier (Levine, 1965; Paul *et al.,* 1966) typical of hapten-specific immune responses induced by hapten-protein conjugates (Gell and Benacerraf, 1961). Responder animals also showed weak delayed hypersensitivity reactions to PLL itself when immunized with DNP-PLL or with PLL in complete adjuvant containing killed human mycobacteria (Green *et al.,* 1966).

By contrast the non-responder animals, comprising about 65–75% of random bred Hartley guinea pigs, showed no evidence of an immune response to identical immunization with DNP-PLL. Thus, delayed skin tests were completely negative, the sera failed to give PCA or ring precipitin tests for DNP-specific antibodies, and non-responders did not develop symptoms of anaphylaxis upon intravenous challenge with DNP-protein conjugates (Kantor *et al.,* 1963). Similar results were found utilizing benzylpenicilloyl-PLL conjugates as the immunogen (Levine, 1964a, b).

It was of interest to determine whether non-responder guinea pigs immunized with hapten-PLL conjugates produce any DNP-specific antibodies at all; i.e., levels of antibody below the level of sensitivity of the methods detailed above. The failure of their sera to elicit PCA reactions indicates that such sera contained less than 1–5 μg γ_1 antibody protein/ml. Sera, assayed by passive hemolysis using DNP-BSA coated tanned RBC, failed to give positive tests (Levine *et al.*, 1963). This test can detect γ_2 antibodies at a minimum concentration of about 5 μg antibody protein/ml. Sera was also assayed for DNP-binding by equilibrium dialysis using dilute solutions of H^3-tagged DNP-ϵ-aminocaproate. There was no statistically significant difference in binding of the DNP-hapten by sera from DNP-PLL immunized non-responders and by sera from control animals immunized with adjuvants alone, although both sets of sera bound quantities of DNP-hapten equivalent to 1–3 μg antibody protein/ml (Green *et al.*, 1966). Also, utilizing a sensitive passive hemagglutination system capable of detecting 0.01 μg antibody protein/ml (Levine and Levytska, 1967), no significant differences were found between the two sets of sera. However, in some animals in each group, DNP-specific hemagglutination reactions at titers of 1/100 to 1/200 were found. Thus, immunization of non-responder guinea pigs with DNP-PLL has not produced clear-cut antibody synthesis, according to our present experience. If some antibody is produced at all, it is produced in concentrations below 1–2 μg antibody protein/ml.

II. GENETIC STUDIES

The all or none character of the immune response to hapten-PLL suggested a simple genetic control. Accordingly, breeding experiments were carried out, and the following findings were made (Levine, Ojeda and Benacerraf, 1963a; Levine and Benacerraf, 1965): Of 26 offspring of 9 pairs of non-responder parents, none were responders; of 22 offspring of 8 pairs of responder parents, 82% were responders. No evidence for sex-linking was found (Table 1). Also, when the highly inbred strains 2 and 13 were studied, 100% of 40 "strain 2" animals and none of

Table 1

IMMUNOGENICITY OF DNP_{24}-PLL_{316} IN OFFSPRING OF RESPONDER PARENTS AND NON-RESPONDER PARENTS *

Parents	Offspring	
Responders * (8 breeding pairs)	Responders * 18; 82 per cent (6 ♂, 12 ♀)	Non-responders ‡ 4; 18 per cent (1 ♂, 3 ♀)
Non-responders † (9 breeding pairs)	0	26 (14 ♂, 12 ♀)

* From Levine, Ojeda and Benacerraf, 1963.

† Responders refer to guinea pigs that showed an immune response to DNP-PLL, evidenced by skin reactivity to DNP_{24}-PLL_{316} and PCA or passive hemolysis of their sera with DNP_{41}-BSA.

‡ Non-responders are animals that were negative by all 3 of these tests.

Table 2

IMMUNOGENICITY OF DNP_{24}-PLL_{316} IN OFFSPRING OF THE MATING OF NON-RESPONDER × HETEROZYGOUS RESPONDER PARENTS *

Family number	Offspring: Responders ‡	Offspring: Non-responders ‡
	NR♀ × HR♂ †	
I	1	1
II	3	1
III	2	2
IV	1	1
V	2	2
VI	0	3
	HR♀ × NR♂	
VII	1	2
VIII	2	1
IX	1	1
X	1	3
Total	14 (45.3%)	17 (54.7%)

* From Levine and Benacerraf, 1965.
† NR = non-responder; HR = heterozygous responder.
‡ See legend, Table 1.

11 "strain 13" guinea pigs were responders. These results suggested that the responder trait is transmitted as a unigenic autosomal Mendelian dominant trait. Further evidence for this view was obtained by breeding non-responders with heterozygous responders (i.e., phenotypic responder guinea pigs whose previous matings had resulted in at least one non-responder offspring). Of the 31 offspring of such matings, 14 (45.3%) were responders. This distribution (45.3:54.7) is not significantly different from the 50:50 distribution expected for the case of a unigenic Mendelian dominant trait (chi square = 0.29; $p > 0.5$). In these experiments, there also was no evidence of sex-linkage (Table 2).

III. RELATIONSHIP OF THE RESPONDER TRAIT TO THE PLL BACKBONE

The breeding experiments indicate that the responder trait (i.e., the ability for a guinea pig to make an immune response to DNP-PLL), is transmitted as a unigenic Mendelian dominant trait, i.e., transmitted by a single gene. It appears likely that this trait depends upon the PLL carrier and not on the hapten since virtually 100% of Hartley guinea pigs can respond immunologically to the DNP or the BPO haptens coupled to homologous or to foreign protein carriers. To study this question, 33 random bred guinea pigs were immunized with a mixture in complete adjuvant of PLL conjugates of 4 structurally dissimilar haptens: DNP, BPO, 5-dimethylamino-naphthalene sulfonyl (DMANS) and p-toluenesulfonyl (tosyl). The 11 responders in this group responded immunologically (by skin and serum tests) to all 4 PLL conjugates, whereas the 22 non-responders failed to show hypersensitivity to any of the 4 conjugates (Table 3). Control experiments excluded the possibili-

Table 3

IMMUNIZATION OF 33 GUINEA PIGS WITH A MIXTURE OF FOUR HAPTEN-POLYLYSINE CONJUGATES (100 μg EACH) *

Guinea Pig No.	Immune Response at Fourteenth Day †			
	Dimethylamino-naphthalene-Sulphonyl-PLL_{316} (10 μg)	$Tosyl_{35}$-PLL_{316} (10 μg)	DNP_{30}-PLL_{316} (10 μg)	Benzyl-$penicilloyl_{25}$-PLL_{316} (10 μg)
1	15 necrosis	20 necrosis	15 necrosis	20 necrosis ‡
2	12 necrosis	13 necrosis	12 necrosis	10 necrosis ‡
3	15 necrosis	17 necrosis	12 necrosis	10 necrosis
4	11 necrosis	15 necrosis	15 necrosis	12 necrosis
5	15 necrosis	17 necrosis	15 necrosis	15 necrosis
6	15 necrosis ‡	20 necrosis	15 necrosis	18 necrosis
7	12 necrosis	16 necrosis	20 necrosis	16 necrosis
8	15 necrosis ‡	20 necrosis	12 necrosis	15 necrosis ‡
9	18 necrosis ‡	20 necrosis	12 necrosis	15 necrosis
10	15 necrosis	16 necrosis	15 necrosis	12 necrosis
11	15 necrosis	18 necrosis	15 necrosis	16 necrosis
12–33 §	negative	negative	negative	negative

* From Levine, Ojeda and Benacerraf, 1963b.

† Average diameters (mm) of skin reactions at 24 h (delayed allergic reactions) are tabulated. Control animals immunized with adjuvants alone gave flat 2–6 mm papular reaction to these conjugates. All 33 animals gave negative skin reactions to PLL.

‡ The sera of the 11 reactors gave positive ring precipitin tests with DMANS-GPA, DNP_{25}-GPA, Tosyl-GPA and BPO_{30}-GPA all at 100 μg/ml., except for the animals and conjugates marked with a double dagger. Their sera were positive for specific anti-hapten antibodies by PCA, titres 1/10 to 1/50.

§ After booster dose of all 4 conjugates, non-reactors failed to show evidence of an immune response by skin and serum tests as well as by intravenous challenge to any of the 4 conjugates.

ties that these results were due to cross-reactivity among the haptens or to non-specific stimulation of immunogenicity (Levine *et al.,* 1963b). Furthermore, responder animals to hapten-PLL were also capable of responding immunologically to a synthetic random linear copolymer of L-glutamic acid and L-lysine (GL) and to DNP-conjugates of GL, whereas non-responders were not capable of making immune responses to these materials (Kantor *et al.,* 1963). The GL random copolymers are likely to contain poly-lysine sequences, although sequence analyses have not been done.

Thus it appears that the physiological effect of the single gene controlling the ability of guinea pigs to respond immunologically to hapten-PLL (and GL copolymer) is in some way related to lysine sequences in the carrier. This gene will hereafter be referred to as the "PLL gene."

IV. ENZYMATIC DEGRADATION OF HAPTEN-PLL

The immune response to a protein immunogen may be considered to involve a sequence of distinct physiological steps, each under separate genetic control. These steps may include enzymatic degradation of the immunogen, coupling of antigen fragments to RNA, synthesis of immunoglobulin chains, and probably others. The effect of the "PLL gene" may be on one of the above steps. As one possibility, we sought to determine whether the PLL gene may act on the enzymatic degradation of hapten-PLL conjugates by tissues involved in immune responses. Degradation of hapten-PLL by responder and non-responder guinea pigs was studied *in vivo* and *in vitro.* Degradation *in vivo* was studied by following the appearance of low molecular weight hapten-lysyl peptides in the urine following an intraperitoneal injection of high molecular weight H^3-DNP-PLL. Degradation *in vitro* was studied by measuring the ability of spleen extracts to degrade a high molecular weight fluorescein-PLL conjugate to smaller peptides. In both test systems, responder and non-responder guinea pigs were capable of degrading hapten-PLL to peptides in an indistinguishable manner (Levine and Benacerraf, 1964) (Figs. 1, 2). Thus, the effect of the "PLL gene" appears not to be on the initial degradation of PLL. The data of Schlossman *et al.* (1965) also support this view. These authors studied the effect of chain length on the ability of α-DNP-oligo-L-lysine peptides to induce an immune response in guinea pigs. They found that the α-DNP peptides containing 4, 5 or 6 lysyl residues were non-immunogenic; the α-DNP-lysyl peptides became immunogenic at the heptamer or the octamer. However, immune responses were obtained only in the responder guinea pigs, demonstrating that the immunogenicity of the smallest effective antigenic unit is under the control of the same gene.

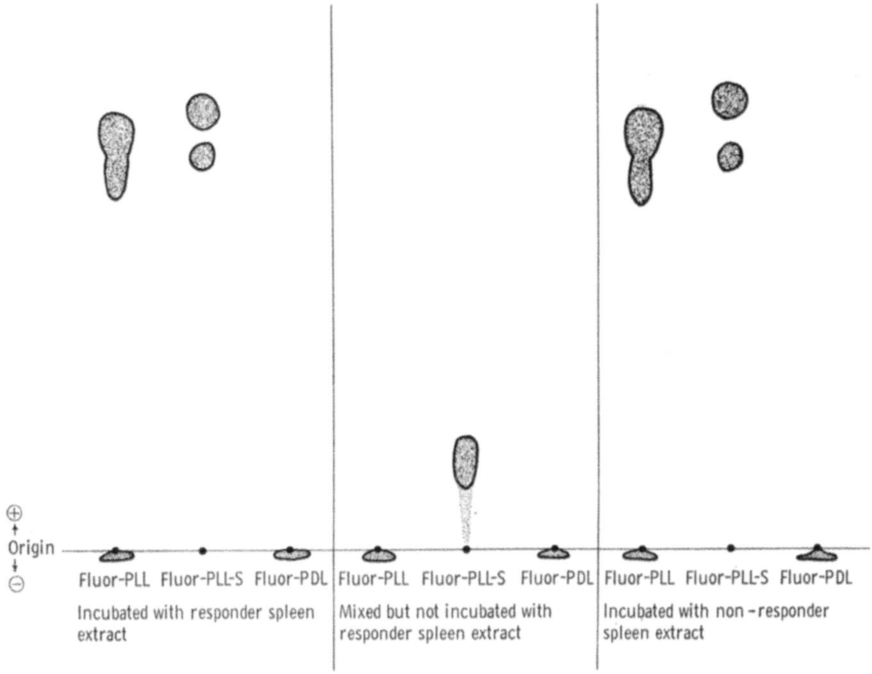

Fig. 1. Drawing of vertical starch gel electrophoresis (pH 8.5) of incubated mixtures of fluorescein-polylysine conjugates with aqueous extracts of responder and non-responder guinea pig spleens. Dark spots are areas of fluorescence seen under ultraviolet light. The fluor-PLL, after incubation with spleen, can penetrate into the gel indicating that it has undergone degradation. The poly-D-lysine conjugate showed no evidence of degradation. PLL-S is succinylated PLL. (From Levine and Benacerraf, 1964.)

Another possibility considered was that DNP-PLL might be taken up by macrophages of immunological tissues differently by responder and non-responder guinea pigs. To study this, 200 μg to 1.0 mg doses of H^3-tagged DNP-PLL was injected into the hind footpads of responder and non-responder animals. Popliteal nodes were excised at various times, and the uptake of H^3 into the node as well as its disappearance was followed by autoradiography. It was found that the uptake of the labelled DNP-PLL into macrophages and its disappearance over the next three weeks occurred indistinguishably in responder and non-responder animals (Vassalli, Levine, and Benacerraf, unpublished). Thus, the "PLL gene" does not appear to control the uptake of PLL-conjugates into phagocyte cells of immunological tissues. Further studies concerned with the fine structural aspects of the distribution of hapten-PLL in lymph nodes of responder vs. non-responder animals are not yet available.

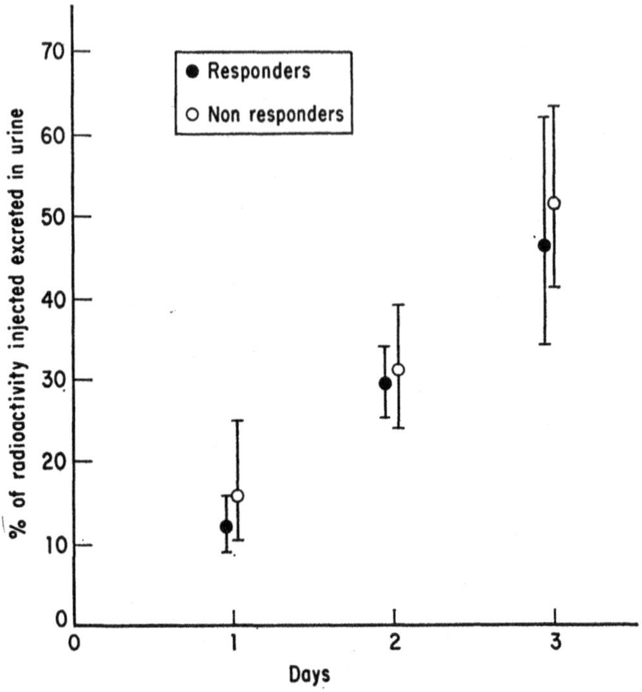

Fig. 2. Cumulative urinary excretions of radioactivity from responder and non-responder guinea pigs injected with 4.3 mg of H^3-DNP-poly-L-lysine intraperitoneally. Average and range of values from 4 guinea pigs are plotted. (From Levine and Benacerraf, 1964.)

V. IMMUNE RESPONSES TO DNP-PLL AGGREGATES WITH BSA

It might be considered that the effect of the "PLL gene" might be to code for a portion of the variable sequence of immunoglobulins concerned with specificity for the PLL portion of the hapten-PLL antigenic determinant unit. However, this possibility is rendered highly unlikely by the observation that non-responder guinea pigs can produce high titers of antibodies specific for DNP-PLL when immunized with aggregates prepared by mixing DNP-PLL with either ovalbumin or BSA at pH 7.5 (Green, Paul and Benacerraf, 1966). Thus even non-responder guinea pigs can produce antibodies with structural specificity for the PLL determinants when properly stimulated. This line of experiments is described and discussed in detail elsewhere in this volume (Green, Levine and Benacerraf, this section).

VI. EFFECT OF SUCCINYLATION OF DNP-PLL ON ITS IMMUNOGENICITY

The foregoing findings indicate that the physiological effect of the PLL gene is not upon the uptake of hapten-PLL by macrophages, nor on the enzymatic degradation of hapten-PLL into smaller peptides by immunological tissues. Nor does it appear likely that the "PLL gene" codes for part of the variable sequence of anti-hapten-PLL antibodies related to their partial specificity for poly-lysine structures. What then is the effect of the "PLL gene"? There is some evidence that, for protein immunogens to induce an immune response, the protein or protein fragments must first become bound or coupled to RNA to form the inducers of antibody synthesis (Fishman and Adler, 1963; Gottlieb *et al.*, 1967). Based on this view, it is conceivable that the "PLL gene" controls a metabolic step in the pathway of formation of the inducer from hapten-PLL. This step may be considered as being a metabolic operation upon hapten-PLL, or upon fragments of this immunogen, such as its coupling to RNA. The role of the "PLL gene" may then be to code for the synthesis of a coupling enzyme which has substrate specificity for a poly-L-lysine sequence. As yet there is no direct experimental data to support such an hypothesis. However, the observation that succinylation of the remaining free-lysine-ϵ-amino groups of hapten-PLL conjugates reduces their immunogenicity virtually to zero (Levine, 1964a) is consistent with this view. Succinylation does not interfere with the enzymatic degradability of hapten-PLL conjugates (cf. Fig. 1) (Levine and Benacerraf, 1964), nor does it greatly interfere with its ability to bind hapten-specific serum antibodies.

VII. FINAL COMMENTS

The hapten-PLL immunogens in guinea pigs emerges as a useful system with which to study certain aspects of the genetic controls and physiological mechanisms of the immune response. Its main advantages are the relative structural simplicity of these immunogens and the applicability of the comparatively clean techniques associated with the use of haptens in immunology. For this system, it has been shown that the ability of guinea pigs to respond immunologically to hapten-PLL is controlled by a single autosomal gene, termed the "PLL gene." Although the physiological effect of this gene in the sequence of steps leading to antibody synthesis is not yet known, it appears reasonably clear that it does not control the uptake of hapten-PLL into phagocytic cells, its initial degradation, or the synthesis of the immunoglobulin chains. For reasons discussed above, it is our present working hypothesis that the "PLL gene" may code for the synthesis of an enzyme with poly-L-lysine substrate specificity which catalyses a metabolic step such as the coupling to RNA of hapten-PLL or of smaller peptides derived from it. To establish clearly the site of control of the "PLL gene" will require biochemical studies on the fate of hapten-PLL conjugates in immunological tissues of responder and non-responder guinea pigs.

References

Arquila, E. R., and J. Finn (1965). Genetic control of combining sites of insulin antibodies produced by guinea pigs. J. Exp. Med., *122*:771.

Carlinfanti, E. (1948). The predisposition for immunity. J. Immunol., *59*:1.

Fishman, M., and F. L. Adler (1963). Antibody formation in vitro. II. Antibody synthesis in x-irradiated recipients of diffusion chambers containing nucleic acid derived from macrophage-incubated antigen. J. Exp. Med., *117*:595.

Fjord-Scheibel, I. (1943). Hereditary differences in the capacity of guinea pigs for the production of diphtheria antitoxin. Acta Path. et Microbiol. Scand., *20*:464.

Gell, P. G. H., and B. Benacerraf (1961). Delayed hypersensitivity to simple protein antigens. Adv. Immunol., *1*:319.

Gottlieb, A. A., V. R. Glisin, and P. Doty (1967). Studies on a macrophage RNA involved in antibody production. Proc. Nat. Acad. Sci., *57*:1849.

Green, I., W. E. Paul, and B. Benacerraf (1966). The behavior of hapten-poly-L-lysine conjugates as complete antigens in genetic responder and as haptens in non-responder guinea pigs. J. Exp. Med., *123*:859.

Ibsen, J. (1959). Differences in primary and secondary immunizability of inbred mice strains. J. Immunol., *83*:448.

Levine, B. B., A. Ojeda, and B. Benacerraf (1963a). Studies on artificial antigens. III. The genetic control of the immune response to hapten-poly-L-lysine conjugates in guinea pigs. J. Exp. Med., *118*:453.

Levine, B. B., A. Ojeda, and B. Benacerraf (1963b). Basis for the antigenicity of hapten-poly-L-lysine conjugates in random bred guinea pigs. Nature, *200*:544.

Levine, B. B. (1964a). Studies on antigenicity. The affect of succinylation of ε-amino groups on antigenicity of benzylpenicilloyl-poly-L-lysine conjugates in random bred and strain 2 guinea pigs. Proc. Soc. Exp. Biol. and Med., *116*:1127–31.

Levine, B. B. (1964b). Antigenicity of hapten conjugates of poly-L-lysine and poly-D-lysine in strain 2 guinea pigs. Nature, *202*:1009.

Levine, B. B. (1965). Inferences on the comparative binding affinities of antibodies mediating delayed and immediate hypersensitivity reactions in the guinea pig. J. Exp. Med., *873*:88.

Levine, B. B., and V. Levytska (1967). A sensitive hemagglutination assay method for DNP-specific antibodies. J. Immunol., *98*:648–52.

Maurer, P. H. (1964). Use of synthetic polymers of amino acids to study the basis of antigenicity. Progr. All., *8*:1–40.

McDevitt, H. O., and M. Sela (1965). Genetic control of the antibody response. I. The demonstration of determinant-specific differences in response to synthetic polypeptide antigens in two strains of inbred mice. J. Exp. Med., *122*:517.

Paul, W. E., G. W. Siskind, and B. Benacerraf (1966). Studies on the effect of the carrier molecules on antihapten antibody synthesis. II. Carrier specificity of anti-2,4-dinitrophenyl-poly-L-lysine antibodies. J. Exp. Med., *123*: 689.

Pinchuck, P., and P. H. Maurer (1965). Antigenicity of polypeptides. XVI. Genetic control of immunogenicity of synthetic polypeptides in mice. J. Exp. Med., *122*:673.

Sang, J. H., and W. R. Sobey (1954). The genetic control of the response to antigenic stimuli. J. Immunol., *72*:52.

Schlossman, S. F., A. Yaron, S. Ben-Efraim, and H. A. Sober (1965). Immunogenicity of a series of α-N-DNP-L-Lysines. Biochemistry, *4*:1638.

Sela, M. (1966). Immunological studies with synthetic polypeptides. Advances in Immunol., *5*:29–129.

Sobey, W. R., J. M. Magrath, and A. H. Reisner (1966). Genetically controlled specific immunological unresponsiveness. Immunol., *11*:511.

THE RELATIONSHIP OF THE HAPTEN TO THE CARRIER IN THE INDUCTION AND SPECIFICITY OF THE IMMUNE RESPONSE

IRA GREEN,* BERNARD B. LEVINE, WILLIAM E. PAUL, AND BARUJ BENACERRAF

New York University School of Medicine, New York, New York

The term hapten was introduced by Landsteiner in 1921 to describe a material (an alcoholic extract of horse kidney) which had the property of being able to combine with antibody, and yet by itself, was unable to initiate an immune response when injected into rabbits. Landsteiner and Simms (1923) were able to demonstrate also that when a foreign protein (pig serum) was added to this material, and this mixture then injected into rabbits, the animals did make an antibody capable of combining with the alcoholic extract of horse kidney. Thus, in this early example the alcoholic extract was the hapten and the pig serum was the carrier.

Forty-six years have now elapsed and this phenomenon has been both extensively studied and made use of to prepare antibodies against many small molecules, but as yet no clear explanation of the phenomenon itself is available. An understanding of the hapten-carrier relationship is closely connected with several other problems, namely: the requirements for immunogenicity, how the immune response is initiated and how antibody specificity is determined, and the relationship between delayed sensitivity (D.S.) and antibody synthesis.

An analysis of the hapten-carrier relationship is complicated by the fact that the usual covalent interaction of the hapten with the protein carrier may cause changes in the tertiary structure of the protein molecule which have important immunologic consequences. For example, guinea pig albumin (GPA) alone does not elicit an immune response in

* Present Address: Laboratory of Immunology, National Institute of Allergy and Infectious Diseases, NIH, Bethesda, Md.

guinea pigs. However, when a hapten such as the dinitrophenyl group is covalently linked to GPA then the conjugate is immunogenic for guinea pigs. Furthermore, after animals are immunized with hapten-protein conjugates, antibodies of many different specificities are produced (Landsteiner, 1945). Some of these antibodies appear to have a specificity directed against determinants on the protein carrier which result from its covalent interaction with the hapten (Haurowitz, 1942; Eisen *et al.*, 1954; St. Rose and Cinader, 1967; Paul *et al.*, 1967). Other antibodies appear to have specificities directed against unaltered portions of the protein carrier. Finally, the antibodies which are directed mainly against the hapten itself, upon close analysis also appear to have some degree of carrier specificity (Levine, 1963; Borek and Silverstein, 1965; Paul *et al.*, 1966).

Another remarkable property of hapten-protein systems is the immunological specificity of immune reactions which depend upon the interaction of actively sensitized cells with antigen such as delayed hypersensitivity reactions (Benacerraf and Levine, 1962), antigen-stimulated DNA synthesis (Dutton and Bulman, 1964) and anamnestic reactions (Ovary and Benacerraf, 1963). The specificity of these reactions, when elicited by hapten-protein conjugates, is characterized by an important contribution of the carrier molecule. Thus, actively sensitized cells not only do not respond to the hapten but also require for optimal responses, the hapten conjugated to the carrier molecule used for immunization.

Recently it has been found that the immune responses of guinea pigs to a simple synthetic poly α amino acid, poly-L-lysine, (PLL) and to hapten conjugates of PLL (H-PLL) is under the control of a single autosomal dominant gene (Kantor *et al.* 1963; Levine *et al.* 1963a). This same gene also controls the response to a linear copolymer of glutamic acid and lysine (GL) and to hapten conjugates of GL. Responder animals carrying this gene form specific antihapten antibodies and display delayed sensitivity (D.S.) after immunization with H-PLL conjugates such as 2,4-dinitrophenyl-PLL (DNP-PLL). It should be noted that responder guinea pigs can, after immunization, make antihapten antibodies to any of the several different hapten-PLL conjugates tried (Levine *et al.*, 1963b). In contrast, guinea pigs lacking the PLL gene do not, after specific immunization, form appreciable serum levels of antihapten antibodies nor display D.S. reactions to DNP-PLL or DNP-GL. Nevertheless, DNP-PLL can behave as a hapten in these genetic non-responder animals, i.e., immunization of these animals with positively charged DNP-PLL complexed to negatively charged foreign albumins elicits the formation of high levels of antihapten antibodies in these animals. However, such animals do not manifest D.S. reactions to DNP-PLL alone (Green *et al.*, 1966).

An analysis of this immune system to study hapten-carrier relationships offers several distinct advantages over studies performed with the hapten conjugates of complex protein antigens. First, the structure of the material used, DNP-PLL, is more precisely defined; second, the same material, DNP-PLL, can behave either as a complete antigen in genetic responder animals or as a hapten in non-responder animals; third, the specificities of the antibodies directed against the hapten and against the carrier can be more exactly determined; and fourth, the electrostatic interaction between the DNP-PLL and the albumin carrier molecule might be expected to produce a minimal amount of distortion of the carrier albumin molecule.

The properties of this system will be summarized first and then they will be related to theories of how the carrier molecule might operate in the immune system.

In contrast to the high levels of serum anti-DNP antibodies produced by non-responder guinea pigs immunized with DNP-PLL complexed with bovine serum albumin (BSA) or ovalbumin, guinea pig albumin (GPA) was only minimally effective as a carrier molecule. Non-responder guinea pigs immunized with DNP-PLL complexed to non-antigenic, non-protein polyanions did not produce anti-DNP antibodies. The anti-DNP antibodies produced in responder animals after immunization with DNP-PLL have partial specificity for PLL as demonstrated by the technique of fluorescence quenching (Paul *et al.,* 1966). The anti-DNP antibodies produced by non-responder guinea pigs after immunization with DNP-PLL complexed to a foreign albumin show exactly the same degree of partial PLL specificity as do the antibodies produced by responder animals immunized with DNP-PLL alone (Green *et al.,* 1966). Furthermore, these antibodies did not show any significant binding for the carrier albumin molecule (Paul and Green, 1966).

Next, the cellular site of synthesis of anti-DNP antibodies and of anti-carrier antibodies was investigated in non-responder animals. It was found that after immunization of such animals with, for example, DNP-PLL.BSA, that even though the BSA is essential for the anti-DNP antibody to be produced, the anti-DNP antibodies and the anti-BSA antibodies were always produced in separate cells (Green *et al.,* 1967).

As mentioned above, guinea pig albumin was found to be only minimally effective as a carrier molecule for DNP-PLL in non-responder animals. In order to investigate the basis for the requirement for an antigenic carrier molecule, we next investigated the effect of tolerance to the foreign carrier (albumin) molecule in non-responder guinea pigs on the ability of the albumin to induce antibodies against the hapten (Green *et al.,* 1968). It was indeed found that non-responder guinea pigs made tolerant to BSA according to the method of Salvin and Smith (1964) did not make anti-DNP antibody after subsequent immunization with DNP-PLL.BSA. These results are shown in Tables 1 and 2.

The *in vitro* reactions to DNP-PLL of lymph node cells from responder and non-responder guinea pigs immunized with DNP-PLL or DNP-PLL.BSA was also investigated to compare the reactivity of these cells with the immune response of the animals (Green *et al.,* 1968). Lymph node cells derived from genetic responder animals immunized with DNP-PLL or DNP-GL proliferated *in vitro* in response to DNP-PLL or DNP-GL, respectively, as measured by the incorporation of ^{3}H thymidine into the DNA of these cells. However, the lymph node cells derived from non-responder animals immunized with DNP-PLL or DNP-GL did not show any proliferation after *in vitro* exposure to DNP-PLL or DNP-GL. When lymph node cells from non-responder animals immunized with DNP-PLL.BSA were exposed to DNP-PLL *in vitro* no stimulation of proliferation was seen in spite of the fact that these cell populations were producing anti-DNP antibodies. However, such cell popula-

Table 1

PRODUCTION OF ANTI-BSA ANTIBODIES IN GENETIC NON-RESPONDER GUINEA PIGS RENDERED TOLERANT TO BSA AND IMMUNIZED WITH DNP-PLL.BSA

		Anti-BSA Antibodies	
Animals	Immunizing Antigens	Gel Diffusion *	Passive Hemagglutination
Tolerant animals	DNP-PLL.BSA		
1	↓	−	<1/20
2		−	<1/20
3		−	<1/20
4		−	<1/20
5		−	<1/20
6		−	1/20
7		+	1/40
8		+	1/40
9		−	1/640
12 control animals not rendered tolerant	DNP-PLL.BSA	All 12 animals +	1/160 to 1/1280 in All 12 animals

* With BSA 0.25 mg/ml.

Table 2

PRODUCTION OF ANTI-DNP-PLL ANTIBODIES IN GENETIC NON-RESPONDER GUINEA PIGS RENDERED TOLERANT TO BSA AND IMMUNIZED WITH DNP-PLL.BSA

Animals	Immunizing Antigens	Anti-DNP-PLL Antibodies: Gel Diffusion †	Anti-DNP-PLL Antibodies: Equilibrium Dialysis ‡ (mM hapten bound/ml serum $\times 10^{10}$)
Tolerant * animals	DNP-PLL.BSA		
1	↓	−	422
2	↓	−	158
3	↓	−	374
4	↓	−	135
5	↓	−	1308
6	↓	−	750
7	↓	−	1500
8	↓	+	1805
9	↓	+	6960
12 control animals not rendered tolerant	DNP-PLL.BSA	All 12 animals +	Mean 5900 Range 2900 to 8200
41 control animals	Complete adjuvant alone	N.D.	Mean 240 ± 205

* These are the same animals as shown in Table 1.
† With DNP-GPA 0.25 mg/ml.
‡ Equilibrium dialysis performed with ^{3}H-DNP-EACA at an initial concentration of 2×10^{-7} M. 7.5 μg of antibody can bind 1000×10^{-10} mM hapten when both sites are occupied, considering the M.W. for guinea pig antibodies to be 150,000.

tion were stimulated to proliferate *in vitro* after exposure to DNP-PLL. BSA or BSA alone. These results are shown in Tables 3 and 4.

Two general hypotheses can be proposed concerning the action of the carrier molecule in inducing antibody synthesis against the hapten it bears. The above data will be interpreted in terms of these two general hypothesis.

(1) The carrier molecule allows antibody to the hapten to be synthesized by forming with the hapten a new and complete antigenic deter-

Table 3

EFFECT OF ANTIGEN ON THE *In Vitro* INCORPORATION OF ^{3}H THYMIDINE IN LYMPH NODE CELLS OF GENETIC RESPONDER GUINEA PIGS IMMUNIZED WITH DNP-PLL, DNP-PLL.BSA OR DNP-GL

			Test Antigens			
No. of Animals	Immunizing Antigen	No. Days After Immunization	DNP-PLL 10 μg/ml	DNP-PLL 1 μg/ml	DNP-PLL.BSA 1 μg/ml	BSA 1 μg/ml
5	Adjuvant & Saline	10–19	1.2 * (1.39–0.70)	1.07 (1.22–0.97)		
3	DNP-PLL	75–85	9.1 (10.5–7.7)	9.5 (20.9–3.4)		
2	DNP-PLL.BSA	15	5.69 (6.6–4.7)	3.05 (4.4–1.75)	6.84 (6.84–6.84)	1.58 (1.6–1.56)
2	DNP-GL	13–19	DNP-GL 100 μg/ml 7.71 (7.65–6.78)	DNP-GL 1 μg/ml 6.74 (7.30–5.64)	DNP-GL 0.01 μg/ml 3.14 (3.72–2.56)	

* This number is the ratio $\frac{\text{counts in DNA from experimental cultures with antigen}}{\text{counts in DNA from control cultures without antigen}}$

Values >1 indicate stimulation of DNA synthesis.

Table 4

EFFECT OF ANTIGEN ON THE *In Vitro* INCORPORATION OF ^{3}H THYMIDINE IN LYMPH NODE CELLS FROM GENETIC NON-RESPONDER GUINEA PIGS IMMUNIZED WITH DNP-PLL, DNP-PLL.BSA OR DNP-GL

No. of Animals	Immunizing Antigen	No. Days After Immunization	Test Antigens			
			DNP-PLL 10 μg/ml	DNP-PLL 1 μg/ml	DNP-PLL.BSA or DNP-PLL.OVA 1 μg/ml	OVA 1 μg/ml
4	DNP-PLL	14–15	0.712 * (0.91–0.40)	1.02 (1.4–0.77)		
2	DNP-PLL.BSA	22		1.24 (1.26–1.22)	7.2 N.D.	
7	DNP-PLL.OVA	11–21	0.707 (1.47–0.438)	1.04 (1.41–0.54)	4.8 (10.2–1.86)	3.0 (5.0–2.3)
2	DNP-GL	13–19	DNP-GL 10 μg/ml 1.13 (1.17–1.09)	DNP-GL 1 μg/ml 1.15 (1.20–1.11)	DNP-GL 0.01 μg/ml 0.93 (1.37–0.50)	

* This number is the ratio $\frac{\text{counts in DNA from experimental cultures with antigen}}{\text{counts in DNA from control cultures without antigen}}$

Values >1 indicate stimulation of DNA synthesis.

minant. This interpretation postulates that the specificity of the anti-hapten antibodies involve both the hapten and adjacent areas of the carrier molecule. According to this view, the energy that will allow binding of a sufficient amount of antigen by cell-bound antibody, to induce the immune response by causing a proliferation of a "specific" cell, is only provided by adding the contribution of the carrier to the immunological specificity of the hapten. In support of this interpretation, there has been ample evidence for carrier specificity in hapten-protein and hapten-polypeptide systems for reactions, such as D.S. and anamnestic reactions, between actively sensitized cells and antigen. Indeed, in the hapten-PLL systems, an unequivocal example of antibody with carrier specificity has been reported in the case of anti-DNP-PLL antibodies produced by genetic responder guinea pigs (Paul *et al.*, 1966). These antibodies show partial specificity for the PLL carrier molecule.

(2) According to an alternative mechanism, the carrier molecule is first recognized as immunogenic by an as yet poorly understood process, the specificity of which may be unrelated to antibodies that may be produced later to determinants on the carrier molecule. Only following this specific recognition and processing step can antibodies be produced to determinants on the molecule.

The first hypothesis, although discussed above in terms of hapten-carrier relationships, also has been considered to be the mechanism of "selection" and then stimulation of proliferation of cells carrying a small amount of the corresponding antibody to ordinary protein antigens. This first hypothesis, in its simplest form, has been, in a sense, central to "selective" theories of antibody formation. The second hypothesis is more indirect, complicated, and ill-defined. The existence of such processing or inducing steps, performed on the antigen, has been hinted at by the work of Fishman and Adler (1963), Askonas and Rhodes (1965), Friedman *et al.* (1965), and Gottlieb *et al.* (1967). These studies suggest that an RNA-antigen moiety produced by macrophages early in the immune response may be a necessary step in the pathway that leads to antibody production. The results of our own studies of the hapten-carrier relationship in the PLL immune system in guinea pig favors the second hypothesis rather than the first hypothesis for the following reasons:

According to the first hypothesis the hapten and carrier, in our case DNP-PLL and BSA, combine to form a complete antigenic determinant which binds an antibody on a cell with a sufficient energy to trigger proliferation of that cell with the subsequent production of large amounts of that antibody. Such an antibody to DNP-PLL produced in non-responder animals immunized with DNP-PLL.BSA therefore should have considerable specificity for the BSA. However, the anti-DNP-PLL antibody made by these animals appears to be identical with anti-DNP-PLL antibodies made by responder animals and the anti-DNP-PLL antibody has not as yet been demonstrated to have any BSA specificity. Indeed, in the non-responder animals the anti-DNP-PLL antibodies and the anti-BSA antibodies are made in separate cells. Thus the data do not fit the first hypothesis.

In addition, the fact that when the immunological machinery is presented with an antigen bearing two determinants DNP-PLL.BSA, or $DNP_{16}BGG$, the antibodies to the different determinants are produced in separate cells (Green *et al.*, 1967c), indicates that the mere "hitting" and "triggering" of multipotential precursor cells by antigens is not consistent with these observations. Another argument against the first hypothesis is that it is very unlikely that the mode of action of the PLL gene operates by controlling the sequence of amino acids in the variable portion of the immunoglobulin molecule which determines antibody specificity. The reason for this is the fact that antibodies can be produced in responder animals against any of several unrelated haptens conjugated to PLL (Levine *et al.*, 1963b); therefore, one would have to postulate, that the PLL gene codes for a segment of immunoglobulin common to all of these diverse antibodies. At present this possibility is not compatible with any known genetic scheme of antibody synthesis. The PLL gene thus can be thought of as acting according to the second hypothesis, to perform some enzymatic processing step on the PLL prior to the recognition of individual antigenic determinants on the molecule.

Another piece of evidence which does not favor the first hypothesis is that in the *in vitro* tissue culture experiments (Green *et al.*, 1967b) the DNP-PLL alone did not stimulate the lymph node cells derived from non-responder animals immunized with DNP-PLL.BSA in spite of the fact that these cell populations contained cells producing anti-DNP-PLL antibodies. Thus, the mere contact of cells producing antibody with a high molecular weight hapten is not sufficient to cause proliferation of such cells. The results of these experiments suggests rather that some type of specific antigen-inducer complex made under the direction of the PLL gene may be necessary for the *in vitro* stimulation of proliferation of lymph node cells and that the mere meeting of an antigenic determinant with a cell bearing an antibody directed against it is not sufficient to cause this cell to proliferate. An alternative interpretation of this particular experiment is that the proliferative response of these cells is not a consequence of contact of these cells with "processed antigen" but rather the processing step itself and the proliferation of cells is very closely related.

The prior establishment of tolerance to the BSA carrier in non-responder guinea pigs abolished the production of anti-DNP-PLL antibodies in these animals when they were immunized with DNP-PLL.BSA.

Thus the carrier must be capable of eliciting an immune response itself in order to induce an antibody to a hapten which it bears.*

According to the first hypothesis the mechanism of action of tolerance to BSA, producing tolerance to DNP-PLL.BSA in non-responder animals, would be the elimination of those cells whose response to DNP-PLL.BSA is dependent on the partial specificity contributed by the portion of the BSA which is postulated to be part of the antigenic determinant of DNP-PLL.BSA. Again, the major difficulty in interpreting this experiment in terms of the first hypothesis is the failure to demonstrate a significant degree of BSA specificity in the anti-DNP-PLL antibodies produced in non-responder animals. According to the second hypothesis, the mechanism by which tolerance to BSA produces tolerance to DNP-PLL.BSA in non-responder animals would be as follows: In untreated non-responder animals, immunized with DNP-PLL.BSA, the required initial inducer step is performed on the BSA rather than on the DNP-PLL; as a consequence the closely associated DNP-PLL determinant is then recognized by the immunological machinery and an antibody is made against it. In an animal tolerant to BSA this initial inducing step may not be able to be performed and therefore no antibody to any determinant on BSA can be synthesized. These experiments also suggest that tolerance can be induced at more than one site along the immunological pathway.

The general nature of the processing step can be thought of as an initial step necessary for the recognition of immunogenicity and the recognition by the immune mechanism of individual antigenic determinants. This initial processing step appears to be one which is genetically controlled in the PLL system; the specificity of this step seems to be different from that of antibodies later to be produced against individual determinants carried on an immunogenic complex. More complex molecules, such as normal protein antigens, might be expected to offer a number of different sites for the operation of genetically controlled processing steps of similar nature but of many different specificities. Thus, in such more complex protein molecules the lack of a site for the operation of a processing step of a single specificity would not be as readily appreciated. Only in a simple homopolymer such as PLL in which a single

* This finding is in apparent contradiction with the observation that hapten-protein immunization of rabbits tolerant to the carrier protein may lead to the production of anti-hapten antibodies. However, these two situations are not completely analogous because it has been shown that the usual covalent conjugation procedure alters the protein carrier molecule sufficiently to produce new antigenic configurations; such configurational changes of the protein molecule may not encompass the hapten. The simple electrostatic aggregation of BSA with DNP-PLL probably does not modify the structure of the BSA molecule to the same extent as does covalent interaction and therefore such new antigenic determinants on the BSA molecule would be less likely to appear.

processing site is offered would this strict type of genetic control of the recognition of immunogenicity be clearly demonstrable. Thus, the role of the carrier in the induction and specificity of the immune response can be viewed as allowing the action of, and providing the site for, this postulated processing step.

The two general hypotheses as to the action of the carrier molecule in the immune response have been discussed as rather separate and mutually exclusive hypotheses in order to delineate more clearly the differences between them, and to stress the fact that the first hypothesis is probably an oversimplification. However, it should be recognized that these theories may not be really separate or mutually exclusive, but rather, that a synthesis of the two hypotheses may be appropriate which would suggest that a sequential series of events may take place. That is, a genetically controlled initial enzymatic processing step may take place on the "antigen," at a later step the products of this first interaction (processed antigenic determinants) would contact specific cell-associated antibody on a thermodynamically driven basis which would then cause a proliferation of these particular cells, with subsequent increased production of the antibody initially associated with them.

SUMMARY

The hapten-carrier relationship has been investigated in a model system in which the immune response to hapten-poly-L-lysine conjugates is under genetic control. In this system one material, dinitrophenylated poly-L-lysine, can behave either as a complete antigen or as a hapten. The data obtained by an analysis of this system suggests that the role of the carrier is to allow and to provide the site for an initial necessary recognition and processing step. This step is genetically controlled and the specificity of this step appears to be unrelated to the specificity of the antibody that is later produced to determinants on the immunogenic molecule.

References

Askonas, B. A., and J. M. Rhodes (1965). Immunogenicity of antigen containing ribonucleic acid preparations from macrophages. Nature, *205*:470–474.

Benacerraf, B., and B. B. Levine (1962). Immunological specificity of delayed and immediate hypersensitivity reactions. J. Exper. Med., *115*:1023–1036.

Borek, F., and A. M. Silverstein (1965). Specificity of guinea pig antibodies and delayed hypersensitivity. Nature, *205*:299–300.

Dutton, R. W., and H. N. Bulman (1964). The significance of the protein carrier in the stimulation of DNP synthesis by hapten-protein conjugates in the secondary response. Immunol., *7*:54–64.

Eisen, H. N., M. E. Carsten, and S. Belman (1954). Studies of hypersensitivity to low molecular weight substances. III. The 2,4 dinitrophenyl group as a determinant in the precipitation reaction. J. Immunol., *73*:296–308.

Fishman, M., and F. L. Adler (1963). Antibody formation initiated in vitro. J. Exper. Med., *117*:595–602.

Friedman, H. P., A. B. Stavitsky, and S. M. Solomon (1965). Induction in vitro of antibodies to phage T_2; antigens in the RNA extract employed. Sci., *149*:1106–1108.

Gottlieb, A. A., J. R. Glišin, and P. Doty (1967). Studies on macrophage RNA involved in antibody production. Proc. Nat. Acad. Sci., *57*:1849–1856.

Green, I., W. E. Paul, and B. Benacerraf (1966). The behavior of hapten poly-L-lysine conjugates as complete antigens in genetic responder and as haptens in non-responder guinea pigs. J. Exper. Med., *123*:859–879.

Green, I., P. Vassalli, and B. Benacerraf (1967). Cellular localization of anti-DNP-PLL and anticonveyor albumin antibodies in genetic non-responder guinea pigs immunized with DNP-PLL albumin complexes. J. Exper. Med., *125*:527–536.

Green, I., W. E. Paul, and B. Benacerraf (1968). Hapten carrier relationships in the DNP-PLL foreign albumin complex system: Induction of tolerance and stimulation of cells *in vitro*. J. Exper. Med., *127*:43–53.

Green, I., P. Vassalli, V. Nussenzweig, and B. Benacerraf (1967). Specificities of the antibodies produced by single cells following immunization with antigens bearing two types of antigenic determinants. J. Exper. Med., *125*:511–526.

Haurowitz, F. (1942). Separation and determination of multiple antibodies. J. Immunol., *43*:331–340.

Kantor, F. S., A. Ojeda, and B. Benacerraf (1963). Studies on artificial antigens. I. Antigenicity of DNP-polylysine and DNP copolymers of lysine and glutamic acid in guinea pigs. J. Exper. Med., *117*:55–69.

Landsteiner, K. (1921). Über heterogenetisches antigen und hapten XV Miteilung über antigene. Biochem. Zeitschrift, *119*:294–306.

Landsteiner, K., and S. Simms (1923). Production of heterogenetic antibodies with mixtures of the binding part of the antigen and protein. J. Exper. Med., *38*:127–138.

Landsteiner, K. (1945). The Specificity of Serological Reactions. Harvard University Press, Cambridge.

Levine, B. B. (1963). Studies on the dimension of the rabbit anti-benzylpenicilloyl antibody-combining sites. J. Exper. Med., *117*:161–183.

Levine, B. B., A. Ojeda, and B. Benacerraf (1963). Studies on artificial antigens. III. The genetic control of the immune response to hapten poly-L-lysine conjugates in guinea pigs. J. Exper. Med., *118*:953–957.

Levine, B. B., A. Ojeda, and B. Benacerraf (1963). Basis for the antigenicity of hapten poly-L-lysine conjugates in random bred guinea pigs. Nature, *200*: 544–545.

Paul, W. E., G. W. Siskind, and B. Benacerraf (1967). A study of the termination of tolerance to BSA with DNP-BSA in rabbits: Relative affinities of the antibodies for the immunizing and paralyzing antigens. Immunol., *13*:147–157.

——— (1966). Studies on the effect of the carrier molecule on antihapten antibody synthesis. II. Carrier specificity of anti-2,4 dinitrophenyl-poly-L-lysine antibodies. J. Exper. Med., *123*:689–705.

Paul, W. E., and I. Green (1966). Unpublished observations.

St. Rose, J. E. M., and B. Cinader (1967). The effect of tolerance on the specificity of the antibody response and on immunogenicity. J. Exper. Med., *125*:1031–1055.

Salvin, S. B., and B. F. Smith (1964). The specificity of allergic reactions. VII. Immunologic unresponsiveness, delayed hypersensitivity and circulating antibody to proteins and hapten protein conjugates in adult guinea pigs. J. Exper. Med., *119*:851–868.

IMPORTANCE OF IMMUNOGENICITY OF THE CARRIER IN INDUCING A RESPONSE AGAINST CARRIER-SYNTHETIC POLYMER AGGREGATES *

PAUL H. MAURER AND PAUL PINCHUCK

Department of Biochemistry, Jefferson Medical College, Philadelphia, Pennsylvania

I. IMMUNOGENICITY OF SYNTHETIC POLYPEPTIDES

During the past decade various types of synthetic polymers of amino acids have been used to study many areas of immunology and immunochemistry. Perhaps the most valuable information obtained has been relative to the requirements for immunogenicity, that is, the ability of a macromolecule to cause an animal to produce antibodies having specificity for some part of the polypeptide.

In our own research we have studied the immune response to linear, essentially random polymers of α-L and α-D amino acids in various species. Table 1 presents a summary of the results obtained which were generally obtained following immunization with polymer emulsified in Freund's complete adjuvant (Maurer, 1964). The responses were evaluated with several *in vivo* and *in vitro* immunological techniques, such as delayed hypersensitivity, active and passive anaphylaxis, passive hemagglutination, gel diffusion and quantitative precipitin techniques. On the whole, the types of immune responses noted were typical of those obtained with classical protein antigens. Responses limited to a single immune manifestation, such as delayed sensitivity, were found only in special situations (Ben-Efraim *et al.,* 1963).

As indicated in Table 1, α-L homopolymers are not immunogenic in

* Paper 27 in the series of Antigenicity of Polypeptides (poly-α-amino acids). This work was supported by USPHS Grants AI-07825 and T1 AI 334 from the National Institute of Allergy and Infectious Diseases and Research Contract No. DADA 17-67-C-7021 from the Department of the Army, Office of the Surgeon General.

Table 1

Response of Various Species to Synthetic Polypeptides

Polymer	Species		
	Rabbit	Guinea Pig	Mouse
α-L Homopolymers	– [1]	–	–
α-L Copolymers	± [2]	±	–
α-L Terpolymers	+ [3]	+	+
α-D (1–4 Amino Acids)	–	–	–

[1] None of animals responded = –.
[2] Some of animals responded = ±.
[3] All of animals responded = +.

these species, nor are they immunogenic in other species studied, such as man and chicken. Polymers of two amino acids (copolymers) are immunogenic in some rabbits and some guinea pigs. This ability to respond has been shown to be genetically transmitted (Levine and Benacerraf, 1965). Mice have not shown any detectable response initiated by copolymers (Pinchuck and Maurer, 1965). In all species, polymers containing three or four α-L amino acids have been excellent immunogens. All polymers composed solely of 1–4 α-D amino acids have not been immunogenic in rabbits, guinea pigs, mice and humans (Maurer, 1965). Recently there have been reports by Janeway and Sela (1967) and Gill *et al.* (1967) that responses to terpolymers of α-D amino acids can be obtained if low doses are used for immunization. This point will be discussed more fully in section IV. However, all are agreed that the responses to α, D-polymers are significantly different from those against α, L-polymers.

II. IMMUNOLOGICAL REACTIVITY OF NON-IMMUNOGENIC POLYPEPTIDES

Although the simplest explanation for non-immunogenicity is that the animal lacks the ability (information) to form antibody-combining sites complementary to these polypeptides, the data presented in this publication and the findings of others indicate that this cannot be the correct explanation. The fact that terpolymers which are very similar to non-immunogenic copolymers (e.g., those in which there is only 4–5 mol% of a third amino acid), are immunogenic and the data on the extensive cross reactions of non-immunogenic α-L homopolymers and copolymers with the antibody induced in response to the related but immunogenic materials would indicate that the animal does not lack the ability to form the appropriate antibody.

A. Induction of Tolerance

In some of our earlier studies we had found that rabbits could be made tolerant to α-L co- and ter-polymers by neonatal injection of the appropriate polymer in solution. The timing and doses required were similar to those involved in inducing tolerance to bovine serum albumin in neonatal rabbits. In addition, as shown in Table 2, neonatal injections into rabbits of high molecular weight polyglutamic acid (PGA) induced tolerance to immunization with the copolymer $Glu_{60}Ala_{40}$ (Maurer *et al.*, 1965). For tolerance to be produced, neonatal injection of the PGA was required. The tolerance was specific, since the response to an immunologically distinct polymer ($Glu_{42}Lys_{28}Ala_{30}$) was not affected. As has been noted with serum proteins, on prolonged immunization with $Glu_{60}Ala_{40}$ some of the animals "escaped" tolerance and produced antibody.

B. Reaction with Antibodies Induced by Related Polymers

1. *Polyglutamic Acid—Rabbit Anti-*$Glu_{60}Ala_{40}$

As might be expected from the tolerance studies reported above, rabbit antisera to $Glu_{60}Ala_{40}$ do cross-react with PGA. Evidence for this, obtained by several immunochemical techniques, is presented in Table 3 (Maurer *et al.*, 1964). Significant inhibition of the homologous precipitin reaction has also been observed with α-L glutamic acid oligomers (4–6 residues) (Gerulat and Maurer, unpublished).

2. $Glu_{90}Tyr_{10}$ *and* $Glu_{60}Ala_{40}$*—Mouse Anti-*$Glu_{60}Ala_{30}Tyr_{10}$

As mentioned in section I, mice do not respond to copolymers, but do respond to terpolymers in which only a small amount of the third amino

Table 2

Effect of Neonatal Injections of Polyglutamic Acid in Rabbits on Subsequent Response to $Glu_{60}Ala_{40}$ Polymer

Polymer Injected			Immune Response	
	Adult			
Neonatally	I.V.[1]	Adjuvant[2]	1st Course No. of Reactors	2nd Course No. of Reactors
Glu_{100}(50–100)[3]	Glu_{100}	$Glu_{60}Ala_{40}$	0/5	0/3
None	Glu_{100}	$Glu_{60}Ala_{40}$	3/3	3/3
Glu_{100}(100)	Glu_{100}	$Glu_{42}Lys_{28}Ala_{30}$	5/5	5/5

[1] Injected on day 75.
[2] Immunized on day 90.
[3] Dose injected, mg.

Table 3

CROSS REACTIONS OF HYPERIMMUNE RABBIT ANTI-$GLU_{60}ALA_{40}$ SERA

Polymer	Per Cent Cross Reaction [1] C.F.	Pptn.	PCA Titer
$Glu_{60}Ala_{40}$ (L,L) [2]	100	100	1000
$Glu_{60}Ala_{40}$ (D,D)	6	0	200
$Glu_{90}Ala_{10}$ (L,L)	65	28	—
Glu_{100} (L)	61	0	200
Glu_{100} (D)	0	0	100

[1] C.F. = Quantitative Complement Fixation; Pptn. = Quantitative Precipitation; PCA = Passive Cutaneous Anaphylaxis.
[2] Optical configuration of amino acids.

acid is present. In this system too, extensive cross reactions with the non-immunogenic copolymers are found. Table 4 summarizes the reactions of a number of sera from inbred mouse strains containing antibody against $Glu_{60}Ala_{30}Tyr_{10}$, with the related non-immunogenic copolymers $Glu_{90}Tyr_{10}$ and $Glu_{60}Ala_{40}$. Neither polyglutamic acid nor polytyrosine have been found to react with the antibody. The apparent inconsistencies (e.g., greater than 100% cross reaction with $Glu_{90}Tyr_{10}$) may be due to the presence of soluble antigen-antibody complexes in the homologous GAT–anti-GAT system (Pinchuck and Maurer, 1968a).

These results indicate that the non-immunogenicity of homo- and copolymers might be due to a lack of ability to "trigger" an immune re-

Table 4

PRECIPITIN CROSS REACTIONS OF ANTI-$GLU_{60}ALA_{30}TYR_{10}$ SERA FROM DIFFERENT INBRED MOUSE STRAINS

Strain	Antibody Precipitated by $Glu_{60}Ala_{30}Tyr_{10}$ µg N/ml	Per Cent Cross Reaction $Glu_{60}Ala_{40}$	$Glu_{90}Tyr_{10}$
CBA/J	102	0	123
BALB/cJ	160	14	134
DBA/2J	249	16	82
C3H/HeJ	421	23	123
129/J	802	40	95

sponse rather than to the inability to produce antibody containing the specific combining sites for the copolymers.

III. NON-IMMUNOGENIC α-L POLYPEPTIDES AS HAPTENS

A consideration of these findings brought to mind the situation found with haptens. Although haptens are immunogenic when coupled to an appropriate protein carrier, they are not immunogenic by themselves. The materials with which we are concerned are much larger than classical haptens; typical molecular weights for a non-immunogenic polypeptide range from 20–50,000, while haptens are in the 100–200 molecular weight range.

It is convenient to divide the immune response into 2 phases: (1) the carrier phase, concerned with triggering of the response, and (2) the hapten phase, concerned with the specificity of the antibody. Non-immunogenic polypeptides would then lack the first property, but would possess the hapten property, in common with such other non-immunogenic materials as classical haptens and macromolecules such as polysaccharides, nucleic acids, and lipids.

A. "External" Carriers

The introduction of the use of "charge" aggregates of methylated bovine serum albumin (MBSA) and DNA suggested that this technique might convert non-immunogenic polypeptides to immunogens (Plescia *et al.*, 1964). Such an aggregate would be analogous to a hapten conjugated protein, and one might consider the charged albumin derivative as an external carrier.

1. *Homopolymers of α-L Amino Acids in Rabbits and Guinea Pigs*

Table 5 presents data obtained when rabbits and guinea pigs were immunized with MBSA aggregates of polyglutamic acid and polyaspartic acid, or poly-L-lysine aggregates formed with acetylated or phosphorylated albumin. In both species the technique worked well with high molecular weight preparations of polyglutamic acid, but not with low molecular weight PGA, polyaspartic, or polylysine (Maurer, 1965). However it is known that if these oligopeptides are directly coupled to a protein (polypeptidyl proteins) antibody formation can be induced.

2. *DNP-Homopolymers in Mice*

The excellence of the response to the 2,4-dinitrophenyl (DNP) group in various species, and the ease of detecting anti-DNP antibodies suggested that non-immunogenic polymer-DNP conjugates would provide a sensitive test of these ideas. In common with copolymers, DNP-poly-L-

Table 5

RESPONSE TO IMMUNIZATION WITH POLYPEPTIDE (α-L AMINO ACID) AGGREGATE COMPLEXES [1]

Polymer	Carrier [2]	Species Injected: Rabbit	Species Injected: Guinea Pig
Glu_{100} (H) [3]	MBSA	13/17	4/55
Glu_{100} (L) [4]	MBSA	1/6	0/9
Asp_{100}	MBSA	0/4	0/18
Lys_{100}	AcBSA	0/12	0/6
Lys_{100}	PBSA	0/6	0/6

[1] $\frac{\text{Number Responding}}{\text{Number Tested}}$.
[2] MBSA = methylated BSA; AcBSA = acetylated BSA; PBSA = phosphorylated BSA.
[3] Molecular weight 50,000.
[4] Molecular weight 5,000.

lysine and DNP-succinylated poly-L-lysine where the ϵ-amino groups are completely modified, are not immunogenic in mice. However, as shown in Table 6, these materials, when aggregated with oppositely charged albumin derivatives, are excellent immunogens, since all the animals produced DNP-specific antibodies (Pinchuck and Maurer, 1968b).

Table 6

RESPONSE OF MICE TO AGGREGATES OF DNP-POLY-L-LYSINE [1]

Polymer [2]	Carrier [3]	Response [4]
DNP-PLL	None	0/40
DNP-PLL	PBSA	11/11
DNP-Succ PLL	None	0/21
DNP-Succ PLL	MBSA	9/9

[1] $\frac{\text{No. Responding}}{\text{No. Tested}}$.
[2] DNP-PLL = Dinitrophenyl coupled poly-L-Lysine; DNP-Succ PLL = Dinitrophenyl coupled succinylated poly-L-Lysine.
[3] PBSA = Phosphorylated BSA.
[4] Determined by passive hemagglutination.

B. "Internal" Carriers

The generally better responses obtained with terpolymers, as compared with copolymers, especially with respect to the consistency of the responses obtained, suggested that these materials had acquired, as a result of their more "complex structure," the carrier property. In this case, however, the carrier would be "internal," i.e., an integral part of the polypeptide. The classical protein antigens, as a result of their complexity, also possess this property.

An example of this is provided by the response of mice and guinea pigs to the terpolymers $Glu_{58}Lys_{38}Tyr_4$ or $Glu_{58}Lys_{38}Phe_4$, where the antibody is predominantly directed to the Glu-Lys copolymer portion of the polymer (Maurer and Pinchuck, 1967).

1. *Internal and External Carriers in* $Glu_{60}Ala_{40}$ *Non-Responder Rabbits*

In collaboration with Dr. Carl Cohen we undertook some studies aimed at learning more about the genetic control of the immune response. With pedigreed rabbits it was possible to demonstrate that the ability to respond to $Glu_{60}Ala_{40}$ was genetically transmitted to the progeny of these rabbits. The availability of such responder and non-responder animals suggested a test system for our ideas about complexity producing an internal carrier. The results are shown in Table 7.

The first part of the table shows that in a relatively genetically heterogeneous population of New Zealand white rabbits non-responsive to

Table 7

Response of $Glu_{60}Ala_{40}$ "Non-Responders" to $Glu_{60}Ala_{40}$-carrier Complexes [1]

Rabbit	Immunogen	Response [2]
NZW	$Glu_{60}Ala_{30}Tyr_{10}$	2/5
	$Glu_{42}Lys_{28}Ala_{30}$	5/5
Pedigreed	$Glu_{60}Ala_{30}Tyr_{10}$	0/12
	Arsanyl-$Glu_{60}Ala_{30}Tyr_{10}$	4/4
	MBSA [3]-$Glu_{60}Ala_{40}$	4/4
	$Glu_{60}Ala_{40}$-Anti-$Glu_{60}Ala_{40}$ ppt (sheep antibody)	4/4

[1] Response evaluated by passive hemagglutination and passive cutaneous anaphylaxis reactions.

[2] $\frac{\text{No. Responding}}{\text{No. Tested}}$.

[3] MBSA = Methylated BSA.

$Glu_{60}Ala_{40}$ only some of the animals could respond subsequently to Glu_{60}-$Ala_{30}Tyr_{10}$, although all responded to $Glu_{42}Lys_{28}Ala_{30}$ which is immunochemically less related to $Glu_{60}Ala_{40}$. These findings are also reminiscent of the earlier observations that those guinea pigs classified as "non-responders" to $Glu_{60}Lys_{40}$ were also non-responders to $Glu_{57}Lys_{38}Ala_{5}$ or $Glu_{58}Lys_{38}Tyr_{4}$ (Maurer, 1963). The pedigreed animals (which were non-responders to both high and low doses of $Glu_{60}Ala_{40}$) did not respond at all to $Glu_{60}Ala_{30}Tyr_{10}$. They did, however, respond to an even more complex derivative, Ars-$Glu_{60}Ala_{30}Tyr_{10}$, producing antibody to both the arsanilic acid and to the polypeptide. The association of $Glu_{60}Ala_{40}$ with a foreign protein, either by ionic linkages (Methylated BSA) or through specific combining sites (sheep anti-$Glu_{60}Ala_{40}$ specific precipitate) led to the formation of anti-$Glu_{60}Ala_{40}$ antibody. It is of some significance that in the latter situation employing the external carriers, the response has been predominantly IgM antibody of a low titer.

It is interesting to compare these observations with the termination of immunologic unresponsiveness to heterologous serum proteins in rabbits. Rabbits that had been made tolerant to bovine serum albumin by neonatal injection could be induced to "break" this tolerance by immunizing them with either cross reacting albumins though distantly related (Weigle, 1961) or with BSA that had been altered by chemical modifications (Weigle, 1962).

In the above situations "tolerance" to the normally immunogenic determinants persisted. However, antibody was produced against newly introduced or previously "non-immunogenic" determinants (Paul *et al.*, 1967). In our studies the association of a non-immunogenic polymer with the appropriate carrier (internal or external) allows the formation of antibody which reacts predominately with the non-immunogenic polymer.

IV. α-D POLYPEPTIDES—EXTERNAL CARRIER COMPLEXES

A. NON-IMMUNOGENICITY OF α-D AMINO ACID POLYMERS

As mentioned in section I, we have failed to detect immune responses to polypeptides composed exclusively of α, D-amino acids in a number of species: rabbits, guinea pigs, mice and man. With the first three species we have used immunization in complete adjuvant, and both high and low doses of immunization. In no instance could we detect a significant response either by *in vitro* or *in vivo* techniques.

In addition to studying the response to α-D amino acid polymers per se, we have attempted to use these polypeptides as carriers for haptens in rabbits and guinea pigs. The conjugates used are as shown in Table 8. None of the α-D-polymers served as carriers for the DNP group or the arsanilic acid group. Here, too, both low and high doses of immunization

Table 8

α-D-AMINO ACID POLYMER-HAPTEN CONJUGATES STUDIED FOR IMMUNOGENICITY

α,D Polymer	Hapten Conjugated to Polymer[1]	Species Studied
Lys_{100}	DNP	Guinea Pig
$Glu_{60}Lys_{40}$	DNP	Guinea Pig, Rabbit
$Glu_{60}Ala_{30}Tyr_{10}$	Ars	Guinea Pig
$Glu_{58}Lys_{38}Tyr_{4}$	DNP	Guinea Pig, Rabbit
$Glu_{58}Lys_{38}Tyr_{4}$	DNP, Ars	Guinea Pig, Rabbit
$Glu_{42}Lys_{28}Ala_{30}$	DNP	Guinea Pig

[1] Ars = arsanilate conjugate; DNP = dinitrophenyl conjugate.

were used, and, in the guinea pigs, we looked for evidence of cellular immunity (delayed sensitivity) as well as humoral antibody.

B. IMMUNOGENICITY OF α-D POLYPEPTIDE AGGREGATES

1. *Anaphylactic Response of Guinea Pigs to α-D Polymer MBSA Aggregates*

In striking contrast to the negative responses obtained with α-D amino acid polymers alone, are the excellent responses obtained when these materials were aggregated with methylated bovine serum albumin (Maurer and Pinchuk, 1967). These responses (Table 9) are based upon the extremely sensitive active systemic anaphylaxis technique. In no case was

Table 9

GUINEA PIG RESPONSE[1] TO D-AMINO ACID POLYMER-MBSA[2] AGGREGATES

Polymer Injected	Response[3]	Polymer Injected	Response[3]
Glu_{100} (α,D)	26/29	$Glu_{60}Ala_{40}$	11/12
Glu_{100} (γ,D)	13/13	$Glu_{58}Lys_{38}Tyr_{4}$	40/48
$Glu_{60}Ala_{40}$	24/30	$Glu_{60}Ala_{30}Tyr_{10}$	29/33

[1] Response determined by systemic anaphylactic reaction.
[2] MBSA = Methylated BSA.
[3] $\frac{\text{No. Responding}}{\text{No. Tested}}$.

Table 10

RABBIT RESPONSE AGAINST MBSA[1] $_{D}GLU_{42D}LYS_{28D}ALA_{30}$ AGGREGATES

	Immunized with		
Course	MBSA	Polymer	MBSA-Polymer
1	0 (6)[2]	0.1–2.9 (8)	4.1–11.3 (8) 0 (2)
2		3.0 (1)	13.5–45.3 (3) 0.5– 2.4 (3)

[1] MBSA = Methylated BSA.
[2] Values in parentheses represent number of animals producing indicated range of plaque-forming cells/10^6 lymph node cells.

delayed sensitivity to the α-D polymers found. Homopolymers, copolymers and terpolymers were all active, as well as a γ-linked polyglutamic acid. Similar results wesre obtained in mice and rabbits using passive hemagglutination to detect the antibody formed.

2. *Plaque-forming Cell Response in Rabbits to MBSA-D Polymer Aggregates*

The Jerne plaque method, which enables one to enumerate directly individual antibody-forming cells, was recently adapted by Dr. Walsh and Miss Egan for studying the response of rabbits against polymers of α-L amino acids. Subsequently Miss Egan studied the response of rabbits to an α-D amino acid polymer, $_{D}Glu_{42D}Lys_{28D}Ala_{30}$, in the presence and absence of MBSA as the aggregating carrier. The data obtained are as shown in Table 10. Immunization with either MBSA alone or the α-D amino acid polymer alone yielded essentially no plaque-forming cells. In contrast, there was an excellent response in most, but not all, NZW rabbits when an α-D-amino acid polymer-MBSA aggregate was used for immunization.

C. "Low Dose" Response and Possible "Spontaneous" Aggregate Formation

As mentioned previously, other groups have reported obtaining responses to α-D amino acid polymers in mice and rabbits. In these experiments, low doses of polymer [10 μg–100 μg] in complete adjuvant produced a weak response, whereas higher doses failed to lead to any antibody formation. A phenomenon analogous to immunological paralysis was postulated to account for this dose dependence. It must be emphasized that in our earlier studies low doses also failed to induce a response.

One possible explanation for these findings is that there might have been *in vivo* "spontaneous" coupling or association of the α-D polypeptide with some protein. The source of the protein might either be the mycobacteria in the complete adjuvant used or some tissue protein in the responding animal. A similar argument has been advanced relative to the immunogenicity of purified polysaccharides in man and mouse.

In no instance where we obtained an anaphylactic response to α-D amino acid polymer aggregates in guinea pigs did we detect any positive immediate or delayed cutaneous reactions. This is of considerable interest when one recalls the recent findings of Schlossman and Levine (1967) that positive delayed cutaneous sensitivity reactions could only be obtained with immunogenic α-DNP-oligo-L-lysines. Indeed, it was found that even a single D-lysine residue abolished both immunogenicity and the ability to elicit a delayed reaction, although this derivative could react with anti-DNP antibody. Further evidence on this point comes from the studies of Jones and Leskowitz (1965). Studying delayed hypersensitivity to arsanillic acid-conjugates, they found α, D-amino acid homo- and co-polymers to be ineffective. However, if the arsanillic acid group was coupled to a free D-amino acid, delayed reactions could be induced. Leskowitz has attributed this effect to an inability of the α-D polypeptide to be metabolized.

V. REQUIREMENT FOR METABOLIZABILITY OF THE CARRIER

As is discussed elsewhere in this volume, there is much evidence ascribing the role of digestion and/or "processing" of the antigen to macrophages. In light of this evidence, and other evidence concerning the importance of metabolizability of the carrier, it would be necessary to demonstrate some enzymatic reaction by macrophages on α-D amino acid polymers before one could consider these materials immunogenic. Thus far no degradation of these polymers has been noted.

That the carrier must be metabolizable is shown in Table 11. Various carriers were used for isomeric α-D and α-L polymers. Materials such as DEAE-cellulose and polyvinylamine, although allowing the polymer to exist in a particulate form, could not act as carriers for haptenic α-D amino acid polymers. Proteins or red cells could act as carriers for polymers aggregated or adsorbed to them. Both homologous (guinea pig albumin and red cells) as well as heterologous carriers (bovine albumin,

Table 11

EFFECT OF CARRIER ON IMMUNOGENICITY OF POLYMERS IN GUINEA PIGS [1]

Carrier [2]	$Glu_{60}Ala_{30}Tyr_{10}$		$Glu_{58}Lys_{38}Tyr_{4}$	
	α-D	α-L	α-D	α-L
Cellex-D	0/7	6/7	0/7	4/7
PVA	0/6	5/6		
Poly-L-Lysine	0/5	0/6		
G. P. rbc			2/7	
Sheep rbc			7/14	
MGPSA	3/5		0/5	
MBSA	10/10	3/10	17/19	0/5

[1] $\frac{\text{No. Responding}}{\text{No. Tested}}$ (Systemic Anaphylaxis).

[2] Cellex-D = DEAE Cellulose; PVA = Polyvinyl amine; G. P. rbc = Guinea pig erythrocytes; MGPSA = Methylated guinea pig serum albumin; MBSA = Methylated bovine serum albumin.

sheep erythrocytes) were satisfactory, although the heterologous materials were more effective carriers.

VI. REQUIREMENT FOR IMMUNOGENICITY OF THE CARRIER

Although metabolizability of the carrier, internal or external, is necessary for immunogenicity, it is not sufficient as shown by the failure of poly-L-lysine to act as a carrier. One is faced with the question of whether the carrier itself must be immunogenic, i.e., if an immune response, either cellular or humoral, to both the carrier and the hapten must be produced.

A. RESPONSE TO A HAPTEN IN THE ABSENCE OF A DETECTABLE ANTI-CARRIER RESPONSE

Regardless of the requirement for immunogenicity of the carrier, there is evidence that the anti-carrier response, if indeed present, may be essentially undetectable. For example, homologous proteins or erythrocytes may serve as carriers (cf. Table 11). In many instances where a heterologous protein, MBSA, was used as a carrier in guinea pigs there was no detectable immediate reaction to the MBSA. However, as mentioned, there was a good correlation between delayed sensitivity to the

Table 12

RESPONSE OF MICE TO HAPTEN α-L POLYMERS [1]

Immunizing Polymer	Response to DNP	Response to Polymer
DNP-$Glu_{60}Lys_{40}$	36/36 [2]	0/36
DNP-$Glu_{57}Lys_{38}Ala_5$	18/18	9/14
DNP-$Glu_{36}Lys_{24}Ala_{40}$	28/28	25/28

[1] Determined by passive hemagglutination.

[2] $\frac{\text{No. Responding}}{\text{No. Tested}}$.

MBSA and antibody to the haptenic polypeptide. In this regard Green *et al.* (1966) have shown that the same guinea pigs responding to DNP-poly-lysine (anti-DNP) also showed a delayed cutaneous reaction to poly-lysine. Unpublished findings of Drs. Plescia and Braun have indicated that MBSA was not immunogenic in mice, and did not "carry" well for DNA. In our own studies in mice, however, MBSA was a good carrier for DNP-poly-L-lysine (cf. Table 6) and for D-polymers. The possibility that undetectable delayed sensitivity against MBSA exists in these mice must be considered.

One may get a response to only part of an immunogenic material with "internal" carriers as well. This is illustrated in Table 12, where all of the mice produced an anti-DNP response when immunized with DNP-α-L copolymers and DNP-α-L terpolymers. An anti-carrier response was detectable in none, some, or all of the animals, depending upon the immunogenicity of the specific polymer acting as carrier (Pinchuck and Maurer, 1968b).

B. RESPONSE IN ANIMALS TOLERANT TO THE CARRIER

An experiment designed to eliminate the carrier function of a metabolizable carrier by induction of specific immunological tolerance is presented in Table 13. Adult guinea pigs were rendered tolerant to MBSA by treatment with Cytoxan for a period of 8 days, wherein MBSA was injected intraperitoneally on day 2. The subsequent response of these animals to an aggregate of MBSA and $_{D}Glu_{58D}Lys_{38D}Tyr_4$ was studied. Although the control groups, treated with cytoxan alone or 0.15 M NaCl made the expected anti-D polymer response, the animals made tolerant to MBSA responded to neither the carrier (MBSA) nor to the haptenic α,D amino acid polymer.

Table 13

EFFECT OF REACTIVITY TO CARRIER ON GUINEA PIG RESPONSE TO D POLYMER—MBSA AGGREGATE [1]

	Response [2, 3] to	
Pretreatment [4]	$_{D}Glu_{58D}Lys_{38D}Tyr_4$	MBSA
Cytoxan + MBSA	1/7	0/7
MBSA	0/8	8/8
Cytoxan + Saline	3/4	
Saline	6/8	

[1] Methylated BSA.
[2] Determined by active systemic anaphylaxis.
[3] $\frac{\text{No. Responding}}{\text{No. Tested}}$.
[4] Cytoxan administered for 8 days; MBSA given on day 2, immunization with polymer-aggregates 2–3 weeks after last cytoxan injection.

Table 13 also shows the response in animals pretreated with MBSA in the absence of cytoxan. None of these animals responded to the α, D-amino acid polymer, and all of them responded to the carrier. This may well represent an example of "antigenic competition." The pre-immunization with the carrier may have pre-empted most of the immuno-competent cells, resulting in a greatly reduced, indeed an undetectable, anti-hapten response. Similarly, if two determinants are presented simultaneously but one is especially efficient in the antigenic competition, antibody might only be detected which is specific for that determinant. An example of this is the data presented in Table 12, where the DNP group was able to block most of the anti-carrier response.

It would appear that the entire complex of polymer and carrier is involved in triggering the response, and that the various potential determinants then compete for pluripotential immunoglobulin-forming cells.

VII. REQUIREMENT FOR COMPLEXITY OF THE CARRIER

Another factor contributing towards immunogenicity is the complexity of the potential immunogen. When one deals with a protein, this complexity is presumably due to the primary structure, the amino acid sequence, and the eventual complexity in determinant groups formed by specific folding of the polypeptide chains. With relatively simple materials, such as predominantly linear copolymers and terpolymers, one

Table 14

RESPONSE OF MICE TO MODIFIED DNP-PLL [1]

Modification	Per Cent ε-NH_2 Groups Modified	DNP Response [2]
None	0	0/10
Deamination	6	16/16
	12	14/14
	100	0/20
Succinylation	60	19/19
	100	0/21

[1] Determined by passive hemagglutination.
[2] $\frac{\text{No. Responding}}{\text{No. Tested}}$.

may speak of complexity in terms of the amino acid composition and/or number of different substituents.

A number of examples of this have already been presented: (1) the incorporation of a limited amount of arsanilate into $Glu_{60}Ala_{30}Tyr_{10}$ converted this material into an immunogen in rabbits unable to respond to $Glu_{60}Ala_{40}$; and (2) the incorporation of a second or third amino acid into polypeptides converted these materials into immunogens, e.g., Glu_{60}-Ala_{40} and $Glu_{90}Tyr_{10}$ were immunogenic in rabbits and guinea pigs, but PGA and polytyrosine were not; or $Glu_{58}Lys_{38}Tyr_4$ and $Glu_{58}Lys_{37}Ala_5$ were immunogenic in mice, but $Glu_{60}Lys_{40}$ was not. The results of a systematic study of this phenomenon is presented in Table 14. As was previously shown (Table 6), mice fail to respond to DNP-coupled homopolymers. If DNP-poly-L-lysine is *partially* deaminated or *partially* succinylated, mice respond by producing an anti-DNP response. However, the completely modified materials are, once more, non-immunogenic (Pinchuck and Maurer, 1968b).

Similar findings of an optimal degree of coupling of a hapten to a carrier have been noted by Kantor *et al.* (1963) for various DNP-poly-L-lysine conjugates in guinea pigs and by Borek and Stupp (1965) with the azobenzenearsonic acid conjugates of poly-L-tyrosine. In all cases studied "overloading" with hapten reduced the immunogenic effect.

VIII. SPECULATIONS ON THE MECHANISM OF ACTION OF THE CARRIER

The evidence for the universality of a carrier function in the immune response, whether internal or external, seems well founded. Some of the properties of the carrier have been described; it must be metabolizable, it must have a certain minimal degree of complexity, and it must be recognized as "foreign." It is not clear, at the present time, what is the relationship between the fragments of antigen produced in the macrophages and the biologically active RNAs extracted from these cells by Fishman and others. In particular it is not clear if the requirement for metabolizability refers only to degradation of the antigen, or to some coupling of antigen and RNA, or to both processes.

Although the chances of producing more "foreign" groupings would obviously increase as a polypeptide becomes more complex, the enhancement of immunogenicity found with increasing complexity of a polymer appears not to be simply due to the formation of additional determinants. For example, the production of anti-gelatin antibodies in rabbits is increased when one uses lightly coupled polytyrosyl gelatin for immunization. This enhancement may be due rather to formation of an appropriate fragment better able to interact with lymphoid cells. Several mechanisms may be hypothesized to account for this altered interaction: (1) the association constant of the processed fragments with macrophage RNA or lymphoid cells may be increased if the fragments contain additional Ala or Tyr; (2) the solubility or, in general, the transport of these fragments in the cell membrane of lymphoid cells may be increased, thus permitting the concentration of the material to become sufficiently high to trigger the proliferation of the cell; (3) the behavior of the fragments as an enzyme substrate may be altered; and (4) the persistence of the altered fragment or fragment-complex may be different from the "unmodified" one.

That antibodies can be produced against determinants in a synthetic polymer, which were uncovered by *in vivo* modification, has been demonstrated (Maurer *et al.*, 1966). The hydroxyl propanolamide derivative of $Glu_{60}Ala_{40}$ does not cross-react with rabbit anti-$Glu_{60}Ala_{40}$. However, some rabbits, when immunized with the modified polymer produced antibody to the derivative and also produced anti-$Glu_{60}Ala_{40}$. In the case of a large, non-degradable hapten, such as a polysaccharide or an α, D-polymer, a "processing" step in the macrophage might be associated with enzymic action on a complex of hapten with some endogenous, immunogenic protein.

Whether one must have a fragment of polymer and a piece of associated, processed protein to participate in the immune response at the gene level is also open for discussion. Much evidence has been advanced for the genetic control of the ability to respond to a polymer, or even a protein, and this may well be at the level of "processing" of the immunogen. Recent findings in our laboratory have indicated that, in contrast to a previous report (Ben-Efraim and Maurer, 1966) that strain-13 guinea pigs do not respond to "lysine containing polymers," definite, although somewhat reduced responses, were obtained against GLAT and

GLA_{20}. Responders and non-responders may be thought to differ in the level or efficiency of their macrophages or other enzymes involved in the processing of the antigen for lymphoid cells.

In summary, based upon our own studies and those of others, it appears that the carrier must be metabolizable and also immunogenic. The recognition is inherent in the specific animals, presumably at an early stage associated with processing of the potential immunogen. Although it is assumed that the carrier must be immunogenic, there are instances where no detectable response to the carrier has been noted in some species. The poor response to DNA-MBSA complexes in mice is in contrast to the good responses obtained with non-immunogenic polypeptide-MBSA complexes in this species. Similarly, MBSA-SIII aggregates are immunogenic in rabbits but not in guinea pigs. It will be necessary to reconcile these differences before a unified theory of the function of a carrier can be developed. Since one observes a greatly diminished anti-hapten response if the carrier is too good an immunogen or if there has been previous response to the carrier, we may be dealing with an antigenic competition. The immunogenic complex consists of many determinants, and a balance must be struck between the immunogenicity of the carrier and that of the carrier-hapten complex if anti-hapten antibody is to be produced.

References

Ben-Efraim, S., S. Fuchs, and M. Sela (1963). Hypersensitivity to a synthetic polypeptide: Induction of a delayed reaction. Science, *139*:1222.

Ben-Efraim, S., and P. H. Maurer (1966). Immune response to polypeptides (poly-α-amino acids) in inbred guinea pigs. J. Immunol., *97*:577.

Borek, F., and Y. Stupp (1965). Specificity of delayed reactions to hapten-polypeptide conjugates. Immunochemistry, *2*:323.

Gill, T. J., H. W. Kunz, and D. S. Papermaster (1967). Studies on synthetic polypeptide antigens. XVIII. The role of composition, charge, and optical isomerism in the immunogenicity of synthetic polypeptides. J. Biol. Chem., *242*:3308.

Green, I., W. E. Paul, and B. Benacerraf (1966). The behavior of hapten-poly-L-lysine conjugates as complete antigens in genetic responder and as haptens in non responder guinea pigs. J. Exp. Med., *123*:859.

Janeway, C. A., Jr., and M. Sela (1967). Synthetic antigens composed exclusively of L- or D-amino acids. I. Effect of optical configuration on the immunogenicity of synthetic polypeptides in mice. Immunology, *13*:29.

Jones, V. E., and S. Leskowitz (1965). Role of the carrier in development of delayed sensitivity to the Azophenyl-Arsonate group. Nature, *207*:596.

Kantor, F. S., A. Ojeda, and B. Benacerraf (1963). Studies on artificial antigens. I. Antigenicity of DNP-polylysine and DNP copolymer of lysine and glutamic acid in guinea pigs. J. Exp. Med., *117*:55.

Levine, B. B., and B. Benacerraf (1965). Genetic control in guinea pigs of immune response to conjugates of haptens and poly-L-lysine. Science, *147:* 517.

Maurer, P. H. (1963). Antigenicity of polypeptides (poly-α-amino acids). IV. Studies in guinea pigs. J. Immunol., *90:*493.

——— (1964). Use of synthetic polymers of amino acids to study the basis of antigenicity. Progr. Allergy, *8:*1.

——— (1965a). Antigenicity of synthetic polymers of L-α amino acids. Fed. Abstracts, *24:*184.

——— (1965b). Antigenicity of polypeptides (poly alpha amino acids). XIII. Immunological studies with synthetic polymers containing only D- or D- and L-α amino acids. J. Exp. Med., *121:*339.

Maurer, P. H., B. F. Gerulat, and P. Pinchuck (1964). Antigenicity of polypeptides (poly-α-amino acids). XI. Quantitative relationships among polymers and rabbit antisera. J. Biol. Chem., *239:*922.

Maurer, P. H., B. F. Gerulat, and P. Pinchuck (1966). Antigenicity of polypeptides (poly-α-amino acids). XIX. Studies with chemically modified polymers. J. Immunol., *97:*306.

Maurer, P. H., and P. Pinchuck (1967). Immunogenicity of synthetic polymers of amino acids; role of carrier and genetic background. In: Immunity, Cancer, and Chemotherapy, E. Mihich, Ed. Academic Press, New York, p. 319.

Maurer, P. H., P. Pinchuck, and B. F. Gerulat (1965). Antigenicity of polypeptides (poly α amino acids). XIV. Studies on immunological tolerance with structurally related synthetic polymers. Proc. Soc. Exp. Biol. and Med., *118:*1113.

Paul, W. E., G. W. Siskind, and B. Benacerraf (1967). A study of the 'termination' of tolerance to BSA with DNP-BSA in rabbits: relative affinities of the antibodies for the immunizing and the paralysing antigens. Immunology, *13:*147.

Pinchuck, P., and P. H. Maurer (1965). Antigenicity of polypeptides (poly-alpha amino acids). XV. Studies on the immunogenicity of synthetic polypeptides in mice. J. Exp. Med., *122:*665.

——— (1968a). Genetic control of aspects of the immune response. In: Regulation of the antibody response, B. Cinader, Ed. Charles C Thomas, Springfield, Illinois, p. 97.

——— (1968b). Antigenicity of polypeptides (poly-α-amino acids). XXVI. Studies of the ability of homo- and copolymers to act as hapten carriers in mice. J. Immunology, *100:*384.

Plescia, O. J., W. Braun, and N. C. Palczuk (1964). Production of antibodies to denatured deoxyribonucleic acid. Proc. Nat. Acad. Sc., *52:*279.

Schlossman, S. F., and H. Levine (1967). Desensitization to delayed hypersensitivity reactions. Fed. Abstracts, *26:*477.

Weigle, W. O. (1961). The immune response of rabbits tolerant to bovine serum albumin to the injection of other heterologous serum albumins. J. Exp. Med., *114:*111.

Weigle, W. O. (1962). Termination of acquired immunological tolerance to protein antigens following immunization with altered protein antigens. J. Exp. Med., *116:*913.

SPECIFICITY OF THE IMMUNE RESPONSE AS A FUNCTION OF THE CARRIER *

O. J. PLESCIA, C. V. RAO, S. CURTIS, V. JOHNSON, AND H. NOLTENIUS

Institute of Microbiology, Rutgers, The State University
New Brunswick, New Jersey

I. USE OF CARRIERS TO PRODUCE ANTIBODIES TO HAPTENS

Some 50 years ago Landsteiner showed the feasibility of the carrier method for eliciting hapten-specific antibodies in suitable hosts. Since then numerous investigators have successfully applied this methodology to obtain antibodies against a vast number of different haptens (Landsteiner, 1945). In every instance, until recently, the haptens were conjugated to the carrier by covalent bonds, and with but one exception the haptens were small molecules. Goebel and Avery (1931) reported that a pneumococcal capsular polysaccharide, which is not immunogenic in rabbits when injected in chemically pure form, became immunogenic once it was conjugated chemically to a carrier protein. Plescia *et al.* (1964) first showed that a carrier need not be conjugated chemically to a hapten, that a carrier need only form a stable complex with the hapten. Methylated bovine serum albumin (MBSA), complexed electrostatically with denatured DNA or type III pneumococcal capsular polysaccharide, proved to be an effective carrier. Rabbits, immunized with these complexes, produced DNA and polysaccharide-specific antibodies. Others, in addition to ourselves (Plescia *et al.*, 1965; Seaman *et al.*, 1965; Maurer, 1965; Green *et al.*, 1966; Levine *et al.*, this volume; Lacour *et al.*, this volume) have since used this methodology to elicit antibodies to synthetic polypeptides, synthetic and natural polynucleotides, chemically altered DNA and non-immunogenic proteins. The use of chemical conjugates of carrier protein with weak antigens has also been reported recently (Sandberg *et al.*, 1967; Nisonoff *et al.*, 1967).

* Supported in part by Contract AT(30-1)-3707 with the U. S. Atomic Energy Commission.

The above studies on the use of carriers were done primarily in rabbits. The overwhelming evidence that carriers are effective in converting inactive antigens into immunogens clearly supports the notion that rabbits have populations of immuno-competent cells[1] capable of synthesizing antibodies specific for determinants of these antigens and that a carrier may be required to promote interactions between these antigens and immuno-competent cells. Presumably the antigens that function as haptens have either too few or no "recognition" groups which immuno-competent cells can recognize as foreign and with which they can react. In all likelihood, these postulated "recognition" groups are antigenic determinants, but there is no a priori reason why they need be.

If the immune response to a given antigenic determinant, whether it belongs to a hapten or an immunogen, depends upon the interaction between immuno-competent cells and the molecule or complex of molecules of which this determinant is a part, then the magnitude of the antibody response to this determinant should depend largely on the nature of the carrier of which it is a part or with which it is associated. There is good evidence to support this prediction (Maurer, this volume; Green *et al.*, this volume). For example, MBSA did not function as a carrier in animals made tolerant to MBSA, and guinea pigs that are non-responders to PLL-DNP produced DNP-specific antibodies after immunization with electrostatic complexes of PLL-DNP and an immunogenic protein. The immune response to a primary antigen would, therefore, appear to depend on the specificity of the interaction between antigen and those immuno-competent cells that "recognize" the antigen as "foreign" and react with it. Similarly, the specificity of a secondary immune response may depend on the specificity of the interaction between the immunogen used to prime the immune response and "memory" cells which are postulated to be primed antibody-forming cells. A question that arises is whether such primed "memory" cells are committed to the original determinants carried by the primary immunogen or whether they can also form antibodies to new determinants that are added on to the previously experienced immunogen. The remainder of this report will deal with this question.

II. SPECIFICITY OF THE SECONDARY RESPONSE

A. Response to BSA-DNP in Rabbits Primed with BSA

Rabbits were hyperimmunized with bovine serum albumin (BSA), to be used later as a carrier for dinitrophenol (DNP), a haptenic determinant. Approximately one year later, when circulating antibodies to

[1] Cells participating in an immune response, either in the processing of the antigen or in the synthesis of antibody.

native BSA could no longer be detected, these rabbits were given a single injection of BSA-DNP as a booster. As controls, normal non-immunized rabbits were also given a single injection of the same BSA-DNP. Both groups of rabbits were bled at days 2, 5, 7 and 10 after this single injection, and the sera were assayed for C′-fixing and precipitating antibodies reactive with native BSA and DNP (in the form of a DNP-conjugate of human serum albumin).

The experimental rabbits, in contrast to the control animals that had not been primed with BSA, showed an immune response to BSA and to DNP typical of a secondary response in terms of the latent period for antibody synthesis and the amount of antibody formed (Fig. 1). A strik-

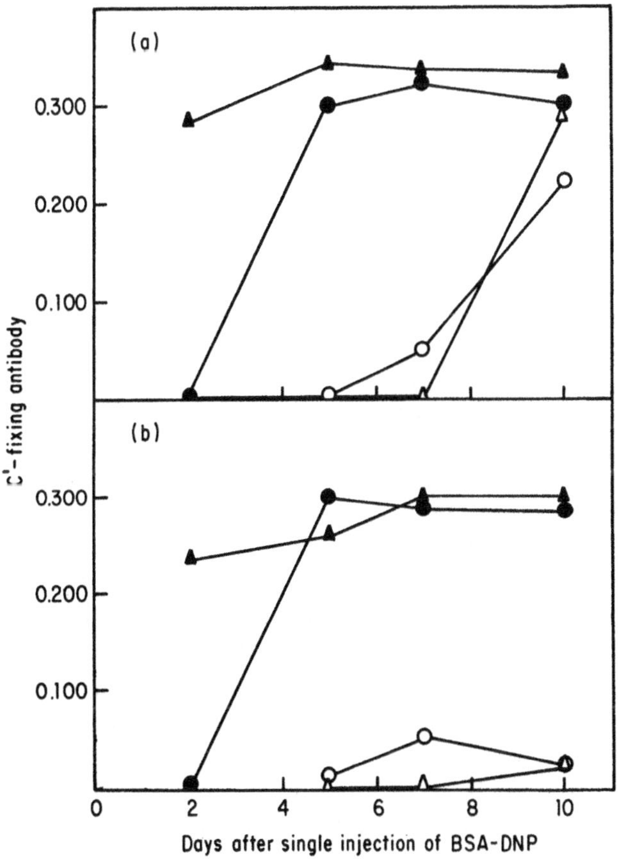

Fig. 1. Antibody response elicited by BSA-DNP in rabbits previously primed to BSA (▲, ●) and in non-primed rabbits (△, ○). Antisera (0.1 ml, 1/200) were assayed, by C′-fixation, for antibodies reactive with BSA, graph (a), and with HSA-DNP, graph (b). C′-fixing antibody is given in O.D. units at 541 mμ, a measure of the hemoglobin liberated from lysed sensitized sheep erythrocytes. A value of 0.340 is equivalent to 100% fixation of C′.

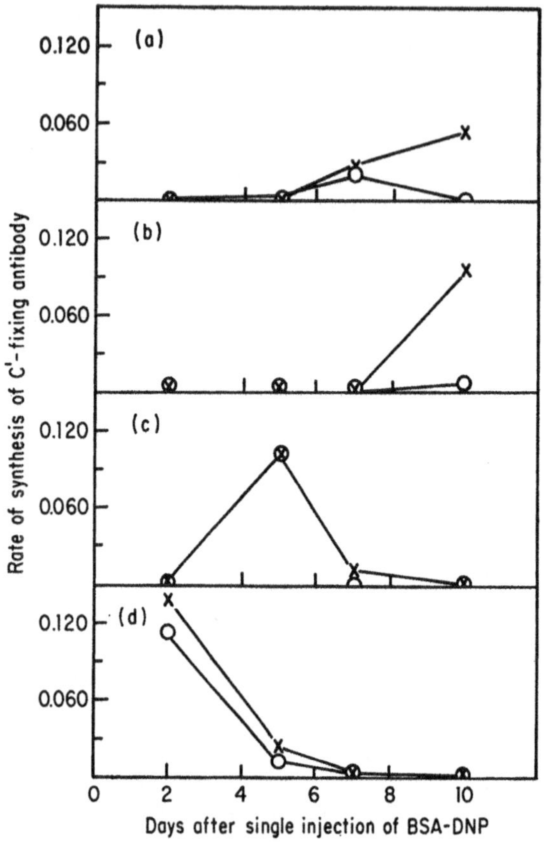

Fig. 2. Rate of synthesis of C′-fixing antibody reactive with BSA (×) and with HSA-DNP (○) in rabbits primed to BSA and boosted with BSA-DNP, graphs (c) and (d), and in rabbits given BSA-DNP only, graphs (a) and (b). The rates were calculated from the data of Fig. 1 and represent the average change in C′-fixing antibody per day.

ing feature of the immune response by primed rabbits was that for a given rabbit the rates of synthesis of antibody reactive with BSA and DNP were virtually identical (Fig. 2). These results suggest that primed cells need not be committed to determinants of the priming immunogen and that the magnitude of the immune response to a given determinant depends on the number and types of cells capable of interacting with molecules or complexes carrying that determinant.

B. Response to Antigens Conjugated to DNP in Rabbits Primed with BSA-DNP

The above experiment was modified to ascertain whether or not a simple hapten can act as a carrier in a secondary response. Rabbits were immunized with BSA-DNP, thus priming them to give a secondary response to DNP on re-injection with DNP conjugated to a different protein. The question was whether the response to DNP-conjugated antigens, different from BSA, would have the characteristics of a primary or secondary response. At a time when circulating antibodies to BSA-DNP could not be detected by C′-fixation, the previously immunized rabbits and normal non-immunized animals were given either one or up to three injections of DNP-conjugates of heterologous or homologous antigens.

1. *Heterologous Antigens*

Conjugates of DNP with human serum albumin (HSA) or bovine γ-globulin (BGG) were injected into primed and control rabbits. Samples of blood were obtained from each rabbit on different days following a single injection, and the sera assayed for C′-fixing antibodies reactive with BSA, HSA or BGG, depending on the DNP-protein conjugate injected.

The responses of the rabbits primed with BSA-DNP were significantly greater, both in terms of the quantity of antibodies formed and the time of onset of detectable antibody formation, than the response of non-primed rabbits. The kinetics of the responses of rabbits injected with HSA-DNP on days 0 and 21 are shown in Figure 3. The primed rabbits produced more antibody to HSA, and at an earlier time, than did a non-primed rabbit. In fact, the control rabbit produced no detectable antibody until after a second injection of HSA-DNP had been given on day 21. There was another striking difference between the responses of the primed and non-primed animals. The former produced an early and substantial quantity of antibody to BSA, the primary carrier protein for DNP; this is typical of a secondary response. In contrast, the control animals produced no detectable C′-fixing antibodies reactive with BSA at any time during the test period.

2. *Homologous Antigens*

Normal tissues (kidney, liver, muscle) of New Zealand white male rabbits were perfused, homogenized and extracted with 0.15 M NaCl at 0°C for several hours according to the method of Witebsky *et al.* (1956). These extracts, dialyzed to remove small molecules ($<$10,000 MW), contain many tissue constituents, of which only very few are likely to be immunogenic in homologous hosts even when they are conjugated to a carrier. Since we had no way of predicting which of these constituents belong to the class of potential immunogens, no attempt was made to isolate pure constituents from these extracts. Instead, they were reacted with a chloro derivative of 2,4 dinitrobenzene, according to the method of Eisen *et al.* (1958) to produce DNP-conjugates of tissue constituents. These DNP-conjugates in complete Freund's adjuvant were injected intramuscularly, once weekly for three weeks, into rabbits primed with BSA-DNP and into normal non-primed rabbits.

The objective of this series of experiments was to ascertain whether or not a hapten, such as DNP, could act as a carrier for homologous antigens and potentiate, or at least enhance, the formation of antibodies to these antigens in animals primed with the hapten. A representative response to rabbit kidney tissue extract, following immunization with

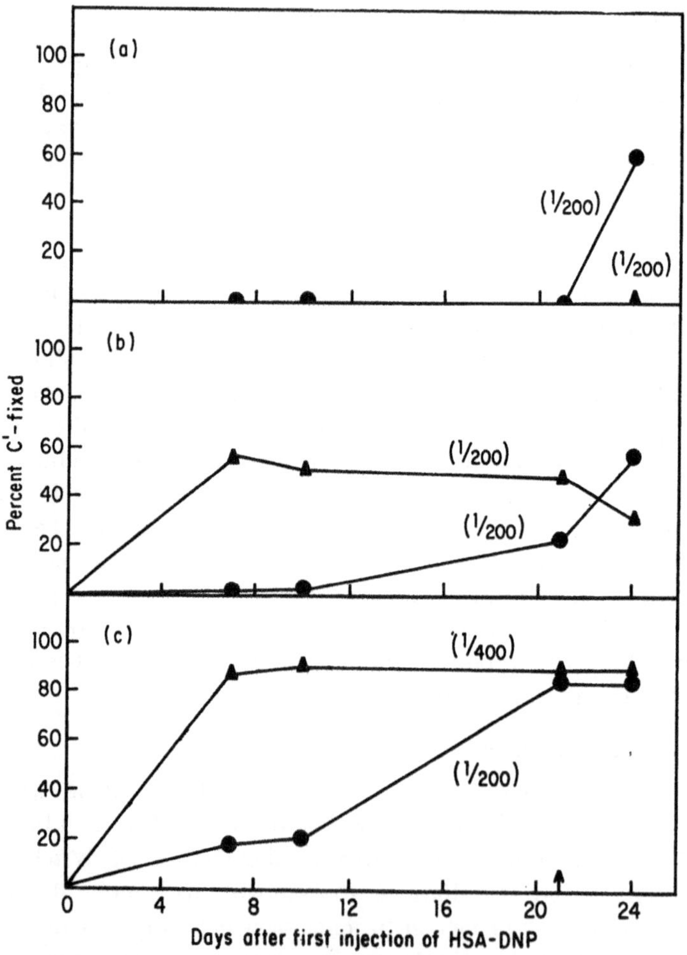

Fig. 3. Antibody response elicited by HSA-DNP in rabbits previously primed to BSA-DNP, graphs (b) and (c), and in a control non-primed rabbit, graph (a). Antisera (0.1 ml, diluted as indicated) were assayed for C′-fixing antibodies reactive with BSA (▲) and with HSA (●). A second injection of HSA-DNP was given on the 21st day, as indicated by the arrow, since the control rabbit had not yet responded to the first injection.

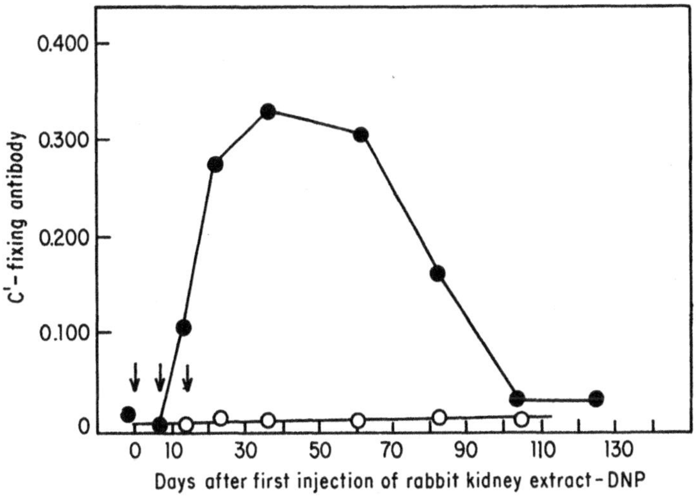

Fig. 4. Antibody response elicited by a DNP-conjugate of rabbit kidney extract in a rabbit previously primed to BSA-DNP (●) and in a control non-primed rabbit (○). Three injections of DNP-kidney extract were given on the days indicated by arrows. The antisera (0.1 ml, 1/100) were assayed for C'-fixing antibodies reactive with rabbit kidney extract. C'-fixing antibody is given in O.D. units at 541 mμ. A value of 0.340 is equivalent to 100% fixation of C'.

DNP-conjugates of such tissue extract, is shown in Figure 4. Clearly, the rabbit primed with BSA-DNP showed a significant and long-lasting response, whereas non-primed rabbits produced little, if any, antibody to homologous tissue antigens.

C. In vitro Stimulation of Spleen Cells from Rabbits Primed with BSA

The influence of the carrier on the specificity of the secondary response was also investigated at the cellular level. After rabbits had been primed with BSA, they were sacrificed, their spleens were removed and suspensions of cells were prepared from them. Aliquots of these cell suspensions were incubated *in vitro* at 37°C for 90 minutes with BSA, derivatives of BSA or a conjugate of BSA and BGG. The cells were then isolated by centrifugation, washed free of the antigen added and transferred into x-irradiated HA/ICR mice. Ten days later the mice were bled, and their sera tested individually for hemagglutinating antibodies by the method of passive hemagglutination in a micro-titer system.

It is evident from the results, summarized in Table 1, that the spleen of a primed rabbit contains cells that are capable of being stimulated to produce antibodies not only by the priming antigen but also by related antigens (conjugates of BSA). The unrelated BGG produced no

Table 1

An Example of the Influence of the Carrier in Stimulating Antibody Formation *in vitro*

Rabbit * Spleen Cells Incubated with	Per Cent of Recipient Mice * Producing Circulating Antibodies † to		
	BSA	BGG	DNP (BGG-DNP)
BSA	70	0	0
BSA-DNP	90	NT (not tested)	70
BSA-BGG (chemical conjugate)	80	90	0
BSA + BGG	60	0	0
BGG	0	0	0
Diluent	10	0	NT

* A rabbit (New Zealand white male) was given a single injection of BSA (5 mg, i.p.) five months before it was sacrificed. Its spleen was removed and a suspension of cells prepared. Aliquots of this suspension (10^7 cells/5 ml) were incubated *in vitro* at 37°C for 90 minutes with above antigens (50 μg N), after which the cells were recovered by centrifugation, washed free of added antigen, and injected i.v. into recipient mice (HA/ICR) that had been given 500 r whole body x-irradiation the day before. Each mouse received 10^6 cells.

† Antibodies were assayed by the microtiter hemagglutination method, using tanned sheep erythrocytes coated with the test antigens shown in the table. Samples of blood were taken from each mouse (10/group) just before and 10 days after the injection of spleen cells. The response to a given antigen was considered positive if the 10-day sample showed at least an 8-fold increase in antibody titer.

stimulation. Furthermore, stimulation by an antigen related to the priming antigen resulted in the production of antibodies not only against determinants of the priming antigen, as expected, but also against new determinants artificially conjugated to the priming antigen. No detectable response to the latter determinants was observed when they were added to the suspension of spleen cells as a simple mixture with the priming antigen, instead of being conjugated to it. These results at the cellular level are in general agreement with those obtained using intact animals (cf. Figs. 1, 3, 4).

III. DISCUSSION

It is difficult to say whether or not the immune response to a series of injections of an immunogen at short intervals, or even to a single injection of an immunogen, is a pure primary response involving only undifferentiated stem cells. There is little question that the response by an animal following reexposure to an immunogen, at a time when circulating antibody is no longer detectable, is primarily of the secondary type with distinct characteristics that differentiate it from a response to an immunogen given for the first time. In both types of response, the antibodies formed react with single determinants, but the immune response is initiated by a complex molecule that generally carries several, and often different, antigenic determinants. One may ask, therefore, whether the type of immune response elicited by an immunogen depends on the overall structure of the immunogen or upon individual determinants which it carries. In other words, what determines the type and specificity of an immune response?

The abundance of data from studies on the use of carriers to produce antibodies to simple haptens, and more recently, the use of carriers to produce antibodies to non-immunogenic polymers (polynucleotides and certain synthetic polypeptides, proteins and polysaccharides) have clearly established that the extent of the response to the hapten depends largely on the immunogenicity of the carrier, or more precisely on the immunogenicity of the carrier-hapten complex, i.e., the ability of the complex to interact with immuno-competent cells, thereby initiating the sequences of steps leading to synthesis of antibody. It is this interaction that presumably determines the specificity of the immune response.

There is general consensus that a secondary immune response involves "memory" cells that become activated either directly or indirectly by the priming immunogen or by a structurally related immunogen.[2] There are, however, apparently conflicting data regarding the specificity of the secondary response provoked by an immunogen related to the priming immunogen, that is whether "memory" cells are committed in their specificity and therefore yield antibody-forming cells able to synthesize antibodies specific for determinants of the priming immunogen only. Data obtained by Dubert (1959), Ashley and Ovary (1965), and B. Levine (1967) support the thesis that "memory" cells are committed. On the other hand, Salvin and Smith (1960) obtained in guinea pigs a superior delayed hypersensitivity reaction to a hapten if it were conjugated to a carrier to which the test animal had been previously sensitized. Similar results were also reported by Dixon and Maurer (1955). Our finding (cf. Figs. 1 and 2) that rabbits give a secondary type response to DNP if it is conjugated to BSA with which the rabbits had been immunized, also indicates that "memory" cells cannot all be committed. Results of our experiments at the cellular level give further support to this conclusion (cf. Table 1). The difference between our results and those of others

[2] No distinction will be made here between "memory" cells as processing cells, antibody-forming cells or cells with both functions.

may be due to differences in the haptens and carriers used. We believe, however, that a more critical factor than differences in materials may be the time interval between the booster injection of related antigen and the last of the injections of priming antigen. The longer the interval, the greater should be the probability that in the population of "memory" cells those that are uncommitted will persist longer and thereby become predominant. A series of experiments to test this prediction is in progress.

A hapten has reportedly been used to enhance the immunogenicity of homologous organ-specific antigens, such as thyroglobulin (Weigle, 1966), but to our knowledge haptens have not been used as carriers in animals first sensitized to the hapten. The results herein reported (cf. Figs. 3 and 4) clearly indicate the feasibility of this approach as a means to obtain even greater enhancement of antibody formation to a weak or inactive immunogen through the use of a hapten.

In essence, the data presented and discussed in this report support the thesis that the specificity of the immune response depends largely on interactions between immunogen and immuno-competent cells which in turn depend upon the number and kind of "recognition" groups carried by the immunogen. It follows from this thesis that the quantity and specificities of antibodies formed depend upon the number and types of immuno-competent cells able to recognize an immunogen as foreign and interact with it. Cells responding to an immunogen for the first time are likely to be uncommitted to any particular determinant; thus, the formation of more antibody to one determinant than to another determinant carried by the same molecule cannot be the result of a selective advantage due to a state of commitment on the part of immunologically responsive cells. There also is no doubt that primed cells, that is "memory" cells, can be committed to determinants of the primary immunogen, but they need not be so committed. Our data can best be explained by assuming the existence of primed uncommitted cells with "memory" for "recognition" groups of the primary immunogen. Such cells should be able to interact efficiently with any immunogen carrying "recognition" groups of the priming immunogen (a structurally related molecule), and because they are members of a cell population that is not totally committed, should be responsive to any potential determinant whether or not it was a determinant of the priming immunogen.

References

Ashley, H., and Z. Ovary (1965). Effect of bovine gamma globulin on subsequent immunization with dinitrophenylated bovine gamma globulin. Proc. Soc. Exptl. Biol. Med., *119*:311–314.

Dixon, F. J., and P. H. Maurer (1955). Specificity of the secondary response to protein antigens. J. Immunol., *74*:418–431.

Dubert, J. M. (1956). Sur la nature de la reponse secondaire. Comptes rendus des seances de l'Academie des Sciences, *243:*1939–1941.

Eisen, H., M. Kern, W. Newton, and E. Helmreich (1959). A study of the distribution of 2,4-dinitrobenzene sensitizers between isolated lymph node cells and extracellular medium in relation to induction of contact skin sensitivity. J. Exptl. Med., *110:*187–206.

Goebel, W. F., and O. T. Avery (1931). Chemo-immunological studies on conjugated carbohydrate-proteins. IV. The synthesis of the p-amino-benzyl ether of the soluble specific substance of type III pneumococcus and its coupling with protein. J. Exptl. Med., *54:*431–436.

Green, I., W. E. Paul, and B. Benacerraf (1966). The behavior of hapten-poly-l-lysine conjugates as complete antigens in genetic responder and as haptens in nonresponder guinea pigs. J. Exptl. Med., *123:*859–879.

Lacour, F., A. M. Michelson, and E. Nahon (1968). Specific antibodies to polynucleotide complexes and their reaction with nucleic acids: Importance of the secondary structure of the antigen. This volume.

Landsteiner, K. (1945). "The Specificity of Serological Reactions." Harvard University Press, Cambridge, Massachusetts.

Levine, B. (1967). Specificity of the anamnestic response to a double hapten conjugate in guinea pigs primed with a single hapten. J. Immunol. *99:*1173–1177.

Levine, L., E. Seaman, and H. Van Vunekis (1968). Immunochemical studies of irradiated DNA. This volume.

Maurer, P. H. (1965). Antigenicity of polypeptides (poly alpha amino acids). J. Exptl. Med., *121:*339–349.

Nisonoff, A., E. Margoliash, and M. Reichlin (1967). Antibodies to rabbit cytochrome c arising in rabbits. Science, *155:*1273–1275.

Plescia, O. J., W. Braun, and N. C. Palczuk (1964). Production of antibodies to denatured deoxyribonucleic acid (DNA). Proc. Natl. Acad. Sci. U. S., *52:* 279–285.

Plescia, O. J., N. C. Palczuk, E. Cora-Figuerra, A. Mukherjee, and W. Braun (1965). Production of antibodies to soluble RNA (sRNA). Proc. Natl. Acad. Sci. U. S., *54:*1281–1285.

Salvin, S. B., and R. F. Smith (1960). Specificity of allergic reactions. II. Azoproteins in the anamnestic response. Proc. Soc. Exptl. Biol. Med., *104:* 584–590.

Sandberg, A. L., M. Liss, and B. D. Stollar (1967). Rabbit antibodies induced by calf thymus histone-serum albumin complexes. J. Immunol., *98:*1182–1189.

Seaman, E., H. Van Vunakis, and L. Levine (1965). Antigenicity of polyribonucleotides. Biochemistry, *4:*1312–1318.

Witebsky, E., N. R. Rose, and S. Shulman (1955). Studies on organ specificity. I. The serological specificity of thyroid extracts. J. Immunol., *75:*269–281.

THE ROLE OF CONFORMATION IN THE STRUCTURE OF ANTIGENIC SITES *

THOMAS J. GILL III, DAVID S. PAPERMASTER, HEINZ W. KUNZ, AND PETER S. MARFEY

Laboratory of Chemical Pathology, Department of Pathology Harvard Medical School, Boston, Mass.

With all you gentlemen, the name's a test
Whereby the nature usually is expressed.
—GOETHE

The role of conformation in the structure of antigenic sites can be explored with synthetic polypeptides and their crosslinked derivatives. By comparing antigenic specificity before and after crosslinking, the effects of stabilization by ordered spatial structure in the construction of antigenic sites can be studied. Three approaches were utilized for the investigation of this new class of antigenic determinants; the study of cross-reactions of L- and D- enantiomorphic polypeptides and their crosslinked derivatives; inhibition of the precipitin reaction by model compounds and small molecular weight polypeptide fragments; and comparison with substituted polypeptides containing groups resembling the structure of the crosslinked sites.

The polymers used were poly $Glu^{52}Lys^{33}Tyr^{15}$ (M = 52,000 or 86,000) and poly $D\text{-}Glu^{48}D\text{-}Lys^{38}D\text{-}Tyr^{14}$ (M = 160,000). Their intramolecularly crosslinked derivatives were prepared by crosslinking with ionic carbodiimide to form an amide bond (Am) between the sidechains of lysine and glutamic acid, or by difluorodinitrobenzene (FFDNB) to form a dinitro-

* This work was supported by grants from the National Science Foundation (GB-940 and GB-3603) and the National Heart Institute (HE 01771). Thomas J. Gill III is the recipient of a Research Career Development Award (K3-AM-5242), and David S. Papermaster was supported by a Training Grant (GM-1235) from the National Institutes of Health.

phenylene crosslink (DPE) between lysine and/or tyrosine sidechains.[1] The structures of the polymers and of the various haptens and inhibitors are shown schematically in Figure 1.

The antibody response to the parent polymer and to its crosslinked derivatives is shown in Table 1. There was no change in the amount of antibody elicited; hence, immunogenicity was not affected by changing the conformation of the polypeptide.

Studies were undertaken to compare the specificity of the antibody elicited by the crosslinked polypeptide with that elicited by a linear polymer and to compare the antibody elicited by polypeptides containing different crosslinks. The results of cross-reaction studies, designed to reveal the overall specificity of the antibodies, are shown in Table 2. The parent linear polypeptide induced antibodies which cross-reacted completely with the crosslinked antigen, so this crosslinking did not cause loss of antigenic sites on the linear portions of the polypeptide chain. In contrast, the antibodies against the crosslinked polypeptides showed decreased cross-reactivity with the linear polymer; hence, antibody formed was specific for the crosslink. This new specificity is clearly evident from the greater cross-reactivity of the anti-DPE-GLT with Am-GLT than with the linear polypeptide. The specificity of the different crosslinks varies and the antibody elicited by Am-GLT showed a greater degree of specificity than that elicited by DPE-GLT. The anti-DPE-GLT antibody cross-reacts 80% with Am-GLT, whereas the anti-Am-GLT antibody cross-reacts only 15% with the DPE-GLT. Thus, the anti-DPE antibody is much more cross-reactive than the anti-Am-GLT antibody. This finding can be accounted for on the basis of a difference in the size of the two crosslinked regions: the Am crosslink is shorter than that of the DPE derivative and thus presents itself as a smaller antigenic site (see Fig. 1). The antibody against the larger DPE crosslinked region can react well with the smaller Am site, whereas the larger DPE region cannot react as well with the antibody to the smaller Am site.

The DPE derivatives have convenient spectral properties and resemble many common haptens, especially DNP; this class of polypeptides was therefore studied in greater detail to probe further the structure of the antigenic site created by crosslinking. The results indicated that the site

[1] The nomenclature of the polypeptides is described in Biopolymers *2*, 283 (1964). The abbreviations used in this paper are:

GLT: poly $Glu^{52}Lys^{33}Tyr^{15}$
ΔGLT: poly $D\text{-}Glu^{48}D\text{-}Lys^{38}DTyr^{14}$
Am: Crosslinked by ionic carbodiimide (Lys-Glu)
DPE: Crosslinked by FFDNB (Lys-DPE-Tyr > Tyr-DPE-Tyr > Lys-DPE-Lys)
(Tyr-DPE): tyrosine-dinitrophenylene hapten
(Lys-DPE): lysine-dinitrophenylene hapten
DNP: dinitrophenyl hapten

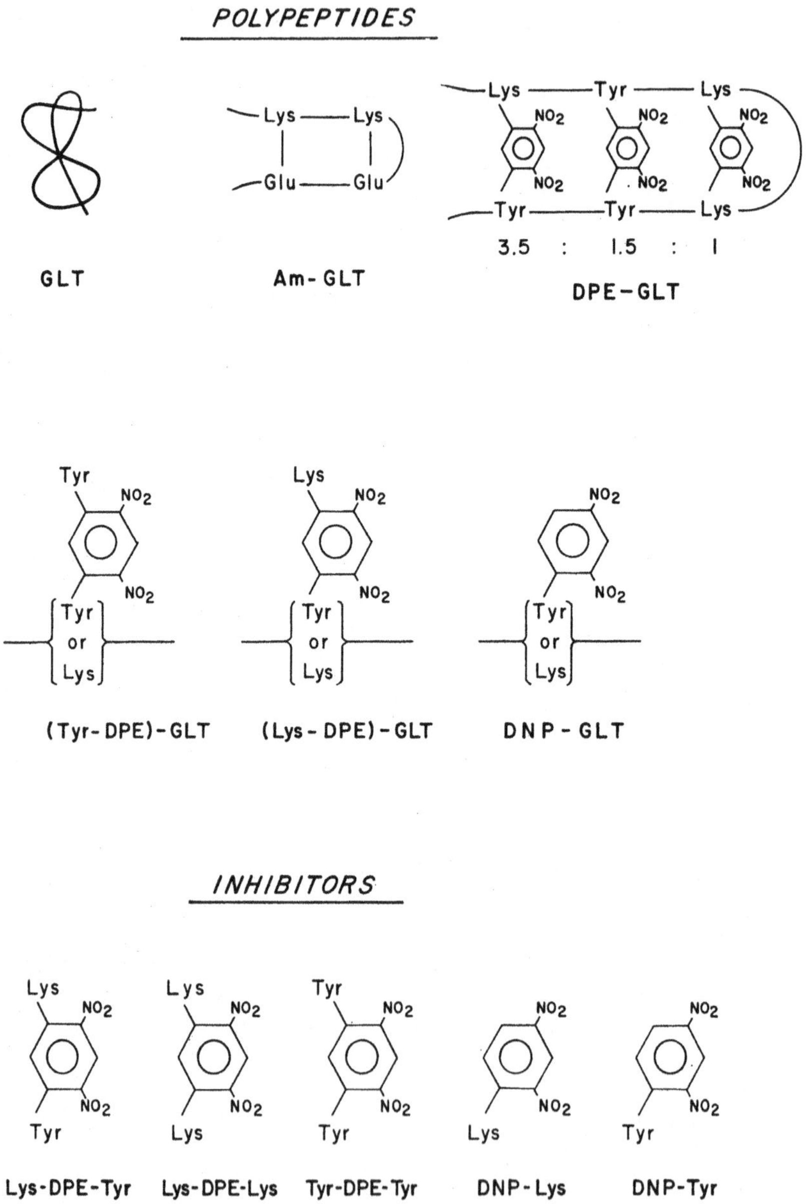

Fig. 1. The schematic structure of the polypeptides, the amino acid-DPE haptens, and the inhibitors used in the immunochemical studies.

Table 1

THE IMMUNOGENICITY OF LINEAR AND INTRAMOLECULARLY CROSSLINKED SYNTHETIC POLYPEPTIDES *

Antigen	Crosslinks		μg AbN/ml ± S.D.
	Per 100 Amino Acids	Per Molecule	
GLT	0	0	123 ± 75
Am-GLT	1.8	6.3	141 ± 55
DPE-GLT	1.5	8.2	139 ± 10

* Gill *et al.,* 1968.

consisted of the crosslink and the peptides to which it is attached. The first approach to this study utilized cross-reactions among the DPE-GLT, DPE-ΔGLT, (Lys-DPE)-GLT, (Tyr-DPE)-GLT, and DNP-GLT antigen-antibody systems (see Fig. 1). The results are summarized in Table 3. Antibody against DPE-GLT cross-reacts well with the DPE-ΔGLT, but not completely, and this finding indicates the importance of the DPE ring. Since there is only 65% cross-reactivity, though, steric constraints due to the peptides attached to both sides of the ring must be important. Further evidence to support the importance of the polypeptide chain in the antigenic sites of the crosslinked polypeptides is the incomplete cross-reactivity of antibodies to (Tyr-DPE)-GLT and (Lys-DPE)-GLT with DPE-ΔGLT. This indicates that the antigenic sites on DPE-ΔGLT must include both the DPE ring and the attached peptides. Antibody to the DNP group reacts to a rather limited extent (30%) with DPE-GLT,

Table 2

THE SPECIFICITY OF ANTIBODY TO LINEAR AND TO CROSSLINKED SYNTHETIC POLYPEPTIDES: THE ROLE OF CONFORMATION IN THE SPECIFICITY OF ANTIGENIC SITES

Antiserum	Per Cent Cross-reaction with Antigens		
	GLT	Am-GLT	DPE-GLT
GLT	100	100	95
Am-GLT	10	100	15
DPE-GLT	60	80	100

Table 3

STRUCTURE OF THE ANTIGENIC SITES INVOLVING SPATIALLY ORDERED (CONFORMATIONAL) STRUCTURE IN CROSSLINKED SYNTHETIC POLYPEPTIDES: CROSS-REACTION STUDIES

	Per Cent Cross-reaction with Antigens	
Antibody	DPE-GLT	DPE-ΔGLT
DPE-GLT	100	65
(Tyr-DPE)-GLT	100	85
(Lys-DPE)-GLT	100	50
DNP-GLT	30	

so that the antibody site encompassed by the DNP ring is much too small to react with the entire DPE antigenic site. This is further evidence that the antigenic site on the crosslinked DPE-GLT must include peptides attached to the DPE ring.

The second approach to a study of the antigenic sites on the crosslinked polypeptides employed inhibition analyses. Tryptic fragments of the parent polymer and bifunctional and monofunctional derivatives of lysine and tyrosine were used (cf. Fig. 1). This series of experiments was designed to probe more closely the fine structure of the antigenic sites. The results are summarized in Table 4. The fact that the fragments of the unlinked GLT polypeptide inhibit the absorbed antiserum against the crosslinked derivative and against the DNP derivative is crucial proof that small peptides are involved in the antigenic sites of these antigens and that the ring, while important, does not comprise the entire antigenic site. None of the bifunctional or monofunctional derivatives inhibits the antibody to GLT with the exception of DNP-tyrosine. This is due to the haptenic nature of the phenol ring which elicits antibody that cross-reacts with the DNP-tyrosine inhibitor; this phenomenon will be discussed later. The lack of inhibition by the bifunctional and monofunctional derivatives in the GLT system demonstrates the specificity of these inhibitors for the antigenic sites involved in the crosslinks. The bifunctional derivative lysine-DPE-tyrosine, which is the structural analogue of the most common crosslink in DPE-GLT, is an extremely good inhibitor of the DPE-GLT antibody. The monofunctional DNP derivatives are less effective inhibitors than the bifunctional DPE derivatives, because they are only partial structural analogues of the crosslinked antigenic sites. The monofunctional DNP compounds are the most effective inhibitors of the DNP antibody, and show a high degree of inhibition, whereas the bifunctional derivatives are less effective because of steric hindrance. In conclusion, the combination of a high degree of inhibition of the DPE-GLT system by Lys-DPE-Tyr and the ability of small GLT peptides to inhibit this system provides the most conclusive evidence that the antigenic site on DPE-GLT includes the DPE ring and the adjacent short peptides attached to it.

The demonstration of a large amount of antibody specific for the conformational structure of the antigen and of the greater potency of these spatially ordered sites compared with that of sites on the linear portions of the polypeptide chains implies that the antigen is structurally intact when it stimulates antibody formation. The role of antigen degradation in regulating antibody formation probably lies in controlling the amount of antigen left intact and capable of stimulating antibody formation over the critical period during which this event takes place. This mechanism can explain the observation of Lapresle (1955) that immunization of rabbits with human serum albumin elicited antibody to the intact molecule and antibody that reacted only with a splenic digest of the albumin: the latter antibody was elicited by the larger degradation products of the albumin formed *in vivo* after injection of the antigen.

From data presented in this study and from various data in the literature (Parker *et al.,* 1966), there is evidence for an inverse relationship

Table 4

STRUCTURE OF THE ANTIGENIC SITES INVOLVING SPATIALLY ORDERED (CONFORMATIONAL) STRUCTURE IN CROSSLINKED SYNTHETIC POLYPEPTIDES: INHIBITION STUDIES *

Inhibitor	μmoles Inhibitor/μmole Specific Ab at 50% Inhibition: GLT	DPE-GLT †	DNP-GLT †
GLT Fragments	18	210	140
Lys-DPE-Tyr	no inhibition	4	no inhibition
Lys-DPE-Lys	no inhibition	11	45
Tyr-DPE-Tyr	no inhibition	14	95
DNP-Lys	no inhibition	22	2
DNP-Tyr	325	36	8

* Gill *et al.,* 1968.

† The antisera were absorbed twice with GLT in order to remove all of the antibody not directed at the antigenic sites involving the crosslink.

RELATIONSHIP BETWEEN SIZE AND POTENCY OF ANTIGENIC SITES

ANTIGEN	SITE	RELATIVE POTENCY
GLT	15-20 Amino acids (linear)	+
DPE-GLT	short peptide } –DPE– { short peptide	+++
DNP-GLT	short peptide } – DNP	++++

Fig. 2.

between the size of an antigenic site and its relative potency: the larger the antigenic site, the less potent it is. This relationship is schematically summarized in Figure 2. The relative potency is a qualitative estimate based on the concentration of inhibitor required for 50% inhibition and on hapten binding data. The explanation for this relationship may lie in the ability of the potent antigenic determinants to provide both antigenic specificity and adequate binding energy to hold the antibody and antigen together. On the other hand, the less potent antigenic determinants provide the specificity, but not adequate binding energy. Hence, the portion of the antigen adjacent to the moiety that determines the specificity is necessary to provide sufficient binding energy to hold the antibody and antigen together through non-specific electrostatic, hydrophobic and van der Waal's interactions.

The role of amino acid sidechains in forming antigenic determinants was investigated further (Gill *et al.*, 1967) by studying the cross-reactivity of isomeric synthetic polypeptides, particularly those containing tyrosine; any cross-reactions between the isomers must be due to sidechain residues, since stereospecificity is confined to the peptide backbone. Using antisera with quite high concentrations of antibody, a relatively small amount (5–10%) of antibody to each isomer was shown to cross-react with the other isomer. As the tyrosine content of the cross-reacting antigen increased, the amount of antibody precipitated increased. This finding of sidechain cross-reactivity may be the explanation for the previously reported (Gill and Matthews, 1963) cross-reactivity between antisera to tyrosine or phenylalanine-containing polypeptides and a variety of proteins: antibody against the aromatic amino acids in the antisera to the synthetic polypeptides cross-reacted with the aromatic residues on the various proteins.

In summary, then, the role of conformation in the structure of antigenic sites of synthetic polypeptides has been delineated thus:

(1) There is no change in immunogenicity with change in conformation.

(2) There is a marked change in potency and specificity of antigenic sites with change in conformation:
   a. Spatially ordered sites are more potent than sites on the linear portions of the polypeptide chain.
   b. Different types of spatially ordered sites differ in their degree of antigenic specificity.

(3) Spatially ordered sites contain two separate portions of the polypeptide chain and the crosslink between them.

(4) The size of an antigenic site varies inversely with its potency.

(5) Amino acid sidechains alone, especially tyrosine, can act as antigenic determinants for a small portion of the antibody elicited by an antigen.

References

Gill, T. J., III, and L. S. Matthews (1963). Studies on Synthetic Polypeptide Antigens. IX. The antigenicity of a new group of linear synthetic polypeptides. J. Biol. Chem., *238*:1373–1379.

Gill, T. J., III, H. W. Kunz, and D. S. Papermaster (1967). Studies on Synthetic Polypeptide Antigens. XVIII. The role of composition, charge, and optical isomerism in the immunogenicity of synthetic polypeptides. J. Biol. Chem., *242*:3308–3318.

Gill, T. J., III, D. S. Papermaster, H. W. Kunz, and P. S. Marfey (1968). Studies on Synthetic Polypeptide Antigens. XIX. Immunogenicity and antigenic site structure of intramolecularly cross linked polypeptides. J. Biol. Chem., *243*: 287–300.

Lapresle, C. (1955). Étude de la Degradation de la Sérumalbumine Humaine par un Extrait de Rate de Lapin. II. Mise en Évidence de Trois Groupements Spécifique Différents dans le Motif Antigénique de l'Albumine Humaine et de Trois Anticorps Correspondents dans la Sérum de Lapin Anti-Albumine Humaine. Ann. Inst. Pasteur, *89*:654–665.

Parker, C. W., S. M. Gott, and M. C. Johnson (1966). The Antibody Response to a 2,4-Dinitrophenyl Peptide. Biochemistry (Wash.), *5*:2314–2326.

DISCUSSION

Chairman: F. HAUROWITZ

HAUROWITZ: Regarding the semantic problem of antigen versus immunogen, I would like to suggest that a substance is an antigen as long as it has not been applied to a living system. If it really brings about a response in a responder, then it becomes an immunogen. If one deals with a non-responder, the substance remains an antigen without being immunogenic.

PLESCIA: Let me suggest that we define an antigen as a substance that contains reactive groups with which antibody can react but for which we do not know whether it can or cannot elicit the formation of antibodies. Once it has been established that it cannot elicit antibody formation, it should be designated a hapten. If you know that it can elicit antibody, it should be referred to as an immunogen.

GREEN: I think we also may have to coin a new word which can cover situations such as DNP-polylysine which is an immunogen for some animals and a hapten for others.

HAUROWITZ: In Dr. Levine's paper the problem of a degrading enzyme was repeatedly mentioned. I don't think it has to be a degrading enzyme. It could be an enzyme that acts as a transferase. We sometimes forget that most of the proteolytic enzymes are such transferases and the picture would be approximately as follows: When an antigen is injected, it may not be degraded but a determinant may be transferred to a factor X and thus become functional. This is not pure speculation. If the antigen were really unchanged in the body, then it should be recoverable from body fluids, but we find, when we use protein antigens with radioactive determinants, that most of the radioactivity appears in the cytoplasmic granules. Therefore, the antigen must have undergone some change, but it need not be a degradation, it may be a transfer to the postulated X factor. In conformance with the data of Campbell and Garvey one could suggest that the X factor may be nucleic acid. However, it seems to me easier from a chemical standpoint to assume that it is a protein because we know that transfer reactions between proteins take place very easily. I wonder if Dr. Levine or any other speakers would have any objections to such a view.

B. LEVINE: As far as the so-called requirement for degradation of large molecular weight materials is concerned, I think it has to be assumed by most of us that such a process does go on in the handling of an antigen prior to the induction of antibody synthesis. However, in reviewing data pertinent to this question, there is really no clear-cut proof that degradation is required. We do know, on the other hand, that materials such as DNP-polylysine undergo degradation *in vivo*, where we do not know. We also know that spleen extracts at pH 4.8 will degrade them into smaller peptides but whether such a process is required for immunogenicity is not known. In this regard it is interesting to point out that DNP-poly-D-lysine of large molecular size, which has been thoroughly dialyzed to remove small molecular weight materials, can induce a good DNP-specific immune response. This is surprising because it is difficult to envision how such a large molecular weight material can undergo processing without being chopped down first. However, as far as I know, there is no enzyme that can degrade poly-D-lysine.

PLESCIA: Perhaps the action of enzymes is only incidental and may only inadvertently influence the production of antibodies. It may be likened to a sword with a double edge. If the action of enzymes changes and modifies the structure of antigenic molecules in such a way that additional groups are exposed, causing the molecules to be recognized as foreign more readily, the altered molecules may become more immunogenic. On the other hand, if these molecules are degraded to fragments that are unable to interact with immuno-competent cells, their immunogenicity may be lost.

At this point I would like to say that one reason for introducing this particular subject in our symposium was the hope that as a result of a thorough discussion of the role of the carrier we might be better able to understand why it has been possible to produce antibodies specific for nucleic acids. You know well that nucleic acids happen to be one class of substances that are easily degraded by extracellular nucleases; therefore, one of the considerations that led to the use of a carrier was the possible need for preventing extensive degradation of polynucleotides by extracellular nucleases.

GILL: To me metabolism means degradation and it may not be required for antibody formation. The role of metabolism may be regulatory and involves the quantity and perhaps to some extent the quality of an antigen. The work of Lapresle on antibodies to bovine serum albumin demonstrated that antibody to internal determinants may be produced. I think that one can interpret this as reflecting a breakdown of the immunogen into several large fragments which in themselves acted as immunogens. Regarding D-polymers of amino acids, it may well be that the usual amount used for immunization paralyzes rather than induces antibody formation. When we used very small amounts of such antigens over a long period of time, we were able to elicit immune responses to D-polymers. I also would like to suggest that the result of complexing of antigens with carriers, such as methylated bovine serum albumin, specific antibody, or polymers of opposite charge, may be a control over the amount of antigen available at any one time to antibody-synthesizing systems, thereby preventing paralysis by making a very small amount of antigen available.

MAURER: In regard to your last point, Dr. Gill, let me state that rabbits which had received tremendous doses of copolymers of glutamate, alanine and tyrosine, enough to paralyze them, did respond when subsequently injected with the copolymer complexed either with methylated bovine serum albumin or specific antibody. Since these rabbits received fairly large amounts of antigen, in repeated injections, and still did not respond they should have been regarded as tolerant, yet they were able to respond when the antigen was presented as a complex.

GILL: Dr. Levine stated that D-amino acid polymers can be degraded *in vivo* very slowly. We have not isolated enzymes capable of doing this but D-amino acid polymers can be broken down. When one chromatographs the urine, one finds peptides of molecular weight of 1000–1200. This shows that degradation of such polymers can occur *in vivo*.

PLESCIA: May I suggest a possible explanation for the apparent paradox that an animal instead of being tolerant to an antigen given in large doses may respond when the antigen is presented with a carrier. The carrier may simply divert the complexed amino acid copolymers to a class of cells that can interact with the carrier or portions of the carrier and hapten.

GILL: We have complexed several good antigens with methylated BSA, for example the copolymer of glutamate, lysine and tyrosine, and also diphtheria toxoid, but we did not affect the extent of antibody formation. In fact, in some cases there was a decrease. How would you explain this?

PLESCIA: I could suggest that in the case of good immunogens there are already enough cells capable of interacting with the immunogen. The response to a good immunogen is probably at a maximum already so that the use of a carrier is not going to be effective.

GILL: I would like to ask Dr. Levine or Dr. Green whether they have tried to immunize non-responders over a prolonged period of time. They indicated the existence of a minimal response and I wonder whether this might indicate the existence of a complex polygenic system with particular penetrance of one gene.

GREEN: I can answer this question partially in the sense that we have immunized guinea pigs over very long periods of time and have never found significant amounts of anti-DNP antibodies. I think what Dr. Levine was trying to say is that if you take normal animals and compare them with non-responders immunized with DNP-polylysine, it is possible to detect small amounts of antibody by the method of equilibrium dialysis or by hemagglutination but this may be due to background levels of anti-DNP antibodies in normal animals. This point has not been settled as yet. I would also like to point out that DNP-polylysine can complex electrostatically with the animal's own albumin. We have shown that guinea pigs injected with complexes of DNP-polylysine and guinea pig albumin will produce low levels of anti-DNP antibodies. It is conceivable that immunization with DNP-polylysine may result in complexing with the animal's own albumin, thus producing a carrier effect.

B. LEVINE: Regarding Dr. Gill's question about the involvement of a multigenic system with incomplete penetrance, I think that it is not possible to rule out this possibility. The only way in which it could be ruled out would be by breeding of homozygous parents and by determining the absence of segregation of non-responders, and this is a very difficult operational problem indeed. The results of the breeding experiments that we have done so far are consistent with the unigenic hypothesis, but I think your point is good and should be recognized as a possibility.

HAUROWITZ: Dr. Green, you mentioned that your hapten was coupled to acidic protein. Have you ever examined the immunogenicity of your hapten coupled to a basic protein like lysozyme or globin?

GREEN: We have tried electrostatic complexing of DNP-polylysine with bovine gamma globulin and under these circumstances nothing happened.

HAUROWITZ: What happens if you bind polylysine covalently to a protein?

GREEN: We did some preliminary studies in which we used glutamyl chloride to bind the DNP-polylysine to bovine gamma globulin. Under these circumstances a good carrier effect was obtained, but we did not pursue these studies since we felt that electrostatic interactions would alter the components to a very minimal degree and thus would provide a more interesting model.

B. LEVINE: Drs. Green and Maurer have described experiments which suggest that the carrier may permit proper metabolism of the unit leading to the formation of an inducer. If one makes animals tolerant to the carrier, the material can no longer lead to antibody production. Does tolerance interfere in this case with metabolism of the antigen?

GREEN: We have a tremendous amount of ignorance about the most fundamental problems of how antibodies are synthesized and therefore we cannot intelligently discuss tolerance. But I think a general statement is possible, namely, what we see as tolerance is the absence of an end-result involving many intermediate steps. There may be interference at any number of these steps and thus tolerance may not always be due to the same cause.

MAURER: I would like to ask Drs. Plescia and Levine how they would reconcile the differences in the results of the NYU and the Rutgers groups regarding the specificity of the secondary response.

B. LEVINE: I think that you are referring to the data on priming with antigen A and boosting with antigen A + B and getting a so-called anamnestic response to both the primed determinant A and also to the new determinant B. I think there are several other groups that have had similar results, for example Maurer and Dixon and also Salvin and Smith. Now, we have had just the opposite type of results; we have done the same experiment and found a secondary response to A and a primary response to B. Ovary and also Dubert have obtained results similar to ours. We did our studies with a polylysine carrier and double hapten conjugates. We primed with DNP-polylysine and boosted with DNP-benzyl peniciloyl polylysine and vice versa, and showed an anamnestic response only to the first hapten and a primary response to the second hapten. I think these two kinds of data can be reconciled if one assumes that there can be more than one method of enhancing an immune response to a second dose of an antigen. One may be through primed cells, the other possible mechanism may involve enhancing effects of antibodies. You will recall that responses to soluble antigens can be enhanced by giving antibody. Therefore, my current view of the two kinds of apparently conflicting data is that they probably both represent enhancement but one may be mediated by primed cells and the other by serum antibodies. I see that Dr. Plescia tried to study this question *in vitro* and I am sure that he has some comments.

PLESCIA: I can offer two possible reasons for the differences in the results. First, I think most of the other work deals with the response to individual artificial haptens and in our work we were simply considering differences in a response to a DNP hapten versus the response to the determinants that are native to the carrier bovine serum albumin. I think this might be important since the results of a number of experiments suggest that the response to a determinant depends on the nature of its carrier. Thus, when two different determinants are carried by molecules that are antigenically unrelated, no competition occurs. On the other hand, if these determinants are carried by the same molecule, competition ensues. Therefore, even if a primed cell is not committed, there is the possibility that already experienced determinants may have, on a purely competitive basis, a selective advantage over a hapten added to the priming antigen. This type of competition would make it appear as if the response were only to the determinant of the priming antigen. Secondly we know that cell populations capable of interacting with immunogenic molecules and capable of synthesizing antibodies are in a dynamic state. We know that the secondary response to a given immunogen will vary somewhat depending on the interval of time between priming and second exposure. Therefore, I would suggest that perhaps there may be a particular time in this interval when there is an appreciable number of primed uncommitted cells capable of participating in the formation of antibodies to new additional determinants as well as to previously experienced determinants, all carried by the same priming antigen.

MAURER: I think it is important to our discussions to recall that we have been dealing with different kinds of animals. All of the work reported here dealt with rabbits, while I have been talking about guinea pigs. When you hyperimmunize guinea pigs with a carrier you cannot get a response at all against carrier-hapten conjugates. Also, rabbits respond very well to methylated albumin-polysaccharide whereas there is no response in guinea pigs to such complexes. Also, the methylated albumin functions poorly as a carrier in mice. Therefore, it is impossible to generalize at the present time.

ABRAMSON: Has anyone coupled or complexed nucleic acid to a hapten or to an antigenic determinant or to any other degradation product of a complete antigen in an effort to determine whether this would by-pass the primary response and result in a secondary response?

GREEN: We did this experiment. DNA is negatively charged and DNP-polylysine is positively charged and therefore we used DNA as a non-antigenic polyanion to which we complexed DNP-polylysine. The complex did not elicit an immune response. Neither did any other large polyanion that we tried, in other words, the carrier itself has to be immunogenic in order to make the reaction go. However, it remains to be seen whether any special RNA, perhaps like that of Dr. Gottlieb or Dr. Fishman, might produce different effects.

GILL: Our experience has been essentially similar. If one puts haptens of a variety of sorts on D-amino acid polymers and uses them under the usual immunizing conditions, one gets no antibody response.

FRANZL: Dr. Levine, you showed data on the degradation of antigen by spleen extracts and reported a lack of difference between non-responders and responders. In view of the likelihood that only a very small amount of antigen is going to participate in the immune response, perhaps less than 1%, I wonder whether you have been able to detect such a small amount in your experiments.

B. LEVINE: Your criticism, Dr. Franzl, is valid and applies to any experiment of this sort.

FRANZL: The next question is whether the macrophage hypothesis has been excluded from your explanation of gene action.

B. LEVINE: Indeed, our working hypothesis is that the effect of the gene may be on some physiological event, possibly in the macrophage. We are not using the word metabolism interchangeably with degradation; however, I think your point is a good one and it may be that a certain kind of degradation is required which cannot be detected by our rather crude methods.

LESKOWITZ: In our studies on hypersensitivity we have obtained data indicating that degradation might play a role in immune responses. We were interested in delayed hypersensitivity to azo-benzene arsanate and found, along with other investigators, that when azo-benzene arsanate is attached to poly D-amino acid polymers, it is nonimmunogenic in delayed hypersensitivity studies. Yet when azo-benzene arsanate is conjugated to D-tyrosine it is immunogenic. I wonder whether Drs. Green, Levine and Gill had comparable experiences using oligopeptides and conjugates thereof.

B. LEVINE: We haven't done this, but I have heard through the grapevine that the introduction of a D-amino acid residue into an octopeptide fragment failed to produce immunogenicity.

GILL: I am sure that you are aware of the work of Sela who showed that hexatyrosine-benzene-arsanate was immunogenic, whereas benzene-arsanate was not. I think this in essence answers your question.

LESKOWITZ: Yes, but these were all L-oligopeptides. I am particularly concerned about the possible immunogenicity of D-oligopeptides.

PINCHUCK: In collaboration with Drs. Fishman and Adler we have taken either normal rabbit or normal mouse macrophages and mixed them with a polymer of glutamate, alanine and tyrosine (GAT). We then extracted the RNA from these macrophages and found that it can induce an immune response to GAT in mice which normally would not be able to respond to GAT. The RNA does seem to contain small but definite amounts of antigen, roughly on the order of what was found in the T2 system.

NUCLEIC ACIDS AS NON-SPECIFIC STIMULATORS OF IMMUNE RESPONSES

STIMULATION OF ANTIBODY-FORMING CELLS BY OLIGONUCLEOTIDES OF KNOWN COMPOSITION *

WERNER BRAUN, MASAYASU NAKANO, LYDIE JARASKOVA, YUKIYOSHI YAJIMA, AND LUIS JIMENEZ

Institute of Microbiology, Rutgers, The State University New Brunswick, New Jersey

The data compiled in this volume illustrate the multiple functions of nucleic acids in antibody formation. Nucleic acid in the form of polydeoxyribonucleotides supplies the basic information for messenger RNA formation and protein synthesis in cells that participate in antibody synthesis. Nucleic acid in the form of messenger RNA transmits this information to RNA-rich cellular sites of polypeptide synthesis, within cells and possibly even among cells. Furthermore, there is some evidence (Braun and Cohen, 1968) suggesting that RNA itself, or something closely associated with it and present in macrophage-released RNA-antigen complexes, may act as an activator or derepressor for stem-cells of antibody-forming lymphocyte populations. In addition, oligomers of nucleic acids influence the rate with which activated stem-cells of antibody-forming lymphocyte populations increase in numbers during a given period of time. This non-specific effect of oligonucleotides on antibody-forming cells will be the subject of this communication.

Our interest in the effects of oligonucleotides on mammalian cells developed as a sequela to earlier studies with bacteria. More than ten years ago we became aware of the ability of oligodeoxyribonucleotides from diverse sources to cause unusual population changes in bacteria (Braun and Whallon, 1954; Braun, Firshein and Whallon, 1957). These effects were produced only by oligonucleotides, such as those present in an enzymatic digest of DNA, but not by mononucleotides, -sides, purines or pyrimidines. Subsequent studies with pneumococci (Firshein and Braun, 1960; Firshein, 1961) revealed that one reason for the unique population changes in bacterial cultures supplemented with oligonucleotides was a

* Supported by NIH grant AM-08742 and NSF grant GB-5887.

selective stimulation of cell multiplication and of rates of DNA synthesis of smooth, virulent mutants; avirulent, rough mutants did not respond. Firshein (see this volume) subsequently demonstrated that oligonucleotides derived either from natural sources or from synthetic homopolymers, can act as inducers of certain nucleotide kinases, i.e., of enzymes involved in the production of nucleoside triphosphates, the building stones of DNA (Kornberg, 1961).

The enhancement of DNA synthesis and cell multiplication by oligonucleotides is not restricted to bacteria. Studies of recent years have shown that mammalian cells, especially those that retain the capacity for multiplication in adult animals, can similarly be stimulated by oligonucleotides. The responding cell populations include tumor cells (Braun, 1958) and cells of the reticulo-endothelial system. Early studies indicative of effects of oligonucleotides on antibody-forming cells included the studies of Talliaferro and Jaroslow (1960) and of Feldman *et al.* (1963) who showed that damage to antibody-forming systems, caused by X-radiation or 6-mercaptopurine, could be repaired by DNA breakdown products. In tests with normal adult mice, Merritt and Johnson (1965), using circulating antibodies as indicators, noted stimulatory effects of enzymatic digests of DNA and RNA, while Braun and his associates (Braun, 1962, 1965; Braun and Nakano, 1965, 1967a, 1967b) observed significant stimulations principally in assays on numbers of antibody-forming spleen cells, and only when antigen was administered in conjunction with DNA digests; digestion of RNA by pancreatic RNAase did not yield stimulatory materials. The effects of DNA digests were shown to be independent of the source of the DNA used but dependent on the presence of oligomers; monomers, deoxyribonucleosides, purines and pyrimidines all were inactive in eliciting stimulatory effects on antibody-forming cell populations.

The literature contains a number of reports referring to stimulatory effects of DNA breakdown products on mammalian cell systems other than the RES. These reports have been reviewed by us recently in conjunction with a more detailed history of the work on stimulatory oligonucleotide effects (Braun and Firshein, 1967), and we should repeat here what we said in that review in regard to the earlier studies by others: it is difficult to assess whether the effects of DNA breakdown products observed in their tests are really comparable to the regulatory effects of oligonucleotides on DNA synthesis and antibody formation. At least some of the effects reported by others may well have been the result of nutritional effects, supplying precursors for DNA synthesis.

I. THE NATURE OF THE STIMULATORY OLIGONUCLEOTIDES

A stimulated rate of increase in the number of antibody-forming cells can be observed when antigenic material, such as sheep red blood cells

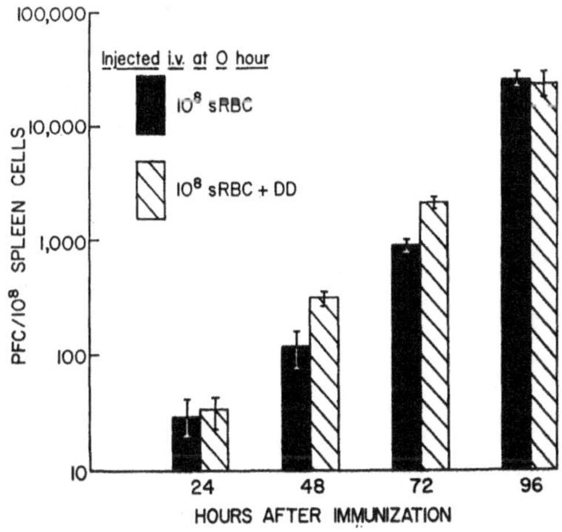

Fig. 1. Influence of DNAase-treated calf thymus DNA (DD) on the number of murine spleen cells (per 10^8 nucleated spleen cells) forming antibodies to sheep red blood cells (sRBC) at various times following immunization.

(sRBC), is injected into mice together with an oligomer-rich enzymatic digest of DNA (Fig. 1 and Table 1), the number of antibody-forming spleen cells being assayed by localized hemolysis in gel according to the method of Jerne *et al.* (1963).* Also, the injection of enzymatic digests of DNA into normal mice will cause an enhancement of phagocytic activity. These effects are obtainable with DNA digests from many different sources; we usually have employed digests prepared from calf thymus DNA. No stimulations result from an injection of mononucleotides (in the form of mononucleoside monophosphates), from mixtures thereof, or from mononucleosides. The size of the active oligomer, producing enhanced immune responses, is difficult to assess in animal tests since the host environment is full of depolymerizing nucleases and thus will alter the size of oligomers *in vivo*. However, tests with defined synthetic oligomers (see below) and tests on strikingly parallel systems in bacteria (cf. Braun, 1965) suggest that active oligomers may contain from two to six nucleotides.

In contrast to the observations by Merritt and Johnson (1965) we have never been able to obtain significant stimulatory effects with enzymatic digests of RNA. This difference may be attributable to several factors: (1) Merritt and Johnson primarily assayed circulating antibodies, thus measuring not only activities dependent on the number of antibody-forming cells releasing antibody over a prolonged period of time, but also possibly altered activities of individual antibody-forming cells; in our studies we have concentrated on assaying effects on numbers of antibody-

* The basic principle of this technique is quite simple. Since red blood cells (e.g., sheep red blood cells) are lysed by specific antibodies in the presence of complement, the number of hemolysin-forming cells in the spleen can be estimated by pouring complement over an agar plate which contains spleen cells (from an sRBC-immunized animal) and sRBC. The red blood cells in the vicinity of an antibody-forming cell will be lysed, and, by counting the number of plaques produced per plate, an estimate of the number of antibody-forming or -releasing cells can be obtained.

Table 1

INFLUENCE OF AN ENZYMATIC DIGEST OF CALF THYMUS DNA ON THE NUMBER OF HEMOLYSIN-PRODUCING SPLEEN CELLS IN CF-1 MICE 40 HOURS AFTER IMMUNIZATION WITH 10^8 sRBC

Treatment of Spleen Donors	Average No. (±S.E.) of Hemolysin-producing Cells Per 10^8 Nucleated Spleen Cells
sRBC	123.3 ± 21.4
sRBC + DD *	278.2 ± 15.8
sRBC + kinetin riboside (2 mg/mouse)	111.5 ± 18.5
sRBC + kinetin riboside + DD	110.9 ± 26.2

* DNA + DNAase. Equivalent to approximately 600 γ DNA.

forming cells present at different periods of time after antigen exposure. (2) Pancreatic RNAase used by us is known to yield RNAase digest containing only oligonucleotides with pyrimidine nucleotide terminals, which may be inactive, whereas oligomers with either purine or pyrimidine nucleotide terminals result from the exposure of DNA to pancreatic DNAase (Steiner and Beers, 1961). (3) The starting RNA preparations employed in the studies of Merritt and Johnson and in our studies may have been sufficiently different to yield non-identical degradation products.

Following our observation that oligodeoxyribonucleotides can stimulate antibody-forming cells, we wondered whether such stimulators might also play a natural role *in vivo* and might be released from intracellular environments under conditions of enhanced antibody synthesis (Braun and Kessel, 1964). Experimental support for this assumption was obtained when we succeeded in identifying a stimulatory fraction rich in oligodeoxyribonucleotides among the materials releasable from spleen cells. We obtained these extracellular materials by exposing antibody-forming spleen cells *in vitro* to antigens to which the spleen donor had been exposed 3 to 20 days earlier (Nakano and Braun, 1967). As illustrated in Figure 2, suspensions of spleen cells from mice immunized with chicken red blood cells were incubated with such cells for several hours in Eagle's medium. After removal of the cells, the supernatant was then injected with an unrelated antigen, e.g., sheep red blood cells, into previously non-immunized mice. Subsequent spleen assays revealed that the supernatant contained a potent non-specific stimulator of antibody-forming cell populations (Table 2). Fractionation of active supernatants on Sephadex G-25 (Braun, Nakano and Plescia, unpublished data) revealed that activity was associated with a large, 260-μ absorbing, molecu-

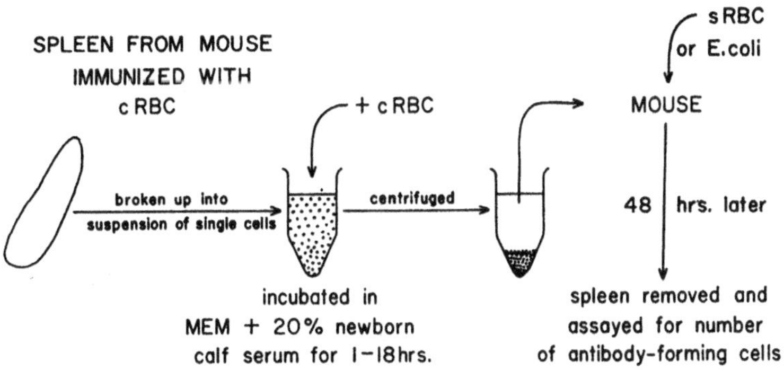

Fig. 2. Procedure for harvesting and use of supernatant from the *in vitro* incubation of immune spleen cells and immunizing antigen (from the J. of Immunology, *99*:571, 1967; used with permission of the publisher).

lar component which yielded active fractions of oligonucleotide size following its exposure to pancreatic DNAase. This finding supported our earlier suggestion (Braun and Kessel, 1964) that the non-specific stimulatory effects of cytotoxic agents, such as bacterial endotoxins (Johnson, 1964), mycobacteria and their fractions (Freund, 1956; Weiss *et al.*, 1964), FUDR, CCl_4 (Braun, 1962), cholchicine (Jaroslow and Talliaferro, 1966), and crystalline silica (Vigliani and Pernis, 1959), may all be the consequence of a release of stimulatory oligonucleotides from intracellular sites. In the case of agents such as endotoxin and mycobacteria the

Table 2

Effects of Supernatants, from Interactions *in vitro* Between Chicken RBC (cRBC) and Normal or Immune Spleen Cells, on the Number of Hemolysin-producing Spleen Cells in Mice 48 Hours After Immunization with Sheep Red Blood Cells (sRBC)

Injected Intravenously into Test Animals	No. of Cells Forming Antibodies to sRBC Per 10^8 Nucleated Spleen Cells
Nothing	26.0 ± 9.4
10^8 sRBC	355.2 ± 39.4
10^8 sRBC + supernatant from (cRBC + cRBC-immune spleen cells)	915.6 ± 141.5
10^8 sRBC + supernatant from (cRBC + normal spleen cells)	350.6 ± 62.5
10^8 sRBC + cell-free medium (minimal essential medium-calf serum)	349.2 ± 69.2

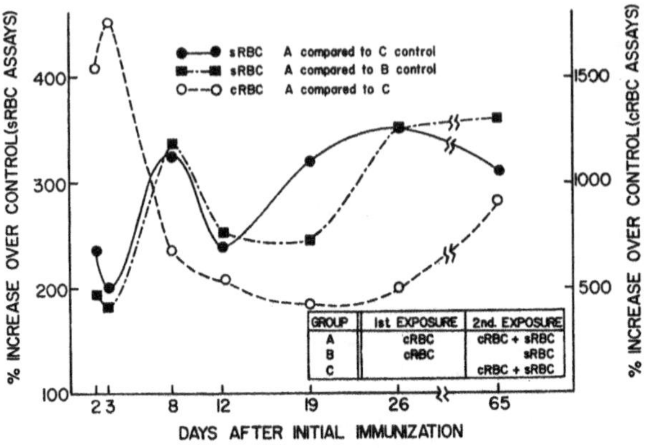

Fig. 3. Influence of reexposure to a previously experienced antigen (10^8 cRBC followed by 10^8 cRBC) on the response to a new antigen (sRBC) at various times following initial immunization. The response is measured in terms of number of antibody-forming spleen cells (anti-sRBC or anti-cRBC) per 10^8 nucleated spleen cells.

release may be triggered by antigen-antibody reactions on cell surfaces which will lead to altered permeability and cell leakage or may cause an actual destruction of the cell.

Oligonucleotide-releasing cell-associated antigen-antibody reactions may also occur *in vivo* as indicated by recent unpublished studies of Lasky and Braun. In these studies an attempt was made to mimic *in vivo* the condition of *in vitro* reactions between antibody-forming cells and a reactive antigen. Accordingly, mice previously exposed to antigen A were reexposed to this antigen at the same time as a new antigen B was administered. Figure 3 illustrates that reexposure to a previously experienced unrelated antigen will enhance the response to a new antigen, as measured in terms of rate of increase of specific antibody-forming cells to the new antigen B. This enhanced response appears to last more than 65 days following the first injection of antigen A, and we assume that it is the consequence of an *in vivo* release of stimulatory oligonucleotides from antibody-forming spleen cells which may release these stimulators after the cells have been in contact with the proper antigen.

Recent studies have indicated that oligonucleotides produced by the *in vivo* depolymerization of double-stranded poly A-poly U molecules can produce a significant and consistent non-specific stimulation of antibody-forming spleen cell populations and of phagocytic cells (Braun and Nakano, 1967c). Examples of the effects on antibody-forming spleen cells are shown in Table 3 and Figure 4. More recently we have determined that the injection of poly C + poly G will produce similar, and usually even stronger, effects (Table 4). Since the activity of enzymatic DNA digests is variable, due presumably to uncontrollable variations in the

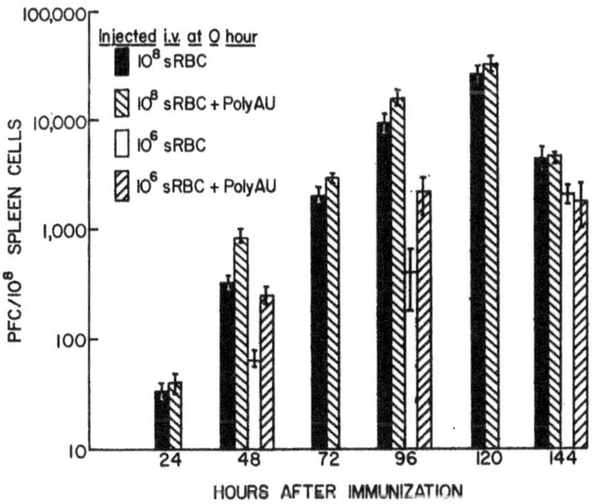

Fig. 4. Influence of poly A-poly U on the number of murine spleen cells (per 10^8 nucleated spleen cells) forming antibodies to sheep red blood cells (sRBC) at various times following immunization with 10^6 or 10^8 sRBC.

course of depolymerization *in vitro,* we now use routinely either complexes of poly A + poly U or of poly C + poly G for stimulation. Poly A, poly C, poly G or poly U, administered as single homopolymers, fail to stimulate (Tables 3 and 4), whereas poly A or poly C complexed to either (a) their complementary homopolymer or (b) methylated bovine serum albumin (MBSA) will produce stimulation (Braun and Nakano, 1967c). The double-stranded molecules of complementary homopolymers, or the homopolymer-MBSA complexes, must be injected as such; injection of the individual components (e.g., poly A, poly U) at different sites fails to elicit stimulation (Braun and Nakano, 1967c). We attribute this requirement, for the administration of homopolymers in the form of a complex, to an excessively rapid depolymerization of single-stranded homopolymers *in vivo.* The double-stranded molecules, as well as the polynucleotide-MBSA complex, are known to be more resistant to nucleases (Plescia *et al.,* 1964).

Some, but not all, preparations of poly-U complexed with MBSA also have shown stimulatory activity (of a lesser magnitude than poly A-MBSA or poly C-MBSA); poly G-MBSA appears inactive. We, therefore, suspect that in mammals as in bacteria (see Firshein *et al.,* this volume) oligonucleotides derived from poly C and poly A are the principal stimulators, but in bacteria the *single*-stranded homopolymers are active (due to a lesser concentration or activity of nucleases?) whereas in mammals the homopolymers have to be administered in a protected *double*-stranded or MBSA-complexed form. Di- to hexanucleotide fractions of poly-A have been tested in mice and have shown activity, particularly when administered as a mixture with poly U.

Firshein *et al.* (1967) have reported that poly A and poly C act as inducers of nucleotide kinases in pneumococci, specifically of dCMP and

Table 3

INFLUENCE OF COMPLEXED AND UNCOMPLEXED HOMOPOLYMERS, AND OF DEOXYNUCLEOSIDE TRIPHOSPHATES, ON THE NUMBER OF HEMOLYSIN-PRODUCING SPLEEN CELLS IN MICE 48 HOURS AFTER IMMUNIZATION

Treatment of Spleen Donors	Average Number (±S.E.) of Hemolysin-producing Spleen Cells Per 10^8 Nucleated Spleen Cells After 48 Hours
CF-1 Mice	
None	15.6 ± 11.1
sRBC (10^8/mouse i.v.)	244.8 ± 29.6
sRBC (10^8/mouse i.v.) + poly A-poly U ‡	992.0 ± 216.1
sRBC (10^8/mouse i.v.) + poly A §	210.6 ± 39.6
sRBC (10^8/mouse i.v.) + poly U §	192.8 ± 50.4
sRBC (10^8/mouse i.v.) + poly A-poly U ‡ + kr ‖	397.0 ± 78.1
AKR Mice	
None	29.8 ± 8.7
sRBC (10^8/mouse i.v.)	466.0 ± 43.6
sRBC (10^8/mouse i.v.) + poly A-poly U *	2287.6 ± 209.4
sRBC (10^8/mouse i.v.) + poly A †	530.0 ± 52.2
sRBC (10^8/mouse i.v.) + poly A-MBSA ‡	1115.0 ± 127.7
sRBC (10^8/mouse i.v.) + MBSA †	437.0 ± 20.7
CF-1 Mice	
None	33.4 ± 31.7
sRBC (10^8/mouse i.v.)	392.6 ± 71.0
sRBC (10^8/mouse i.v.) + poly C †	325.2 ± 84.8
sRBC (10^8/mouse i.v.) + poly C-MBSA ‡	694.6 ± 100.5
sRBC (10^8/mouse i.v.) + MBSA †	230.2 ± 51.3
CF-1 Mice	
None	17.3 ± 10.6
sRBC (10^8/mouse i.v.)	282.8 ± 87.5
sRBC (10^8/mouse i.v.) + poly A-poly U *	794.7 ± 108.7
sRBC (10^8/mouse i.v.) + dATP, dGTP, dCTP, dTTP *	479.8 ± 52.2
sRBC (10^8/mouse i.v.) + dATP, dGTP, dCTP, dTTP * + kr ‖	242.6 ± 55.8

* 150 γ/mouse of each i.v.
† 750 γ/mouse i.v.
‡ 750 γ/mouse of each i.v.
§ 1500 γ/mouse.
‖ 2 mg/mouse.

Table 4

Influence of Complexes Between Poly G and Poly C on the Number of Hemolysin-producing Spleen Cells in CF-1 Mice 48 Hours After Immunization with 10^8 sRBC

Treatment of Spleen Donors *	Average No. (±S.E.) of Hemolysin-producing Spleen Cells Per 10^8 Nucleated Spleen Cells After 48 Hours
None	6.7 ± 2.5
sRBC	310.5 ± 48.2
sRBC + poly G-poly C	1697.4 ± 152.9
sRBC + poly A-poly U	1085.9 ± 38.7
sRBC + poly G	254.7 ± 33.3
sRBC + poly C	282.5 ± 14.8
sRBC − poly G-poly C	13.3 ± 2.6
sRBC − poly G	12.8 ± 2.6
sRBC − poly C	15.7 ± 1.7

* Where homopolymers were used, 150 γ/mouse of each was injected i.v.

dGMP kinases. This means that these polymers stimulate the production or activity of two of the four enzymes that are concerned with the formation of deoxynucleoside triphosphates and -diphosphates from deoxynucleoside monophosphates. The deoxynucleoside triphosphates are, of course, known to be the immediate precursors of DNA (Kornberg, 1961). If poly A-poly U, or poly C-poly G, similarly affect nucleotide kinase activities in mammalian cells, one would anticipate that the administration of the products, i.e., of all four deoxynucleoside triphosphates would result in a stimulation of antibody-forming cells. This was found to be the case (Table 3). In contrast, administration of all four deoxynucleoside monophosphates fails to produce such effects.

We thus may conclude that A- or C- (and possibly also U-) rich oligonucleotides, acting as inducers of enzymes concerned with DNA synthesis, can stimulate DNA synthesis, and with it rates of cellular activity and division of certain cells of the RES. Other agents and conditions that are known to stimulate such activities may in many instances do so by triggering the release of the actual stimulators, that is the release of A- or C-rich oligonucleotides. This conclusion is further supported by the finding that kinetin riboside (the riboside of 6-furfuryl amino purine), and also N^6-(Δ^2-isopentenyl) adenosine, a natural component of transfer RNA (Hall *et al.*, 1966), abolish the stimulatory effects of all of the stimulators discussed above, namely the effects of poly A-poly U (Table 3), poly C-poly G, mixtures of deoxynucleoside triphosphates (Table 3), supernates from antigen-exposed immune spleen cells, readministration of antigens, and of cytotoxic agents such as FUDR (Braun and Nakano, 1966a, 1967). Some endotoxin effects also are antagonized by kinetin riboside (Kessel and Braun, 1966). It is noteworthy that kinetin riboside and N^6-(Δ^2-isopentenyl) adenosine interfere only with the stimulated immune response but not with the normal immune response. We shall return to this point again later on.

It is also noteworthy that a known inhibitor of interferon release, namely urethan (DeMaeyer-Guignard and DeMaeyer, 1967), acts like kinetin riboside, interfering with the non-specific stimulation of antibody-forming cells by oligonucleotides. A possible relationship between the stimulation of interferon-release or -production and the stimulation of antibody-forming cells also is suggested by the recent report of enhancement of interferon levels after exposure of animals to heterologous double-stranded RNA (Lampson *et al.*, 1967) and multistranded synthetic polynucleotide complexes (Field *et al.*, 1967). Single-stranded RNA proved inactive.

II. THE UPTAKE OF STIMULATORY OLIGONUCLEOTIDES BY CELLS OF THE RES

DNA digests, poly A-poly U, or poly C-poly G will evoke a stimulation of the rate of increase in numbers of specific antibody-forming cells only when administered either with specific antigen or with substances that are suspected of altering the permeability of lymphocyte membranes (Braun and Nakano, 1965, 1967a, 1967b; Braun, Nakano and Freedman, 1966). Thus, the sole administration of DNA digest or of poly A-poly U to mice will cause no alteration in the background number of spleen cells forming antibodies to any of various antigens that have been tested, e.g., sRBC or cRBC. However, when these materials are injected i.p. or i.v. just prior to, with, or shortly after, the i.p. or i.v. injection of a specific antigen, they will enhance the rate of increase in spleen cells forming antibodies to the specific antigen. For example, poly A-poly U administered with sRBC will increase the number of spleen cells forming anti-sRBC 48 hours later (Table 3) but it will not affect the background number of spleen cells forming anti-cRBC or anti-*E. coli.* On the other hand, when permeability-altering agents such as chlorpromazine, phenoxybenzamine, streptolysin C, or low concentrations of cortisone, are given in conjunction with DNA digests or poly A-poly U, small but significant increases in the number of many types of antibody-forming spleen cells can be observed. For example, chlorpromazine administered at the time of injection of poly A-poly U will result, 48 hours later, in an above-background number of spleen cells that form antibodies to sRBC, cRBC, and *E. coli*. We have interpreted these findings as follows: stimulatory oligonucleotides ordinarily cannot get into lymphocytes, but they can get into these cells after membrane permeability has been altered. Thus, after administration of a specific antigen, a reaction between the antigen and antigen-sensitive cells (Kennedy *et al.,* 1966), or possibly also between antigen and antibody on the surface of antibody-forming cells, may open up these particular cells, but no other lymphocytes, for entrance of the stimulators; consequently, stimulation is restricted to those cells making antibody to the antigen administered together with the stimulatory material. In contrast, when the permeability of *all* lymphocytes is altered by agents such as chlorpromazine or cortisone, stimulatory oligonucleotides may enter into all members of an already activated clone of lymphocytes. The attendant dilution of available stimulator into many cells, however, will cause a relatively weak response in comparison to responses obtained when only cells forming antibodies to one antigen are stimulated.

When mice are injected with DNA digests or poly A-poly U and nothing else, a significant increase in the rate of clearance of carbon particles can be observed (Freedman and Braun, 1965). This enhancement of phagocytosis does not appear to require the presence of a specific antigen or of a non-specific modifier of cell permeability. Nevertheless, the possibility that antigen present on the surface of carbon particles may participate in the stimulation of phagocytic cells by oligonucleotides cannot be ruled out.

III. THE RESPONSE TO STIMULATORY OLIGONUCLEOTIDES

Assays on antibody-forming or -releasing spleen cells at different times following immunization (Figs. 1 and 4) indicate that oligonucleotides influence principally the early rate of increase in the number of such cells, the optimum number eventually obtained is about the same in stimulated and non-stimulated animals. It is not known whether in addition to effects on cell numbers, the antibody-forming activity per cell may also be affected; oligonucleotides do not increase the overall number of cells per spleen.

Dosage and route of administration can influence the magnitude of the oligonucleotide effect (Braun and Nakano, 1965). Active materials remain in the circulation of mice treated with DNA digests for relatively long periods of time. Thus, passive transfer of serum, from a mouse exposed to DNA digest 48 hours earlier, to a recipient animal at time of immunization, will produce stimulation. However, seventy-two hours after DNA digest treatment, the stimulator is no longer in the circulation in appreciable amounts (Braun and Nakano, 1967b). These data are quite similar to those described in the case of passive transfer of the activity elicited by Freund's adjuvant (Dawe *et al.*, 1965).

Both primary and secondary responses can be stimulated by oligonucleotides but the stimulation affects principally 19S antibody-forming cells, the effect on 7S antibody-forming cells is of questionable significance (Table 5). The extent of stimulation is greater in newborn animals than in adult animals (Hechtel *et al.*, 1965), and this finding has suggested that in older animals the stimulatory effects may be superimposed upon those already produced by a background level of oligonucleotides which may be present as the result of immunological and physical injuries to cells (Braun and Nakano, 1967b).

Now let us return to an observation cited earlier, namely the ability of kinetin riboside (kr) and of N^6-(Δ^2-isopentenyl)adenosine to block stimulated responses but not the normal immune response (cf. Tables 1 and 3). One explanation for this selective suppression of only the excess of antibody-forming cells usually present after stimulation by oligonucleotides or cytotoxic adjuvants may be a physiological difference between the stimulated cells and the cells participating in the normal immune response. The latter might be resistant to kr, and to agents with similar activity, whereas the former might be sensitive. In an attempt to detect such possible heterogeneity among spleen cells forming antibodies to one group of antigenic determinants (e.g., those of sRBC), we initiated studies with Dr. B. Jaroslow regarding the recovery of hemolysin-forming spleen cells following centrifugation in a sucrose density

Table 5

INFLUENCE OF POLY A-POLY U ON THE SECONDARY RESPONSE

Treatment of Spleen Donor at Time of 2nd Antigen Exposure *		Average No. of Hemolysin-producing Cells Per 10^8 Spleen Cells After				
		0 hr	24 hr	48 hr	72 hr	96 hr
sRBC (10^8/mouse)	19S	–	1895.2	*14355.4* †	50094.0	32107.6
	7S	–	2866.8	12533.8 ‡	18152.0	28878.4
sRBC (10^8/mouse) +	19S	–	1043.4	*24903.2* †	31603.6	45662.8
Poly A-poly U (150 γ each)	7S	–	2489.6	16018.6 ‡	13799.2	31457.6
None	19S	1186.6	–	263.7	–	340.4
	7S	5738.5	–	793.8	–	1025.3

* All mice (CF-1) received 10^8 sRBC/mouse 14 days earlier.
† $t = 2.648$ $P < 5\%$.
‡ $t = 1.5120$ $P < 20\%$.

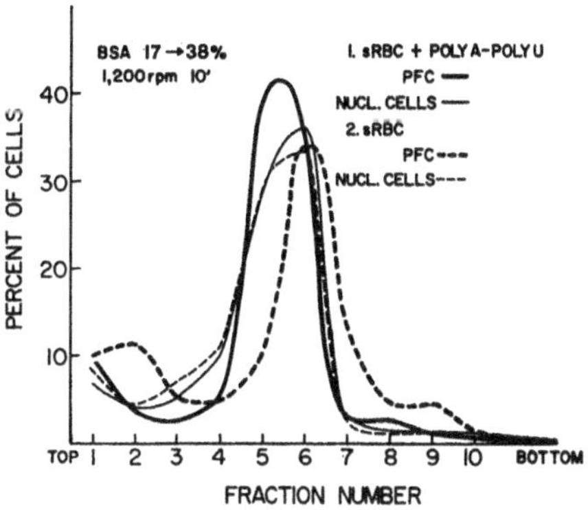

Fig. 5. Recovery of plaque-forming (anti-sRBC) spleen cells from different fractions of a BSA gradient, 17–38 percent. Pooled spleen cell suspensions, prepared from three spleens of immunized mice, were floated on top of a BSA gradient prepared with the aid of a gradient maker. After centrifugation at 1,200 rpm for 10′ at 5°C, fractions were collected by siphoning and assayed according to the Jerne technique.

gradient (Jaroslow, Rahman and Nakano, unpublished data). Recovery of hemolysin-forming cells from different parts of the gradient was obtained, and there was some suggestion that after stimulation of the spleen donors with poly A-poly U more cells were recovered from a fraction of light density. We recently extended these studies to centrifugation in a bovine serum albumin gradient. Spleen cells from mice immunized 3 days earlier with sRBC or sRBC + poly A-poly U were carefully separated from each other by teasing and straining through a stainless steel mesh; they were centrifuged for a brief period in minimal essential medium to remove cell clumps, and then centrifuged for 10 minutes at 1200 rpm in a 17% to 38% BSA gradient. Fractions were then collected, checked for spleen cell numbers, and assayed by the Jerne technique. A typical result is shown in Figure 5. It can be seen that the spleens of stimulated animals contain many hemolysin-forming cells that band in a part of the gradient of low density. Thus it appears that in stimulated animals, cells of different density are recruited for antibody formation. These cells may either represent cells that ordinarily fail to participate in antibody formation, or more likely they may be younger cells resulting from more rapid cell multiplication after stimulation. In any event, the results indicate a heterogeneity of the antibody-forming cell population in stimulated animals, and agents such as kinetin riboside seem to affect antibody-forming cells recoverable from the lighter part of the density gradient. It is noteworthy that kr can also inhibit a certain proportion of antibody-forming cells *in vitro* when added to Jerne plates; the actual proportion of spleen cells that is inhibited shifts from day to day after immunization. This again suggests a heterogeneity of the spleen cell population participating in 19S antibody formation to a given antigen, but the basis for this heterogeneity is obscure. Similarly, there is no explanation for the heterogeneity of cell populations forming IgG antibodies to a single hapten as recently reported by Mäkela (1967). However, it is unlikely that the inhibition by kr obtained *in vitro* is related to that observed *in vivo,* since the inhibition *in vitro* can affect a high proportion of hemolysin-forming spleen cells from animals that are engaged in a normal non-stimulated immune response, whereas, it will be recalled, the *in vivo* inhibition affects only the stimulated response.

Finally, one may ask: how are antibody-forming spleen cells distributed after stimulation of the spleen donor by poly A-poly U, by DNA digest or by cytotoxic adjuvant? To get an answer to this question, one of us (L.J.) has employed a technique previously utilized by Nakano and Braun (1966) to demonstrate the non-random distribution of antibody-forming cells in the spleen. The procedure employed by Nakano and Braun was to cut a spleen from a mouse, immunized with sRBC 3–4 days earlier, into approximately 20 fragments of near-equal size and to assay the number of hemolysin-forming cells in each of the fragments. From the earlier results with this technique it had become apparent that there were micro-colonies (clones?) of cells forming anti-sRBC immuno-

Table 6

EFFECTS OF AN ENHANCER (POLY A-POLY U) AND A SUPPRESSOR (6-MERCAPTOPURINE) OF IMMUNE RESPONSES ON THE DISTRIBUTION OF HEMOLYSIN-FORMING CELLS IN SPLEEN FRAGMENTS OF CF-1 MICE IMMUNIZED 96 HOURS EARLIER WITH 10^8 sRBC. EACH SET OF 24 FIGURES SHOWS THE ORIENTATION OF THIS FRAGMENT IN THE SPLEEN OF THE DONOR. PROBABLE EXTENT OF INDIVIDUAL CLONES ARE INDICATED BY BRACKETING

Hemolysin-forming spleen cells (×1/25) in the spleen of a mouse immunized with								
sRBC + 6-Mercaptopurine (5 mg/mouse)			sRBC alone			sRBC + Poly A-poly U (150 γ of each)		
20	140	35	390	290	140	640	144	100
45	35	190	90	270	390	140	220	180
195	102	90	110	370	180	160	600	400
26	176	270	140	180	120	220	480	260
15	40	50	270	220	60	480	680	160
20	260	240	125	144	200	240	180	120
70	70	180	200	350	340	320	800	560
180	240	40	295	120	70	700	760	240
Average: 111			Average: 211			Average: 354		

globulins in the spleen, producing fragments with high number of hemolysin-forming cells and other fragments with relatively low numbers of such cells. It has now been observed that the number of fragments with high numbers of hemolysin-forming cells does not change materially after administration of sRBC together with a stimulatory dose of poly A-poly U; neither does it change when responses are partially suppressed by 6-mercaptopurine. All that changes in the spleen of stimulated or suppressed animals is the absolute number of hemolysin-forming cells in the fragments with high numbers (see example in Table 6). From this one may conclude that following stimulation of antibody formation by oligonucleotides the number of stem-cells ("antigen-sensitive" cells) responding to an antigenic stimulus, and thus the number of resulting antibody-forming clones, remains the same. What is altered in stimulated animals is the rate of increase of antibody-forming cells in activated clones. Oligonucleotides thus seem to influence the rate of division of activated (derepressed), committed, antibody-forming cells. In addition, if some of the increases in antibody-forming clones really do occur by means other than cell division (cf. Braun and Nakano, 1967b; Eidinger and Pross, 1967), oligonucleotides might also exert their effects by influencing the rate of spreading of something like "activated information" from one cell to a neighboring cell. We hope that more information regarding the latter possibility will become available from studies on antibody-forming cells in tissue cultures. Such studies are now in progress.

References

Braun, W. (1958). Cell population dynamics and somatic change. J. Cellular Comp. Physiol. Suppl. 1, *52*:337–369.

Braun, W. (1962). Effects of cell components on cell multiplication and changes in cell populations. *In* Biological interactions in normal and neoplastic growth. Boston: Little, Brown and Co.

Braun, W. (1965). Influence of nucleic acid degradation products on antibody synthesis. *In* Molecular and cellular basis of antibody formation. New York: Academic Press.

Braun, W., and E. P. Cohen (1968). On the role of nucleic acids in antibody formation. *In* Regulation of the antibody response. B. Cinader (ed.). Springfield, Ill.: Charles C Thomas.

Braun, W., and W. Firshein (1967). Biodynamic effects of oligonucleotides. Bacteriol. Rev., *31*:83–94.

Braun, W., W. Firshein, and A. Whallon (1957). Effects of deoxyribonucleic acid breakdown products on bacterial population changes and virulence. Science, *125*:445–447.

Braun, W., and R. W. I. Kessel (1964). Cytotoxicity of endotoxins in relation to their effects on host resistance. *In* M. Landy and W. Braun (ed.), Bacterial endotoxins. New Brunswick, New Jersey: Rutgers University Press.

Braun, W., and M. Nakano (1965). Influence of oligodeoxyribonucleotides on early events in antibody formation. Proc. Soc. Exptl. Biol. Med., *119:*701–707.

Braun, W., and M. Nakano (1967a). The role of cell breakdown products in the stimulation of immune responses. *In* International Symposium on Adjuvants of Immunity. Symposia Series in Immunobiological Standardization, vol. 6. Basel/New York: S. Karger Publishing Co.

Braun, W., and M. Nakano (1967b). The reinforcement of immunological stimuli by oligodeoxyribonucleotides. *In* R. Smith and R. Good (ed.), Ontogeny of Immunity. University of Florida Press.

Braun, W., and M. Nakano (1967c). Antibody formation: Stimulation by polyadenylic and polycytidylic acids. Science, *157:*819–821.

Braun, W., M. Nakano, and H. Freedman (1966). On the role of antigen in the stimulation of multiplication of antibody-forming cells. Federation Proc., *25:*370.

Braun, W., and J. Whallon (1954). The effects of DNA and enzyme-treated DNA on bacterial population changes. Proc. Natl. Acad. Sci. U. S., *40:*162–165.

Dawe, D. L., D. Segre, and W. L. Myers (1965). Passive transfer of the action of Freund's adjuvant by serum of rabbits injected with the adjuvant. Science, *148:*1345–1347.

DeMaeyer-Guignard, J., and E. DeMaeyer (1967). Depression of circulating interferon response in Balb/c mice after urethan treatment. Science, *155:*482–484.

Eidinger, D., and H. F. Pross (1967). The immune response to sheep erythrocytes in the mouse. I. A study of the immunological events utilizing the plaque technique. J. Exptl. Med., *126:*15–33.

Feldman, M., A. Globerson, and D. Nachtigel (1963). The reactivation of the immune response in immunologically suppressed animals, p. 427–449. *In* Conceptual advances in immunology and oncology. New York: Hoeber Medical Division, Harper & Row, Publishers.

Field, A. K., A. A. Tytell, G. P. Lampson, and M. R. Hilleman (1967). Inducers of interferon and host resistance, II. Multistranded synthetic polynucleotide complexes. Proc. Nat. Acad. Sci. U. S., *58:*1004–1010.

Firshein, W. (1961). Effects of deoxyribonucleic acid products on deoxyribonucleic acid synthesis of virulent and avirulent pneumococci. J. Bacteriol., *82:* 169–180.

Firshein, W. (1968). This volume.

Firshein, W., C. Benson, and M. Sease (1967). Deoxycytidylate and deoxyguanylate kinase activity in pneumococci after exposure to known polyribonucleotides. Science, *157:*821–822.

Firshein, W., and W. Braun (1960). Effects of enzymatic deoxyribonucleic acid digests on population changes and deoxyribonucleic acid synthesis of pneumococci. J. Bacteriol., *79:*246–260.

Freedman, H. H., and W. Braun (1965). Influence of oligodeoxyribonucleotides on phagocytic activity of the reticuloendothelial system. Proc. Soc. Exptl. Biol. Med., *120:*222–225.

Freund, J. (1956). The mode of action of immunologic adjuvants. Adv. Tuberc. Res., *7:*130–148.

Hall, R., M. J. Robins, L. Stasiuk, and R. Thedford (1966). Isolation of N^6-

(γ, γ-dimethyl allyl)adenosine from soluble ribonucleic acid. J. Amer. Chem. Soc., *88*:2614–2615.

Hechtel, M., T. Dishon, and W. Braun (1965). Influence of oligodeoxyribonucleotides on the immune response of newborn AKR mice. Proc. Soc. Exptl. Biol. Med., *119*:991–993.

Jaroslow, B. N., and W. H. Taliaferro (1966). The effect of colchicine on the hemolysin response in unirradiated and irradiated rabbits. J. Infect. Diseases, *116*:139.

Jerne, N. K., A. A. Nordin, and C. Henry (1963). The agar plaque technique for recognizing antibody-producing cells, p. 109–122. *In* Cell-bound antibodies. Philadelphia: Wistar Institute Press.

Johnson, A. (1964). *In* "Bacterial Endotoxins," M. Landy and W. Braun (eds.). New Brunswick, New Jersey: Rutgers University Press.

Kennedy, J. C., J. E. Till, L. Siminovitch, and E. A. McCulloch (1966). The proliferative capacity of antigen-sensitive precursors of hemolytic plaque-forming cells. J. Immunol., *96*:973–980.

Kornberg, A. (1961). Enzymatic synthesis of DNA. Ciba Lectures Microbial Biochem. New York: John Wiley & Sons, Inc.

Lampson, G. P., A. A. Tytell, A. K. Field, M. M. Nemes, and M. R. Hilleman (1967). Inducers of interferon and host resistance. I. Double-stranded RNA from extracts of *Penicillium funiculosum*. PNAS, *58*:782–789.

Makela, D. (1967). Cellular heterogeneity in the production of an antihapten antibody. J. Exptl. Med., *126*:159–170.

Merritt, K., and A. G. Johnson (1965). Studies on the adjuvant action of bacterial endotoxins on antibody formation. VI. Enhancement of antibody formation by nucleic acids. J. Immunol., *94*:416–422.

Nakano, M., and W. Braun (1966). Fluctuation tests with antibody-forming spleen cell populations. Science, *151*:338–340.

Nakano, M., and W. Braun (1967). Cell-released non-specific stimulators of antibody-forming spleen cell populations. J. Immunol., *99*:570–575.

Plescia, O., W. Braun, and N. Palczuk (1964). Production of antibodies to denatured deoxyribonucleic acid (DNA). Proc. Natl. Acad. Sci. U. S., *52*: 279–285.

Steiner, R. F., and R. F. Beers (1961). Enzymatic probes of polynucleotide structure, p. 65–72. *In* Polynucleotides. Amsterdam: Elsevier Publishing Co.

Taliaferro, W. H., and B. N. Jaroslow (1960). The restoration of hemolysin formation in x-rayed rabbits by nucleic acid derivatives and antagonists of nucleic acid synthesis. J. Infect. Diseases, *107*:341–350.

Vigliani, E. C., and B. Pernis (1959). An immunological approach to silicosis. J. Occupational Med., *1*:319–328.

Weiss, D. W., R. S. Bonhag, and J. A. Parks (1964). Studies on the heterologous immunogenicity of a methanol-insoluble fraction of attenuated tubercle bacilli (BCG). J. Exp. Med., *119*:53–70.

INFLUENCE OF OLIGONUCLEOTIDES OF KNOWN COMPOSITION ON RATES OF DNA SYNTHESIS IN PNEUMOCOCCI *

W. Firshein and R. C. Benson

Department of Biology, Wesleyan University, Middletown, Conn.

For some time, we have been interested in the ability of virulent strains of pneumococci and streptococci to synthesize DNA in excess of their normal content in resting cell suspensions provided that a mixture of oligodeoxyribonucleotides and a substrate pool of deoxyribonucleosides and -tides is present (Firshein, 1961; Firshein and Zimmermann, 1964). In the case of pneumococci, this excess DNA has been shown to be genetically competent (Firshein, 1965). Comparable stimulatory effects are not produced in cell suspensions of avirulent strains of these genera. Experimental evidence has indicated that the oligodeoxyribonucleotides (derived from a variety of deoxyribonuclease treated DNAs) induce the synthesis of high levels of two deoxyribonucleotide kinases (deoxycytidylate and deoxyguanylate) which are present in very low amounts in unsupplemented cell suspensions. The two remaining kinases (deoxyadenylate and thymidylate), which are present in sufficient amounts throughout the incubation period, are unaffected by the supplementation of cell suspensions with oligodeoxyribonucleotides. The net result of this apparent induction is an increase in the intracellular pool of all four deoxyribonucleoside triphosphates required for DNA synthesis (Firshein, 1965).

In recent experiments, it was observed first, that the greater the concentration of cytosine (C), and to a limited extent adenine (A), in the oligodeoxyribonucleotide fraction, the greater the stimulatory effects on kinase activity, and second, that polyribonucleotides of C, and to a lesser extent of A, exert greater effects than the oligodeoxyribonucleotides in enhancing deoxyribonucleotide kinase activity. The active

* This research was supported, in part, by a grant from the National Cancer Institute of the Public Health Service, CA-06343. William Firshein is a Career Development awardee of the Public Health Service.

homopolymers (of A and C) had previously been observed by Braun and Nakano (1967) to enhance antibody formation in mammalian cell systems in a manner similar to that of the oligodeoxyribonucleotides.

The present study was undertaken first, to determine whether the effects of the polyribonucleotides and the oligodeoxyribonucleotides were related, second, to characterize the effects of the polyribonucleotides to a greater extent, third, to establish whether the synthesis of a specific kinase-protein was involved in the enhancement by the homopolymers, and if so, fourth, to ascertain whether the active polyribonucleotides interacted with cell components that might control the rate and extent of kinase synthesis, such as DNA and RNA. All of the experiments to be described were carried out with a virulent strain of type III *Diplococcus pneumoniae,* A66, which responds maximally to the oligodeoxyribonucleotides and polyribonucleotides, with respect to stimulation of deoxycytidylate (dCMP) and deoxyguanylate (dGMP) kinase activity. The methods for growth of cells, preparation of resting cells, enzyme extraction, and assay of deoxyribonucleotide kinase activity were described in detail previously (Firshein, 1961, 1965). Other methods will be described in the text or in the legends of the tables and figures.

Initial experiments surveyed the effects of various polyribonucleotides on the activity of dCMP and dGMP kinases partially purified from cell suspensions of pneumococci after 25 minutes of incubation (Table 1). It can be seen that the addition of poly C enhanced the activity of both dCMP and dGMP kinases, whereas poly A exerted stimulatory effects on dGMP kinase activity only. Poly U, poly I, or poly G were ineffective in stimulating kinase activity. Poly C and poly A had no effects on dCMP and dGMP kinase activity when added to isolated enzyme extracts. Additional experiments showed that molecular complexes between poly C + poly I or poly A + poly U were unable to exert any effects on dCMP and dGMP kinase activity, suggesting that a single-stranded structure is required for stimulation (Table 2).

When 2.5 mg/ml of poly C was treated with a concentration of pancreatic ribonuclease (10 μg/ml) that caused extensive degradation of poly C to cytidine monophosphate (CMP), no stimulatory activity could be detected. However, when poly C was treated with small amounts of RNAase (from 0.03 to 0.003 μg/ml), stimulatory activity for dCMP and dGMP kinases was equivalent or greater than that of the intact homopolymer. This result, suggesting that C-oligomers were more active or as active as the homopolymer was supported by the finding that the addition of (C^{14}) poly C to cell suspensions resulted in the recovery of approximately 80% of the cell-associated radioactivity in the cold trichloroacetic acid-soluble fraction rather than in the fraction soluble in hot trichloroacetic acid (nucleic acid fraction) (Table 3). The extent of degradation of the (C^{14}) poly C in the acid-soluble fraction was ascer-

Table 1

EFFECTS OF POLYRIBONUCLEOTIDES ON dCMP AND dGMP KINASE ACTIVITY OF PNEUMOCOCCI

Polyribonucleotide Added to Bacterial Suspensions	Concentration (μg/ml)	Specific Activity After 25 min. (mμmoles C^{14} Product/hr/mg Protein) $\times 10^3$	
		dCMP Kinase	dGMP Kinase
	25	9.2	4.2
Poly-C	50	11.8	9.4
	100	15.8	17.5
	25	3.2	9.8
Poly-A	50	3.5	10.5
	100	4.8	8.9
	25	2.9	3.1
Poly-U	50	3.5	3.0
	100	2.8	2.5
	25	4.1	3.5
Poly-I	50	3.2	3.5
	100	2.8	3.0
	25	3.2	4.5
Poly-G	50	3.0	4.8
	100	3.0	5.1
None	—	3.1	3.8

Pneumococci (type III, strain A66) were cultured in 2–3 L of CAT medium (Firshein, 1961) for 18 hr at 37°C. Resting cells (prepared by centrifuging and washing the cells 3 times with 0.02M sodium-potassium phosphate buffer) were suspended in the same buffer containing glucose (1%), Difco casitone (1%), and catalase (0.005%). The suspensions (2–4×10^9 cells/ml) were shaken in a New Brunswick gyrotory water bath shaker at 37°C for various periods of time in the presence and absence of the supplements. Deoxyribonucleotide kinases were extracted from these cells as described in detail previously (Firshein, 1965). The assay mixture contained the following (in 0.6 ml): Tris (hydroxymethyl aminomethane), pH 7.5, 20.0 μmoles; ATP, 5.0 μmoles; $MgCl_2 \cdot 6H_2O$, 9.0 μmoles; (C^{14}) deoxyribonucleotide, 0.07 μmoles (1.4 μc); kinase, 20–100 μg (in potassium phosphate buffer, 0.01M, pH 7.5). Incubation of the assay tubes was at 37°C for 30 minutes. Determination of specific activity was described previously (Firshein, 1965).

Table 2

EFFECTS OF MOLECULAR COMPLEXES BETWEEN POLY C + POLY I AND POLY A + POLY U ON dCMP AND dGMP KINASE ACTIVITY

Addition to Bacterial Suspensions	Concentration (μg/ml)	Specific Activity After 25 min. (mμmoles C^{14} Product/hr/mg Protein) × 10^3	
		dCMP Kinase	dGMP Kinase
poly C + poly I	50 each	3.9	4.1
	100 each	2.0	2.4
poly A + poly U	25 each	2.0	4.2
	50 each	1.7	4.8
poly C	50	12.0	9.6
	100	14.8	16.2
poly A	25	3.0	9.9
	50	3.3	11.2
none	—	3.5	4.1

See Table 1 for assay conditions. The molecular complex was formed prior to addition to the cell-suspensions by incubation for 30 minutes at 25°C as described by Davies and Rich (1958).

Table 3

DISTRIBUTION OF RADIOACTIVITY IN PNEUMOCOCCAL EXTRACTS AFTER INCORPORATION OF (C^{14}) POLY C

State of Cells	Counts/min Over Background	Fraction of Total Extract Counted	Total Counts/min
unwashed cells	36,034	1/27	972,910
thoroughly washed cells	10,436	1/6	62,616
cold trichloroacetic acid extract	9,862	1/5	49,310
ethanol-ether extracts	14	1/2	28
hot trichloroacetic acid extract	5,928	1/2	11,856
protein extract	22	1/4.5	99
Per cent recovery from washed cells			97

Methods for extraction were taken from Firshein (1961). Incubation time and temperature was 25 minutes at 37°C. Specific activity of added (C^{14}) poly C was 0.8 μc/18 μg.

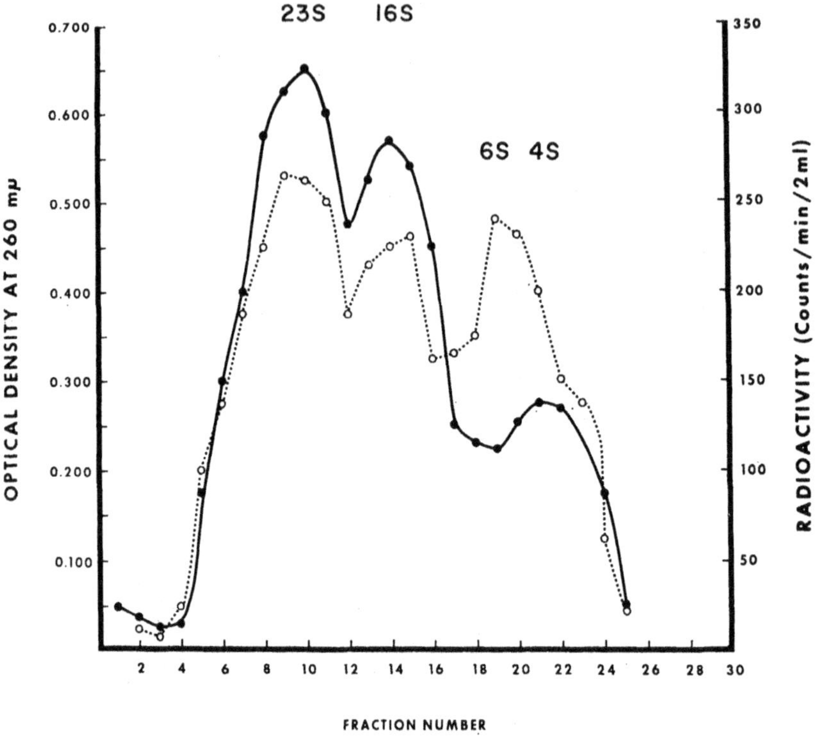

Fig. 1. Sucrose density fractionation of RNA extracted from pneumococcal cells exposed to (C^{14}) poly C. After incubation of cell suspensions RNA and any undegraded (C^{14}) poly C were extracted by a modification of the method of Lodish and Zinder (1966). Bentonite (8 mg/ml, final concentration) was added to inhibit RNAase activity and sodium deoxycholate (1 ml of 5% solution per 25 ml cell suspension) was used to lyse the bacteria. After extraction, 10 optical density units (at 260 mμ) in 0.5 ml were layered on 55 ml of a sucrose gradient (5–20%) prepared in 0.01M sodium acetate buffer (pH 5.2) containing 0.1M NaCl. The tubes were centrifuged at 24,500 rev/min for 20 hours in a SW 25.2 rotor of a Spinco model L-2 ultracentrifuge. The bottom of the tube was punctured, and 2.0 ml fractions were collected in a Vanguard fraction collector. Optical density readings were made on each fraction at 260 mμ in a Zeiss PMQ spectrophotometer. Radioactivity measurements were performed on solubilized (with 1.5N NH_4OH) nucleic acid precipitates obtained by treatment of each fraction with 6.0 ml of cold 0.6N perchloric acid, bovine serum albumin serving as a carrier (0.1 ml of solution containing 10 mg/ml). Specific activity of added (C^{14}) poly C was 1.2 μc/400 μg. ______ = optical density at 260 mμ; .... = radioactivity. Poly C sediments at 6S.

tained by chromatographing an aliquot of the extract in a solvent consisting of propanol, NH_4OH, and H_2O (55:10:35, v/v/v) as described by Nirenberg and Leder (1964) for 48 hours together with a reference standard of CMP. The radioactivity of the separated products was scanned in a Vanguard chromatogram scanner, Model 880. Eighty-seven per cent of the acid-soluble fraction consisted of oligomers of C, while 13% of the initial poly C was degraded to CMP.

The radioactivity in the nucleic acid fraction was found to be, in part, undegraded (C^{14}) poly C and either RNA labelled in the cytosine moiety from incorporated (C^{14}) CMP, and/or (C^{14}) poly C (or its oligomers—see Table 7) complexed with a specific fraction of cellular RNA (Fig. 1).

An important question to answer was whether an optimum extracellular chain-length of C was required for stimulatory activity, or whether more than one length would suffice for activity. Accordingly, various chain-lengths of C-oligomers were prepared as described in Table 4 and assayed for their ability to affect kinase activity. It can be

Table 4

EFFECTS OF VARIOUS OLIGOMERS OF DIFFERENT CHAIN LENGTHS ON dCMP KINASE ACTIVITY AND dGMP KINASE ACTIVITY

Length of C-oligomer (nucleotides) Added to Bacterial Suspensions	Specific Activity After 25 min. (mμmoles C^{14} Product/hr/mg Protein) $\times 10^3$	
	dCMP Kinase	dGMP Kinase
2	4.5	5.0
3	8.8	10.5
4	10.1	14.0
5	14.0	16.4
6	13.4	16.8
7	13.8	18.0
8	15.0	18.3
poly C	15.4	18.3
none	3.2	4.0

Assay mixtures were described in Table 1. Concentration of each oligomer was adjusted to the same concentration of poly C (100 μg/ml) used in the other experiments. To prepare the C-oligomers, 30 mg poly C were dissolved in 8.0 ml H_2O and treated with a solution of phosphate-free pancreatic ribonuclease A (2.25 μg/ml final concentration) for 30 minutes at 37°C. The entire digest (in 1 ml aliquots) was streaked across the bottom of a Whatman #3 MM sheet and the various oligomers separated by chromatography in the propanol solvent described in the text (Nirenberg and Leder, 1964) together with a reference standard of CMP. After the paper was dried, seven ultraviolet quenching areas were detected by ultraviolet light (2538 A). Each area was cut into small strips and eluted in 10 ml of propanol-H_2O (55:35, v/v). The eluates were evaporated to dryness *in vacuo* and the residues dissolved in H_2O to yield final concentrations of 800 μg/ml. The chain length of each oligomer was estimated by comparison with the Rf position of the CMP standard and by phosphate analysis of the oligomers after treatment with an *Escherichia coli* alkaline phosphatase (Lipsett, Heppel, and Bradley, 1961).

seen that maximum stimulation occurred when tetramers of C were added and no significant difference was observed between tetramers and the higher oligomers in this respect. Di- and trinucleotides were only one-quarter and three-quarters as effective, respectively, as the tetramers.

It is important to note that not every batch of poly C purchased from commercial suppliers has been active in stimulating kinase activity, particularly when a viscous solution is produced after the homopolymer is dissolved in water. The reasons for this are unknown, but they may reflect either a permeability problem or the presence of a secondary structure which is inactive. However, if inactive batches of poly C are treated with an amount of RNAase (less than 0.5 μg/30 mg poly C/8.0 ml) that produces only a slight degradation of the homopolymer, stimulatory activity is restored. Such RNAase treatment is now performed routinely with every batch of poly C. RNAase alone has no effect on kinase activity.

The enhancement of dCMP and dGMP kinase activity by C-oligomers was found to be dependent on an external source of amino acids and was inhibited by the addition of chloramphenicol, suggesting that, as in the case of stimulation by oligodeoxyribonucleotides (Firshein, 1965), protein synthesis, rather than enzyme activity, is involved in the enhancement (Table 5). Further proof for the involvement of protein synthesis was obtained by the finding that the specific activity of dCMP kinase purified 110 fold over crude extracts from cells exposed to poly C was approximately 5–6 fold greater than that extracted from control cells (Table 6).

An attempt was made to determine whether poly C, or its oligomers, could complex with cell-components that might control the rate and extent of kinase synthesis *in vivo* such as DNA and RNA. Annealing experiments, carried out with cell-extracted nucleic acids and poly C, or its oligomers, using the methods of Ritossa *et al.* (1966) and Kubinski *et al.* (1966) for DNA and Nonoyama and Ikeda (1966) for RNA, showed a significant complexing between poly C and the nucleic acids, but not between C-oligomers and nucleic acids (Fig. 2A, B; Fig. 3A, B). The experiments were divided into two parts, the first part determined the optimum RNA or DNA concentration to use with a specific concentration of poly C (Figs. 2A and 3A, respectively), and the second part assayed the total amount of poly C bound to RNA and DNA as a function of the poly C concentration (Figs. 2B and 3B, respectively).

Although C-oligomers did not exhibit any complexing with RNA or DNA, the *in vivo* situation may be different. It will be recalled that most of the poly C taken up by the cells was degraded to oligomers after incubation (see Table 3). The following experiment suggests that such C-oligomers may have complexed with RNA *in vivo*. RNA was extracted from cells exposed to the oligomers or poly C after a 25 minute incubation, and the same annealing experiments were performed as de-

Table 5

EFFECTS OF CHLORAMPHENICOL AND DEFICIENCY OF AMINO ACIDS ON dCMP AND dGMP KINASE ACTIVITY IN THE PRESENCE AND ABSENCE OF POLY C

Additions to Bacterial Suspensions				
Poly C	Chloram-phenicol	Casitone	Kinase Assayed	Specific Activity After 25 min. (mμmoles (C^{14}) Product/hr/mg Protein) × 10^3
+	−	+		14.3
+	+	+	dCMP	4.4
+	−	−		6.4
−	−	+		3.2
−	+	+		1.8
−	−	−		2.9
+	-	+		19.0
+	+	+		5.0
+	−	−	dGMP	7.3
−	−	+		3.9
−	+	+		2.8
−	−	−		4.1

Reaction mixtures and concentrations of poly A and poly C were as described in Table 1. Cells were washed four times with 0.85% saline adjusted to pH7.2 with dilute NaOH, and allowed to "starve" in the solution for 30 minutes before inoculation into the suspending medium to deplete as much of the internal amino acid pool as possible. Casitone (enzymatic digest of Difco casein used as the external amino acid source), a normal constituent of the suspending medium, was omitted and added as shown in the table at a final concentration of 1.0%. Chloramphenicol concentration was 100 μg/ml in the suspending medium.

scribed above. Table 7 shows that the complexing between poly C and RNA extracted from cells previously exposed to the oligomers or poly C was reduced considerably as compared to RNA extracted from unexposed cells. In fact, exposure to oligomers elicited a greater inhibition on a concentration basis than exposure to poly C.

Finally, analysis of the oligomer effects on DNA synthesis in cell suspensions of pneumococci in the presence of a substrate pool of deoxyribonucleosides and deoxyribonucleotides revealed that a selective enhancement of DNA synthesis occurred in comparison to RNA and protein synthesis in a manner similar to that observed with oligodeoxyribonucleotides (Table 8).

Table 6. Purification of dCMP Kinase from Poly-C-supplemented and Control Cells

Fraction	Total Volume		Total Activity		Total Protein		Yield		Specific Activity	
	Poly C	Control	Poly C	Control	Poly C	Control	Poly C	Control	Poly C	Control
	[ml]		[(C^{14}) μmoles/hr]		[mg]		[%]		[(C^{14}) μmoles/hr/ mg protein]	
A) Crude Extract	85	85	255	52	228	254	100	100	1.11	0.20
B) Streptomycin fraction	100	100	225	50.8	198	214	88	97	1.13	0.23
C) Calcium Phosphate Gel fraction	90	90	185	47	135	172	72	87	1.31	0.27
D) Alumina Cγ Gel fraction	16	16	26	6.5	15.5	17.1	9.4	13.3	1.60	0.40
E) Aγ-2	16	16	64	13	8.6	10.7	25	24	7.5	1.2
F) Aγ-3	14	14	68	12.4	5.8	6.3	27	23	12.0	2.0
G) DEAE cellulose fraction	16	16	22	4.0	0.61	0.68	9.0	7.6	34.0	5.8
H) DEAE Sephadex fraction	10	10	20	3.5	0.18	0.21	8.0	6.7	111.0	16.6

Assay mixture is described in Table 1. dCMP kinase was extracted from 25 liters of resting cells prepared as described previously (Firshein, 1961, 1965). The extraction procedure was taken from Lehman *et al.* (1958) for steps A through D, and from Sugino *et al.* (1966) for steps E through H. The Alumina Cγ-gel fraction is obtained by washing the gel containing the adsorbed kinases with 0.005 M potassium phosphate buffer (pH 7.5) containing 2-mercaptoethanol. Aγ-2 is extraction of the gel with 0.01M phosphate buffer, and Aγ-3 is extraction of the gel with 0.025M phosphate buffer. Fractions D, E, and F were pooled for chromatography on DEAE cellulose. After the last step, the fractions were dialyzed against 0.01M potassium phosphate buffer (pH 7.5) and stored at −20°C.

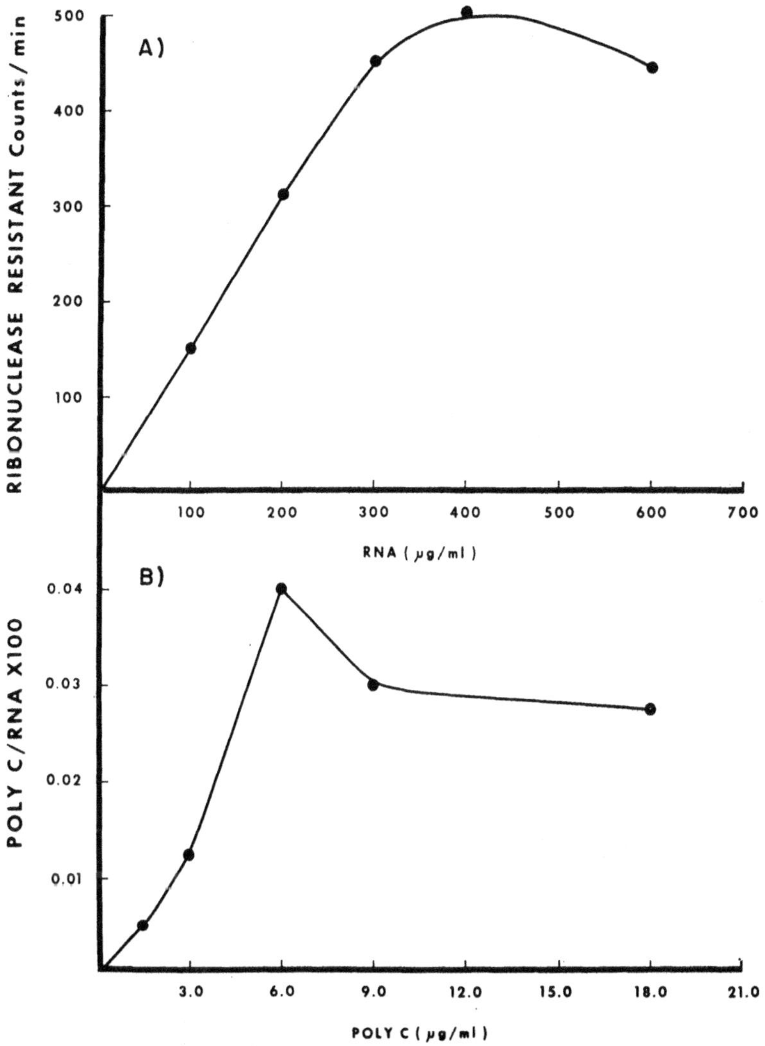

Fig. 2. Complexing of (C^{14}) poly C with pneumococcal RNA. (C^{14}) poly C, or its (C^{14}) oligomers, were diluted in 2 × S.S.C. (S.S.C. is 0.15M NaCl + 0.015M sodium citrate buffer, pH 7.0) to a concentration of 1–9 μg/ml poly C and 75–600 μg/ml RNA. The total volume was 2.0 ml. Controls containing (C^{14}) poly C or its oligomers were diluted in 2.0 ml of 2 × S.S.C. without RNA. The mixtures and controls were incubated for 2 hours at 25°C, after which time phosphate-free pancreatic ribonuclease was added at a final concentration of 40 μg/ml. The solution was incubated for 30 minutes at 25°C, and the radioactivity of the RNAase-resistant fraction (complexed material) was assayed by treatment with cold perchloric acid as described in Fig. 1. The addition of RNAase was omitted when the oligomers were being assayed, since they were not precipitated by perchloric acid. **A.** Complexing of (C^{14}) poly C to RNA as a function of the concentration of RNA. Specific activity of (C^{14}) poly C was 1.3 μc/9.0 μg. **B.** Complexing of (C^{14}) poly C to RNA as a function of the concentration of (C^{14}) poly C. Concentration of RNA was 400 μg/ml.

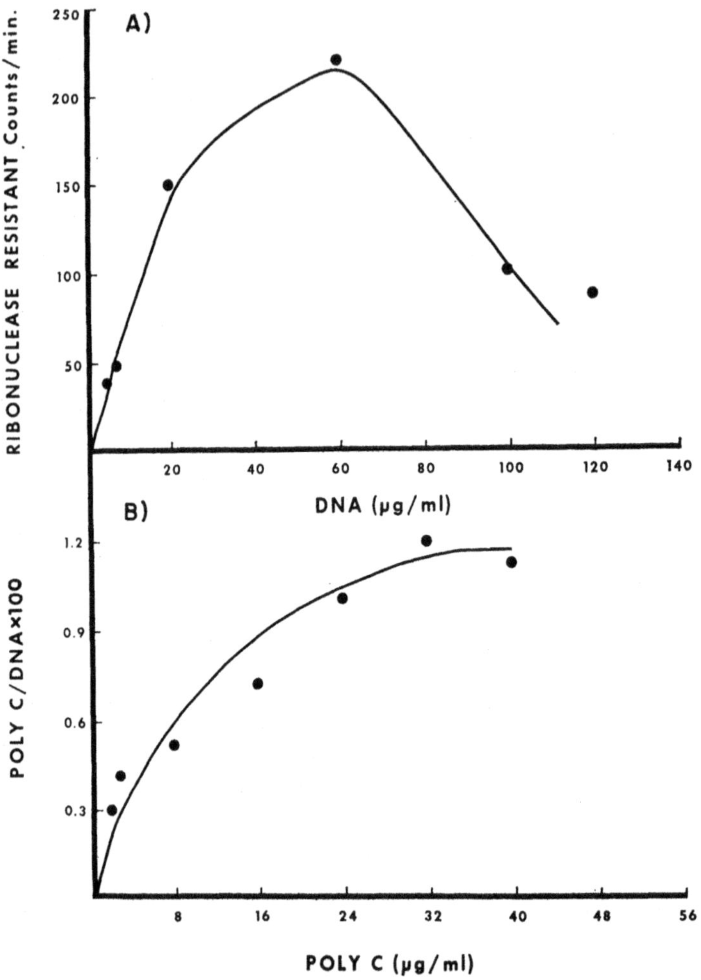

Fig. 3. Complexing of (C^{14}) poly C with denatured pneumococcal DNA. DNA was extracted as described by Marmur (1963) except that sodium deoxycholate was used as the lysing agent, and sodium perchlorate addition was omitted. For annealing experiments with DNA, it was necessary first to denature the DNA by heat treatment as described by Marmur and Doty (1961). Annealing was carried out in 1.2 ml of 2 × S.S.C. (see Fig. 2) at pH 5.0 (Kubinski *et al.,* 1966) containing 2–50 μg of (C^{14}) poly C or its oligomers and 10–100 μg of denatured DNA. The mixture was incubated for 2 hours at 57°C with appropriate controls. After incubation, the pH was adjusted to 6.5 with dilute alkali and phosphate-free RNAase was added at a final concentration of 40 μg/ml. The mixture was incubated at 30°C for 1 hour and the radioactivity of the ribonuclease-resistant fraction assayed as described in Fig. 2. **A.** Complexing of (C^{14}) poly C to DNA as a function of the DNA concentration. Specific activity of (C^{14}) poly C was 1.3 μc/9.0 μg. **B.** Complexing of (C^{14}) poly C to DNA as a function of the concentration of (C^{14}) poly C. Concentration of DNA was 60 μg/ml.

Table 7

In Vitro Complexing of (C^{14}) Poly C with Pneumococcal RNA Extracted from Cells Exposed to Poly C or Its C Oligomers

RNA Extracted from Pneumococci Exposed to	Ribonuclease-resistant Counts/min Over Background
Poly C	91 ± 9
C-oligomers	36 ± 5
None	540 ± 20

Poly C or its oligomers (at final concentrations of 100 μg/ml each in the suspending medium) were added to cell suspensions and incubated for 25 min at 37°C. After incubation, RNA was extracted as described in Fig. 1 and a complexing experiment performed as described in Fig. 2a, b. A ratio of 400 μg of RNA to 9.0 μg of (C^{14}) poly C/ml in 2 × S.S.C. was used. The solution was incubated for 2 hours at 25°C after which time the extent of complexing was determined as ribonuclease-resistant counts/min. The experiment was repeated four times.

Table 8

Effects of Poly C, Poly A or Both in Combination on Nucleic Acid and Protein Synthesis in the Presence of Deoxyribonucleotides and -sides

Additions to Bacterial Suspensions			Percentage Increase Over Zero Time After 70 min. of Incubation		
Poly C	Poly A	Deoxyribonucleosides and Deoxyribonucleotides	DNA	RNA	Protein
+	−	+	85	14	15
−	+	+	53	16	19
+	+	+	91	17	17
−	−	+	35	20	20
+	−	−	18	14	13
−	+	−	20	20	19
+	+	−	23	15	18
−	−	−	20	16	18

Methods for extraction and measurements of nucleic acids and protein were described previously (Firshein, 1961, 1965). Concentration of poly C and poly A were 100 and 50 μg/ml respectively. Deoxyribonucleoside concentration was 800 μg/ml (200 each of deoxyadenosine, deoxyguanosine, deoxycytidine, and thymidine); deoxyribonucleotide concentration was 800 μg/ml (200 each of the phosphorylated derivatives described above).

DISCUSSION

The above results suggest that oligoribonucleotides of C (and to a lesser extent of A) and oligodeoxyribonucleotides derived from enzymatic DNA digests affect the same biosynthetic process, i.e., dCMP and dGMP kinase metabolism. The lack of activity of other homopolymers (of U, of I, and of G) suggests that the stimulatory effects are not due simply to the addition of polyanions, but to the specific bases present in the homopolymer. Enzyme synthesis, rather than enzyme activity, is affected by the active polyribonucleotides. This is shown by the fact that (1) chloramphenicol inhibits the stimulatory effects of the oligomers; (2) omission of amino acids from the suspending medium reduces the stimulatory effects considerably; and (3) the specific activity of dCMP kinase purified from stimulated pneumococci is 5–6 fold higher than dCMP kinase extracted from a similar number of unstimulated pneumococci.

It is likely that C and A oligomers affect the same biosynthetic processes in the bacterial system and in the mammalian cell system described by Braun and Nakano (1967) despite the differences in requirement for single-stranded homopolymers and double-stranded molecular complexes, respectively. These differences may merely reflect a rapid degradation of homopolymers in the circulation of the animal, thus necessitating protection from nuclease action by the injection of a molecular complex which is more resistant to nuclease degradation. However, once the complex is incorporated into the proper cells, it may be degraded over a period of time to active oligomers. In contrast, in the bacterial system, nuclease activity may not be as pronounced in the extracellular environment, and single-stranded polyribonucleotides may remain relatively intact until they are taken up by the cells and degraded to active oligomers by intracellular nucleases. A molecular complex would probably be more difficult to degrade by the pneumococci during the short time period (25 minutes) of the assay period, and little or no active oligomers would result.

Although the actual mechanism leading to the enhancement of kinases is unknown, a suggestion of possible mechanisms of action of the oligomers can be obtained by considering the dependence of their effects on protein synthesis together with two additional observations. First, the failure of a poly C + poly I complex, or of a poly A + poly U complex, to affect kinase activity could be due to RNAase effects *in vivo,* thus preventing the formation of active complex with an intracellular kinase-controlling component. Second, the apparent ability of the active oligomers to complex with RNA *in vivo* (see Table 7) shows that such oligomers can interact with cell components that control the rate and extent of kinase synthesis. Thus, the active oligomers could induce higher levels of dCMP and dGMP kinases through interaction between repressor and the oligomer. Alternatively, the active supplements could, in some manner, stimulate the process by which the kinases are released from the polysomes, or facilitate the transcription process itself. Although it is not certain which, if any, of these mechanisms are involved, recent results of Kubinski *et al.* (1966) suggest that the effects of the active oligomers may be related to an interference with repressor action. These investigators have found that there are extensive sequences of C in the DNA of a variety of organisms, including bacteria, and that these regions may bind the DNA-dependent RNA polymerase. They have postulated that such clusters may correspond to specific initiation points for transcription of multicistronic messages. However, if such C-clusters were also capable of binding repressor molecules to prevent attachment of the polymerase, the presence of poly C or its oligomers, could conceivably relieve the inhibition preferentially by interfering with the binding of the repressor. It is interesting to note in this respect that poly C can interfere with the binding between denatured DNA and poly G, poly IG, and ribosomal RNA, while poly A prevents binding of DNA with poly U (Kubinski *et al.*, 1966).

In conclusion, the direct or indirect effects of poly C and poly A (or their oligomers) on kinase synthesis may be an important factor in elucidating certain metabolic aspects of control of DNA synthesis in pneumococci and other Gram-positive cocci. In addition, these studies may help in understanding certain aspects of the control of antibody synthesis in mammals.

References

Braun, W., and M. Nakano (1967). Antibody formation: Stimulation by polyadenylic and polycytidylic acids. Science, *157*:819–821.

Davies, D., and A. Rich (1965). The formation of a helical complex between polyinosinic acid and polycytidylic acid. J. Am. Chem. Soc., *80*:1003–1005.

Firshein, W. (1961). Effects of deoxyribonucleic acid products on DNA synthesis of virulent and avirulent pneumococci. J. Bacteriol., *82*:169–180.

Firshein, W. (1965). Influence of DNA degradation products and orthophosphate on deoxynucleotide kinase activity and DNA synthesis in pneumococcus type III. J. Bacteriol., *90*:327–336.

Firshein, W., and E. M. Zimmerman (1964). *In vivo* and *in vitro* effects of DNA degradation products on virulent and avirulent group A streptococci. J. Gen. Microbiol., *36*:237–248.

Kubinski, H., Z. Opara-Kubinska, and W. Szybalski (1966). Patterns of interaction between polyribonucleotides and individual DNA strands derived from several vertebrates, bacteria and bacteriophages. J. Mol. Biol., *20*:313–328.

Lehman, I. R., M. J. Bessman, E. S. Simms, and A. Kornberg (1958). Enzymatic synthesis of DNA I. Preparation of substrates and partial purification of an enzyme from *Escherichia coli. J. Biol. Chem., 233:*163–170.

Lipsett, M. N., L. A. Heppel, and D. F. Bradley (1961). Complex formation between oligonucleotides and polymers. J. Biol. Chem., *236:*857–863.

Lodish, H. F., and N. D. Zinder (1966). Replication of the RNA of bacterio phage F_2. Science, *152:*372–378.

Lowry, O. H., N. J. Rosebrough, A. L. Farr, and R. J. Randall (1951). Protein measurement with the Folin phenol reagent. J. Biol. Chem., *193:*265–275.

Marmur, J. (1963). A procedure for the isolation of DNA from microorganisms. *In* Methods of Enzymology, Vol. VI. Ed. by S. P. Colowick and N. O. Kaplan. Academic Press Inc., N. Y., pp. 726–738.

Marmur, J., and P. Doty (1961). Thermal renaturation of deoxyribonucleic acids. J. Mol. Biol., *3:*585–594.

Nirenberg, M., and P. Leder (1964). RNA codewords and protein synthesis. Science, *145:*1399–1407.

Nonoyama, M., and Y. Ikeda (1966). Mode of phage RNA replication studied by the annealing method. Biochim. et Biophys. Acta, *129:*548–553.

Ritossa, F. M., K. C. Atwood, and S. Spiegleman (1966). On the redundancy of DNA complementary to amino acid transfer RNA and its absence from the nuclear organizer region of *Drosophila melanogaster.* Genetics, *54:*663–676.

Sugino, Y., H. Teraoka, and H. Shimono (1966). Metabolism of deoxyribonucleotides. Purification and properties of deoxycytidine monophosphokinase of calf thymus. J. Biol. Chem., *241:*961–969.

ENHANCEMENT OF ANTIBODY FORMATION BY NUCLEIC ACIDS AND THEIR DERIVATIVES

A. G. JOHNSON, J. SCHMIDTKE, K. MERRITT, AND I. HAN
University of Michigan Medical School, Ann Arbor, Michigan

It is common knowledge today that antibody forming cells can be either suppressed or enhanced by various manipulative procedures. Our objective in this presentation is to describe experiments in which nucleic acids and their derivatives were found to act as adjuvants by decreasing the induction period and increasing the concentration of circulating antibody following a single injection of antigen.

The experiments which guided us into the study of nucleic acids evolved from our seeking to understand the means by which endotoxins derived from Gram negative bacteria act as potent adjuvants to antibody synthesis (Johnson *et al.*, 1956). It was postulated on the basis of an observed cytotoxicity *in vivo* on small lymphocytes, located perifollicularly near antibody-forming cells (Ward *et al.*, 1959), that endotoxin released, from lymphocytes, nucleic acids which stimulated division of the initial, or early, antibody-forming cells. In order to obtain evidence that injection of endotoxin resulted in enhancement of cell division, experiments were designed to determine the capability of endotoxin to overcome the effects on antibody formation of a specific inhibitor of DNA synthesis and cell division, 5 fluoro-2-desoxy-uridine (FUDR).

It was found (Merritt and Johnson, 1965) that a single injection of FUDR was able to inhibit antibody formation, provided the injection was given 18 hours or more after administration of antigen. These data were interpreted as revealing that there was an 18 hour latent period with respect to cell division after injection of protein antigen and, with the knowledge that the effects of FUDR are over rapidly, that injection of FUDR did not inhibit antibody formation unless it was injected at a time when potential antibody cells were ready to divide.

When this experiment was repeated in mice given endotoxin with

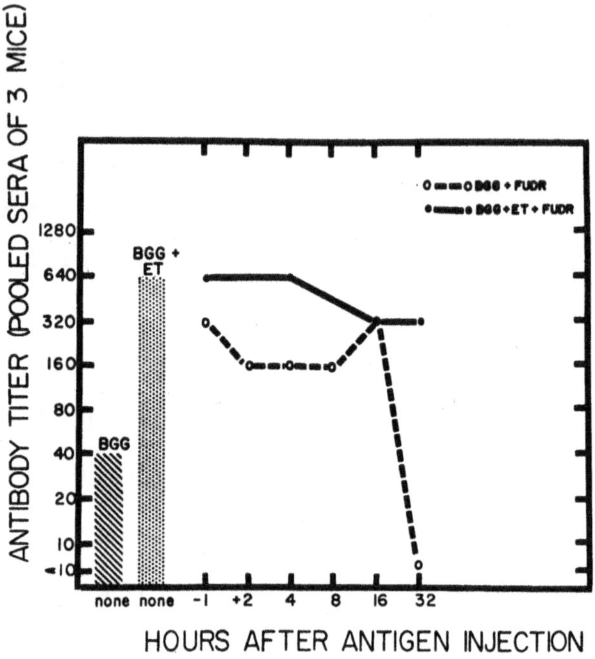

Fig. 1. The ability of bacterial endotoxin to enhance antibody formation in mice given a single injection of 5-fluoro-2-deoxy-uridine (FUDR) 32 hours after bovine gamma globulin. Each point represents the passive hemagglutination antibody titer on day 12 of a group of mice given a single injection of FUDR varying with respect to the time after antigen as shown on the abscissa.

antigen (Fig. 1), antibody formation was found to occur in spite of the injection of FUDR at 32 hours, at which time FUDR inhibited antibody synthesis in non-endotoxin stimulated mice. Thus, it is evident that cell division is a prerequisite for antibody formation after a single injection of antigen and that endotoxin, by overcoming FUDR inhibition, probably affected cell division in a positive manner.

If our hypothesis were correct that endotoxin increased division of antibody-forming cells through the mediation of released nucleic acids, then nucleic acids, per se, should act as adjuvants. Table 1 summarizes the results of some pertinent experiments. It may be seen that both DNA and RNA were capable of shortening the induction period and increasing antibody titers to bovine gamma globulin (BGG). This adjuvant action was not abolished by treatment of the nucleic acids with their respective nucleases, unless the DNA-DNAase or RNA-RNAase mixture was dialyzed. Thus, it appeared that low molecular weight products, oligonucleotides, were the functional components of nucleic acids. Lyophilization, concentration and subsequent enhancement of antibody titers following injection of the dialysate together with BGG into mice confirmed this supposition.

Table 1

THE ADJUVANT ACTION IN MICE OF DNA, RNA, AND THEIR BREAKDOWN PRODUCTS IN TERMS OF ANTIBODY SYNTHESIS

Products Injected: BGG (3 mg) +	Antibody Titer on Day 7	Antibody Titer on Day 12
–	<10	20
DNA	40	320
DNA + DNAase	80	320
DNA + DNAase (dialyzed)	<10	40
DNA + DNAase (dialysate)	40	320
RNA	<10	160
RNA + RNAase	<10	640
RNA + RNAase (dialyzed)	<10	40
RNA + RNAase (dialysate)	160	320

Fractionation of the DNA-DNAase mixture or the dialysate by DEAE ion exchange chromatography using 7M urea-0.3M NaCl as eluant resulted in multiple peaks, reflecting oligonucleotides of varying base composition (Johnson and Hoekstra, 1967). Each of these peaks excepting the tetranucleotides, when pooled and freed of the eluant, possessed adjuvant activity. Thus, it appears that this property may be independent of the base composition of the oligonucleotide. Inasmuch as 25–55% protein was present in 5 of the 8 active oligonucleotide preparations, DNA from rabbit spleens was treated for 10 minutes with 0.5 *N* H_2SO_4 to remove histones. As is seen in Table 2, this material also was capable of adjuvant action.

Table 2

THE ADJUVANT ACTION OF DNA TREATED TO REMOVE HISTONES

Product Injected: BGG (3 mg) +	Antibody Titer Day 9
–	320
DNA *	2560

* Rabbit spleen DNA treated 10 min. with 0.5 *N* H_2SO_4. Collected as sediment after washing 3× with saline. 3 mg injected/mouse.

Table 3

ADJUVANT ACTION IN MICE OF POLYADENYLIC-POLYURIDYLIC ACID COMPLEXES

Product Injected * BGG (3 mg) +	Antibody Titer Day 11
—	160
Poly A-U, 300 μg	2560
Poly A, 300 μg	160
Poly U, 300 μg	320
Endotoxin, 10 μg	5120

* All products injected intraperitoneally.

We also investigated the adjuvant effect of mixtures of commercially available homoribopolymers, namely polyadenylic acid and polyuridylic acid, as reported by Braun and Nakano (1967). That such mixtures were indeed potent adjuvants in our system is documented in Table 3. Complex, or helix formation, resulting from admixture of poly A and poly U appeared to be necessary for elevated titers inasmuch as neither poly A nor poly U, when given alone, was active in this respect.

It is clear from the results in Table 4 that binding of antigen and its subsequent slow release was not the means by which enhancement of antibody formation took place inasmuch as poly A-U complexes, given by a route different from that of the antigen, retained their adjuvant property.

Varying the time of injection of poly A-U complexes prior to antigen

Table 4

EFFECT OF ROUTE ON ADJUVANT ACTION OF POLYADENYLIC-POLYURIDYLIC ACID COMPLEXES

Product Injected: BGG (3 mg) +	Route	Antibody Titer Day 10
—	Intravenous	80
Poly A-U, 300 μg	Intravenous	1280
Endotoxin	Intravenous	640
—	Intraperitoneal	40
Poly A-U, 300 μg	BGG, IP; Poly A-U, IV	320
Endotoxin	Intraperitoneal	640

Table 5

LACK OF ADJUVANT ACTION OF POLYADENYLIC-POLYURIDYLIC ACID COMPLEXES INJECTED PRIOR TO ANTIGEN

Product Injected: BGG (3 mg) +	Time Injected	Antibody Titer Day 11
Poly A-U *	−48 hr	10
Poly A-U	−24 hr	10
Poly A-U	−12 hr	10
Poly A-U	−6 hr	10
Poly A-U	−1 hr	640
Poly A-U	with antigen	1280
Endotoxin	with antigen	640

* 150 μg polyadenylic acid + 150 μg polyuridylic acid mixed together and injected intravenously into mice in total volume of 0.2 ml with antigen.

established the ineffectiveness of this adjuvant unless it was given close to the time of antigen injection, within less than 6 hours (Table 5).

For future studies, in order to compare the fate or processing of antigen with and without adjuvant, it was important to determine the minimal dose of BGG which could elicit antibody under the influence of the poly A-U complexes. In Table 6, it may be seen that the poly A-U complexes permit the expression of antigenicity of as little as 0.01 micrograms of this protein antigen.

Experiments designed to determine the biochemical or morphologic

Table 6

ADJUVANT ACTION OF POLYADENYLIC-POLYURIDYLIC ACID ON MINUTE AMOUNTS OF ANTIGEN

Product Injected: BGG (0.01 μg) +	Antibody Titer on Days 6	8	10	12
—	0	0	0	0
Poly A-U *	80	160	160	160
Endotoxin	40	40	80	160

* 150 μg polyadenylic acid + 150 μg polyuridylic acid mixed together and injected intravenously into mice in total volume of 0.2 ml with antigen.

Table 7

EARLY APPEARANCE OF ANTIBODY IN MICE TREATED WITH POLYADENYLIC-POLYURIDYLIC ACID COMPLEXES

Product Injected: BGG (2 mg) +	Antibody Titer on Days				
	3	5	7	9	11
—	0	0	20	80	80
Poly A-U *	20	40	160	640	1280
Endotoxin	20	40	160	1280	1280

* 150 μg polyadenylic acid + 150 μg polyuridylic acid mixed together and injected intravenously into mice in total volume of 0.2 ml with antigen.

phase or phases of the immune system in which the oligonucleotides and homoribopolymers act are being initiated. Suggestive evidence that poly A-U complexes might function by increasing division of potential antibody forming cells is seen in Table 7, where a comparison of the shortening of the induction period achieved with both the poly A-U complex and endotoxin is seen. Trace titers are apparent in 3–5 days in both treated groups, whereas significant amounts of antibody did not appear until 7–9 days in control animals receiving only antigen.

With respect to the cell type affected by the adjuvant, our early evidence suggests that it may not be the small lymphocyte. When the % transformation of human small lymphocytes to large blast cells was measured following their exposure to poly A-U complexes, either with or without specific sensitizing antigen, no real effect was noted (Friedman and Johnson, unpublished results).

In summary, isologous or heterologous DNA and RNA, as well as their breakdown products and commercially available homoribopolymers have been shown to be capable of increasing the activity of antibody forming cells thereby causing them to synthesize more antibody in a shorter period of time. This suggests that the potential or early stage of antibody forming cell is functioning suboptimally with respect to antibody synthesis and that this deficiency can be remedied by supplying oligonucleotides either exogenously by injection or endogenously, through the mediation of cytotoxic substances such as endotoxin.

References

Braun, W., and M. Nakano (1967). Antibody formation: Stimulation by polyadenylic and polycytidylic acids. Science, *157*:819–821.

Johnson, A. G., S. Gaines, and M. Landy (1956). Studies on the O antigen of Salmonella typhosa. V. Enhancement of antibody response to protein antigens by the purified lipopolysaccharide. J. Exp. Med., *103*:225–246.

Johnson, A. G., and G. Hoekstra (1967). Acceleration of the primary antibody response. Chapter 22, pgs. 187–190 in *Ontogeny of Immunity,* University of Florida Press, Gainesville.

Merritt, K., and A. G. Johnson (1965). Studies on the adjuvant action of bacterial endotoxins on antibody formation. VI. Enhancement of antibody formation by nucleic acids. J. Immunol., *94*:416–422.

Ward, P. A., A. G. Johnson, and M. R. Abell (1959). Studies on the adjuvant action of bacterial endotoxins on antibody formation. III. Histologic response of the rabbit spleen to a single injection of a purified protein antigen. J. Exp. Med., *109*:463–474.

ACKNOWLEDGMENTS

Table 1 and Figure 1 are reprinted with permission of the Journal of Immunology.

RESTORATION OF IMMUNOLOGIC CAPACITIES IN IRRADIATED ANIMALS BY NUCLEIC ACIDS AND THEIR DERIVATIVES *

MIROSLAV M. SIMIĆ AND DUŠAN T. KANAZIR

Boris Kidrič Institute of Nuclear Sciences, Beograd, Yugoslavia

During the last decade restorative effects of nucleic acids and their derivatives have been extensively studied in different irradiated biological systems. Although the results of these investigations have often varied markedly, and sometimes even have been contradictory or confusing, it seems now well established that nucleic acids and their derivatives may restore radiation injury. Thus, the proportion of surviving irradiated bacteria capable of forming colonies is significantly increased after addition of exogenous DNAs into the growth medium (Kanazir *et al.*, 1959a; Igali, 1964). Exogenous nucleic acids or their derivatives display similar restorative effects in irradiated mammalian cells growing *in vitro* and may, in addition, lead to some cytological and biochemical improvements (Djordjević *et al.*, 1962; Miletić *et al.*, 1963; Petrović *et al.*, 1963; Horikawa *et al.*, 1963; Kostić *et al.*, 1965; Petrović and Nias, 1967). When nucleic acids or their derivatives are injected into irradiated animals, one can observe (a) a better survival in lethally or sublethally irradiated animals (Detre and Finch, 1958; Kanazir *et al.*, 1959b; Maisin *et al.*, 1960; Paoletti *et al.*, 1964; Wilczok and Mendecki, 1965; Sugahara *et al.*, 1966; Libinzon *et al.*, 1967); (b) an improvement of reproductive capacities (Savković, 1964; Savković and Hajduković, 1965); (c) a morphologic improvement of some of their tissues and organs (Pantic *et al.*, 1962); and (d) enhanced metabolism of nucleic acids in some organs (Bećarević *et al.*, 1962; Petrović *et al.*, 1962).

There are also numerous investigations showing that nucleic acids and

*Supported by the Yugoslav Federal Nuclear Commission.

their derivatives may restore immunologic capacity in irradiated animals (Taliaferro and Jaroslow, 1960; Jaroslow and Taliaferro, 1962, 1966; Simić *et al.*, 1959, 1965a; Kornfeld and Miller, 1961; Feldman *et al.*, 1962; Hammond *et al.*, 1962; Friedman, 1964; Simić and Šljivić, 1966). In the present report these investigations will be reviewed and evaluated with special reference to (a) the known effects of radiation on antibody formation and (b) the repair of radiation damage observed in other biological systems. Some possible restorative mechanisms will be discussed because of their importance for a better understanding of processes involved in antibody formation.

I. EFFECTS OF RADIATION ON ANTIBODY FORMATION

Much work has been published on various effects of X-radiation on immune response but there is still no unequivocal explanation of radiation-induced injury of antibody synthesis. From the radiobiological point of view, several important features of this complex injury might be explained by assuming that radiation affects a synchronously growing cell population which develops through proliferation and differentiation of antigenically stimulated progenitor cells. Thus, it is quite certain that radiation-induced injury can result in a loss of cell reproduction, cellular death, and may also lead to mitotic delay and inhibition of mitosis in proliferating competent cells. Variations in radiosensitivity during the mitotic cycle and as a function of the stage of differentiation may influence the rate of expression of radiation-induced damage. However, not all radiation effects on an immune response can be completely understood in terms of the above mechanisms. For example, although a cell exposed to ionizing radiation must undergo a finite dose-dependent loss of reproductive integrity, it can be shown that this is not so in the case of immunologically competent progenitor cells (Simić *et al.*, 1965b). Thus, two quite different radiation doses may produce identical damage to immunologic responses when given to competent progenitor cells at different times during a 24-hour period prior to antigenic stimulation. In both cases *similar* numbers of antibody (Ab)-forming cells arise although similar numbers of their progenitors were irradiated with *different* radiation doses. On the other hand, the *same* radiation dose given to progenitor cells at various times during a 24-hour period prior to immunization produces *different* degrees of impairment of immunologic capacity. Obviously, the known radiation effects are not sufficient to explain these observations satisfactorily.

In general, there are three main categories of radiation-induced modifications of a primary antibody response (Simić *et al.*, 1965b). Modifications belonging to the first category, involving various degrees of a generalized depression of immunologic capability, are a result of radiation injury to progenitor cells before they are stimulated by antigen. The second category encompasses rather uniform modifications which may vary from a moderate depression to a moderate enhancement of immunologic capacity; these modifications occur when radiation affects processes involved in the initial events of an immune response. Finally, modifications belonging to the third category appear when radiation affects antigenically stimulated proliferating and differentiating cells. Considering the specific purpose of this report, modifications belonging to the first category will be discussed in more detail.

Irradiation of progenitor cells prior to their stimulation with antigen results in an impairment of all main parameters of an immune response. Latent phase, rate of antibody rise, and peak titer are always altered simultaneously indicating that a decreased number of functional Ab-forming cells appears at a slower rate following a prolonged lag. The extent of such a complex injury of immunologic capacity depends upon two variables: (a) *dose* of radiation, and (b) the *time-period* separating exposure and antigen injection. Thus, on the one hand, injury is proportional to the dose of radiation when the time-period is kept constant (Taliaferro and Taliaferro, 1964; Simić *et al.,* 1965b); on the other hand, injury is proportional to the time-period for the first 12–18 hours after irradiation when the dose of radiation is held constant, as demonstrated by injecting antigens at various hours after exposure (Simić *et al.,* 1965b). In other words, additional impairment of the capacity to respond to an antigen can be prevented if an antigenic stimulus is given at some time before a 24-hour period following irradiation, i.e., before maximal depression for a radiation dose has been reached. Thereafter, immunologic recovery is very slow and usually takes several weeks. These findings indicate a possible involvement of the following factors in the development of radiation injury when irradiation occurs prior to immunization: (a) early interphase death; (b) loss of "inducibility" of progenitor cells; and (c) loss of reproductive integrity, i.e., loss of the normal proliferative potential of "induced" precursors of Ab-forming cells. Because of the well-known high radiosensitivity of lymphocytes, one could easily imagine that interphase death of progenitor cells may be the main reason for the rapid loss of immunologic capacity during the 12–18-hour period following irradiation. However, in this case one should also anticipate that an antigenic stimulus would be able to "protect" the responding progenitor cells from interphase death. Although such a possibility was suggested by the observation that increased resistance of lymphocytes to X-rays followed their incubation with phytohemagglutinin (Schrek and Stefani, 1964), direct evidence is still lacking. In fact, histological observations (Keuning *et al.,* 1963) indicate that some lymphocytes, especially those in immunologically responsive regions of lymphoid organs, tend to survive sublethal radiation doses while losing their responsiveness to antigen 12 to 24 hours after irradiation. One can only speculate about the nature of such a functional impairment involving loss of capacity of progenitor cells to initiate rapid proliferation and differentiation following antigenic stimulation. However, it appears to us that it might involve a specific radiation injury interfering with the two major functions of DNA, i.e., with the regulation of self-replication and with DNA transcription, that is, with the DNA-dependent synthesis of RNA. Antibody synthesis, which involves cellular proliferation and differentiation as well as the synthesis of specific antibody protein and thus is dependent on DNA synthesis, must play an essential role. It is well established that X-irradiation affects the structure and synthesis of DNA (Beltz *et al.,* 1957; Ord and Stocken, 1958; Berry *et al.,* 1961; Huntley and Lajtha, 1962). The inhibition of DNA synthesis may be due to: inhibition of phosphorylation (Creasey and Stocken, 1959); injury to DNA template activity (Walwick and Main, 1962; Weiss and Wheeler, 1967); increased activity of DNAases (Pierucci and Regelson, 1965); or an inhibition of enzymes needed for DNA synthesis (Bollum *et al.,* 1960; Sugino *et al.,* 1963; Berg and Goutier, 1967). In addition, radiation affects the synthesis of rapidly labeled nuclear RNAs (Petrović *et al.,* 1962; Thompson *et al.,* 1966; Uchiyama *et al.,* 1966; Berg and Goutier, 1967) presumably because of a damage to the priming ability of DNA for RNA polymerization (Zimmermann *et al.,* 1964; Weiss and Wheeler, 1967). Also, it is probable that radiation injury to mechanisms of transcription and translation may produce delayed death in those progenitor cells that were "inducible," i.e., capable to enter into a rapid cell cycle at the moment of antigenic stimulation. This loss of reproductive integrity probably is accompanied by mitotic delay or inhibition of mitosis thus leading not only to a decrease in the number of potential Ab-forming cells, but also altering the normal rate of their development. Finally, it should be stressed that the ultimate expression of late effects probably depends not only on the initial injury to progenitor cells, but also on their ability to repair the radiation-induced injury since repair mechanisms seem to be operating in mammalian cells. Thus, the non-S-phase incorporation of C^{14}-thymidine into DNA of mammalian cells suggests the existence of such repair mechanisms (Rassmussen and Paintair, 1966). Furthermore, from experiments on incorporation of 2-C^{14}-thymidine into thymus DNA after 400 r total-body irradiation, results were obtained indicating "synthesis" of DNA without a detectable net increase in the total amount of DNA (Sugino *et al.,* 1963). These observations support the assumption that in cells of higher organisms there exist enzymatic mechanisms capable of protecting informational functions of DNA through repair of alterations or defects in the DNA structure.

From this it may be postulated that the number of Ab-forming cells which will appear in the course of an immune response when radiation is given before antigen, will depend on (a) the number of progenitors at the moment of exposure; (b) the number of "inducible" progenitors at the moment of antigenic stimulation; (c) the probability of survival of antigenically induced competent cells; and (d) the capacity of repair processes. Consequently, depression of immunologic capacity under these conditions may be overcome (1) by increasing the number of progenitors before radiation exposure, e.g., through preimmunization (Makinodan and Albright, 1962) or nonspecific stimulation of the lymphoreticular system (Stender *et al.,* 1962); (2) by increasing the number of "inducible" progenitors present at the moment of antigen injection, e.g., by preventing early death and loss of "inducibility" with chemical protectors (Simić *et al.,* 1960) or by repairing "inducibility"; (3) by increasing the probability of survival of antigenically induced proliferating cells; (4) by stimulating such proliferation and differentiation; and (5) by enhancing spontaneous repair processes. However, and this should be stressed specifically, under conditions of antigen administration to previously untreated animals at the time of maximal immunologic depression, i.e., 1 or 2 days after radiation exposure, an improved development of Ab-forming cells, due to post-irradiation treatment, should be a consequence of (a) a repair of "inducibility"; or (b) a repair of survival probability; or (c) a stimulation of proliferation and differentiation of induced cells; or (d) maintenance and stimulation of spontaneous repair processes.

It must also be mentioned that because of the possible involvement of an initial processing of antigens by macrophages as a first stage in antibody formation, one should consider the possibility that radiation injury to the function of macrophages may play an important role in the depression of immunologic ability when antigen is injected after radiation exposure. However, morphological studies of cells comprising the reticuloendothelial system (Bloom, 1948), studies on the effects of X-radiation on the functional capacity of phagocytes *in vivo* (Benacerraf, 1960) and *in vitro* (Perkins *et al.,* 1966), as well as studies on antigen localization in X-irradiated animals (Jaroslow and Nossal, 1966), strongly indicate that it is very unlikely that radiation injury to the processes of engulfing, degradation and antigen processing could be an important cause of impairment of the immune response in animals immunized after irradiation. This conclusion is certainly true for sublethal doses of X-radiation since even after kiloroentgen doses *in vitro* (Perkins *et al.,* 1966) or *in vivo* (Simić *et al.,* 1965c) functional capacities of macrophages seem not to be altered significantly at least during the first two days after irradiation.

Finally, it should be stressed that antigenically stimulated competent cells proliferate and differentiate normally when transferred to X-irradiated recipients (Perkins *et al.,* 1961). This indicates strongly that there is nothing in irradiated animals that inhibits development of a population of Ab-forming cells once their progenitors have been properly stimulated by antigen.

II. EFFECTS OF NUCLEOPROTEINS, NUCLEIC ACIDS AND THEIR DERIVATIVES ON THE ANTIBODY-FORMING CAPACITY OF IRRADIATED ANIMALS

As already mentioned, data obtained from experiments on restoration of radiation injury by exogenous nucleic acids and their derivatives have been contradictory or confusing at times mainly because (a) processes leading to recovery are still unknown and most probably complex, and (b) because the experimental conditions were often not comparable. Considering the latter reason, in attempting to explain possible restorative mechanisms we shall refer mainly to those data which have been obtained from investigations characterized by similar experimental conditions; this includes (a) experiments in which a sublethal dose of X-radiation (400–600 r) was given 1 or 2 days before immunization and (b) experiment in which response to a particulate antigen (red cells or bacteria) was measured during a prolonged period of time following immunization. Table 1 summarizes the main results from such experiments, showing the extent of restorative ability of nucleoproteins, nucleic acids and their derivatives. Both deoxyribonucleoprotein (DNP) and ribonucleoprotein (RNP) extracted from lymph nodes of specifically preimmunized donors may almost completely restore radiation-damaged immunologic responses, as indicated by the experiments of Friedman (1964). These experiments showed that there are practically no differences between responses of non-irradiated and irradiated (425 r) recipients of DNP or RNP administered 1 day after radiation exposure, because both nucleoproteins appear to have marked restorative potency for the immunogenic material present in the preparations used. However, one should not neglect the possibility that the observed responses could be due to the transfer of a few viable competent cells which escaped destruction during extraction of the nucleic acids from lymphoids. Similar results were also obtained with nucleic acids (DNA and RNA together) isolated from spleens of specifically preimmunized rats and administered during 3 days following irradiation. In these experiments (Simić and Šljivić, 1966), however, the first of four injections of nucleic acids was given immediately after radiation exposure and the antigen one day thereafter. The peak titer in animals thus treated was almost completely restored but the response was delayed almost to the same extent as in non-treated controls. Since preparations of nucleic acids may contain

Table 1

EFFECTS OF NUCLEOPROTEINS, NUCLEIC ACIDS AND SOME OF THEIR DERIVATIVES ON ANTIBODY-FORMING CAPACITIES IN X-IRRADIATED ANIMALS

Investigators	Species	X-ray Dose (r)	Responses Tested	Materials Used	Source	Effect
			A. Nucleoproteins			
Friedman, 1964	Rabbit	425	Agglutinins to *Shigella paradysenteriae*	DNP [1] (430–920 μgN)	Homologous lymph node [2]	Marked restoration
Friedman, 1964	Rabbit	425	Agglutinins to *Shigella paradysenteriae*	RNP [1] (142–680 μgN)	Homologous lymph node [2]	Marked restoration
			B. Both Nucleic Acids Together			
Simić and Šljivić, 1966	Rat	400	Hemolysins to SRC	DNA + RNA [3] (7 + 5 mg)	Homologous spleen [4]	Marked restoration
Simić *et al.*, 1965	Rat	450	Hemolysins to SRC	DNA + RNA [3] (11 + 8 mg)	Homologous spleen	Partial restoration
			C. Deoxyribonucleic Acid			
Taliaferro and Jaroslow, 1960	Rabbit	400	Hemolysins to SRC	DNA [6] (100 mg)	Heterologous (bovine) thymus	Insignificant restoration
Simić *et al.*, 1959	Rat	600	Hemolysins to SRC	DNA [7] (19 mg)	Homologous spleen	Partial restoration
Feldman *et al.*, 1962	Rat	500–550	Agglutinins to SRC	DNA [8] (8 mg)	Homologous spleen	Partial restoration
Simić *et al.*, 1965	Rat	450	Hemolysins to SRC	DNA [5] (12 mg)	Homologous spleen	Partial restoration

			D. Ribonucleic Acid			
Jaroslow and Taliaferro, 1956	Rabbit	400	Hemolysins to SRC	RNA [6] (100 mg)	Heterologous (yeast)	No restoration
Simić and Šljivić, 1966	Rat	450	Hemolysins to SRC	RNA [5] (11 mg)	Homologous spleen	Partial restoration
Simić and Šljivić, 1966	Rat	450	Hemolysins to SRC	RNA [5] (20 mg)	Homologous liver	No restoration
			E. DNAase Treated DNA			
Taliaferro and Jaroslow, 1960	Rabbit	400	Hemolysins to SRC	Partially degraded DNA [9]	Heterologous (bovine) thymus	Partial restoration
Feldman *et al.*, 1962	Rat	500–550	Agglutinins to SRC	Partially degraded DNA [10]	Homologous spleen	Partial restoration
Simić *et al.*, 1965	Rat	450	Hemolysins to SRC	Totally degraded DNA [11]	Homologous spleen	Insignificant restoration
Simić and Šljivić, 1966	Rat	450	Hemolysins to SRC	Totally degraded DNA [11]	Homologous liver	Insignificant restoration
			F. RNAase Treated RNA			
Taliaferro and Jaroslow, 1960	Rabbit	400	Hemolysins to SRC	Partially degraded RNA [12]	Yeast	Partial restoration
Simić and Šljivić, 1966	Rat	450	Hemolysins to SRC	Totally degraded RNA [13]	Homologous spleen	Insignificant restoration
Simić and Šljivić, 1966	Rat	450	Hemolysins to SRC	Totally degraded RNA [13]	Homologous liver	No restoration

[1] One i.v. injection *without* antigen 1 day after exposure to radiation.
[2] From donors sacrificed 72 hours following injection with *Shigella paradysenteriae*.
[3] Four i.p. injections through 3 post-irradiation days; SRC injected i.v. 1 day following irradiation.
[4] From donors sacrificed 72 hours following injection with SRC.
[5] Five i.p. injections through 4 post-irradiation days; SRC injected i.v. 2 days following irradiation.
[6] One i.v. injection at the same time as SRC i.v., i.e., 1 day following irradiation.

[Table continues on following page]

Table 1 (*Continued*)

EFFECTS OF NUCLEOPROTEINS, NUCLEIC ACIDS AND SOME OF THEIR DERIVATIVES ON ANTIBODY-FORMING CAPACITIES IN X-IRRADIATED ANIMALS

Investigators	Species	X-ray Dose (r)	Responses Tested	Materials Used	Source	Effect
			G. Nucleotides, Nucleosides and Bases			
Taliaferro and Jaroslow, 1960	Rabbit	400	Hemolysins to SRC	Mixture of nucleosides and -tides (100 mg); nucleotide diphosphates (20 mg); guanosine (57 mg); adenylic acid (10 mg); adenosine (10 mg); adenine (10 mg) [14]		No restoration

[7] Six i.v. and six i.p. injections through 5 post-irradiation days; SRC injected i.v. 1 day following irradiation.

[8] One i.p. injection at the same time as SRC i.v., i.e., 1 day following irradiation.

[9] 100 mg calf thymus DNA incubated with DNAase at 37°C and injected i.v. with antigen 1 day following irradiation.

[10] DNA incubated for 2 hours at 37°C with 1:100 DNAase and injected i.v. with SRC 1 day following irradiation.

[11] Ethanol non-precipitable products of DNA that had been digested with DNAase (1:50) for two hours at 37°C; the material was administered in five i.p. injections through 4 post-irradiation days and SRC were injected i.v. 2 days following irradiation.

[12] 100 mg yeast RNA incubated with RNAase at 37°C and injected with SRC 1 day following irradiation.

[13] Ethanol non-precipitable products of RNA that had been digested with RNAase for 1 hour at 37°C; the material was administered in five i.p. injections through 4 post-irradiation days and SRC were injected i.v. 2 days following irradiation.

[14] The nucleosides and nucleotides were deoxyadenosine, deoxyguanosine, deoxycytidine, thymidine and their monophosphates. The nucleotide diphosphates were uridine, cytidine, adenosine and guanosine. They were given i.v. at the same time as SRC; i.e., 1 day following irradiation.

some immunogenic material, the marked restoration could be due, at least in part, to antigenic stimulation at a time when there is still a marked immunologic responsiveness present. Under similar experimental conditions, DNA and RNA isolated from spleens of normal, non-immunized animals gave only a partial restoration when injected together for 4 days following irradiation into animals that received antigen 2 days after exposure (Simić *et al.*, 1965a). Furthermore, restoration obtained by simultaneous administration of both spleen DNA and RNA did not differ either quantitatively or qualitatively from that obtained by administration of spleen DNA or RNA alone (Simić *et al.*, 1965a).

Restorative abilities of DNA should be considered in respect to (a) its origin (in terms of organ and species specificity), (b) its molecular size, and (c) in respect to the radiation dose given to the recipient animals. Thus, although homologous spleen DNA when injected into irradiated whole animals possesses the ability to restore both hemolysin (Simić *et al.*, 1959, Simić *et al.*, 1965a) and hemagglutinin (Feldman *et al.*, 1962) responses to sheep red blood cells in rats, heterologous thymus DNA was unable to restore hemolysin responses in rabbits under similar experimental conditions (Taliaferro and Jaroslow, 1960). These data may indicate a possible importance of the origin of DNA in the restoration phenomenon. Numerous investigations on restoration in other experimental systems, such as survival of bacteria, cultured mammalian cells and animals, also have led to questions regarding the importance of the origin of restorative DNA. However, for the time being, the answer to this question is still quite unclear because under some experimental conditions restorative ability could be correlated with the origin of administered DNA both in terms of organ and species specificity (Kanazir *et al.*, 1959a and 1959b; Djordjević *et al.*, 1962; Libinzon *et al.*, 1967), while other investigations have strongly indicated that the origin of the DNA does not play any important role (Savković, 1964; Savković and Hajduković, 1965; Wilczok and Mendecki, 1965).

A more decisive answer is available in regard to the question regarding a correlation between molecular size of DNA and its ability to restore immunologic reactivity. Partially digested homologous spleen DNA was able to restore hemagglutinin-forming capacity in sublethally irradiated rats as effectively as non-digested high molecular weight DNA (Feldman *et al.*, 1962). Ethanol non-precipitable products of enzymatically digested homologous spleen DNA, however, were markedly less effective in restoring hemolysin response in the same species (Simić *et al.*, 1965a). Finally, deoxyribonucleotides and deoxyribonucleosides, given separately or in mixture, proved to be quite ineffective in restoring hemolysin responses in sublethally irradiated rabbits (Taliaferro and Jaroslow, 1962). However, it was also found that heterologous thymus DNA, although ineffective when administered in non-digested form, significantly restored hemolysin responses in irradiated rabbits when used after incubation with DNAase (Taliaferro and Jaroslow, 1960). From such data it may be concluded that the restorative property is not associated with high molecular weight of DNA, but the fact that DNA loses this property when completely degraded, may indicate that there is probably a critical size for the DNA fragment below which it is devoid of any restorative potency. It should also be remembered that any DNA injected in animals is degraded by host nucleases. Direct evidence that the restorative ability of administered DNA may be correlated with its molecular weight has been obtained recently in survival experiments with lethally irradiated animals (Wilczok and Mendecki, 1965; Libinzon *et al.,* 1967).

Evaluation of restorative effects of homologous spleen DNA on hemolysin responses in rats irradiated with different doses of X-rays (Simić *et al.,* 1965a) has indicated that DNA has the same proportional capacity at each X-ray dose, and suggested that DNA would probably be ineffective in animals with totally suppressed immunologic abilities. An inability to restore antibody-forming capacity in animals rendered anergic by high doses of radiation was clearly shown in experiments with the cytotoxic agent colchicine (Jaroslow and Taliaferro, 1966) which restores depressed immunologic abilities in sublethally irradiated animals presumably by a release of nucleic acid degradation products (Jaroslow and Taliaferro, 1962). On the basis of the above data it would follow that nucleic acids and their derivatives are able to restore immunologic capacity in irradiated animals only in cases where this capacity has not been totally depressed by radiation. Similar conclusions also can be reached on the basis of experiments on the restoration of survival of mammalian cells grown in culture (Djordjević, unpublished observations).

Homologous spleen RNA partially restored hemolysin responses in sublethally irradiated rats, but restorative effects were insignificant when ethanol non-precipitable degradation products of the same RNA were administered. Under similar experimental conditions both high molecular weight and totally degraded homologous liver RNA were ineffective (Simić and Šljivić, 1966). In contrast, yeast RNA was ineffective in restoring hemolysin responses in sublethally irradiated rabbits when administered in a non-degraded form; a partial restoration was obtained by injecting the same RNA after incubation with RNAase (Taliaferro and Jaroslow, 1960). Effects of yeast RNA or yeast ribonucleotides have been observed in regard to increasing survival of irradiated animals (Detre and Finch, 1958; Maisin *et al.,* 1960; Sugahara *et al.,* 1966), restoring serum bactericidal activity *in vivo* after irradiation (Kornfeld and Miller, 1961), and reducing mortality from experimental infection in irradiated animals (Hammond *et al.,* 1962).

III. SOME POSSIBLE EXPLANATIONS FOR THE RESTORATIVE EFFECTS OF NUCLEIC ACIDS AND THEIR DERIVATIVES ON RADIATION-DEPRESSED ANTIBODY-FORMING CAPACITIES

An important characteristic of the restored immune response in the investigations cited above, as compared with the immune response of irradiated controls, is the attainment of a higher peak titer after a slightly shorter or unchanged lag period and during a usually unchanged exponential phase of the response. This finding strongly indicates that in animals given nucleic acids or their derivatives, antigen caused a greater number of progenitor cells to proliferate and to differentiate. The findings also suggest that an increased proportion of induced cells survived irradiation, but that the rate of proliferation of induced cells was probably very similar to that in non-treated controls. As discussed in more detail before, the following four processes could lead to a restoration of antibody-forming capacities in irradiated animals given antigen at the time of maximal immunologic depression, i.e., 1–2 days after exposure: (1) repair of "inducibility," (2) repair of survival probability, (3) stimulation of proliferation, and (4) facilitation of spontaneous repair processes.

Jaroslow and Taliaferro (1966), on the basis of their extensive studies, reached the conclusion that digests of nucleic acids restore antibody formation in animals immunized after irradiation mainly by increasing the number of progenitor cells receptive to the antigen. Postulating that progenitor cells respond to antigen only at a specific stage of their mitotic cycle, and assuming therefore that radiation makes these cells nonresponsive by damaging the synthesis of DNA, they have suggested that restorative agents act by repairing DNA synthesis. Consequently, more progenitor cells would be able to reach the specific antigen-sensitive stage in their cell-cycle and would thus become susceptible to antigenic stimulation. The explanation proposed by Jaroslow and Taliaferro (1966) involves repair of DNA synthesis as the main cause of restoration and is based upon two assumptions: (a) that competent progenitor cells must be dividing relatively rapidly and must be synthesizing DNA prior to antigenic stimulation; and (b) that radiation damages this synthesis. Recent investigations, however, rather strongly indicate that progenitor cells are not synthesizing DNA actively prior to stimulation (Syeklocha *et al.*, 1966). On the other hand, it was also clearly shown that DNA synthesis itself, once it has begun, is not inhibited by relatively moderate (sublethal) radiation doses, but the same doses are quite effective in inhibiting synthesis when given before the beginning of the synthetic period (e.g., Beltz *et al.*, 1957). It is therefore reasonable to postulate that progenitor cells do not synthesize DNA prior to antigenic stimulation and have a very long generation time, but that they must be able to change this steady state and enter into a rapid cell cycle when DNA synthesis is initiated as the result of antigenic stimulation. It would follow (Simić *et al.,* 1965a) that radiation applied *prior to* antigenic stimulation inflicts damage to some cellular mechanisms of the progenitor cells, namely a mechanism necessary to initiate antigenically induced DNA synthesis. This damage would result in a non-responsive state which we called earlier a "loss of inducibility." If, however, radiation acts on "induced" cells, i.e., cells in which antigenically induced DNA synthesis has already been initiated, synthesis would continue more or less unhampered, and radiation effects on antibody formation would be the complex result of changes in growth rate, mitotic delay, inhibition of mitosis and delayed death presumably due to an injury to regulatory genetic mechanisms involved in the development of a new population of Ab-forming cells. If this is correct, then any repair of radiation-damaged processes affecting the triggering and support of DNA synthesis and replication in irradiated progenitor cells would increase the number of antigen-responsive, "inducible," progenitor cells; the result would be a restoration of antibody formation. In this sense, a repair of DNA synthesis and its functions could be a central point in the restoration of radiation-damaged antibody synthesis (Jaroslow and Taliaferro, 1960; Simić *et al.,* 1965a; Jaroslow and Taliaferro, 1966).

The nature of processes leading to a repair of radiation-damaged cellular mechanisms, involved in the triggering and support of DNA synthesis in antigen-exposed progenitor cells, is completely unknown. However, there are some possibilities which might be envisaged. Investigations on the effects of radiation on DNA synthesis indicate, as mentioned before, that phosphorylation of nucleotides to triphosphates may be affected and that this injury, accompanied by an inhibition of polymerase synthesis, prevents formation of polynucleotide chains. A remarkable decrease in the activity of dCMP deaminase and TMP synthetase, which may limit the endogenous supply of TMP for DNA synthesis, was also demonstrated in thymus after irradiation (Sugino *et al.,* 1963). Since Firshein (1965) has shown that exogenous oligodeoxyribonucleotides stimulate DNA synthesis of bacteria by increasing levels of nucleotide kinases, it appears conceivable that a radiation-induced block of nucleotide synthesis in damaged progenitor cells could be repaired by a similar mechanism. Such an explanation, involving stimulation of induction of nucleotide kinases, is indirectly supported by the finding that the rate of increase of Ab-forming cells in non-irradiated animals is significantly influenced by exogenous oligodeoxyribonucleotides (Braun and Nakano, 1965; Hechtel *et al.,* 1965), and possibly also by oligoribonucleotides (Merritt and Johnson, 1965). This mechanism could explain the restorative ability of

DNA digests of relatively high molecular weight. On the other hand, the finding that radiation interferes with the synthesis of rapidly labeled nuclear RNA (Uchiyama *et al.*, 1966; Berg and Goutier, 1967) suggest that the inability of progenitor cells to synthesize DNA after antigenic stimulation might be due to an impaired synthesis of some proteins that are indispensable for the triggering and support of DNA synthesis. This impairment may be the result of a block in the synthesis of m-RNA or may be the consequence of a synthesis of abnormal m-RNAs. Thus, the finding that in thymus, spleen and liver of irradiated animals enhanced RNA synthesis precedes the peak of DNA synthesis strongly suggests that synthesis of RNA is a necessary prerequisite for the production of enzymes and/or for other factors required for the synthesis of DNA and the recovery of cells and organs. The possibility of a repair of radiation-damaged RNA synthesis by exogenous DNA has been suggested by the finding (Bećarević *et al.*, 1962; Petrović *et al.*, 1962) that exogenous DNA, administered to irradiated animals, stimulates incorporation of labeled precursors into RNA. It was also demonstrated that exogenous DNA incorporated into irradiated cells may influence the incorporation rate of H^3-thymidine into the DNA of these cells (Djordjević *et al.*, 1965). Finally, it should be pointed out that preparations of nucleic acids appear to be most effective in stimulating normal, or restoring radiation-damaged "inducibility" when they act simultaneously with injected antigen (Braun and Nakano, 1965; Jaroslow and Taliaferro, 1966), and that their restorative capacities might be even greater when they are derived from lymphoid organs of preimmunized donors (Friedman, 1964; Simić and Šljivić, 1966) presumably because they contain some bound antigen under these conditions.

Recovery of immunologic capacities in animals treated with nucleic acids may also be due to a repair of reproductive integrity, i.e., of the proliferative capacity of antigenically "induced," proliferating precursors of Ab-forming cells. In fact, the appearance of an increased number of functional Ab-forming cells during the response indicates that exogenous nucleic acids may increase the probability of survival of proliferating antigenically "induced" cells by protecting them from delayed death. Such a possibility is indirectly suggested by the finding that exogenous nucleic acids or their derivatives may repair the reproductive capacity of irradiated mammalian cells *in vivo* and *in vitro* (Miletic *et al.*, 1963; Horikawa *et al.*, 1963; Petrović *et al.*, 1963; Sugahara *et al.*, 1966). Processes leading to reproductive death are complex and still unknown but since chromosome breaks and mutations, among others, have been invoked as possible causes, a primary damage to the genetic apparatus, i.e., to the chromosomal structure and/or to the DNA structure, to synthetic pathways and to functions may be involved. The possibility of a repair of chromosomal damage by nucleic acids has been suggested by the finding that the percentage of chromosome aberrations was decreased significantly if irradiated cells were treated with isologous DNA after irradiation; however, post-irradiation treatment with heterologous DNA was followed by an increased number of chromatide aberrations (Karpfel *et al.,* 1963). Exogenous DNA might repair damage to the structure of DNA by replacing a damaged part of the host DNA molecule with a corresponding part of the donor DNA, as proposed by Kanazir (Kanazir *et al.,* 1959a). Such a possibility is suggested by several findings. Thus, restorative effects on survival of animals has been observed under some conditions only when polymerized and homologous DNA was administered (Kanazir *et al.,* 1959b). Furthermore, it has been found that the restorative effect was closely related to the molecular weight of the injected DNAs (Wilczok and Mendecki, 1965; Libinzon *et al.,* 1967). Finally, it has been shown that large molecules of DNA may actually penetrate into mammalian cells and that such incorporated DNA may be active in the host cell under some conditions (for review see Ledoux, 1965). Since in the restoration of antibody formation by DNA some indications have been obtained suggesting that the effect could be correlated with the molecular size and origin of the administered DNA, the possibility should be envisaged that a radiation-induced structural impairment of DNA might be repaired by a replacement of the damaged part of the DNA molecule by a corresponding part of administered DNA. Such a mechanism alone, or together with the possible restorative effects on the synthetic pathways of DNA, may result in an increased survival of proliferating competent cells.

It has been shown that the rate of increase of progenitor cells, after stimulation by an antigen, can be increased significantly by oligodeoxyribonucleotides (Braun and Nakano, 1965). However, it seems improbable that merely a stimulation of proliferation of "induced" competent cells could account for the recovery of radiation-damaged immunologic capacities. First of all, such an effect without an increased number of "inducible" progenitors and without a repair of survival of "induced" cells, should manifest itself primarily in a shortening of the exponential phase of the response accompanied by an increased rate of antibody rise and without any increase in peak titers. However, this is usually not the case. Second, a stimulation of proliferation alone should not increase the number of developing Ab-forming cells, since there is no reason to believe that a rapid proliferation could increase the probability of survival of proliferating radiation-damaged cells. It may therefore be concluded that an enhancement of proliferation of antigenically stimulated cells by exogenous nucleic acids does not play an important role in the restoration phenomenon.

Since the ultimate expression of a radiation effect on proliferating cell populations is influenced substantially by spontaneous repair mechanisms, which depend essentially upon growth conditions and cell metabolism, one cannot exclude the possibility that exogenous nucleic acids restore immunologic abilities by sustaining the already existing repair processes inside the cell (Miletić *et al.*, 1964). Such a possibility is strongly supported by the finding that with increasing radiation doses the restorative potential parallels the loss of immunologic capacities and is probably nil when immunologic capacities are nil (Simić *et al.*, 1965; Jaroslow and Taliaferro, 1962). Thus, exogenous nucleic acids are ineffective in restoring antibody-forming capacities when such capacities are *completely* inhibited by high radiation doses. However, a certain amount of damage can be repaired spontaneously when the amount of radiation-induced damage does not exceed the capacity of existing intracellular repair processes, and exogenous nucleic acids may increase this capacity of repair mechanisms and may thus lead to an even better survival.

Finally, it is possible that the restitution of immunologic capacity by nucleic acids could also be due, at least in part, to a repair of radiation-induced injury to the functional capacity of antigen-trapping and -processing cell systems, because it was shown that deoxyribonucleotides produce a significant stimulation of the phagocytic activity of the reticuloendothelial system (Freedman and Braun, 1965). However, such a mechanism, even if operative, should not be important because, as already discussed, a radiation-induced injury of the phagocytic activity is probably not a major cause of depression of the immunologic capacities of animals immunized 1–2 days after exposure to sublethal doses of irradiation.

IV. SUMMARY

Investigations on the restoration of antibody-forming capacities in X-irradiated animals treated with nucleic acids and their derivatives have been reviewed. The results have been discussed with respect to possible causes of radiation injury to antibody formation, and also in the light of restoration effects observed in other irradiated biological systems. The following four major processes are suggested to be involved in the restoration of immunologic abilities when radiation affects competent progenitor cells *prior to* their stimulation by antigen: (a) repair of "inducibility"; (b) repair of reproductive potential; (c) stimulation of proliferation; and (d) enhancement of spontaneous repair. Some biochemical mechanisms that may be involved have been discussed with emphasis on the role of nucleic acids in antibody synthesis.

References

Bećarević, A., S. Petrović, V. Janković, D. Kanazir, and G. Jovicki (1962). In *Biological Effects of Ionizing Radiation at the Molecular Level,* IAEA (Vienna), p. 421.

Beltz, R. E., J. van Lancker, and V. R. Potter (1957). Cancer Res., *17*:688.

Benacerraf, B. (1960). Bacteriol. Rev., *24*:35.

Berg, T. L., and R. Goutier (1967). Arch. Internat. Physiol. Biochim., *74*:909.

Berry, R. J., E. Hell, and L. G. Lajtha (1961). Internat. J. Radiation Biol., *4*:61.

Bloom, W. (1948). In *Histopathology of Irradiation from External and Internal Sources*. McGraw-Hill, New York.

Bolum, F. J., J. W. Anderegg, A. B. McElya, and V. R. Potter (1960). Cancer Res., *20*:138.

Braun, W., and M. Nakano (1965). Proc. Soc. Exp. Biol. Med., *119*:701.

Creasey, W. A., and L. A. Stocken (1959). Biochem. J., *72*:519.

Detre, K. D., and S. C. Finch (1958). Science, *128*:656.

Djordjević, O., Lj. Kostić, and D. Kanazir (1962). *Nature, 195*:614.

Djordjević, O., Lj. Kostić, and D. Kanazir (1965). Photochem. Photobiol., *4*:511.

Feldman, M., A. Globerson, and D. Nachtigal (1962). In *Mechanisms of Immunological Tolerance*, ed. by Hašek, Lengerova and Vojtiškova, Publishing House of the Czechoslovak Academy of Sciences (Prague), p. 305.

Firshein, W. (1965). J. Bacteriol., *90*:327.

Freedman, H. H., and W. Braun (1965). Proc. Soc. Exp. Biol. Med., *120*:222.

Friedman, H. (1964). Ann. N. Y. Acad. Sci., *114*:444.

Hammond, C. W., S. K. Anderle, and C. P. Miller (1962). Proc. Soc. Exp. Biol. Med., *109*:690.

Hechtel, M., T. Dishon, and W. Braun (1965). Proc. Soc. Exp. Biol. Med., *119*:991.

Horikawa, M., Y. Doida, and T. Sugahara (1963). Exptl. Cell Research, *32*:404.

Huntly, G. M., and Z. G. Lajtha (1962). Internat. J. Radiation Biol., *5*:447.

Igali, S. (1964). Acta Biol. Acad. Sci. Hung., *15*:28.

Jaroslow, B. N., and G. J. V. Nossal (1966). Aust. J. Exp. Biol. Med. Sci., *44*:609.

Jaroslow, B. N., and W. H. Taliaferro (1962). In *The Effects of ionizing radiations on immune processes*, ed. C. A. Leone, New York, Gordon and Breach, p. 301.

——— (1966). J. Infect. Dis., *116*:139.

Kanazir, D. T., O. Ž. Čečuk, B. N. Krajinčanić (1959a). Bull. Boris Kidrič Inst. Nucl. Sci. (Beograd), *9*:133.

Kanazir, D. T., A. Bećarević, B. Panjevac, M. Simić, and G. Ristić (1959b). Bull. Boris Kidrič Inst. Nucl. Sci. (Beograd), *9*:145.

Karpfel, Z., J. Šlotova, and E. Paleček (1963). Exptl. Cell. Res., *32*:147.

Keuning, F. J., J. van der Meer, P. Nieuwenhuis, and P. Oudendijk (1963). Lab. Invest., *12*:156.

Kornfeld, L., and C. P. Miller (1961). J. Immunol., *86*:215.

Kostić, Lj., O. Djordjević, and D. Kanazir (1965). Photochem. and Photobiol., *4*:515.

Ledoux, L. (1965). In *Progress in Nucleic Acid Research and Molecular Biology*, ed. Davidson and Cohn. Academic Press, New York and London.

Libinzon, R. E., S. A. Rogachova, and G. G. Rusinova (1967). Radiobiologiya, *7*:455.

Maisin, J., P. Dumont, and A. Dunjić (1960). Nature, *186*:487.

Makinodan, T., and J. F. Albright (1962). J. Cell. Comp. Physiol., *60*, Suppl. 1, 129.

Merritt, K., and A. G. Johnson (1965). J. Immunol., *94*:416.
Miletić, B., D. Petrović, Lj. Zajac (1963). Nature, *197*:90.
Miletić, B., D. Petrović, A. Han, and Lj. Sasel (1964). Radiation Res., *23*:94.
Ord, M. A., and L. A. Stocken (1958). Nature, *182*:1787.
Pantić, V., N. Stošić, D. Kanazir, and G. Jovicki (1962). Nature, *193*:83.
Paoletti, C., J. P. Rebière, G. Aubin, and M. Tubiana (1964). Rev. Franc. Etud. Clin. Biol., *9*:950.
Perkins, E. H., P. Nettesheim, and T. Morita (1966). J. Reticul. Soc., *3*:71.
Perkins, E. H., M. A. Robinson, and T. Makinodan (1961). J. Immunol., *86*:533.
Petrović, D., and A. H. W. Nias (1967). Internat. J. Radiation Biol., *11*:609.
Petrović, D., B. Miletić, and B. Brdar (1963). Internat. J. Radiation Biol., *7*:131.
Petrović, S., D. T. Kanazir, and A. Bećarević (1962). In *Biological Effects of Ionizing Radiation at the Molecular Level,* IAEA (Vienna), p. 429.
Pierucci, O., and W. Regelson (1965). Radiation Res., *24*:619.
Rasmussen, R. E., and R. B. Painter (1966). J. Cell. Biol., *29*:11.
Savković, N. V. (1964). Nature, *203*:1297.
Savković, N. V., and S. J. Hajduković (1965). Internat. J. Radiation Biol., *9*:361.
Schrek, R., and S. Stefani (1964). J. Nat. Cancer Inst., *32*:507.
Simić, M. M., and V. S. Šljivić (1966). In *Recovery of Irradiated Cells and Organisms by Nucleic Acids.* Bull. Boris Kidrič Inst. Nucl. Sci. (Beograd), *17,* Suppl. 1, p. 147.
Simić, M. M., M. Ž. Petković, and D. D. Mančić (1959). Bull. Boris Kidrič Inst. Nucl. Sci. (Beograd), *9*:155.
Simić, M. M., V. S. Šljivić, and M. Ž. Petković (1960). Bull. Boris Kidrič Inst. Nucl. Sci. (Beograd), *10*:149.
Simić, M. M., V. S. Šljivić, M. Ž. Petrović, and D. M. Čirković (1965a). In *Antibody Formation in Irradiated Rats.* Bull. Boris Kidrič Inst. Nucl. Sci. (Beograd), *16,* Suppl. 1, p. 125.
——— (1965b). In *Antibody Formation in Irradiated Rats.* Bull. Boris Kidrič Inst. Nucl. Sci. (Beograd), *16,* Suppl. 1, p. 17.
——— (1965c). In *Antibody Formation in Irradiated Rats.* Bull. Boris Kidrič Inst. Nucl. Sci., *16,* Suppl. 1, p. 87.
Stender, H. S., D. Strauch, and H. Winter (1962). In *Effects of Ionizing Radiations on Immune Processes,* ed. C. A. Leone, New York, Gordon and Breach, p. 315.
Sugahara, T., H. Nagata, and T. Tanaka (1966). Radiat. Res., *29*:516.
Sugino, J., E. Frenkel, and R. L. Potter (1963). Radiation Res., *19*:682.
Syeklocha, D., L. Siminovitch, J. E. Till, and E. A. McCulloch (1966). J. Immunol., *96*:472.
Taliaferro, W. H., and B. N. Jaroslow (1960). J. Infect. Dis., *107*:341.
Taliaferro, W. H., and L. G. Taliaferro (1964). J. Infect. Dis., *114*:285.
Thompson, J. S., Ch. D. Severson, and R. W. Reilly (1966). Radiation Res., *29*:537.
Uchiyama, T., N. Fausto, and J. L. Van Lancker (1966). J. Biol. Chem., *241*:991.
Walwick, E. R., and R. K. Main (1962). Biochim. Biophys. Acta, *55*:225.
Weiss, J. J., and C. M. Wheeler (1967). Biochim. Biophys. Acta, *145*:68.
Wilczok, T., and J. Mendecki (1965). Internat. J. Radiation Biol., *9*:201.
Zimmermann, F., H. Kröger, U. Hagen, and K. Keck (1964). Biochim. Biophys. Acta, *87*:160.

NUCLEIC ACIDS AND INDUCTION OF ANTIBODY SYNTHESIS IN INHIBITED SYSTEMS *

Bernard N. Jaroslow

Division of Biological and Medical Research
Argonne National Laboratory
Argonne, Illinois

The effects of radiation (Taliaferro *et al.,* 1964) and various drugs (Gabrielsen and Good, 1967) on the immune response have demonstrated that the mechanism of antibody synthesis follows the same course as most of the other proteins in that the *induction* of antibody synthesis is associated with nucleic acid synthesis.

Molecular events associated with induction of antibody synthesis can be inhibited by radiation or chemical agents. The inhibition of induction can be overcome by specific agents which restore † antibody-forming capacity. Thus, the effects of radiation on induction of the primary response in rabbits were investigated in collaboration with W. H. Taliaferro (Jaroslow and Taliaferro, 1958; Taliaferro and Jaroslow, 1958, 1960) and the effects of Actinomycin D on the induction of the secondary response *in vitro* were investigated in collaboration with K. Shortman (Jaroslow and Shortman, 1966).

Restoration of antibody-forming capacity of the primary response in irradiated rabbits was studied by the following procedures:

(1) Rabbits were exposed to 400 r whole-body irradiation.

(2) One day later they were injected into the ear vein with sheep red cells and various nucleic acid or tissue preparations.

(3) The rabbits were bled, from the ear vein, 12 times in 30 days and

* Work supported by the U. S. Atomic Energy Commission.

† We use the terms "restorative agents" and "restoration of antibody forming capacity" (Jaroslow and Taliaferro, 1956) in the sense that something was given to bypass a radiation-induced lesion in contrast to the use of the word "restoration" in discussions of recovery from radiation injury by Bacq and Alexander, 1955.

the serum was titrated by the 50% hemolysin end-point method of Taliaferro and Taliaferro (1950, 1956).

(4) The antibody curve was drawn on semi-log paper and the peak titer was used as the primary indicator of restorative activity.

Antibody-forming capacity, as measured by peak titer, was partially restored after radiation injury had fully developed, by injecting a variety of preparations (Table 1) at the same time as antigen. Specific enzymatic digests of DNA and RNA, and ribonuclease had restorative activity (Fig. 1). On the other hand, the purines, a mixture of nucleosides and nucleotides, nucleotide-diphosphates (Fig. 2), deoxyribonuclease (Fig. 1), and polymerized heterologous nucleic acids (Fig. 1) were not restorative. The role of kinetin (6-furfurylaminopurine) (Fig. 2), an analog of adenine, will be discussed later. Indoleacetic acid (Fig. 2), a plant growth hormone, is active through stimulation of DNA synthesis in the inhibited animal (see discussion in Jaroslow, 1960).

Table 1

THE COMPARATIVE RESTORATIVE ACTIVITIES OF VARIOUS MATERIALS
(Based on the data of Jaroslow, B. N. and Taliaferro, W. H., 1956, and Taliaferro, W. H. and Jaroslow, B. N., 1960)

	Restoration		
	Complete *	Partial *	None *
Tissue Preparations	HeLa Cells	Rabbit Spleen, Mouse Spleen	Rabbit Kidney, Rabbit Muscle
Nucleic acids, digests, and derivatives	—	DNA inc. with DNAase	Polym. DNA (Calf thymus)
		RNA inc. with RNAase	Polym. RNA (yeast)
		RNAase	DNAase
		Kinetin	Adenine, adenosine adenylic acid
			Guanosine
			Mixture of nucleosides, nucleotides and nucleotide-di-P
Other	Colchicine Yeast Extract	3-Indoleacetic acid	—

* Complete restoration means that the mean log peak titer was not significantly different from the non-irradiated controls ($P > .05$). Partial restoration means that the mean log peak titer was significantly different from both the non-irradiated and irradiated controls ($P < .01$). None means that the mean log peak titer was not significantly different from the irradiated controls ($P > .05$).

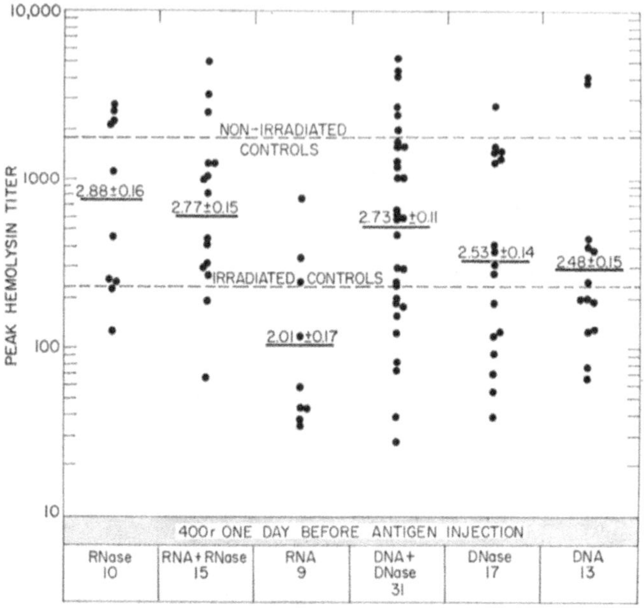

Fig. 1. The distribution and mean log peak titer ±S.E. for each treatment group compared with the mean log peak titer of the non-irradiated and irradiated control groups (broken lines). RNAase, RNA + RNAase, and DNA + DNAase were significantly different from both control groups ($P < .01$). RNA, DNA and DNAase were not significantly different from the irradiated controls ($P < .05$). (Based on data from Taliaferro and Jaroslow, 1960.)

In essence, restorative activity depends upon the presence of oligonucleotides, whether supplied directly in enzymatic digests or from nucleic acid pools of cells and tissues actively synthesizing DNA, or as the result of the increase of circulating endogenous nucleic acid degradation products after the subcutaneous injection of cytotoxic agents (Table 1).

In the second series of experiments, the induction of the secondary response in rat lymph node cultures was inhibited by the addition of Actinomycin D to the medium before antigen was added (Jaroslow and Shortman, 1966). The protocol for these experiments was:

(1) Rats were injected into the foot pads with polymerized flagellin (PF), prepared from *Salmonella adelaide.*

(2) Two to six months after injection, the popliteal lymph nodes were removed and cut into 1 mm³ pieces.

(3) They were then added to Eagle's Minimum Essential Medium, with supplements, in a Leighton culture tube.

(4) Actinomycin D (1 μg/ml) and other desired preparations were added to the appropriate cultures (Table 2).

(5) After four hours, the medium was removed, the culture was washed

with the culture medium and fresh medium, 1 mg of PF/ml, and various nucleic acid derivatives were added as needed (Table 2).

(6) The culture medium was removed every three days and fresh culture medium (without antigen) was added.

(5) The medium, after removal, was frozen and stored until it could be titrated by the bacterial immobilization technique of Ada *et al.,* 1964.

The induction of the secondary response was completely inhibited in cultures treated with Actinomycin D prior to the addition of antigen. If RNA and RNAase were added with the antigen, antibody-forming capacity was restored. If DNA and DNAase or a mixture of the four ribonucleosides were used, there was no restoration. There was no enhancement by the RNA digest in the non-inhibited culture, nor was there any discernible competition between the RNA digest (10 mg RNA plus 1 μg RNAase/ml) and the Actinomycin D (1 μg/ml) when both were present in the culture together (Table 2).

These two experiments are similar in the manner in which inhibition

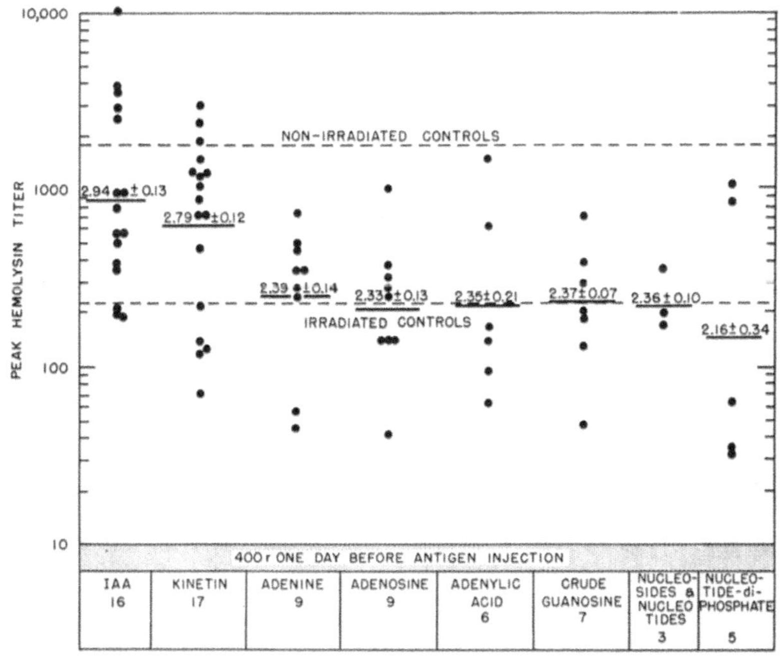

Fig. 2. The distribution and mean log peak titer ±S.E. for each treatment group compared with the mean log peak titers of the non-irradiated and irradiated control groups (broken lines). IAA and kinetin were significantly different from both control groups ($P < .01$). The various purine, nucleoside and nucleotide mixtures were not significantly different from the irradiated controls ($P < .05$). (Based on data from Taliaferro and Jaroslow, 1960.)

Table 2

TREATMENT DURING THE FIRST 24 HOURS OF ORGAN CULTURE
(Taken from Jaroslow, B. and Shortman, K., 1966)

Group No.	Hours After Start of Culture 0–4 *		4–24 *		Antibody † Titer
		Experiment #2			
1	AD ‡	wash	PF	wash	nil (3)
2	AD	wash	PF, ribonucleosides	wash	nil (5)
3	AD	wash	PF, RNA, RNAase	wash	40 (5)
4		wash	PF, RNA, RNAase	wash	59 (5)
5		wash	PF	wash	41 (5)
6		wash		wash	3 (2)
		Experiment #3			
7	AD	wash	PF	wash	nil (4)
8	AD, RNA, RNAase	wash	PF	wash	nil (5)
9	AD, DNA, DNAase	mash	PF	wash	nil (5)
10	AD, DNA, DNAase	wash	PF, DNA, DNAase	wash	nil (5)
11	AD, RNA, RNAase	wash	PF, RNA, RNAase	wash	26 (4)
12		wash	PF	wash	28 (4)
13	RNA, RNAase	wash	RNA, RNAase	wash	6 (2)
14	DNA, DNAase	wash	DNA, DNAase	wash	nil (2)

* Additions were made in the following amounts/ml: AD, 1 μg; PF, 1 μg; RNA, 10 mg; RNAase, 0.5 μg; DNA, 3 mg; DNAase, 10 μg; ribonucleosides, 10 mg each of adenosine, guanosine, cytidine and uridine.

† Titers shown are the geometric mean titers for the group on day 11 in Experiment #2 and on day 10 in Experiment #3. Numbers in parentheses are the number of cultures in each group.

‡ AD = Actinomycin D.

was overcome. In the first case a radiation block to DNA synthesis was overcome by the activity of oligonucleotides of DNA; in the second case, an Actinomycin D block to mRNA synthesis was overcome by the activity of oligonucleotides of RNA but not of DNA. Polymerized homologous DNA and oligonucleotides have also restored antibody-forming capacity in other irradiated animals immunized with various antigens (Simić *et al.*, 1959; Kind and Johnson, 1959; Feldman *et al.*, 1963). It is likely that oligonucleotides of DNA and RNA operate in the same manner to overcome a block to the synthesis of DNA and RNA, respectively.

These observations have led to the formulation of a hypothesis for the role of oligonucleotides in overcoming a block to nucleic acid synthesis, a hypothesis that correlates the restoration of nucleic acid synthesis with the process of induction of the immune response.

We assume that the restorative activity of the oligonucleotides enhances nucleic acid polymerase activity so that available nucleosides and nucleotides can be incorporated more efficiently into the nucleic acid being synthesized. In addition, the oligonucleotides are also incorporated into the new nucleic acids.

To make the proposed mechanism feasible, it is necessary: (1) That oligonucleotides get into the cell; (2) That nucleic acid synthesis is inhibited at some stage between phosphorylation and polymerization; and (3) That oligonucleotides stimulate nucleic acid synthesis in the presence of nucleosides and nucleotides.

Oligonucleotides get into the cell. Hill (1967) demonstrated that large tritium-labeled polynucleotides were incorporated into the nucleolus of non-dividing cells. Bensch and King (1961) demonstrated that oligonucleotides from *E. coli* were incorporated without degradation into nuclei of L cell fibroblast cultures. Other supportive evidence is presented in the proceedings of the Conference on Recovery of Irradiated Cells by Nucleic Acids (1967).

The inhibition of DNA synthesis in irradiated cells, at some stage between phosphorylation and polymerization, is demonstrated by the intracellular accumulation of nucleosides and nucleotides after irradiation (Stocken, 1959; Potter, 1959). This accumulation is probably the result of an observed decrease, in irradiated rat livers and irradiated rat kidneys, of the amount of intracellular kinase and polymerase (Bollum *et al.,* 1960; Main *et al.,* 1963). These enzymes catalyze the phosphorylation of nucleosides and the polymerization of nucleotides, respectively.

A stimulation of nucleic acid synthesis by oligonucleotides was observed by Keir (1962) in a cell-free system for primed DNA synthesis and by Gros *et al.* (1963) for DNA-primed RNA synthesis. Han (personal communication) found that DNA synthesis was stimulated in X-irradiated L cells after addition of deoxynucleotides to the tissue culture medium.

Firshein and Braun (1958, 1960) observed a selective stimulation of S cells of Pneumococcus, but not of R cells, in cultures supplemented with oligonucleotides. This stimulation was further enhanced by the addition of nucleosides and nucleotides to the medium. Without the oligonucleotides, however, the nucleosides and nucleotides had no effect. Firshein (1965) demonstrated that the addition of oligodeoxynucleotides to a culture of Pneumococcus type III stimulated the synthesis of deoxycytidylic and deoxyguanylic kinases, and that the addition of deoxynucleosides and deoxynucleotides preferentially enhanced DNA synthesis to an even greater degree. In the absence of the digest, the nucleosides and nucleotides had no effect.

According to these findings, the role of oligodeoxynucleotides is to increase the synthesis of DNA through stimulation of kinase synthesis and subsequent utilization of the nucleoside and nucleotide pool. Therefore, it is expected that oligonucleotides would be active in radiation-inhibited systems, in which there is a decrease in kinase and polymerase synthesis (Bollum *et al.*, 1963) with a resultant intracellular accumulation of nucleosides and nucleotides (Stocken, 1959; Potter, 1959). With the association of oligonucleotides, kinases, nucleosides and nucleotides in the enhancement of nucleic acid synthesis in irradiated animals, it becomes easier to relate the increased amount of antibody induction that follows the administration of oligonucleotides with antigen.

Antigen initiates changes that result in maturation and proliferation of stimulated cells into large populations of antibody-forming cells but all the cells that are genetically suitable are not in a physiological state susceptible to antigenic stimulation. Thus, the stimulation of nucleic acid synthesis, by oligonucleotides, at the time of antigen injection, may result in the development of more cells into a state responsive to induction by antigen. This conclusion from the analysis of molecular events agrees with the ideas of Jaroslow and Taliaferro, 1966, arrived at by an analysis of dose responses to x-ray, antigen and restorative agent (colchicine).

The non-inhibited systems, in which oligonucleotides enhance the immune response (Merritt and Johnson, 1965; Braun, 1965; Braun and Nakano, 1965), in all likelihood operate in the same manner, i.e., the efficiency of the antigen is greatly increased by the presence of the oligonucleotide supplement because more of the stimulated cells can respond to stimulation. The work of Braun and Nakano, 1965, affords a particularly good demonstration of the oligonucleotide effect in normal animals. The injection of DNA digest at the same time as sheep red cells induced a threefold increase in the production of plaque-forming cells. A combination of sheep red cells, DNA digest and kinetin or kinetin riboside resulted in a reduction of plaque-forming cells to the level of response obtained after immunization with sheep red cells alone. Kinetin and kinetin riboside in large amounts (2 mg/20 gm mouse) did not alter the number of plaque-forming cells to sheep RBC. It appears, in this instance, that kinetin competes with nucleosides, as an analog, to abort the stimulation induced by oligonucleotides. The same relationship between DNA digest and kinetin was also observed by Freedman and Braun (1965) in a study of phagocytic activity; by Braun and Nakano (1967) when polyadenine plus polyuridine were substituted for DNA digest. Low doses of kinetin (10 mg/3 kg rabbit), on the other hand, restored antibody-forming capacity in irradiated rabbits (Fig. 2) and increased the number of plaque-forming cells in mice in the absence of oligonucleotide supplements (Braun and Nakano, 1965).

In summary, I propose that only a minor percentage of the population of cells that are genetically suitable for antigenic stimulation are in a receptive physiological state. Oligonucleotides can induce this receptive state by stimulating nucleic acid synthesis. The specific role of the oligonucleotides appears to be to increase the level of kinase and to stimulate polymerization of the nucleotides.

References

Ada, G. L., G. J. V. Nossal, J. Pye, and A. Abbot (1964). Antigens in immunity. I. Preparations and properties of flagella antigens from *Salmonella adelaide*. Aust. J. Exp. Biol. Med. Sci., *42*:267–282.

Bacq, A. M., and P. Alexander (1955). Fundamentals of radiobiology. London: Butterworths Scientific Publications.

Bensch, K. G., and D. W. King (1961). Incorporation of heterologous deoxyribonucleic acid into mammalian cells. Science, *133*:381–382.

Bollum, F. J., J. W. Anderegg, A. B. McEllya, and V. R. Potter (1960). Nucleic acid metabolism in regenerating rat liver. VII. Effect of X-radiation on enzymes of DNA synthesis. Cancer Research, *20*:138–143.

Braun, W. (1965). Influence of nucleic acid degradation products on antibody synthesis. In *Molecular and Cellular Basis of Antibody Formation* (J. Šterzl, Ed.), pp. 525–534. Publishing House of the Czechoslovak Academy of Science, Prague.

Braun, W., and M. Nakano (1965). Influence of oligodeoxyribonucleotides on early events in antibody formation. Proc. Soc. Exper. Biol. Med., *119*:701–707.

——— (1967). Antibody formation: Stimulation by polyadenylic and polycytidylic acids. Science, *157*: 819–821.

Feldman, M., A. Globerson, and D. Nachtigal (1963). The reactivation of the immune response in immunologically suppressed animals. In *Conceptual Advances in Immunology and Oncology* (R. W. Cumley, D. M. Aldridge, J. Haroz, and J. McCay, Eds.), pp. 427–449. Harper and Row, Inc., New York.

Firshein, W. (1965). Influence of deoxyribonucleic acid degradation products and orthophosphate on deoxynucleotide kinase activity and deoxyribonucleic acid synthesis in pneumococcus type III. J. Bacteriol., *90*:327–337.

Firshein, W., and W. Braun (1958). On the nature of the selective effects of deoxyribonucleic acid digests upon pneumococci of different virulence. Proc. Natl. Acad. Sci. U. S., *44*:918–923.

Firshein, W., and W. Braun (1960). Effects of enzymatic deoxyribonucleic acid digests on population changes and deoxyribonucleic acid synthesis of pneumococci. J. Bact., *79*:246–260.

Freedman, H. H., and W. Braun (1965). Influence of oligodeoxyribonucleotides on phagocytic activity of the reticuloendothelial system. Proc. Soc. Exp. Biol. Med., *120*:222–225.

Gabrielsen, A., and R. Good (1967). Chemical suppression of adaptive immunity. Advances in Immunology, *6*:91–229, ed. F. J. Dixon and J. H. Humphrey. New York and London: Academic Press.

Gros, R., J. Dubert, A. Tissières, S. Bourgeois, M. Michelson, R. Soffer, and L. Legault (1962). Regulation of metabolic breakdown and synthesis of messenger RNA in bacteria. Cold Spring Harbor Symposia on Quantitative Biology, *28*:299–313.

Hill, M. (1967). The uptake of deoxyribonucleic acid released from damaged cells in tissue cultures. Exp. Cell Res., *45*:533–549.

Jaroslow, B. N. (1960). Factors associated with initiation of the immune response. J. Infect. Diseases, *107*:56–64.

Jaroslow, B. N., and K. Shortman (1966). Restoration of actinomycin D-inhibited antibody synthesis by ribonucleic acid digests. J. Infect. Diseases, *116*:601–605.

Jaroslow, B. N., and W. H. Taliaferro (1956). The restoration of hemolysin-forming capacity in X-irradiated rabbits by tissue and yeast preparations. J. Infect. Diseases, *98*:75–81.

——— (1958). Effect of nucleic acid digests in restoration of hemolysin production in irradiated rabbits. Fed. Proc., *17*:519.

——— (1966). The effect of colchicine on the hemolysin response in unirradiated and irradiated rabbits. J. Infect. Diseases, *116*:139–150.

Keir, H. M. (1962). Stimulation and inhibition of deoxyribonucleic acid nucleotidyltransferase by oligodeoxyribonucleotides. Biochem. J., *85*:265–276.

Kind, P., and A. G. Johnson (1959). Studies on the adjuvant action of bacterial endotoxins on antibody formation. I. Time limitation of enhancing effect and restoration of antibody formation in X-irradiated rabbits. J. Immunol., *82*:415–427.

Main, R. K., L. J. Cole, and E. R. Walwick (1963). Inhibition of thymidine-kinase-DNA-polymerase activity in the kidney by X-radiation before or after uninephrectomy. Int. J. Rad. Biol., *7*:45–51.

Merritt, K., and A. G. Johnson (1965). Studies on the adjuvant action of bacterial endotoxins on antibody formation. VI. Enhancement of antibody formation by nucleic acids. J. Immunol., *94*:416–422.

Potter, R. L. (1959). The effect of ionizing radiation on deoxyribonucleotide metabolism *in vivo*. Fed. Proc., *18*:304.

Proceedings of the conference on recovery of irradiated cells by nucleic acids (1967). Publication of the Boris Kidrić Institute for Nuclear Research, Belgrade. In press.

Simić, M. M., M. Ž. Petković, and D. D. Mančić (1959). Partial restoration of hemolysin-forming capacity in X-Irradiated rats by homologous spleen desoxyribonucleic acid. Bull. Inst. Nucl. Sci. "Boris Kidrich" (Belgrade), *9*:155–164.

Stocken, L. A. (1959). Some observations on the biochemical effects of X-radiation. Radiation Res. Suppl., *1*:53–72.

Taliaferro, W. H., and B. N. Jaroslow (1958). Restoration of antibody-forming capacity in x-irradiated rabbits. 2nd Int. Conf. Peaceful Uses of Atomic Energy, *23*:79–83.

Taliaferro, W. H., and B. N. Jaroslow (1960). The restoration of hemolysin formation in X-rayed rabbits by nucleic acid derivatives and antagonists of nucleic acid synthesis. J. Infect. Diseases, *107*:341–350.

Taliaferro, W. H., L. G. Taliaferro, and B. N. Jaroslow (1964). Radiation and Immune Mechanisms. Academic Press, New York and London.
Taliaferro, W. H., and L. G. Taliaferro (1950). The dynamics of hemolysin formation in intact and splenectomized rabbits. J. Infect. Diseases, *87*:37–62.
Taliaferro, W. H., and L. G. Taliaferro (1956). X-ray effects on hemolysin formation in rabbits with spleen shielded or irradiated. J. Infect. Diseases, *99*:109–128.

INDUCTION OF AN ALLOGENEIC-LIKE INHIBITION BY MEANS OF DNA OBTAINED FROM UNTREATED CELLS *

J. Leslie Glick

Roswell Park Memorial Institute, New York State Department of Health
Buffalo, New York

I. INTRODUCTION

Several laboratories have recently presented evidence of an apparently nonimmunological phenomenon which bears some resemblance to an immunological reaction (Ginsburg and Sachs, 1965; Holm and Perlman, 1965; Bergheden and Hellström, 1966; Hellström, 1966; Möller *et al.*, 1966; Fialkow, 1967; Hellström *et al.*, 1967; Möller, 1967). Hellström (1966) has applied the term "allogeneic inhibition" to this phenomenon, in which cells of one strain are destroyed by cells of another strain because of a difference with respect to certain isoantigens. Such inhibition has been obtained both *in vivo* and *in vitro* by means of normal or even lethally X-irradiated cells from nonimmunized animals. Extracts of the allogeneic cells have also been found to cause cell destruction. However, isoantibodies reacting against certain antigenic determinants of the allogeneic cell-line negate the latter's inhibitory property (Möller, 1967). The mechanism by which "allogeneic inhibition" occurs thus does not appear to be mediated by antibodies. Instead, special antigenic constituents of the inhibitory cell presumably attach to specific antigens of the target cell, thereby causing subtle configurational changes and inducing cell death.

At about the same time that the above studies were reported, my laboratory was engaged in an investigation on the destruction of tumor cells by means of certain kinds of DNA (Glick and Goldberg, 1965, 1966; Glick, 1967a, 1967b; Glick and Salim, 1967a). It now appears that

* This work was supported in part by U. S. Public Health Service Grant, CA-08763.

the mechanism by which DNA causes cell death may be similar to that of "allogeneic inhibition," except that the DNA, after entering the target cell, must first utilize its genetic information in order to synthesize RNA. The RNA may then cause the formation of non-antibody protein, which in turn may interact with the specific target antigens and cause cell death (Glick, 1967c; Glick and Sahler, 1967).

II. SPECIFICITY OF DNA-INDUCED CELL DESTRUCTION

The method we use to prepare DNA involves treatment of the appropriate tissue or cell-line with sodium lauryl sulfate, followed by extraction in 1.0M NaCl, deproteinization in a mixture of chloroform and butanol, and precipitation of the DNA in ethanol (Szybalska and Szybalski, 1962). Except where indicated, we routinely expose the DNA to RNAase and repurify it according to the above procedure. The resulting DNA preparations have been tested mainly on L1210 mouse leukemia cells, although other cell-lines have been employed. The L1210 cells grow equally well in culture medium RPMI #1634 or in DBA_2 or BDF_1 mice. In all cases the doubling time is about 12 hours. When injected intraperitoneally, the cells will proliferate in the ascitic form. When injected subcutaneously, these cells will form solid tumors. Table 1 outlines the procedure for working with DNA-treated L1210 cells.

Table 2 summarizes the effects of different DNA preparations on L1210 tumor growth. In each experiment the cells were incubated with the DNA, resuspended in DNA-free saline, and then injected into BDF_1 mice. Every time a group of cells was treated with DNA, a corresponding batch of cells was treated with the buffer alone. The results demonstrate

Table 1

EXPERIMENTAL PROCEDURE FOR TREATING L1210 CELLS WITH DNA

1. Suspend 6×10^6 cells in 3.5 ml of phosphate-buffered saline (pH 7.5) + 5.5 mM glucose + 0.2 mM spermine.
2. Add 0.5 ml of saline-citrate (pH 7.5) with or without 360 μg of DNA.
3. Incubate for 30 min at 37°C.
4. Centrifuge and resuspend in DNA-free, phosphate-buffered saline + glucose.
5. Reincubate at 37°C or inject subcutaneously into 10 BDF_1 or DBA_2 mice.

Table 2

SPECIFICITY OF INHIBITION OF L1210 TUMOR GROWTH BY BDF_1 MOUSE THYMUS DNA

Substances Incubated with Cells	Per Cent of Control Values: Mean Wt. of 8-Day-Old Tumors		Incidence of Palpable Tumors When 70% of Control Mice Were Positive		Incidence of Deaths When 70% of Control Mice Had Died	
Saline-citrate (control)	100	(10)	100	(100)	100	(80) [a]
Untreated BDF_1 DNA	40 [c]	(10)	41 [b]	(89)	46 [b]	(59)
RNAase-treated BDF_1 DNA	43 [d]	(10)	43 [d]	(20)		
DNAase-treated BDF_1 DNA	144	(10)	100	(20)	121	(20)
100°C-heated BDF_1 DNA	117	(10)	114	(10)		
60°C-heated BDF_1 DNA			0 [c]	(10)		
UV-irradiated BDF_1 DNA			86	(20)		
Untreated L1210 DNA			90	(60)	96	(60)
Untreated *E. coli* DNA			119	(30)		

[a] Numbers in parentheses represent the numbers of mice injected.
[b] $P < 0.001$.
[c] $P < 0.01$.
[d] $P < 0.02$.

that native, BDF_1 mouse thymus DNA was significantly effective in inhibiting the growth of L1210 tumors. Such inhibition was particularly striking, since 3 types of measurements were employed. The appearance of palpable tumors was delayed, as was the occurrence of deaths. BDF_1 DNA also inhibited the size of tumors, although complete inhibition was only occasionally observed (Glick and Goldberg, 1965). To date, the only other type of DNA tested which inhibits L1210 tumor growth is the closely related DBA_2 DNA (Glick and Goldberg, 1966). In all cases, degradation of the DNA by means of DNAase (2 μg/ml), denaturation by means of heating to 100° (followed by quick cooling at 0°C), or UV-irradiation (35 ergs/mm^2/sec for 10 min) completely prevented the DNA from exerting its biological effect. Intact L1210 DNA or *E. coli* DNA also were ineffective. These results thus indicate that the inhibitory property of the thymus DNA was not only dependent upon its high molecular weight and helical structure, but that it may also have been dependent upon its specific primary structure (Glick, 1967a).

Tables 3–5 report the effects of DNA on cell viability, when DNA-

Table 3

EFFECTS OF DIFFERENTLY TREATED DNA PREPARATIONS FROM BDF_1 OR DBA_2 MOUSE THYMUS ON L1210 CELL VIABILITY

Treatment of DNA	Number of Experiments	Inhibition of Viability at Various Times (min) *					
		−30	0	30	60	90	120
Untreated	11	0	0	+	++		
RNAase	16	0	0	+	++	++	++
DNAase	4	0	0	0	0	0	0
100°C heat	5	0	0	0	0	0	0
60°C heat	4	0	0	+	++	++	++
Fractionated: MW, 5×10^4	2	0	0	0	0	0	0
Fractionated: MW, 5×10^5	2	0	0	0	0	0	0

* Cells were incubated with or without DNA from −30 to 0 min and were reincubated in DNA-free medium from 0 to 120 min. The plus signs indicate that the viability of DNA-treated cells was inhibited to a mean value of 10–19% (+) and 20–29% (++) below the corresponding control values. Zero (0) indicates that the viability was inhibited to a mean value ≦5%. In all cases s.e.m. was between ±2 and ±8%.

Table 4

SPECIFICITY OF INHIBITION OF L1210 CELL VIABILITY BY BDF_1 OR DBA_2 MOUSE DNA

Source of DNA			Inhibition of Viability *
Species	Strain	Tissue	
Mouse	BDF_1	Thymus	++ (10)
Mouse	DBA_2	Thymus	++ (22)
Mouse	DBA_2	Liver	+++ (8)
Mouse	DBA_2	Spleen	++ (8)
Mouse	DBA_2	L1210 Leukemia	0 (14)
Mouse	Swiss	Thymus	0 (8)
Mouse	Swiss	S180 Sarcoma	0 (4)
Rat	W. D. Fisher	Thymus	0 (8)
Hamster	Syrian	SB1 Carcinoma	0 (4)
Calf	–	Thymus	0 (4)
Human	–	Spleen	0 (4)

* See Table 3 for explanation of symbols. Only those results are shown which were obtained at 90 and 120 minutes after reincubation of DNA-treated cells in DNA-free medium. Numbers in parentheses indicate the numbers of determinations.

Table 5

INHIBITION OF VIABILITY OF SPECIFIC TUMOR LINES BY VARIOUS DNA PREPARATIONS

Source of DNA	Tumor line	Inhibition of Viability *	
DBA_2 mouse thymus	L1210 mouse leukemia	++	(22)
	S180 mouse sarcoma	++	(6)
	S37 mouse sarcoma	++	(6)
	Ehrlich mouse carcinoma	0	(4)
	Human leukemia RPMI #1245	0	(4)
	Human mesothelioma RPMI #212	0	(4)
Swiss mouse thymus	S180 mouse sarcoma	++	(8)
	L1210 mouse leukemia	0	(8)
	Ehrlich mouse carcinoma	0	(4)
Human spleen	Human leukemia RPMI #1245	+++	(14)
	Human lymphoma P-3J	++	(4)
	L1210 mouse leukemia	0	(4)
Human leukemia RPMI #5630	Human lymphoma P-3J	+++	(8)
	Human leukemia RPMI #1245	0	(14)

* See Table 4 for details.

treated cells were reincubated in DNA-free medium. Loss of cell viability was detected by measuring uptake of trypan blue and cell lysis (Glick and Goldberg, 1966). Table 3 indicates that the inhibitory effect of BDF_1 or DBA_2 mouse thymus DNA on L1210 cell viability *in vitro* was dependent upon the structural integrity of the DNA molecule. Even after the DNA had been chromatographed on centrifuged DEAE-cellulose paper pulp (Davila *et al.,* 1965), fractions of molecular weight 5×10^4 and 5×10^5 were still not large enough to cause cell damage. The association between the genetic information of a specific DNA and its ability to induce cell death *in vitro* is suggested in Tables 4 and 5. Only BDF_1 and DBA_2 DNA were capable of destroying L1210 cells. This effect was not organ-specific but was found to be species-specific and possibly even strain-specific. Thus, Swiss mouse DNA, which inhibited S180 cell viability, did not inhibit L1210 cell viability. Even DBA_2 DNA was not inhibitory to all of the mouse tumor lines tested. Moreover, DNA obtained from one of the human cell lines was toxic to a second human cell line but not to a third.

Table 6

UPTAKE OF DIFFERENT TYPES OF DNA BY L1210 CELLS *

DNA Concentration ($\mu g/10^6$ Cells)	Source of DNA †	Uptake at 30 min ($\mu g/10^6$ Cells)
60	DBA_2 mouse thymus	0.100
	Swiss mouse thymus	0.095
	L1210 mouse leukemia	0.107
70	DBA_2 mouse thymus	0.122
	Swiss mouse thymus	0.119
	L1210 mouse leukemia	0.120

* Incorporation of DNA was terminated by exposure of cells to DNAase, as previously described (Glick, 1967c).

† Tritiated DNA was prepared from 11 day old mice or 6 day old ascites cells, 24 hours after intraperitoneal injections of thymidine-^3H (25 μc in 0.1 ml of saline).

Current data indicate that the unique ability of a given DNA to cause cell death is not simply due to an unusually large absorption of the DNA by the recipient cell. Table 6 demonstrates that there is no difference between uptake of the inhibitory DBA_2 DNA and that of 2 non-inhibitory types of DNA into L1210 cells. Unlike the cytotoxic effect of DNA, uptake of DNA would appear to be relatively independent of its genetic information.

III. MOLECULAR EFFECTS OF EXOGENOUS DNA ON L1210 CELLS

Table 3 demonstrates that the cell destruction caused by BDF_1 or DBA_2 DNA was preceded by a latent period longer than the time required for DNA to penetrate the cells. The following experiments were performed in order to determine if specific metabolic changes were induced by exogenous DNA, which may have preceded or accompanied the loss of L1210 cell viability. Details of this study have been presented elsewhere (Glick and Sahler, 1967). Table 7 summarizes the results of treating L1210 cells with various types of DNA and then incubating the cells in DNA-free medium containing uridine-^3H (1.7 μc/ml). Incorporation of uridine-^3H into the DNA and RNA fractions was stimulated by BDF_1 DNA but not by L1210 DNA or *E. coli* DNA. The apparently marked elevation of DNA synthesis in the BDF_1 DNA-treated cells occurred within 30 minutes following the addition of isotope. This pattern was also observed when thymidine-^{14}C was used instead of uridine-^3H (Glick and Sahler, 1967). What was perhaps even more interesting was

Table 7

EFFECTS OF DNA PREPARATIONS OBTAINED FROM DIFFERENT SOURCES ON THE INCORPORATION OF URIDINE-^{3}H INTO DNA AND RNA OF L1210 CELLS

Pretreatment of Cells	Number of Samples	DNA Synthesis (Corrected CPM/10^7 Cells)		RNA Synthesis (Corrected CPM/10^7 Cells)	
		30 min	60 min	30 min	60 min
Saline-citrate	8	72	169	862	996
BDF_1 DNA	8	364 (+405) *	529 (+212) *	1042	1981 (+99) *
L1210 DNA	4	100	147	793	688
E. coli DNA	2	112	104	611	883

* Values in parentheses represent only the % changes which were statistically significant. In each case, $P < 0.05$.

the apparent increase of RNA synthesis in the BDF_1 DNA-treated cells, since this activity was not observed until 30 minutes after the rise of newly labeled DNA.

We now had to determine if the increased synthesis of nucleic acids was responsible for the subsequent loss of cell viability, or if it was involved in a repair mechanism in those cells which survived. Table 8

Table 8

INFLUENCE OF METABOLIC INHIBITORS ON THE INHIBITION OF L1210 CELL VIABILITY BY DBA_2 MOUSE THYMUS DNA *

1st Incubation	2nd Incubation	3rd Incubation	Inhibition of Viability Caused by DNA at Various Times (min) of the Final Incubation		
			60	90	120
Cells ± DNA	Cells only	—	+	++	++
Cells ± DNA	FUDR + Cells	—	+	0	0
FUDR + Cells ± DNA	FUDR + Cells	—	0	0	0
ACT + Cells	Cells ± DNA	Cells only	++	++	++
Cells ± DNA	ACT + Cells	Cells only	0	0	0
Cells ± DNA	PUR + Cells	—	0	0	0

* See Table 2 for explanation of symbols. Results were obtained from typical experiments. Each experiment was performed 3 times. FUDR = 5-fluorodeoxyuridine (10^{-6} M); ACT = actinomycin D (0.25 μg/ml); PUR = puromycin dihydrochloride (5×10^{-6} M).

demonstrates that the addition of 5-fluorodeoxyuridine (FUDR) to cells which had been exposed to DBA_2 DNA almost completely blocked the cytotoxic effect of the DNA. Since FUDR is a specific inhibitor of DNA synthesis, the above result indicated that the stimulation of DNA synthesis in the DNA-treated cells was required in order to cause cell destruction. Moreover, the addition of FUDR after the DNA treatment permitted some initial cell damage, so that the stimulation of DNA synthesis caused by the exogenous DNA must have occurred soon after its uptake. As was expected, a double exposure of cells to FUDR, both during and after the DNA treatment, completely prevented any DNA-induced cell damage.

Table 8 also indicates that the burst of RNA synthesis in the DNA-treated cells must have been necessary for subsequent cell death. Actinomycin D, an inhibitor of DNA-dependent RNA synthesis, was found to block the cytotoxic effect caused by DNA when applied to the cells for only 15 minutes after the DNA treatment. The cells were then thoroughly washed and reincubated. Presumably any actinomycin D remaining in the cells would have been bound to both the host DNA and the donor DNA (Kirk, 1960; Goldberg *et al.*, 1962). However, when the cells were pretreated with actinomycin D for 15 minutes, washed, and then exposed to DNA, the actinomycin had no effect on DNA-induced cell death (Table 8). In this case, the actinomycin should have been bound primarily to the host DNA but hardly at all to the donor DNA (see Glick and Sahler, 1967). These results suggest that at least some of the newly formed RNA which was involved in causing cell death was probably transcribed by the donor DNA or by a replicated form of the donor DNA, instead of the host DNA. The results in Table 9 support this hypothesis,

Table 9

EFFECT OF BDF_1 MOUSE THYMUS DNA ON RNA SYNTHESIS OF ACTINOMYCIN D-PRETREATED CELLS

1st Incubation	2nd Incubation	3rd Incubation (30 min)	Number of Samples	Incorporation of Uridine-^{3}H in Terms of % Control
Cells only	Cells only	Cells + Uridine-^{3}H	8	100 (Control)
Cells only	Cells + DNA	Cells + Uridine-^{3}H	8	111
Cells + ACT *	Cells only	Cells + Uridine-^{3}H	6	20 †
Cells + ACT	Cells + DNA	Cells + Uridine-^{3}H	6	65

* ACT = actinomycin D.
† $P < 0.05$, compared to any other group.

since the pretreatment of cells with actinomycin D caused a marked decrease in RNA synthesis. Addition of BDF_1 DNA to actinomycin-pretreated cells resulted in a partial restoration of rapidly labeled RNA. In fact, incorporation of uridine-^3H into RNA of actinomycin-pretreated cells was stimulated 3-fold by BDF_1 DNA, even before DNA stimulation of RNA synthesis was observed in the actinomycin-untreated cells. As discussed elsewhere, it is unlikely that the addition of DNA, after the actinomycin D treatment, merely permitted a shift in the binding of actinomycin from host DNA to donor DNA in order to stimulate RNA synthesis (Glick and Sahler, 1967).

Finally, BDF_1 DNA also appeared to stimulate the incorporation of lysine-^{14}C into protein of L1210 cells (Glick and Sahler, 1967). The newly synthesized protein was apparently required in order for cell death to occur, since the addition of puromycin, an inhibitor of protein synthesis, completely suppressed the loss of viability of DNA-pretreated cells (Table 8).

IV. DESTRUCTION OF L1210 CELLS BY MULTIPLE GENETIC SITES

By means of an appropriate equation (Table 10A), one can estimate the per cent of cells inactivated by DNA, when DNA-treated cells are injected into mice. The length of time required for palpable tumors to develop is converted into the corresponding number of cell generations, where there are 2 L1210 cell generations per day. The greater the number of cells destroyed by DNA, the greater the number of cell divisions required for the remaining unharmed cells to form palpable tumor (Glick, 1967a, 1967c). Thus, at a concentration of 100 $\mu g/10^6$ cells, BDF_1 DNA usually causes a 3-day delay in the appearance of palpable L1210 tumors. This delay corresponds to 6 extra cell generations, and the per cent of cells inactivated by the DNA is calculated to be 98% (Glick and Goldberg, 1965). By determining this per cent for various temperatures and time periods of exposure of L1210 cells to BDF_1 DNA, one can calculate the temperature coefficient, Q_{10}, for the resulting inhibition of tumor growth. A Q_{10} of 3.1 has in fact been reported, thereby indicating that the inhibition caused by DNA involves an energy-dependent process (Glick, 1967a). However, this procedure alone does not signify whether the energy-requiring process is related to the uptake of DNA or to the intracellular events following uptake.

A recent study has demonstrated that the rate of incorporation of radioactively labeled DNA into L1210 cells is optimal during the first 5 minutes of incubation, and that about 50% of the total amount of DNA to be incorporated penetrates the cells within this time period (Glick, 1967c). The Q_{10} during this initial 5-minute period was found to be 3.2. Because of the striking similarity between the Q_{10} values for

Table 10

MATHEMATICAL ANALYSIS OF DNA-INDUCED CELL DEATH

A. Estimation of R, the per cent of cells destroyed by DNA

1. Let T_s, T_d = No. of injected cells that form tumor after saline or DNA treatment, respectively
   G_s, G_d = No. of cell generations from injection of saline-treated or DNA-treated cells, respectively, till appearance of palpable tumor
2. $2^{G_s}(T_s) = 2^{G_d}(T_d)$
3. $R = \frac{(T_s - T_d)}{T_s} = 1 - 2^{(G_s - G_d)}$

B. Equation for P(Y = O), the probability that no lethal DNA molecule enters a cell

1. Let M = Total no. of DNA molecules surrounding a cell
   N = Total no. of DNA molecules entering a cell
   X = No. of lethal DNA molecules surrounding a cell
2. $P(Y = O) = \frac{C(X, O) \times C(M\text{-}X, N\text{-}O)}{C(M, N)}$
3. In general, $C(A, B) = \frac{A!}{B!(A-B)!}$
4. $\text{LN } P(Y = O) = -(X/M) \times N$

C. Estimation of X/M, the fraction of lethal molecules in the DNA preparation

1. Obtain R for various values of M
2. Let P(Y = O) = 1-R
3. Obtain N for various values of M
4. Plot LN P(Y = O) *v.* N, and determine the slope as (−X/M)

uptake of DNA and DNA-induced inhibition of tumor growth, the possibility was likely that such inhibition was strongly dependent on uptake of the BDF_1 or DBA_2 DNA.

All of the results cited above, concerning DNA-induced cell death, are consistent with an hypothesis involving genetic expression of the specific, exogenous DNA. However, the per cent of cells destroyed by the DNA is much too large for the results to be explained on the basis of genetic transformation for a specific genetic marker. Therefore, the possibility was considered that any one of a number of genetic sites present in the BDF_1 or DBA_2 DNA was capable of directing the formation of a foreign substance, within the L1210 cells, which was inhibitory to these cells. Cell destruction would thus result, if at least one of the "lethal" genetic sites entered the cell.

An expression can now be derived for P(Y = O), the probability that no lethal DNA molecule enters a cell. This probability would correspond to the per cent of cells not harmed by the exogenous DNA. To obtain

P(Y = O), we must first consider the exogenous DNA population to consist of 2 subpopulations: lethal and nonlethal. We can then determine P(Y = O), by calculating all of the ways for a cell to incorporate O lethal molecules from X, the lethal subpopulation; (N-O) nonlethal molecules from (M-X), the nonlethal subpopulation; and N total molecules from M, the entire DNA population. If one assumes that N and X are much less than M, P(Y = O) is simply a function of N and X/M, where X/M is the fraction of lethal molecules in the exogenous DNA preparation (Table 10B; see also Glick, 1967c).

The hypothesis can now be tested, that any one of a number of genetic sites in the genome of BDF_1 or DBA_2 mice is capable of producing a substance against L1210 cells. One merely determines a number of values for the per cent of cells inactivated by various DNA concentrations. The per cent of cells not inactivated is then calculated and is plotted against the amount of DNA incorporated by the cells, for the appropriate DNA concentration (Table 10C). If the hypothesis is correct, a negative semilogarithmic slope should result, which should extend straight back to a value of 100% unharmed cells at 0 uptake of DNA. As described in detail elsewhere, the above relationship does indeed hold when tested empirically (Glick, 1967c). Table 11 lists the results of 4 separate experiments. In each case a fresh preparation of BDF_1 DNA was employed. The average molecular weight of the DNA was assumed to be 4×10^6

Table 11

ESTIMATED FRACTION OF LETHAL MOLECULES IN DNA PREPARATIONS FROM BDF_1 MOUSE THYMUS

DNA Preparation	Groups of Mice Injected with DNA-Treated or Saline-treated L1210 Cells *	Lethal Molecules per 10^6 Total Molecules †
A	6	63
B	5	88
C	4	160
D	4	90
Mean	—	100 ± 21

* Each group of mice, consisting of 10 animals, were injected with cells exposed to a concentration of DNA varying from 0–116 $\mu g/10^6$ cells. The per cent of cells inactivated by each concentration of DNA was calculated for each group of mice (see Table 10A).

† Incorporation of DNA into L1210 cells determined for various concentrations of DNA (Glick, 1967c). The fraction of lethal molecules was calculated, as in Table 10C.

(Glick, 1967c). An average of 100 per 10^6 molecules was estimated to be lethal. Presumably if only one of these molecules entered an L1210 cell, the cell would die. The actual number of lethal genetic sites normally present in a BDF_1 mouse cell was also calculated, on the assumption that there was only 0 or 1 lethal genetic site per DNA molecule. Since BDF_1 mouse thymus contains approximately 8 $\mu\mu$g of DNA per cell, the minimum number of genetic sites presumed to be lethal against L1210 cells was estimated to be 120 per BDF_1 genome (Glick, 1967c).

As a check on the validity of the above mathematical analysis, the equation for $P(Y = O)$ has been shown to hold for data obtained from experiments of genetic transformation in bacteria (see Glick, 1967c). When $P(Y = O)$ and X/M were defined in terms of transforming molecules instead of lethal molecules, the equation indicated what was expected, that not more than one gene per bacteria genome could have been responsible for transformation.

V. CONCLUSIONS

The mechanism by which a specific DNA may act against a particular cell line appears to involve at least the following three steps: (a) entry of the lethal DNA molecule(s) into the target cell; (b) expression of specific genetic information of the donor DNA, resulting in the formation of the actual allogeneic or xenogeneic inhibitor; (c) recognition and destruction of some host antigenic constituent(s) by the newly synthesized inhibitor. Evidence from a number of other laboratories supports the plausibility of each of the above steps. With respect to uptake of DNA by mammalian cells, various studies have demonstrated that the donor DNA enters the nucleus in a highly polymerized form and may even be integrated into the host DNA (Gartler, 1960; Borenfreund and Bendich, 1961; Rabotti, 1963; Meizel and Kay, 1965; Ledoux *et al.*, 1967). Concerning the ability of exogenous DNA to manufacture new gene products, several laboratories have already obtained results suggesting that mammalian cells can be genetically transformed (Kraus, 1961; Bradley *et al.*, 1962; Szybalska and Szybalski, 1962; Podgajetskaja *et al.*, 1964; Roosa, 1966; Glick and Salim, 1967b). As for the concept of allogeneic inhibition, whereby antigens of one cell strain can recognize and destroy another cell strain, evidence for this interaction has been cited in the Introduction.

The data presented here, on the specificity of DNA-induced cell death, are consistent with the results of others. Halpern *et al.* (1966) have reported mouse and rat DNA preparations to inhibit the growth of mouse S180 and rat W256 tumors, respectively. DNAase-degraded DNA was ineffective, whereas DNA subjected to proteolytic enzymes retained its activity. Intact DNA prepared from calf thymus, salmon sperm, and bacteria did not affect tumor growth. Floersheim (1962) has also found a specificity involved in the destruction of CBA mouse bone marrow cells by exogenous DNA. However, both Halpern *et al.* (1966) and Floersheim (1962) have observed that isologous DNA was inhibitory to cell growth. Those results are not necessarily inconsistent with the data reported here, since one might conclude that genetic information normally contained in some cell lines is potentially capable of destroying these cells, whereas such genetic information might be lacking in other cell lines. Finally, other investigators have reported that not only mammalian DNA but also bacterial DNA can inhibit the growth of homologous cells (Smith and Cress, 1961; Gimlin *et al.,* 1963; Zahn and Tiesler, 1963; Smith, 1964; Piechowska and Shugar, 1967).

Although the results which I have reported demonstrate a reproducible, inhibitory effect of BDF_1 or DBA_2 DNA on L1210 cells, it is important to note that variation of the incubation conditions can seriously alter this effect. If L1210 cells are exposed to DNA for too long a period or if the DNA-treated cells are incubated in a growth medium, there will be little or no inhibitory effect of the DNA (Glick and Goldberg, 1966; Glick, 1967a). A recent study also demonstrates that preincubation of L1210 cells in saline medium, prior to DNA treatment, enables the DNA to protect the cells against loss of cell viability (Glick, 1967d). This phenomenon, which also occurs in DNA-treated cells that are shaken too rapidly, is independent of the type of DNA employed. Unlike the cytotoxic effect of DNA, the protective effect presumably does not result from expression of specific genetic information.

In conclusion, the ability of a given DNA to destroy a specific cell line is probably dependent on the following factors: (a) specific nucleotide sequences in the DNA; (b) the size of the isolated DNA molecules; (c) the extent of uptake of the DNA; (d) the ability of the target cell to degrade the DNA; (e) the availability of the appropriate enzymes necessary for decoding the DNA; (f) the possibility that the target cell could either repress transcription of information initiated by the DNA or inactivate end products or proteins resulting from the translation of this information; (g) the absence or concealment of the appropriate antigenic constituents of the target cell, which would preclude the recognition and destruction of the target cell by the newly synthesized gene products.

References

Bergheden, C., and K. E. Hellström (1966). Studies on allogeneic inhibition. II. Inhibition of mouse tumor cell growth by *in vitro* contact with cells containing foreign H-2 antigens. Intern. J. Cancer, *1*:361–369.

Borenfreund, E., and A. Bendich (1961). A study of the penetration of mammalian cells by deoxyribonucleic acids. J. Biophys. Biochem. Cytol., *9*:81–91.

Bradley, T. R., R. A. Roosa, and L. W. Law (1962). DNA transformation studies with mammalian cells in culture. J. Cell. Comp. Physiol., *60*:127–137.

Davila, C., P. Charles, and L. Ledoux (1965). The chromatography of nucleic acid preparations on DEAE-cellulose paper. I. Fractionation of deoxyribonucleic acid on paper strips or on centrifuged paper pulp. J. Chromatog., *19*:382–395.

Fialkow, P. J. (1967). Chromosomal breakage induced by extracts of human allogeneic lymphocytes. Science, *155*:1676–1677.

Floersheim, G. L. (1962). Cytotoxic action of nucleic acids. Experientia, *18*:328–329.

Gartler, S. M. (1960). Demonstration of cellular uptake of polymerized DNA in mammalian cell cultures. Biochem. Biophys. Res. Commun., *3*:127–131.

Gimlin, D. L., E. Farquharson, and F. R. Leach (1963). Effect of bromouracil-labelled transforming deoxyribonucleic acid of *Bacillus subtilis* on cell growth. Nature, *197*:407–408.

Ginsburg, H., and L. Sachs (1965). Destruction of mouse and rat embryo cells in tissue culture by lymph node cells from unsensitized rats. J. Cell. Comp. Physiol., *66*:199–219.

Glick, J. L. (1967a). The specificity of inhibition of L1210 tumor growth by mouse thymus deoxyribonucleic acid. Cancer Res., *27*:175–180.

——— (1967b). The specificity of inhibition of tumor cell viability by DNA. Cancer Res., *27*:2338–2341.

——— (1967c). Quantitative studies concerning the uptake of mouse thymus DNA by L1210 leukemia cells. Cancer Res., *27*:2350–2356.

——— (1967d). Incubation conditions which influence contrasting effects of thymus DNA on L1210 cell viability. Exptl. Cell Res., *45*:690–693.

Glick, J. L., and A. R. Goldberg (1965). Inhibition of L1210 tumor growth by thymus DNA. Science, *149*:997–998.

——— (1966). The action of thymus DNA on L1210 leukemia cells. Trans. N. Y. Acad. Sci., *28*:741–753.

Glick, J. L., and C. Sahler (1967). Specific metabolic changes in DNA-treated L1210 leukemia cells. Cancer Res., *27*:2342–2349.

Glick, J. L., and A. P. Salim (1967a). Inhibition of tumour cell viability by DNA. Nature, *213*:676–678.

——— (1967b). DNA-induced pigment production in a hamster cell line. J. Cell Biol., *33*:209–212.

Goldberg, I. H., M. Rabinowitz, and E. Reich (1962). Basis of actinomycin action, I. DNA binding and inhibition of RNA-polymerase synthetic reactions by actinomycin. Proc. Natl. Acad. Sci. U. S., *48*:2094–2101.

Halpern, R. M., B. C. Halpern, G. Ciliv, and R. A. Smith (1966). Inhibition of tumor transplantation by DNA. Biochem. Biophys. Res. Commun., *24*: 559–563.

Hellström, K. E. (1966). Studies on allogeneic inhibition. I. Differential behavior of mouse tumors transplanted to homozygous and F_1 hybrid hosts. Intern. J. Cancer, *1*:349–359.

Hellström, K. E., I. Hellström, and C. Bergheden (1967). Studies on allogeneic inhibition. III. Inhibition of mouse tumor cell colony formation *in vitro* by contact with lymphoid cells containing foreign H-2 antigens. Intern. J. Cancer, *2*:286–296.

Holm, G., and P. Perlman (1965). Phytohemagglutinin-induced cytotoxic action of unsensitized immunologically competent cells on allogeneic and xenogeneic tissue culture cells. Nature, *207*:818–821.

Kirk, J. M. (1960). The mode of action of actinomycin D. Biochim. Biophys. Acta, *42*:167–169.

Kraus, L. M. (1961). Formation of different haemoglobins in tissue culture of human bone marrow treated with human deoxyribonucleic acid. Nature, *192*:1055–1057.

Ledoux, L., G. B. Gerber, P. Charles, J. Remy, and J. Remy-Defraigne (1967). Uptake of labelled DNA by isolated perfused organs. Experientia, *23*:16–18.

Meizel, S., and E. R. M. Kay (1965). Studies of the incorporation *in vitro* of deoxyribonucleic acid by cells of the Ehrlich-Lettré ascites carcinoma. Biochim. Biophys. Acta, *103*:431–439.

Möller, E. (1967). Cytotoxicity by nonimmune allogeneic lymphoid cells. J. Exptl. Med., *126*:395–405.

Möller, G., V. Beckman, and G. Lundgren (1966). *In vitro* destruction of human fibroblasts by non-immune lymphoid cells. Nature, *212*:1203–1207.

Piechowska, M., and D. Shugar (1967). Inhibitory and lethal effects of DNA on transformable streptococci. Biochem. Biophys. Res. Commun., *26*:290–295.

Podgajetskaja, D. J., V. M. Bresler, I. M. Surikov, T. N. Ignatova, and J. M. Olenov (1964). The transforming action of deoxyribonucleic acid isolated from sarcolysine-resistant tumors on sarcolysine-sensitive tumors. Biochim. Biophys. Acta, *80*:110–115.

Rabotti, G. F. (1963). Incorporation of DNA into a mouse tumor *in vivo* and *in vitro*. Exptl. Cell Res., *31*:562–565.

Roosa, R. A. (1966). DNA-mediated transformation in mammalian lymphoma cells. Ninth International Cancer Congress (Tokyo), Abstracts, p. 189.

Smith, A. G. (1964). Cytologic and cytotoxic changes induced by nucleic acids. Cancer Res., *24*:603–613.

Smith, A. G., and H. R. Cress (1961). Effect of tumor deoxyribonucleic acid upon tissue cultures of fibroblasts. Lab. Invest., *10*:898–907.

Szybalska, E. H., and W. Szybalski (1962). Genetics of human cell lines, IV. DNA-mediated heritable transformation of a biochemical trait. Proc. Natl. Acad. Sci. U. S., *48*:2026–2034.

Zahn, R. K., and E. Tiesler (1963). Die beeinflussung des wachstums des Ehrlich-ascites-tumors der mous durch native und mit ultraviolettem licht bestrahlte heterolge desoxyribonucleinsäure. Z. Naturforschung, *18b*:523–525.

DISCUSSION

Chairman: M. LANDY

LANDY: Dr. Jaroslow, have you done any work on restoration of antibody formation under conditions of limited concentrations of antigen, and, if so, what were the effects? Was the magnitude increased?

JAROSLOW: Both irradiated (400 r) and non-irradiated rabbits were injected with different amounts of sheep red cells and a constant amount of colchicine, a restorative agent. There was no effect on the peak titer reached by rabbits given antigen in amounts that fall in the logarithmic portion of the antigen dose-response curve. At antigen doses in the asymptotic portion of the dose-response curve, however, colchicine enhancement of the peak titer of antibody increased with the dose of antigen. This was true for the response of both irradiated and non-irradiated rabbits.

BRAUN: Dr. Jaroslow, I have a question in regard to the *in vitro* experiments with actinomycin and oligonucleotides. How have you ruled out the possibility of a direct interaction between actinomycin and oligonucleotides?

JAROSLOW: Such an interaction is possible. The RNA digest was not restorative in the presence of an excess of actinomycin D. It was only active after the excess of inhibitor was removed, leaving bound actinomycin D behind. On the other hand, this occurred only in the presence of antigen. Further work is required to clarify the results.

JOHNSON: I would like to ask Dr. Braun or Dr. Jaroslow what mechanism they envision for opening up of cells by an antigen so that oligonucleotides can penetrate.

BRAUN: A reaction between antigen and antibody at cell membrane sites is known to alter permeability. There are many data in the literature documenting this; for example, I can refer you to some of the work which was done by Dr. Kessel in our laboratory, but there are many other data regarding the ability of cell-associated antigen-antibody reactions to alter cell permeability.

JOHNSON: This, of course, requires preexisting antibody. You are willing to accept that?

BRAUN: No, in the case of the stimulation of a specific response I am assuming that there is enough antigen around to react with antibody on antibody-forming cells, some of which may preexist, while others may have been activated. Reactions on the membrane of antibody-forming cells may then alter the permeability sufficiently to permit the entrance of the stimulator.

May I now address a question to Dr. Johnson. He has looked at circulating antibody titers and has shown a stimulation by poly A-poly U. We have shown a comparable stimulation on the basis of number of antibody-forming cells and we have pointed out that this stimulation involves principally an alteration of the rate of increase in the number of such cells. What nobody has done as yet, as far as I know, is to determine to what extent the synthetic capacities of individual cells has been altered by the stimulators. I think this would be an important question. Have you looked at this?

JOHNSON: If you give me about a month, I hope I will be able to answer this question. We have these experiments going on in our laboratory now.

GLICK: I would like to ask Dr. Johnson how he explains the lack of an effect of tetranucleotides. Does this always occur with such oligomers?

JOHNSON: We have repeated such tests three times. In each case the tetranucleotides did not work. I do not have an explanation for it, unless the fact that the tetranucleotides are coming off the column at about five o'clock when I prepare columns for an overnight run has something to do with this.

BRAUN: Using commercially available tetra-A we found that it will stimulate.

LANDY: The results of Drs. Braun's and Johnson's studies imply that stimulators may exist in limiting concentrations *in vivo*. It seems to me that this is somewhat at odds with the current work on *in vitro* systems by Dutton and Mishell. Their *in vitro* systems do not contain the types of materials that are the subject of this discussion, yet they can get a commitment in terms of plaque-forming cells far exceeding that which can be detected *in vivo*. This would imply that we have, instead of a shortage, perhaps a repression *in vivo* that you are overcoming with the aid of your stimulators.

BRAUN: In regard to the *in vitro* system, the response to the types of stimulators that Dr. Johnson, Dr. Jaroslow and I have studied remains to be determined. It may well be that conditions prevail *in vitro* that are not comparable to those that can occur *in vivo*.

JAROSLOW: I would like to ask Dr. Firshein regarding his ideas on how oligonucleotides might stimulate kinase synthesis, whether he thinks this is strictly an energy release or a phosphorylation or something of that nature.

FIRSHEIN: Since you need sequences of at least three cytidine residues which cannot be interrupted by any other base, since you need single-stranded material, and since protein synthesis is involved, I think that there could be some type of derepression. You find long sequences of Cs in the DNA of a variety of microorganisms and higher organisms, some of them 100–120 nucleotides long, which presumably play a role in the initiation of multi-cistronic messages. Now, in our system, where we test under conditions of limited protein synthesis, it is possible, as Stent has suggested, that the ribosomes that are not participating in protein synthesis, or uncharged transfer RNA, might combine with such regions to prevent transcription, perhaps by preventing binding of the polymerase. By adding C oligomers or A oligomers, both of which bind with ribosomal RNA and transfer RNA, a relief from this so-called universal repression might be obtained and the polymerase might now be able to attach. However, if this were the case, then other enzymes should be affected. Accordingly, we are now making an exhaustive study of a variety of enzymes involved in DNA metabolism. The fact that we do get a greater percentage of polysomes in the stimulated population does suggest that some of the ribosomes have been converted to a system permitting increased protein synthesis.

BRAUN: Dr. Glick, do you believe that your effects may be dependent on the ability of a cell to depolymerize DNA? You showed some strain specificity. Are possible restriction phenomena involved in what you have seen?

GLICK: You mean whether or not DNAases degrade differentially depending on the type of DNA? We have tested several cells. To some extent certain cells will degrade DNA and the inability of DNA to affect those cells may be due to the fact that DNA is degraded. But with cells which are inhibited by DNA, there is no differential effect of DNAase activity on inhibitory DNA and non-inhibitory DNA.

NOVELLI: I have a comment regarding effects of oligonucleotides. Bernheimer has shown that the synthesis of streptolysin S by certain strains of hemolytic streptococci can be greatly stimulated by yeast nucleic acid. After the discovery of polynucleotide phosphorylase Ochoa and associates made synthetic polymers and these acted in a similar manner to stimulate the synthesis of streptolysin S. However, some of the polymers made in New York by Ochoa's group did not work, whereas others made at the same time by some Japanese investigators worked. An exchange of polymers between laboratories confirmed the difference and showed that the polymers made in Japan were enriched with respect to guanylic acid. Subsequently it was shown that polyriboguanylic acid was the actual stimulator and in fact Japanese workers subsequently crystallized streptolysin S and found that guanylic acid is covalently bound to the protein. I have forgotten the length of the nucleotide chain but I think it is about 9 nucleotides.

FIRSHEIN: Spiegelman has reported recently that the polymerase that synthesizes $Q\beta$ RNA consists of two parts, one of which is a poly G-polymerase dependent on poly C. Thus oligomers of C appear to play an important role under at least some natural conditions.

ORTIZ-ORTIZ: It has been reported that kinetin as well as oligodeoxyribonucleotides can stimulate antibody production. I would like to ask Dr. Braun if he has attempted to determine the effect of mixed administration of kinetin and oligodeoxyribonucleotides.

BRAUN: We only have tested such mixtures at concentrations at which kinetin inhibits the effects of oligonucleotides. It must be remembered that kinetin stimulates at low concentrations and inhibits at high concentrations; kinetin riboside is always inhibitory.

LESKOWITZ: Feldman and co-workers have recently shown that X-irradiation affects the ability of macrophages to process antigen, an effect that perhaps is at least as important as the effect of radiation on lymphoid cells; also, Nossal and co-workers have suggested that neonatal animals have ineffective or inefficient macrophage systems. I wonder, in view of these findings and in view of Dr. Jaroslow's demonstration that oligoribonucleotides can restore antibody formation in irradiated animals, whether the effect of the oligonucleotides may be at the macrophage processing level rather than at the antibody synthesis level.

JAROSLOW: I did look into antigen localization in irradiated animals that were given colchicine and thereby had experienced a restoration of antibody production, but the localization of antigen in these animals was identical with that in irradiated animals. This is not particularly surprising since the process of phagocytosis is relatively radioresistant; I found that it took doses of up to 16,000 r to stop this follicular localization, but such observations do not preclude the possibility that the effect might not be at the macrophage level.

BRAUN: Dr. Hechtel has published studies on the effects of stimulatory oligonucleotides on the responses of newborn mice. She demonstrated that stimulatory effects can only be observed after the mouse has attained the ability to produce a detectable antibody response to the particular antigen under study. When such a response occurs in a newborn mouse, the stimulatory effects of oligonucleotides are very much greater than in adult animals. Subsequently we were able to demonstrate that it is possible to initiate very early antibody responses in newborn mice by providing them with macrophages from adult mature animals. On the basis of such observations one may suggest that as long as the processing system is inoperative, the stimulators cannot do anything to affect it. Once the processing system operates, the stimulator may stimulate and may do so perhaps both at the level of processing cells as well as of antibody-forming cells.

REGELSON: We have been working with synthetic polyanions of polysulphonites, polyphosphates and polycarboxylates. These compounds are inhibitory for transplanted tumors and impede viral induction through interferon production. One of these compounds, a divinyl ether maleic anhydride copolymer, Pyran copolymer, induces interferon in both the mouse and man. Recently we gave some of this material to Dr. Braun to have him determine whether it has enhancing activity on antibody formation, and I wonder whether he would comment on his observations.

BRAUN: Your materials, Dr. Regelson, do indeed have a stimulatory effect on the rate of increase of antibody-forming cells, in a fashion similar to the oligonucleotides that have been discussed here. Since the polymers that you have employed are known to affect membranes, one is tempted to suggest that the Pyran effect may be another instance of triggering the release of stimulatory oligonucleotides from injured cells.

ALEXANDER: The comment by Dr. Regelson prompts me to ask Dr. Glick whether he has ruled out the possibility that the killing effects of DNA which he is observing *in vitro* may be cytotoxic effects merely due to polyanions. At a hundred micrograms per ml, which is the concentration he is using, polyanions tend to be very toxic to cells particularly if there is little protein in the medium. Perhaps he is dealing with a nonspecific polyelectrolyte effect, and perhaps the difference between those DNAs that are not cytotoxic and those that are may be due to their protein content.

GLICK: The amount of protein present in our DNA preparations is pretty much the same for all DNA preparations. There is hardly any, certainly no more than about ½ per cent of the DNA. With respect to the concentration of DNA, an effect can be observed with as little as 10 micrograms per ml which is about $\frac{1}{10}$ the order of magnitude that you just mentioned. As far as the incubation conditions are concerned, they are extremely important. If you incubate too long with DNA, before reincubation of the cells, the degree of inhibition becomes very slight. Furthermore, if you preincubate your cells before you add the DNA, you may obtain an apparent enhancement of cell viability. This protection is independent of the type of DNA used and occurs only with certain cell lines. Why, I do not know. It does occur with L1210 cells, but it doesn't occur with some other cell lines.

ALEXANDER: Dr. Braun, the compound made by Dr. Regelson has, as you stated a little while ago, an effect very similar to your poly A and poly U. I am wondering whether you are trying to impose the same explanation on the two systems and if so whether this may not be forcing the issue.

BRAUN: My current prejudice is that the materials that can damage cells can release the actual stimulators of antibody-forming cells. I believe that the oligonucleotides may be the actual stimulators and can act directly on the cells involved. There are many ways in which such stimulators can be made available and I would say that the majority of materials which have been used in the past, including such polymers as Dr. Regelson's, endotoxins and many other cytotoxic materials, all may provide identical stimulators by triggering their release.

ALEXANDER: Dr. Braun, I wonder to what extent the oligonucleotide effects that you have studied might be of a pharmacological nature. You have stressed that the effect is purely one of rate. However, the antigens involved have to get to the node or lymphoid tissue to induce antibody, and they have to penetrate through blood vessels and other organs. We know that the rate of cell division once antigens get into lymph nodes is extremely rapid. These cells divide with a division time of about 12 hours, so I am wondering how far your effect, Dr. Braun, is on rate of transport. Has, in fact, any work been done to study the pharmacology of these agents?

BRAUN: We know that oligodeoxyribonucleotides as well as poly A plus poly U are non-pyrogenic. We don't know anything more about possible pharmacological effects of these materials. However, we have not found it necessary to assume any indirect effects since the evidence presented by Firshein for the direct stimulation of nucleotide kinases, and the finding that you can get the same effect with antibody-forming cells by providing the products of such kinase activity, suffice to explain the results we have obtained. This does not rule out the possibility that there may be an indirect effect as well.

REGELSON: The question regarding the pharmacological effects prompts me to make some comments. It is well known that stimulation of the reticuloendothelial system (RES) affects the metabolism of a variety of drugs. This has been most strikingly demonstrated by Wooles and others regarding the detoxification of hexabarbital and other drugs. Hexabarbital sleeping time increases remarkably following RES stimulation. RES phagocytic function is measured by a variety of methods including clearance of chromate-tagged red cells, carbon particles and a lipid emulsion. These effects of the RES on drug metabolism are mediated via nonspecific hydrolyzing enzymes. Most recently, Wooles has shown that steroid metabolism is dramatically affected by RES stimulation and interestingly, a substantial portion of the adrenal cortex has phagocytic capacity. Thus, it may be that one has to consider the action of oligonucleotides, and of other agents that have immunologic enhancing activity, as possibly involving effects on drug metabolizing enzymes, with particular reference to effects on adrenal steroid metabolism. The latter is known to have profound action on lymphocytic populations and function. If RES stimulation is involved in drug-metabolizing effects, then the other side of the coin is also pertinent: what do drugs do to RES function?

BEERS: I would like to make a comment that may have some relationship to the question of direct versus indirect activities of enhancing agents. Three years ago we attempted to determine the fate of polyadenylic acid in mammals. We injected radioactively labelled polyadenylic acid intracardially and did not observe any endonuclease or any polynucleotide phosphorylase action on the synthetic polymer. However, after seven minutes the material was completely cleared from the plasma of the animals. We subsequently found that much of the radioactivity appeared in the urine, probably in the form of ADP. This has suggested to us that at least for single-stranded material such as poly A, clearance through the kidney may be a principal factor. One other comment that I should like to make is in respect to the frequent implication that mixing of poly A and poly U will result in a double-stranded helix. It is quite likely that under certain circumstances a triple-stranded structure may result since, particularly in the presence of divalent cations, the triple-stranded structure is favored over the double-stranded form.

FIRSHEIN: Is this true for poly C and poly I as well?

BEERS: No, I was referring only to poly A and poly U.

DUBISKI: I have a question for Dr. Braun. Were the enhancing effects observed only on cells making antibodies of high hemolytic capacity or also on cells making antibodies of low hemolytic capacity? In studies with Drs. Cinader and Chou we found that no matter what the change in the immunizing conditions, the proportion of cells making antibodies of low hemolytic capacity was always substantial. In most cases it was higher than the proportion of cells making antibodies of high hemolytic capacity.

BRAUN: We have tested the stimulatory effects both by the conventional Jerne technique in which you have 19S hemolytic antibody as well as after addition of antibody to detect 7S antibodies. The stimulation affects principally 19S-producing cells but to a limited extent it affects also 7S-producing cells.

COLE: I have two questions for Dr. Glick. First, do you think that the phenomenon that you have described is related in any way to the allogeneic inhibition phenomenon studied by the Swedish workers? Secondly, have you investigated DNA extracted from irradiated cells or the effects of irradiating DNA itself? Will this lead to inactivation?

GLICK: With respect to your last question, I haven't tried X-irradiated DNA but I have tried UV-irradiated DNA which was not effective. In regard to your first question, the answer is yes and no. I may be inducing the formation of some non-antibody proteins or other inhibitory peptide which can then react with constituents of the cell.

SIGEL: Dr. Glick, could you possibly have a virus or viral DNA involved in your effect? Or if you don't have a direct viral DNA effect, do you perhaps have something similar to what Roger Weil recently reported at a Gordon Conference, namely that injection of mice with polyoma virus can lead to the production of an endonuclease? This enzyme then chops up the host DNA into pieces which can be packed into capsules of polyoma virus. What is the molecular weight of the DNA that you are working with?

GLICK: About 4 or 5 million. Yes, it is possible that I might have a virus. But let me point out that the effect occurs very rapidly. The cells start to die within two hours after *in vitro* exposure. I don't know of any virus that has such rapid effects.

LACOUR: Dr. Glick, to what extent do you think your phenomenon may differ from that observed by Niu with RNA?

GLICK: Niu has interpreted his results dealing with tumor inhibition by RNA as being due to a transformation into liver cells. I don't claim anything so dramatic. However, in all fairness to Niu, there have been a number of reports that have claimed an inhibition of tumor growth by RNA. I also have tested the effect of RNA on tumor growth and I have never reported the results because I don't know what to make of them. I do find positive results but they are not very reproducible from one RNA preparation to another.

LACOUR: Dr. Glick, our experience has been that the RNA effects are very reproducible at least when you use intact RNA centrifuged in a gradient.

GLICK: I don't wish to dispute that. However, most people that have reported RNA inhibition of tumor growth were not very careful about the way in which they prepared their RNA.

ROLE OF NUCLEIC ACIDS IN SPECIFIC ANTIBODY FORMATION (PART I)

ANTIBODY FORMATION INITIATED *IN VITRO* WITH RNA AND RNA-ANTIGEN COMPLEXES

M. Fishman, F. L. Adler, and M. Holub
Public Health Research Institute, New York City

The idea that macrophages may convert antigens into "active molecules" (Ehrich, 1956) which then incite antibody formation in other cells has recently engaged the attention of several laboratories. Experiments conducted *in vitro* (Fishman, 1959; Friedman *et al.*, 1965; Gottlieb *et al.*, 1967; Jacherts *et al.*, 1966) as well as *in vivo* (Askonas and Rhodes, 1965; Gallily and Feldman, 1967; Mitchison, 1965; Pribnow and Silverman, 1967) have yielded evidence that implicates the macrophage as a necessary though not necessarily sufficient cellular component of the antibody response. There is reason to believe that the macrophage in question may belong to a subclass of morphologically similar cells which is distinguished by its cellular associations (Nossal, 1965; Thiery, 1959), its ability to retain antigen (R. Spiers and E. Spiers, 1964), and by other yet poorly defined attributes. That macrophages in themselves may produce antibodies has been shown by some (Holub *et al.*, 1964), and that they contain ribonucleic acid relevant to the formation of immunoglobulins has been demonstrated by others (Adler *et al.*, 1966).

Our work has led us to conclude that antibody formation by lymph node fragments from normal animals can be initiated *in vitro* by ribonucleic acid (RNA) fractions from macrophages which have been incubated with the antigen in question, but neither by the antigen alone nor by a mixture of antigen and RNA from macrophages that had not been exposed to the antigen. At least two fractions of RNA have been implicated, one free of demonstrable antigen as the mediator of 19S antibody formation, the other complexed with antigen as the agent responsible for 7S antibody synthesis. In the present paper the evidence for this concept is briefly reviewed and data bearing upon the cellular origins of the active RNA fractions and the antibodies are presented.

Early indications for the dual nature of RNA from specifically stimulated macrophages came from the finding that RNA from macrophages which had been exposed to T2 phage doses ranging from 10^6 to 10^{12} plaque-forming units (PFU) per 10^9 cells evoked similar 19S antibody responses whereas 7S antibody was formed only in response to RNA from macrophages which had been incubated with at least 10^{11} PFU (Fishman and Adler, 1967). It has also been possible to show that the 19S antibody formation is not greatly enhanced by an increase in the amount of RNA that is added to the lymph node fragments. The pertinent data, shown in Table 1, indicate that the use of 2 mg of RNA resulted in a poor response, probably because of toxicity of the RNA or traces of phenol. Though similar studies on the 7S response have yet to be made, the present data suggest that 19S antibody formation is relatively dose-independent, both in its macrophage phase and in its productive stage in the lymph node cells.

One of several possible explanations for the greater antigen requirement of the 7S antibody response, which appeared to parallel *in vivo* findings, was the assumption that this response depends on the presence of antigen in the RNA, possibly as a complex with RNA. Indirect support for this concept came from the finding that this RNA was more resistant to inactivation by ribonuclease than the RNA responsible for the 19S response (Fishman *et al.,* 1964). These observations led to a series of experiments in which answers were sought to the following questions: (1) Are the RNA fractions responsible for 19S and 7S antibody forma-

Table 1

THE IgM ANTIBODY FORMED IN LYMPH NODE FRAGMENTS IN RESPONSE TO VARYING AMOUNTS OF RNA FROM ANTIGENICALLY STIMULATED MACROPHAGES

Amount of RNA Added/Plate	Per Cent Neutralization (5-day culture fluids) 1:2	1:4 Dilutions
2000 μg	9	0
1000 μg	42 *	19
500 μg	36 *	14
300 μg	31 *	9
100 μg	25 *	8
50 μg	0	0

* Sign. 1% level.

Table 2

SUPPRESSION OF THE IMMUNOGENICITY OF RNA FROM MACROPHAGES THAT HAVE BEEN INCUBATED WITH T2 PHAGE

RNA Treated with	Effect on Antibody Response 19S	7S
RNAase (1:100) *	–	+
RNAase (1:15)	–	–
Anti-T2 serum	+	–
Pronase	+	–
Trypsin, chymotrypsin	+	+
DNAase	+	+

* Enzyme-substrate ratio.
+ Response present.
– Response absent.

tion distinct entities? (2) Does the RNA which mediates synthesis of 7S antibody contain antigen and, if so, is the antigen component of the complex relevant and essential to the immunogenic function of the complex? (3) Is the RNA instrumental in 19S antibody formation free of antigen and, if so, how does it function?

With regard to the first question it has been possible to show that the two RNA fractions can be separated from each other chromatographically (Fishman and Adler, 1967). It has also been demonstrated that the presence of Actinomycin D during the incubation of macrophages with antigen selectively inhibits formation of the RNA responsible for 19S antibody production (Fishman *et al.,* 1964). An affirmative answer to both parts of the second question came from several lines of investigation. It could be shown that treatment of the RNA with potent antisera against T2 phage resulted in the precipitation of a small amount (less than 1 per cent) of RNA which, upon re-extraction, evoked only 7S antibody while the RNA that remained in the supernatant gave rise solely to 19S antibody (Fishman and Adler, 1967). It also was found that addition of trace amounts of purified anti-T2 antibody to the RNA specifically quenched the 7S antibody response. Finally, it was observed that treatment of the RNA with pronase yielded a product which evoked only 19S antibody formation. These findings, summarized in Table 2, show that an antigenic product of the phage is present in the RNA fraction responsible for 7S antibody formation and that both this fragment and the RNA are essential for the initiation of the 7S response.

Technical limitations preclude definitive proof of the absence of antigen; thus the answers to the third question must be based on indirect evidence. Among such indirect lines of evidence is our observation that the 19S antibody carries allotypic markers characteristic of the immunoglobulins of the macrophage donor in their light polypeptide chains (Adler *et al.*, 1966) and, as found in preliminary experiments, also in their heavy polypeptide chains. In the light of current concepts this can most readily be explained by attributing informational or messenger activity to the RNA and by assuming that the specific 19S antibody is a product of cells that have received multiple "hits" with such RNA. The size of the RNA in question, and its ready acceptance of tritiated uridine in pulse experiments (Fishman and Adler, 1967) are compatible with this notion.

The data shown in Table 3 introduce a current line of investigation from which we hope to obtain further insight into the problems just discussed. In this experiment two antigens, namely *E. coli* phage T2 and *B. subtilis* phage SP 32, were mixed and incubated with peritoneal exudate cells. The RNA extracted from these cells stimulated specific 19S and 7S antibody formation against each of the two antigens. It will be of interest to see whether individual cells in the lymph node fragments make antibody against both antigens simultaneously either during the 19S or the 7S phases of the response. Also, by using macrophages from animals with different allotypes it will be possible to incubate each antigen with one set of cells, extract and mix the RNA, and study allotypic specificities of the antibodies engendered by these preparations.

Table 3

ANTIBODY FORMATION IN LYMPH NODE FRAGMENTS STIMULATED BY RNA FROM MACROPHAGES WHICH HAD BEEN INCUBATED WITH A MIXTURE OF TWO UNRELATED PHAGES

Days of Culture	Per Cent Neutralization *	
	Anti-T2	Anti-SP †
3	56 *	33 *
6	33 *	0
9	18	20
13	33 *	35 *
16	24 *	48 *

* Sign. 1% level.
† *B. subtilis* phage (SP 32).

Table 4

ABILITY OF DIFFERENT CELL TYPES TO PRODUCE IMMUNOGENIC RNA UPON INCUBATION WITH T2 PHAGE

RNA Source	Early Response (4–6 days)	Late Response (10–14 days)
Peritoneal cells (PE) *	++	++
Monolayer cells (macrophages-PE)	++	++
Polymorphonuclear cells	–	–
Lymph node cells	–	–
PE cells + lymph node cells (1:1)	++	++
Alveolar macrophages	–	–

* Based on the addition of 100 μg RNA/plate. The cells were incubated with 100 plaque-forming units of phage/cell.

The data shown in Table 4 summarize our findings from experiments which were performed to determine the cellular source(s) of the RNA involved in 19S and in 7S antibody production. It will be noted that peritoneal exudate cells, or macrophages isolated from such exudates by selection of cells that attach to glass, were the only competent cells. The RNA extracted from such "purified" peritoneal macrophages transferred the allotypic markers, and a deliberate enrichment of the peritoneal exudate with lymphocytes had no detectable effect. It seems unlikely, therefore, that the active RNA comes from lymphocytic cells in the peritoneal exudate which have become stimulated either directly by antigen or by a product of the reaction of macrophages with the antigen. Additional support for the view that the macrophage is the source of the active RNA came from experiments in which peritoneal exudate cells were treated with tridymite (crystalline silica) either before or immediately after their incubation with antigen. Only in the latter case was an active RNA obtained. If tridymite indeed acted specifically on the macrophages (Pernis and Paronetto, 1962) this cell once again appears to be the most likely source of the active RNA.

Our studies on the nature and number of cells in the lymph node fragments that produce the two classes of antibodies are still in a preliminary stage. Histological studies of frozen sections from lymph node fragments, employing methyl green-thionin and periodic acid-Schiff staining techniques, show that most of the lymphocytes disappear during the first days of culture. A proliferation of basophilic reticulum cells follows, and a zone of large macrophages containing lymphocytic debris is seen on the periphery. Growing out from the fragment are fibroblastic strands with occasional lymphocytes attached to the fibroblastic cells. Starting on the third day of culture newly formed centers of blast cells and activated lymphocytes appear, and these persist for about 8 days. These centers are clearly set off from the surrounding connective tissue which turns necrotic after 9–12 days of culture. At that time a loosely organized lymphatic tissue forms in the outgrowth from the fragment but it contains neither centers of proliferation nor activated lymphocytes. Rare and isolated plasma cells are found in the strands of reticulum cells.

Preliminary results of attempts to identify antibody-producing cells by the fluorescent antibody technic show that such cells can be detected from the 2nd day of culture. Fluorescence is consistently confined to isolated cells and these cells are usually of histiocytic or activated reticulum cell morphology and occasionally are activated lymphocytes or blast cells. Thus far we have not been able to detect a consistent difference in the type of cells making 19S antibody early during the culture and those making 7S antibody on day 13. These studies are being continued.

Of interest in the light of the histological background just described are the results of experiments shown in Table 5. Here lymph node fragments were placed into culture and RNA was added after varying periods of incubation. It will be noted that the 19S response could be elicited in cultures which had been pre-incubated for up to 7 days, i.e., 19S antibody was found in such cultures on day 12. It is apparent, therefore, that the rise of 19S antibody to a peak on day 5, and its subsequent disappearance in cultures to which the RNA was added immediately, cannot be attributed to the disappearance of a responsive cell type.

At the present stage of our studies we feel that the competent perito-

Table 5

The Synthesis of Antibody in Lymph Node Fragments in Culture for Varying Periods Prior to Addition of the Immunogenic RNA

Day RNA Added to Cultures	Early Response (IgM) (5 days)	Late Response (IgG) (12 days)
0	++	++
2	++	+++
5	+++	–
7	++	–
9	–	–
12	–	–

neal macrophage plays a unique role in the 19S antibody formation. We think that this cell has the potential for immunoglobulin and thus for antibody production which, however, is not fully expressed. Whether the proposed block exists at the level of polypeptide chain assembly or antibody secretion is of course not known. It may be supposed that such cells are uniquely adapted toward recognition of specific antigens and that a result of their exposure to antigen would be the induction of increased messenger RNA synthesis. Transfer of such RNA to other cells would then account for the synthesis of immunoglobulin with the allotypic marker and antibody specificities specified by the RNA. We cannot be certain whether the RNA that forms the complex with antigenic fragments and elicits 7S antibody comes from such macrophages. This pre-formed RNA could well be from other macrophages with a less developed potential for immunoglobulin synthesis, or it could possibly be ribosomal RNA from the cells that yield the messenger RNA in which bound polypeptide chains form the link between antigenic fragment and RNA.

References

Adler, F. L., M. Fishman, and S. Dray (1966). Antibody formation initiated *in vitro.* III. Antibody formation and allotypic specificity directed by ribonucleic acid from peritoneal exudate cells. J. Immun., *97*:554–558.

Askonas, B.A., and J. M. Rhodes (1965). Immunogenicity of antigen-containing ribonucleic acid preparations from macrophages. Nature, *205*:470–474.

Ehrich, W. E. (1956) in Handbuch der allgemeinen Pathologie. F. Buchner, E. Letterer, and F. Roulet, Editors, Vol. 7, Part 1, page 137. Springer, Berlin.

Fishman, M. (1959). Antibody formation in tissue culture. Nature, *183*:1200–1201.

Fishman, M., J. J. Van Rood, and F. L. Adler (1964). The initiation of antibody formation by ribonucleic acid from specifically stimulated macrophages. Molecular and cellular basis of antibody formation. Proc. of a Symposium, Prague, pp. 491–501.

Fishman, M., and F. L. Adler (1967). Macrophage RNA and antibody synthesis. Immunity, Cancer, Chemotherapy, Vol. 1, Academic Press, pp. 177–185.

——— (1967). The role of macrophage-RNA in the immune response. XXXII Cold Spring Harbor Symposium on Quan. Bio. "Antibodies," *32*:343–348.

Friedman, H. P., A. B. Stavitsky, and J. M. Solomon (1965). Induction *in vitro* of antibodies to phage T2: Antigens in the RNA extract employed. Science, *149*:1106–1107.

Gallily, R., and M. Feldman (1967). The role of macrophage in the induction of antibody in X-irradiated animals. Immun., *12*:197–206.

Gottlieb, A. A., V. R. Glisin, and P. Doty (1967). Studies on macrophage RNA involved in antibody production. Proc. Natl. Acad. Sci., *57*:1849–1856.

Holub, M., I. Riha, and V. Kamarytova (1964). Immunological competence of different stages of the lymphoid cell. Molecular and cellular basis of antibody formation. Proc. of a Symposium, Prague, pp. 447–458.

Jacherts, D., and H. Noltenius (1966). Antikorpe-Synthese *in vitro*. V. Die Informationsübertragung von der antigen-verarbeitenden Zelle auf die antikörperproduzierende Zelle. Z. med. Mikrobiol. u. Immunol., *152*:112–133.

Mitchison, N. A. (1965). Quantitative aspects of immunological tolerance. Int. Arch. Allergy, *27*:364.

Nossal, G. J. V. (1965). Self-recognition. Annals of N. Y. Acad. Sci. Autoimmunity-experimental and clinical aspects, *124*:37–49.

Pernis, B., and F. Paronetto (1962) . Adjuvant effect of silica (tridymite) on antibody production. Proc. Soc. Exp. Biol. Med., *110*:390–392.

Pribnow, J. F., and M. S. Silverman (1967). Studies on the radiosensitive phase of the primary antibody response in rabbits. I. The role of the macrophage. J. Immun., *98*:225–229.

Spiers, R. S., and E. E. Spiers (1964). Cellular reactions to reinjection of antigen. J. Immun., *92*:540–549.

Thiery, J. P. (1959). Microcinematographic contributions to the study of plasma cells. Ciba Foundation Symposium on Cellular Aspects of Immunity, p. 59. (G. E. W. Wolstenholme and M. O'Connor, ed.)

ROLES OF DNA AND RNA IN THE ANTIBODY RESPONSE *

ABRAM B. STAVITSKY

Department of Microbiology, Case Western Reserve University School of Medicine, Cleveland, Ohio

It now is evident that DNA and RNA play major roles in the antibody response. This paper presents studies from this laboratory which aim at a better definition of these roles. Already published experiments are summarized only briefly, while current work is treated more completely. Finally, on the basis of some of the evidence available from this and other laboratories, a broad picture of the antibody response is drawn to further discussion, speculation, and experimentation.

I. EXPERIMENTAL

A. ANTIGEN PROCESSING AND PRODUCTION OF IMMUNOGENIC RNA-ANTIGEN COMPLEXES (GUSDON AND STAVITSKY, 1967)

Recent studies utilizing an *in vitro* model system (Fishman, 1961; Adler, Fishman and Dray, 1966), suggested that RNA extracted from peritoneal cells exposed to phage contains information for the synthesis of 19S antibody to the phage. In other experiments (Askonas and Rhodes, 1965; Friedman *et al.*, 1965) antigenic fragments were detected in the RNA derived from peritoneal cells exposed to antigens. It was, therefore, proposed that the RNA induces antibody synthesis because of its antigen content; that the RNA serves only as carrier or adjuvant which promotes induction. More recently, Gottlieb *et al.* (1967) have adduced evidence of another kind to support the ancillary role of the RNA in the induction of antibody synthesis.

* This investigation was supported by U. S. Public Health Service grant AI-01865. The experiments done in collaboration with Drs. Moore and Schoenberg were supported in part by U. S. Public Health Service grant AM-07161.

The experiments to be discussed employed still another approach to the question of the role of RNA in the induction of antibody synthesis. RNA obtained from lymph node cells of rabbits which were synthesizing γG antibody (Bauer *et al.,* 1963) was injected repeatedly into nonimmunized rabbits. The capacity of the recipients to synthesize antibody to the antigen with which the donors of the RNA was injected, as well as the presence of immunological memory, were determined. The results, summarized in Table 1, indicate that injection of RNA from the cells of 7S antibody-producing rabbits usually induced the appearance of low levels of 19S hemagglutinating antibody, occasionally of low titers of 7S hemagglutinating antibody, and, less frequently, immunological memory for 7S antibody production. It was found by microcomplement fixation (Wasserman and Levine, 1961) that each RNA preparation, whether further fractionated on methylated albumin columns or not, contained hemocyanin or bovine γ globulin antigen.

This experiment was open to the criticism that not only was RNA utilized but two other adjuvants, methylated bovine albumin and precitated alum. Moreover, it was conceivable that the antibody which appeared in the recipients was not synthesized but was associated with the RNA and subsequently released. The experiment summarized in Table 2 was, therefore, done. RNA was isolated from the spleens of hemocyanin-immunized rabbits. This RNA was injected without further adjuvant into the left popliteal lymph node of a rabbit. One week later the rabbit was bled out and its two popliteal lymph nodes and spleen cultured as cell suspensions in C^{14} glycine. *De novo* synthesis of antibody was assayed by the incorporation of radioactive glycine into antibody (Stavitsky and Gusdon, 1966). The results indicate that *de novo* synthesis of antibody had occurred.

B. Early Antibody Response (Roszman and Stavitsky, 1967)

1. *Antibody Synthesis*

Ideal systems for study of the mechanisms of the antibody response should exhibit early synthesis of antibody so that analyses of the preceding antigen-processing and inductive stages could be confined to a few hours or days. Since several studies indicated that the injection of key hole limpet hemocyanin induces the early appearance of antibody in the serum of rabbits, experiments were done to determine how soon after the injection of this antigen antibody synthesis could be demonstrated in *in vitro* cultures of lymph node cells. Both intracellular and extracellular antibody were assayed in some experiments.

Table 3 presents the results of a number of these experiments. It was found that lymph node cells removed from rabbits 1 day after the inoculation of antigen synthesized antibody (5250). Initial experiments with

Table 1

ANTIBODY RESPONSE OF RABBITS TO THE INJECTION OF RNA-ANTIGEN COMPLEXES AND TO SUBSEQUENT INJECTION OF ANTIGEN

Rabbit No.	Response to Primary Injection			Response to Second Injection		
	RNA Inoculum	Titer Day 7	Antibody Class	Antigen Inoculum	Titer Day 5	Titer Day 10
1	500 μg unfr. KLH	0	–	KLH	160	*2560* *
2	1500 μg unfr. KLH	80	19S	KLH	160	*2560*
3	500 μg unfr. KLH	80	19S, 7S	KLH	*2560*	5120
4	500 μg KLH-I	80	19S	-(dead)		
5	500 μg KLH-II	160	19S	KLH	160	*1280*
6	265 μg KLH-III	80	19S	KLH	160	*2560*
7	500 μg unfr. BGG	40	19S, 7S	BGG	*2560*	5120
8	1500 μg unfr. BGG	0		(dead)		
9	500 μg unfr. BGG	0		BGG	160	*2560*
10	500 μg BGG-I	0		-(dead)		
11	500 μg BGG-II	0		-(dead)		
12	500 μg BGG-III	0		KLH	160	*2560*
13	None	0		KLH	160	*2560*

* Italicized titers indicate first appearance of 7S antibody.

Rabbits weighing 3 kilograms were injected into each hind foot pad with 5 mg of alum-precipitated bovine γ globulin or 2 mg alum-precipitated key hole limpet hemocyanin. At least 4 injections were given with an interval of 21–50 days between injections. Three days after the last injection of antigen, the popliteal lymph nodes were removed and placed in Eagle's medium. Cells were teased out of the tissue and free of aggregates and debris by passage through gauze, centrifuged and washed in Warner's buffer A (Warner, Rich and Hall, 1962) at 4°C. The packed cells from 6 lymph nodes were then extracted with phenol by a method already described (Hiatt, 1962). The RNA finally prepared was dissolved in buffer (sodium acetate, 0.01 M, pH 5.0; lithium chloride, 0.14 M) for further treatment. This RNA was further fractionated on a methylated albumin column (Comb, Brown and Katz, 1964). Three separate peaks were eluted, the first presumably sRNA (I), the second 16–18S RNA (II) and the third 25–29S RNA (III) (Phillipson, 1961). Each peak was desalted by passage through P2 Bio-Gel (Bio-Rad Laboratories, Richmond, California) columns and the effluent peaks were concentrated by lyophilization.

The RNA fractions were complexed with methylated albumin for injection (Plescia, Braun and Palczuk, 1965). The RNA-albumin complex was alum-precipitated and injected into the hind foot pads of rabbits on days 1, 43 and 64. Eight months after the initial series of injections, at which time no hemagglutinating antibody was detectable, the rabbits were injected into the hind foot pads with 5 mg alum-precipitated bovine γ globulin or 2 mg alum-precipitated hemocyanin and then bled daily for 12 days. Antibody was assayed by the passive hemagglutination method (Stavitsky, 1954). The class of antibody was determined by assaying hemagglutinating antibody activity of fractions obtained by sucrose density gradient ultracentrifugation (Edelman, Kunkel and Franklin, 1958). KLH–key hole limpet hemocyanin; unfr.-RNA not fractionated on methylated albumin column; I, II, III–fractions I, II and III eluted from methylated albumin column, I being eluted first, etc. BGG–bovine γ globulin.

Table 2

De Novo ANTIBODY SYNTHESIS FOLLOWING THE INTRANODAL INJECTION OF RNA FROM SPLEENS OF HEMOCYANIN-IMMUNIZED RABBITS

Tissue Cultured	CPM in Specific Antibody
Left node	234
Right node	146
Spleen	294

Immunization of rabbits with key hole limpet hemocyanin and preparation of RNA were carried out (as described in legend to Table 1) from spleens of hemocyanin-immunized rabbits. One hundred and fifty micrograms of this RNA was injected into the left popliteal lymph node of an anesthetized rabbit which 4 days before had been injected with 0.5 ml of a mixture of equal volumes of sheep blood and Alsever's solution. The blood was injected to cause hypertrophy of the homolateral lymph node so that it could readily be grasped manually, exteriorized through the incision and held for the injection of RNA. Seven days later the rabbit was bled out and the left and right popliteal lymph nodes and spleen removed. Cells were teased out of these organs into Eagle's medium and 2×10^7 cells per ml of medium incubated for 24 hours in the presence of 1 uC Glycine-1-C^{14} (15 uC/uM). Determination of radioactive antibody to hemocyanin was then made by the co-precipitation method (Stavitsky and Gusdon, 1966). To each sample of culture medium 0.2 ml of rabbit anti-egg albumin and an equivalent amount of egg albumin were added, incubated at 37°C for 1 hour and then overnight at 4°C. The precipitates which formed were washed with saline and plated onto a Millipore (HA) filter, dried and counted in a low background gas flow counter. It was usually necessary to do two or more non-specific co-precipitations to bring the non-specific counts to a low level. Then one specific co-precipitation was done with rabbit anti-key hole limpet hemocyanin and an equivalent amount of hemocyanin. This precipitate was prepared and counted as above. Only the specific CPM are shown as the non-specific CPM were at background (5 CPM) in this experiment.

several lymph nodes from unimmunized rabbits failed to disclose the production of hemagglutinating antibody (5249, 5257) or the incorporation of C^{14} amino acids into antibody (5261). However, subsequent attempts employing hemagglutination and/or the incorporation assay revealed that lymph node cells from unimmunized rabbits (5249, 5257, 5324–5327) may produce antibody to hemocyanin. In two instances (5257, 5261) the injection of antigen increased the production of antibody compared to the uninjected node from the same animal.

Although the amounts of antibody produced were small, occasionally it was possible to determine the class of antibody produced. As shown

Lymph Node Cells from Normal and Hemocyanin-stimulated Rabbits

Rabbit No.	Cell Source	Immunization	CPM-Incorporation into: Total Protein	Non-specific Total	Non-specific Final	Antibody Protein	HA Titer	Class of Antibody	Class of Protein
5249	LLN + RLN	none			101	560	0	γM	
5257	LLN	none			74	159	0		γM
	RLN	1 day			98	265	0	γM	γM
5261	LLN	none	19,275	3,870	75	59	0		
	RLN	1 day	43,390	17,984	52	165	20		γM
5270	LLN	1 day	26,122	17,683	119	229	20	γM, γG	
	RLN	2 days	14,422	5,070	87	294	80	γM, γG	
5250	LLN + RLN	1 day			126	448		γM, γG	
5324	LLN	none					40		
	RLN	1 day					20		
5325	LLN	none					20		
5326	LLN	none					20		
5327	LLN	none					20		

One times 10^7 lymph node cells from unimmunized rabbits or from rabbits immunized 1 or 2 days previously in the hind foot pads with 1.5–2.0 mg alum-precipitated key hole limpet hemocyanin were incubated at 37°C for 19 hours in Eagle's medium containing 1–2 μc glycine-1-C^{14} (15 μc/μ M) or 2.5 μc mixed amino acids (UL-C^{14}-Volk Radiochemical Co., catalogue no. 1709). At the end of the incubation, the medium was separated by centrifugation. In some experiments an aliquot was subjected to precipitation with an equal volume of 10% trichloracetic acid, and the precipitates collected on an HA Millipore filter for drying and then counting in a Nuclear Chicago low background flow counter. In other experiments the medium was utilized for non-specific and specific co-precipitation as described in the footnote to Table 2.

LLN-left popliteal lymph node; RLN-right popliteal lymph node; Total protein-precipitated by TCA; Non-specific, total-sum of CPM of all egg albumin-anti-egg albumin precipitates; Final-last egg albumin-anti-egg albumin precipitate; Antibody protein-CPM in hemocyanin-anti-hemocyanin precipitates; Class of antibody-results of hemagglutination or incorporation assays of antibody in fraction obtained by zone electrophoresis (Müller-Eberhard, 1960) and/or by sucrose density gradient ultracentrifugation (Edelman *et al.*, 1958). Class of protein refers to determinations of radioactivity of TCA-precipitates made from zone electrophoretic and/or density-gradient-prepared fractions.

in Table 3, only γM antibody was detected in cultures from rabbits 5249 and 5257, whereas both γM and γG antibodies were observed in cultures from 5270 and 5250.

2. *Protein Synthesis*

In some experiments the capacity of lymph node cells to synthesize protein (TCA-precipitable radioactivity) and protein which was co-precipitated non-specifically by egg albumin-anti-egg albumin precipitates (non-specific radioactivity) as well as the class(es) of protein synthesized were determined. Table 3 also summarizes these data. Rabbit 5261 is typical of the findings to date. The left lymph node cells from the unimmunized animal incorporated 19,275 counts per minute into TCA-precipitable material, presumably protein, and about 20% as many counts into protein that is non-specifically co-precipitable. The right lymph node cells taken from this animal 1 day after immunization incorporated 43,000 counts into protein and almost 42% of these counts were non-specifically co-precipitable. A much smaller number of counts were incorporated into antibody. Further analysis revealed that much of the protein synthesized and secreted by the unimmunized and immunized lymph node cells was γM globulin and that the immunized cells incorporated 2–3 times as many counts into this fraction as the unimmunized cells.

3. *Effect of Inhibitors Upon Nucleic Acid, Protein and Antibody Synthesis*

Previous experiments in this and other laboratories (summarized in Stavitsky and Gusdon, 1966) indicated that, once induced, antibody synthesis was somewhat resistant to inhibition by Actinomycin D. These observations suggested that this synthesis was dependent upon stable messenger RNA. Other observations (summarized in Stavitsky and Gusdon, 1966) suggested that Actinomycin did inhibit antibody synthesis when injected into animals before antigen was given. It was, therefore, of interest to determine whether the early antibody response was inhibited by Actinomycin *in vitro*. Because of the continuing argument about whether DNA synthesis is required for antibody synthesis (Dutton and Mishell, 1967), the relation between DNA synthesis and the early antibody response was explored in experiments in which DNA synthesis was inhibited by 5-bromodeoxyuridine. Table 4 summarizes the results of a number of experiments. The following observations were made:

1. Although the early antibody response occasionally was inhibited completely (5270–1 day), more commonly this response was inhibited only partially (5270–2 days; 5291) or even stimulated (5283; 5292) by Actinomycin.

2. The synthesis of non-antibody protein usually was much more susceptible to inhibition than the synthesis of antibody (5270–2 days;

Table 4. Effects of Actinomycin and 5-Bromodeoxyuridine on Nucleic Acid, Protein and Antibody Synthesis During Early Primary Response

Rabbit No.	Immunization	Source of Cells	Time of Incub.	Inhibitor-	μg	CPM-Incorporation into DNA	RNA	Protein	Non-specif	Antibody
5270	1 day	LLN	19 hr	0				26,122	119	229
		LLN	19 hr	Actino	0.5			530	32	33
	2 days	RLN	19 hr	0				14,422	69	294
			19 hr	Actino	0.5			6,262	87	108
5282	1 day	Spleen	19 hr	0				80,696	166	344
			19 hr	Actino	5.0			596	38	129
5283	1 day	Spleen	19 hr	0				200,800	161	299
			19 hr	Actino	5.0			66,888	175	348
5291	1 day	LLN + RLN	19 hr	0				14,127	52	264
			19 hr	Actino	5.0			12,285	119	134
5292	2 days	LLN + RLN	19 hr	0				18,488	102	215
			19 hr	Actino	5.0			14,004	84	292
5299	1 day	LLN + RLN	8 hr	0		450	354	1,668	36	59
				Actino	5.0	140	22	758	25	69
				BRUD-	200.0	109		658	31	96
			23 hr	0		6700	2425	57,764	123	317
				Actino	5.0	142	145	113,776	19	67
				BRUD-	200.0	160		52,148	26	83

These experiments were carried exactly like those shown in Table 3, with the exception that 0.1 μc uridine-2-C^{14} (specific activity 0.2 $\mu c/\mu M$) was added to some tubes as a precursor of RNA and 1 μc thymidine-2-C^{14} (specific activity 1.0 $\mu c/\mu M$) to other tubes as a precursor to DNA. Actinomycin or 5-bromodeoxyuridine were added to some tubes as indicated. DNA was determined by the application of Kahan's method (Kahan, 1960) to the well-washed cells, and RNA by the extraction of the cells by the method of Hiatt, 1962. The CPM in the nucleic acid precipitates were determined following plating on Millipore HA, drying and counting in a Nuclear-Chicago low background flow counter.

5282; 5283), although this difference was not always striking (5291 and 5292).

3. When cells from rabbits immunized 1 day before were incubated for 8 hours *in vitro* with sufficient Actinomycin to inhibit DNA synthesis partially and RNA synthesis completely, antibody synthesis was unaffected (5299). When incubation was continued for another 15 hours, DNA synthesis was still inhibited, a slight amount of RNA synthesis occurred, but antibody synthesis was inhibited completely.

4. When the cells from rabbits immunized 1 day before were incubated for 8 hours with enough 5-bromodeoxyuridine to inhibit DNA synthesis substantially, antibody synthesis was not inhibited and was, in fact, stimulated. Upon incubation for another 15 hours, some DNA synthesis occurred, but antibody synthesis was completely inhibited.

4. *Polyribosomes in Lymph Node Cells of Unimmunized and of Immunized Rabbits (Folds and Stavitsky, 1967)*

Since the experiments just presented indicated that lymph node cells from both unimmunized and immunized rabbits synthesized non-antibody and antibody proteins, it was of interest to determine whether immunization affected the polyribosome content of lymph node cells. Experiments such as that illustrated in Figure 1 indicated that 1 day after the injection of hemocyanin there was an increase in the number of heavy and light polyribosomes in the cells from the draining lymph nodes.

C. Secondary Antibody Response (Gusdon, Stavitsky, Schoenberg and Moore, 1967)

1. *Duration of Antibody Synthesis*

Previous studies (summarized in Stavitsky, 1961) demonstrated that the secondary antibody response may be short-lived. We have reexamined this question with the rabbit anti-hemocyanin system, especially with a view to explaining the short duration of antibody synthesis. Table 5 presents typical results. The general trend can best be gauged by comparing the ability to incorporate the label by cells from left and right lymph node removed from individual rabbits at different times after anamnestic antigenic stimulation. Compare the lymph node of rabbit 5048 removed before challenge and that removed on day 3; the nodes of 5044 removed on days 2 and 4; the nodes of 5043 removed on days 3 and 4; the nodes of 5037 removed on days 4 and 5; the nodes of 5038 removed on days 4 and 6. Newly labeled, nascent antibody was first detected on day 2. Antibody synthesis increased on day 3, was maximal on day 4 and dropped thereafter. Immunofluorescence studies gave corresponding results: immunofluorescence per cell for specific intracellular

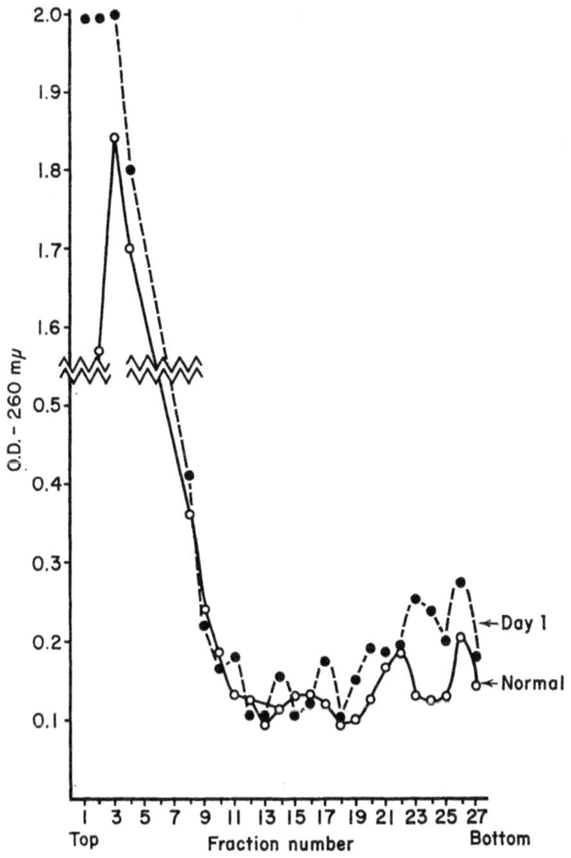

Fig. 1. Polyribosome profiles obtained from popliteal lymph node cells before immunization and one day after immunization with hemocyanin. The two left popliteal lymph nodes of two unimmunized rabbits were removed under anesthesia. 1.0 mg of alum-precipitated key hole limpet hemocyanin was then injected into the right hind foot pads of each of these rabbits and the right popliteal lymph nodes removed 1 day later. The methods for preparing the polyribosomes were essentially those of Becker and Rich (1966). The nodes were removed into sterile ice-cold Eagle's medium and washed once in this medium. The tissue was minced with scissors in this medium and the fragments were teased with wire gauze. The expressed cells were filtered through sterile gauze, washed once in ice-cold medium, resuspended in 5 ml medium and counted. One times 10^8 cells were employed for each preparation. The cells were lysed in cold hypotonic buffer (pH 7.4, containing 0.01 *M* Tris-HCl, 0.015 *M* $MgCl_2$, 0.01 *M* KCl) in the presence of 0.5% sodium desoxycholate and 50 μg deoxyribonuclease (Worthington Biochemical Co., Freehold, New Jersey). 3.5 ml of this suspension was layered onto a 5–30% sucrose density gradient and centrifuged for 2 hours at 22,000 rpm in a Spinco Model L ultracentrifuge. Twenty-seven 1 ml fractions were collected using an Isco gradient fractionator (model 180). The OD at 260 mμ of each fraction was determined with a Zeiss spectrophotometer. Fraction 1 represents the top and fraction 27 the bottom of the gradient.

Table 5. Incorporation of C^{14} Amino Acids into Antibody at Various Times During Anamnestic Response

Number of Rabbit	Source of Cells	Time of Removal of Node–Days	Counts Per Minute	
			Non-specific	Specific
5048	left node	0	102	120
	right node	3	172	613
5044	left node	2	205	291
	right node	4	367	1236
5043	left node	3	178	615
	right node	4	171	645
5037	left node	4	120	516
	right node	5	75	241
5038	left node	4	116	738
	right node	6	110	322

Each rabbit was injected into the hind foot pads 3 times with 2 mg alum-precipitated key hole limpet hemocyanin at one month intervals and then rested for 8–10 weeks. Each was then injected with 2 mg alum-precipitated KLH into the hind foot pad. The left and right popliteal lymph nodes were removed on the days after challenge indicated. The nodes were teased to yield individual cells. Two times 10^7 cells in Eagle's medium were incubated on a roller drum for 30 minutes at 37°C. Ten microcuries of purified C^{14} labelled protein hydrolysate (New England Nuclear Corp., Boston, Mass.) were added to each tube for 2–3 minutes. Incorporation was stopped by plunging the tube into ice water and adding 5 ml of cold buffer. The cells were recovered by centrifugation and washed with 5 ml of buffer and centrifuged. One ml of post-nuclear fraction from Novikoff cells was added to the cell pellet to insure the stability of the polyribosomes (Scharff and Uhr, 1965). One ml of hypotonic buffer, pH 7.4, containing 0.01 M Tris-HCl, 0.015 M $MgCl_2$, 0.01 M KCl and 0.1 ml of a 10% solution of sodium deoxycholate was also added. When lysis of the cells was complete, the contents of several tubes (4–5 ml) from the same lymph node were layered on a 25 ml cold 5 to 30% linear sucrose gradient (Britten and Roberts, 1960). After centrifugation for 2 hours at 25,000 rpm in an SW 25 rotor in a Spinco Model L ultracentrifuge, 30–33 fractions were collected from the bottom of the tube (Martin and Ames, 1962), and the optical density of each fraction determined at 260 mμ. The polyribosomal, ribosomal and post-ribosomal fractions were pooled separately, dialyzed against 0.15 M NaCl overnight at 4°C, and then concentrated by ultrafiltration to a volume of approximately 4 ml.

The determination of incorporation of C^{14} amino acids into anti-KLH antibody was made by the co-precipitation method as described in the legend to Table 2.

EA-anti-EA non-specific precipitate; specific = non-specific = last/KLH-anti-KLH precipitate. These counts represent the sum of determinations on the polyribosomal, ribosomal and post-ribosomal fractions. For example, for Rabbit 5044, right lymph mode removed on day 4, the data in counts/minute were as follows: *Polyribosomes:* non-specific–155, specific–418; *Ribosomes:* non-specific–48, specific–114; *Postribosomal fraction:* non-specific–165, specific–704.

anti-hemocyanin was most intense on days 3 and 4 and present in the largest number of cells on day 4 and decreased subsequently. These observations paralleled the increment in plasma cells and the capacity of the cells to incorporate the C^{14} amino acids during the pulse period.

2. *Morphologic Observations*

Although there was a rapid development and growth of germinal centers and an increase in number and size of lymphocytes in the cortex and corticomedullary junction, most of the plasma cells containing antibody were in the medulla. Therefore, the electron microscopic studies were limited to the cells in this site.

The nodes from rabbits that received the three monthly injections of antigen, but not the anamnestic stimulus, had a larger number of plasma cells in the medullary areas of the nodes than had those from nonimmunized animals, but few compared to those from animals that were stimulated. In the absence of anamnestic stimulation, the majority of the cells in the medullary region were lymphocytes with a small amount of cytoplasm, few mitochondria, scattered single or aggregated ribosomes and some short segments of endoplasmic reticulum.

Each lymph node from a challenged animal showed variability in the stages of differentiation of the plasma cells. However, a general pattern of development could be determined. The changes described illustrate a sequence of development of individual cells and should not be construed to indicate that the cells required 4 days to mature into plasma cells. The prominence of one cytoplasmic pattern at each time reflects the larger number of cells differentiating into plasma cells after each mitotic cycle.

Within a few hours after challenge with antigen many lymphocytes in the medulla were increased in size. There was an increase in cytoplasm and ribosomes were more numerous. Some mitochondria were enlarged. The nuclei gradually enlarged. After this mitochondria were more common and endoplasmic reticulum with associated polyribosomes more extensive in many lymphocytic cells (Fig. 2). The nuclei attained their greatest size during this stage of development. Mitoses were common. Soon the channels between the lamellae were filled and distended with slightly electron-dense material (Fig. 3).

After reaching this stage there was a release of material from the cisternae of the endoplasmic reticulum, causing their collapse and a reduction in cytoplasmic volume (Fig. 4). The cell margins were irregular and possessed delicate cytoplasmic extensions. Some of these extensions were detached from the cell as shed cytoplasm.

The return of the plasma cells toward a lymphocytic form was most apparent at 5 days. At this time most of the plasma cells in the medulla of the nodes had less cytoplasm and the nuclei were small. The cell mar-

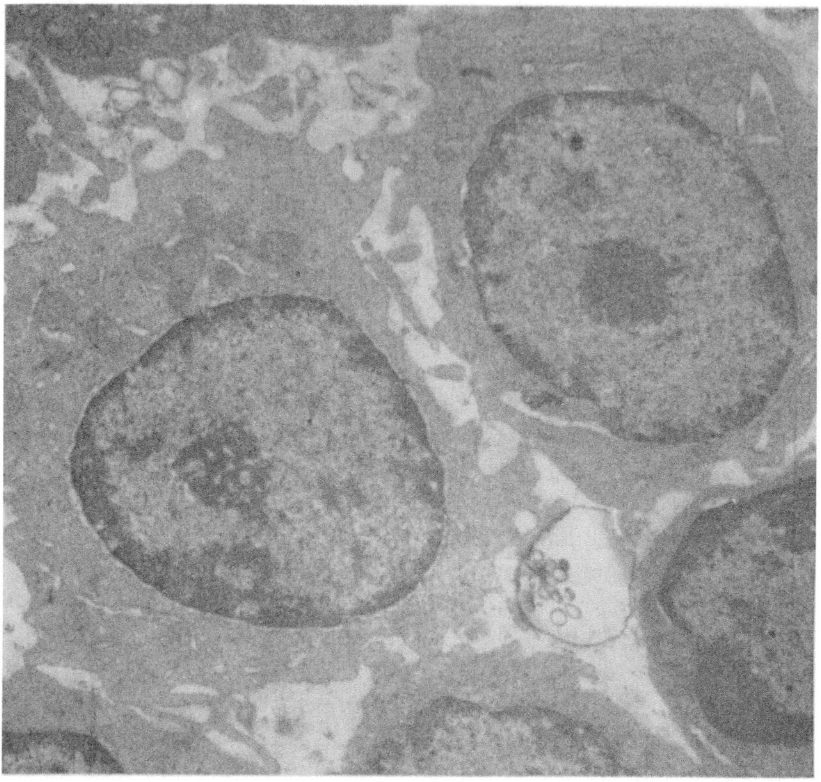

Fig. 2. Stages in maturation process of medullary lymphocytes. The cells have abundant cytoplasm, numerous ribosomes, polyribosomes and a few segments of endoplasmic reticulum with associated ribosomes. ×12,000.

gins had many empty vesicles and fine processes that interdigitated with neighboring cells, creating a sponge-like appearance in the intercellular spaces (Fig. 5). The cytoplasm contained numerous ribosomes and polyribosomes, a few collapsed profiles of endoplasmic reticulum and fewer mitochondria. By six days the cells more closely resembled lymphocytes, though their margins were still somewhat irregular and their ribosomes and polyribosomes more numerous than in the original lymphocyte population (Fig. 6). By light microscopy these cells could not be distinguished from small and medium lymphocytes. Only rarely did the cells show changes suggestive of death.

II. DISCUSSION

Certain features of these experiments should be stressed. First, most of them have been conducted *in vivo,* although the assays of antibody, protein and nucleic acid syntheses and observations of morphologic changes have been performed *in vitro*. Secondly, where possible, antibody synthetic capacity has been assayed by the incorporation of C^{14} amino acids into antibody. This evidence of *de novo* synthesis of a specific protein is more rigorous than the assay of antibody activity—such as hemagglutination or virus neutralization—because antibody activity conceivably may result from the release of antibody from a cell-, RNA- or antigen-bound state. Moreover, sudden changes in synthetic capacity can be registered more promptly by the incorporation assay. Thirdly, all of the experiments involved the foot pad or intravenous injection of 1.0–2.0 mg of alum-precipitated key hole limpet hemocyanin. For this reason, the observations and conclusions may not always apply to other immunogens. Nevertheless, it is felt that these experiments provide much useful information and methodology applicable to other systems. Finally, the data presented are not highly detailed so that the best that can be hoped for is a broad, general picture which hopefully can be used as a basis for more incisive and detailed analyses at many points.

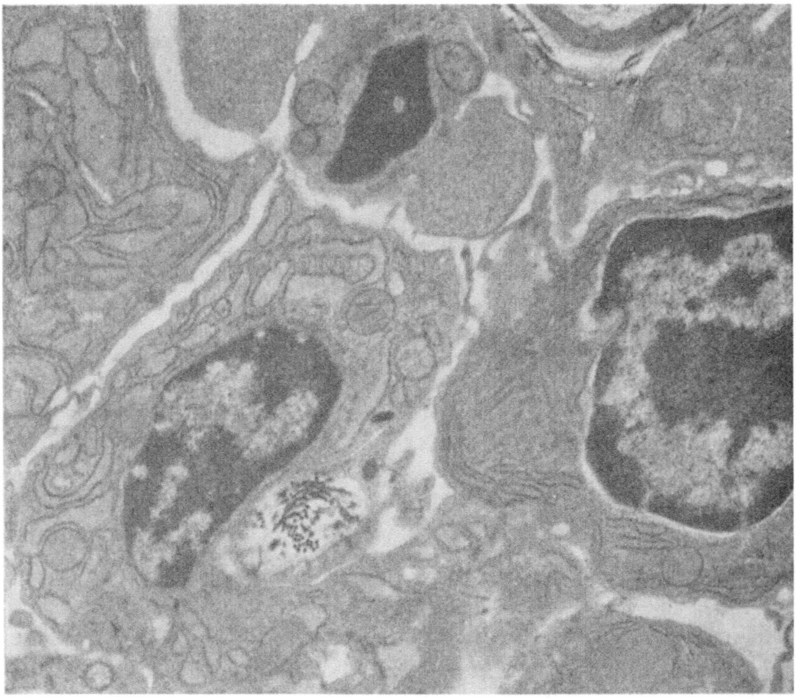

Fig. 3. Several mature plasma cells. The cisternae of these cells are markedly distended with slightly electron-dense material, presumably antibody. ×12,000.

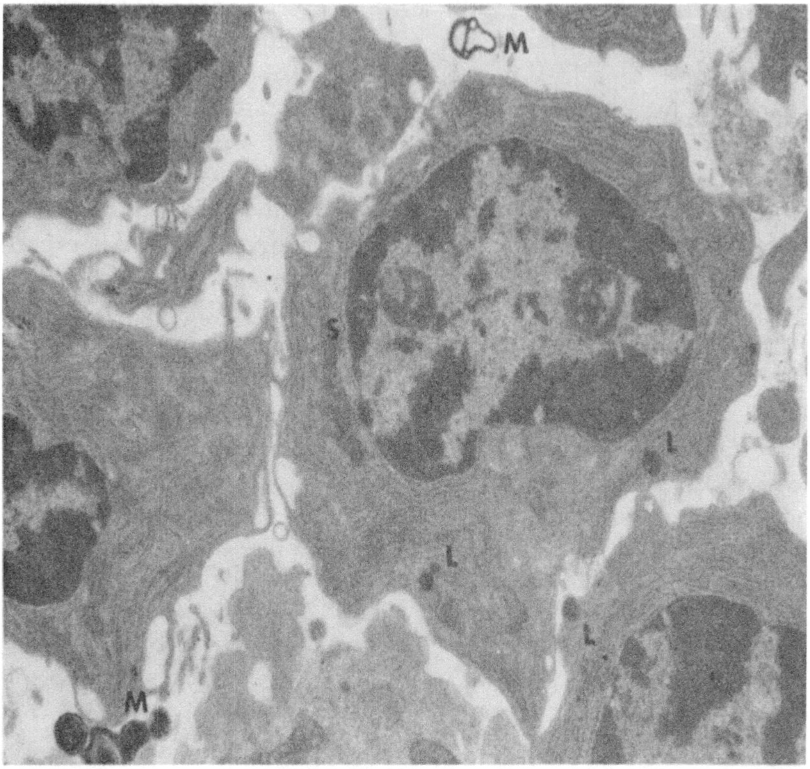

Fig. 4. Plasma cells assuming a lymphocytic form. The material in the cisternae is markedly reduced. The margins of the cells are more irregular with delicate processes. Myelin figures (M) and lysosomal-like bodies (L) are present. One cell has a partially sequestered area (S) of cytoplasm. The latter changes are important in distinguishing the plasma cells returning toward the lymphocytic form from those that are differentiating. ×12,000.

A. Cellular Uptake and Degradation of Antigen in Relation to Induction and Persistence of Antibody Response

The view that the antibody response is initiated by the uptake and degradation of antigen by macrophages now is supported by a rapidly growing body of evidence (Ford *et al.,* 1966; Gallily and Feldman, 1967; Frei *et al.,* 1965). The uptake and degradation of hemocyanin (Askonas and Rhodes, 1965) and T_2 phage (Friedman *et al.,* 1965; Uhr and Weissman, 1965) by macrophages *in vitro* and *in vivo* has been reported. RNA or RNA-antigen complexes which are immunogenic *in vivo* (Askonas and Rhodes, 1965; this paper) and *in vitro* (Fishman, 1961; Friedman *et al.,* 1965; Gottlieb *et al.,* 1967) have been isolated from macrophages exposed to antigens *in vitro.* Immunogenic RNA-antigen complexes have also been isolated from lymph nodes and spleen (this study) and from liver (Garvey and Campbell, 1957). Adler *et al.* (1967) described two types of RNA isolated from peritoneal cells exposed to T_2 phage *in vitro*. One type, which is free of demonstrable antigen and appears to be induced by the phage, induces the formation of γM antibody upon addition to lymph node cells *in vitro*. The second type apparently is preformed in the peritoneal cells, induces the synthesis of γG antibody in

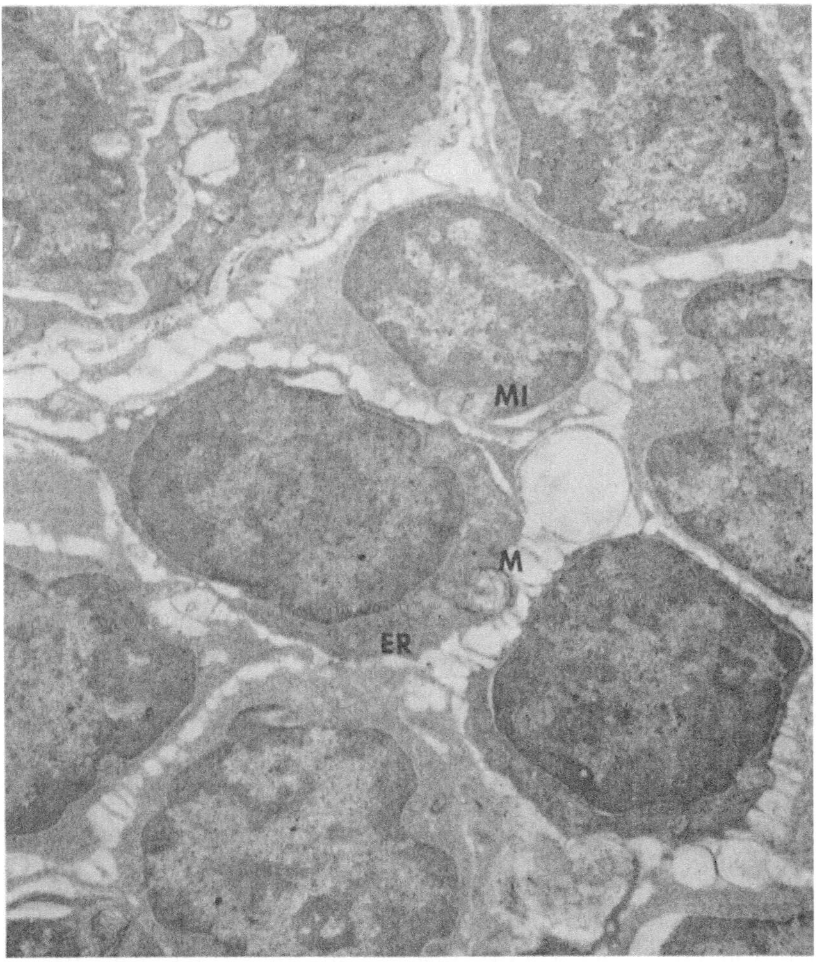

Fig. 5. Cells in medulla on day 5 after anamnestic stimulation. Many cells show sparse cytoplasm. The margins of the cells have many fine processes interdigitating with neighboring cells. Ribosomes are numerous and there are a few segments of collapsed endoplasmic reticulum (ER) and disrupted mitochondria (MI). A few myelin figures (M) are still present. Though the nuclei are smaller, there is nothing to suggest their degeneration. ×12,000.

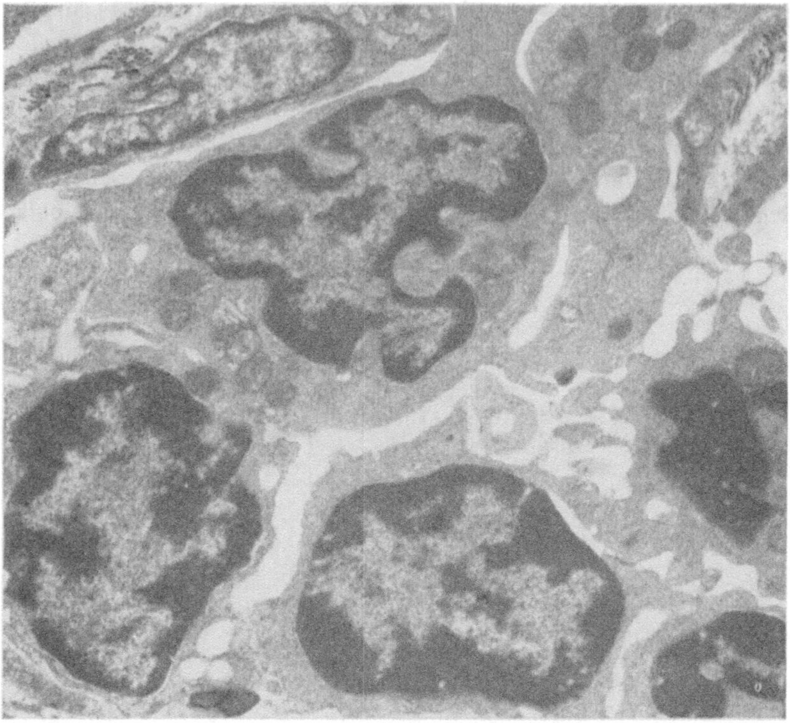

Fig. 6. Return of cells toward lymphocytic form. The cells more closely resemble lymphocytes, but their margins are still irregular and ribosomes more numerous than in unstimulated animals. Compare with Fig. 2. ×12,000.

lymph node cells and, because it is precipitated by anti-T_2 serum, appears to contain T_2 antigen. The second type may well be analogous to the RNA-antigen complexes described by Askonas and Rhodes (1965) and Friedman *et al.* (1965). Adler *et al.* (1967) presented data on the allotypic specificity of the antibodies directed by the RNA to support the conclusion that macrophage-derived RNA contains information for the synthesis of the light chains of the γM antibody which appears in lymph node cells. The experiments of Gottlieb *et al.* (1967) also suggest that pre-formed macrophage RNA combines with antigen to yield a complex which is immunogenic for lymph node cells *in vitro*. Jacherts and Noltenius (1966) reported that *in vitro* stimulation of macrophages by bacterial antigens results in the production of two types of informational RNA. One of them acts as messenger RNA in the macrophage and causes the rapid formation of a cell-bound α-macroglobulin antibody. The other RNA is ineffective within the macrophage but initiates the production of γ globulin antibody in cultures of normal spleen cells. Both types of antibody appear upon addition of RNA to a cell-free system. The finding of antibody within 10 minutes in this system is most surprising and calls for more convincing evidence that *de novo* synthesis of antibody rather than release of preformed antibody is occurring.

The system employed by Fishman (1961) and Adler *et al.* (1967) is very complicated, involving heterogeneous cell populations, at least two species of RNA and two molecular species of antibody. A particularly vexing point to this author is the lack of rigorous evidence that *de novo* synthesis of antibody occurred. Much more direct evidence is desirable that the RNA which induces γM antibody is indeed messenger RNA. It may be feasible to decide whether this RNA is encoded for antibody synthesis by adding it to a cell-free system containing polyribosomes from cells other than lymphoid cells and determining whether these polyribosomes synthesize antibody and/or non-antibody γM globulin. Such a transfer of information has been reported with another mammalian protein-synthesizing system (Weisberger and Armentrout, 1966).

Assuming that intracellular immunogenic RNA and/or RNA-antigen complexes are formed by macrophages as a result of interaction with antigen, several other questions follow:

1. How is the information for antibody synthesis transferred from macrophage to lymphocytes? Macrophage RNA has been transferred *in vitro* to lymph node cells (Fishman *et al.,* 1963) and morphologic evidence of adhesion and intercellular bridges between lymphocytes and macrophages in immunized lymph nodes (Schoenberg *et al.,* 1964) has been reported. Recently, Ford *et al.* (1966) have done experiments which suggest that living macrophages which have ingested sheep erythrocytes can prime lymphoid cells for antibody formation *in vitro* whereas lymphoid cells cannot be induced by native antigen directly.

2. What is the role of circulating antibody and of pre-formed cellular antibody in the uptake of antigen? It is not known whether the macrophage is capable of developing into an antibody-forming cell or whether it merely takes up cytophilic antibody (Berken and Benacerraf, 1966). Experiments summarized in Table 3 indicate that many rabbits possess low levels of hemagglutinating antibody to hemocyanin. It is not known whether sufficient antibody is produced to promote antigen uptake. It is also not known how frequently antibody to other antigens is present in unimmunized animals; certainly so-called normal antibody to bacteriophages, erythrocytes and bacterial lipo-polysaccharides has been observed (Boyden, 1966). There is considerable evidence for the function of cellular as distinct from circulating receptors in the secondary response (Mitchison, 1966), but none for the primary response.

3. What is the role of RNA or RNA-antigen complexes in the induction of immunological memory and of tolerance? Table 1 illustrates the induction of immunological memory by RNA-antigen complexes but the

level of memory is certainly very low in these rabbits. No evidence is available concerning the induction of tolerance.

B. Characteristics of Inductive Phase of Antibody Response

Based on the length of time between antigenic stimulation and appearance of antibody, until recently the inductive phase has been assumed to be at least a week in length (Stavitsky, 1961). However, as more sensitive assays have been applied to the search for antibody, this phase was reduced to a few days or less. Moreover, detailed studies have revealed that many events occur within hours or days after contact of antigen with tissues. These include not only the formation of RNA-antigen complexes described in the previous section but striking morphologic changes in subcellular architecture of lymphoid cells (Feldman, 1964; Gusdon, Stavitsky, Schoenberg and Moore, 1967); increased RNA, DNA and protein synthesis (Mach and Vassalli, 1965a and b); appearance of new species of RNA detectable by hybridization with DNA (Cohen, 1967); increase in specific activity of enzymes of nucleotide metabolism (Râska and Cohen, 1967); increase in RNA possessing template activity (Mach and Vassalli, 1965a and b); appearance of increased numbers of polyribosomes (Norton, *et al.*, 1965; Folds and Stavitsky, 1967); and early appearance of antibodies (Svehag and Mandel, 1964). The observations of an early increase in levels of γM globulin and antibody to hemocyanin in this study (Roszman and Stavitsky, 1967) are also in accord with this picture.

It is not known which of these events are specifically induced, i.e., caused only by specific antigens and which may be stimulated by other substances such as phytohemagglutinin, endotoxin, Freund's adjuvant (Mellman and Rawnsley, 1966) and polyadenylic and polycytidylic acids (Braun and Nakano, 1967). Recent evidence suggests that many genes are activated by phytohemagglutinin (Kleinsmith *et al.*, 1966).

Popliteal lymph node cells from some unimmunized rabbits produced low levels of hemagglutinating antibody to hemocyanin and/or incorporated C^{14} amino acids into this antibody *in vitro* (Table 3). This finding complicates the attribution of the early antibody response solely to the reaction of cellular receptors with antigen. It seems clear that the secondary response involves the reaction of antigen with cellular receptors (Mitchison, 1966) and it would be reasonable to assume that such receptors also operate in the primary response. Perhaps this dilemma can be resolved by attempting to induce primary antibody formation by addition of antigen to cells *in vitro* in the absence of demonstrable circulating antibody. Such an *in vitro* primary induction of antibody to erythrocytes has been described recently (Mishell and Dutton, 1967), but its applicability to other antigens has not yet been demonstrated.

The data shown in Tables 3 and 4 suggest that there is a small amount of information in lymph node cells, presumably messenger RNA, which can be utilized or mobilized quickly for antibody synthesis. This information seems to last at least 8 hours after which renewed RNA and DNA synthesis are required for the resumption of antibody synthesis. Recent observations with an *in vitro* antibody induction system (Dutton and Mishell, 1967) also point up the importance of continued DNA synthesis for induction. Other observations indicate that DNA synthesis may be required generally for cellular differentiation (Sobel, 1966) and virus-induced cell transformations (Bader, 1965).

The question of the role of antigen in induction and maintenance of antibody synthesis has been raised again recently because investigators using sensitive methods have not found antigen in antibody-forming cells (Nossal *et al.,* 1965). However, if the simplistic view is taken that antigen serves to increase the level of messenger RNA synthesis or to reactivate messenger RNA from an inactive form and that some of this RNA is stable for long periods of time (Miller, 1964; Stavitsky and Gusdon, 1966), antibody synthesis could continue on polyribosomes encoded with this RNA in the absence of antigen. Once mRNA was degraded, antigen would again be required for induction. Whether the feedback inhibition of antibody synthesis by passively administered antibody (Dixon *et al.,* 1967) may be exerted at this level remains to be determined.

In several experiments it was found (Table 3) that information is present in lymph node cells for the early synthesis of both γM and γG globulin antibodies. This observation is in line with previous indications (summarized in Wei and Stavitsky, 1967) that a separate line of cells has differentiated for the synthesis of each of these immunoglobulins. Whether a common precursor for these two lines exists is not known.

It may be speculated that the lag period between injection of antigen and appearance of antibody as well as the amount of antibody made may be related to the amount of basal antibody synthesis occurring in animals *before* immunization. The basal synthesis would be expected to reflect the number of pre-existing cells in the population which contain cellular receptors for a particular antigen, whether these receptors were made by the cells themselves and/or picked up passively. There is already considerable evidence of basal synthesis of antibody to erythrocytes, phages and bacterial lipopolysaccharides (Boyden, 1966) in unimmunized animals.

It is clear from the data in Table 3 that most of the protein synthesized by lymph node cells is not antibody and that much of this non-antibody protein is non-specifically co-precipitated by egg albumin-anti-egg albumin precipitates. The relationship of synthesis of this non-antibody protein to synthesis of antibody is not understood and is under investigation in this laboratory.

C. Characteristics of Synthetic Phase of Antibody Response

Antibody synthesis during the anamnestic response usually is short-lived, lasting only a day or two, although synthesis may go on at a low level for months or years (Miller, 1964). The data presented in Table 5 are consistent with this picture. The electron microscopic evidence presented in Figures 2–6 suggests that about the time antibody synthesis declines in the lymph node the plasma cells are replaced with lymphocytoid cells. This observation is reminiscent of observations by light microscopy of the appearance of many lymphocytoid cells in the spleens of immunized rats after the disappearance of the plasma cells (Wissler *et al.*, 1957). Alternative explanations of the disappearance of plasma cells and hence of the decline in antibody synthesis include cell death and emigration of cells from the node. Little evidence of cell death was observed in this study. Cell migration is known to occur (Gowans and McGregor, 1965), but probably is insufficient to account for the almost total loss of recognizable plasma cells. The evidence thus suggests that transformation of plasma cells into other cell type(s) is the most likely explanation. The decline in antibody synthesis may be related to a disaggregation of polyribosomes owing to loss of messenger or transfer RNA, or to interference with the binding of ribosomes to the membranes of the endoplasmic reticulum (Blobel and Potter, 1967). These and other explanations are under investigation.

Previous (Stavitsky, 1961) and current (Folds and Stavitsky; Ortiz-Muniz and Sigel, 1967; Mishell and Dutton, 1967) experiments indicate that antibody synthesis persists for a much longer time when antibody-forming tissues are maintained *in vitro* than when they are maintained *in vivo*. Presumably, some regulatory controls which restrict antibody synthesis *in vivo* either do not operate or operate inefficiently *in vitro*. It is likely that study of this *in vitro* system and, especially, of mechanisms whereby it can be restricted will shed light on the nature of these regulatory controls.

The primary and secondary responses usually are distinguished primarily on the basis of the longer lag period of the primary, and the types and amounts of antibodies synthesized (Bauer *et al.*, 1963). However, recent findings of the early onset of antibody synthesis in the primary in this study and others (Svehag and Mandel, 1963) as well as of the production of both γM and γG antibodies during the primary response (Wei and Stavitsky, 1967) prompt a reexamination of the question of the distinction between responses: is it not possible that they are mechanistically similar and only quantitatively distinguishable? Further study should reveal other attributes of these responses, such as the role of macrophages and the susceptibility to feedback inhibition by antibody (Dixon *et al.*, 1967).

III. SUMMARY

This paper considers the antigen-processing, inductive and synthetic phases of the antibody response. Current evidence and thoughts concerning the role of RNA and RNA-antigen complexes were summarized. It was found that lymph node cells from unimmunized rabbits synthesized low levels of antibody to key hole limpet hemocyanin. Lymph node cells from rabbits immunized 1 day before produced low levels of γM and γG antibodies to this antigen. These syntheses were not inhibited initially by Actinomycin D or 5-bromodeoxyuridine, suggesting that there is a small amount of information which can be mobilized quickly for antibody synthesis. This information appears to last at least 8 hours after which renewed RNA and DNA synthesis seem to be required for resumption of antibody synthesis. It was speculated that the lag period between injection of antigen and the appearance of antibody as well as the quantity of antibody made are related to the amount of basal antibody synthesis occurring in animals before immunization. Antibody synthesis was found to be short-lived *in vivo*. At about the time antibody synthesis declined, the plasma cells in the lymph node appeared to be replaced by lymphocytoid cells derived from them. Some of the mechanisms which might control the level of antibody synthesis *in vivo* were considered.

References

Adler, F. L., M. Fishman, and S. Dray (1966). Antibody formation initiated *in vitro*. III. Antibody formation and allotypic specificity directed by ribonucleic acid from peritoneal exudate cells. J. Immun., *97*:554–558.

Askonas, B. A., and J. M. Rhodes (1965). Immunogenicity of antigen-containing ribonucleic acid preparations from macrophages. Nature, *205*:470–474.

Bader, J. P. (1965). Transformation by Rous sarcoma virus: a requirement for DNA synthesis. Science, *149*:757–758.

Bauer, D. C., M. J. Mathies, and A. B. Stavitsky (1963). Sequences of synthesis of γ-1 macroglobulin and γ-2 globulin antibodies during primary and secondary responses to proteins, salmonella antigens and phage. J. Exp. Med., *117*:889–907.

Becker, M. J., and A. Rich (1966). Polyribosomes of tissues producing antibodies. Nature, *212*:142–146.

Berken, A., and B. Benacerraf (1966). Properties of antibodies cytophilic for macrophages. J. Exp. Med., *123*:119–144.

Blobel, G., and V. R. Potter (1967). Studies on free and membrane-bound ribosomes in rat liver. II. Interaction of ribosomes and membranes. J. Mol. Biol., *26*:293–301.

Boyden, S. V. (1966). Natural antibodies and the immune response, vol. 5, New York and London: Academic Press, pp. 1–29.

Braun, W., and M. Nakano (1967). Antibody formation: stimulation by polyadenylic and polycytidylic acids. Science, *157*:819–821.

Britten, R. J., and R. B. Roberts (1960). High resolution density gradient sedimentation analysis. Science, *131*:32–34.

Cohen, E. P. (1967). The appearance of new species of RNA in the mouse spleen after immunization as detected by molecular hybridization. Proc. Nat. Acad. Sci., *57*:673–680.

Comb, D. G., R. Brown, and S. Katz (1964). The nuclear DNA and RNA components of the aquatic fungus *Blastocladiella emersonii.* J. Mol. Biol., *8*:781–789.

Dixon, F. J., H. Jacot-Guillarmod, and P. J. McConahey (1967). The effect of passively administered antibody on antibody synthesis. J. Exp. Med., *125*:1119–1136.

Dutton, R. W., and R. I. Mishell (1967). Cell populations and cell proliferation in the *in vitro* response of normal mouse spleen to heterologous erythrocytes. Analysis by the hot pulse technique. J. Exp. Med., *126*:443–454.

Edelman, G. M., H. G. Kunkel, and E. C. Franklin (1958). Interaction of the rheumatoid factor with antigen-antibody complexes and aggregated gamma globulin. J. Exp. Med., *108*:105–120.

Feldman, J. D. (1964). Ultrastructure of Immunologic Processes, Advances in Immunology, vol. 4, New York and London: Academic Press, pp. 175–248.

Fishman, M. (1961). Antibody formation *in vitro.* J. Exp. Med., *114*:837–856.

Fishman, M., R. A. Hammerstrom, and V. P Bond (1963). *In vitro* transfer of macrophage RNA to lymph node cells. Nature, *198*:549–551.

Folds, J., and A. B. Stavitsky, unpublished observations, 1967.

Ford, W. L., J. L. Gowans, and P. J. McCullagh (1966). The origin and function of lymphocytes. Ciba Symposium, The Thymus, Boston: Little, Brown and Co., pp. 58–78.

Frei, P. C., B. Benacerraf, and G. J. Thorbecke (1965). Phagocytosis of the antigen, a crucial step in the induction of the primary response. Proc. Nat. Acad. Sci., *53*:20–23.

Friedman, H. P., A. B. Stavitsky, and J. M. Solomon (1965). Induction *in vitro* of antibodies to phage T2: antigens in the RNA extract employed. Science, *149*:1106–1107.

Gallily, R., and M. Feldman (1967). The role of macrophages in the induction of antibody in X-irradiated animals. Immunology, *12*:197–206.

Garvey, J. S., and D. H. Campbell (1957). The retention of S^{35}-labeled bovine serum albumin in normal and immunized rabbit liver tissue. J. Exp. Med., *105*:361–372.

Gottlieb, A. A., V. R. Glisin, and P. Doty (1967). Studies on macrophage RNA involved in antibody production. Proc. Nat. Acad. Sci., *57*:1849–1856.

Gowans, J. L., and D. D. McGregor (1965). The immunological activities of lymphocytes. Progr. Allergy, 9. Basel: Karger, pp. 1–78.

Gusdon, J. P., and A. B. Stavitsky (1967). Fed. Proc., *26*:642 (abstract).

Gusdon, J. P., A. B. Stavitsky, M. D. Schoenberg, and R. D. Moore, unpublished observations, 1967.

Hiatt, H. A. (1962). Rapidly labeled RNA in rat liver nuclei. J. Mol. Biol., *5*:217–229.

Jacherts, D., and H. Noltenius (1966). Antikörper-Synthese *in vitro*. V. Die informations übertragung von der antigen-verarbeitenden zelle auf die antikörper-produzierende zelle. Z. med. mikrobiol. u. immunol., *152*:112–133.

Kahan, F. M. (1960). Purification and measurement of microgram amounts of radioactive nucleic acids and proteins from animal cells in tissue culture. Anal. Biochem., *2*:107–126.

Kleinsmith, L. J., V. G. Allfrey, and A. E. Mirsky (1966). Phosphorylation of nuclear protein early in the course of gene activation in lymphocytes. Science, *154*:780–781.

Mach, B., and P. Vassalli (1965a). Template activity of RNA from antibody-producing tissues. Science, *150*:622–626.

Mach, B., and P. Vassalli (1965b). Biosynthesis of RNA in antibody-producing tissues. Proc. Nat. Acad. Sci., *54*:975–982.

Martin, R. G., and B. N. Ames (1961). A method for determination of the sedimentation behavior of enzymes: application to complete Freund's adjuvant. J. Biol. Chem., *236*:1372–1379.

Mellman, W. J., and H. M. Rawnsley (1966). Blastogenesis in peripheral blood lymphocytes in response to phytohemagglutinin and antigens. Fed. Proc., *25*:1720–1722.

Miller, J. J. (1964). An autoradiographic study of plasma cell and lymphocyte survival in rat popliteal lymph nodes. J. Immun., *92*:673–681.

Mishell, R. I., and R. W. Dutton (1967). Immunization of dissociated spleen cell cultures from normal mice. J. Exp. Med., *126*:423–442.

Mitchison, N. A. (1966). Recognition of antigen by cells. Progress in Biophysics and Molecular Biology, vol. 16, London: Pergamon Press.

Müller-Eberhard, H. J. (1960). A new supporting medium for preparative electrophoresis. Scand. J. Clin. and Lab. Inv., *12*:33–37.

Norton, W. L., D. Lewis, and Morris Ziff (1965). The effect of progressive immunization on polyribosomal size in lymphoid cells. Proc. Nat. Acad. Sci., *54*:851–856.

Nossal, G. J. V., G. L. Ada, and C. M. Austin (1965). Antigen in immunity. IX. The antigen content of single antibody-forming cells. J. Exp. Med., *121*:945–954.

Ortiz-Muniz, G., and M. M. Sigel (1967). Long-term synthesis of antibody *in vitro*. Proc. Soc. Exp. Biol. Med., *124*:1178–1182.

Phillipson, L. (1961). Chromatographic separation and characteristics of nucleic acids from HeLa cells. J. Gen. Physiol., *44*:899–910.

Plescia, O. J., W. Braun, and N. C. Palczuk (1965). Molecular and cellular basis of antibody formation. Prague: Czechoslovak Academy of Science Publishing House, pp. 61–69.

Râska, K., and E. P. Cohen (1967). Specific activity of several enzymes of nucleotide metabolism in mouse spleen after immunization. Clin. and Exp. Immunol., *2*:in press.

Roszman, T., and A. B. Stavitsky, unpublished observations, 1967.

Scharff, M. D., and J. W. Uhr (1965). Functional ribosomal unit of γ-globulin synthesis. Science, *148*:646–648.

Schoenberg, M. D., V. R. Mumaw, R. D. Moore, and A. A. Weisberger (1964). Cytoplasmic interaction between macrophages and lymphocytic cells in antibody synthesis. Science, *143*:964–965.

Sobel, J. S. (1966). DNA synthesis and differentiation in embryonic kidney mesenchyme *in vitro*. Science, *153*:1387–1389.

Stavitsky, A. B. (1961). *In vitro* studies of the antibody response. Advances in Immunology, vol. 1, New York and London: Academic Press, pp. 211–261.

Stavitsky, A. B. (1954). Micromethods for the study of proteins and antibodies. I. Procedure and general applications of haemagglutination and haemagglutination-inhibition reactions with tannic acid and protein-treated blood cells. J. Immun., *72*:360–367.

Stavitsky, A. B., and J. P. Gusdon (1966). Symposium on *in vitro* studies of the immune response. IV. Role of nucleic acids in the anamnestic antibody response. Bacteriol. Rev., *30*:418–426.

Svehag, S. E., and B. Mandel (1964). The formation and properties of poliovirus-neutralizing antibody. I. 19S and 7S antibody formation: differences in kinetics and antigen dose requirement for induction. J. Exp. Med., *119*:1–20.

Uhr, J. W., and G. Weissmann (1965). Intracellular distribution and degradation of bacteriophage in mammalian tissues. J. Immun., *94*:544–550.

Wasserman, E., and L. Levine (1961). Quantitative micro-complement fixation and its use in the study of antigenic structure by specific antigen-antibody inhibition. J. Immun., *87*:290–295.

Wei, M. M., and A. B. Stavitsky (1967). Molecular forms of rabbit antibody synthesized during the primary response to human albumin. Immunology, *12*:431–444.

Weisberger, A. S., and S. A. Armentrout (1966). Directed protein synthesis by messenger ribonucleoprotein and ribosomes from different mammalian species. Proc. Nat. Acad. Sci., *56*:1612–1620.

Wissler, R. W., F. W. Fitch, M. F. La Via, and C. H. Gunderson (1957). The cellular basis for antibody formation. J. Cell. and Comp. Physiol., *50*:265–302 (supplement 1).

THE ANTIGEN-RNA COMPLEX OF MACROPHAGES *

A. ARTHUR GOTTLIEB

Department of Chemistry, Harvard University Cambridge, Mass.†

I. INTRODUCTION

Previous studies from this laboratory have demonstrated the existence of a minor fraction of macrophage RNA which has a banding density in cesium sulphate solution lighter than that of the bulk of the cellular RNA (Gottlieb *et al.*, 1967). If rat macrophages are exposed to T2 bacteriophage, nearly all of the immunogenic activity attributed to the RNA of the macrophage is found in the minor light density fraction. The immunogenic activity of the minor band, as well as its ability to band in a cesium sulphate density gradient, is destroyed by incubation with pronase. This report presents additional information regarding the purification of this minor RNA-protein complex of macrophages and the determination of some of its properties.

II. ISOLATION AND PURIFICATION OF THE MINOR BAND

A. PREPARATIVE DENSITY GRADIENT ULTRACENTRIFUGATION

Rat macrophages were obtained by intraperitoneal injection of mineral oil and incubated in phosphate-free MEM Eagle's solution for 18–20 hours in the presence of 1.0 mc of P^{32}. RNA was prepared by phenol

* I am indebted to Prof. Paul Doty for encouragement and support during the course of this research and for counsel in the preparation of this manuscript. My appreciation is also due Dr. Vlada Glisin, Prof. Alec Sehon, Dr. Helga Doty and Mr. Olke Uhlenbeck for helpful discussion about various aspects of this work.

This research was supported by a research grant from the National Institutes of Health (HD-01229) and by a special fellowship from the National Institute of General Medical Sciences (5-F3-GM 28, 302-02).

† Present address: Department of Medicine, Harvard Medical School, Robert B. Brigham Hospital, Boston, Mass.

extraction of the cells as previously described (Gottlieb *et al.*, 1967). The RNA obtained in several experiments had a specific activity of 40,000–55,000 cpm per μg. The separation of the light density protein-RNA fraction from the bulk of the cellular RNA in a cesium sulphate density gradient is shown in Figure 1.

In order to exploit this separation for preparative purposes we turned to the use of the fixed angle rotor technique described by Flamm, Bond and Burr (1966) for density gradients which allows the banding of larger amounts of RNA than is possible using the swinging-bucket technique. Using the fixed angle rotor, the gradient is much steeper towards the bottom of the tube than in the middle of the tube. This permits the minor band to be spread out over a larger section of the tube. By recycling the minor band from one gradient, it is possible to effect a relatively clean separation of the minor band from the bulk of the cellular RNA. This is shown in Figure 2. Its buoyant density is 1.585 ± .005. Approximately 5% of the cellular RNA appeared in the minor band.

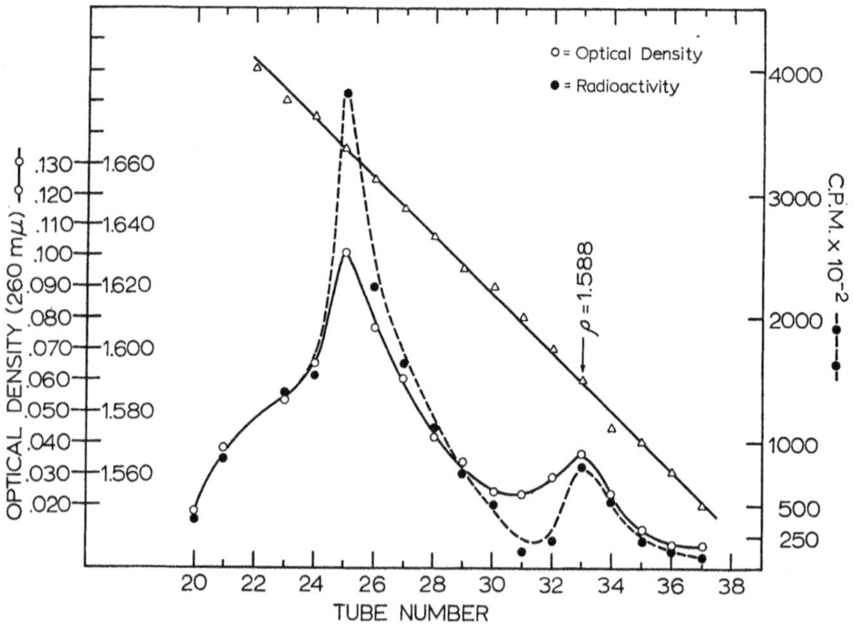

Fig. 1. Cesium sulphate gradient distribution of RNA from bacteriophage-infected macrophages. Macrophages (5.5×10^8) were incubated in 5.0 ml of MEM Eagle's medium containing calf serum but deficient in inorganic phosphate for 18 hours in the presence of 1.0 mc of P^{32}-orthophosphate. Bacteriophage T2 was added and the incubation continued for an additional 30 minutes. RNA was extracted as described in *Methods* and 50 μg placed in cesium sulphate and centrifuged in a swinging-bucket rotor at 33,000 rpm for 72 hours at 25°. (Reprinted with permission from Proceedings of the National Academy of Sciences.)

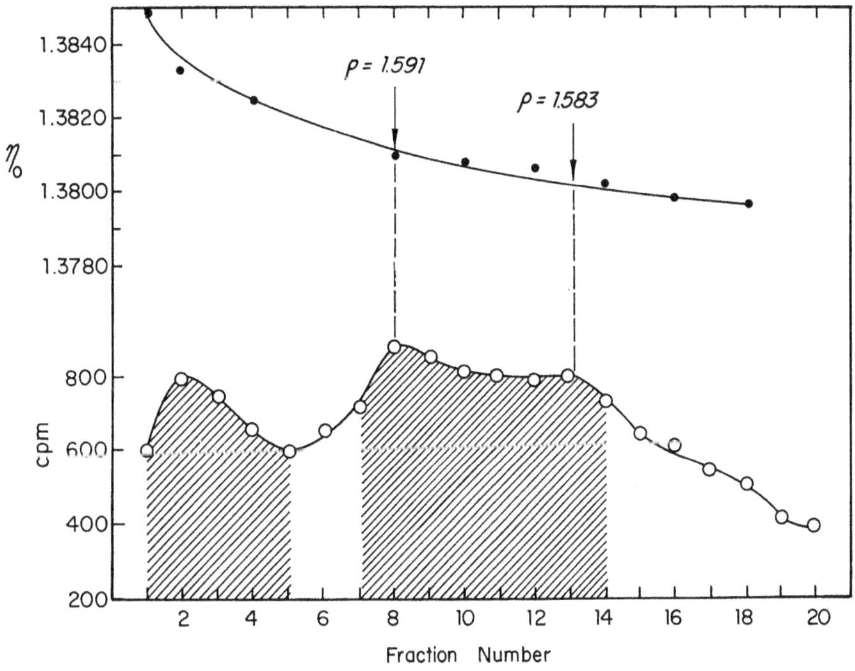

Fig. 2. Partial purification of the minor fraction of macrophage RNA by rebanding in cesium sulphate. The minor peak obtained from a cesium sulphate density gradient of P^{32}-labelled macrophage RNA was banded in a second cesium sulphate density gradient. Fifteen drop fractions were collected through a No. 23 needle. The radioactive and refractive index measurements along the gradient are shown.

B. Chromatography on TEAE-Cellulose in 7.5 M Urea

To purify and analyze the minor band further, column chromatography on TEAE-cellulose was employed. The minor component from a preparative density gradient was rebanded in cesium sulphate solution to remove major band material. The minor band from this second density gradient was collected, pooled, passed on Biogel P-2 to remove cesium sulphate and transferred to a column of TEAE-cellulose in 7.5 M urea. Elution was carried out with a salt gradient from 0.15 M NaCl to 0.5 M NaCl. The elution profile is shown in Figure 3. A small amount of label remained on the column after elution to 0.45 M NaCl and could only be recovered by elution with 1N NaOH. The bulk of the radioactivity elutes at a salt concentration of 0.33 M NaCl. If this peak is recovered and rebanded in Cs_2SO_4, it is found that the label bands principally at the density of the minor band as shown in Figure 4. Thus, the purification of the minor band is complete, and it is characterized by a buoyant density of 1.588 in Cs_2SO_4 solution.

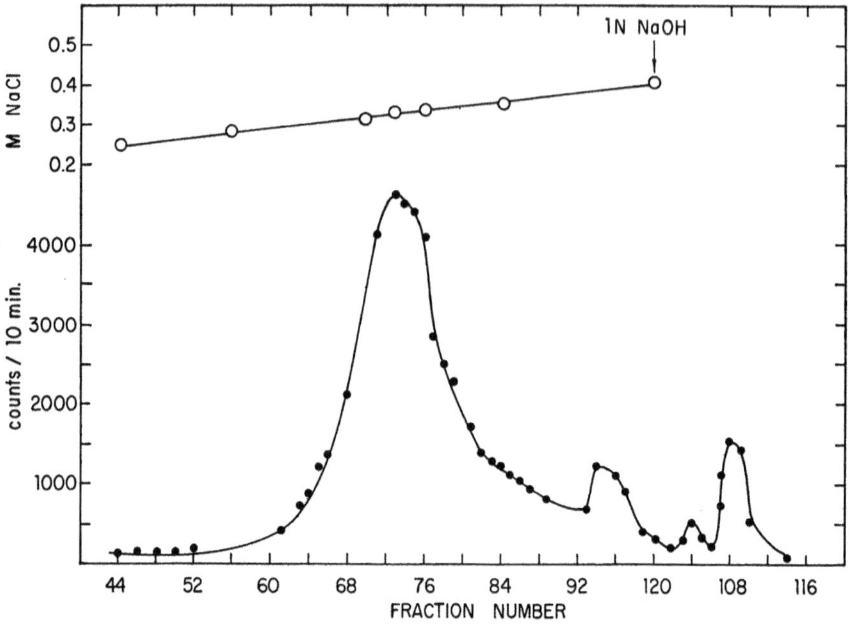

Fig. 3. Salt gradient elution profile of minor fraction of macrophage RNA on TEAE-cellulose in 7.5 *M* urea. A P^{32} labelled minor band preparation was recycled through a cesium sulphate density gradient, collected and desalted through a column of Biogel P-2 polyacrylamide gel. 10,000 cpm were placed on a 10 ml column of TEAE-cellulose in 7.5 *M* urea buffered with 0.01 *M* Tris (pH 8.2). Elution was carried out from 0.15 *M* NaCl to 0.45 *M* NaCl. 4 ml fractions were collected and 1 ml aliquots assayed for radioactivity. The radioactivity and salt gradient profiles are shown.

The base composition of this material was determined by alkaline hydrolysis followed by electrophoresis on paper. The results are shown in Table 1.

III. PROPERTIES OF THE "MINOR" BAND

A. Effect of Actinomycin on the Formation of Immunogenic RNA from Antigen-exposed Macrophages

The observation that the minor band, indicative of the RNA-protein complex, was present in macrophages irrespective of exposure to a specific antigen suggested that the RNA backbone of the complex was not synthesized in response to antigen exposure. Since Actinomycin-D inhibits RNA synthesis, it provided a means of testing this hypothesis.

Pretreatment of the macrophage population with Actinomycin-D (Dactinomycin, Merck) for 2.5 hours at a concentration of 0.3 μg/ml failed to impair the ability of the macrophages to form an immunogenic

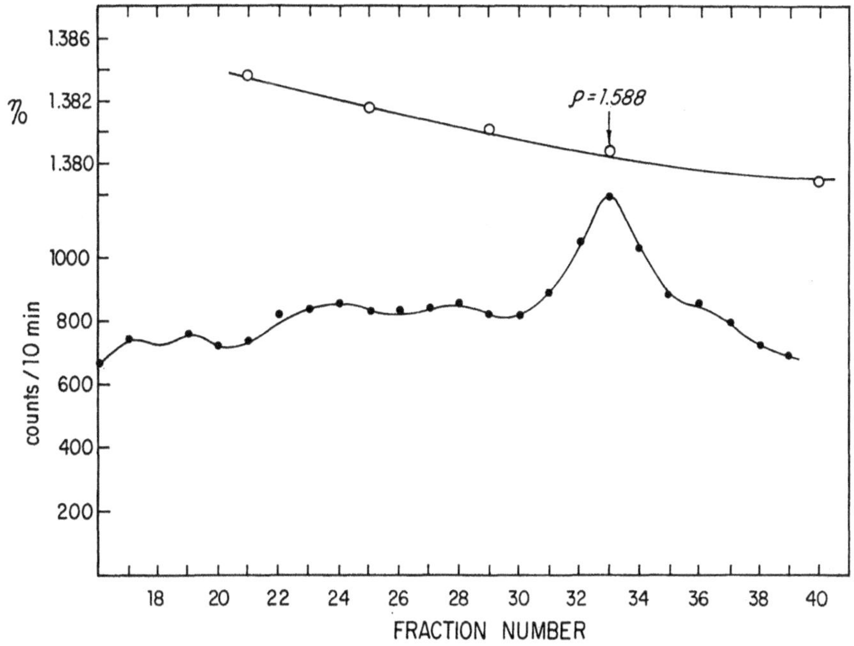

Fig. 4. Rebanding in cesium sulphate of the 0.33 *M* NaCl fraction from TEAE-cellulose. An aliquot of the 0.33 *M* NaCl fraction from the profile shown in Fig. 3 was desalted on Biogel P-2, concentrated and rebanded in a cesium sulphate preparative gradient. The radioactive profile along the gradient is shown.

Table 1

BASE COMPOSITION ANALYSIS OF PURIFIED MINOR BAND

	Mole Per Cent			
Peak	CMP	AMP	GMP	UMP
Minor band	26.1	20.8	32.5	20.6
Major band	24.2	27.8	28.2	20.6
Total Macrophage RNA	22.6	24.6	30.0	23.0

P^{32} labelled macrophage RNA was carried through two cycles of Cs_2SO_4 density gradient ultracentrifugation prior to chromatography on TEAE-cellulose in 7.5M Urea. An aliquot of the 0.33M NaCl fraction was hydrolyzed in 0.3M KOH for 24 hours and electrophoresed on paper. Each of the bases, identified by standard markers, was eluted from the electrophoretogram and counted in Breys-Cabosil solution. Aliquots of total RNA and of major band were analyzed in a similar fashion.

T2 antigen-RNA complex. These results are presented in Table 2. Thus the RNA fraction in question is not being synthesized at this time.

To demonstrate that the Actinomycin was indeed inhibiting RNA synthesis the base composition of the total cellular RNA synthesized in the presence and absence of Actinomycin D was determined (Table 3). Substantial changes in base composition of macrophage RNA are seen in macrophage populations which have been exposed to a concentration

Table 2

Sample	Input	Plaque Ct. at 40 min	Per Cent Plaque Reduction
	Set A		
25 μg T2-Mϕ-RNA *	352	210	40.3
25 μg T2-Mϕ-RNA from cells pretreated with actinomycin	352	226	35.8
50 μg T2-Mϕ-RNA from cells pretreated with actinomycin	352	210	40.3
Control (no RNA)	352	337	4.3
	Set B		
10 μg T2-Mϕ-RNA	301	230	23.5
10 μg T2-Mϕ-RNA from cells pretreated with actinomycin	301	223	25.9
Control (no RNA)	301	286	5.0

* T2-Mϕ-RNA refers to RNA from T2 infected macrophages.

1.38×10^8 macrophages were incubated in 5.0 ml of phosphate buffered saline with 0.3 μg/ml (Set A) or 0.9 μg/ml (Set B) of Actinomycin-D for 2.5 hours at 37°C at which time 2.7×10^9 bacteriophage T2 were added and the incubation continued for an additional 30 minutes. RNA was then extracted and directly added to cultures of spleen fragments. The medium was removed at six days and assayed for antibody activity. A control culture consisting of spleen fragments to which saline alone was added was carried along simultaneously. Neutralizing antibody was assayed by a modification of the bacteriophage plaque reduction technique (Gottlieb, 1966). Input refers to the number of plaque forming units added per 0.1 ml. of test mixture. At 40 minutes, 0.1 ml. aliquots were removed in triplicate, plated on *E. coli* B and the percentage plaque reduction computed.

Table 3

EFFECT OF ACTINOMYCIN-D ON THE BASE COMPOSITION OF P^{32}-LABELLED RNA FROM MACROPHAGES

	Mole Per Cent			
	CMP	AMP	GMP	UMP
Non-treated cells	21.8	21.6	31.4	25.2
Actinomycin-treated cells	31.8	23.0	23.0	22.2

Two sets of macrophages were exposed to 2.0 mc of P^{32} orthophosphate in 5.0 ml of normal saline for three hours. To one set, Actinomycin-D was added to a final concentration of 0.3 μg/ml. At the end of the incubation period, RNA was extracted from the cells and base ratio analyses carried out as described in Table 1.

of 0.3 μg/ml of the drug for 3 hours. Thus the relative stability of the RNA of the complex is established. It is of interest to note that the concentration of Actinomycin-D used in these experiments has been previously shown by Uhr to inhibit antibody fonmation to T2 bacteriophage (1963).

B. COMPARATIVE SPECIFIC ACTIVITIES OF THE RNA FROM MAJOR AND MINOR BANDS

If the formation of the RNA backbone of the minor band is not dependent on antigen exposure, it would be expected that the specific activity of the RNA in the minor band should be less than that of the major band.

When the macrophages are incubated for 18 hours in the presence of P^{32} orthophosphate and the RNA is then extracted and banded in a cesium sulphate density gradient, it is found that the specific activity of the minor light-density band is always less than that of the major RNA band (Table 4). In these experiments, it was not possible to determine the pattern of labelling of the RNA *in vivo* because of the difficulty in preparing undegraded preparation of RNA from macrophages.

These observations support the idea that the RNA backbone of the minor band is preformed.

C. DISTRIBUTION OF I^{125}-LABELLED ANTIGEN IN A Cs_2SO_4 DENSITY GRADIENT DISTRIBUTION OF MACROPHAGE RNA

Since direct testing of the major and minor fractions of macrophage RNA reveals immunogenicity only in the minor band, one would expect

Table 4

COMPARATIVE SPECIFIC ACTIVITIES OF THE RNA FROM MAJOR AND MINOR BANDS

	Specific Activity (cpm/μg)	
Expt.	Major	Minor
1	43,500	18,700
2	55,000	15,450
3	38,750	13,000

Macrophages were incubated with P^{32} orthophosphate for 18 hours in Eagle's MEM medium deficient in cold phosphate as described in the text. RNA was prepared and banded in a cesium sulphate density gradient. The major and minor bands were collected, desalted on columns of Biogel P-2 and the specific activities of the RNA in each fraction determined.

that the antigen would be selectively concentrated within the minor band.

By use of I^{125}-labelled T2 bacteriophage, it was possible to demonstrate directly the presence of T2 bacteriophage protein in the RNA-protein complex of macrophages. When 2.1×10^{10} I^{125} labelled bacteriophages, having a specific activity of 3.7×10^6 cpm per μg, were incubated with 2.3×10^8 macrophages for thirty minutes at 37°C in buffered phosphate saline, the total cellular RNA which was subsequently extracted had a specific activity of 26 cpm/μg. This represents 7.2×10^{-5} μg phage protein per μg of total RNA. 35 μg of this RNA were banded in a cesium sulphate density gradient, ten drop fractions were collected through a No. 23 needle, the optical density was determined for each fraction, and the entire fraction was dissolved in Brays-Cabosil solution and counted in a liquid scintillation counter. The profile of the I^{125} label as a function of position in the gradient is shown in Figure 5. I^{125} label is found to be concentrated in the region corresponding to the minor component. Additional radioactivity is found distributed in the gradient as material of low molecular weight, but no radioactivity over this background is observed in the region of the major peak.

D. ANALYTICAL DENSITY GRADIENT STUDIES

In order to obtain additional information regarding the response of the minor band to certain agents, studies were carried out using Cs_2SO_4 density gradients in the analytical centrifuge.

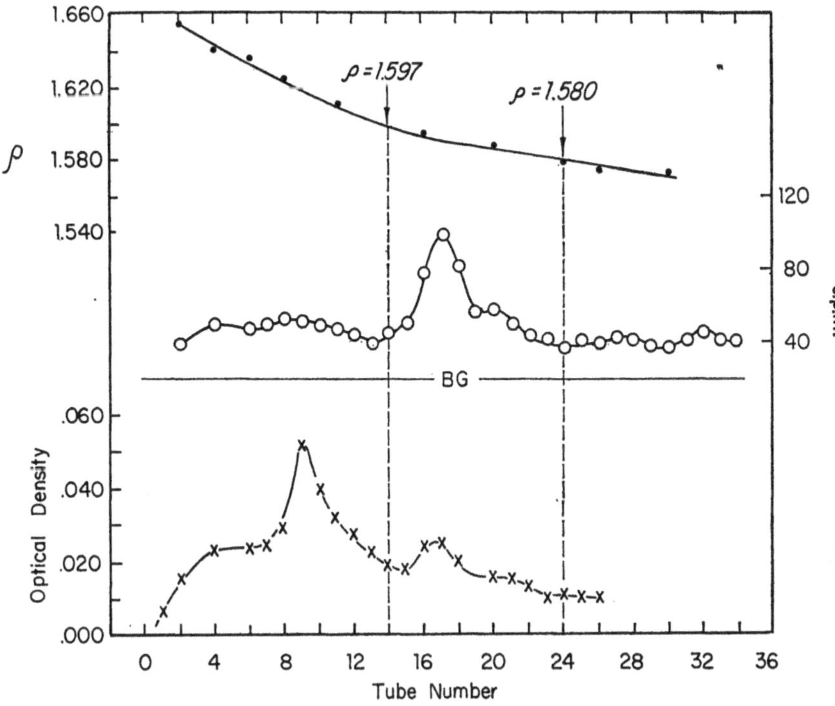

Fig. 5. Selective uptake of I^{125}-labelled antigen by the minor fraction of macrophage RNA. 2.3×10^8 macrophages were incubated with 2.1×10^{10} I^{125}-labelled T2 bacteriophage for 30 minutes at 37°C in phosphate buffered saline. The total I^{125} input was 1.7×10^7 cpm. The RNA was extracted from the macrophages and had a specific activity of 26 cpm/μg RNA. 35 μg was banded in a cesium sulphate density gradient and fractionated as described in *Methods*. The I^{125} radioactivity, optical density and solution density profiles are indicated. Counter background was 31 counts per minute.

1. *Effect of Selected Enzymes on the "Minor" Band*

When 10 μg of RNA from macrophages was incubated with 2 μg of RNAase-A (Sigma, 2700 U/mg) for one hour at 25°C and then banded in a cesium sulphate analytical gradient, the profile shown in Figure 6B was obtained. It is seen that the minor component is essentially preserved while the major RNA peak is substantially reduced. It required 10 μg of RNAase-A to eliminate the minor band completely. The minor band appeared to be more sensitive to the action of Taka-Diastase T1 RNAase (15 Units/μg). Five units of this enzyme were equivalent to 0.5 μg of RNAase-A as judged by the ability of the two preparations to yield acid-soluble fractions from P^{32} labelled rat liver RNA. When 10 μg of macrophage RNA was treated with 3 units of T1-RNAase for one hour at 25°C and banded in a similar analytical gradient the situation shown in Figure 6C was obtained. Again the minor component is pre-

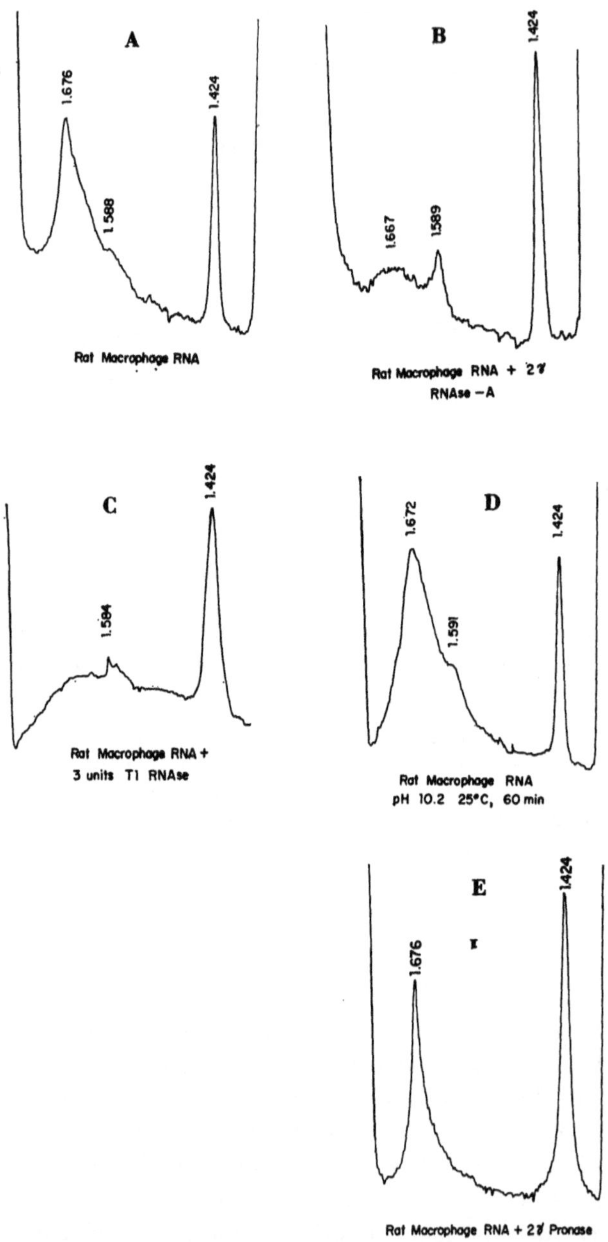

Fig. 6. Effects of RNAase-A, RNAase-T1, pronase and alkali on the minor fraction of macrophage RNA. **A.** 9.5 μg of RNA from T2 infected macrophages was banded in a cesium sulphate density gradient of initial density 1.54 g/cc and spun at 52,640 rpm for 21 hours after addition of dAT (ρ = 1.424) as a standard reference marker. **B.** 11.0 μg of macrophage RNA were treated with 2 μg of RNAase-A in 0.01 *M* Tris buffer (pH 7.4) for 1 hour at 25°C prior to banding in cesium sulphate. **C.** 11.0 μg of macro-

served, but was eliminated by 10 units of T1-RNAase (equivalent to 1.0 gamma of RNAase-A). Incubation of macrophage RNA at pH 10.3 for 2 hours at 25°C gave the profile shown in Figure 6D. Here the minor component is accentuated and the major component modestly reduced. Similar treatment at pH 11.3 and 12.3 eliminated both bands. Treatment of macrophage RNA with pronase eradicates the minor band (Fig. 6E).

In summary, it appears that limited treatment of whole RNA preparations from macrophages with RNAase-A or RNAase-T1 is quite selective in that the minor band, the immunogenic RNA-protein complex, remains unaffected while the cellular RNA is degraded. Moreover, the minor band RNA does not become enriched at the expense of the degradation of the major peak. Furthermore, treatment of RNAs from tissues free of the minor band with these nucleases did not lead to the formation of material banding in the minor region. It should be noted that the macrophage RNA preparations used in these studies were all obtained as degraded material sedimenting below 5S.

In all respects, the action of RNAase-A, T1 RNAase, pronase and alkali on the minor band was very similar to the action of these agents on the immunogenicity of total RNA preparations from T2 infected macrophages. We previously reported that pronase and large amounts of RNAase-A destroyed the immunogenicity of macrophage RNA preparations (Gottlieb *et al.,* 1967). Now it is seen that these enzymes also eliminate the minor band in cesium sulphate density gradients. In parallel with the effect of higher enzyme concentration mentioned above we found that the immunogenicity of 10 μg of RNA from T2 infected macrophages was not affected by 1.0 unit of T1-RNAase but was destroyed by 10 units of this enzyme. Similarly, treatment of macrophage RNA at pH 11.2 for two hours destroyed the immunogenicity. Again, the behavior of the minor band parallels the immunogenic potential of the RNA preparation.

2. *A Search for the Minor Band in Other Tissues of the Rat*

RNA was prepared from various rat tissues and banded in cesium sulphate gradients. As shown in Figure 7 it was possible to demonstrate the existence of a minor band in reduced amount, relative to macro-

---

phage RNA was treated with 3 units of RNAase-T1 in 0.01 *M* Tris buffer (pH 7.4) for 1 hour at 25°C before banding in cesium sulphate. **D.** 10.0 μg of macrophage RNA was adjusted to pH 10.1 with NaOH and held at 25°C for one hour. The solution was neutralized with NaH_2PO_4 prior to addition of cesium sulphate. **E.** 7.3 μg of macrophage RNA was treated with 2 μg of self-digested pronase in 0.01 *M* Tris buffer (pH 7.4) for 36 hours at 37°C prior to banding in cesium sulphate.

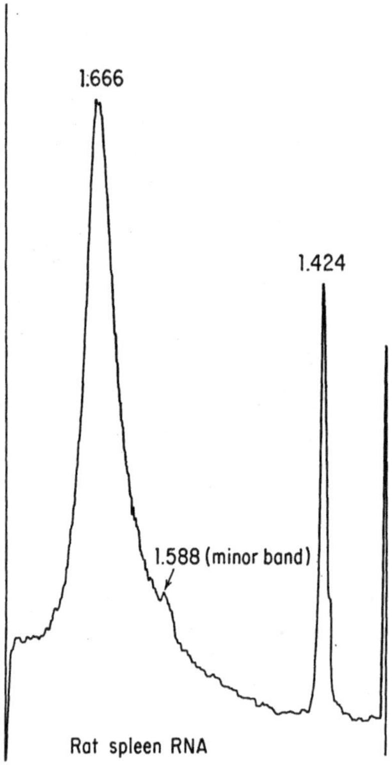

Fig. 7. Demonstration of a "minor" band in rat spleen. 20 μg of rat spleen RNA was banded in a cesium sulphate density gradient. The densitometer tracing was prepared from a picture taken at 280 mμ.

phages, in rat spleen. A preparative cesium sulphate gradient of 700 μg of rat liver RNA is shown in Figure 8 and the minor fraction is again evident. By contrast, as shown in Figure 9, the minor band was not found by these techniques in RNA from rat thymus, kidney, thyroid or mouse plasma cell myeloma MOPC-21 of Potter *et al.* (1965).[1] Treatment of these RNAs with RNAase-A did not lead to the formation of the "minor" band. Thus the light-density RNA-protein complex appears to reside only in tissues which contain macrophages or macrophage-like cells.

3. *Composition of the RNA-Protein Complex*

An estimate of the amount of protein in the minor band was made by calculating the difference in banding densities of the major and minor bands relative to the banding density of T2 bacteriophage protein. Bacteriophage T2 was subjected to osmotic shock and banded in a cesium sulphate density gradient, after addition of SP 82 DNA as a

[1] This tumor was kindly supplied by Dr. M. Potter.

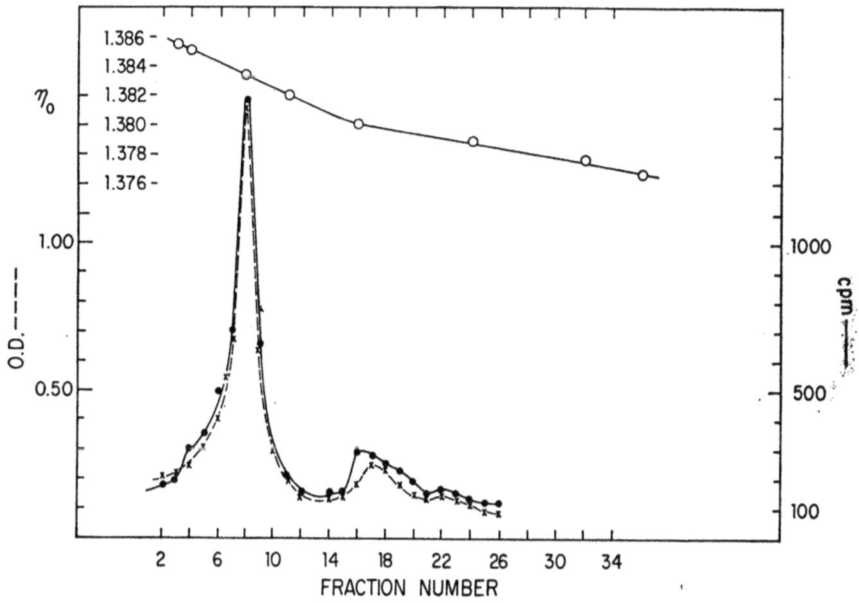

Fig. 8. Band profile of rat liver RNA in a cesium sulphate preparative density gradient. RNA was extracted from a rat labelled *in vivo* with 1.0 mc of P^{32} orthophosphate for twenty-four hours before sacrifice. 700 μg of the RNA was banded in a type 40 fixed angle rotor and fractionated as described in *Methods*. The fractions were diluted to 0.5 ml and the optical density of each fraction determined. 10 μl of each fraction was counted in Brays-Cabosil scintillation medium. The optical density, radioactivity and solution density profiles are shown.

standard reference marker. The protein peak was found to have a density of 1.376. By using the formula

$$\frac{\rho_{\text{major}} - \rho_{\text{minor}}}{\rho_{\text{major}} - \rho_{\text{protein}}} \times 100$$

an estimate of 28% was obtained for the protein content of the minor band.

IV. SUMMARY AND DISCUSSION

The experiments reported here, employing T2 bacteriophage as antigen, provide additional evidence for the idea that the immunogenicity of RNA preparations from antigen-exposed macrophages is related to a light density RNA component which appears to be a protein-RNA complex.

The failure of Actinomycin-D to affect the linkage of T2 bacterio-

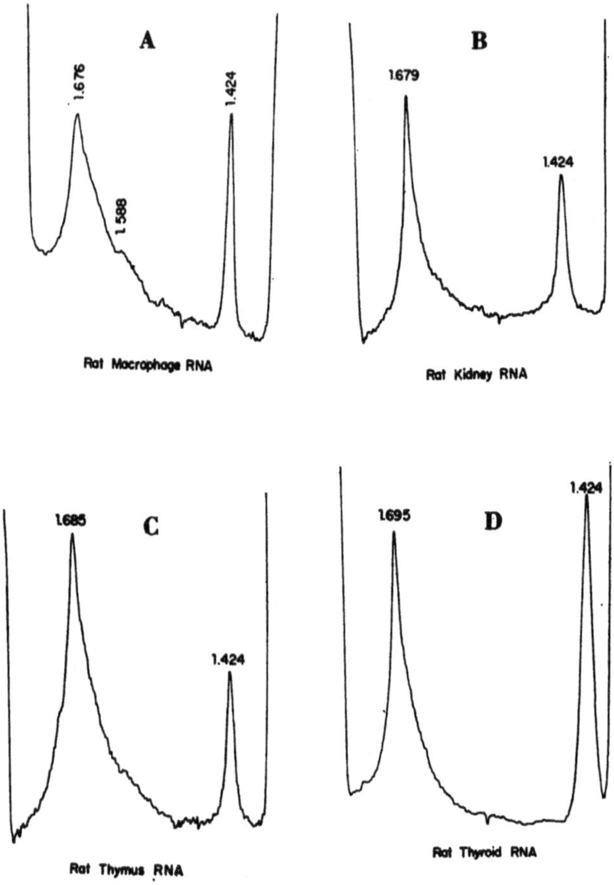

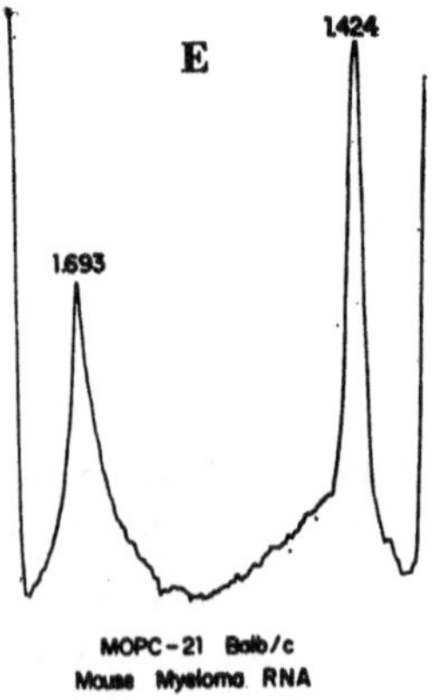

Fig. 9. Absence of light-density, minor fraction in rat kidney, thymus, thyroid and mouse myeloma RNA. RNA from the various tissues indicated was banded in a cesium sulphate density gradient after addition of dAT standard. **A.** 9.5 μg macrophage RNA. **B.** 9.0 μg rat kidney RNA. **C.** 10 μg rat thymus RNA. **D.** 8.8 μg rat thyroid RNA. **E.** 5.0 μg mouse myeloma RNA.

phage antigen to macrophage RNA and the observation that the specific activity of the minor band is consistently less than that of the major band are compatible with the existence of an antigen-carrier RNA in macrophages which is not newly formed on antigenic challenge and is complexed with a substantial amount of protein. The existence of the minor band in RNA preparations from macrophages not exposed to antigen, and the direct demonstration of antigen in the region of the minor band implicate this light-density RNA-protein complex as the immunogenic locus of the macrophage. Moreover, this fraction has so far been observed only in macrophages or tissues containing significant numbers of macrophages or macrophage-like cells such as the fixed tissue histiocytes (Kuppfer cells) of the liver.

Treatment of macrophage RNA with RNAase-A or RNAase-T1 appears to unmask the minor component and demonstrates its resistance to these enzymes relative to the RNA of the major band. There appeared to be no conversion of the major band to minor band material by treatment with these nucleases. In the case of RNAs which were shown to lack the minor RNA component, it was not possible to form the minor band by treatment with these enzymes. In all respects, the immunogenicity of macrophage RNA parallels the behavior of the light density, minor-RNA component.

Although antigen appears to be selectively concentrated within the minor band, the nature of the protein attached to the RNA of the minor band of macrophages not experimentally exposed to antigen remains unknown. These peptides may have arisen from previous exposure of the macrophage cells to unknown antigens during the life history of the cell. This would be compatible with the storage of immunological information in the macrophage, a possibility which seems not unlikely in view of the established demonstration of residual antigen with fixed tissue elements and macrophages of the reticuloendothelial system.

References

Fishman, M. (1961). Antibody Formation *in vitro*. J. Exptl. Med., *114*:837.

Fishman, M., and F. L. Adler (1963). Antibody Formation Initiated *in vitro*. J. Exptl. Med., *117*:595.

Flamm, W. G., H. E. Bond, and H. E. Burr (1966). Density Gradient Ultracentrifugation of DNA in a Fixed-Angle Rotor. Biochim. Biophys. Acta, *129*:310.

Gottlieb, A. A. (1966). Assay of Neutralizing Antibody: Increased Sensitivity of Assay by Plating on Salt-Deficient Agar. Nature, *212*:513.

Gottlieb, A. A., V. R. Glisin, and P. Doty (1967). Studies on Macrophage RNA Involved in Antibody Production. Proc. National Acad. Sci., *57*:1849.

Greenwood, F. C., W. M. Hunter, and J. S. Glover (1963). The Preparation of

I^{131} Labelled Human Growth Hormone of High Specific Activity. Biochemical J., *89*:114.

Potter, M., E. Appella, and S. Geisser (1965). Variations in the Heavy Polypeptide Chain Structure of Gamma Myeloma Immunoglobulins from an Inbred Strain of Mice. J. Mol. Biol., *14*:361.

Uhr, J. W. (1963). Actinomycin-D: Its Effect on Antibody Formation *in vitro*. Science, *142*:1476.

ON THE ROLE OF ANTIGEN FRAGMENTS AND RNA IN THE IMMUNE RESPONSE OF RABBITS TO A SOLUBLE ANTIGEN *

JUSTINE S. GARVEY

California Institute of Technology, Pasadena, California

I. INTRODUCTION

The purpose of some early studies from this laboratory was to isolate, by mild chemical procedures, antigen material † that had been retained

* Discussions with Drs. Henry Rinderknecht and Dan H. Campbell have been extremely helpful to the progress of these studies, and the author is appreciative of the very capable technical assistance of Mrs. Berta Weliky, Mr. Roger Smith and Miss Betty Aalseth.

This investigation was supported by the National Institute of Allergy and Infectious Diseases. Contribution No. 3591.

† *Explanation of Terms*

Antigens (Ag): S^{35}BSA:	Tracer antigen obtained from diazotization-coupling reactions of S^{35}-sulfanilate with bovine serum albumin.
S^{35}KLH:	Tracer antigen obtained from diazotization-coupling reactions of S^{35}-sulfanilate with key hole limpet hemocyanin.
BSA:	Bovine serum albumin referred to as carrier protein or native, unlabelled protein to distinguish it from S^{35}BSA.
KLH:	Key hole limpet hemocyanin referred to as carrier protein or native, unlabelled protein to distinguish it from S^{35}KLH.
Soluble antigen:	Non-particulate, but colloidal, rather than a true solution. (Term applies to the above antigens.)
Secondary injection:	Application of term varies, referring to *in vivo* or *in vitro* stimulation and to a tracer antigen or a native carrier protein depending upon experimental conditions.
Antigen (Ag) material:	Metabolized antigen that has been isolated from tissues of an injected rabbit, usually on the basis of radioactivity, for the purpose of further identification. (Term "material" often omitted in text.)
RNA:	Antigen-associated material, assayed either by the orcinol color reaction or by 260 mμ absorption and quantitated by comparison with a known amount of purified yeast RNA similarly assayed.
Ag-RNA complex:	Orcinol-positive, U.V.-absorbing, antigen material.

in the tissues of an immunized rabbit. Liver tissue was chosen for the technical reason that the quantity of retained antigen was greater in this tissue than in any other tissue. When liver brei was fractionated with sucrose, antigen material was concentrated in the soluble fraction. Further fractionation to remove non-antigen material was effected by absorption on Dowex-2 resin and subsequent elution with sodium salicylate. Ribonucleic acid was present with the concentrated antigen, and when further fractionation was carried out with 50% ammonium sulfate, the nucleic acid and antigen remained in the soluble fraction with an S_w value of 3.6 to 4.3 (Garvey and Campbell, 1957). The fraction that contained antigen and RNA was at least 200-fold more active for *in vitro* anaphylactic sensitization, i.e., in the Schultz-Dale reaction than the original antigen and was tentatively regarded as a complex since neither component was dialyzable unless denatured by heat or alkali. Studies carried out over the past dozen years have amplified and extended these initial findings but determining their significance in the immune reaction still requires much work.

II. RESULTS

A. The Secondary Response

1. *Aspects of the Secondary Response:*

From the finding of a degraded form of antigen, i.e., an antigen fragment that had enhanced biological activity compared with the original antigen it was postulated that such a form of antigen might account for the rapidity of the secondary response (Campbell and Garvey, 1961). Studies such as the following have strengthened this hypothesis.

a. *Loss of Primary Injection, an Effect of Antibody.* Antigen retention, even in the liver, is relatively small when expressed as a percentage of the injected dose, and decreases with time after injection. When only the first injection in a series of injections was labelled, the detection of radioactivity could be used to indicate antigen material remaining from the initial injection (Garvey and Campbell, 1958a). Quantitative assays indicated that retention in the liver varied inversely with circulating precipitating antibody, i.e., a linear plot was obtained that showed antigen retention to be high when the serum antibody was low. Further investigation showed that the essential factor was intracellular antibody-forming activity and not antibody per se in the circulation since passive transfer of antibody failed to affect the rate of loss of antigen from liver (Garvey and Campbell, 1958b).

b. *Cellular Events.* More recently, cellular aspects of these findings have been demonstrated by radioautography (Garvey and Campbell,

1966). The absence of antigen material that is usually present after primary injection of labelled antigen is obvious when histological sections of a liver are compared from a rabbit given only the radioactive antigen and one that was given secondary injections of the corresponding *unlabelled* carrier protein (Fig. 1). When the time after secondary injection is reduced from days to minutes and only one reinjection is made instead of several, then the initial stages of antigen loss may be observed (Fig. 1). The derangement of parenchymal liver cells with cytoplasmic alteration and antigen release is a striking observation within one hour following a secondary injection of the native carrier protein into a rabbit that has received an injection of radioactive antigen one to two weeks earlier. Parallel studies have been made of histological reactions occurring after the secondary injection in tissues other than the liver, e.g., lymph nodes and spleen. It is often possible to detect an increase of antigen material in these latter tissues; this increase seems attributable to transfer of antigen from the liver. A common finding is a conspicuous localization of antigen material in the secondary nodules of the spleen after secondary injection of KLH into a rabbit that has received a primary injection of S^{35}KLH (see Fig. 2).

From radioautographic observations on tissue sections it was concluded that the antigen material was predominantly in the parenchymal liver cells at the times observed following the initial injection of antigen. Confirmation of this point was gained through the isolation and examination of preparations that were principally either parenchymal or R. E. cells as shown in Figure 3. When comparable volumes of cells were counted to determine radioactivity present at 1 day following injection of S^{35}BSA, at least a 30-fold greater value was obtained for the parenchymal cell preparation compared with the R. E. cell preparation. It may be seen from the photographic reproduction in Figure 3 that the cell volume for individual cells of the two types is vastly different and, furthermore, the observation of grains is difficult in freshly isolated cells of either type. When cells are cultured, grains indicating the presence of antigen are better visible in the stretched parenchymal cells compared to freshly isolated cells. Furthermore, the assay of culture medium shows radioactivity, thus indicating the loss of antigen material *in vitro* similar to the slow, continuous loss *in vivo*.

c. *Circulation of Antigen and RNA*. Another finding that is associated with the tissue studies is the detection of an increased amount of antigen material in the circulation following secondary stimulation with antigen. Data relative to this study have been plotted in Figure 4. The quantity of RNA material that is detectable by the orcinol reaction is likewise increased after the secondary injection and this rise is more rapid, greater, and more sustained than that observed after the primary injection.

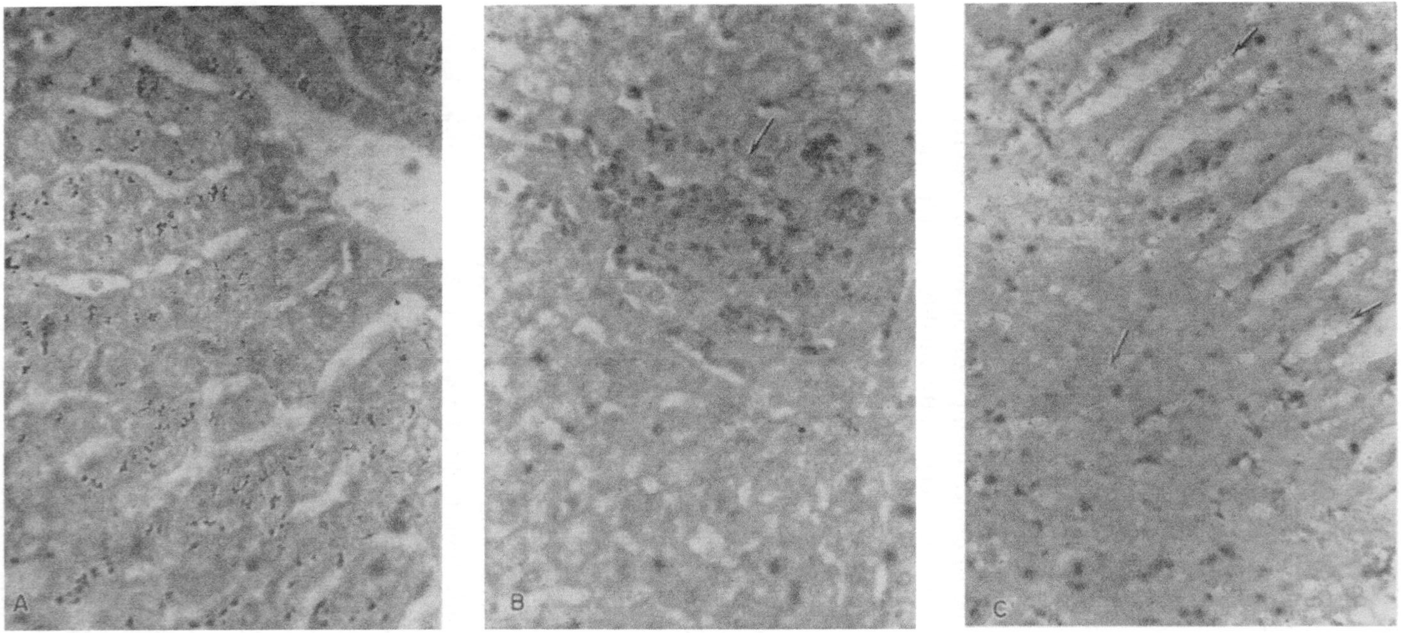

Fig. 1. Photomicro-radioautographs prepared with 5μ paraffin tissue sections as described by Garvey and Campbell (1966). In **A,** radioautographic grains reveal S^{35}Ag retention in parenchymal liver cells at 21 days following a single injection of S^{35}BSA. With 4 secondary injections of the *unlabelled* carrier BSA into another rabbit **B,** which had been injected initially with S^{35}BSA, a negligible number of grains is observed and this would indicate the absence of S^{35}Ag although the interval of time since the initial injection of S^{35}BSA and the histological preparation was 21 days, as in **A.** (The reinjections were begun 2 days after the initial injection of S^{35}BSA and 2–3 day intervals separated the reinjections, with the last of the four injections being 9 days after the initial injection of S^{35}BSA; thus there was a lapse of 12 days from the last injection to the time when tissue was obtained for histological studies.) The field in **B** shows a concentrated localization of lymphoid cells (see arrow), but these lack radioactive antigen material as do the parenchymal liver cells.

When instead of multiple secondary injections, only a single secondary injection of *unlabelled* carrier BSA is made, at an interval of at least 7 days following the injection of S^{35}BSA, and when the tissues are obtained for histology within one hour after the reinjection of unlabelled antigen, the findings are as demonstrated in **C.** Arrows indicate released antigen as observed throughout the field of deranged parenchymal cells and "migrant" lymphoid cells. It is to be noted that with reference to time and procedure of immunization, there is a close similarity to the findings with sera. ×800.

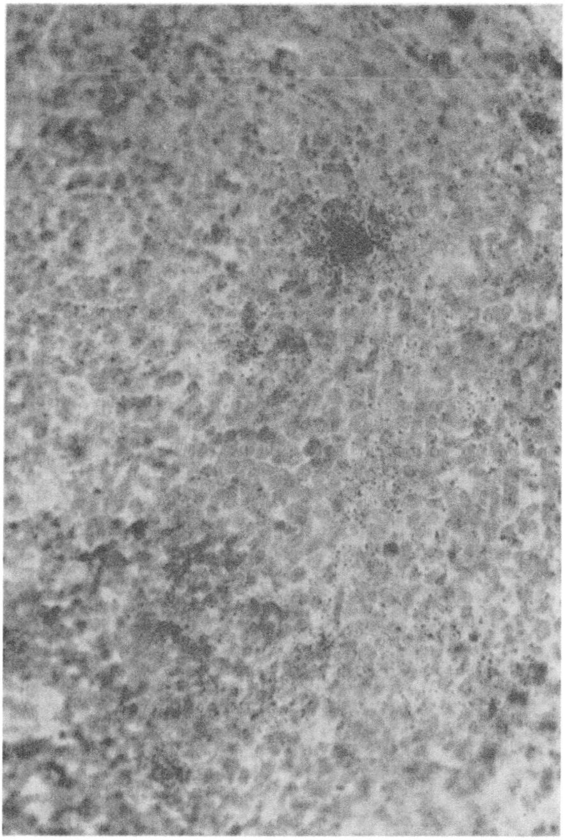

Fig. 2. After immunization as described in Fig. 1 for A and B, marked localization of material from the initial injection can be noted in lymphoid tissue, particularly in the spleen; localization is particularly marked when the initial antigen is S^{35}KLH and the secondary antigen is KLH. ×1100.

d. *Excretion of Antigen and RNA.* A similar increase in antigen and RNA material can be observed in urine following secondary injection. Since details of the excretion studies have been published (Garvey *et al.*, 1967) the major findings need only to be mentioned, these being an increased amount of antigen material in the urine following secondary stimulation and a corresponding increase in nucleic acid material.

2. *Summary of Findings Relative to Role of Antigen Material*

Experimental findings such as those mentioned led to the diagram in Figure 5. A retention of antigen material from a primary injection of antigen may occur at a significant molecular level in some cells for the life span of a rabbit; this is plausible despite continuous loss of antigen

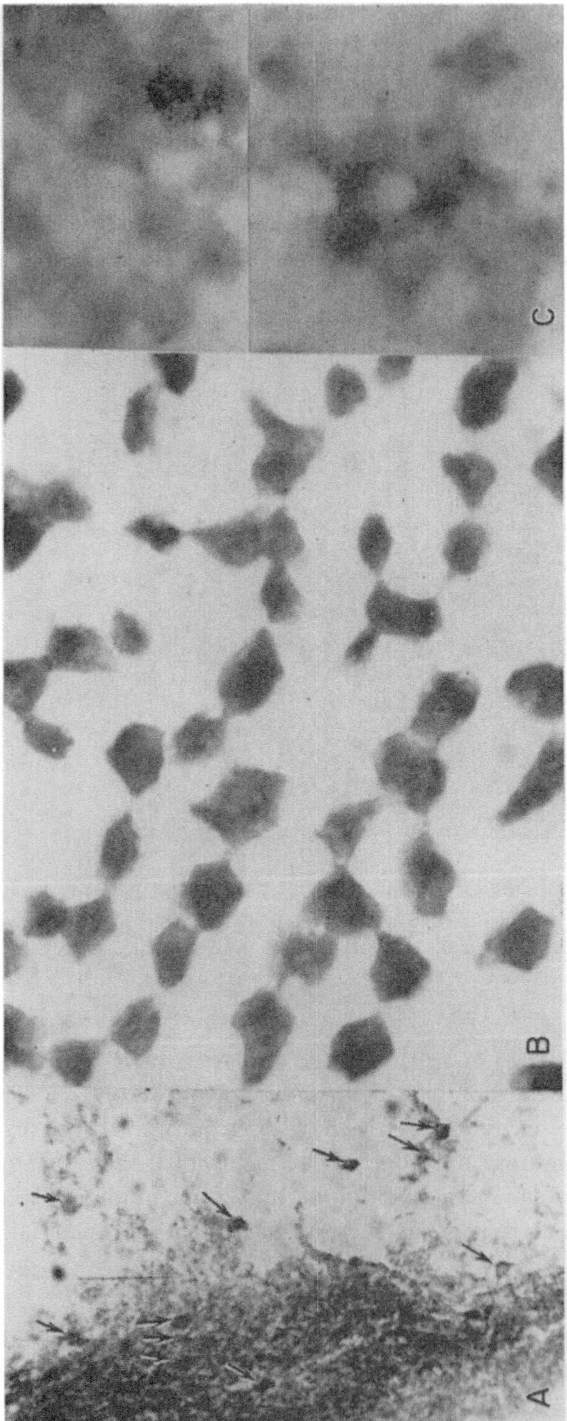

Fig. 3. Separation and study of parenchymal and R. E. cells from liver tissue. In order to determine more conclusively which of the two main groups of liver cells, the parenchymal or the R. E., contain antigen material, cells have been separated according to Garvey (1961). This method yields distinctly different preparations of cells as shown by comparison of **A** (R. E. cells) and **B** (parenchymal cells). **C** is a photomicro-radioautograph of an *in vitro* culture of parenchymal cells. ×1100.

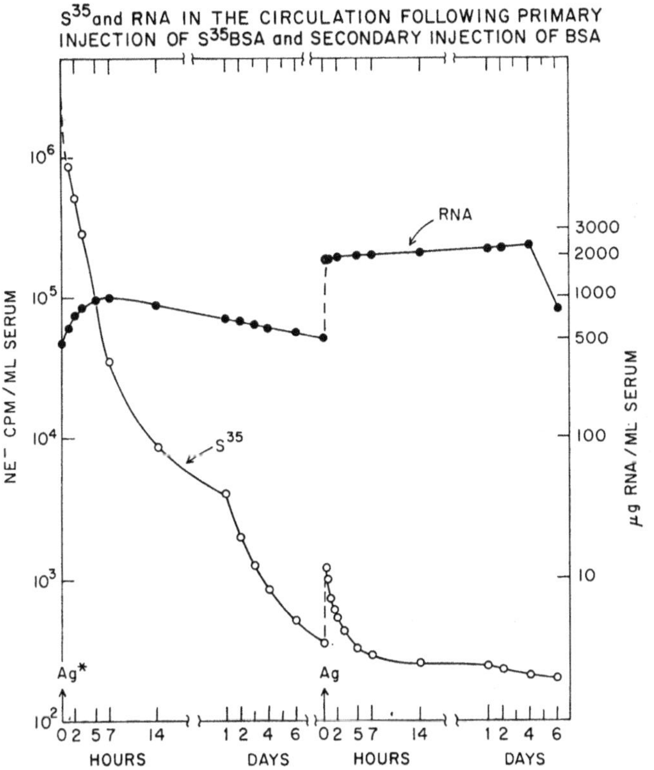

Fig. 4. Determination of radioactivity and RNA in sera obtained from consecutive bleedings of a rabbit injected initially with 2 ml S^{35}BSA, containing 20 mg protein and 100 μc S^{35}, and 7 days later with 1 ml BSA, containing 5 mg *unlabelled* carrier BSA. Both injections were by intravenous route, but only the initial injection contained radioactivity. The time of the bleedings, which is shown on the abscissa, was expanded for the plotting of data obtained for bleedings during the first day after each injection. The time scale was reduced for the plotting of data from samples obtained at daily intervals beginning one day after each injection. Radioactivity, the curve with open circles, refers to the scale shown on the left ordinate and RNA, the curve with solid circles, to the scale on the right ordinate.

material from cells. When antigen is introduced into the circulation a second time and cells become reexposed to new antigen, the loss of antigen material that is already present from the previous injection is accelerated. Cells (parenchymal liver cells) show a cytotoxic reaction (see photomicro-radioautographs in Fig. 3), i.e., a hypersensitivity due to the presence of antibody. There is some evidence for the latter, as indicated in the serum fractionation (Fig. 8), and in prior studies (Garvey and Campbell, 1956). The released antigen is lost to the vascular system and subsequently encounters other tissue cells that are efficient antibody-forming cells. During this course of events antigen material is also excreted.

B. Studies to Elucidate the Nature of the Antigen-RNA Complex

The information obtained from several sources of materials, from varied conditions of immunization, and from a variety of techniques suggested that ribonucleic acid was involved in the metabolism of antigen and that a stable, chemical linkage may be formed between nucleic acid and antigen material. Nonetheless the question about the nature of the association between antigen and RNA was not answered and conclusive evidence was sought by further investigation. One of the most logical forms of antigen material to investigate was that representing a maximal degree of degradation of the original antigen, yet minimal association with cellular constituents, particularly with antibody. The nucleopeptide material referred to as *n-p* by Saha *et al.* (1964) seemed

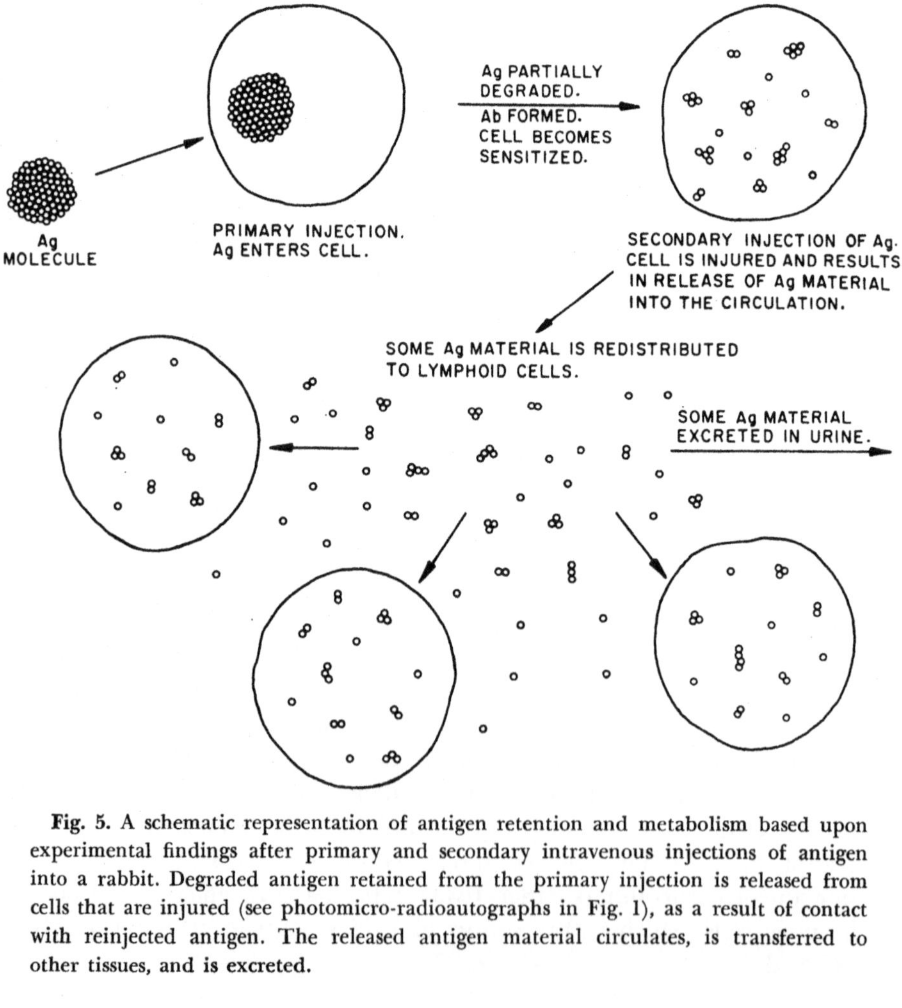

Fig. 5. A schematic representation of antigen retention and metabolism based upon experimental findings after primary and secondary intravenous injections of antigen into a rabbit. Degraded antigen retained from the primary injection is released from cells that are injured (see photomicro-radioautographs in Fig. 1), as a result of contact with reinjected antigen. The released antigen material circulates, is transferred to other tissues, and is excreted.

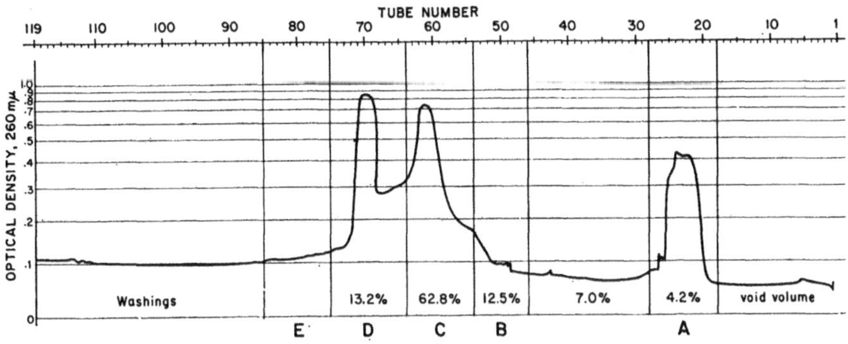

Fig. 6. Fractionation of liver *n-p* on Sephadex G-50 with recording of 260 mμ absorption, and of the radioactivity in peaks. A tracing is shown for the fractionation, recorded with 260 mμ optics for a 19 hr., 40 min. run. The tube number in the collection series is shown at the top of the figure. The contents of the tubes were pooled according to absorption peaks and, as indicated by the vertical lines, these pools were designated as A, B, C, D and E. The remaining tubes were pooled and designated as indicated. The distribution of eluted radioactivity is indicated for the various peaks.

especially worthy of further investigation. A second form of antigen material that previous studies had ear-marked for further investigation was that released from liver at the time of reinjection of antigen. The latter was of particular interest since it may play an initial role in the stimulus leading to increased antibody production that characterizes the secondary response. From the previous investigations it appeared that either the urine or the serum might be used as a source of the released antigen material.

Parts of the investigations * on the liver *n-p* and on serum, representing primary and secondary antigen respectively, will be described.

1. *Liver as a Source of Antigen Material from the Primary Injection of Antigen*

a. *Preparation and Fractionation of n-p.* A rabbit, injected intravenously with 2.5 ml S^{35}BSA, containing 30 mg protein and 400 μc S^{35}, was killed 24 hours later. Liver brei was treated with perchloric acid and the soluble *n-p* fraction was obtained essentially as described by Saha *et al.* (1964). A sample of 500 mg lyophilized *n-p* was reconstituted in water and the solution was fractionated on a column of G-50 Sephadex, using water for elution. The effluent was scanned for U.V. absorption at 260 mμ and the recording of this has been traced as shown in Figure 6. The corresponding fraction (tube) number is given at the top of the figure. Contents of the various tubes were pooled according to the absorption

* A detailed report of these investigations will be published in collaboration with Dr. Henry Rinderknecht and Dr. Dan H. Campbell.

peaks and as indicated by the vertical lines; these pools were assigned a letter designation, *A, B, C, D* and *E*. The remaining tubes were pooled and designated as indicated. Radioactivity was determined in each pooled material and the relative percentage eluted in each pool was calculated. The void volume, *E*, and washings together accounted for less than 1% of the radioactivity. There was a significant amount of radioactivity in all absorption peaks and also in the pool prepared from the tubes collected between *A* and *B*. The material *C* that showed a high absorption peak and contained the most radioactivity was used in additional studies.

b. *Physical-chemical Properties of C.* The heterogeneous composition of *C* was obvious after an initial separation by descending chromatography. The observed U.V. absorption, traced on the chromatogram, is shown and described in Figure 7. The strip of paper that contained the chromatogram was subsequently stitched on paper that was prepared for high voltage electrophoresis. An additional separation was obtained by the latter technique, relying initially on visual observation of U.V. absorption to identify individual components. Components that possessed a peptide bond were identified by the chlorination color reaction described by Rydon and Smith (1952) but performed according to the modification used in Dr. William J. Dreyer's laboratory. Further identification was made of components that were radioactive, and the spectral identification of the components is in progress.

c. *Biological Properties of C.* The components, identified by number in Figure 7, have been extracted with water, with extreme precaution being exercised to avoid extraneous contamination during lyophilization and subsequent handling. The *in vitro* lymph node culture technique, as described by Halliday and Garvey (1964), has been used to identify which individual components, among those numbered in Figure 7, have specific immunogenic activity as indicated by "triggering" the secondary response. The passive hemagglutination method, using BDB cells prepared as described by Arquilla (1967) and sensitized with BSA, has been applied to this initial screening for immunogenic properties. The results have shown that the anodic components 2, 3, and 4 and components 5, 8, and 10, which are neutral to slightly cathodic, produce a positive response. The results seem particularly significant in view of the finding that the peak *in vitro* antibody titer of these particular isolated fractions approximates that obtained with the optimal concentration of original antigen used in a control culture, but more important is the finding that a positive titer for antibody in the culture fluid is obtained more than 30 days after stimulation by these particular antigen materials, whereas the titer of antibody in response to the original antigen is lower or even negative at this time. Isolation of these components on a preparative scale seems feasible now that an extremely immunogenic form of antigen material can be isolated.

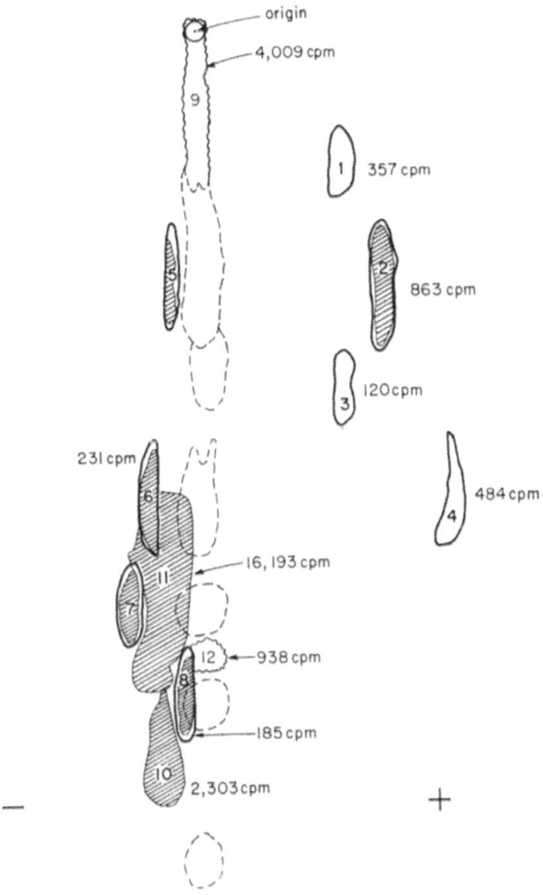

Fig. 7. Two-dimensional separation of constituents in C (refer to Fig. 6), i.e., mapping by chromatography followed by electrophoresis. Findings are summarized for U.V.-absorbing, peptide-staining and radioactive components. An aqueous sample equivalent to 3.5 O.D. units at 260 mμ was separated initially by descending chromatography in solvent consisting of 2 parts of 0.5 *M* ammonium acetate and 5 parts of absolute ethanol, adjusted with glacial acetic acid to pH 3.8. Examination by a short wave U.V. lamp source revealed the absorption spots that are indicated by discontinuous lines and the fluorescence spots that are indicated by wavy lines.

The strip of paper that contained the chromatogram was stitched on No. 3 Whatman paper and a further separation of components was obtained by electrophoresis in 7% formic acid for 2 hours, at 3300 volts and a temperature of 11–14°C. The paper was examined for U.V. absorption and the respective areas were traced with solid lines.

Peptides (cross-hatched areas) were detected by staining. Areas that are numbered on the map were extracted with water, and the indicated radioactivity is that found in 16 unstained maps. The pooled extracts of individual components were used in a tissue culture assay method in which the results were used to screen biologically active components.

2. *Serum as a Source of Antigen Material Released as a Result of Secondary Stimulation with Antigen*

a. *Screening of Sera.* During the initial investigation, which revealed findings presented in Figure 4, a number of bleedings were made and the sera were assayed for radioactivity and for orcinol-positive material. On the basis of these two assays it appeared that an animal could be "bled out" within one hour after the second injection for a maximal amount of released antigen in the serum. When, in addition, hemagglutination titers were also obtained on the individual sera, certain correlations were noted that led to the grouping of "bleed out" sera on the following basis: (1) When the orcinol-reaction showed a very significant rise after the primary injection (at least a two-fold increase over the normal, preinjection level) a positive hemagglutination titer was obtained in the absence of a secondary injection of antigen. In such a rabbit the level of released antigen was not as high on secondary injection of antigen as in a group #2 serum. (2) When neither a marked increase in the orcinol value nor a positive serum antibody titer was obtained after primary injection alone, the release of both antigen material and RNA material was very marked after the reinjection of antigen.

From a series of daily bleedings after the secondary injection into other animals that were not bled out but were used for background information, it became known that antibody would be detected after a time lag of 2–4 days whenever the release of antigen and RNA material was at a very significant level such as in the example given in Table 1. Thus, a "bleed out" serum described as (2) in the table characterizes a secondary response. An occasional animal that failed to show an increased level of RNA-positive material and an increased rate of antigen release at any time after the primary or secondary injection was still giving a negative reaction in the passive hemagglutination assay as long as the test period of 7 days that followed the secondary injection. It seems probable that the rate at which antigen is released in the interval between the two injections also differs for the two groups, i.e., if there is a very significant rise in RNA material in the circulation after the primary injection alone, there probably occurs an increased rate of antigen release. An increased release of antigen is more obvious after secondary injection of antigen, i.e., at a time when the level of radioactivity is low compared to early after the primary injection, a point that is obvious from the comparison of two sets of sera, each set representing serial bleedings from a single rabbit (Table 1). Such observations are mentioned simply because they emphasize the importance of examining individual sera before pooling, and also because they have been useful guide lines in the selection of material for characterization.

b. *Alcohol Fractionation of Sera.* In addition to the orcinol reaction

Table 1

SCREENING OF SERA BY RADIOACTIVITY, ORCINOL AND HEMAGGLUTINATION ASSAYS

Antigen Injection		(1) Time	Net cpm	μg RNA	HA		(2) Time	Net cpm	μg RNA	HA
		−1 d., −1 hr.*	0	560			−1 d., −1 hr.*	20	510	
$S^{35}BSA$	→	0				→	0			
		1d	9,263	1240			1d	11,143	520	
		2d	4,672	1290			2d	4,898	600	
		5d	1,584	1295			5d	1,462	505	neg
		7d †	1,044	975	80		7d †	1,209	635	neg
BSA	→	0				→	0			
		1 hr.‡	1,065	950	neg		1 hr.‡	3,144	1050	neg

* Average of 2 separate normal, pre-injection sera.
† Bleeding 1 hr. before reinjection.
‡ After reinjection "bleed out."
HA = passive hemagglutination.

and radioactivity assay, sera collected from individual rabbits, bled at different times after injection and reinjection of antigen (see Figure 4), were fractionated with ethanol. The procedure was the same as described by Garvey *et al.* (1967) for a crude separation of urine into fractions that were assayed for radioactivity and RNA material. The findings differed from those obtained for urine and lend strong support to the probability that intracellular degradation is a significant mechanism of antigen metabolism, rather than an enzymatic degradation per se in the circulation. Furthermore, the distribution of radioactivity within the various fractions seems to be related to antibody formation. Throughout the first hour after primary injection of antigen, essentially all the radioactivity and RNA material were precipitated by 2 volumes of ethanol. The precipitated radioactivity could be separated into two fractions on the basis of solubility in water, and unlike a control of S^{35}BSA in serum that was about equally divided in radioactivity between the two fractions, the sample serum had 6–8 times more radioactivity in F-1 (water insoluble) than in F-2 (water-soluble). By the third hour there was a shift in the relative percentage of radioactivity, from F-1 to F-2 and F-4 (soluble at high alcohol concentration) and this continued, so that by 14 hrs after injection F-2 and F-4 together equalled F-1. F-3, the insoluble fraction obtained at high alcohol concentration, contained less than 2% of the radioactivity at any time. Between 1 and 3 days there was a relative loss in radioactivity from F-4, which was reflected as a gain in either F-1 *or* F-2; which of these two fractions gained, appeared related to whether or not antibody appeared after the primary injection. The F-1 radioactivity was higher in an antibody-forming animal than in an animal that failed to form antibody. As far as the reinjection of antigen is concerned, an animal that continued to be negative showed essentially no shift in the relative percentages of radioactivity in the various fractions before and after reinjection of antigen. If an animal had produced antibody during the time intervening between the two injections of antigen, a shift in the relative percentage of radioactivity from F-1 to F-2 occurred, the change being particularly significant between 4–6 days when the increased antibody production, related to the reinjection of antigen, became evident in the circulation. In all samples from different animals the RNA material was precipitated at low alcohol concentration, thus differing from the findings with urine. Qualitative information regarding these shifts in radioactivity will be obtained when this particular method is combined with column fractionation.

c. *Sucrose Density Gradient Centrifugation.* The method described by Kunkel (1960) has been used with "bleed out" sera of both groups (1) and (2), described in Table 1, and also with two sera from the same animal, one of which was collected at 15 minutes following an initial injection of S^{35}BSA and the other at 15 minutes following a secondary injection

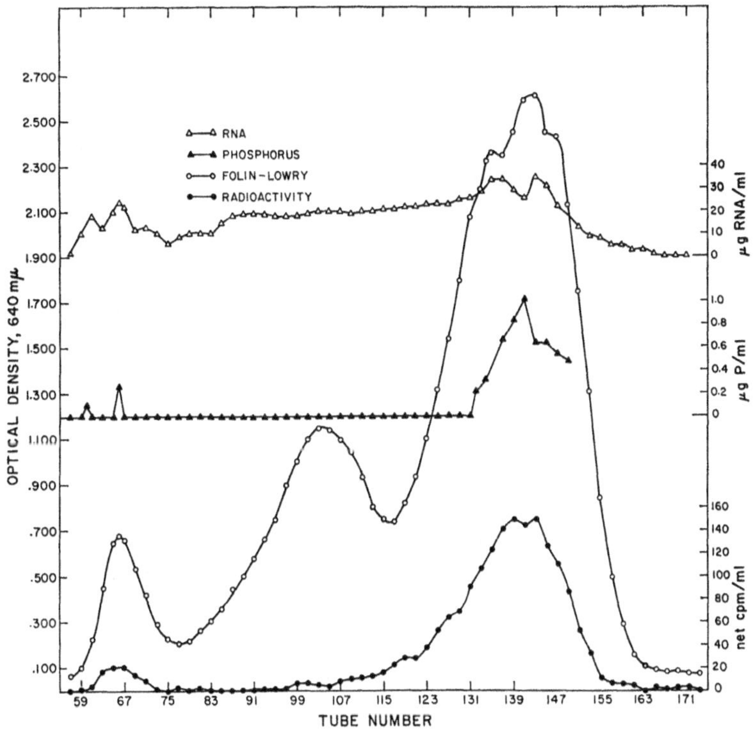

Fig. 8. Sephadex G-200 fractionation of a serum obtained one hour after injection of *unlabelled* carrier BSA into a rabbit that had received a primary injection of S^{35}BSA 7 days previously. (See text for data on the radioactivity in this whole serum and for previous serum samples from the same animal.) A 10-ml sample was layered between the buffer (0.1 *M* Tris-HCl, pH 8.0 in 0.2 *M* NaCl) in a column, 5 × 52 cm, using a procedure similar to Flodin and Killander (1962). Fractions, 5 ml/16 min., were eluted with Tris-NaCl buffer and collected over a 40-hour period. The distribution of protein in the fractions was determined by the Folin-Lowry test as described by Lowry *et al.* (1951). The O.D. units read at 640 mμ were plotted as open circles, the values of which may be found on the left ordinate. Each fraction was also assayed for radioactivity for which a curve with solid circles has been plotted and to which the bottom scale along the right ordinate applies. The phosphorus content of fractions, determined by the method of Allen (1940), are read from the solid triangles with reference to the middle scale on the right ordinate. Orcinol-positive values for the fractions have been obtained by the method of Mejbaum (1939), standardized with yeast RNA, and the resulting values have been plotted as open triangles with reference to the top scale along the right ordinate.

of BSA. The separation of protein into a minor and a major component matched the original description by Kunkel (1960). Radioactivity was found distributed in all fractions of the primary sample but fractionation of radioactivity into a minor and major peak, corresponding to the protein peaks, occurred in the reinjection sample. The exception to these findings was the absence of radioactivity in the minor protein peak of a "bleed out" serum of group (1) in Table 1, i.e., from an animal that had demonstrated antibody before secondary injection. The fractionation obtained by this method (including the detection of phosphorus and RNA) was very similar to that resulting from Sephadex G-200 column chromatography as described in the next section, but the latter type of fractionation had distinct technical advantages.

d. *Sephadex G-200 Fractionation.* This method of fractionation has been adopted on a semi-preparative scale for whole sera as described by Flodin and Killander (1962), and was utilized, more recently, following a preliminary ethanol fractionation as described above. The results obtained on a whole serum of group (2) (Table 1) are presented in Figure 8. The two peaks, characterized by a varying amount of radioactivity, contain small amounts of phosphorus and have a significant concentration of orcinol-positive material. These findings have been confirmed by replicate fractionations and assays on 10 ml samples of whole serum. To date, the results have been less significant with these fractions than with *n-p* components similarly assayed *in vitro* for biological activity. For this reason enzymatic treatment is being carried out with the material from both peaks of radioactivity. In the case of the higher molecular weight peak the enzymatic treatment is being preceded by acid dissociation because of the suspicion that antibody may be present together with antigen material.

III. DISCUSSION AND SUMMARY

This report has emphasized experimental findings and has attempted to suggest how these facts might be helpful in understanding why the secondary response to antigen results in such pronounced antibody formation. The physical-chemical findings of our study give additional support to the existence of complexes of RNA and antigen material. After electrophoretic separation, an association of the two materials still existed. The high level of immunogenicity of these materials was demonstrated by an *in vitro* method which is used at present for the screening of immunogenic material. The presence of radioactivity, derived from S^{35}BSA, was detected in all biologically active fractions with the exception of component 5 on the "finger-print" map reproduced in Figure 7. The fact that peptide was not detected in some components despite the presence of radioactivity emphasizes the greater sensitivity of the radiolabelling procedure; the peptide stain procedure has a minimal sensitivity of detection of about 0.3 μg for glycine and 0.5 μg for BSA in a 12 mm diameter spot (Rydon and Smith, 1952).

Related to the question of detectability (both physical-chemical and biological) is the question of whether undegraded antigen may be present in the material. The calculation of protein based on radioactivity gives extremely low values; a maximum value was obtained for component *10* (Fig. 7), namely 0.15 μg for the total extraction from 16 maps (=the number of replicate maps for which the radioactivity is given in Fig. 7) and 0.025 μg protein/ml as antigen concentration in the biological test. The areas involved in the map suggest a concentration insufficient for detection of some components by the peptide bond spray, and in regard to the antigenic stimulation in the *in vitro* system, the calculated amounts of unaltered protein fail to meet the required level of antigen for a secondary response *in vitro.* Other facts, such as the entire procedure of isolation and separation of the components, and also their absorption characteristics, conflict with the possibility that unaltered protein may be present. However, it is obvious that additional separation of component *10,* the biologically active component free of nucleic acid material, is required prior to further characterization of active materials. More detailed analysis of the separated components must be made in order to find a basis for the varying degrees of biological activity that was observed. From such studies, some inference may become possible regarding the nature of the antigen involved in immune reactions *in vivo.*

The evidence gained from the serum studies, as well as the photomicroradioautography of tissues, implicated cells as responsible for antigen metabolism. The radioautographic studies also suggests cellular interactions, as indicated by the migratory activity of lymphoid cells into the area of parenchymal cells from which antigen is being released after secondary injection of antigen. The tissue reactions are also suggestive of an involvement of antibody. Since nucleotide material is present in the first peak from Sephadex chromatography, and also in the high density peak from density gradient centrifugation, nucleotide material and/or antibody may account for the small amount of antigen material that exists as a component with higher molecular weight than the major amount of antigen material. At this stage in the isolation and characterization of antigen material from the serum, it is reasonable to conclude that, as previously indicated by studies on liver materials, nucleic acid is an inseparable companion of antigen in the early stages of the immune response.

References

Allen, R. J. L. (1940). The estimation of phosphorus. Biochem. J., *34*:858–865.

Arquilla, E. R. (1967). Antigen-red cell conjugates prepared with bisdiazobenzidine *in* Methods in Immunology and Immunochemistry, M. W. Chase and C. A. Williams (editors), Academic Press, Inc.

Campbell, D. H., and J. S. Garvey (1961). The fate of foreign antigen and speculations as to its role in immune mechanisms. Lab. Investigation, *10*:1126–1150.

Flodin, P., and J. Killander (1962). Fractionation of human serum proteins by gel filtration. Biochim. Biophys. Acta, *63*:403–410.

Garvey, J. S. (1961). Separation and *in vitro* culture of cells from liver tissue. Nature, *191*:972–974.

Garvey, J. S., and D. H. Campbell (1956). Studies of the retention and properties of S^{35}-labeled antigen in livers of immunized rabbits. J. Immunol., *76*:36–45.

——— (1957). The retention of S^{35}-labelled bovine serum albumin in normal and immunized rabbit liver tissue. J. Exper. Med., *105*:361–372.

——— (1958a). The *in vivo* identification in liver tissue of an S^{35}-azo label with its protein carrier. Arch. Biochem. Biophys., *73*:507–508.

——— (1958b). Effect of secondary injections of antigen upon the retention in liver of a primary injection. J. Exper. Med., *107*:497–506.

——— (1966). Autoradiographic investigation of tissues after primary and multiple antigenic stimulation of rabbits. Nature, *209*:1201–1202.

Garvey, J. S., D. H. Campbell, and M. L. Das (1967). Urinary excretion of foreign antigens and RNA following primary and secondary injections of antigens. J. Exper. Med., *125*:111–126.

Halliday, W. J., and J. S. Garvey (1964). Some factors affecting the secondary immune response in tissue cultures containing hydrocortisone. J. Immunol., *93*:757–762.

Kunkel, H. G. (1960). Macroglobulins and high molecular weight antibodies. Chapter 8 *in* The Plasma Proteins, F. W. Putnam (ed.), Academic Press, Inc., New York, Vol. I, 279–307.

Lowry, O. H., N. J. Rosebrough, A. L. Farr, and R. J. Randall (1951). Protein measurement with the Folin phenol reagent. J. Biol. Chem., *193*:265–275.

Mejbaum, W. (1939). Über die Bestimmung kleiner Pentosemengen, insbesondere in Derivated der Adenylsäure. Z. Physiol. Chem., *258*:117–120.

Rydon, H. N., and P. W. G. Smith (1952). A new method for the detection of peptides and similar compounds on paper chromatograms. Nature, *169*:922–923.

Saha, A., J. S. Garvey, and D. H. Campbell (1964). The physicochemical characterization of the ribonucleic acid-antigen complex persisting in the liver of immunized rabbits. Arch. Biochem. Biophys., *105*:179–192.

THE NATURE OF IMMUNOGENIC RNA-ANTIGEN COMPLEXES IN IMMUNE AND TOLERANT MICE *

HERMAN FRIEDMAN

Departments of Microbiology, Albert Einstein Medical Center and Temple University School of Medicine Philadelphia, Pennsylvania

There is currently widespread interest concerning the role of subcellular fractions and nucleic acids in antibody-formation. During the late 1950's it was reported that nucleoprotein-rich extracts from donor rabbits immunized with bacterial antigens produced specific agglutinins in nonimmune individuals (Sterzl and Hrubesova, 1956; Friedman, 1959). More recent studies concerning molecular aspects of antibody formation have been concerned with the apparent conversion of "normal" nonimmune lymphoid tissues or cells *in vitro* to specific immunologic competence following treatment with RNA-rich substances extracted from lymphoid cells or macrophages exposed to a specific antigen, either *in vitro* or *in vivo*. For example, RNA-rich extracts from rat macrophages exposed to bacteriophage antigen *in vitro* appear to stimulate specific antibody formation by normal lymph node cultures (Fishman, 1961; Fishman and Adler, 1963). The nature and mode of action of such nucleic acid-rich extracts is under investigation by a number of groups. Although much of the original work in this area implicates low molecular weight RNA as a possible "code" for antibody formation (Cohen, 1967; Fishman and Adler, 1963; Campbell and Garvey, 1963) some recent evidence suggests that an antigen, or an antigenic determinant, possibly closely associated with RNA, may be involved in such transfer reactions either *in vivo* or *in vitro* (Askonas and Rhodes, 1965; Friedman *et al.*, 1965; Adler *et al.*, 1966).

* Supported by grants from the National Science Foundation and the American Cancer Society. The capable technical assistance of Mrs. Gertrude Mellof, Mr. Leonard Silverman, Mrs. Leony Mills, Mr. Barry Silvers, Miss Joyce Jenderowski, and Mr. Steven Sylk during various portions of this study is acknowledged.

Most studies concerning transfer of immune reactivity by subcellular extracts *in vitro* and *in vivo* have measured resulting antibody increases by serologic methods. Data obtained by several groups indicate that appropriately obtained and prepared nucleic acid-rich extracts stimulate specific antibody formation and other immune responses in normal animals and tissue cultures (Fong *et al.,* 1961; Nester *et al.,* 1961; Mannick and Egdahl, 1962; Campbell and Garvey, 1963; Michelazzi *et al.,* 1965; Esposito and Ligniere, 1966; Wilson and Wecker, 1966; Thor, 1967). Such procedures detect only the end product of immunologic reactivity, not cellular events. However, during the past few years studies utilizing localized hemolytic plaque techniques have also indicated that RNA-rich extracts from immune donors can induce active antibody formation measurable on the cellular level. Normal spleen cells incubated with nucleic acid-rich extracts from donor mice immunized with sheep red blood cells have produced an increase in the number of specific hemolysin-forming cells *in vitro* (Cohen and Parks, 1964; Friedman, 1964b and c, 1966a; Cohen *et al.,* 1965, 1967).

In a number of earlier studies we had noted that nucleic acid-rich extracts prepared by salt extraction procedures from spleens of donor rabbits immunized with soluble antigens of *Shigella paradysenteriae* induced specific agglutinin formation in nonimmune recipient rabbits, both in normal and in X-irradiated recipients (Friedman, 1959, 1963a, 1964a). Also, in studies with mice it was found that nucleic acid-rich material from donor mice immunized either with Shigella antigen or sheep red blood cells induced specific agglutinin formation in nonimmune recipient mice (Friedman, 1963b, 1966a, b). The activity of the extract could be abolished by pre-treatment of the extracts *in vitro* with RNAase or with specific antiserum to the immunizing antigen (Friedman, 1966a, b).

For the experiments to be described here, nucleic acid-rich extracts were prepared by phenol extraction and alcohol precipitation methods from mice at various times after immunization with Shigella antigen (Fig. 1). In addition, similar extracts were prepared from mice that were tolerant to Shigella antigen as a result of injection of the specific antigen at birth. The extracts were injected into groups of recipient mice, either nonimmune or previously primed with specific antigen (Fig. 1). The immunologic response of the recipients was determined by serologic titration of serum specimens and by an indirect hemolytic plaque assay with spleen cell suspensions (Fig. 1). In additional experiments to be described here, RNA-extracts were added to spleen cell cultures, either from normal or antigen-primed mice, and the cellular response *in vitro* was determined by the plaque technique. Information obtained from these studies suggests that persisting antigenic determinants in the RNA-rich extracts may be important components of the systems in which antibody production is elicited by the transfer of RNA-rich extracts.

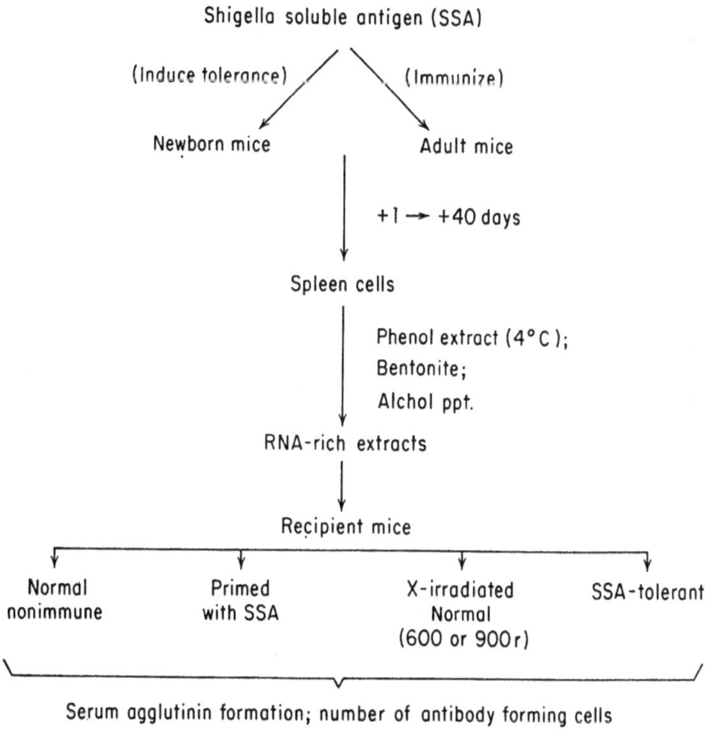

Fig. 1. Experimental scheme for detection of immunogenic activity in RNA-rich extracts from spleens of Shigella-immune or -tolerant mice.

I. METHODS AND MATERIALS

A. Experimental Animals

NIH albino A mice, obtained from a closed colony maintained by random breeding by a local dealer in the Philadelphia area, were used for all experiments reported here. The animals were housed in groups of six in plastic cages and were fed water and mouse pellets *ad libitum*. On occasion, antibiotics were added to the drinking water.

B. Antigens

A soluble antigenic extract derived by trypsin digestion of washed, alcohol killed suspensions of overnight cultures of *Shigella paradysenteriae* was used for immunization, for induction of immunologic tolerance, and for coating of sheep erythrocytes for indirect hemolysis (Friedman, 1960, 1962, 1964a, 1965). Sheep erythrocytes, purchased commercially, were washed and suspended in physiological saline solution for injection.

C. Immunization

Adult mice were injected intraperitoneally (i.p.) with 0.1 ml of the soluble Shigella antigens (SSA) containing 20 μg N per ml (0.2 to 0.4 mg dry weight). Mice immunized with sheep RBC were inoculated i.p. with 0.5 ml of a 10 per cent suspension of freshly washed erythrocytes (4×10^8 RBC).

D. Induction of Tolerance

Litters of newborn mice were injected i.p. on the day of birth with 0.1 ml SSA, as described previously (Friedman and Gaby, 1960; Friedman, 1962, 1965).

E. Serology

Blood samples were obtained from experimental and control mice from the retro-orbital plexus prior to and following immunization. Serum agglutinin titers were determined by micro-titer techniques using serial two-fold dilutions in 0.025 ml volumes physiological saline. Hemolysin titers were determined following addition of 0.025 ml of guinea pig serum, containing four units of complement, to each dilution prior to incubation at 37°C for one hour. An 0.05 per cent suspension of alcohol-killed Shigella, or an 0.5 per cent suspension of sheep erythrocytes coated with Shigella antigen (Friedman, 1967) was used for the determination of serum antibody titers. For serologic inhibition tests, 0.025 ml of a 1:800 to 1:1600 dilution of hyperimmune mouse or rabbit anti-Shigella serum was incubated with 0.025 ml of serial dilutions of the test specimen being assayed for Shigella antigen. Following one hour of incubation at 37°C, 0.025 ml of an 0.5 per cent suspension of antigen-coated erythrocytes was added to each titration cup to determine the degree of inhibition of passive hemagglutination. In duplicate tests, 0.025 ml of an 0.05 per cent suspension of alcohol-killed Shigella was added to the serum dilutions to determine inhibition of bacterial agglutination by the test samples. Serologic tests were performed in duplicate on at least two separate occasions.

F. Nucleic Acid Extracts

Spleens from 10 or more control or experimental mice were rapidly frozen with dry ice and extracted with cold phenol and phosphate buffer, as described elsewhere (Friedman, 1964b and c). RNA-rich material was precipitated by alcohol. The RNA content of the extracts was estimated by the orcinol reaction and by UV spectrophotometry.

G. RNA Transfer

One half ml of the nucleic acid-rich extract, containing 0.1 to 0.5 mg RNA, was injected i.p. into recipient animals, or added to tissue culture suspensions consisting of 2 to 5×10^6 nucleated spleen cells per ml. Medium 199 containing 20 per cent mouse or calf serum was used. The tissue culture tubes were incubated in a slowly rotating drum at 37°C and the medium was usually changed every two to three days. Inactivation by enzymes was determined under the following conditions: RNA-rich extracts were incubated with 1 to 5 μg RNAase for 30 minutes at 25°C prior to injection into animals or prior to addition to tissue culture suspensions. Aliquots of RNA extracts were similarly incubated with 1 to 10 μg DNAase per ml or with 100 μg trypsin per ml. In other experiments 1.0 ml of RNA-extracts were incubated with 0.1 ml of a 1:100 dilution of high-titered rabbit or mouse anti-Shigella serum, or, as a control, with normal serum, for 30 minutes at 37°C.

H. Detection of Antibody-forming Cells

The number of antibody-plaque-forming cells (PFC) per spleen or per million nucleated cells was determined by an indirect localized hemolytic plaque technique in agar gel, as described previously (Friedman, 1964d, 1966c). In brief, 0.1 ml of a cell suspension of lymphoid cells was added to 2 ml warm (48–52°C) melted Noble agar to which had been added a dilute suspension of sheep erythrocytes coated with Shigella antigen, or, as a control, uncoated RBCs. This mixture was rapidly and carefully overlaid onto a Petri dish containing a base layer of clear agar. Following solidification of the upper layer and incubation for 1 hour at 37°C, guinea pig complement, diluted 1:15, was added and the plates were incubated again for 30 minutes. Clear zones of hemolysis (plaques) appeared around individual antibody-forming spleen cells. Enumeration of the number of PFC per plate indicated the proportion of detectable antibody-forming cells to non-immune cells in the suspensions. In all instances cell suspensions were tested in duplicate or triplicate with several cell concentrations so as to obtain approximately 50 to 300 plaques per plate. Cell counts and viability were determined with a hemocytometer and by Trypan Blue dye exclusion technique.

II. RESULTS

A. Serologically Detectable Shigella Antigen in RNA Extracts

Phenol extracts prepared from spleens of mice immunized with 20 μg SSA six to ten days previously did not contain serologically detectable

Table 1

PERSISTENCE OF SEROLOGICALLY DETECTABLE SHIGELLA ANTIGEN IN SERA AND SPLEEN CELL FRACTIONS OF MICE INJECTED AS ADULTS OR NEONATES

Preparation	Serologically Detectable Antigen [a] (Day after donor injection with SSA)				
	+1	+3	+5	+7	+14
Immune Adult: [b]					
Serum	+++	±	0	0	0
Spleen Homogenate	+++	±	0	0	0
Spleen Nuclei	++	0	0	0	0
Spleen RNA-Extract	+++	±	±	0	0
Tolerant: [c]					
Serum	++	+	0	0	0
Spleen Homogenate	+++	+	±	0	0
Spleen Nuclei	++	±	0	0	0
Spleen RNA-Extract	++	±	0	0	0
Normal Spleen Homogenate:	0	0	0	0	0

[a] Agglutination-inhibition or passive H.A.-inhibition assay for antigen; detectable level 0.01 μg SSA per ml. At least 5 mice per group per day.

[b] 5- to 7-week-old mice injected i.p. with 20 μg SSA.

[c] Newborn mice injected i.p. with 20 μg SSA.

antigen as determined by hemagglutination-inhibition or direct agglutination-inhibition tests (Table 1). Sufficient antigen to inhibit a positive agglutination reaction was present in the phenol extracts prepared the first day after immunization of donors. However, by the third to fifth day there was only a trace amount of antigen, if any, detectable in the extracts. Likewise, no serologically detectable Shigella antigen could be found in phenol extracts prepared from spleens of newborn mice five to seven days after injection of a tolerance-inducing inoculum of SSA. Antigen was detected in the RNA-rich extracts during the first 24 hours after inoculation at birth. Particulate fractions prepared from spleens of either adult immune or newborn tolerant mice generally had no detectable SSA within a few days after inoculation. Cell-free spleen homogenates and nuclei did not have detectable antigen five days after antigen inoculation (Table 1). Similar results were obtained with serum specimens from these mice (Table 1).

B. Antibody Responses in Mice Treated with RNA-Rich Extracts from Shigella-immune Donors

Most normal mice, inoculated i.p. with about 100 μg RNA-rich extract from mice immunized with SSA, responded with a detectable agglutinin titer (Tables 1 and 2). The highest titers, usually 1:32 to 1:64, were observed in mice injected with RNA extracts from donors immunized 2 to 7 days previously. Agglutinin titers in the recipients usually reached a peak level by six to nine days, similar to that observed in actively immunized animals. When recipients were primed with specific Shigella antigen several months previously, more mice responded with higher titers (Table 2). Usually there were several recipient mice with titers as high as 1:256 or 1:512. Extracts obtained from donor mice immunized one or two days to a week prior to sacrifice were most immunogenic (Tables 2 and 3).

In several experiments prospective nonimmune recipient mice were X-irradiated 24 hours prior to RNA inoculation with either 600 or 900 r (Friedman, 1964). The animals treated with 900 r rarely had any detectable agglutinin response following transfer of the RNA extracts. On the other hand, many recipients treated with 600 r generally had low but detectable agglutinin titers following injection of RNA extracts. However, these mice also responded with similar titers to active immunization (Table 3).

Table 2

Agglutinin Responses of Normal Recipient Mice Injected with RNA Extracts Obtained at Various Time Intervals After Injection of Shigella Antigen into Donor Mice of Different Ages

Age in Days of RNA-Donor Mice at Time of SSA Injection [b]	Positive Agglutinin Titers Following RNA Transfer [a] (Day after donor injection with SSA)				
	+2	+8	+15	+30	+60
<1	6/8	5/6	4/5	2/6	2/8
30–50	3/5	4/6	2/4	2/5	0/6
75–110	3/4	3/5	2/4	0/5	1/7
Nonimmune Donor	0/7	0/5	0/5	–	–

[a] 50–100 μg donor RNA extract injected per recipient; positive titer = 1:32 or greater within 5 to 8 days after injection.

[b] 20 μg SSA injected i.p. per donor mouse.

Table 3

AGGLUTININ RESPONSES IN ADULT NORMAL, SHIGELLA-IMMUNE, SHIGELLA-TOLERANT, OR X-IRRADIATED MICE INJECTED WITH RNA EXTRACTS OBTAINED AT VARIOUS TIMES FROM ADULT IMMUNE OR TOLERANT DONORS

Donors and Days After SSA Injection [a]		Donor Mean Peak Titer	Recipient Mean Peak Agglutinin Titers [d]				
			Non-Immune Mice	SSA-Primed [e] Mice	X-Irradiated Mice		SSA-Tolerant Mice
					600 r	900 r	
Adult Immune [b]	+1	<1:4	1:58	1:270	1:11	1:20	1:21
	+3	1:58	1:46	1:310	1:16	1:11	1:10
	+10	1:1792	1:28	1:141	1:4	1:4	1:18
	+20	1:964	1:14	1:75	<1:4	<1:4	1:30
	+40	1:114	1:11	1:21	<1:4	<1:4	1:17
Tolerant [c]	+1	<1:4	1:66	1:226	1:31	1:16	1:30
	+3	1:4	1:58	1:280	1:20	1:20	1:12
	+10	1:11	1:31	1:215	1:4	1:7	1:26
	+20	1:19	1:20	1:65	1:4	1:4	1:25
	+40	1:12	1:8	1:24	<1:4	<1:4	1:11
Normal Adult	–	<1:4	<1:4	<1:4	<1:4	<1:4	1:15
SSA (2 μg)	–	–	1:465	1:2094	1:281	1:12	1:30

[a] At least 5 donor mice per group per day.
[b] 5- to 7-week-old mice injected i.p. with 20 μg SSA.
[c] Injected i.p. at birth with 20 μg SSA.
[d] 50–200 μg donor RNA extract transferred to 3–4 mice per group.
[e] 10–15-week-old recipient mice injected 5–8 weeks previously with 20 μg SSA.

C. Transfer of RNA from Shigella-tolerant Donor Mice

RNA extracts prepared from spleens of donor mice injected at birth with "tolerizing" concentrations of Shigella antigen stimulated detectable agglutinin formation in most recipient mice, both in primed as well as in normal mice (Tables 1, 2, and 3). The agglutinin titers of such recipients were generally similar to those obtained following transfer of RNA from adult immunized mice (Tables 1, 2, and 3). Figure 1 indicates the comparative agglutinin responses of 20 groups of recipient mice, in one experiment, as follows: (a) control mice injected with several concentrations of SSA only; and (b) experimental mice injected with RNA-rich extracts prepared from spleens of donor mice, either adult or newborn, inoculated with 20 μg SSA 7, 14, or 21 days prior to sacrifice. The 6- to 8-week donor mice were "immune" and the 12-hour mice were "tolerant." The highest agglutinin titers occurred in SSA-primed recipient mice, in comparison to the normal recipients. Primed mice injected with RNA-rich extracts from donor animals treated at birth with a tolerance-inducing SSA inoculum had an agglutinin response comparable to that observed in other animals injected with 10^{-6} to 10^{-8} mg of SSA (Fig. 1).

D. Transfer of RNA Extracts to Shigella-tolerant Recipients

When recipient mice were specifically tolerant to Shigella following neonatal inoculation of SSA, no significant agglutinin response occurred following the injection of RNA-rich extracts obtained from adult immune mice or from neonatal tolerant mice, or following the injection of SSA (Table 3); tolerant recipients were equally unresponsive to all three preparations.

E. Effect of treatment of RNA-Extracts with RNAase or Antiserum

In vitro incubation of RNA-rich extracts with a small quantity of RNAase for 30 minutes abolished the immunogenic activity of the extracts in both normal and primed recipients (Table 4). Treatment of RNA extracts, with 0.1 ml of a 1:100 dilution of hyperimmune mouse or rabbit anti-Shigella serum for 30 minutes at 37°C, also abolished the immunogenic activity (Table 4). Similar treatment of these extracts with DNAase, trypsin, or with normal mouse or rabbit serum had no effect.

F. Antibody-forming Cells in RNA-Treated Recipients

In the previous experiments the immune response of recipient mice was determined by assaying serum agglutinin titers. Such titers could reflect either (a) a few cells stimulated to produce relatively large amounts

Table 4

AGGLUTININ FORMATION IN NORMAL AND SSA-IMMUNE MICE RECEIVING RNA FROM TOLERANT OR ADULT IMMUNE DONORS TREATED *in vitro* WITH NUCLEASES, TRYPSIN OR ANTISERUM

RNA Donor [a]	RNA Treatment *in vitro*	Recipient Agglutinin Titer: Normal, No. Posit.[d]/ No. Inject.	Normal, Mean Peak	SSA Primed, No. Posit.[d]/ No. Inject.	SSA Primed, Peak Mean
Adult Immune [b]	None	6/9	1:53	8/8	1:198
(+3–5 days)	RNAase	0/6	1:4	0/5	1:13
	DNAase	5/5	1:47	8/8	1:181
	Trypsin	4/4	1:58	6/6	1:165
	Anti-Shigella Serum	1/10	1:11	1/9	1:16
	Normal Serum	5/5	1:40	5/5	1:170
Tolerant [c]	None	3/4	1:65	4/4	1:155
(+3–6 days)	RNAase	0/5	1:4	1/6	1:8
	DNAase	4/4	1:58	3/4	1:190
	Trypsin	3/3	1:65	3/3	1:158
	Anti-Shigella Serum	0/5	1:4	0/7	1:4
	Normal Serum	4/4	1:48	4/4	1:175
SSA (2 μg)		6/6	1:364	10/10	1:3065

[a] Phenol extracts from 10–20 donors.
[b] Injected i.p. with 20 μg SSA at 5–8 weeks of age.
[c] Injected at birth with 20 μg SSA i.p.
[d] Serum titer of 1:16 or greater.

of antibody, or (b) a large number of antibody-forming cells each secreting a very small quantity of antibody. The indirect plaque assay, with Shigella, previously used to demonstrate antibody-forming cells to Shigella in normal immune or tolerant mice (Friedman, 1966), was utilized to enumerate the number of antibody-forming cells in spleens of mice treated with RNA extracts from immune and tolerant donors. There was a rapid increase in the number of specific PFC in spleens of mice actively immunized with 20 μg SSA, with 60,000 or more PFC appearing within 5 days after immunization. Primed mice had a more rapid increase in PFC response. Normal mice injected with 0.1 to 0.2 mg RNA extract from either SSA immune or tolerant donors also had an increase in the number of specific spleen PFC within 3 to 6 days. The number of PFC in spleens of primed recipients was generally three- to eight-fold greater than that in normal recipient spleens. The maximum number of PFC in both groups usually occurred when RNA extracts were obtained, from either adult or tolerant donor mice, two to seven days after SSA inoculation (Tables 5 and 6). Treatment of RNA extracts, either from adult

Table 5

AGGLUTININ TITERS AND NUMBERS OF ANTIBODY-FORMING CELLS OF NORMAL, IMMUNE AND TOLERANT MICE INJECTED WITH RNA OBTAINED AT DIFFERENT TIMES FROM IMMUNE OR TOLERANT DONOR MICE

	Donor			Immune Response of Recipients [d]					
				Normal		SSA Primed [a]		SSA Tolerant [c]	
Source of RNA Extract	Day After SSA Inj.	Aggl. Titer	PFC Per Spleen (×10³)	Aggl. Titer	PFC Per Spleen (×10³)	Aggl. Titer	PFC Per Spleen (×10³)	Aggl. Titer	PFC Per Spleen (×10³)
Adult Primary [a]	+3	1:185	7.4	1:38	3.8	1:170	17.3	1:15	0.9
	+6	1:665	24.6	1:47	4.4	1:135	11.5	1:30	1.7
	+10	1:575	18.6	1:30	4.1	1:201	15.0	—	—
Adult Secondary [b]	+3	1:3110	35.8	1:29	4.1	1:150	21.3	—	—
	+6	1:4840	69.3	1:55	3.5	1:198	14.8	1:28	1.5
Tolerant [c]	+3	1:11	1.6	1:64	5.4	1:111	14.5	1:17	1.1
	+6	1:15	2.8	1:28	5.9	1:89	19.3	1:24	2.0
	+10	1:18	2.7	1:50	6.5	1:230	11.2	—	
Nonimmune	—	<1:4	0.6	1:4	0.8	1:18	3.5	1:18	1.8
SSA (20 μg)	—	—	—	1:296	34.5	1:4075	84.6	1:29	2.1

[a] 5–8-week-old normal mice injected i.p. with 20 μg SSA.
[b] 15–16-week-old mice injected with 20 μg SSA 7 to 12 weeks after primary injection with Shigella.
[c] Newborn mice injected i.p. with 20 μg SSA.
[d] 3 to 4 recipients per group; injected i.p. with donor RNA and tested 5–8 days later.

Table 6

ANTIBODY RESPONSES, IN TERMS OF SPLEEN CELLS AND AGGLUTININ TITERS, OF NORMAL AND SHIGELLA-IMMUNE MICE TREATED WITH RNA EXTRACTS FROM ADULT IMMUNE OR TOLERANT DONOR MICE WITH OR WITHOUT *in vitro* INCUBATION WITH RNAASE OR ANTISERUM

Donor		Immune Response of Recipients							
		Normal				SSA Primed			
RNA Extract Source	RNA Treatment *in vitro* [a]	PFC/Spleen on Day +3	+5	+10	Peak Aggl. Titer	PFC/Spleen on Day +3	+5	+10	Peak Aggl. Titer
Adult Immune [b]	None	1486	1965	1143	1:72	3186	9461	2743	1:214
(+3–6 days)	RNAase	763	645	731	1:12	1135	1011	1265	1:38
	Antiserum	905	1160	1005	1:8	1750	9102	1160	1:19
Tolerant [c]	None	2015	2410	–	1:60	2506	8340	7515	1:185
(+3–6 days)	RNAase	1184	1200	1156	1:8	1950	1564	1760	1:22
	Antiserum	–	916	890	1:14	1109	978	1340	1:42
Normal	None	1258	1740	1800	1:4	1865	1175	1530	1:35
SSA (20 μg)	–	16,584	58,430	43,560	1:345	47,650	93,450	85,860	1:4198

[a] 150–200 μg RNA extract, in 1.0 ml medium, treated *in vitro* with 10 μg RNAase or 0.1 ml 1:10 antiserum, titer = 1:1024.
[b] 5–8-week-old mice injected i.p. with 20 μg SSA.
[c] Newborn mice injected i.p. with 20 μg SSA.

immune or neonatal tolerant donor spleens with RNAase or with specific anti-Shigella serum prior to inoculation, markedly suppressed the appearance of plaque-forming cells in recipient spleens. RNA extracts obtained from normal donors had little if any effect on the number of cells forming antibodies to Shigella in recipient mice.

G. *In vitro* Stimulation of Plaque-forming Cells with RNA Extracts

When spleen cells from mice primed with SSA one to three months previously were cultured *in vitro,* addition of a small quantity of SSA, but not of sheep erythrocytes, stimulated a detectable increase in plaque-forming cells (Table 7). There was no significant increase in specific plaque-forming cells in normal spleen cell cultures following treatment with SSA *in vitro.* The maximum PFC count after SSA exposure *in vitro* of cells from primed mice generally occurred by the second or third day.

Table 7

Antibody Forming Cells in Spleen Cell Cultures of Normal or Shigella-primed Mice Maintained *in vitro* for 48 Hours After Exposure to RNA Extracts from Immune or Tolerant Mice

Source of Donor RNA	Day After Donor Immunization	Antibody-forming Cells Per 10^6 Spleen Cells: After Exposure of Normal Cells	After Exposure of Cells from Primed Mice [a]
Adult Immune [a]	+3	10.8	33.8
	+6	11.4	45.1
	+10	18.3	40.5
	+10-RNAase [c]	<1.0	15.4
	+10-Anti-SSA Serum [d]	<1.0	11.0
Tolerant [b]	+3	10.3	48.0
	+6	7.4	55.9
	+10	11.6	50.5
	+10-RNAase [c]	<1.0	7.8
	+10-Anti-SSA Serum [d]	<1.0	11.5
SSA (1–2 μg)	—	<1.0	178.6
SSA + Normal RNA	—	<1.0	184.5
Sheep RBC-	—	<1.0	10.5

[a] 5–8-week-old mice injected i.p. with 20 μg SSA.
[b] Newborn mice injected i.p. with 20 μg SSA.
[c] RNA extract treated *in vitro* for 30 minutes with RNAase.
[d] RNA extract treated *in vitro* for 30 minutes with specific antiserum.

There was no evidence that increased numbers of PFC could be detected when plates were treated with anti-mouse gamma globulin, a procedure reported to permit the detection of 7S plaque-forming cells to sheep erythrocytes (Riha and Sterzl, 1965; Wortis and Dresser, 1965). There was a detectable increase in the number of PFC in most cultures of spleen cells from normal or primed mice treated with RNA-rich extracts from donor animals that had been inoculated with SSA either as adults or as neonates three to ten days previously (Table 8). There were similar increases in the number of PFC in cultures of primed mice treated with RNA-extracts from either adult immune or neonatal tolerant mice (Table 7). The stimulatory effect of the RNA-rich extracts from either adult or tolerant donors was suppressed by prior incubation with either RNAase or anti-Shigella serum. RNA extracts from normal mice had no effect on plaque-forming cells *in vitro*. Similarly, the cultivation of either normal or primed cells *in vitro* following stimulation with a mixture of SSA and RNA extract from spleens of normal mice resulted in no greater plaque response than occurred with SSA alone (Table 7).

III. DISCUSSION

Normal and Shigella antigen-primed mice, but not specifically tolerant or X-irradiated recipient mice, responded to injection of RNA extracts from either immune or tolerant mice with detectable antibody formation to Shigella on both the humoral and cellular level. The response was similar to that occurring following active immunization with specific antigen (Table 8).

These results suggest that the antibody activity observed in recipient mice, as well as in spleen cell cultures treated with RNA extracts from

Table 8

SUMMARY OF IMMUNOLOGIC RESPONSES TO SHIGELLA ANTIGEN OF VARIOUS GROUPS OF MICE AFTER INJECTION WITH RNA-RICH EXTRACTS FROM VARIOUS DONOR MICE, AS COMPARED TO INJECTION WITH SSA

	Recipient Mice				
				X-Irradiated	
RNA Donor	Normal	SSA-Primed	SSA-Tolerant	600 r	900 r
Adult Immune–Primary	+	++	0	±	0
Secondary	+	++	0	±	0
Tolerant	+	++	0	±	0
Normal, Non-Immune	0	0	0	±	0
SSA only	++	++++	0	0	+

immune and tolerant donors, may be related, at least partially, to persisting antigen. Since the immunogenic activity of the extracts from Shigella-immune donors persists for several weeks, it seems plausible that a postulated RNA-antigen complex may be an important factor in antibody formation throughout immunogenesis, not only during the inductive period. However, demonstration of a similar immunogenic activity with spleen extracts from Shigella-tolerant animals, which are *not* forming significant amounts of detectable antibody, raises the question as to the biologic importance of this material.

The bacterial system used here offers a very sensitive model for demonstrating immunologic reactivity in animals to small quantities of antigen. It is likely that all mice, prior to specific immunization, are naturally "exposed" to Gram-negative bacterial antigens. However, the administration of one microgram or less of SSA into normal mice results in a brisk immune response, marked by rapid appearance of specific antibody forming cells and serum agglutinins. Following a second injection of SSA, several weeks or months later, there is a more rapid and sustained appearance of specific antibody-forming cells and agglutinins.

Although there was no consistent or direct straight line relationship between the amount of antigen injected and mean titers of various control groups, a dose response was apparent when normal or primed mice were injected with graded concentrations of SSA (Fig. 2). The results of this study have indicated that most nonimmune, as well as previously primed mice, injected with RNA-rich extracts from spleens of Shigella-immunized donor mice also form readily detectable serum agglutinins and plaque-forming cells. This response was similar to that which occurred when a small quantity of free SSA was injected directly into control mice.

Within a few days after active immunization of donor animals with SSA, there is obviously a rapid dilution or metabolism of the antigen so that it is undetectable by classic serologic techniques (Table 1) (Friedman and Gaby, 1960; Friedman, 1962, 1963b). Thus, if RNA extracts from these donors contain antigen, or antigenic determinants, the quantity present is probably extremely small, possibly no more than a few hundred molecules. If the immunologic activity of such RNA-rich extracts is directly related to this amount of retained antigen, as postulated by ourselves and others, it would suggest that the complex may act as a "super antigen." Such enhanced immunogenicity could be related directly to the RNA moiety, which could act either as a nonspecific adjuvant (Merritt and Johnson, 1965; Braun, 1965) or as a metabolic stimulator of antibody formation (Braun, 1967). On the other hand, the entire activity of the extracts could be a direct function of an "immunologic code" mediated by RNA. Any antigen molecules, if present, would then be merely nonimportant contaminants. However, if this were the case, it seems difficult

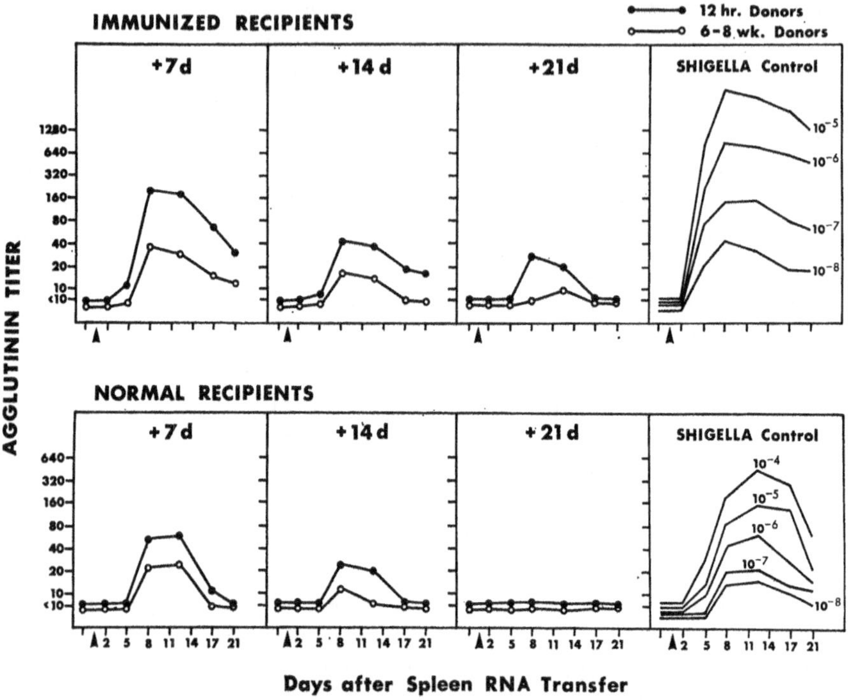

Fig. 2. Agglutinin titers in sera of normal (lower) and Shigella-primed (upper) mice following injection with either graded concentrations of Shigella antigen, in mg (right panel), or 0.1–0.6 mg of RNA extract from newborn or adult mice injected with 20 μg Shigella antigen either 7, 14, or 21 days previously.

to explain why RNA-extracts from mice that are tolerant to Shigella can also stimulate antibody formation in normal and primed recipients. It is possible that *antigen* associated with RNA-rich extracts from tolerant donors may be a direct inducer of antibody-formation, whereas similar extracts from immune, non-tolerant animals would act by a different mechanism whereby the *RNA* is a direct immunologic code and the antigen only an insignificant contaminant.

The studies with nucleases indicated that RNA extracts prepared both from immune and tolerant donors could be inactivated by RNAase treatment *in vitro*. This suggests that the ribonucleic acid in the extract is biologically important. In addition, incubation of extracts, prepared either from immune or tolerant donors, with high-titered anti-Shigella serum also inactivated the stimulating activity. Other agents tested, such as normal serum, DNAase, or trypsin, were ineffective. Thus, it seems that both RNA and an antigenic determinant may be important components of the phenol extracts which induce immunologic reactivity in normal animals or cell cultures.

It seems noteworthy that Shigella-tolerant recipient mice did not respond to RNA extracts, prepared from either normal immune or tolerant donors. Since there was no "breaking" of tolerance by such extracts, in this or previous experiments (Friedman, 1965a), it appears that the tolerant mice may not be able to respond either to an RNA "code," to antigen, or to an RNA-antigen complex. The mechanism of tolerance is still poorly understood. It is generally agreed that tolerance is manifested by a lack of detectable antibody forming cells. However, it has not yet been possible to distinguish between (a) absence of specific reactive cells, which may be eliminated during tolerance induction, and (b) presence of potential antibody-forming cells which are suppressed, possibly by "excess" intracellular antigen. The failure to induce agglutinin formation in Shigella-tolerant mice with RNA-rich extracts, as occurred in normal or primed recipients, may be due to either a lack of immunologically receptive cells for an RNA "code," or to a lack of cells which can respond to a "super antigen." The experiments reported here do not permit differentiation between these two suggested mechanisms. The results do indicate that RNA extracts from tolerant mice share the capacity with similar extracts from normal immune mice to induce specific antibody in non-tolerant recipients.

The experiments with lethally irradiated recipients also indicated that mice incapable of forming antibody to a specific antigenic stimulation cannot produce detectable antibody, either in the serum or on the cellular level, when injected with RNA extracts capable of inducing specific antibody formation in normal or antigen-primed recipients. Mice treated with 600 r had a low level of response, but many could also respond to direct antigenic stimulation with SSA.

The studies with antibody-forming cells complemented those concerned with serum agglutinin titers. Mice injected with RNA extracts from donor animals, either immune or tolerant to Shigella, generally formed increased numbers of specific antibody plaques to Shigella, as compared to the background PFC level. The peak plaque response usually occurred five to eight days after injection of RNA extracts, similar to that observed in normal mice actively immunized with SSA. The number of PFC in spleens of non-primed recipients was much lower than that in spleens of mice previously injected with SSA. Recipient mice injected with antigen several weeks or months prior to administration of RNA extracts had a plaque response comparable to primed mice challenged with very small quantities of Shigella, such as 1 or 2 μg or less. Such a plaque response did not occur in recipients when the RNA extracts, either from immune or tolerant donors, were pre-treated *in vitro* with either RNAase or anti-Shigella serum. These results extend the findings based on agglutinin titers and indicate that both an RNA moiety and Shigella antigen determinants may be important components of the extracts.

Stimulation of PFC formation in tissue cultures of spleen cells from normal, non-immune mice suggests that a "conversion" to immunity occurred *in vitro.* Although such cultures rarely formed increased numbers of PFC when stimulated *in vitro* with SSA alone, it should be noted that the background PFC count to Shigella antigen is relatively high, suggesting prior immunologic exposure of the mice to this or related antigens. Stimulation of primed spleen cells *in vitro* with RNA-extracts was similar to that which occurred upon addition of free SSA, in small quantity, to the cultures. However, the RNA moiety appeared to be a necessary component since the *in vitro* stimulatory activity could be abolished by prior treatment of the extracts with RNAase as well as with antiserum, similar to that observed with the *in vivo* test system (Table 7). Addition of RNAase to control cultures treated with SSA only did not interfere with the *in vitro* increase in PFC. Addition of anti-Shigella serum to either RNA extracts or to free SSA, prior to incubation with spleen cell cultures, blocked the subsequent PFC response.

The specific cellular source of the active RNA-antigen material in the phenol extract is not known. Although all of the results presented here were obtained with extracts prepared from spleens, essentially similar results have been obtained in a few preliminary experiments with lymph node and liver extracts from immune donors. Extracts from peritoneal cell exudates from the same donors have been ineffective in this system. In addition, we have failed in a number of experiments to obtain stimulatory RNA-rich extracts from spleen, lymph node or peritoneal exudate cell suspensions from normal mice incubated with SSA *in vitro,* at various concentrations, and for varying lengths of time. In other control experiments, SSA which had been incubated *in vitro* with RNA-rich extracts from normal donors, followed by phenol extraction, did not stimulate antibody formation either *in vivo* or *in vitro.* Also, SSA injected into recipient mice together with 0.1 to 0.4 mg RNA extract from normal donor mice resulted, in general, in no higher titer than that which appeared in other mice injected with SSA alone.

If a specific RNA-antigen complex is indeed an important factor in immunogenesis, and not an experimental artifact, the actual mechanism of action is still obscure. It is certainly possible, as indicated above, that an immunologic "code" is responsible for most of the "conversion" activity reported here as well as by other investigators. Investigations with allotypic markers strongly suggest that at least one portion of the antibody response in tissue culture systems to bacteriophage antigens may be due to information directly coded in RNA (Adler *et al.,* 1966). It is possible that such an RNA "code" may be stabilized by small amounts of antigen. It is also possible that an RNA molecule codes for one molecular form of antibody and the antigen or antigen-RNA complex for another form (Adler *et al.,* 1966). It also seems possible, as suggested by

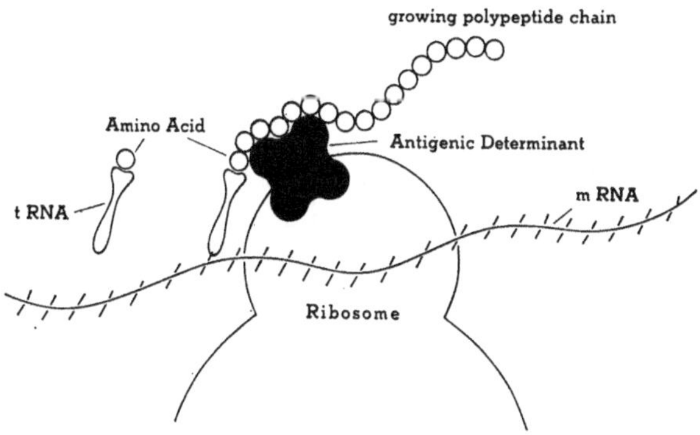

Fig. 3. Schematic representation of a possible mode of action of an antigen-RNA complex as a "direct" template in the translation of messenger RNA coded for a gamma globulin chain on a ribosome (modified from Haurowitz, 1966).

others, that an antigenic determinant associated with RNA acts merely as a steering mechanism to direct the nonspecific RNA stimulator into the proper cell (Braun, 1968). This would presume that only certain cells are genetically competent to form antibodies to a specific antigenic determinant. The receptors on such specific cells, formed by an unknown mechanism, would permit specific "recognition" of an RNA-antigen complex.

An alternative hypothesis is that an RNA-antigen complex may be closely related to a functional antibody-forming template, in a manner analogous to that suggested by Haurowitz (Haurowitz, 1965, 1966). Although it is necessary to assume certain qualifications, such that only a few ribosomes per cell are actively engaged in synthesis of antibody-reactive sites at a single time, it may be possible that an RNA-antigen complex, isolated either from macrophages or other lymphoid cells which have processed the original antigen, may direct the "coding" of specific amino acid sequences in a small portion of the gamma globulin chains, possibly by a built-in translational "misreading" mechanism on specific ribosomes (Fig. 3). In tolerant animals, excess RNA-antigen complexes on or at such a site would then specifically block synthesis, since no translation could occur.

IV. SUMMARY

Phenol extracts prepared from spleens of mice immunized with soluble antigen of Shigella (SSA) induced a detectable and specific immune response in normal and primed mice. There was a specific increase in serum antibody formation as well as in the number of specific antibody-forming cells both in recipient mice as well as in tissue culture suspensions of spleen cells from normal and antigen-primed mice. The greatest activity was obtained with ribonucleic acid extracts prepared from mice within a few days after antigen injection. The largest amount of antibody appeared in recipient mice which had been primed with specific antigen. However, a detectable increase in antibody formation, both on the serum and cellular level, occurred in a majority of non-immune recipients treated with RNA-extracts. Extracts prepared at various time intervals from spleens of mice injected at birth with a tolerance-inducing concentration of Shigella antigen also stimulated antibody-formation in recipient mice, both in normal mice as well as in mice previously injected with antigen. Similar extracts from control donors not injected with antigen did not stimulate antibody formation. RNA-rich extracts prepared from immune or tolerant mice did not elicit a significant immune response in recipient mice treated with 900 r 24 hours prior to transfer, or when injected into young adult mice specifically tolerant to Shigella antigen following neonatal administration of SSA. The active RNA-rich extracts from immune and tolerant donor mice stimulated significant increases in the number of cells forming antibody to SSA in tissue culture suspensions of spleen cells from antigen-primed donor mice, similar to the effect of free SSA. The stimulatory activity of these extracts was abolished by prior treatment *in vitro* with RNAase, as well as by incubation with specific anti-Shigella serum, but not by exposure to normal serum. Treatment with DNAase or trypsin was ineffective. These results suggest that both an RNA moiety and an antigenic determinant, having serologic properties of Shigella antigen, may be involved as an immunogenic complex in spleens of mice injected either with an immunizing or a tolerance-inducing concentration of antigen.

References

Adler, F. L., M. Fishman, and S. Dray (1966). Antibody formation initiated *in vitro.* III. Antibody formation and allotypic specificity directed by ribonucleic acid from peritoneal exudate cells. J. Immunol., *97*:554.

Askonas, B. A., and J. M. Rhodes (1965). Immunogenicity of antigen-containing ribonucleic acid preparations from macrophages. Nature, *205*:470.

Braun, W. (1965). Influence of nucleic acid degradation products on antibody synthesis, in *Molecular and Cellular Basis of Antibody Formation,* p. 525. Academic Press, N. Y.

Braun, W., and E. P. Cohen (1968). The role of nucleic acids in antibody formation, in *Regulation of the Antibody Response,* Charles C Thomas.

Campbell, D. H., and J. S. Garvey (1963). Nature of retained antigen and its role in immune mechanisms. Adv. in Immunol., *3*:261.

Cohen, E. P., and J. J. Parks (1964). Antibody production by nonimmune spleen cells incubated with RNA from immunized mice. Science, *144*:1012.

Cohen, E. P., R. W. Newcomb, and L. K. Crosby (1965). Conversion of nonimmune spleen cells to antibody-producing cells by RNA: strain specificity of the response. J. Immunol., *95*:583.

Cohen, E. P. (1967). Conversion of nonimmune cells into antibody producing cells by RNA. Nature, *213*:462.

Dresser, D. W., and Wortis (1965). Use of an antiglobulin serum to detect cells producing antibody with low hemolytic efficiency. Nature, *208*:859.

Esposito, E., and E. L. C. Ligniere (1966). Transfer of the immunization to a bacterial antigen by RNA. Experientia, *22*:603.

Fishman, M. (1961). Antibody formation *in vitro*. J. Exp. Med., *114*:837.

Fishman, M., and F. L. Adler (1963). Antibody formation initiated *in vitro*. II. Antibody synthesis in X-irradiated recipients of diffusion chambers containing nucleic acid derived from macrophages incubated with antigen. J. Exp. Med., *117*:595.

Fong, J., D. Chin, and E. S. Elberg (1963). Studies of tubercle bacillus-histiocyte relationships. VI. Induction of cellular resistance by ribosomes and ribosomal RNA. J. Exp. Med., *118*:371.

Friedman, H. (1959). Transfer of agglutinin formation by immune lymph node cell fraction. Fed. Proc., *18*:567.

——— (1962). Transfer of antibody formation by spleen cells from immunologically unresponsive mice. J. Immunol., *89*:257.

——— (1963a). Appearance of anti-bacterial agglutinins in X-irradiated rabbits receiving immune spleen nucleoproteins. Experientia, *19*:537.

——— (1963b). Persistence of antigen in nucleoprotein fractions of mouse spleen cells during antibody formation. Nature, *199*:502.

——— (1963c). Detection of persisting Shigella antigen in subcellular fractions of spleens from immunologically unresponsive mice. Proc. Soc. Exp. Biol. Med., *113*:1040.

——— (1963d). Persistence of Shigella antigen in tissues of immunologically unresponsive mice. Fed. Proc., *22*:440.

——— (1964a). Specific restoration of agglutinin formation in X-irradiated rabbits by nucleoprotein fractions from immune donors. Ann. N. Y. Acad. Scies., *114*:444.

——— (1964b). Acquisition of antibody plaque forming activity by normal mouse lymphoid cells incubated *in vitro* with RNA from immune mouse spleens. Biophys. Biochem. Res. Comm., *17*:272.

——— (1964c). Antibody plaque formation by normal spleen cells maintained in tissue culture following *in vitro* incubation with immune RNA. Science, *146*:934.

——— (1964d). Distribution of antibody plaque forming cells in various tissues of several strains of mice. Proc. Soc. Exp. Biol. and Med., *117*:526.

——— (1965a). Transfer of RNA extracts from immune donor spleen cells to Shigella tolerant recipient mice. Nature, *207*:1315.

——— (1965b). Absence of anti-Shigella antibody plaque forming cells in spleens of immunologically tolerant mice. Bacter. Proc., p. 255.

——— (1966a). Immune responsiveness and unresponsiveness to Shigella: the role of antigen containing RNA extracts. Bacteriol. Proc., p. 35.

——— (1966b). Antibody formation and nucleic acids: Induction of immune reactivity in normal animals and tissue cultures with RNA-rich extracts from immune donors. Proc. IX. Internat'l Congr. Microbiol., p. 554, Moscow, U.S.S.R.

——— (1966c). Immunologic tolerance to microbial antigens. II. Suppressed antibody plaque formation to Shigella antigen by spleen cells from tolerant mice. J. Bacteriol., *92*:820.

Friedman, H., and W. L. Gaby (1960). Immunological unresponsiveness in mice following neonatal exposure to Shigella antigens. J. Immunol., *84*:441.

Friedman, H. P., A. B. Stavitsky, and J. M. Solomon (1965). Inhibition *in vitro* of antibodies to phage T_2: Antigens in the RNA extract employed. Science, *149*:1106.

Haurowitz, F. (1965). Antibody formation. Physiol. Rev., *45*:1.

——— (1966). Discussion of paper by Porter and Weiss. J. Cell Physiol., *67*: Suppl. 1:61.

Mannick, J. A., and R. H. Egdahl (1962). Ribonucleic acid in "transformation" of lymphoid cells. Science, *137*:976.

Merritt, K., and A. G. Johnson (1965). Studies on the adjuvant actions of bacterial endotoxins. VI. Enhancement of antibody production by nucleic acids. J. Immunol., *94*:416.

Michelazzi, L., A. Novelli, I. Baldini, U. M. Marinari, and G. Nanni (1965). Induction of antibody response in lymphoid cells *in vivo* and *in vitro* by RNA-immune carrier from immune serum. Experientia, *21*:585.

Nester, M., O. Makela, and G. J. V. Nossal (1961). Studies on the transfer of nucleoprotein fractions from immunized spleens. Trans. Bull., *28*:98.

Sterzl, J., and M. Hrubesova (1956). Transfer by nucleoprotein fractions of antibody forming capacity into nonimmunized recipients. Folia Biol., *2*:21.

Sterzl, J., and Riha (1965). Detection of cells producing 7S antibodies by the plaque technique. Nature, *208*:858.

Thor, D. E. (1967). Delayed hypersensitivity in man. A correlation *in vitro* and transfer by an RNA extract. Science, *157*:1567.

Wilson, D. B., and E. E. Wecker (1966). Quantitative studies on the behavior of sensitized lymphoid cells *in vitro*. III. Conversion of "normal" lymphoid cells to an immunologically active status with RNA derived from isologous lymphoid tissues of specifically immunized rats. J. Immunol., *97*:512.

STIMULATION OF ANTI-TUMOR ACTIVITY OF THE HOST WITH RNA FROM IMMUNE LYMPHOCYTES *

P. ALEXANDER, E. J. DELORME, L. D. HAMILTON AND J. G. HALL

Chester Beatty Research Institute
Clifton Avenue, Sutton, Surrey

I. INTRODUCTION

Serological procedures involving heterologous systems showed that many tumors contain substances which were not present in "comparable" normal tissues, and such components are frequently referred to as tumor-specific antigens. In this connection the term "specific" may be misleading since the phenomenon depends on the choice of the normal tissue used for comparison. Thus hepatomas are rich in so-called tumor-specific antigen when the reference organ is adult liver but the majority disappear when the comparison is made with embryonic liver. The observation (Foley, 1953) that the transplantation of primary, chemically induced sarcomata into syngeneic animals could be prevented by first immunizing the prospective recipients with the tumor, showed that these sarcomata contained on their plasma membrane a component, capable of evoking transplantation immunity, which is totally absent (or not expressed) in any of the normal cells of the animal. Subsequent studies (cf. reviews by Old and Boyse, 1964, and Alexander and Hamilton-Fairley, 1967) revealed transplantation-type tumor-specific antigens (TTSA) in most experimentally induced tumors whether the causative agent was chemical, physical or viral. The TTSA of chemically induced tumors appear to be unique to each tumor and there is no cross-reaction between, for example, two sarcomata induced by the same carcinogen in the same

* This investigation has been supported by grants to the Chester Beatty Research Institute (Institute of Cancer Research: Royal Cancer Hospital) from the Medical Research Council and the British Empire Cancer Campaign for Research, and by a U. S. Public Health Service research grant from the National Cancer Institute.

strain of rodent. Thus, immunologically, each of the primary tumors used in the experiments reported in this paper is presumably antigenically unique.

There are several possible ways in which the presence of TTSA can be used as a point of attack for immunotherapy of primary tumors (cf. review Alexander, 1968) so long as the host has not become tolerant to the TTSA. For experiments of this type we have used sarcomata induced in rats—both outbred and pure-line—by subcutaneous implantation of a pellet of 3:4 benzpyrene. In spite of the fact that these tumors have never been observed to regress spontaneously the host is neither tolerant to the TTSA nor immunologically crippled and a lymphocyte-mediated reaction of the primary host against the tumor has been demonstrated in a number of different ways (Mikulska *et al.,* 1966; Alexander *et al.,* 1968). The reason why the established tumor grows relentlessly is due to the fact that the immune response is overwhelmed by the actively growing tumor. In this system, therefore, immunotherapy need not initiate a new response but need only augment or strengthen an existing one. Some anti-tumor effect was produced by immunization with irradiated autografts of tumor (Haddow and Alexander, 1964) but the use of immune lymphocytes proved to be much more effective.

II. "INDIRECT" ACTION OF IMMUNE LYMPHOCYTES

Our original working hypothesis was that the injection of anti-tumor lymphocytes, obtained from donor animals immunized with a piece of the tumor to be treated, might augment the host's anti-tumor action in a purely passive sense by increasing the total number of "cytotoxic" cells that are available to attack the primary tumor. While neither spleen nor lymph node cells from immunized donors had any influence on the growth rate of the primary tumors, lymphocytes obtained from the thoracic duct between 5 to 10 days after immunization (with biopsy fragments taken from the tumor to be treated) and injected intravenously exerted a marked growth-inhibitory effect and occasionally cured the primary tumors (Delorme and Alexander, 1964). Only the tumor used for immunization responded in this way and this indicated that the action was directed against the TTSA since these are unique for each tumor. This applied equally whether the injected lymphocytes were syngeneic or allogeneic with respect to the tumor-bearing animal.

A number of studies (Alexander, Delorme and Hall, 1966; Jeejeebhoy *et al.,* 1966; Alexander *et al.,* 1968) indicated that the injected lymphocytes did not interact immediately with the tumor and that the mode of action was less direct than originally envisaged. While therapy by "passive cellular immunity" can be demonstrated against ascites cells (Alexander *et al.,* 1966) in the case of the primary rat sarcomata, the injected lymphocytes appear to initiate a reaction which is taken to completion by the host.

The possibility that the anti-tumor effect might be achieved by the administration of a cell-free extract was suggested by the finding that heterologous (i.e., sheep) lymphocytes obtained by cannulating the efferent lymphatic of a sheep's node that had been previously stimulated with rat tumor, had a pronounced and specific anti-tumor action (Alexander, Delorme and Hall, 1966). Using this system it became apparent that the anti-tumor effect was probably brought about by the large immuno-blasts which leave the node between 3 to 7 days after antigenic stimulation. These cells do not survive for any length of time in the rats and are rapidly sequestered in the spleen following i.v. injection.

III. ANTI-TUMOR ACTIVITY OF A NUCLEIC ACID FRACTION OF THE LYMPHOCYTES

The idea that the RNA from the immuno-blasts might be responsible for the activity observed was suggested by the fact that these lymphocytes are very rich in RNA (on average they contain 6 times as much RNA as do small lymphocytes), and by the fact that RNA had been shown to have a role in the genesis of the immune response by Fishman, Mannick and others (see this symposium). To be frank, when we came to test this approach we were not very sanguine of success and for this reason the choice of the method of isolation of the RNA was determined by the requirement that all the high molecular weight RNA present in the cell should be extracted. Accordingly, a method of Kirby was modified in which the lymphocytes were homogenized in 4-aminosalicylic acid and then treated with phenol containing 8-hydroxyquinoline (see Alexander *et al.*, 1967). The end-product was far from pure and contained a considerable and variable amount of DNA as well as some protein. While the intravenous route was the most effective with cells, the RNA was not administered in this way because of the high RNAase activity of blood. Working on the assumption that to be effective the RNA must enter lymphoid cells of the rat, the nucleic acid extract was injected into the footpads so that it might reach lymph nodes directly by lymphatic channels.

The general experimental procedure was as follows: Two rats with primary tumors of between 1.5 to 2.0 cms were selected. About half the tumor mass was excised from each rat. One of the tumor masses was used for immunization and the other discarded. Between 5×10^9 to 3×10^{10} lymphocytes were obtained from the immunized donor animals (either several rats or one sheep); the yield of RNA was approximately 2 mg/gm of packed wet cells. RNA in doses of 0.5 mg per treatment (number of treatments was determined by the amount of RNA available) was injected into both rats; the one whose tumor had been used for immunization received so-called "specific" treatment whereas the other one constituted the "control." The rate of tumor growth was determined by daily diameter measurements. Tumors in rats which had received no treatment—but had only been biopsied—required an average of 11 days (and in no instance more than 18 days) to increase in volume by a factor of 4 (i.e., a diameter increase of 1.6 times).

In our original publication (Alexander *et al.,* 1967) we reported the results of the experiments—fifteen pairs of tumors in all—in which an adequate number of lymphocytes were obtained. In none of the "control" rats was there any effect of the RNA on the rate of tumor growth whereas there was a temporary shrinkage of the tumor and an overall delay in growth in 10 out of 15 and a complete "cure" of one tumor. Since this publication, 9 further pairs of animals have been treated. In this later series, one cure has been obtained and 6 significant growth delays, i.e., the time for the volume of the tumor to quadruple was respectively 24, 26, 29, 29, 40, 49 and greater than 210 days. Eight of the nine control animals required between 8 to 17 days for a four-fold increase of volume but one control animal had its tumor delayed by 120 days. Whether this constitutes a true cross-reaction between the TTSA of the two tumors chosen as the pair can unfortunately not be determined as these were tumors in non-inbred rats.

From these experiments we conclude that the nucleic acid extracts simulate the effects of the intact lymphocytes in inducing a highly specific—i.e., restricted to one tumor per preparation—growth-retarding effect on primary rat sarcomata.

While from a practical point of view an effect on primary tumors is of interest, such tumors constitute a very difficult experimental system for the elucidation of the mechanisms that are involved and the identification of the active principle in the cell free preparation. Each primary tumor is to some extent unique and to compare the effectiveness of different procedures therefore requires a series of tests. These experiments are extremely time-consuming; the treatment of one "pair" of rats per week requires the almost full-time work of two scientists since complex surgery of the sheep and prolonged biochemical extraction procedures have to be carried out. As a consequence, all we can claim at the present is to have shown that the procedure described retards the growth rate of the one class of tumors studied.

A. Grafted versus Primary Tumors

The sarcomata if induced in pure-line rats can be transplanted readily into syngeneic hosts and such grafted tumors constitute stable and reproducible biological test systems. Unfortunately established grafted sarcomata have in our hands failed to respond to a significant extent either to treatment with intact immune cells—whether syngeneic, allogeneic or heterogeneic—or the nucleic acid extract prepared from such cells. The failure to respond cannot be ascribed to a loss of TTSA on transplantation since the capacity to induce immunity is retained. A possible explanation is that the grafted tumors grow between 2 to 3 times as rapidly (based on increase in volume) as did the primary tumors from which they were derived. There is much evidence to indicate that this difference in growth rate arises from the fact that in the primary host the tumor is slowed by an active reaction against it which has been built up over a long period. On transplantation the tumor is in an "immunologically virgin" environment in which it can grow more rapidly and in which any immuno-therapeutic procedure is much less likely to produce a detectable effect. Injection of the nucleic acid extract two days before implanting of sarcoma resulted in the failure of the graft to take in a few sporadic instances. The irregularity of the phenomenon, however, makes it difficult to know if any significance can be attached to the observation.

We have failed to influence either the growth or, by a pretreatment, the take of the mouse lymphoma L5178 Y in its syngeneic strain DBA/2 with nucleic acid extracts from lymphocytes sensitized to the lymphoma.

B. Failure of a Purer RNA Preparation to Be Active *in vivo*

A major difficulty in the RNA extraction procedure is the formation in the early stages of the preparation of a DNA-protein gel, the presence of which complicates all subsequent steps. At the suggestion of the late Professor Kirby a detergent—1% of tris-isopropyl-naphthalene sulphonate—was added to the solution of 4-amino salicylic acid in which the lymphocytes were homogenized. This addition greatly facilitated the subsequent stages and preparations containing 90% RNA and only 10% DNA were obtained in high yield. The final material gave non-viscous solutions which were quite unlike the gel-like suspensions that were obtained by the original method. However, in repeated attempts these RNA preparations failed to arrest the growth of the primary sarcomata and they were also quite ineffective in a series of *in vivo* experiments in which transfer of immunity to bacterial antigens and B.C.G. was looked for (see below).

This relatively pure RNA was, however, active in an *in vitro* test based on the system developed by Mannick (see this symposium). The L5178 Y lymphoma grows readily *in vitro* as a suspension culture with a cloning efficiency which exceeds 50%. The growth of these cells was inhibited *in vitro* by spleen cells from mice that had been immunized against the tumor and in this system activity both against the normal transplantation antigens as well as the TTSA was demonstrated (Alexander *et al.*, 1966). Non-immune spleen cells were rendered cytotoxic to the L5178 Y cells by exposure for 30 min to nucleic acids (at a concentration of 0.1 mg/ml) extracted from the lymphocytes of sheep that had been immunized with L5178 Y.

Why the addition of the detergent should abolish the *in vivo* activity of the preparation has not been established. There is no indication that the quality of the RNA is altered and the most prominent effect of the detergent is to remove DNA-gel from the final product. It may be that the presence of DNA—particularly in a gel form—facilitates the uptake of the RNA by the lymphoid cells or that it protects against degradation by RNAase in interstitial tissue. On the other hand the possibility that the anti-tumor effect is brought about not by RNA but by a contaminant cannot be excluded, although it would appear to be unlikely.

IV. ATTEMPTS TO TRANSFER IMMUNITY TO BACTERIAL ANTIGENS WITH NUCLEIC ACIDS

Either *Brucella abortus,* or *S. typhi* "O" or "H" antigen was injected into the drainage area of the prefemoral nodes of sheep, the efferent duct was cannulated and the cells, including a high proportion of blast cells, were collected. Rats were then injected with either the immune lymphocytes or a nucleic acid fraction derived from them by one of several isolation procedures. The serum of the rats was tested for agglutinating antibodies. While in general the titre was raised by one or two dilutions over that present before injection of the sheep cells or their nucleic acids, the increase did not appear to be statistically significant.

V. POSSIBLE MECHANISMS

To attribute the anti-tumor activity of the nucleic acid extract to the presence of a "super antigen" requires that such a material is present in lymphocytes since the cells from which the extract was made contained no macrophages. While the cell suspensions used contained small lymphocytes there are many reasons for believing that the large lymphoblasts provide the active material. These cells are the direct precursors of plasma cells (Birbeck and Hall, 1967) and possibly also of lymphocytes involved in graft rejection (Gowans and McGregor, 1965). These large RNA-rich cells probably contain the messenger RNA required for the synthesis of the proteins of circulating antibody or cell-bound immunity. It is tempting to speculate that RNA from such cells can enter uncommitted lymphoid cells and bestow on to them the capacity to make specific antibody, free or cell-bound. Phenotypic transformation by RNA has been observed in bacteria and in spleen cells by Fishman (see this symposium).

One might speculate that such a reaction could also cause an increase in the host reaction to the tumor (via the TTSA) and in this way account for the specific anti-tumor activity seen by us with the RNA-containing extracts from immune lymphoblasts. We face the problem why the capacity to form agglutinating antibodies to bacterial antigens cannot be transferred with RNA-containing extracts. A possible explanation could be that circulating antibodies are made up of H and L chains and their synthesis requires that within each antibody-producing cell there is a correctly balanced complement of the two types of messenger RNA responsible for the H and L chains respectively. Even if messenger RNA can enter cells the probability of introducing the correct proportion so that the phenotypically transformed cell can make 19S antibody is very small. If the effector mechanism responsible for the reaction directed against the tumor antigens involves a protein having one polypeptide chain only, then entry of a single molecule of messenger RNA into a suitable cell might cause its production. Some such mechanism might make it possible for extraneously administered RNA to induce cell-mediated immunity.

References

Alexander, P. (1968). Immunotherapy of Experimental Tumours *in* "Progress in Experimental Tumour Research." Basel: S. Karger, ed. F. Homburger, vol. 10, in the press.

Alexander, P., and G. H. Fairley (1967). Cellular Resistance to Tumours. Brit. Med. Bull., *23*:86–92.

Alexander, P., D. I. Connell, and Z. B. Milulska (1966). Treatment of a Murine Leukaemia with Spleen Cells or Sera from Allogeneic Mice Immunized Against the Tumour. Cancer Research, *26*:1508–1515.

Alexander, P., E. J. Delorme, and J. G. Hall (1966). The Effect of Lymphoid Cells from the Lymph of Specifically Immunized Sheep on the Growth of Primary Sarcomata in Rats. Lancet, i, 1186–1189.

Alexander, P., E. J. Delorme, L. D. G. Hamilton, and J. G. Hall (1967). Effect of Nucleic Acids from Immune Lymphocytes on Rat Sarcomata. Nature, London, *213*:569–572.

Alexander, P., J. Bensted, E. J. Delorme, J. G. Hall., L. D. G. Hamilton, and J. Hodgett (1968). Treatment of Primary Sarcomata by Enhancing Host Defense with Immune Lymphocytes or their RNA *in* 21st Annual Symposium on Fundamental Cancer Research of M. D. Anderson Hospital and Tumor Institute. Pub. by Williams and Wilkins, Baltimore, in the press.

Birbeck, M. S. C., and J. G. Hall (1967), Transformation *in vivo* of Basophilic Lymph Cells into Plasma Cells; an Electronmicroscope Autoradiographic Study. Nature, *214*, page 183.

Delorme, E. J., and P. Alexander (1964). Treatment of Primary Fibrosarcoma in the Rat with Immune Lymphocytes. Lancet, ii, 117–120.

Foley, E. J. (1953). Antigenic Properties of Methyl-cholanthrene Induced Tumours in Mice of the Strain of Origin. Cancer Res., *13*:835–837.

Gowans, J. L., and D. D. McGregor (1965). The Immunological Activities of Lymphocytes. Progr. in Allergy, *9*:1–29.

Haddow, A., and P. Alexander (1964). An Immunological Method of Increasing the Sensitivity of Primary Sarcomas to Local Irradiation with X-rays. Lancet, i, 452–457.

Jeejeebhoy, H. F., E. J. Delorme, and P. Alexander (1966). The Anti-tumour Effect of Lymphoid Cells Placed in Millipore Diffusion Chambers. Transplantation, *4*:397–403.

Mikulska, Z. B., C. Smith, and P. Alexander (1966). Evidence for an Immunological Reaction of the Host Directed Against Its Own Actively Growing Primary Tumour. J. Nat. Cancer Inst., *36*:29–35.

Old, L. J., and E. A. Boyse (1964). Immunology of Experimental Tumours. Ann. Rev. Med., *15*:167–188.

DISCUSSION

Chairman: M. Cohn

BRAUN: In starting this discussion, I would like to present to you, for your consideration and criticism, a scheme of initiation of antibody formation that differs slightly from those proposed and discussed by others. Our scheme is based not only on our own experimental data and ideas but borrows heavily from those of others. Like many others we assume the participation of a two-cell system, as shown below (Fig. 1), and we picture the initial events as involving an uptake of potential immunogens by macrophages and a processing of such material by some members of the macrophage population. There may be recognition mechanisms at this level. We have proposed that the resulting RNA-rich, antigen-containing complex, which is known to be able to initiate antibody formation in lymphocytes of non-immunized animals, may represent a complex of a non-specific activator and antigen which acts as a sort of pilot-boat permitting the entrance of the activator into appropriate preexisting stem-cells of antibody-forming populations. We assume that the non-specific activator, which may be either RNA or something closely associated with it, ordinarily cannot enter preexisting stem-cells of antibody-forming lymphocyte populations (clones). However, an antibody-like recognition site on these lymphocytes, by reacting with the antigenic determinant complexed to the macrophage-elicited RNA, may permit entrance of the activator into those cells that already possess the capacity of making antibody-protein capable of reacting with that particular antigenic determinant. The entrance of the activator into these members of the preexisting stem-cell population of lymphocytes will activate these cells so that they will perform and increase in numbers, giving rise to antibody-forming clones. The activator cannot get into the other stem-cells.

This scheme has a number of interesting consequences, some of which we have subjected to experimental tests. First of all, you could predict that if our idea is correct that antigen merely represents the pilot boat guiding a non-specific activator into preexisting stem-cells, you may ruin this guidance system by opening up, so to speak, all of the stem-cells in a non-specific manner, thereby misguiding an available activator (which is produced after exposure to antigen) into the wrong stem-cells. In recent months we have succeeded in producing such misguidance or dispersion of the postulated activator. We have been able to demonstrate that if you give an antigen together with agents that are known or suspected of altering the permeability of lymphocytes, false, or better—unexpected, responses result. This means you activate clones which should not be activated. So far the production of every type of antibody that we have looked for, for example anti-sRBC, anti-cRBC, anti-*E. coli*, has proved to be turned on after the administration of either sRBC or key hole limpet hemocyanin in the presence of chlorpromazine, phenoxybenzamine, epinephrine, streptolysin or culture filtrates of staphylococci given to us by Dr. W. R. Chesbro. Also, endotoxin given to endotoxin-tolerant animals at time of injection of sRBC produces increases in spleen cells making antibody to antigens that have not been administered. Thus all of these agents seem to permit a dispersion of an activator into the wrong stem-cell. It should be pointed out that under these conditions one obtains a limited, but nevertheless very significant, activation of antibody formation to antigens not presented to the animal.

The scheme shown in Figure 1 also can explain all known phenomena of non-responsiveness. For example, one may assume that the processing capacities of macrophages are not yet mature in the early stages of an

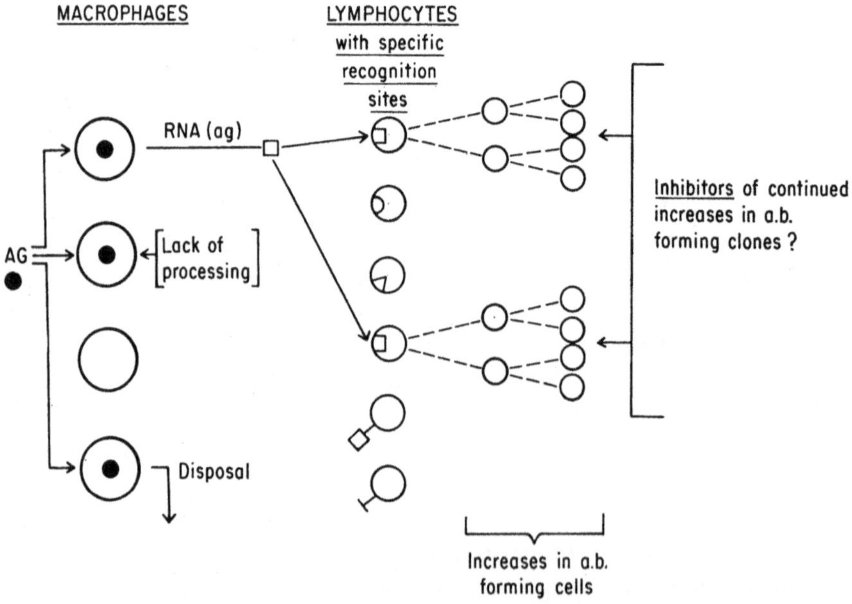

Fig. 1. Scheme illustrating a possible mode of initiation of antibody formation.

individual's development, whereas the assortment of lymphocytic stem-cells is already present; therefore, any antigen present during this period, either self-antigens or antigens introduced into the animal, would fail to be combined with an activator and would instead go as "naked" antigen, devoid of activator, to the antibody-like recognition sites blocking the entrance door, so to speak, and preventing the subsequent entrance of the activator. If this idea is correct, it could be predicted that newborn animals may lack the ability to process many potential antigens and therefore lack the capacity to form antibody to these antigens. We have obtained data indicating that this may indeed be the case. It is well known that if you give an antigen such as sheep red blood cells to newborn mice no antibodies to sRBC are formed; such mice are unable to respond to this antigen for many days after their birth. However, if you present this same antigen to newborn mice together with mature macrophages from adult mice, the newborn mice will make antibodies to sRBC very beautifully. Next, one could suggest that the known phenomenon of non-responsiveness after injury to the processing system of adult animals could again represent a by-passing of the processing system, thereby bringing naked antigen to available recognition sites without activator; again this would prevent subsequent uptake of the activator. The fact that injury to the processing system will result in a prolonged non-responsiveness to antigens administered at such a time was shown by Feldman. He used X-irradiation to injure the processing system. We have shown similar effects by "constipating" the macrophage system with the help of thorotrast; an antigen given to such blocked animals will indeed cause a prolonged specific non-responsiveness, presumably by causing a blocking of the recognition site with naked antigen. And finally, the phenomenon of immune paralysis can also be explained very easily on the basis of our proposal because in any normal antibody response there may be a competition between antigen that has been hooked on to the activator and naked antigen, both of which may compete for the recognition site on the stem-cell. If the processing capacities are overloaded by the administration of too much antigen, naked antigen, devoid of activator, may occupy the recognition sites and interfere with activation because the activator now can no longer enter into the cell. Also, antibodies or enzymes that ruin responses, as was discussed earlier in this symposium, may do so by either ruining the activator or the "pilot boat," i.e., by neutralizing the guiding antigen which now reacts with free circulating antibody instead of with antibody-like recognition sites.

COHN: In assessing the nature of antibody formation, we must take account of the satisfying finding, which we owe to Fishman, that there is a two-cell system involved. We are beginning to believe that an intermediate cell, suspected to be a macrophage or something related to it, may process antigen and that resulting processed material, rich in RNA, from that point on acts in a way that is special and different from the free antigen that we inject. Dr. Braun has presented to you a theory covering these aspects. Now it doesn't seem to me that this is where the real problem lies. The problem that is difficult to understand is the recognition of the antigen by the macrophage. This poses the problem of enzymatic systems of a special kind which can distinguish between what has been injected from what is normally inside the cell. It also poses a problem of concern to chemists, namely that when you fragment a substance into small pieces the chances that these fragmented pieces are going to carry antigenic determinants that are related to the injected molecule are very low. It is true that you can break an antigen down and obtain fragments that are identical, but in general when substances are broken down to the level of nucleopeptides that we have heard about, they would lose their antigenicity. Therefore, we would have to postulate two kinds of special enzymatic systems: (1) one that would link an RNA to an antigenic determinant, and it is incumbent upon those working in the field to find out what that linkage is, and (2) one that processes the antigen in such a manner that it maintains its antigenic configuration. In addition to such complex events Dr. Fishman has thrown another bombshell into the story. He has introduced a messenger RNA which phenotypically transforms cells and which passes from one cell to another. He has shown this by demonstrating that the antibody protein produced carries the genetic marker of the donor of the transforming RNA. Here again, as he himself has pointed out, we are faced with a structural problem since we know that antibodies consist of a light and heavy chain. We know that they are made under the direction of two separate messenger RNAs and therefore both of these RNAs would have to enter into the same cell. Since the population of potential antibody-forming cells is extremely heterogeneous, the probability that the two RNAs would go into a single cell is very low. Thus, we are faced with findings which, on the surface, present an obvious paradox to what we know. I think we should concentrate in our discussion on the chemistry of the RNA-rich complex, the enzymology involved in its formation and its function in the biology of this system.

STAVITSKY: I am wondering whether any other assays have been applied in the allotypic system employed by Fishman and Adler, looking, for example, for A4 globulins in addition to antibody specific for the antigen under test. One would expect that there would indeed be only a very small amount of A4 globulin produced and I am wondering whether you have any evidence regarding this.

FISHMAN: Actually these allotype experiments are preliminary and I hope that this is rather obvious from what has been said. All we have on hand is the appearance of the allotype marker on specific antibody as revealed by inhibition analysis in our neutralization tests. I would like to comment on Dr. Cohn's postulation that what we see in a tissue culture environment may actually be going on in the animal. The question is how much of the information obtained in tissue culture experiments can be correlated with events taking place in the animal. Now, we can show, for instance, an association of macrophage and lymphocyte. We can actually show passage of material from one cell to another. We don't know what this material is. We can also show that there is passage of some material from macrophage to macrophage. Therefore, getting material from one cell to another appears to be no problem if we accept some of the data that come from *in vitro* studies. But we do not know anything about the RNA-antigen complex and how it functions *in vivo*. The only data on this that I am aware of are those of Dr. Garvey and those of some Italian investigators who have claimed to have picked up RNA-antigen complexes *in vivo*. I don't see any dilemma in the allotype transfer experiments since RNA was applied in high concentrations so that the chances of getting two RNAs into the same cell do not appear too remote. I don't know whether this can happen in the animal.

COHN: I still believe that the critical question is whether or not the RNA-antigen complex is a necessary intermediate in the synthesis of antibody.

ALEXANDER: There certainly is one situation in which it would not appear to be necessary and that is in the graft versus host reaction when one puts parental lymphocytes obtained from thoracic duct into F1 hybrids. One then gets a graft versus host reaction and yet no macrophages capable of recognizing the other parental strain have been put into the animal.

GOTTLIEB: I would like to make an additional point regarding the necessity for a processing step. It should be remembered that the Fishman type of experiment dealing with antigen-RNA interactions has really only been done with rather large antigens, certain bacterial antigens, phage antigens and hemocyanin. We are faced, therefore, with the problem of generality. Do the available data necessarily imply that every antigen will have to go through the postulated pathway?

FISHMAN: I think one has to remember that one is never dealing with a primary response. When an antigen is injected, both primary and secondary responses may be involved. Perhaps at this point Coons' classification of cells may become important. If you recall, he postulated that there are three types of cells which he called X, Y and Z, and they are supposed to be different in terms of their reactivity with antigen and their needs for induction. One may postulate that at any given time an animal may possess a varied proportion of X, Y and Z cells. Now, what is required to stimulate X to Y might be quite different from what it takes to stimulate Y to Z. Therefore, we may have many experiments in which you can bypass the "processing" if Y cells exist, and I think we have some evidence for this from data such as those reported by Dr. Landy, namely that the induction period required for appearance of antibody can be extremely short. Therefore, one is not only dealing with different cells in terms of their differentiation towards antibody synthesis, but one is also dealing with different stimuli that can act at different cell levels. Accordingly, it seems difficult to talk about one specific material affecting the whole general immune system.

GARVEY: I believe that in respect to soluble antigens the role of the macrophage may be incidental while it may be needed as a processing agent for insoluble antigens. The evidence that we have accumulated indicates that soluble antigens really have no encounter with macrophages; instead parenchymal cells seem to be the primary contact and the second contact seems to be with lymphoid cells.

GOTTLIEB: With regard to the chemistry of the RNA-antigen complex, this is something that is currently being investigated from two points of view. One is to determine the base sequence of the oligonucleotides involved in this minor fraction of macrophage RNA, and second, and perhaps more important, is to determine to which base or bases antigen is linked. Furthermore, we are trying to determine whether different antigens attach to different base regions of the oligonucleotide. Unfortunately I cannot as yet give you any definitive information on this. I would agree that it is quite likely that there is an enzyme in macrophages which takes degraded antigen and links it to the RNA carrier.

STAVITSKY: With respect to the character of this RNA-antigen complex, you indicated, Dr. Gottlieb, that the minor component was ribonuclease-resistant. Does this mean that it can still function as an immunogen after treatment with ribonuclease and why is it relatively ribonuclease-resistant?

GOTTLIEB: Ribonuclease resistance is a relative matter. If one compares the degradation of major and minor components after exposure to the same amount of ribonuclease, one finds that the minor component is preserved. This is also true with T1 ribonuclease. If one wants to think somewhat teleologically about this, one could suppose that the relative resistance to ribonuclease may serve a rather direct physiological purpose, namely to permit this material to be biologically active even after exposure to serum nucleases.

FISHMAN: A comment about the formation of the RNA-antigen complex. We have started experiments in which we have tried to solubilize the antigen and hook it on to some RNA. So far mere mixing of these components does not suffice. We are now testing cell sap materials to determine whether indeed it may contain enzymes capable of coupling the two components.

DUBISKI: I would like to make some comments which relate to all of the papers and to Dr. Braun's theory. In association with Drs. Chou and Cinader we made some experiments in which by cell transfer we produced antibody formation in newborn rabbits. We took lymphoid cells from adult normal rabbits, incubated them with antigen *in vitro,* injected them into newborn normal rabbits and observed antibody formation in these rabbits. These antibodies had allotype specificity of the recipient. In addition to these antibodies we found other immunoglobulins which were formed very early and were rather short-lived. These latter immunoglobulins did not have antibody activity against human albumin which was the antigen in our experiments and these particular immunoglobulins had the donor allotype. We used three kinds of cells in these experiments, and we observed that all of them were almost equally efficient in transferring the antibody-producing capacity, namely peritoneal exudate cells, lymph node cells and thymus cells. However, only peritoneal exudate cells and lymph node cells were usually able to stimulate the donor type immunoglobulin formation in the recipient rabbits. After transfer of thymus cells we only occasionally observed a very long-lived donor type immunoglobulin formation. A few of the rabbits were transformed into chimeras in which the donor type immunoglobulin formation persisted for a very long time and we were even able to transfer this ability for donor type immunoglobulin formation into a second generation of rabbits.

SEHON: In Dr. Fishman's experiments species barriers were never crossed, whereas in Dr. Alexander's experiments they were. What is the role of species barriers?

ALEXANDER: Species barriers are not a problem in our system. We have also been able to use rat thoracic duct cells and extracted the RNA from these. However, our main interest has been for obvious practical reasons in the use of a heterologous system.

FISHMAN: An inability to cross species barriers has been reported in much of the *in vivo* work. Mitchison believes that there is strain specificity and so does Feldman. In our studies with synthetic copolymers which we are carrying out in cooperation with Drs. Pinchuck and Maurer, we have crossed species barriers and I think that the RNA-antigen complex can indeed pass these barriers.

GARVEY: We also have demonstrated the ability of an antigen-RNA complex to cross species barriers. Our complexes elicited good Schultz-Dale reactions in the guinea pig.

GLICK: Dr. Gottlieb, are there any rare bases or methylated bases in the minor band of the macrophage RNA?

GOTTLIEB: The answer is we don't know as yet.

GLICK: Has anybody attempted to alter the structure of the antigen-RNA complex? For example, have there been any attempts to treat the antigen-RNA complex with radiation or antibiotics to see if one can alter the RNA? This would be of interest in relation to the question whether or not RNA is acting merely as a carrier or is also doing something else of its own inside the recipient cell.

GOTTLIEB: UV-irradiation at wavelengths around 250 $m\mu$ inactivates the immunogenic activity of the complex. This was interpreted to be of an effect on the protein component, and not on the RNA.

SIGEL: I am not clear whether Dr. Alexander has done experiments with RNAase. I mean what evidence is there that there is RNA present in his extracts?

ALEXANDER: Yes, RNAase abolishes the activity but we have as yet not done any dose-dependent experiments.

SIGEL: I noticed in Dr. Friedman's presentation that the RNA-antigen complex derived from tolerant animals caused better immunization than similar material from immune animals. I would like to have him comment on this.

FRIEDMAN: We consider this difference to be due to quantitation. 20γ of antigen into a newborn mouse is a lot more antigen per tissue than 20γ into an adult animal. We are actually transferring the same quantity of RNA from both sources and also, as Nossal and associates have shown, antigen does localize in the newborn animal and persists very much longer.

ABRAMSON: I would like to ask Dr. Braun about the system in which he immunizes the immunologically incompetent newborns with the help of macrophages from competent adult animals. He suggests that this results in a primary response. If these animals are challenged with the same antigen later on in life when they are immunologically competent, do they show a primary response or an anamnestic response?

BRAUN: This has not been tried as yet.

PANIJEL: I have two questions. The first is related to the report of Dr. Garvey. Two years ago we found in the sera of germ-free animals some anti-RNA antibodies able to give soluble complexes. These were natural antibodies and we have suggested they may function normally as a carrier of RNA. Do you think that this is possible and may your nucleopeptide be a degradation product of some complex between such natural antibodies and RNA? If this should be so, it is possible to consider that one may not only be dealing with pathological auto-antibodies but also with useful auto-antibodies that play a carrier role in the organism. The second question is very brief and general. It is surprising that during the discussion of the RNA-antigen complex not much has been stated regarding the possibility that there may be activity of the RNA alone or of the antigen alone. It seems to me that this question is particularly important in order to reconcile the results of Fishman, Friedman and Gottlieb.

GARVEY: You are asking whether the peptides we found might be part of natural antibodies. The answer must be: this is unlikely in view of the way in which we have isolated the complexes. To your question about the RNA versus the antigen and the importance of each as entities unto themselves, I agree that this is an important question to get at next. But first one must analyze the complex to determine the linkage between the two components and then consider the question that you raised.

ADLER: Dr. Stavitsky raised the question whether we had independent evidence for the allotype transfer aside from the data that we presented, namely the inhibition of T2 neutralization by allotype antisera. The answer is yes as reported by us in a recent *Journal of Immunology* paper. The presence of the allotype was tested by an inhibition of tanned cell hemagglutination.

DREYER: Dr. Braun, are the macrophages in a very young mouse unable to carry out the production of RNA-antigen complexes? The reason for asking this question is related to the question Mel Cohn asked, namely whether this process is necessary for antibody formation. Since we know that many kinds of antibodies can be made in the prenatal animal, I would be inclined to agree with Justine Garvey and say that perhaps some kinds of antigens, for example very large ones, may require a garbage disposal mechanism or a degradation, but other kinds of antigens may not need RNA or macrophages at all. Otherwise you could not make antibodies in a prenatal mouse or prenatal fetal pig.

BRAUN: My answer to this problem would be that the processing capabilities for different antigens mature at different times.

L. LEVINE: The size of the antigenic determinant or the size of the antigenic fragment bound to RNA appears to be, from Dr. Gottlieb's published papers, roughly 25–30 amino acids; it could be even smaller. It should be pointed out that it is difficult to see how 25 amino acids could retain the original conformation of a polypeptide in its native state. If one adds to this the constraint due to the covalent linkage between RNA and peptide, it becomes even more difficult to understand how one can get antibodies specific for native determinants of a globular protein.

PLESCIA: Apropos of this I would like to add another thought which also could be disturbing. We tend to talk about determinants which are peptide in nature, but we also know that determinants come from polysaccharides and polynucleotides. This, therefore, requires some scheme in which RNA can be linked with equal ease to oligonucleotides and oligosaccharides.

ROLE OF NUCLEIC ACIDS IN SPECIFIC ANTIBODY FORMATION (PART II)

FRACTIONATION OF RNA CAPABLE OF TRANSFERRING TRANSPLANTATION IMMUNITY *

JOHN A. MANNICK, M.D.

Department of Surgery, Boston University Medical Center
Boston, Mass.

I. INTRODUCTION

Previous work from this laboratory (Mannick and Egdahl, 1962, 1964) indicated that the capacity to react immunologically against specific transplantation antigens may be transferred to previously unsensitized lymphoid cells by incubation of these cells *in vitro* with ribonucleic acid (RNA) extracted from the lymph nodes of specifically immunized donors. The assay for this apparent transfer of transplantation immunity was based on the transfer reaction (Brent, Brown and Medawar, 1958) and on the accelerated rejection of specific skin allografts, both assays being carried out in rabbits. Similar observations have been reported by others who also studied the transfer of transplantation immunity in rabbits, rats and mice (Clarke and Wilson, 1963; Wilson and Wecker, 1967; Sabbadini, 1967; Rigby, 1967).

Working along similar lines, Fishman and Adler (1963) and Cohen and Parks (1964) showed that the capacity for conventional antibody formation may be similarly transferred by means of RNA-containing material extracted from specifically sensitized cells. More recently Cohen *et al.* (1965) found that the active RNA material sedimented with a 8–12S fraction in a sucrose density gradient. Because it seemed important to determine whether or not similar mechanisms were involved in the transfer of transplantation immunity by lymphoid RNA and because it was desirable to characterize the RNA fraction responsible for the transfer of transplantation immunity, active RNA preparations in the present study were fractionated by sucrose density gradient ultracentrifugation.

* Supported in part by USPHS grants CA 08091 and AM 10824.

II. MATERIALS AND METHODS

Adult rabbits of both sexes and of three breeds (New Zealand White, California White, and Dutch) were used as experimental animals. The animals were caged individually and fed a standard laboratory diet with water *ad libitum*. The basic experimental protocol, illustrated in Figure 1, has been described previously (Mannick and Egdahl, 1964). A New Zealand White recipient was immunized against a California donor by the application of full thickness skin allografts to the lower hind legs and to the anterior chest wall, and by the injection of a suspension of donor spleen cells into the 4 foot pads. After 7 or 8 days, at the time of rejection of the skin allografts, the stimulated axillary and popliteal lymph nodes were removed from the New Zealand recipient. RNA was extracted from these lymph nodes and then, with or without an intervening fractionation step, was incubated for 20 minutes at 37°C in 10 ml of Hank's solution with spleen cells obtained from a second New Zealand rabbit. Following incubation the spleen cells were collected by centrifugation, washed, resuspended in Hank's solution, counted in a clinical hemocytometer, and subjected to a viability determination utilizing 0.1% trypan blue. Cell viability was greater than 80% in all experiments reported here. The suspensions of autologous spleen cells, varying in number from 100 to 300×10^6, were returned by intravenous infusion

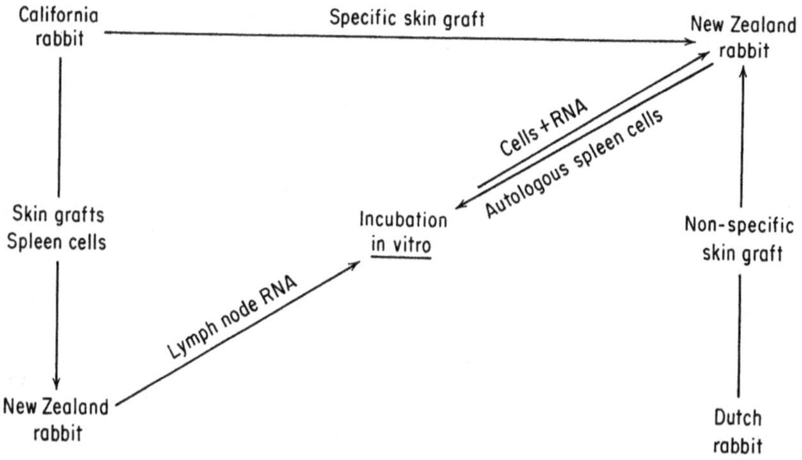

Fig. 1. Basic experimental protocol used in the present studies. A New Zealand recipient rabbit is sensitized against a California donor by skin allografts and injection of donor spleen cells. Stimulated reigonal lymph nodes are excised from the recipient. RNA is extracted from these nodes and incubated *in vitro* with spleen cells from a second New Zealand rabbit. After incubation the spleen cells are reinfused into the same rabbit. This animal is then challenged with a specific skin allograft from the original California donor and a nonspecific allograft from an indifferent Dutch rabbit.

into the New Zealand rabbit from which they came. This rabbit, serving as the test animal, had received 24–48 hours previously a full thickness skin allograft from the original California donor, applied to the right ear and a full thickness skin allograft from a Dutch rabbit, applied to the left ear. Skin allograft survival was determined by gross inspection. The day of graft rejection was taken to be the first day on which unequivocal infarction of the graft occurred or at least 75% of the epidermis was clearly necrotic.

All surgical procedures were performed under sterile conditions and under intravenous pentobarbital anesthesia supplemented with local procaine infiltration.

RNA was extracted from lymph nodes by the hot phenol method as described by Cohen and Parks (1964). The freshly excised lymph nodes were trimmed of all fat and connective tissue and quickly frozen with dry ice. The frozen tissue was pulverized with a mortar and pestle and homogenized in a mixture containing equal parts of 0.01 M acetate buffer, to which was added 0.5% sodium dodecyl sulfate and 0.2% bentonite, and freshly distilled phenol that was saturated with buffer and contained 0.1% 8-hydroxyquinoline. Following homogenization, the mixture was heated to 55°C in a water bath with continuous stirring and quickly cooled to 12°C in an alcohol-dry ice bath. The phases were separated by centrifugation at 8000 × g for 10 minutes at 12°C. The aqueous phase was removed and the interface and phenol phase were reextracted with an additional ⅔ volume of fresh buffer. The aqueous phase was again separated and the combined aqueous phases were reextracted twice with ½ volume of fresh phenol, repeating the heating and cooling cycle each time. After the final phenol extraction, the RNA was precipitated from the aqueous phase at 2°C by adding sodium acetate to a final concentration of 0.3 M and adding 2.5 volumes of absolute ethanol. The precipitate was recovered by centrifugation, washed 6 times with 66% ethanol at 2°C and redissolved before use.

For sucrose density gradient ultracentrifugation, the RNA was dissolved in a small amount of Hank's solution and up to 600 μg of RNA in 0.5 ml were layered onto 4 ml of sucrose ranging in concentration from 12.5 to 30% to give a continuous density gradient. Centrifugation was carried out at 5°C for 5 hours at 38,000 rpm in a Spinco SW-39 rotor in a model L preparative ultracentrifuge. Alternatively, up to 1 mg of RNA in .75 ml of Hank's solution was layered onto 10 ml sucrose gradients and centrifuged in the 969 swinging bucket head of an International BD-2 preparative ultracentrifuge at 25,000 rpm for 18 hours at 4°C. Following centrifugation, fractions were collected from the bottom of the tube and the optical density of each fraction was determined at 260 mμ in a Beckman DU spectrophotometer. RNA from fractions obtained from the gradients was precipitated with acetate and ethanol as before and was collected by centrifugation. The RNA was dissolved in Hank's solution for biological testing.

Spleen cell suspensions were prepared in these experiments by pressing the freshly excised spleen through a coarse stainless steel wire screen into a small volume of Hank's solution. The cells were washed twice with Hank's solution and dispersed by gentle pipetting prior to incubation with RNA.

III. RESULTS

A. Control Skin Allografts

The survival of California and Dutch skin allografts, simultaneously placed on opposite ears of the same New Zealand recipient rabbit, was determined in a series of control studies performed concomitantly with the experimental series reported below. The survival of these control skin allografts is listed in Table 1. Control California allografts had a

Table 1

Skin Allograft Survival in Control New Zealand Rabbits

Allograft Donor	No. Animals	Allograft Survival Time (Days)	
		Range	Mean
California	22	6–12	8.0 ± 1.7 S.D.
Dutch	22	6–12	8.3 ± 1.8 S.D.

mean survival time of 8.0 ± 1.7 * days, and control Dutch skin allografts had a mean survival time of 8.3 ± 1.8 days. Prior work from this laboratory (Mannick and Egdahl, 1964) had demonstrated that the intravenous administration of autologous spleen cells had no effect upon the survival time of skin allografts.

B. Hot Phenol RNA Controls

Since the RNA utilized in previously reported attempts to transfer transplantation immunity had been extracted by the cold phenol technique, lymphoid RNA, extracted by the hot phenol technique, was tested for its capacity to transfer transplantation immunity. In these experiments, the protocol followed is outlined in Figure 1. The total RNA

* S.D.

Table 2

SKIN ALLOGRAPH SURVIVAL IN HOT PHENOL RNA CONTROLS

Allograft Donor	Conc. RNA (μg/ml)	No. Animals	Allograft Survival Time (Days)	
			Range	Mean
Specific California	250–1200	8	4–8	5.4 ± 1.5 S.D.
Nonspecific Dutch	250–1200	8	5–11	7.6 ± 2.0 S.D.

extracted from the lymph nodes of a sensitized New Zealand recipient was incubated *in vitro* with spleen cells from a New Zealand test rabbit. These autologous spleen cells were reinfused into the test animal and the survival time of specific California and non-specific Dutch skin allografts, simultaneously applied to the test animal, was determined. The results of this experiment are listed in Table 2. The mean survival time of the specific California skin allografts in this group of New Zealand test animals was 5.4 ± 1.5 days, whereas the mean survival time of the non-specific Dutch skin allografts on the same test animals was 7.5 ± 2.0 days. The survival of the specific California skin allografts was significantly shorter than the survival of control California allografts, $p < 0.001$. The survival time of the specific California allografts was also significantly shorter than the survival of the nonspecific Dutch allografts, $p < 0.05$.

The yield of RNA obtained by the hot phenol technique appeared to be somewhat greater than that previously reported when the cold phenol technique was utilized (Mannick and Egdahl, 1964). As noted in Table 2, the concentration of RNA in the 10 ml incubation medium ranged from 250–1200 μg/ml.

C. DENSITY GRADIENT FRACTIONATION

In these experiments the same protocol was followed except that the RNA extracted from the lymph nodes of the sensitized New Zealand recipient was subjected to sucrose density gradient ultracentrifugation. The optical density at 260 mμ of the RNA fractions obtained by this method is shown in Figure 2. Peaks sedimenting at approximately 28S, 18S, and 4S were seen.

The RNA sedimenting in the top half of the gradient was separated from that sedimenting in the bottom half of the gradient, as illustrated in Figure 2, and each was incubated with spleen cells obtained from one

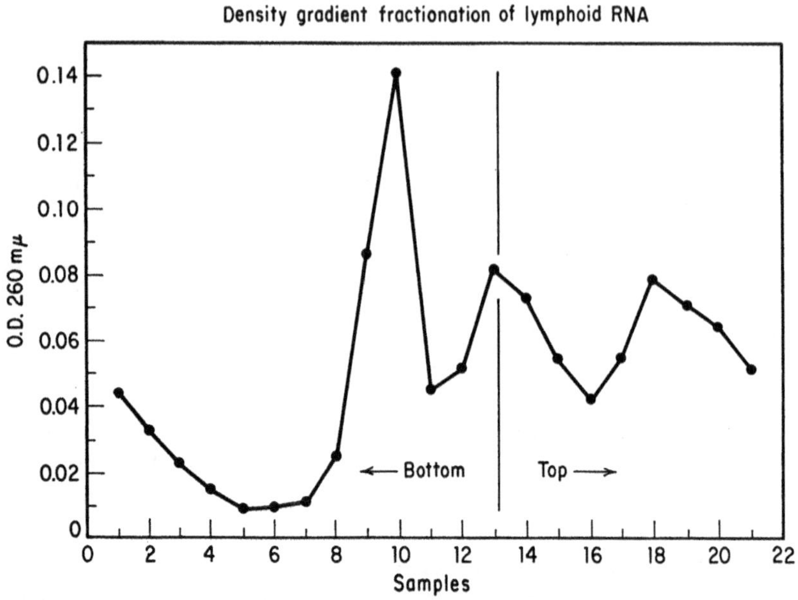

Fig. 2. Optical density at 260 mμ of lymphoid RNA fractions obtained after ultracentrifugation in sucrose gradient for 5 hours at 38,000 rpm. First fraction from the bottom of tube is at left. Division of RNA sample into top and bottom halves for biological testing is indicated.

of two New Zealand test rabbits. After incubation, the spleen cells were reinfused back into the animals from which they came and as before the survival of specific California and non-specific Dutch allografts applied to opposite ears of each of the test animals was determined.

As shown in Table 3, the mean survival time of the specific California skin allografts in test rabbits, which had received autologous spleen cells incubated in RNA obtained from the bottom half of the density gradient, was 7.8 ± 1.2 days, a time not significantly different from the survival of control California skin allografts, $p > 0.6$. The mean survival time of the nonspecific Dutch skin allografts applied to these same New Zealand test animals was 8.2 ± 0.9 days. The mean survival times of the specific California and nonspecific Dutch skin allografts in this group of test animals were not significantly different, $p > 0.4$.

The mean survival time of the specific California skin allografts in test animals, which had received autologous spleen cells incubated with RNA from the top half of the density gradient, was 5.7 ± 1.8 days, a time significantly shorter than the survival of control California skin allografts, $p < 0.01$. The mean survival time of the nonspecific Dutch skin allografts applied to the same New Zealand test animals was 7.9 ± 1.0 days. In this experimental group the mean survival time of the

Table 3

Skin Allograft Survival. RNA from Bottom Half of Gradient

Allograft Donor	Conc. RNA (μg/ml)	No. Animals	Allograft Survival Time (Days) Range	Mean
Specific California	35–220	10	6–10	7.8 ± 1.2 S.D.
Nonspecific Dutch	35–220	10	7–10	8.2 ± 0.9 S.D.
RNA from Top Half of Gradient				
Specific California	40–140	10	4–9	5.7 ± 1.8 S.D.
Nonspecific Dutch	40–140	10	6–9	7.9 ± 1.0 S.D.

specific California grafts was significantly shorter than that of the nonspecific Dutch grafts, $p < 0.01$. It thus appeared that the active RNA fraction was present in the top half of the sucrose density gradient.

Accordingly, the RNA in the top half of the density gradient was further divided into two fractions in another series of experiments, a lighter fraction (0–8S) and a heavier fraction (8–18S) as shown in Figure 3. As before, RNA from these two fractions was separately incubated with spleen cells obtained from two New Zealand test rabbits. Following incubation, the cells were reinfused back into the New Zealand animals from which they came. As before, these rabbits were challenged with specific California and nonspecific Dutch skin allografts.

The results of this experiment are illustrated in Table 4. The mean survival time of the specific California allografts in New Zealand test animals, receiving autologous cells incubated with RNA from the 0–8S fraction, was 8.8 ± 3.0 days, a time not significantly different from the survival time of control California allografts, $p > 0.5$. The mean survival time of the nonspecific Dutch allografts in the same New Zealand test animals was 7.4 ± 2.0 days. The mean survival times of the specific California and nonspecific Dutch skin allografts did not significantly differ from one another, $p > 0.3$.

The mean survival time of the specific California allografts in New Zealand test animals, receiving autologous spleen cells incubated with RNA from the 8–18S fraction, was 5.9 ± 1.5 days, a time significantly different from the survival time of control California allografts ($p < 0.01$). The mean survival time of the nonspecific Dutch allografts in the same

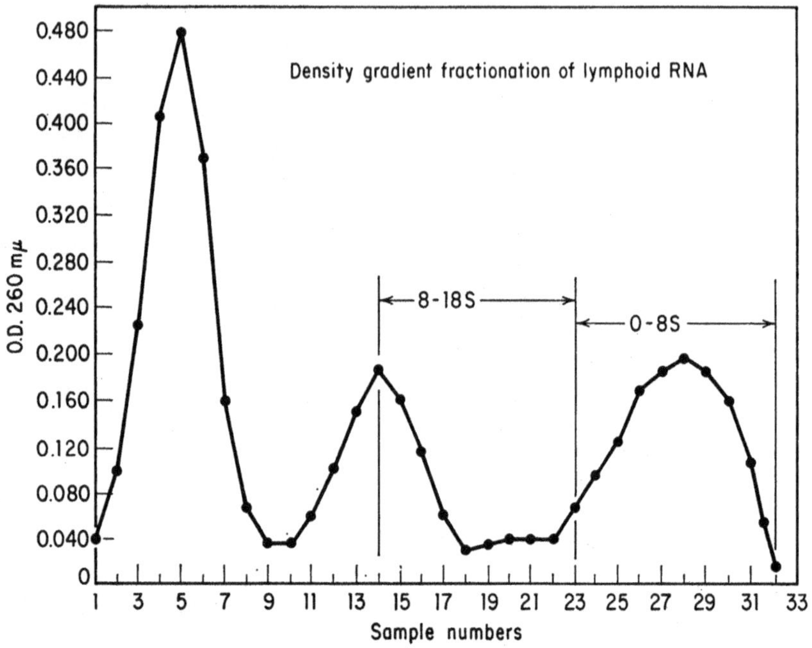

Fig. 3. Optical density at 260 mμ of lymphoid RNA fractions obtained after ultracentrifugation in sucrose gradient for 18 hours at 25,000 rpm. Division of RNA sample into 0–8S and 8–18S fractions for biological testing is indicated.

Table 4

SKIN ALLOGRAFT SURVIVAL

Allograft Donor	Conc. RNA (μg/ml)	No. Animals	Allograft Survival Time (Days) Range	Mean
		0–8S RNA		
Specific				
California	35–180	8	6–13	8.8 ± 3.0 S.D.
Nonspecific				
Dutch	35–180	8	6–12	7.4 ± 2.0 S.D.
		8–18S RNA		
Specific				
California	35–175	9	4–8	5.9 ± 1.5 S.D.
Nonspecific				
Dutch	35–175	9	6–11	8.1 ± 1.6 S.D.

New Zealand test animals was 8.1 ± 1.6 days. The survival time of the specific California allografts was significantly shorter than the survival time of the nonspecific Dutch skin allografts ($p < 0.02$). It thus appeared that the RNA fraction responsible for the transfer of transplantation immunity sedimented within the 8–18S fraction.

D. Ribonuclease Controls

In a final series of experiments, RNA extracted from the lymph nodes of a sensitized New Zealand recipient was subjected to density gradient fractionation as before. Again the 8–18S fraction was separated. In this series of experiments, however, bovine pancreatic ribonuclease, five times recrystallized, was added at a concentration of 10 μg/ml to the medium containing the 8–18S RNA immediately prior to incubation with spleen cells from the New Zealand test rabbit. Following incubation the autologous spleen cells were reinfused back into the New Zealand rabbit from which they came. This test rabbit was challenged with a specific California skin allograft and a nonspecific Dutch skin allograft.

The results of this experimental series are shown in Table 5. The mean survival time of the specific California skin allografts in the New Zealand test animals, which had received autologous cells incubated with ribonuclease-treated RNA from the 8–18S fraction, was 8.0 ± 0.6 days, a time not significantly different from the survival time of control California skin allografts ($p = 1$). The mean survival time of the nonspecific Dutch skin allografts applied to the same New Zealand test animals was 8.5 ± 1.4 days. The mean survival times of the specific and nonspecific allografts did not differ from one another significantly ($p > 0.5$). Thus, a brief exposure to ribonuclease at low concentrations apparently had entirely destroyed the ability of the 8–18S RNA to transfer transplantation immunity.

Table 5

Skin Allograft Survival. RNAase-Treated 8–18S RNA

Allograft Donor	Conc. RNA (μg/ml)	No. Animals	Allograft Survival Time (Days)	
			Range	Mean
Specific California	80–180	6	7–9	8.0 ± 0.6 S.D.
Nonspecific Dutch	80–180	6	7–10	8.3 ± 1.4 S.D.

IV. DISCUSSION

The present results indicate that the RNA fraction responsible for transfer of the capacity to cause accelerated rejection of specific skin allografts sediments between 8 and 18S in a sucrose gradient. It is of interest that Cohen *et al.* (1965) previously found that the RNA capable of transferring conventional antibody forming ability lies in the 8–12S fraction. The fact that the RNA fractions capable of transferring the two types of immune reactivity lie in similar portions of a sucrose density gradient suggests that these molecules are of similar size. It is of further interest that Mach and Vassalli (1965) have found that RNA sedimenting in the 6–12S fraction is far more active than other lymphoid RNA fractions in stimulating the formation of new polypeptides in an *in vitro* protein synthesizing system and is, therefore, presumably RNA with messenger activity.

It is possible that the 8–18S RNA of the present experiments and the 8–12S RNA of Cohen *et al.* is messenger RNA and that this message is transferred to previously uncommitted lymphoid cells during incubation with the RNA. In the experiments involving transfer of conventional antibody forming capability, it is logical to suggest that such a transferred messenger might code for the synthesis of a portion, presumably the combining site, of a specific immunoglobulin molecule. In the case of the transfer of transplantation immunity, however, the protein whose synthesis the messenger RNA might direct is not apparent.

Transplantation immunity ordinarily can be transferred with sensitized cells but not with serum; allografts protected from lymphoid cells but not the serum of an allogeneic host do not undergo conventional rejection, and lymphocytes, in the absence of detectable specific antibody or complement, are capable *in vitro* of attacking and killing allogeneic cells against which they have been sensitized. All of these facts provide strong evidence that the transplantation immune response is a form of cellular immunity. While it is entirely possible that transplantation immunity, like other cellular immune responses, may be mediated by means of cell-bound antibody attached to the effector cells, the presence of such antibody in or on the lymphocytes engaged in carrying out a typical transplantation rejection response has never been demonstrated convincingly. On the other hand, lymphocytes stimulated *in vitro* by the isoantigens of allogeneic cells respond with increased synthesis of DNA and protein (Bain *et al.*, 1964; Cooperband *et al.*, 1967). Wilson (1965) has also shown that an inhibitor of RNA and protein synthesis will prevent the destructive action of sensitized lymphocytes upon specific allogeneic target cells. Nevertheless, whether or not the synthesis of any new immunoglobulin or, in fact, any new protein, is necessary for lymphoid cells to manifest transplantation immunity cannot be determined at present with any certainty.

If the 8–18S RNA isolated in the present experiments and the 8–12S RNA separated by Cohen, *et al.*, is indeed a messenger molecule it is difficult to explain how this messenger can continue to direct effectively the synthesis of new immunoglobulin, or other protein, once the cellular proliferative response, thought to be associated with the initiation of both antibody formation and cellular immunity, has taken place. The added messenger presumably would be diluted progressively by multiple cell divisions unless the messenger molecule were also endowed with the property of self-replication. The fact that the fate of the added RNA, in experiments involving RNA-mediated transfer of immunity, is entirely unknown adds to the difficulty of providing a satisfactory answer to this problem.

A second possibility is that the RNA capable of transferring immune capability acts as a derepressor once it has penetrated cells genetically capable of producing a specific immune response. With respect to transplantation immunity, evidence either for or against this hypothesis is entirely lacking. However, in experiments dealing with the transfer of conventional antibody forming capability Adler *et al.* (1966) have recently reported that lymphoid cells, stimulated by RNA from a donor of foreign gammaglobulin allotype, initially produce immunoglobulin M of donor rather than recipient allotype. This evidence, particularly if it can be confirmed, would argue strongly against the idea of derepression and for the possibility that the added RNA is indeed a messenger molecule.

A third possibility is that the RNA preparation capable of transferring transplantation immunity contains a highly active form of transplantation antigen, and that it is the antigen rather than the RNA which is responsible for the apparent transfer of immunity. Several reports have indicated that specific antigen is contained within the RNA preparations capable of immune transfer (Askonas and Rhodes, 1965; Friedman *et al.*, 1965). Since the isoantigenic molecules responsible for the induction of transplantation immunity have not been clearly defined chemically, a decision as to whether or not such material is present in the 8–18S RNA isolated in the present experiments cannot be made.

Certain features of the RNA-mediated transfer of transplantation immunity make the presence of antigen as a sole explanation for this phenomenon unlikely. First, previous work from this laboratory (Mannick and Egdahl, 1964) and the present experiments have indicated that the RNA-mediated transfer of transplantation immunity is effective when the RNA-treated cells are administered to the test animal as long as 48 hours after the application of the test allografts. The RNA-treated cells are ineffective in shortening specific allograft survival time when administered 48 hours prior to the application of the test allografts. If antigen contained within the RNA preparation were responsible for the effect one would expect a greater degree of sensitization by the antigen when the RNA-treated cells were administered 2 days prior to grafting than when the cells were administered 2 days after application of the grafts. The opposite effect was observed. Furthermore, unpublished observations in this laboratory have shown that large doses of foreign transplantation antigen in the form of whole lymphoid cells or subcellular antigenic material administered to recipient animals 48 hours after grafting have no detectable effect on the survival time of specific skin allografts taken from the antigen donor. Certainly, if an antigen-RNA complex is important in mediating the observed transfer of immunity, the RNA itself must play an important role in the production of the effect since small amounts of ribonuclease have consistently prevented the transfer of immunity. It is now well known that transplantation antigens in most mammalian species, including the rabbit (Mannick *et al.*, 1964), are resistant to the action of ribonuclease.

It thus seems clear that a satisfactory explanation of the biochemical mechanisms involved in the transfer of transplantation immunity mediated by the 8–18S RNA isolated in the present experiments cannot be made from the facts currently available. Further characterization of the RNA species responsible for the immune transfer, and of the metabolic events which follow exposure of lymphoid cells to this molecule, will be necessary before the phenomenon may be understood more clearly.

References

Adler, F. L., M. Fishman, and S. Dray (1966). Antibody formation initiated *in vitro*. III. Antibody formation and allotypic specificity directed by ribonucleic acid from peritoneal exudate cells. J. Immunology, *97*:554–558.

Askonas, B. A., and J. M. Rhodes (1965). Immunogenicity of antigen-containing ribonucleic acid preparations from macrophages. Nature, *205*:470–474.

Bain, B., R. V. Magdalene, and L. Lowenstein (1964). The development of large immature mononuclear cells in mixed leukocyte cultures. Blood, *23*: 108–116.

Brent, L., J. Brown, and P. B. Medawar (1958). Skin transplantation immunity in relation to hypersensitivity. Lancet, *2*:561–563.

Clarke, S. D., and R. E. Wilson (1963). Localization of RNA responsible for direct transfer activity. Surg. Forum, *14*:583–590.

Cohen, E. P., and J. J. Parks (1964). Antibody production by nonimmune spleen cells incubated with RNA from immunized mice. Science, *144*:1012–1013.

Cohen, E. P., R. W. Newcomb, and L. K. Crosby (1965). Conversion of nonimmune spleen cells to antibody-forming cells by RNA: Strain specificity of the response. J. Immunology, *95*:583–590.

Cooperband, S. R., H. Bondevik, K. Schmid, and J. A. Mannick (1968). Transformation of human lymphocytes: inhibition by homologous alpha globulin. Science, *159*:1243–1244.

Fishman, M., and F. L. Adler (1963). Antibody formation initiated *in vitro*. II. Antibody synthesis in x-irradiated recipients of diffusion chambers containing nucleic acid derived from macrophages incubated with antigen, *117*:595–602.

Friedman, H. P., A. B. Stavitsky, and J. M. Solomon (1965). Induction *in vitro* of antibodies to phage T2: Antigens in the RNA extract employed. Science, *149*:1106–1107.

Mach, B., and P. Vassalli (1965). Biosynthesis of RNA in antibody-producing tissues. Proc. Nat. Acad. Sci., *54*:975–982.

Mannick, J. A., and R. H. Egdahl (1962). Ribonucleic acid in "Transformation" of lymphoid cells. Science, *137*:976–977.

——— (1964). Transfer of heightened immunity to skin homografts by lymphoid RNA. J. Clin. Invest., *43*:2166–2177.

Mannick, J. A., J. T. Graziani, and R. H. Egdahl (1964). A transplantation antigen recovered from cell culture medium. Transplantation, *2*:321–333.

Rigby, P. (1967). Correction of homograft immunity in tumor bearing animals with lymphocytic RNA. Clinical Research, *15*:339.

Sabbadini, E., and A. H. Sehon (1967). Acceleration of allograft rejection induced by RNA from sensitized donors. Internat. Arch. Allergy, *32*:55–63.

Wilson, D. B. (1965). Quantitative studies on the behavior of sensitized lymphocytes *in vitro*. J. Exper. Med., *122*:167–172.

Wilson, D. B., and E. E. Wecker (1966). Quantitative studies on the behavior of sensitized lymphoid cells *in vitro*. III. Conversion of "Normal" lymphoid cells to an immunologically active status with RNA derived from isologous lymphoid tissues of specifically immunized rats. J. Immunology, *97*:512–516.

INDUCTION OF ACCELERATED GRAFT REJECTION AND ENHANCED GRAFT-VERSUS-HOST REACTION WITH RNA FROM LYMPHOID ORGANS OF ANIMALS IMMUNIZED WITH ALLOGENEIC TISSUES

E. SABBADINI * AND A. H. SEHON

Departments of Experimental Surgery and Chemistry
McGill University, Montreal, Quebec, Canada

As previously reported, accelerated rejection of skin allografts in adult rabbits (Sabbadini and Sehon, 1967a) and in mice (Sabbadini and Sehon, 1967b), and enhanced graft-versus-host reaction (GVHR) in newborn F_1 hybrid mice (Sabbadini and Sehon, 1967b), were induced with spleen cells incubated with RNA preparations extracted from lymphoid organs of immunized mice (I-RNA). The underlying feature of these two immunological manifestations was their specificity with respect to the transplantation antigens of the animals used to immunize the donors of the I-RNA; therefore, both types of reaction were considered to be an expression of the transfer of immunity by I-RNA.

The present report surveys some of the observations leading to the conclusion that transplantation immunity can be transferred by I-RNA and some experiments suggesting the possible absence of transplantation antigens in I-RNA preparations.

I. MATERIALS AND METHODS

Tissue culture medium 199 (TCM) (Morgan *et al.*, 1950) and Earle's balanced salt solution (BSS) (Earle, 1943) were obtained from Difco Laboratories, Detroit, Michigan.

Phosphate-buffered saline (PBS) was prepared according to Dulbecco

* Fellow of the Medical Research Council of Canada, 1966–67.

and Vogt (Dulbecco and Vogt, 1954); for incubation of spleen cells with RNA, PBS containing 0.7 M sucrose (PBS-sucrose) was used.

Radioactive chromium (^{51}Cr) was supplied by Atomic Energy of Canada Ltd., Ottawa, Ont., as sodium chromate in NaOH; portions containing 60 μc were neutralized with HCl and sterilized by autoclaving prior to use. Radioactivity was measured with a 1070 Auto/Gamma Well Scintillation counter (Nuclear Chicago Co., Des Plains, Ill.), with a window width of 10 volts.

A. Animals

The following inbred strains of mice, 8–10 weeks old, listed with their H_2-alleles in brackets, were purchased from Jackson Laboratories, Bar Harbor, Maine: A/J (a), CBA/J (k), C3H/HeJ (k), C57B1/6J (b). For production of F_1 hybrids the animals were bred in the Animal House of the Donner Building on McGill Campus.

B. Immunization

For the immune stimulation of lymph nodes, two procedures of immunization were used: skin grafting and subcutaneous implantation of fragments of spleen, lymph nodes, thymus and liver. These procedures were previously described in detail (Sabbadini and Sehon, 1967b); the three strain combinations used were CBA → A, C57B1 → A and A → CBA.

For production of cytotoxic sera the recipients received 3–4 i.p. injections, 4 weeks apart, of 5, 10, 20 and 50 million allogeneic lymphoid cells representing a mixture of cells from spleens, lymph nodes and thymi. Sera were collected 7 days after the last injection. The sera from 10–20 mice were pooled, heated at 56°C for 30 minutes, absorbed with the appropriate lymphoid cells and stored in a frozen state in portions of 0.1 ml. For absorption of cross-reacting antibodies, 5×10^7 lymphoid cells were suspended in 1 ml of serum, incubated at room temperature for 30 minutes and centrifuged at 1500 r.p.m. for 15 minutes. The absorption was repeated until no further cytotoxicity was observed in the cross-reactivity test.

Two pools of sera were used for the cytotoxic inhibition test: Pool 1 was an anti-CBA serum produced in mice A and absorbed with C57B1 lymphoid cells; Pool 2 was an anti-C57B1 serum produced in mice A and absorbed with CBA lymphoid cells.

C. Cytotoxic Test and Cytotoxic Inhibition Test

The cytotoxic test was performed according to Sanderson (Sanderson, 1964) using cells from axillary lymph nodes labelled by incubation with radioactive chromium (40 μc of ^{51}Cr per ml for 30 minutes at 37°C), followed by five washes with BSS.

To a series of tubes containing 0.1 ml of serum in twofold serial dilution, 10^5 labelled cells suspended in 0.1 ml were added, mixed thoroughly and left for 5 minutes at room temperature. To each tube 0.1 ml of fresh fourfold diluted guinea pig serum was added as a source of complement and the tubes were incubated for 30 minutes at 37°C with continuous mechanical agitation. To every tube 2.0 ml of saline was added and the tubes were centrifuged at 2000 r.p.m. for 5 minutes. The radioactivity released into the medium was measured for 1 ml of the supernatant of each tube.

The number of lysed cells was calculated using the expression

$$P = \frac{(S_e - S_o)}{(S_m - S_o)} \times 100,$$

where P = percentage of lysed cells and S_e, S_o and S_m represent counts in the supernatants for the experiment, before lysis and on complete lysis, respectively (Sanderson, 1964).

The cytotoxic inhibition test consisted of incubating for 15 minutes at room temperature 0.1 ml portions of the cytotoxic serum, in a dilution capable of inducing 50% lysis, with 0.1 ml of the RNA preparation to be tested. Portions of 0.1 ml of the labelled cell suspension were then added and incubated for 5 minutes. The mixture was diluted with 2 ml of BSS, centrifuged and resuspended in diluted fresh guinea pig serum. The rest of the procedure was the same as for the cytotoxic test.

D. Preparation of RNA

The RNA was extracted with hot phenol from lymph nodes of mice; the details of the procedure of extraction and purification are reported elsewhere (Sabbadini and Sehon, 1967b). The RNA preparations from immune and non-grafted animals will be referred to as I-RNA and NI-RNA, respectively.

E. Incubation of Spleen Cell Suspensions with RNA

The spleen cell suspensions were generally prepared in BSS and washed three times with the same medium by resuspension and centrifugation at 150 g for 5 minutes. They were then resuspended in PBS-sucrose and mixed with the RNA solution in the same medium. The final concentration of the cells was 1–2 × 10^7/ml (only in some experiments, when larger numbers of cells were needed was this concentration exceeded) and the RNA concentration was 200–600 μg/ml. The cells were incubated with RNA for 15 minutes at 37°C and the suspension was diluted with cold PBS, centrifuged and resuspended in BSS and injected as soon as possible.

F. Experiments with RNA in Skin-Grafted Adult Mice

Pools of 5–12 × 10^8 spleen cells were incubated with varying amounts of RNA from isologous mice and 10^8 cells were injected intravenously into each isologous recipient; when an animal was to receive a higher dose of the cells, a series of injections of 10^8 cells each was given on consecutive days. The animals received one skin graft on each side of the chest 24 hours after the first injection, one from the strain used for immunizing the donors of the RNA (specific graft) and the other from a different strain (control graft). Six days after grafting, the skin grafts were removed and examined histologically. The grafts were classified as rejected if more than 95% of the epithelium had sloughed off and as non-rejected in every other instance.

G. Assay for GVHR

To induce GVHR, parental strain spleen cells were injected intraperitoneally into newborn F_1 hybrids. The intensity of the resulting GVHR was evaluated by the spleen assay of Simonsen (1962) using the statistical evaluation according to Finney (1952). The two strain combinations, (AxCBA)F_1 and (AxC57Bl)F_1, were chosen.

Spleen cells obtained from normal A mice were incubated with the I-RNA preparation (test cell suspension) and another portion of A cells was incubated under the same conditions without RNA (standard cell suspension).* The activity of the two cell suspensions was tested in the same hybrid litter, using three different logarithmically spaced doses of the test cell preparation and three equivalent doses of the standard cell preparation; in addition, at least one animal in each litter receiving isologous hybrid cells was used as a control. All the cell preparations to be tested were made up to the same total number of cells by adding the appropriate amount of hybrid cells, isogenic with the recipients; in this way each animal received an identical number of cells equal to the highest number of A cells (further details can be found elsewhere, Sabbadini and Sehon, 1967b).

For interpretation of the experiments with RNA, the results were expressed as the ratio of activities due to cells incubated with RNA and with medium alone; this ratio according to Simonsen (1962) was designated as the factor of immunization (F.I.).

* The use of NI-RNA was omitted from the standard cell suspensions since, as will be shown later, it had no effect on the activity of spleen cells.

II. RESULTS

A. RNA Preparations

The RNA preparations from lymph nodes or spleens of immune animals did not present any physical or chemical characteristics significantly different from those of RNA preparations from non-immune animals. The protein and the DNA content were in the range of 10–20 μg and 2.4–7.6 μg per mg of RNA, respectively. In sucrose density gradients the RNA preparations were resolved into 3 components with sedimentation coefficients of 28S, 18S and 4–5S, respectively, and the faster sedimenting component normally being present at a higher concentration than the 4–5S component. When this pattern was reversed, i.e., when the concentration of the 4–5S component was higher, the RNA was considered degraded. Such partially degraded preparations were usually discarded; occasionally, however, they were used in order to establish if the integrity of the RNA molecules was critical for the biological reaction.

B. Skin Grafts

Skin grafts applied to untreated recipients occasionally showed initial signs of rejection on the 6th day after grafting, but no complete rejec-

Table 1

Rejection of Skin Grafts After Treatment of Recipient with Spleen Cells Incubated with I-RNA [a]

Donor Strain of Specific Graft	Donor Strain of Control Graft	Recipient Strain	Specific Grafts [b]	Control Grafts [b]
CBA	C57Bl	A	9/52	1/52
C57Bl	CBA	A	5/10	0/10
A	C3H	CBA	6/24	0/24

[a] Mice of the recipient strain received 1–3 I.V. injections of 1×10^8 isologous spleen cells, incubated for 15 minutes at 37°C with I-RNA (400–600 μg/ml in PBS-sucrose), extracted from lymph nodes of mice of the same strain immunized with tissues from animals of the specific donor strain. The injections were given on consecutive days. One skin graft from the specific donor strain and one from the control donor strain were applied 24 hours after the first injection. The skin grafts were removed on the 6th day after grafting for histological examination.

[b] The numerator represents the number of skin grafts completely rejected on the 6th day; the denominator represents the number of animals in the group.

Table 2

EFFECT OF VARIOUS CONTROL TREATMENTS ON REJECTION OF SKIN GRAFTS [a]

Treatment	CBA Grafts [b]	C57Bl Grafts [b]
Isologous cells incubated with PBS-sucrose [c]	0/18	0/18
Isologous cells incubated with "non-immune" RNA [d]	0/22	0/22
Isologous cells incubated with "immune" RNA which had been treated with RNAase [e]	0/16	0/16

[a] Mice of strain A received on consecutive days 3 I.V. injections of preparations c, d and e. A skin graft from CBA mice and a skin graft from C57Bl mice, were applied 24 hours after the first injection. The grafts were removed on the 6th day after grafting for histological examination.

[b] See footnote b, Table 1.

[c] Spleen cells from normal A mice were incubated at 37°C for 15 minutes in PBS-sucrose and then injected I.V. (1×10^8 cell per injection to each recipient).

[d] RNA was extracted from lymph nodes of untreated A mice, dissolved in PBS-sucrose (500 μg/ml) and used for incubation of spleen cells of A mice for 15 minutes at 37°C; 1×10^8 cells per injection were administered to each recipient.

[e] RNA from lymph nodes of A mice, immunized with CBA tissues, was dissolved in PBS (without Ca^{++} and Mg^{++} ions) to a final concentration of 300 μg/ml and RNAase was added to a final concentration of 4 μg/ml. After incubation for 1 hour at 37°C, the RNA was re-extracted with phenol. The RNA was then used for incubation of spleen cells as described in footnote d and 1×10^8 of these cells were injected I.V. into each recipient.

tion was observed in more than 200 experiments with different strain combinations.

As shown in Table 1, the treatment of the grafted animals, with spleen cells which had been incubated with I-RNA, induced an accelerated rejection of 18–50% of the "specific" grafts in all the three strain combinations tested. With the exception of one out of 86, the "control" grafts were not rejected.

From the results shown in Table 2, it is obvious that all the controls were negative. Thus, no accelerated rejection was observed after transfer of isologous spleen cells incubated with NI-RNA or with PBS-sucrose in the absence of RNA, or with I-RNA which had been pre-incubated with RNAase.

To examine the possibility that transplantation immunity might be

Table 3

EFFECT OF INJECTIONS OF I-RNA ON SKIN GRAFTS [a]

Route of Injection	CBA Grafts [b]	C57Bl Grafts [b]
Intravenous [c]	0/12	0/12
Intraperitoneal [d]	0/6	0/6
Subcutaneous [e]	0/6	0/6
Foot pads [f]	0/6	0/6

[a] I-RNA obtained from lymph nodes of mice immunized with CBA tissues was dissolved in PBS and injected into mice A at the doses listed in c–f.

[b] See footnote b, Table 1.

[c] One intravenous injection of 500 μg of I-RNA was administered per mouse A; the recipients were immediately grafted and the skin grafts were removed 6 days later for histological examination.

[d] Three intraperitoneal injections of 200 μg of I-RNA at weekly intervals; the recipients were grafted 1 week after the last injection.

[e] Three subcutaneous injections of 200 μg of I-RNA at weekly intervals; the recipients were grafted 1 week after the last injection.

[f] Three injections into the hind foot pads of 200, 400 and 600 μg were given 1 week apart; the recipients were grafted 1 week after the last injection.

induced by direct injection of I-RNA, without preincubation with spleen cells, different routes of administration and different schedules were used. Since these experiments were performed with the main purpose of determining if any antigen-like factor could induce active immunization of the recipients, the injections were repeated at weekly intervals, as indicated in Table 3, and the animals were grafted 7 days after the last injection. From the results listed in this table, it would appear that no immunity could be conferred by direct administration of I-RNA, i.e., without pre-incubation with spleen cells.

C. SPLEEN ASSAY

As shown in Table 4, incubation of spleen cells with NI-RNA did not modify the capacity of these cells to induce GVHR in newborn F_1 hybrids, as compared to cells incubated in the absence of RNA. In contrast, as is evident from data of Tables 5 and 6, incubation of spleen cells with I-RNA resulted in a stronger GVHR in a significant number of ex-

Table 4

SPLEEN ASSAY IN F_1 HYBRIDS, TREATED WITH A SPLEEN CELLS, INCUBATED WITH NI-RNA [a]

Source of the RNA [b]	Hybrid Combination	No. of Litters Tested	Concentration of the NI-RNA (μg/ml)	F.I.[c]
Lymph nodes	AxCBA	6	500	0.97
Lymph nodes	AxCBA	5	700	0.96
Lymph nodes	AxC57B1	6	600	1.04
Spleen	AxCBA	5	350	1.07
Spleen	AxC57B1	6	500	1.02

[a] Two suspensions, from the same pool of normal A spleen cells, were incubated for 15 minutes at 37°C with NI-RNA in PBS-sucrose or with PBS-sucrose without RNA. In each litter a group of 3 animals was treated I.P. with cells incubated with NI-RNA (each animal received a different dose), and a second group of 3 animals was treated with the cells incubated without RNA (these animals received doses corresponding to those used for the three animals of the first group). One or two animals in each litter received isogenic hybrid cells for control. The animals were sacrificed 8 days later and the spleens were weighed.

[b] NI-RNA was extracted from lymph nodes and spleens of normal A mice.

[c] The factor of immunization (F.I.) represents the ratio of the activity of the cells incubated with RNA to that of the cells incubated in PBS-sucrose only.

periments. This effect was not obtained with all RNA preparations, and the reasons for the lack of activity in some cases are not clear. However, it is to be stressed that when the same active I-RNA preparations were tested in different experiments, the results were quite reproducible.

The results reported in Table 7 show that the increase of GVHR induced by I-RNA was higher in the case of specific strain combinations than in F_1 hybrids of control strain combinations, the effect in the latter strain combinations being non-significant. The results of the experiments listed in Table 8 demonstrate that while RNAase abolished the activity of the I-RNA preparations, treatment of these preparations with DNAase or pronase did not reduce significantly their activity.

D. CYTOTOXIC INHIBITION TEST

The cytotoxic inhibition test was employed in an attempt to determine directly the possible presence of transplantation antigens in the active I-RNA preparations.

Table 5

SPLEEN ASSAY IN (AxCBA)F_1 HYBRIDS, TREATED WITH A SPLEEN CELLS, INCUBATED WITH I-RNA FROM LYMPH NODES [a]

I-RNA Preparation [b]	No. of Litters Tested	Concentration of the I-RNA (μg/ml)	F.I. [c]
LN-9	8	600	1.77 †
	6	400	1.67 *
LN-10	5	350	1.12
	5	700	1.08
LN-12	6	500	1.47 *
LN-13	5	500	1.11
	6	800	0.95
LN-14	6	500	1.62 *

[a] The experimental conditions were identical to those described in footnote a of Table 4 except that I-RNA was used instead of NI-RNA.

[b] Each I-RNA preparation was extracted from lymph nodes of A mice immunized with CBA tissues.

[c] See footnote c, Table 4.

* $p < 0.05$

† $p < 0.01$

Table 6

SPLEEN ASSAY IN (AxC57BC)F_1 HYBRIDS, TREATED WITH A SPLEEN CELLS, INCUBATED WITH I-RNA FROM LYMPH NODES [a]

I-RNA Preparation [b]	No. of Litters Tested	Concentration of the I-RNA (μg/ml)	F.I. [c]
LN-11	6	600	1.68 *
LN-15	6	500	0.98
LN-16	6	500	1.51 *
LN-18	7	600	1.06

[a] See footnote a, Table 5.

[b] I-RNA was extracted from lymph nodes of A mice immunized with tissues from C57Bl mice.

[c] See footnote c, Table 4.

* $p < 0.05$

Table 7

SPECIFICITY OF I-RNA IN THE GVHR ASSAY [a]

RNA Preparation [b]	Tissues Used for Immunization	Strain Combination Used for Test	No. of Litters	F.I. [c]
LN-14	CBA	AxCBA	6	1.62 *
		AxC57Bl	6	1.31
LN-16	C57Bl	AxC57Bl	6	1.51 *
		AxCBA	5	1.18

[a] The experiment described in footnote a, Table 4, was performed by injecting spleen cells of the same pool, after incubation with I-RNA, into newborn (AxCBA) F_1 and (AxC57Bl)F_1 hybrids.

[b] Preparation LN-14 was obtained from lymph nodes of A mice, immunized with CBA tissues. Preparation LN-16 was obtained from lymph nodes of A mice, immunized with C57Bl tissues.

[c] See footnote c, Table 4.

* $p < 0.05$

Table 8

EFFECT OF TREATMENT WITH DIFFERENT ENZYMES ON THE ACTIVITY OF I-RNA IN GVHR [a]

I-RNA Preparation [b]	Enzyme Used	Strain Combination	No. of Litters Tested	F.I. Prior to Enzymatic Treatment	F.I. After Enzymatic Treatment
LN-9	RNAase [c]	AxCBA	5	1.77 †	1.09
LN-11	RNAase [d]	AxC57Bl	6	1.68 *	1.06
LN-14	RNAase [d]	AxCBA	6	1.62 *	0.99
LN-14	DNAase [d]	AxCBA	5	1.62 *	1.56 *
LN-14	pronase [d]	AxCBA	6	1.62 *	1.67 *

[a] The experiment described in footnote a, Table 4, was performed using I-RNA preparations which had been incubated with the enzyme indicated in column 2.

[b] I-RNA preparations LN-9 and LN-14 were obtained from lymph nodes of A mice immunized with CBA tissues, preparation LN-11 was obtained from lymph nodes of A mice immunized with C57Bl tissues.

[c] After incubation the I-RNA + RNAase mixture was used for incubation of the spleen cells without purification.

[d] Incubation and purification were performed prior to incubation with the spleen cells.

* $p < 0.05$

† $p < 0.01$

Three I-RNA preparations, previously found capable of enhancing GVHR in the specific F_1 hybrid combinations, were incubated in concentrations of 1.5–3 mg/ml with the appropriate antisera (cytotoxic to cells of the strains used for immunization of the donors of the RNA) and with a control antiserum possessing cytotoxic activity for cells of an unrelated strain. The cytotoxic activity of the sera was then tested with the appropriate lymph node cells, as described. No detectable inhibition of the cytotoxic activity of these sera was observed.

III. DISCUSSION

The experimental data reported in this paper demonstrated that a state of transplantation immunity could be induced in spleen cells by incubation with I-RNA. Moreover, spleen cells so treated conferred adoptive immunity on adult isologous animals and induced enhanced GVHR in newborn F_1 hybrids.

The main feature which emerges from this study is the specificity of the resulting immunological response with respect to transplantation antigens, as demonstrated on the one hand by the accelerated rejection of the "specific" grafts and by the stronger GVHR in the F_1 hybrids of the "specific strain combination," and on the other by the lack of any such effect of I-RNA on "non-specific grafts" and on F_1 hybrids of "non-specific strain combinations." Therefore, both the accelerated rejection of skin grafts and the enhanced GVHR were considered to be expressions of the same basic phenomenon of transfer of transplantation immunity with cells activated by I-RNA.

The exquisite specificity of the reactions might be attributed to the presence in the active I-RNA preparations of some factor(s) normally involved in the immune reaction to allografts. In analogous studies (Askonas and Rhodes, 1965; Fishman *et al.,* 1965; Friedman *et al.,* 1965), RNA preparations extracted from peritoneal cells, which had phagocytosed different antigens were shown to be capable of inducing antibody formation on incubation with lymphoid cells, the presence of antigenic determinants of high immunogenicity in these RNA preparations was demonstrated.

Therefore, it seemed reasonable to postulate the presence of transplantation antigens in the I-RNA preparations and that these determinants were responsible for the transfer of transplantation immunity. However, the experimental data presented do not support this hypothesis. Thus, pronase, which is a potent proteolytic enzyme, did not affect the activity of the I-RNA preparations, although the mouse transplantation antigens have so far been shown to possess properties of proteins (Davies, 1967). Consequently, this evidence alone would suggest that the activity of I-RNA was not due to transplantation antigens or that these antigens might have been complexed with RNA in a special manner making them resistant to attack by proteolytic enzymes. Moreover, the search for transplantation antigens by the cytotoxic-inhibition test did not demonstrate the presence of such antigens in the I-RNA preparations tested; however, it is to be pointed out that the sensitivity of this test may have been too low for the detection of very small amounts of transplantation antigens.

The direct injection of I-RNA preparations without previous incubation with spleen cells failed to induce immunity in the recipients. This result could be due to degradation of the RNA by nucleases present in the body fluids of the recipients. This result by itself cannot be considered to strengthen any argument for or against the possible presence of transplantation antigens in the I-RNA preparations, since the RNA portion of a hypothetical RNA-antigen complex might be required as a carrier of the antigenic determinants to the immunocompetent cells capable of being activated into a state of acquired transplantation immunity.

One may, therefore, conclude that the exact physicochemical definition of the active factor in the I-RNA, as well as the elucidation of the mechanism responsible for activation of the spleen cells to a state of transplantation immunity, must await further experimentation.

References

Askonas, B. A., and J. M. Rhodes (1965). Immunogenicity of antigen-containing ribonucleic acid preparations from macrophages. Nature, London, *205*:470.

Davies, D. A. L. (1967). Properties of soluble histocompatibility antigens studied mainly in mice. Proceedings of the 1st International Congress of the Transplantation Society.

Dulbecco, R., and M. Vogt (1954). Plaque formation and isolation of pure lines with poliomyelitis viruses. J. Exp. Path., *45*:398.

Earle, W. R. (1943). Production of malignancy *in vitro*. IV. The mouse fibroblast cultures and changes as seen in living cells. J. Nat. Cancer Inst., *4*:165.

Finney, D. J. (1952). Statistical methods in biological assay. Ch. Griffin and Co., London.

Fishman, M., J. J. van Rood, and F. L. Adler (1964). The initiation of antibody formation by ribonucleic acid from specifically stimulated macrophages. Proc. Symp. on Molecular and Cellular Basis of Antibody Formation, J. Sterzl, ed., Prague, p. 491.

Friedman, H. P., A. B. Stavitsky, and J. M. Solomon (1965). Induction *in vitro* of antibodies to phage T2: Antigens in the RNA extract employed. Science, *149*:1106.

Morgan, F. J., H. J. Morton, and R. C. Parker (1950). Nutrition of animal cells in tissue culture. I. Initial studies on a synthetic medium. Proc. Soc. Exp. Biol. Med., *73*:1.

Sabbadini, E., and A. H. Sehon (1967a). Acceleration of allograft rejection induced by RNA from sensitized donors. Int. Arch. Allergy, *32*:55.

——— (1967b). Transfer of transplantation immunity with RNA from lymphoid organs of immunized animals. Proceedings of the 1st International Congress of the Transplantation Society, Paris, p. 56.

Sanderson, A. R. (1964). Cytotoxic reactions of mouse iso-antisera: preliminary considerations. Brit. J. Exp. Path., *45*:398.

Simonsen, M. (1962). Graft-versus-host reactions; their natural history and applicability as tool of research, in "Progress in Allergy." P. Kallos and B. H. Waksman, eds., vol. 6, p. 349.

UNIQUE SPECIES OF RNA IN PERITONEAL CELLS EXPOSED TO DIFFERENT ANTIGENS *

EDWARD P. COHEN † AND KAREL RASKA, JR.‡

Rutgers Medical School, New Brunswick, N. J.

A suitable inducer of cell differentiation in the mammalian host stimulates a complex series of events which often includes the division of cells and the synthesis of specialized protein. According to current concepts specialized cell function requires the activation of information stored in a latent form in the genome of the cell for the synthesis of new RNAs and new protein. Thus, if the differentiated response is of sufficient magnitude and if the assay is of adequate sensitivity, one might detect new RNA synthesis in differentiating cells. Cells which differentiatc to form proteins which are distinguishable from one another may form detectably different species of RNA.

Antibody formation has been considered a suitable model for the study of induced cell differentiation (Watson, 1965). The injection of antigen triggers the onset of lymphoid cell division and the synthesis and release of highly specialized protein.

I. RESPONSE OF LYMPHOID CELLS TO ANTIGEN

The development of methods for quantitating the cellular response to specific antigens revealed that a 500-fold increase in the number of antibody forming cells in the spleen of the mouse followed a single intraperitoneal injection of sheep RBC. These cells were thought to arise through division of a relatively few precursors which responded to the antigen (Koros *et al.*, 1966). As a reflection of these mitotic events, the

* Supported by U. S. Public Health Service grant AI 06852-03.

† Present address: Larabida—University of Chicago Institute, Chicago, Ill.

‡ Present address: Institute of Organic Chemistry and Biochemistry, Czechoslovak Academy of Sciences, Prague, Czechoslovakia.

specific activity of thymidine kinase in the spleen of infected mice rose to three and one-half times control values within 48 hours (Raska and Cohen, 1967).

Accelerated biosynthesis of RNA in the spleens of mice also began within two hours after the injection of antigen. The specific activity of RNA in the mouse spleen exposed to pulses of tritium-labeled uridine after the injection of sheep RBC rose to two and one-half times the specific activity of RNA from the spleens of noninjected control animals (Table 1). Also, the measurement of the specific activity of uridine kinase in the spleens of mice that were injected with sheep RBC revealed significant increases within hours (Fig. 1). These changes could have been predicted because the average RNA content per cell was found to be increased in the spleens of immunized animals (Juhasz and Rose, 1963).

The "resting" spleen synthesizes RNA at a basal rate. Thus, a large proportion of the newly labeled material is formed in response to the injection of antigen and likely is either directly or indirectly involved in the events which occur as cells in the spleen prepare for the synthesis of antibody.

The cells found in the normal mouse peritoneal cavity appear to respond to antigen in a manner similar in magnitude to the response by the spleen. Bussard (1967) reported that as many as 1% of mouse peritoneal lymphocytes responded *in vitro* to sheep RBC by forming antibody. It may be that not every cell which might have responded *in vivo* did so under artificial conditions. Thus, the true proportion of responding cells may have been even larger in this population of cells. Bussard con-

Table 1

RNA SYNTHESIS IN THE MOUSE SPLEEN AFTER ANTIGEN

Hours After Antigen	Specific Activity Immune/Control	Antibody-forming Cells Per 10^6 Spleen Cells
2	1.64	0.9
24	1.12	0.8
72	2.47	54
96	1.02	470
120	1.02	712

B_6AF_1 mice received 0.1 ml of 25% sheep RBC intravenously, control mice received isologous mouse cells intravenously. Sixty minutes before they were killed, each mouse received 50 μc of tritium-labeled uridine in 0.1 ml and the specific activity of the RNA extracted from the spleens was determined (Cohen, 1967a).

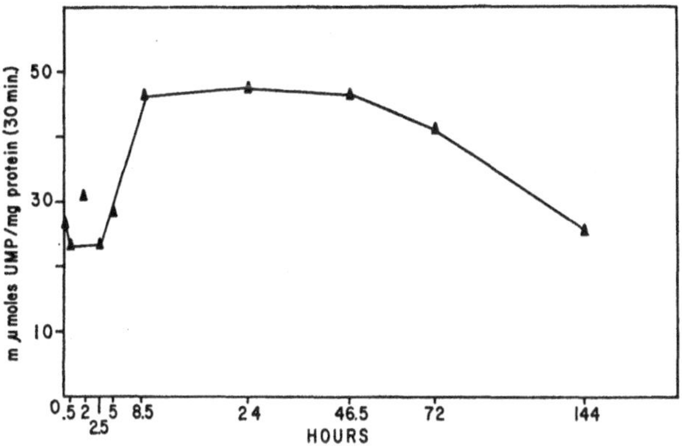

Fig. 1. Specific activity of uridine kinase in mouse spleen after injection of sheep RBC. B_6AF, mice (Jackson Labs, Bar Harbor, Maine) were injected i.p. with 10^9 sheep RBC. At varying times thereafter the specific activity of uridine kinase in their spleens was determined (Raska and Cohen, 1967a).

cluded that the proportion of cells in the peritoneal cavity with the latent potential to respond to sheep RBC by forming antibody was far greater than was expected previously. This proportion was more than could be accounted for by theories which postulate that the origin of the diversity of the immune response is the result of somatic mutation (Lederberg, 1959; Burnet, 1959). According to these theories, only a relatively few cells which had mutated to the potential to form antibody to sheep RBC would have responded.

Because of the magnitude of the RNA response to specific antigen, we applied techniques developed in other systems for the *in vitro* formation of molecular hybrids between RNA and DNA in an attempt to answer questions related to the mechanism of the immune response.

II. THE APPLICATION OF MOLECULAR HYBRIDIZATION TECHNIQUES TO THE IMMUNE RESPONSE

Recently the techniques of molecular hybridization previously applied to the study of microbial and viral gene expression were applied to study questions in mammalian systems. Hoyer *et al.* (1965) constructed tables of evolutionary relationships which were based on homology between RNAs and DNAs of various species. Church and McCarthy (1967a, 1967b) detected the appearance of new species of RNA in regenerating and embryonic mouse liver and concluded that these RNAs were the expression of latent gene potentials in those cells.

In molecular hybridization studies a saturation plateau results upon the addition of increasing amounts of labeled RNA to a fixed amount of DNA under conditions which allow the formation of RNA:DNA

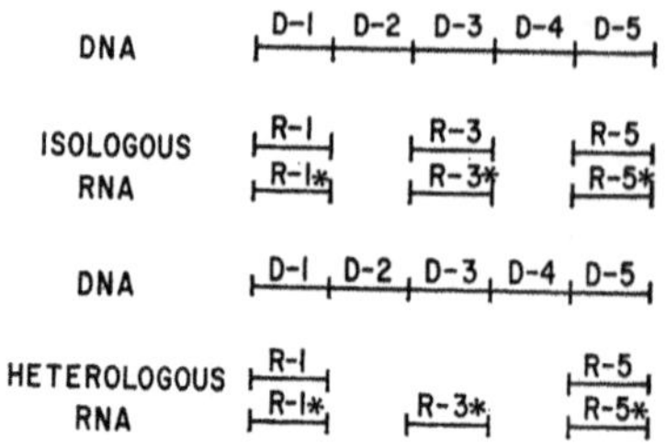

Fig. 2. Diagrammatic representation of molecular hybridization illustrating the principle of competition by either isologous or heterologous RNA.

hybrids. This plateau is reached after all the available sites for hybridization in DNA become occupied with complementary species of RNA. The number of counts which remain complexed with the DNA after removal of non-hybridized RNA is an index of the number of different molecular species of RNA which are present in the preparation. It is independent of the specific activity of the preparation of RNA used for hybridization studies. (As a representative member of each species of RNA hybridizes to its complementary site, other members of that species are washed away. The specific activity of each molecular species may differ. Therefore, the specific activity of the whole preparation may vary over a wide range without affecting the cpm which remain with the filter after incubation with saturating amounts of labeled RNA.)

The basis of an important control rests on the ability of unlabeled RNA to compete with labeled RNA for sites of hybridization. This capacity for competition between two RNAs is an indication of the number of molecular species of RNA in common in the two preparations. In a similar manner the relative capacity of two unlabeled RNAs to compete with one preparation of labeled RNA is an index of the number of molecular species in common between the two unlabeled preparations (Fig. 2).

In applying techniques of molecular hybridization to the study of the immune response, we labeled RNA of lymphoid cells during the period of new RNA synthesis which occurs after the cells are exposed to antigen. Initially we confirmed Hoyer's finding of the specificity of the hybridization reaction in the mouse system (Cohen, 1967b). Further, we found that new [1] species of RNA appeared in the spleens of mice that were injected with antigen (Cohen, 1967b) and in mouse peritoneal cells that were incubated *in vitro* with antigen (Fig. 3).[2]

The presence of different species of RNAs may be detected by recipro-

[1] We realize that some species of RNA may be present in amounts too small to be detected by molecular hybridizations and that "new" species may represent an increase in such species. The term "new" is used for convenience only.

[2] In general terms, a foreign material becomes an antigen only if it stimulates the formation of antibody. In these experiments we did not determine whether the cells exposed to "antigen" eventually would form antibody; hence the term "antigen" is used for convenience.

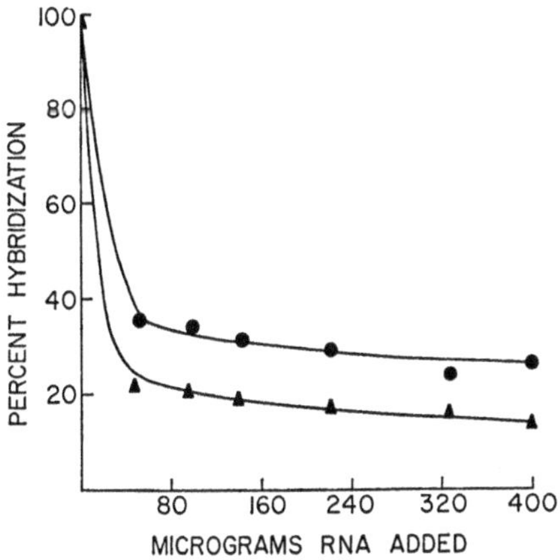

Fig. 3. Competition for sites of hybridization by unlabeled RNA from p.e. cells incubated with sheep RBC or by unlabeled RNA from control p.e. cells incubated without added sheep RBC. Peritoneal exudate cells (p.e. cells) were obtained from CD-1 mice (Charles River) two days after the i.p. injection of 3 ml thioglycollate broth. Labeled RNA was prepared from approximately 10^9 cells (Scherrer and Darnell, 1962) after incubation for 60 minutes at 37°C in 25 ml Eagle's medium containing 250 μc/ml tritium-labeled uridine and 0.3 ml of a 10% suspension of washed sheep RBC. Two pools of unlabeled RNA were prepared from approximately 10^9 p.e. cells after incubation for 60 minutes at 37°C in Eagle's medium with or without the addition of 0.3 ml sheep RBC. Forty μgm of tritium-labeled RNA (3300 cpm/μgm) from p.e. cells exposed to sheep RBC was incubated for 24 hours at 65°C in 5 ml of 2 × SSC (SSC is .15 *M* sodium chloride, 0.15 *M* sodium citrate) with membrane filters containing 20 μgm DNA (Marmur, 1961). After incubation, the filters were washed, treated with ribonuclease and washed again (Cohen, 1967b). Other filters were incubated with the same concentration of tritium-labeled RNA from p.e. cells exposed to sheep RBC and with varying amounts of unlabeled RNA from either control cells or cells exposed to sheep RBC. After incubation the filters were treated as described previously. Triangles: unlabeled RNA from p.e. cells incubated with sheep RBC. Closed circles: unlabeled RNA from p.e. cells incubated without sheep RBC.

cal competition experiments with preparations of unlabeled RNA. We began a series of reciprocal competition experiments between labeled and unlabeled RNAs from cells exposed to four particulate antigens (sheep RBC, chicken RBC, rabbit RBC, *E. coli*) and one soluble antigen of differing specificities (human gamma globulin). The results indicated that cells exposed to different antigens formed species of RNA which were distinguishable from one another by these methods. Two reciprocal competitions between RNAs from cells exposed to different antigens are presented (Figs. 4, 5).

An additional control was performed. Filters containing mouse DNA were saturated with unlabeled RNA from peritoneal cells incubated with either sheep RBC or chicken RBC. Increasing amounts of labeled RNA from cells incubated with sheep RBC were added to saturation.

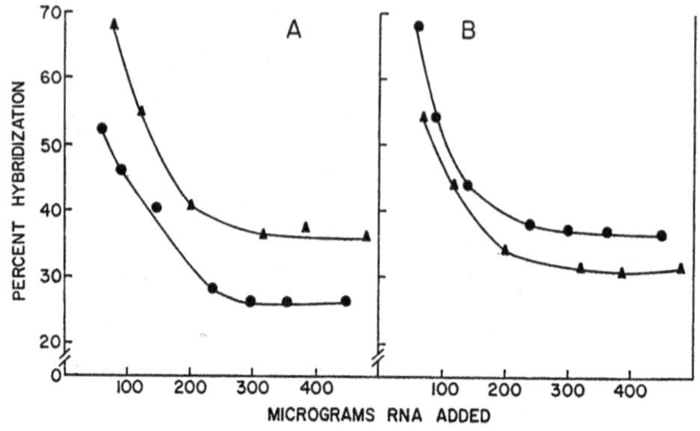

Fig. 4. Competition for sites of hybridization by unlabeled RNA from cells exposed to either sheep RBC or chicken RBC. **A.** Forty μgm of tritium-labeled RNA from p.e. cells exposed to sheep RBC were incubated with filters containing 20 μgm DNA and varying amounts of unlabeled RNA from p.e. cells exposed to either sheep RBC or chicken RBC. **B.** Other filters containing 20 μgm DNA were incubated with 40 μgm tritium-labeled RNA from p.e. cells exposed to chicken RBC and varying amounts of unlabeled RNA from either of the *same pools* of RNA used in **A.** Triangles: unlabeled RNA from p.e. cells incubated with chicken RBC. Closed circles: unlabeled RNA from p.e. cells incubated with sheep RBC.

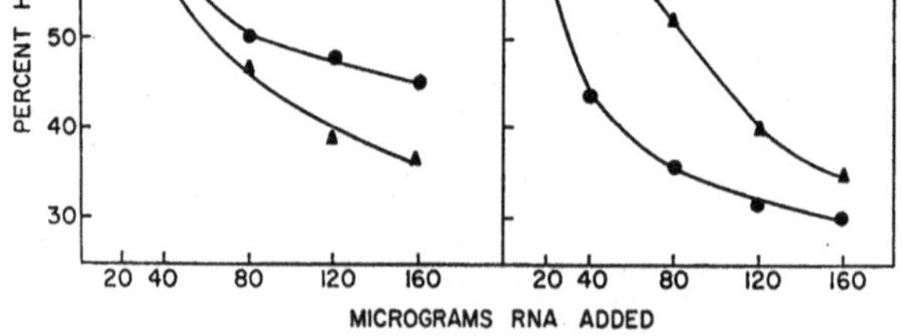

Fig. 5. Competition for sites of hybridization by unlabeled RNA from cells exposed to either sheep RBC or *E. coli.* **A.** Fifty μgm of tritium-labeled RNA from p.e. cells exposed to sheep RBC were incubated with filters containing 20 μgm DNA and varying amounts of unlabeled RNA from cells exposed to either sheep RBC or *E. coli.* **B.** Other filters were incubated with 50 μgm labeled RNA from p.e. cells exposed to *E. coli* and varying amounts of unlabeled RNA from either of the *same pools* of RNA used in **A.** Triangles: unlabeled RNA from p.e. cells incubated with sheep RBC. Closed circles: unlabeled RNA from p.e. cells incubated with *E. coli.*

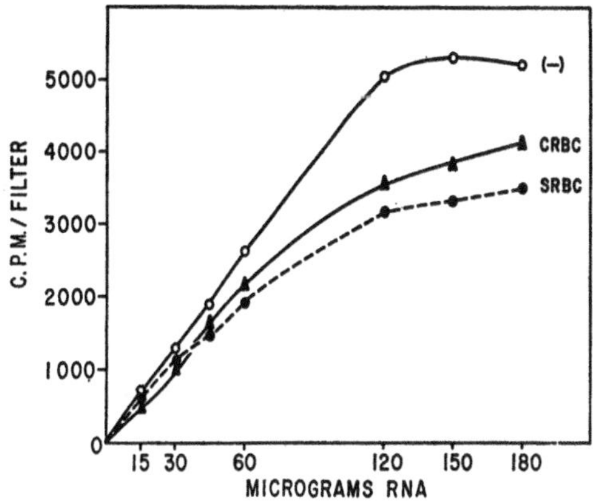

Fig. 6. Presaturation of the sites of hybridization of labeled RNA from cells exposed to sheep RBC by unlabeled RNA from cells exposed to either sheep RBC or chicken RBC. One hundred twenty μgm of unlabeled RNA from cells exposed to either sheep RBC or chicken RBC were incubated for 24 hours at 65°C in 2 × SSC with filters containing 20 μgm mouse DNA. Control filters were incubated without added RNA. After incubation, the filters were washed and varying amounts of labeled RNA from p.e. cells exposed to sheep RBC (12,600 cpm/μgm) were added. The filters were then incubated for an additional 24 hours and washed again before the cpm were determined.

Unlabeled RNA from cells exposed to sheep RBC competed more effectively for sites of hybridization than did unlabeled RNA from cells incubated with chicken RBC (Fig. 6). This result is consistent with our conclusion that cells exposed to different antigens formed species of RNA which differed from one another. That RNA from cells exposed to chicken RBC competed at all suggests that many RNAs in the two preparations were the same.

III. THE ROLE OF ANTIGEN-UNIQUE SPECIES OF RNA IN THE IMMUNE RESPONSE

Many specialized cell functions are activated as cells respond to antigen. Among such effects are the degradation of the antigen and the synthesis of antibody. Conceivably, the species of RNA we detected may be devoted entirely or in part to one of these functions.

The mouse peritoneal cell population consists primarily of lymphocytes and macrophages. Macrophages are cells with the capacity to ingest and degrade antigen. Lymphocytes synthesize antibody and are thought to differentiate further to plasma cells. We separated these cell classes and tested each for the presence of different species of RNA after incubation with different antigens. The cells were separated by allowing the macrophages to adhere to the bottom of glass bottles. The supernatant consisted of lymphocytes and small proportions of other cells as well as 1% to 2% macrophages which failed to adhere. The results indicated that new species of RNA appeared in the cells which adhered to glass after they were incubated with sheep RBC; however, we were unable to detect different species of RNA when the adhering cells were incubated with different antigens (Table 2). Lymphocytes, on the contrary, formed both new and different species of RNA when exposed to different antigens (Fig. 7). (Our failure to detect different species of RNA in macrophages incubated with antigens of different specificity does not necessarily prove that macrophages do not discriminate between antigens in the species of RNA they synthesize. Quite obviously, our ability to detect differences between preparations of RNA is limited by the resolution of the method.)

One function of lymphocytes is the synthesis of antibodies of varying specificity. These cells contain few lysosomes, and it is unlikely that the more generalized responses, such as DNA synthesis and the onset of cell division, would require the synthesis of different species of RNA for different antigens. Thus, the appearance of unique species of RNA in these cells suggests that this RNA plays an essential but as yet unknown role in antibody formation. These RNAs may act as specific messengers for the formation of immunoglobulins. This conclusion must be made with extreme caution. Certain cell functions may be involved which are

Table 2

HYBRIDIZATION OF RNA FROM ADHERING CELLS IN THE PRESENCE OF COMPETING RNA

Labeled RNA from Cells Exposed to	Percentage Hybridization with Competing RNA from Cells Exposed to					
	Sheep RBC			Chicken RBC		
Sheep RBC	(1) 35%	(2) 44%	(3) 32%	(1) 36%	(2) 42%	(3) 32%
Chicken RBC	(1) 26%	(2) 45%	(3) 37%	(1) 27%	(2) 46%	(3) 36%

Peritoneal exudate cells from CD-1 mice were incubated in Eagle's medium for 3 hours at 37°C in glass bottles. After incubation the nonadhering cells were removed and the adhering cells were incubated for an additional hour with either sheep RBC or chicken RBC. Labeled RNA from such cells was prepared in a similar manner except that 0.1 mc/ml tritium-labeled uridine was added to the incubation medium. The results of three reciprocal competition experiments are presented. Twenty μgm of labeled RNA and at least 180 μgm unlabeled RNA was used in each experiment. The filters contained 20 μgm DNA.

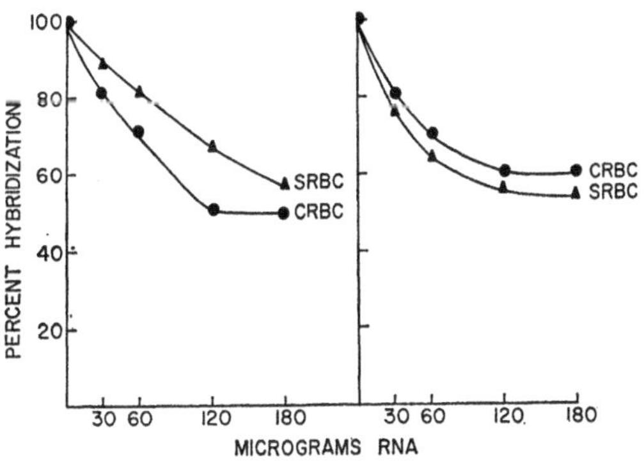

Fig. 7. Competition between unlabeled RNAs from peritoneal cells which did not adhere to glass. P.e. cells were incubated at 37°C in Eagle's medium three hours in glass bottles. At the end of the incubation the cells which did not adhere to the glass were removed and incubated for one hour with either sheep RBC or chicken RBC as described above. *Left:* Thirty μgm of labeled RNA from non-adhering cells exposed to chicken RBC were incubated with membrane filters containing 10 μgm DNA from the spleens of mice and with varying amounts of unlabeled RNA from non-adhering cells exposed either to sheep RBC or chicken RBC. *Right:* Thirty μgm labeled RNA from non-adhering cells exposed to sheep RBC were incubated with membrane filters containing 10 μgm DNA, and with varying amounts of unlabeled RNA from either of the same pools used in the left graph.

only indirectly involved in antibody formation and which presently are obscure. These hypothetical responses may require different RNAs for different stimuli. The precise function of a nuclear RNA found in duck erythrocytes (Attardi, 1966) and HeLa cells (Scherrer *et al.*, 1963; Mach and Vassalli, 1965) is unknown. It is of large molecular weight, has a rapid synthesis and breakdown and is thought to be a precursor of ribosomal RNA. Part of the RNAs we detected may be represented by such materials.

IV. RESPONSE OF PERITONEAL CELLS TO "NON-ANTIGENS"

Some forms of proteins are "non-antigens," that is, they fail to provoke immune responses when injected into hosts which are otherwise fully immunocompetent. The synthetic polymer, DNP_{15} poly-L-lysine is one such compound in mice (Pinchuck, 1967). If its primary structure includes enough alanine, as the polymer DNP_{15} $lysine_{70}$ $alanine_{30}$, it becomes an antigen.

We exposed peritoneal cells from mice *in vitro* to both DNP_{15} poly-L-lysine and DNP_{15} $lysine_{70}$ $alanine_{30}$. RNA was extracted from these cells and tested by means of molecular hybridizations. The antigen, DNA_{15} $lysine_{70}$ $alanine_{30}$ stimulated the appearance of RNAs distinguish-

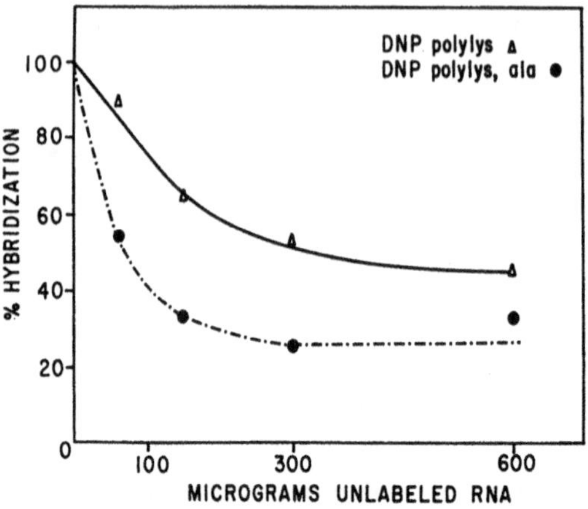

Fig. 8. Competition for sites of hybridization by unlabeled RNA from p.e. cells exposed to either DNP_{15} poly-L-lysine or DNP_{15} poly-lysine$_{70}$ alanine$_{30}$. Approximately 10^9 p.e. cells were incubated in 25 ml Eagle's medium at 37°C for 45 minutes with the addition of 50 μgm of either DNP_{15} poly-L-lysine or DNP_{15} poly-lysine$_{70}$ alanine$_{30}$. Both polymers were dissolved in 1% normal mouse serum. Labeled RNA was prepared from p.e. cells exposed to 50 μgm DNP_{15} poly-lysine$_{70}$ alanine$_{30}$ in 25 ml Eagle's medium containing 250 μc/ml tritium-labeled uridine. For hybridization 30 μgm tritium-labeled RNA was incubated with filters containing 20 μgm DNA and varying amounts of unlabeled RNA from cells exposed to either polymer.

able from the response by the cells to DNP_{15} poly-L-lysine (Fig. 8). The "non-antigen," DNP_{15} poly-L-lysine, however, failed to stimulate the formation of species of RNA which were detectably different than those formed by control cells (Fig. 9).

Thus, the *in vitro* addition of a "non-antigen" to peritoneal cell suspensions failed to lead to the formation of new species of RNA. This strengthens our suspicion that these RNAs may be essential for the immune response.

V. PREPARATION OF AN ENRICHED MESSENGER RNA FRACTION

As described above, the addition of a suitable antigen to a responding population of cells stimulates the rapid formation of a new fraction of RNA. The rate of synthesis of this population may exceed the rate of synthesis of other species of cellular RNA, transfer RNA, ribosomal RNA and non-antibody messenger RNAs. The fractions which form in response to antigen may represent a greater proportion of the newly formed cellular RNA than any of the others. Therefore, pulse-labeling of RNA during the phase of active synthesis should result in a labeled population which has a higher specific activity than other RNAs formed

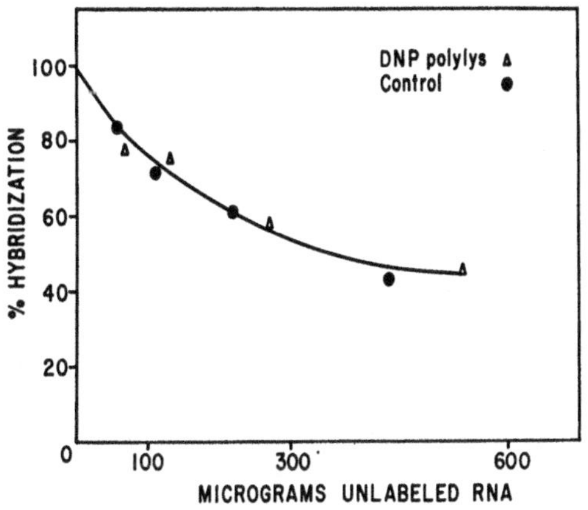

Fig. 9. Competition for sites of hybridization by unlabeled RNA from p.e. cells exposed to DNP_{15} poly-L-lysine or by unlabeled RNA from p.e. cells incubated without added polymer. Conditions were as described in the legend which accompanies Fig. 8 except that unlabeled RNA was prepared from cells incubated with DNP_{15} poly-L-lysine or without added polymer. Labeled RNA was prepared from p.e. cells incubated with DNP_{15} poly-L-lysine. Thirty μgm of labeled RNA was added to filters containing 20 μgm DNA and varying amounts of unlabeled RNA from either group.

at a slower rate during the same period of labeling. (The participation of nuclear RNA, mentioned above, is unknown.) If we asume that the proportion of genes coding for these newly formed messengers is relatively deficient when compared to the number of copies of the gene product (RNA), then it may be possible to isolate these RNAs by hybridization techniques. The principle of this isolation (enrichment) technique is based upon the preparation of columns containing DNA which is immobilized in nitrocellulose but still available for hybridization with complementary strands of RNA (Bautz, 1966). Species of RNA which are in excess should saturate their complementary sites and the remainder will then pass through the column. Other species of RNA, present in lesser amounts, should remain behind.

Columns containing DNA immobilized in nitrocellulose were prepared. RNA was prepared from peritoneal cells which were labeled with tritium-labeled uridine during a period of incubation with antigen. It was then applied to the column and allowed to pass through under conditions suitable for the formation of molecular hybrids. Preliminary results indicate that the first fractions to be collected were of higher specific activity than those which followed (Fig. 10). This indicated that a partial separation of the various species of RNA had occurred. The role of these highly labeled RNAs in the immune response remains to be determined.

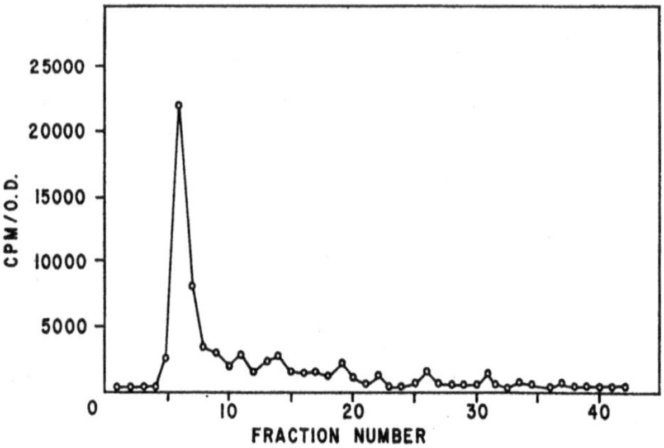

Fig. 10. Fractionation of tritium-labeled RNA from p.e. cells exposed to sheep RBC. A 22 ml column containing denatured mouse DNA was prepared according to techniques described by Bautz *et al.* (1966). Tritium-labeled RNA from p.e. cells exposed to sheep RBC was prepared as described previously and 193 μgm in 0.5 ml of the labeled RNA containing 550,000 cpm were added to the top of the column. The RNA was passed through under conditions found to be optimal for the formation of molecular hybrids. The optical density of each 3.0 ml fraction was determined before the RNA was precipitated with 0.5 ml of a 30% TCA solution. The precipitate was collected on nitrocellulose filters (0.45 μ-porosity millipore) and after drying was counted in a liquid scintillation counter.

VI. HYBRIDIZATION OF RNA WITH DNA FROM IMMUNIZED ANIMALS

The DNA used in the experiments described above was obtained from the spleens of animals that were not specifically immunized. This suggested that the sites for hybridization of the various species of RNA we detected were present in a large proportion of normal cells. If the different species of RNA that formed in response to different antigens are indeed messengers for immunoglobulins, then we are forced to conclude that at least some of the information for the synthesis of these RNAs preexisted in many cells before immunization contrary to what would be predicted by clonal selection theories. The technique for molecular hybridization we used is unable to detect the presence of the genes for specific antibody synthesis which may exist in a few stem cells only.

If part of the cellular potential to form antibodies of varying specificity arose through a process akin to somatic mutation (Lederberg, 1959; Burnet, 1959), or somatic recombination (Edelman and Gailly, 1967; Smithies, 1967), then specific immunization which stimulates division of antibody-forming cells might be expected to result in an increase in the proportion of sites for hybridization for specific messengers. As a test of this postulate, we immunized mice with either sheep or chicken RBCs and four days later, when the proportion of antibody forming cells in the spleen was the highest, extracted DNA from the spleens of these animals for hybridization studies. RNA was prepared from peritoneal exudate cells exposed to either of the same antigens. In these experiments, we hybridized RNA from cells exposed to either antigen and with its specific or nonspecific DNA. (RNA from normals was hybridized with DNA from immunized animals as a baseline control on the preparation of the DNAs.) We were unable to detect differences in DNAs from either specifically or nonspecifically immunized mice. RNA from cells exposed to sheep RBC hybridized as well with DNA from animals immunized with chicken RBC (Table 3). RNA from normals formed more hybrids with both DNAs than with DNA from nonimmunized mice.

VII. DISCUSSION

The results of this series of experiments suggest that different species of RNA appear in cells exposed to different antigens and that the information for the synthesis of at least part of these species of RNA existed in the normal genome.

The role of these different species of RNA in the immune response has not been established. There are several reasons to suspect that they may play a direct role in antibody synthesis: (1) they were found in the cell population which failed to adhere to glass, consisting of more than 95% lymphocytes (lymphocytes are not primarily phagocytic cells but include cells which begin antibody synthesis upon appropriate stimulation); (2) we were unable to detect the presence of different species of RNA when such cells were exposed to non-antigens; (3) the proportion of genes coding for these newly formed RNAs appeared to be relatively deficient when compared to the number of copies of the gene product. The preparation of columns containing immobilized DNA with sites available for hybridization allowed the isolation of a high specific activity fraction. RNA synthesis after antigen occurs at a rapid rate suggesting that some species of RNA labeled during this period would be of higher specific activity than others synthesized at a basal rate during the same period.

At least part of the information for the synthesis of these RNAs appeared to be present in cellular DNA from normal mice. If such information was entirely clonally distributed in precursor cells, as by a process of somatic mutation, then we would have been unable to detect the presence of different species of RNA in cells exposed to different antigens. This conclusion is further supported by experiments with DNA from immunized mice. Immunization before extraction of DNA failed to increase the proportion of sites available for the formation of hybrids with RNA from cells exposed to the same antigen.

Table 3

Hybridization of RNA with DNA from the Spleens of Immunized Mice

Experiment	RNA (μgm)	cpm Remaining with DNA from Mice Immunized with	
		Rabbit RBC	Sheep RBC
1	R-RBC 287	4200	3310
	S-RBC 287	4850	3130
		Chicken RBC	Sheep RBC
2	C-RBC 180	2500	2800
	S-RBC 180	2700	3500
3	C-RBC 244	1200	1900
	S-RBC 244	2400	3700

DNA was extracted from the spleens of mice four days after they were injected i.p. with either sheep RBC, chicken RBC, or rabbit RBC. Filters containing 10 μgm DNA were prepared as described previously. Labeled RNA was prepared from peritoneal cells of mice exposed for one hour *in vitro* to either sheep RBC, chicken RBC, rabbit RBC as described in the legends which accompany the figures. The results of three experiments are presented.

C-RBC: RNA from peritoneal cells exposed to chicken RBC.
S-RBC: RNA from peritoneal cells exposed to sheep RBC.
R-RBC: RNA from peritoneal cells exposed to rabbit RBC.

Therefore, if these different species of RNA represent messengers for the synthesis of immunoglobulins, then it may be that at least a part of the potential to form antibodies of varying specificity is inherited and not acquired by mutational processes which are thought to occur throughout embryonic development and the life of the animal.

Plasma cell myelomas appear to be normal globulins produced in excess. Amino acid sequencing of these proteins and data on the genetic code in *E. coli* have suggested that the potential to form antibodies of varying specificity arose through processes akin to mutation and recombination (Smithies, 1967). The question is whether these events occurred throughout evolution of the species or throughout the lifetime of the individual.

The problem of explaining the broad diversity of the immune response may be responsible for theories which attempt to explain this potential by processes of somatic mutation.

Our purpose is not to explain why these theories may be wrong. However, we may be reminded of several observations which may make it easier to accept an alternative concept. Immunoglobulins have a four chain structure. This evolutionary development is not accidental. Such polymorphism allows maximum flexibility in protein specificity with minimal requirements for genetic information. That is, as few as one hundred separate genes for L chains and one hundred separate genes for H chains would provide the animal with the potential to form antibodies with 10,000 different specificities (Smithies, 1963). Theoretically, the number of antigens with which 10,000 antibodies can combine is quite large if one assumes that families of immunoglobulins act in concert toward common antigenic determinants (Talmage and Cohen, 1965). Polymorphism in the construction of functional proteins is well known in the formation of hemoglobin (Ingram, 1961) and several enzymes (Jukes, 1966).

Bussard (1967) concluded that the proportion of peritoneal cells which respond *in vitro* to sheep RBC by forming hemolysins is too large to be explained by theories of antibody formation which rely on somatic mutational events. Dreyer and Bennett (1965) and Putnam and his associates (1967) postulate an evolutionary origin of the immune potential. Our observations are consistent with the concept that part of the potential to form antibodies of varying specificity is inherited.

References

Attardi, E., H. Parnas, M-I. H. Hwang, B. Attardi (1966). Giant-size rapidly labeled nuclear ribonucleic acid and cytoplasmic messenger ribonucleic acid in immature duck erythrocytes. J. Mol. Biol., *20*:145.

Bautz, E. K. F., and E. Reilly (1966). Gene-specific messenger RNA: Isolation by the deletion method. Science, *151*:328.

Burnet, M. (1959). "The clonal selection theory of acquired immunity." Nashville: Vanderbilt University Press.

Bussard, A. E., and M. Lurie (1967). Primary antibody response *in vitro* in peritoneal cells. J. Exptl. Med., *125*:873.

Church, R. B., and B. J. McCarthy (1967a). Ribonucleic acid synthesis in regenerating and embryonic liver. I. The synthesis of new species of RNA during regeneration of mouse liver after partial hepatectomy. J. Mol. Biol., *23*:459.

——— (1967b). Ribonucleic acid synthesis in regenerating and embryonic liver. II. The synthesis of RNA during embryonic liver development and its relationship to regenerating liver. J. Mol. Biol., *23*:477.

Cohen, E. P. (1967a). Conversion of nonimmune cells into antibody-forming cells by RNA. Nature, *213*:462.

——— (1967b). The appearance of new species of RNA in the mouse spleen after antigen. Proc. Nat. Acad. Sci., *57*:673.

Dreyer, W. J., and J. C. Bennett (1965). The molecular basis of antibody formation: A paradox. Proc. Nat. Acad. Sci., *54*:864.

Edelman, E. M., and J. A. Gailly (1967). Somatic recombination of duplicated genes: a hypothesis on the origin of antibody diversity. Proc. Nat. Acad. Sci., *57*:353.

Hood, L. E., W. R. Gray, W. J. Dreyer (1966). On the mechanism of antibody synthesis: A species comparison of 2-chains. Proc. Nat. Acad. Sci. (U. S.), *55*:826.

Hoyer, B. H., E. T. Bolton, B. J. McCarthy, and R. B. Roberts (1965). The evolution of polynucleotides *in* Evolving Genes and Proteins, eds. V. Bryson and H. J. Vogel, New York: Academic Press, p. 581.

Ingram, V. (1961). Gene evolution and the hemoglobins. Nature, *189*:704.

Juhasz, A., and B. Rose (1963). Studies on antibody formation. I. Relationship between nucleic acids and antibody synthesis in the primary response in the rat. Can. J. Biochem., *41*:2983.

Jukes, T. H. (1966). Molecules and evolution. Columbia University Press, New York and London, pp. 153–190.

Koros, A. M., H. Fuji, and N. K. Jerne (1966). Kinetics of proliferation of clones of antibody-producing cells. Abstract number 650 in Abstracts of the 50th annual meeting of the Federation of American Societies for Experimental Biology.

Lederberg, J. (1959). Genes and antibodies. Science, *129*:1649.

Mach, B., and P. Vassalli (1965). Biosynthesis of RNA in antibody-producing tissues. Proc. Nat. Acad. Sci., *54*:975.

Marmur, J. (1961). A procedure for the isolation of deoxyribonucleic acid from microorganisms. J. Mol. Biol., *3*:208.

Pinchuck, P. (1967). Personal communication. We thank Dr. Pinchuck for supplying us with the polymers.

Putnam, F. W., T. Shinoda, K. Titani, and M. Wikler (1967). Immunoglobulin structure: Variation in amino acid sequence and length of human lambda light chains. Science, *157*:1050.

Raska, K., Jr., and E. P. Cohen (1967). The specific activity of several enzymes of nucleotide metabolism in the mouse spleen after antigen. Clin. Exper. Immunol., in press.

Scherrer, K., and J. E. Darnell (1962). Sedimentation characteristics of rapidly-labeled RNA from HeLa cells. Biochem. Biophys. Res. Comm., *7*:486.

Scherrer, K., H. Latham, and J. E. Darnell (1963). Demonstration of an unstable RNA and of a precursor to ribosomal RNA in HeLa cells. Proc. Nat. Acad. Sci., *49*:240.

Smithies, O. (1963). Gamma-globulin variability: A genetic hypothesis. Nature, *199*:1231.

—— (1967). Antibody variability. Science, *157*:267.

Talmage, D. W., and E. P. Cohen (1965). Antibody production and specificity in "Immunological Diseases." M. Samter, ed. Little, Brown and Company, Boston, p. 87.

Watson, J. (1965). "The molecular biology of the gene." New York: Benjamin (Bennett).

STUDIES *IN VITRO* ON IMMUNOGLOBULIN BIOSYNTHESIS *

Alan R. Williamson

National Institute for Medical Research, Mill Hill, London N.W.7. England

I. INTRODUCTION

Study of the biosynthesis of immunoglobulin in mice has been greatly facilitated by the use of neoplastic plasma cells. *In vitro* experiments with plasma cells from 5563 myeloma tumor and also with hyperimmune lymph node tissue have furnished evidence on the synthesis and assembly of the subunits of immunoglobulin-G. In the first part of this article these findings will be summarized and a model of IgG biosynthesis proposed. The second section will deal with the construction of a cell-free system, with which to study IgG biosynthesis in more detail; preliminary studies on the functioning of this system show that at least some of the stages in IgG synthesis are carried out and point to the potential of the system.

The results of *in vitro* studies with intact cells and with cell-free systems indicate that, despite the unique nature of immunoglobulins, translation and the subsequent steps in immunoglobulin synthesis do not differ essentially from the mode of biosynthesis of other mammalian proteins.

II. INTACT CELL STUDIES

Plasma cells of myeloma tumor 5563 synthesize and secrete a single molecular species of immunoglobulin-G. This has been shown by *in vitro* labelling and characterization of the newly formed radioactive G-mye-

* The author wishes to thank Dr. B. A. Askonas for much helpful discussion and encouragement during the course of these studies and for critically reading the manuscript. He also wishes to thank Mrs. J. Henderson for her excellent technical assistance.

loma globulin both before and after secretion (Awdeh *et al.*, 1967); the electrophoretic heterogeneity of 5563 myeloma protein as it is found in serum is due to modifications of changed residues after synthesis and secretion of the homogeneous globulin. It follows that in these monoclonal cells only a single heavy chain variant and a single light chain variant are synthesized, thus assuring symmetrically assembled IgG molecules $[(LH)_2]$.

In a series of collaborative experiments, with Dr. B. A. Askonas, a study was made of the synthesis of G-myeloma globulin by 5563 plasmacytoma cells *in vitro*. From our findings, which we reviewed recently (Askonas and Williamson, 1967a), we can construct a picture of the various stages as illustrated in Figure 1 and outlined below.

1. Heavy (H) and light (L) chains appear to be synthesized as separate, complete polypeptide chains. This is deduced from immunological characterization of the different size classes of polyribosomes engaged in synthesis of H and L chains (Williamson and Askonas, 1967); the data suggested the existence of two separate species of m-RNA with the expected sizes to code for complete H and L chains. It follows that H and L chains are the likely subunits from which IgG is assembled.

2. Polyribosomes engaged in synthesis of H and L chains appear from electron microscope studies to be confined to the surface of the endo-

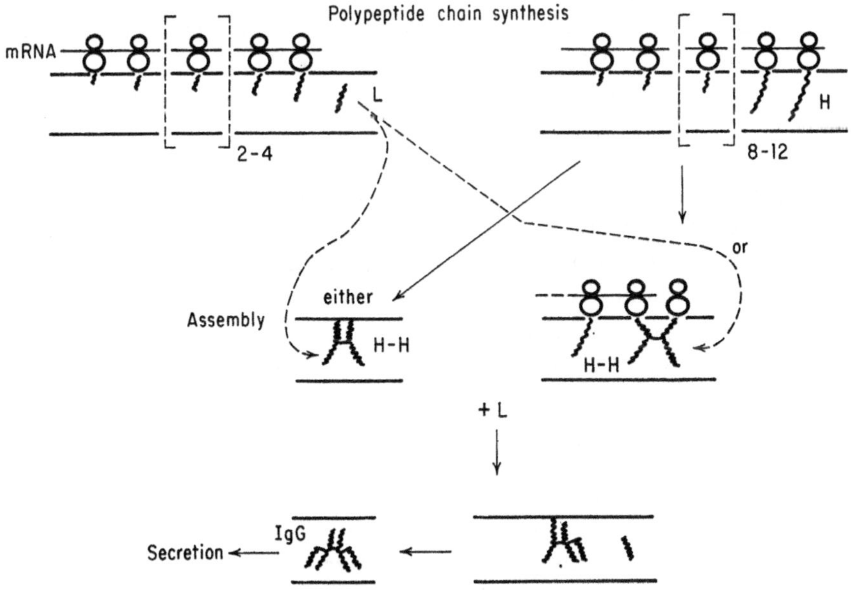

Fig. 1. Proposed scheme for immunoglobulin-G biosynthesis and assembly based on studies with mouse 5563 plasmacytoma cells.

plasmic reticulum (de Petris, 1967); also newly synthesized and assembled G-myeloma globulin appears to be retained within the cisternae of the endoplasmic reticulum (E.R.) prior to secretion (de Petris, unpublished observations; see also Swenson *et al.,* 1967). These observations suggest that IgG synthesis closely resembles the synthesis of other secretory proteins with vectorial release of completed polypeptide chains into the cisternae of the E.R. (Redman *et al.,* 1966). This proposal and the mode of membrane-ribosome interaction described by Sabatini *et al.* (1966) have been incorporated into Figure 1.

3. Light chains are autonomously released from their site of synthesis into a small pool of free light chains.

4. Free light chains function as a rapidly turning over pool of intermediates in IgG assembly: there is a balanced synthesis of H and L chains and *no* net excess synthesis or secretion of L chains. Evidence for these points has been obtained using both 5563 plasmacytoma cells and hyperimmune mouse lymph node tissue (Askonas and Williamson, 1966a, b; 1967b).

5. A short lag in the appearance of labelled H chain in G-myeloma protein (Williamson and Askonas, 1968) and the detection of labelled H chain dimers (H-H) (Askonas and Williamson, 1967) suggests that assembly proceeds stepwise as illustrated in Figure 1. Heavy-heavy chain dimerization through an interchain disulphide bridge could well involve intrapolyribosomal interaction. Association between L chains and H-H most probably takes place in the vicinity of H chain synthesis; L chain attachment may precede and assist release of H-H dimer from polyribosomes or the adjacent membrane (alternatives shown in Fig. 1) or H-H may have a brief free existence.

A number of problems remain unanswered by the intact cell studies although the overall picture permits some of the outstanding questions to be more clearly defined. It was with these questions in mind that work was initiated on the setting up of a cell-free system in which biosynthesis of immunoglobulin molecules could be studied in more detail. The cell-free system described below has not yet answered many questions but its properties suggest its potential usefulness.

III. CELL-FREE SYSTEM STUDIES

The preparation of a subcellular system for studying protein biosynthesis can be approached in two ways: either to utilize the liberated cellular contents in as crude (i.e., unmanipulated) a manner as possible, or to isolate the necessary subcellular components and then reassemble them. The latter approach was chosen in the present work, but it was simplified by using plasma cells as a source of ribosomal material only, supplementing with ancillary soluble components from more convenient sources.

The polyribosomes engaged in IgG synthesis appear to be membrane-bound and most previous subcellular systems from IgG-producing tissue have been based on microsomal particle fractions (Ogata *et al.,* 1961; Eisen *et al.,* 1961; Askonas, 1961; Van der Meer & Koningsberger, 1965; Stenzel & Rubin, 1966; Nezlin & Kulpina, 1967). Wust and Novelli (1964) described a cell-free amino acid incorporating system which utilized rat spleen ribosomes and soluble components from rat liver. More recently Talal (1966) obtained amino acid incorporation in a cell-free system containing both ribosomes and soluble components from rat spleen.

A. Preparation of Ribosomes from Neoplastic Plasma Cells

The present system was prepared using plasma cell polyribosomes detached from membranes in order to determine what functions can be retained. The solid form of the plasma cell tumor 5563 was used as the source of ribosomes. Tumor tissue was homogenized in buffered sucrose in the presence of Macaloid and a post-mitochondrial supernatant was prepared (see Table 1). Analysis of the particulate fraction of post-mitochondrial supernatant on a sucrose gradient shows the presence of both membrane-bound ribosomes and free polyribosomes (Fig. 2). To this supernatant was added DOC to a final concentration of 0.5% and then ribosomes were pelleted through 30% sucrose. Ribosomes prepared in this way contain a broad spectrum of intact polyribosomes (Fig. 3), similar to that observed previously in 5563 ascitic tumor cells (Williamson and Askonas, 1967).

Table 1

Preparation of Ribosomes from 5563 Tumor Tissue

1. Homogenize tissue in 1 volume 0.15 *M* sucrose-solution E * containing 0.1% Macaloid.†
   Glass-teflon homogenizer clearance 0.1 mm.
2. Centrifuge at 12,000 r.p.m. for 5 mins.
3. Supernatant plus ⅙th volume 5% DOC layered on sucrose (15 ml 15% over 15 ml 30%) in solution E. Centrifuge at 30,000 r.p.m. for 2 hrs.
4. Pellet suspended in 0.25 *M* sucrose.
5. Ribosome suspension centrifuged at 10,000 r.p.m. for 5 mins.

* Solution E is 10^{-2} *M* Tris Cl, 10^{-2} *M* KCl, 10^{-3} *M* $MgCl_2$, pH 7.5.

† Macaloid is sodium magnesium lithofluorosilicate.

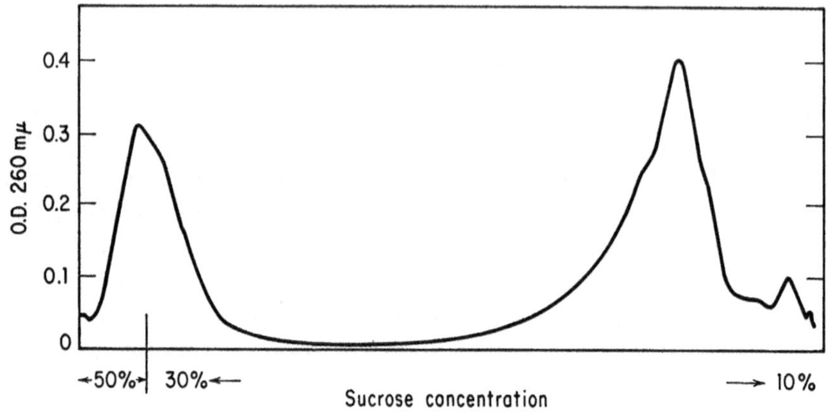

Fig. 2. Membrane-bound and free ribosomes from 5563 plasma cells. Total microsomes pelleted from post-mitochondrial supernatant (see Table 1) and analyzed on sucrose gradient, 4.2 ml linear 10–30% sucrose over 0.5 ml 50% sucrose (in solution E) centrifuged in Spinco SW 39 rotor for 20 mins. at 38,000 r.p.m. Optical density was monitored automatically using ISCO density gradient fractionator.

B. Cell-free System

Isolated plasma cell polyribosomes were assayed for ability to incorporate amino acids in a cell-free system supplemented with transfer RNA (t-RNA) from rabbit liver (Cantoni *et al.*, 1962), and a partially purified enzyme fraction from rabbit reticulocytes (Hardesty *et al.*, 1963); this fraction contains transfer enzymes and aminoacyl RNA synthetases and is almost free of nuclease activity. The ability of plasma cell ribosomes to function for peptide bond formation in this mixed system is shown by the synthesis of polyphenylalanine when poly U is added as exogenous

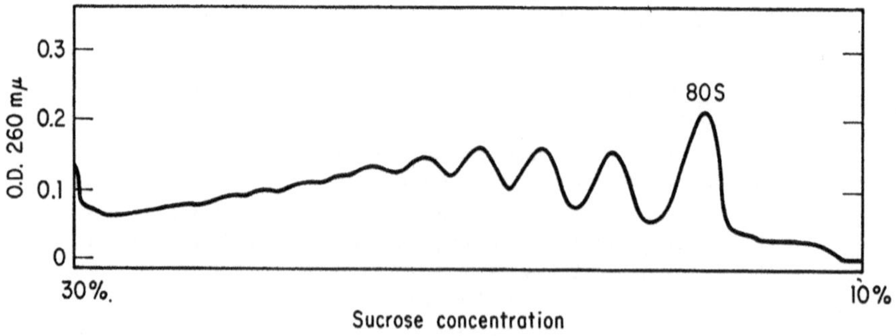

Fig. 3. Polyribosomes from 5563 plasma cells. Polyribosomes prepared as described in Table 1 and analyzed on linear 10–30% sucrose gradient (in solution E) centrifuged in Spinco SW 25 rotor for 150 mins. at 24,000 r.p.m.

Table 2

ENZYME AND RNA REQUIREMENTS FOR POLY U STIMULATION OF PLASMA CELL RIBOSOMES

Assay Conditions	Incorporation of ^{14}C-Polyphenylalanine (mμmoles/mg Ribosomes)
1. Complete system	3.1
2. – minus poly U	0.15
3. – minus energy	0.04
4. – minus enzyme	1.1
5. – minus t-RNA	0.72
6. Complete zero time	0.04

The complete system was as described by Williamson *et al.* (1967) for reticulocyte ribosomes containing rabbit reticulocyte enzyme fraction, rabbit liver transfer RNA and an energy source comprising ATP, GTP, creatine phosphate and creatine kinase. The mixture was in a final volume of 0.5 ml and contained 200 μg poly U (Miles Laboratories). Incubation was at 37° for 120 mins. and was terminated by chilling, followed by addition of ⅙th volume 6 *N* NaOH containing 10^{-2} *M* ^{12}C-phenylalanine. After 5 mins. at room temperature samples were precipitated with 5% trichloracetic acid and filtered on Oxoid membranes for measurement of radioactivity in a Nuclear-Chicago gas-flow counter.

messenger (Table 2). The requirements for enzyme fraction and t-RNA are similar to those found with reticulocyte ribosomes (Williamson *et al.*, 1967).

Mouse plasma cell ribosomes thus appear to function adequately for peptide bond formation with t-RNA and enzymes from rabbits. In the absence of added poly U, incorporation of amino acids into polypeptides in this system should represent the translation of mRNA contained in the plasma cell polyribosomes. Some of the parameters of the endogenous incorporation are shown in Table 3. Incorporation was absolutely dependent on a source of energy (ATP and generating system). The stimulations by enzyme fraction and t-RNA were measured with optimal amounts of each; optimal Mg^{++} concentration was 3.5*mM* and there was no incorporation in the absence of Mg^{++}. Amino acid incorporation was found to be proportional to the amount of added polyribosomes (Fig. 4).

Table 3

REQUIREMENTS FOR ^{14}C-VALINE INCORPORATION BY PLASMA CELL RIBOSOMES

Reaction Mixtures	Incorporation ($\mu\mu$moles/mg Ribosomes)
1. Complete system	114
2. – minus energy	11
3. – minus enzyme	69
4. – minus t-RNA	75
5. – minus amino acid mixture	113
6. Complete zero time	8

The complete system contained, in a volume of 0.5 ml: 1.0 mg enzyme fraction; 50 μg t-RNA; 10 μg creatine kinase; 0.33 μmole dipotassium ATP (pH 7.5); 3.3 μmoles creatine phosphate (pH 7.0); 0.08 μmoles GTP (pH 7.0); 6.7 μmoles GSH; 17 μmoles Tris Cl buffer (pH 7.5); 17 μmoles KCl; 1.7 μmoles $MgCl_2$; .01 μmole ^{14}C-Valine; .02 μmole each ^{12}C-amino acid except valine; and 0.5 mg ribosomes. Incubation at 37° for 60 mins. Samples precipitated and counted as described under Table 2.

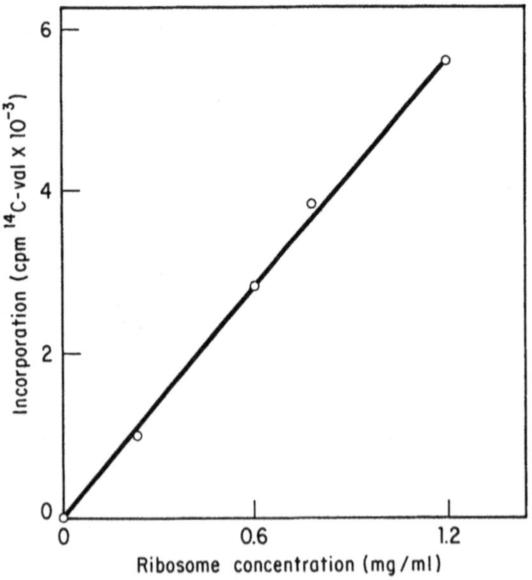

Fig. 4. Dependence of ^{14}C-Valine incorporation on ribosome concentration. The cell-free system was as described under Table 3 with ^{14}C-Valine (specific activity 34 mc/mmole) and varying amounts of plasma cell ribosomes.

C. Properties of Plasma Cell Ribosomes in a Cell-free System

1. *Products of Endogenous Amino Acid Incorporation*

It was next essential to check whether any of the incorporation directed by plasma cell polyribosomes represents synthesis of IgG determinants. In order to obtain the labelled nascent peptides in soluble form for immunological analysis, puromycin was added to the cell-free system, after incorporation had essentially ceased, and incubation continued for 10 minutes. This resulted in the release of 64% of the labelled nascent chains which were separated from ribosomes by ultracentrifugation (Table 4). The larger labelled polypeptides were selected by gel filtration on a Sephadex G100 and then tested for co-precipitation with 5563 G-myeloma protein and rabbit antiserum to various myeloma protein determinants; as control, co-precipitation with rabbit IgG—goat anti-rabbit

Table 4

Immunological Characterization of Polypeptides Synthesized in Cell-free System

Procedure	Radioactivity in Polypeptides (cpm ^{14}C-leu)
1. Cell-free system 40 min × 37° then plus 2×10^{-4} *M* puromycin 10 min × 37°.	
Soluble fraction	7,740
Ribosomal pellet	14,100
2. Gel filtration of soluble fraction (12,690 cpm) on G-100 Sephadex. Frontal peak	6,090
3. Labelled peptides from step 2. (Aliquots: 1,690 cpm each)	
(a) Plus 50 μg 5,563 myeloma protein and 50 μl antiserum to myeloma protein. Ab/Ag precipitate	670
(b) Plus 50 μg rabbit IgG and 110 μl goat anti-rabbit IgG serum Ab/Ag precipitate	290
4. Specific precipitation (Difference 3a–3b)	380

Cell-free system was prepared as described under Table 3 except that it contained 1.8 mg ribosomes, ^{12}C-amino acids other than leucine and ^{14}C-leucine (specific activity 155 mc/mole). After incubation the mixture was diluted to 2 ml with 0.25 *M* sucrose and centrifuged at 105,000 g for 60 mins. G-100 column was equilibrated with phosphate buffered saline. Incubation with antisera and preparation of antibody/antigen precipitates for counting in Packard liquid scintillation counter were as described previously (Askonas and Williamson, 1966b).

IgG was measured. The results (Table 4) show that about 20% of the labelled polypeptides tested were specifically co-precipitated with myeloma protein by antiserum to myeloma protein.

2. *Coding Fidelity of Plasma Cell Ribosomes*

Aminoglycoside antibiotics have been shown to cause misreading of the genetic code by bacterial ribosomes (Davies, 1964). Similar translational errors have not been observed directly using mammalian ribosomes in the presence of these antibiotics (Schweet *et al.*, 1965; Weinstein *et al.*, 1966). However an effect of streptomycin on the synthesis of specific antibody by plasma cells *in vitro* has been interpreted as possibly being due to the insertion of erroneous amino acids into the polypeptide chains of immunoglobulins (Krueger, 1965). Since the susceptibility to induced coding errors has been shown to be a property of ribosomes (Davies, 1964; Cox *et al.*, 1964), it was of interest to check the effect of aminoglycoside antibiotics on the fidelity of translation by plasma cell ribosomes in the heterologous cell-free system. To this end poly U was used as added messenger RNA and the incorporation of amino acids other than phenylalanine, which is the normal amino acid coded by poly U, was examined. The technique employed was to use ^{14}C-labelled protein hydrolysate as a source of radioactive amino acids and to measure incorporation in the presence of a large excess of ^{12}C-phenylalanine which damps out incorporation of ^{14}C-phenylalanine. Under these conditions incorporation should represent a gross measure of coding errors. It was found that there is a low level of incorporation stimulated by poly U at 12–16 *mM* magnesium ion concentrations (Fig. 5). Addition of either streptomycin or neomycin effected no increase in the level of incorporation directed by poly U and indeed reduced incorporation to, or below, the endogenous level seen in the absence of poly U (Fig. 5). Thus we do not see any antibiotic-induced coding errors using plasma cell ribosomes in a cell-free system.

Inhibition of peptide bond formation by streptomycin and neomycin was checked using poly U directed polyphenylalanine synthesis on plasma cell ribosomes. In the presence of streptomycin, incorporation was inhibited only slightly at levels up to 10^{-3} *M*; neomycin, however, inhibits incorporation $> 99\%$ at only 2×10^{-4} *M*. Plasma cell ribosomes thus respond to these antibiotics in a similar manner to reticulocyte ribosomes (Schweet *et al.*, 1965; Weinstein *et al.*, 1966).

3. *Functioning of Polyribosomes*

Polyribosomes are thought to function in a dynamic way with motion of ribosomes relative to mRNA. In a cell-free system both attachment and detachment of ribosomes from mRNA can be demonstrated (see review by Schweet *et al.*, 1964). In the reticulocyte cell-free system polyribo-

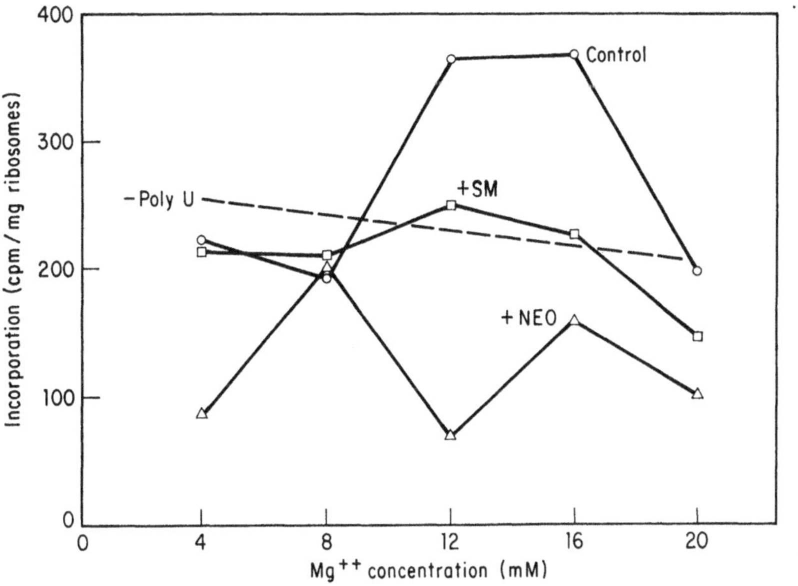

Fig. 5. The effect of streptomycin and neomycin on the fidelity of translation of poly U by plasma cell ribosomes. The cell-free system for poly U stimulation of plasma cell ribosomes was used (see Table 2) but with 0.5 μmoles ^{12}C-phenylalanine and 0.04 μc ^{14}C-labelled amino acids (protein hydrolysate, specific activity 600 μc/mg). Streptomycin (2×10^{-4} *M*) and neomycin (10^{-5} *M*) were added where indicated.

somes break down to monomeric ribosomes during incubation and this detachment from mRNA can be correlated with release of completed polypeptide chains (Hardisty *et al.*, 1963).

When plasma cell polyribosomes were incubated in the heterologous cell-free system breakdown to monomeric ribosomes occurs as incorporation proceeds (Fig. 6); this breakdown is also accompanied by some release of labelled polypeptides (Fig. 6). The extent of this release is at present variable from one preparation of polyribosomes to another and conditions of preparation and incubation to give optimal release are being sought. The time course of incorporation into ribosomes and into soluble released polypeptides was measured (Fig. 7); these curves are consistent with the idea that the ribosomal incorporation is into nascent chains which are precursors of the released polypeptides.

IV. DISCUSSION

The essential features of the scheme shown in Figure 1, based on the study of 5563 myeloma protein, are consistent with the findings on IgG biosynthesis in other systems. There is, however, scope for differences in the details. For instance, balanced synthesis of heavy and light chains does not obtain in a number of neoplastic plasma cell lines. Indeed, the known products of such tumors include free L chain (Bence-Jones protein) either alone, or in addition to 7S globulin, half molecules (L-H) and fragments resembling Fc (heavy chain disease). Excess L chain production and secretion has also been reported in rabbit immune tissue (Shapiro *et al.*, 1966). The order of formation of interchain disulphide bridges might also vary from one species of IgG molecule to another. This order could be a function of the relative stabilities, in a reducing environment, of the H-H and L-H disulphide bonds; this property varies amongst species of IgG (Askonas and Williamson, 1967c).

The *in vitro* studies both in intact cells and the subcellular system suggest that the components and processes involved in IgG biosynthesis do not vary significantly from the general pattern of mammalian protein biosynthesis. The functioning of ribosomes from IgG-synthesizing cells in the cell-free system shows that they closely resemble ribosomes from other mammalian cells with respect to translation of either endogenous mRNA or exogenous synthetic messenger; with poly U as messenger, plasma cell ribosomes were found to show similar resistance to induced

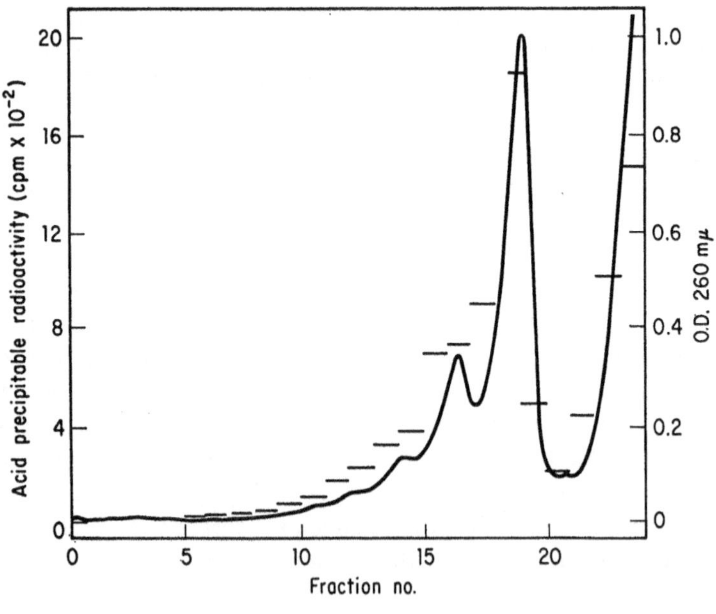

Fig. 6. Sucrose gradient analysis of amino acid incorporation in cell-free system. The complete cell-free system was as described in Table 3 with 1 mg plasma cell ribosomes and ^{14}C-Valine (specific activity 34 mc/mmole). Incubation was for 60 minutes at 37°. The total mixture was analyzed on 30 ml linear 10–30% sucrose gradient (in solution E), centrifuged in Spinco SW 25 rotor for 120 mins. at 24,000 r.p.m. Optical density was measured automatically (continuous line). Fractions were collected, 0.5 mg BSA added as carrier and acid-precipitable radioactivity measured (see Table 2).

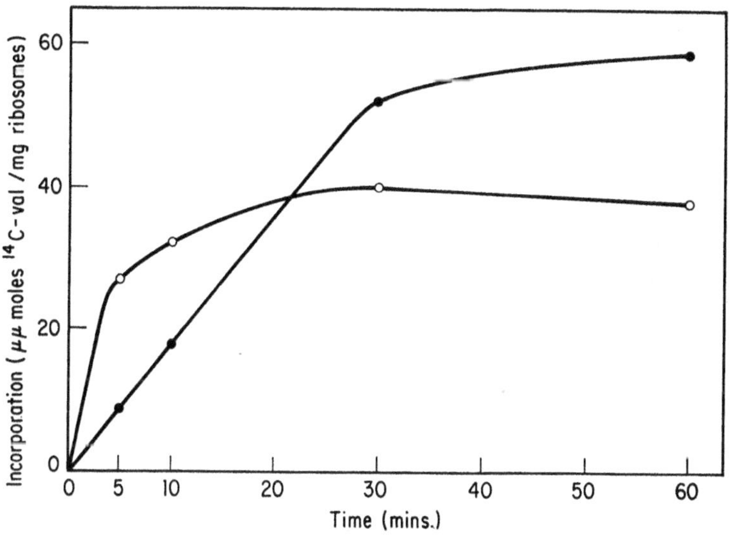

Fig. 7. Time course of incorporation of ^{14}C-Valine into plasma cell ribosomes and soluble protein in cell-free system. The complete cell-free system was as described in Table 3 containing 720 μg ribosomes in 0.5 ml assay. Incubation was at 37° and 0.1 ml aliquots were withdrawn at indicated times, diluted to 1.5 ml with 0.25 *M* sucrose and centrifuged at 105,000 g for 60 mins. Incorporation into soluble supernatant protein (solid circles) and pelleted ribosomes (open circles) was measured as described in Table 2 using 0.5 mg BSA carrier protein.

miscoding, i.e., similar fidelity of translation to other mammalian ribosomes.

In view of the number of reports, quoted earlier in this article, of amino acid incorporating systems from IgG producing tissue, it would seem to be worthwhile discussing the need for another cell-free system. In a number of the systems using microsomal particulate fractions evidence has been offered for incorporation of amino acids into IgG determinants (Ogata *et al.,* 1961; Eisen *et al.,* 1961; Stenzel and Rubin, 1966; Neslin and Kulpina, 1967). The complete synthesis of IgG molecules has not been demonstrated in these systems and the microsomal system would be less suitable than a ribosomal system for some cell-free studies. The systems based on ribosomes freed from membranes (Wust and Novelli, 1964; Talal, 1966) incorporate amino acids into ribosome-bound nascent polypeptides which were not characterized.

The major difficulty encountered in the previous studies, and in the present system, is the small extent of amino acid incorporation relative to incorporation in other mammalian cell-free systems; this seems to reflect the difficulties of isolating intact polyribosomes from plasma cells which have a high nuclease content. In the present studies emphasis is being placed on increasing the extent of incorporation; some progress has been made in this respect but the best incorporation yet obtained is still only 25% of that in an optimal reticulocyte cell-free system. Moreover in the reticulocyte system *de novo* synthesis and release of complete chains of hemoglobin can be demonstrated (see review by Schweet *et al.*).

One of the reasons for choosing to set up a cell-free system using ribosomes, rather than microsomes, was to study the functions of polyribosomes from the rough endoplasmic reticulum in the absence of membranes. The properties of the mixed cell-free system, even at its present stage of development, show that some functons of membrane-bound polyribosomes can be preserved after removal of the phospholipid membranes. Part of the amino acid incorporation directed by the endogenous messenger of plasma cell polyribosomes appears to represent synthesis of antigenic determinants of IgG; it should be stressed that this does *not* constitute cell-free synthesis of IgG but is only offered as evidence that the plasma cell ribosome preparations include some of the polyribosomes which were involved in IgG synthesis in the cell. The most encouraging aspect of the present system is that it is apparently possible to preserve natural release of completed polypeptide chains. The variability of release from one ribosome preparation to another may reflect the difficulty of removing polyribosomes from membranes in an undamaged condition. Release of completed chains is necessary for completion of synthesis of a multichain protein such as IgG and in the cell-free system observation of the assembly steps would be possible only if release functions normally.

References

Askonas, B. A. (1961). A study on globulin formation by plasma-cell neoplasm (5563) transplantable in mice. Biochem. J., *79*:33–43.

Askonas, B. A., and A. R. Williamson (1966a). Biosynthesis of immunoglobulins on polyribosomes and assembly of the IgG molecule. Proc. Roy. Soc. B, *166*:232–243.

——— (1966b). Biosynthesis of immunoglobulins. Free light chain as an intermediate in the assembly of γG-molecules. Nature, *211*:369–372.

——— (1967a). Biosynthesis and assembly of immunoglobulin-G. Cold Spring Harbour Symp. on Quant. Biol., *32*:223–232.

——— (1967b). Balanced synthesis of light and heavy chains of immunoglobulin-G in mouse lymph nodes. Nature, *216*:264–267.

——— (1967c). Balanced heavy and light chain synthesis in immune tissue and disulphide bond formation in IgG assembly. Nobel Symposium III, Stockholm: Almquist and Wiksell, pp. 369–383.

Awdeh, Z. L., B. A. Askonas, and A. R. Williamson (1967). The homogeneous γG-immunoglobulin produced by mouse plasmacytoma 5563 and its subsequent heterogeneity in serum. Biochem. J., *102*:548–553.

Cantoni, G. L., H. V. Gelboin, S. W. Luborsky, H. H. Richards, and M. F. Singer (1962). Studies on soluble ribonucleic acid of rabbit liver. III. Preparation and properties of rabbit liver soluble RNA. Biochim. Biophys. Acta, *61:* 354–367.

Cox, E. C., J. R. White, and J. G. Flacs (1964). Streptomycin action and the ribosome. Proc. Natl. Acad. Sci., *51:*703–709.

Davies, J. E. (1964). Studies on the ribosomes of streptomycin-sensitive and resistant strains of *Escherichia coli.* Proc. Natl. Acad. Sci., *51:*659–664.

Eisen, H. N., E. S. Simms, E. Helmreich, and M. Kern (1961). Incorporation of amino acids into a γ-globulin-like protein by cell-free extracts from lymph nodes. Trans. Assoc. American Physicians, *74:*207–215.

Hardesty, B., R. Miller, and R. Schweet (1963). Polyribosome breakdown and hemoglobin synthesis. Proc. Nat. Acad. Sci., *50:*924–931.

Krueger, R. G. (1965). The effect of streptomycin on antibody synthesis *in vitro.* Proc. Natl. Acad. Sci., *54:*144–152.

Nezlin, R. S., and L. M. Kulpina (1967). Incorporation of radioactive amino acids into γ-globulin in a cell-free system from rat spleen. Biochim. Biophys. Acta, *138:*654–656.

Ogata, K., S. Omori, and R. Hirokawa (1961). Studies on the biosynthesis of antibody and other γ-globulins in microsomes of immunized spleen cells. J. Biochem., *49:*660–672.

de Petris, S. (1967). Polyribosomes in thin sections of 5563 plasmacytoma cells. J. Mol. Biol., *23:*201–216.

Redman, C. M., P. Siekevitz, and G. E. Palade (1966). Synthesis and transfer of amylase in pigeon pancreatic microsomes. J. Biol. Chem., *241:*1150–1158.

Sabatini, D., Y. Tashiro, and G. E. Palade (1966). On the attachment of ribosomes to microsomal membranes. J. Mol. Biol., *19:*503–524.

Schweet, R. S., R. Arlinghaus, J. Shaeffer, and A. Williamson (1964). Studies on haemoglobin synthesis. Medicine, *43:*731–745.

Shapiro, A. L., M. D. Scharff, J. V. Maizel, and J. W. Uhr (1966). Synthesis of excess light chains of gamma globulin by rabbit lymph node cells. Nature, *211:*243–245.

Stenzel, K. H., and A. L. Rubin (1966). Biosynthesis of γ-globulin: studies in a cell-free system. Science, *153:*537–539.

Swenson, R. M., and M. Kern (1967). The synthesis and secretion of γ-globulins by lymph node cells. 1. The microsomal compartmentalization of γ-globulins. Proc. Nat. Acad. Sci., *57:*417–422.

Talal, N. (1966). Polyribosomes and protein synthesis in the spleen. J. Biol. Chem., *241:*2067–2074.

Van der Meer, J. H. H., and V. V. Koningsberger (1965). The synthesis of bacteriophage φX174-neutralizing protein in a subcellular system obtained from the spleen of φX174 immunized rats. Biochim. Biophys. Acta, *103:*136–145.

Weinstein, I., M. Ochoa, Jr., and S. Friedman (1966). Fidelity in the translation of messenger RNAs in mammalian subcellular systems. Biochemistry, *5:*3332–3339.

Williamson, A. R., and B. A. Askonas (1967). Biosynthesis of immunoglobulins: The separate classes of polyribosomes synthesizing heavy and light chains. J. Mol. Biol., *23:*201–216.

Williamson, A. R., and B. A. Askonas (1968). Time course of labelling of light and heavy chains of immunoglobulin-G and turnover of the free light-chain intermediates. Arch. Biochem. Biophys., in press.

Williamson, A. R., E. Hausmann, R. Heintz, and R. Schweet (1967). Homopolynucleotide inhibition of poly U-dependent polyphenylalanine synthesis. J. Mol. Biol., *26*:267–278.

Wust, C. J., and G. D. Novelli (1964). Cell-free amino acid incorporation by rat spleen ribosomes. Arch. Biochem. Biophys., *104*:185–191.

SYNTHESIS AND ASSEMBLY OF IMMUNOGLOBULIN *

JONATHAN W. UHR AND CHAYA MOROZ

Irvington House Institute and Department of Medicine
New York University School of Medicine, New York

Two aspects of antibody formation have been recently studied using concepts and techniques that have emerged from molecular biology:

1. The informational problem. The capacity of a single animal to form antibody molecules to a seemingly unlimited number of antigens has led to a search for uniqueness somewhere in the DNA-RNA-protein sequence. It is now generally agreed that the information for the specificity of antibody resides in the DNA of antibody forming cells (Lederberg, 1950; Smithies, 1963; Dreyer *et al.*, 1965; Brenner and Milstein, 1966; Edelman and Gally, 1966; Cioli and Baglioni, 1966). The central question is whether each clone of antibody-forming cells has a different DNA coding for the variable portion of the light (L) or heavy (H) chain of immunoglobulin or whether each clone has the same DNA but with derepression of a different set of cistrons. This question has been approached but not yet answered by DNA-RNA hybridization studies (Cohen and Raska, 1967; Greenberg and Uhr, 1967).

2. The problem of synthesis and assembly of immunoglobulin which is a multi-chain disulfide-bonded protein with attached carbohydrate moieties.

In this paper, we will discuss the latter problem, i.e., the transcription and translation leading to synthesis of gamma-globulin polypeptide chains, their assembly into the immunoglobulin molecule and the addition of the carbohydrate moiety.

* This work was done under the sponsorship of the Commission on Immunization of the Armed Forces Epidemiological Board, and was supported in part by the U. S. Army Medical Research and Development Command, Department of the Army, under Research Contract No. DA-49-193-MD-2043, and in part by U. S. Public Health Service Grant No. AI-01821.

I. TRANSCRIPTION AND TRANSLATION

Earlier studies suggested that immunoglobulin synthesis depended upon DNA-dependent RNA synthesis. This conclusion was based on the finding that actinomycin D at low concentrations prevents antibody formation induced *in vitro* (Uhr, 1963; Smiley *et al.*, 1964) or *in vivo* (Jerne *et al.*, 1963). It was further shown that lymph node cell suspensions removed from hyperimmunized rabbits were similarly sensitive to actinomycin D but that such Actinomycin D-treated cells could synthesize RNA and protein at rapid rates if provided with new messenger RNA (mRNA) by means of infection with Newcastle Disease Virus (Tawde *et al.*, 1966). This observation argued against a secondary effect of antinomycin D, such as destruction of the cellular machinery necessary for protein synthesis.

It was also shown that gamma-globulin was formed on polyribosomes (Scharff and Uhr, 1965). Thus, nascent gamma-globulin chains could be identified on polyribosomes from rabbit lymph node cells. This finding, in contrast to an earlier one by Stenzel *et al.* (1964), confirmed that the polyribosome was the active unit of immunoglobulin synthesis. The implication is that a mRNA molecule presumably containing the information for the amino acid sequence of the immunoglobulin in question is being translated within the cytoplasm of the antibody-forming cell.

II. ASSEMBLY

The first two questions considered were whether or not light (L) and heavy (H) chains are made on separate polyribosomes and whether assembly takes place on or off polyribosomes. One approach to the first question was based on the fact that the size of polyribosomes is in general proportional to the length of the mRNA, that is, messengers are usually fully "loaded" with ribosomes (Warner *et al.*, 1962). Thus, if separate messenger molecules are formed, L and H chains should be associated with polyribosomes of different and appropriate sedimentation constants. Alternatively, L and H chains might be formed on the same polyribosome with a high sedimentation constant that accommodates the longer polycistronic mRNA.

These possibilities were tested by determining whether L and H chains could be identified on different-sized polyribosomes. Preliminary studies were first performed on the "polyribosomal profiles" of 2 mouse plasma cell tumors, one of which formed both L and H chains and the other only L chains. These studies suggested that L chains were formed on polyribosomes of approximately 190 S and H chains on heavier polyribosomes of approximately 270 S. Using the tumor, MPC_{11}, that makes both H and L chains, an attempt was made to determine if H and L chains were present on the 270 S and 190 S polyribosomes respectively.

MPC_{11} tumor cells were labeled with a ^{14}C-amino acid mixture for 90 seconds to label fully nascent chains (chains on polyribosomes), and then cold amino acids were added in excess as a chase (Shapiro *et al.,* 1966). Polyribosomes were separated on a sucrose gradient, and the polyribosomes in the fractions corresponding to 270 S and 190 S were concentrated by centrifugation. The nascent polypeptide chains were released from the polyribosomes, reduced by urea and mercaptoethanol, and the chains analyzed by acrylamide gel electrophoresis.

Figure 1 shows the results of electrophoresis of the nascent chains associated with 190 S and 270 S polyribosomes after chases of 15 and 30 seconds. In the 190 S fraction, an L chain peak was present 15 seconds after the beginning of the chase and had almost disappeared by 30 seconds. No H chain peak was discernible. In the 270 S fraction, an H chain peak was found at 15 and 30 seconds which persisted for 60 to 75 seconds but not thereafter (not shown in Fig. 1). In addition, an L chain peak was found in this fraction and the peak persisted for at least 90 seconds.

The results of this experiment suggest that (1) L chains are formed

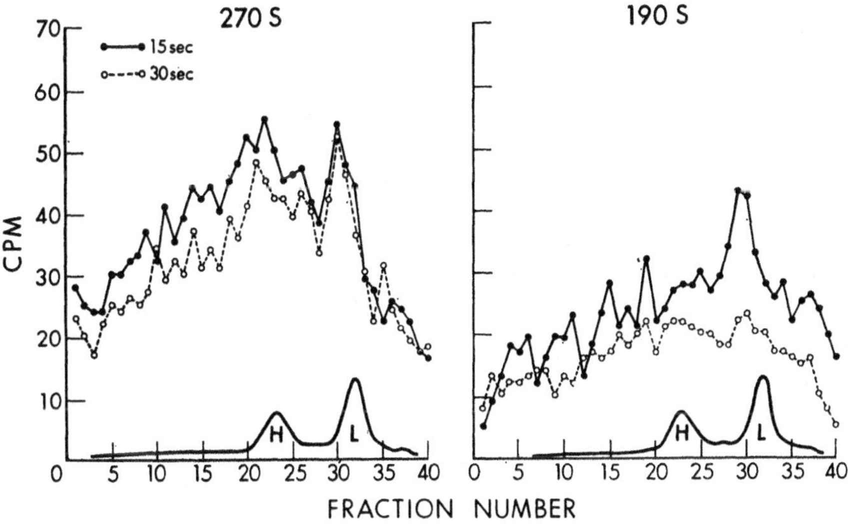

Fig. 1. Electrophoretograms of nascent polypeptides associated with 270S and 190S polyribosomes of the MPC_{11} tumor. Cells were labeled with ^{14}C-amino acids and chased as described in the text. Polyribosomes of selected sizes were concentrated from pooled fractions of three gradients by centrifugation. The nascent polypeptides were released by 100 γ/ml of ribonuclease at 37°C for 30 min., treated with SDS, urea, and ME, and analyzed electrophoretically. H and L chains were run on additional gels simultaneously and the optical densities of eluted protein are shown in the bottom curve of each panel (Shapiro *et al.,* 1966).

on the smaller polyribosomes in about 30 seconds and H chains on the larger polyribosomes in about 75 seconds and (2) assembly occurs on the larger polyribosomes where completed and released L chains become attached to nascent H chains.

The persistence of the L chain peak on the 270 S polyribosomes after nascent H chains have been "chased off" is probably due to an excess of labeled L chains in the intracellular pool which continue to attach to nascent unlabeled H chains on the larger polyribosomes.

A different approach to this problem (Shapiro *et al.,* 1966) was to examine the rate of appearance of released, completed chains rather than of nascent polypeptide chains. If L chains are less than half the length of H chains and are completed in less than half the time needed for completion of H chains, it follows that more radioactivity should be "chased off" the H chain than the L chain polyribosomes after nascent chains on both classes of polyribosomes are fully labeled. For example, if, as indicated by the previous experiment, it takes 30 seconds to make an L chain and about 75 seconds an H chain, after 30 seconds of chase, all radioactivity will be chased off the L chain polyribosome but considerable radioactivity (approximately one-third) would still remain on the H chain polyribosomes. Hence, with a further period of chase the remaining H chain radioactivity would be chased off H chain-polyribosomes. The prediction therefore is that the ratio of L/H radioactivity of released, completed chains should decline as a function of time of chase.

This was tested by a pulse-chase experiment similar to that described previously, in which aliquots were sedimented on sucrose gradients and the *supernatants* were analyzed by acrylamide gel electrophoresis for their content of labeled L and H chains. As can be seen in Figure 2, the prediction was verified in that the L/H ratio of acid precipitable radioactivity, which was high initially, declined over a period of 90 seconds. The additional finding that between 90–120 seconds there was a slight

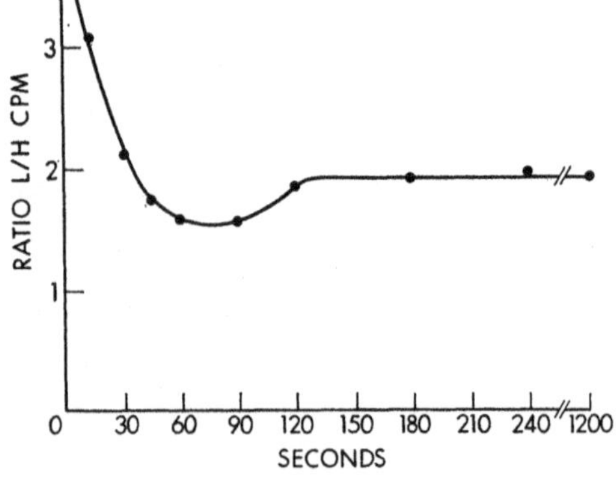

Fig. 2. Ratio of radioactivity in completed L and H chains after chasing. Cells were pulse-labeled for 1½ min. with five ^{14}C-labeled amino acids and then chased by a 200-fold excess of unlabeled amino acids. Equal aliquots were taken at intervals from 15 to 240 sec. after chasing, and analyzed on a sucrose gradient. The relative radioactivity in completed and released chains was determined electrophoretically on the material from the tops of each gradient (Shapiro *et al.,* 1966).

increase in the ratio of L to H chain radioactivity is most easily explained by the continued release from the 270 S polyribosomes of a few residual labeled L chains from the intracellular pool which are associated with unlabeled H chains. This experiment was consistent with the results of the previous one, i.e., that L and H chains are formed on different polyribosomes (of approximate sedimentation constants of 190 and 270 S, respectively) and that assembly occurs on the 270 S polyribosomes where completed and released L chains bind to nascent H chains.

Williamson and Askonas (1967) have obtained independent evidence that H and L chains are made on polyribosomes of different size. They did not obtain evidence, however, for L chains on the H chain polyribosomes. This difference in results may be due to differences in methodology.

The molecular composition of the LH aggregates released from the H chain polyribosomes has not been ascertained. The finding that HL is secreted by this and other plasma cell tumors (Cohn, 1967) suggests that HL is a precursor. Thus, an HL dimer may be released from the H chain polyribosome and then bind to another HL dimer to form the tetrameric immunoglobulin. Scharff *et al.* (1967) have obtained considerable, but inconclusive, evidence for the HL dimer being the precursor of the final molecule. Askonas and Williamson (1967), however, obtained evidence that supports another possibility, i.e., that the last 2 nascent H chains dimerize before attachment of L chains.

It is not known whether or not assembly of L to nascent H chain is obligatory for H chain release. The only evidence to support this possibility is the failure to find myeloma tumors that secrete only isolated H chains whereas tumors that secrete "free" L chains are common. The role of assembly in release of H chains is an important point in elucidating regulatory mechanisms in immunoglobulin synthesis. There are, of course, many other sites at which regulatory mechanisms might act; for example, rates of transcription and translation of the two messenger RNAs (for L and H chains respectively), half-lives of the messengers; binding of the mRNAs at sites on the endoplasmic reticulum or in the cell sap, assembly of the tetramer, addition of carbohydrate moiety (to be discussed), etc.

An attempt was made to find the approximate sedimentation values of the mRNAs associated with the two classes of polyribosomes forming L and H chains (Shapiro *et al.*, 1966). For this purpose, aliquots of tumor cells were exposed to tritiated uridine for 10–15 minutes in order to label only the polyribosome associated "messenger" RNA and none of the 28 S and 16 S ribosomal RNA. Polyribosomes from the 190 S and 270 S fractions of the gradient were centrifuged and the pellets dissolved in 1% SDS. The extract was then resedimented on a second sucrose gradient.

With the MPC_{11} tumor, which synthesizes both H and L chains, the rapidly labeled RNA associated with 190 S polyribosomes peaked between 9 and 11 S. A similar result was obtained using the 190 S fraction from a tumor (BJ) which produces only L chains. The rapidly labeled non-ribosomal RNA isolated from 270 S polyribosomes of MPC_{11} was more heterogeneous in size and peaked at more than 14 S. Similar RNA was not isolated from the 270 S fraction of the BJ tumor.

This rapidly labeled non-ribosomal RNA obtained from polyribosomes is presumably mRNA. Since 30–50% of the protein synthesizing activity of the cell population is committed to immunoglobulin synthesis, it is reasonable to assume that the mRNA for L and H chains are contained in the corresponding RNAs that were isolated and probably accounts for the peaks. The half-lives of the various mRNAs in these cells are not known, however, and it could be successfully argued that the labeling is primarily of mRNAs of short half-life for synthesis of cellular protein whereas the mRNA for immunoglobulin synthesis has a comparatively long half-life. The results of studies using actinomycin D, referred to previously, argue against this point of view. Formal proof that the above RNAs are the mRNAs for synthesis of immunoglobulin requires the use of such messengers in a cell-free system and identification of the polypeptide chains as L or H chains or, possibly, specific precipitation of the messenger on its polyribosomes by use of antiserum directed to the completed nascent chains in question.

If we assume that the above RNAs code for immunoglobulin, their coding capacities can be calculated (Hall and Doty, 1958). For example, the mRNA from the 190 S polyribosomes codes for a chain of approximately 25,000 molecular weight. This finding is consistent therefore with the translation of the L chain from one starting point as indicated by the studies of Lennox *et al.* (1967).

III. ADDITION OF CARBOHYDRATE MOIETY

In contrast to protein synthesis, addition of the carbohydrate moiety is not directly involved with the DNA-RNA-protein pathway and occurs presumably as a result of precisely regulated enzymatic activities. Questions of concern are whether the carbohydrate moiety is added entirely at the polyribosomal level and whether or not its addition depends upon continued protein synthesis. These questions were explained by exposing MPC_{11} cells to a sufficient concentration of puromycin to inhibit completely protein synthesis and then determining the extent of incorporation of carbohydrate precursors into immunoglobulin (Moroz and Uhr, 1967). It can be seen in Figure 3 that incorporation of ^{14}C-mannose into acid-precipitable material continued at an approximately linear rate after

Fig. 3. Incorporation of ^{14}C-amino acids and ^{14}C-mannose into TCA precipitable material by MPC_{11} cells. ●—● 5×10^7 cells were exposed to 10 μc of ^{14}C amino acids for 5 min. and were then treated with 2×10^{-4} *M* puromycin. ○—○ 5×10^7 cells were treated with 2×10^{-4} *M* puromycin for 5 min. and were then exposed to 10 μc of ^{14}C mannose. (Moroz and Uhr, 1967.)

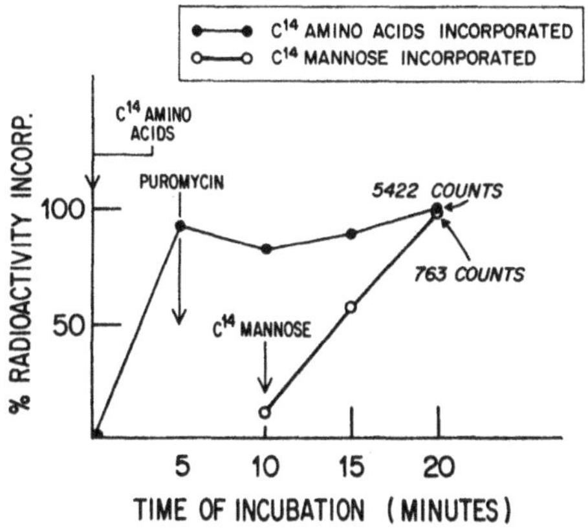

puromycin treatment. Examination of the intracellular protein during this ten-minute period revealed a progressive increase in immunoglobulin and LH dimer.

Other experiments using ^{14}C-mannose and ^{14}C-galactose have revealed that the addition of carbohydrate can continue for even longer periods of time after treatment of cells with cyclohexamide or puromycin. Thus, protein synthesis is not essential for carbohydrate synthesis, and, in contrast to the rapid assembly of polypeptide chains, the addition of the carbohydrate moiety occurs in large part after release of nascent chains from polyribosomes and much later in the intracellular life of the gamma-globulin molecule. Indeed, carbohydrate addition may not be completed until just prior to secretion. Thus, it has been shown with many glycoproteins, including immunoglobulin (Melchers and Knopf, 1967), that the secreted material has a much higher average percentage of carbohydrate compared to the intracellular glycoprotein (Eylar, 1965). This observation has led Eylar to postulate that the carbohydrate moiety is an intracellular signal for secretion across the cell membrane.

It was still considered possible, of course, that the addition of the carbohydrate moiety begins at the polyribosomal level even though completion of the process does not occur until after release of the chains. We investigated therefore whether glucosamine, the first hexose residue of the carbohydrate, was added onto nascent chains. For this purpose, MPC_{11} cells were exposed to ^{14}C-glucosamine for 35 minutes and polyribosomes were then obtained from the cell fractions. The polyribosomes were pelleted and treated with puromycin to release nascent chains. The optical density E_{280} and TCA-precipitable radioactivity were determined on the nascent chains and also on completed chains at the top of the gradient.

The results indicate that 80% of the label was released by puromycin and that the specific activity of the nascent chains from polyribosomes labeled with ^{14}C-glucosamine did not differ remarkably from that of the completed and released chains.

This experiment suggests therefore that glucosamine is incorporated on polyribosomes; whereas, as indicated before, mannose and galactose are incorporated primarily after completion and release of polypeptide chains from polyribosomes. It should be stressed, however, that when the nascent chains from the above experiment were electrophoresed on acrylamide gel, we were unable to identify radioactive peaks in the HL, H or L chain areas. We lack proof, therefore, that glucosamine is incorporated into gamma-globulin polypeptide chains rather than into other polypeptide chains being synthesized by this tumor.

IV. SUMMARY

The transcription and translation mechanisms responsible for immunoglobulin synthesis appear similar to those for other proteins. However, L and H chains are made on separate polyribosomes, of approximately 190 S and 270 S, respectively. Assembly occurs by attachment of free released L chains onto nascent H chains. Glucosamine appears to be added onto nascent chains whereas completion of the carbohydrate moiety appears to occur primarily on released and completed chains.

References

Askonas, B. A., and A. R. Williamson (1967). Balanced heavy and light chain synthesis in immune tissue and disulphide bond formation in IgG assembly. Nobel Symposium III, Stockholm: Almquist and Wiksell, pp. 369–383.

Brenner, S., and C. Milstein (1966). Origin of antibody variation. Nature, *211:* 242–243.

Cioli, D., and C. Baglioni (1966). Origin of structural variation in Bence Jones proteins. J. Mol. Biol., *15*:385–388.

Cohen, E. P., and K. Raska (1967). Antigen-unique species of RNA in peritoneal cells of mice which do not adhere to glass. Cold Spring Harbor Symp., XXXII: 349–351.

Cohn, M. (1967). Natural history of the myeloma. Cold Spring Harbor Symp., XXXII: 211–221.

Dreyer, W. J., and J. C. Bennett (1965). The molecular basis of antibody formation: A paradox. Proc. Nat. Acad. Sci., *54*:864–869.

Edelman, G. M., and J. A. Gally (1967). Somatic recombination of duplicated genes: An hypothesis on the origin of antibody diversity. Proc. Nat. Acad. Sci., *57*:353–358.

Eylar, E. H. (1965). On the biological role of glycoproteins. J. Theoretical Biol., *10*:89–113.

Greenberg, L. J., and J. W. Uhr (1967). DNA-RNA hybridization studies on gamma globulin synthesizing tumors in mice. Cold Spring Harbor Symp., XXXII: 243–248.

Hall, B. D., and P. Doty (1958). Microsomal Particles and Protein Synthesis, pp. 27–35. London: Pergamon Press.

Jerne, N. K., A. A. Nordin, and C. Henry (1963). The agar plaque technique for recognizing antibody-producing cells. Cell-Bound Antibodies, pp. 109–116. Philadelphia: The Wistar Institute Press.

Lederberg, J. (1959). Genes and antibodies. Science, *129:*1649–1653.

Lennox, E. S., P. M. Knopff, A. J. Munro, and R. M. E. Parkhouse (1967). A search for biosynthetic subunits of light and heavy chains of immunoglobulins. Cold Spring Harbor Symp., XXXII: 249–254.

Melchers, F., and P. M. Knopff (1967). Biosynthesis of the carbohydrate portion of immunoglobulin chains: Possible relation to secretions. Cold Spring Harbor Symp., XXXII: 255–262.

Moroz, C., and J. W. Uhr (1967). Synthesis of the carbohydrate moiety of γ-globulin. Cold Spring Harbor Symp., XXXII: 263–264.

Scharff, M. D., A. L. Shapiro, and B. Ginsberg (1967). The synthesis, assembly and secretion of gamma globulin polypeptide chains by cells of a mouse plasma cell tumor. Cold Spring Harbor Symp., XXXII: 235–241.

Scharff, M. D., and J. W. Uhr (1965). Functional ribosomal unit of gamma-globulin synthesis. Science, *148:*646–648.

Shapiro, A. L., M. D. Scharff, J. V. Maizel, Jr., and J. W. Uhr (1966). Polyribosomal synthesis and assembly of the H and L chains of gamma globulin. Proc. Nat. Acad. Sci., *56:*216–221.

Smiley, J. D., J. G. Heard, and M. Ziff (1964). Effect of actinomycin D on RNA synthesis and antibody formation in the anamnestic response *in vitro.* J. Exp. Med., *119:*881–893.

Smithies, O. (1963). Gamma-globulin variability: A genetic hypothesis. Nature, *199:*1231–1236.

Stenzel, K. H., W. D. Phillips, D. D. Thompson, and A. L. Rubin (1964). Functional ribosomes in antibody-producing cells. Proc. Nat. Acad. Sci., *51:*636–642.

Tawde, S., M. D. Scharff, and J. W. Uhr (1966). Mechanisms of γ-globulin synthesis. J. Immunol., *96:*1–7.

Uhr, J. W. (1963). Actinomycin D: Its effect on antibody formation *in vitro.* Science, *142:*1476–1477.

Warner, J. R., A. Rich, and C. E. Hall (1962). Electron microscope studies of ribosomal clusters synthesizing hemoglobin. Science, *138:*1399–1403.

Williamson, A. R., and B. A. Askonas (1967). Biosynthesis of immunoglobulins: The separate classes of polyribosomes synthesizing heavy and light chains. J. Mol. Biol., *23:*201–214.

ON THE ROLE OF NUCLEIC ACIDS AS GENES CONFERRING PRECISE CHEMOSPECIFICITY TO DIFFERENTIATED CELL LINES *

WILLIAM J. DREYER AND WILLIAM R. GRAY

Norman W. Church Laboratory of Chemical Biology, Division of Biology California Institute of Technology, Pasadena, California

If you will be persuaded by me, pay little attention to Socrates, but much more to the truth, and if I appear to you to say anything true, assent to it; but if not, oppose me with all your might, taking care that in my zeal I do not deceive both myself and you, and like a bee depart, leaving my sting behind.

Socrates, in Plato's *Phaedo,* 360 BC

We shall focus attention today on two types of cellular differentiation in which highly specific cells are produced. The first example is the development of the brain. Now that we know the genetic code and the dogma of protein synthesis one can express the process with the "simple" equation:

$$\text{CHROMOSOMAL DNA} + \text{GENETIC CODE} \xrightarrow[\boxed{+?}]{} \text{BRAIN}$$

The second example of cellular differentiation is the development of chemospecificity in cells of the immune system—the antibody producing cells. In this second case the fundamental problem may be expressed more clearly in molecular terms: How can our genetic mechanisms generate perhaps hundreds of thousands of cell lines, each committed to the production of a different type of antibody molecule? We believe that this question can be expressed by the equation:

* Portions of this paper have been adapted from a presentation given at the Cold Spring Harbor Symposium on Antibodies, June 1967. This work was supported by grant USPH GM-06965.

CHROMOSOMAL DNA

$$+ \text{GENETIC CODE} \underset{\boxed{+?}}{\longrightarrow} \text{COMMITTED LYMPHOCYTES (memory cells)}$$

Even this simple (or simple-minded) expression is open to controversy. Others prefer hypotheses that may be expressed as follows:

NEWLY CODED SOMATIC DNA (OR RNA)

$$+ \text{GENETIC CODE} \underset{\boxed{+?}}{\longrightarrow} \text{COMMITTED LYMPHOCYTES (memory cells)}$$

There is, strangely enough, an equivalent school of thought in brain research:

NEWLY CODED SOMATIC RNA

$$+ \text{GENETIC CODE} \underset{\boxed{+?}}{\longrightarrow} \text{MEMORY}$$

Another feature common to both fields at this point in time is a high degree of controversy.

Dr. Melvin Cohn has introduced these questions in his talk entitled "The Molecular Biology of Expectation" (see p. 671). He pointed out that the immune system and the brain both represent relatively late evolutionary adaptations which are utilized by vertebrates to deal with the unexpected in the environment. This, in itself, is not the subject of controversy. Rather, it is the question of heritable vs. new genetic information, and mechanisms for using these genes, that is debatable. Since Dr. Cohn dwelled rather extensively on the subject of antibody formation we shall start today with some comments on the brain.

I. THEORY OF THE MOLECULAR BASIS OF DIFFERENTIATION

A. Surface Displays of Chemospecific Molecules in the Brain

Figure 1 is taken from the work of Sperry and his associates (Sperry, 1963). This diagrammatic representation of the brain of a goldfish serves to illustrate the pathway which the optic nerve tract takes from the retina of the eye to the tectum in the brain. Approximately 500,000 individual nerve fibers are used to carry information through the optic nerve tract to very precise locations in the tectum. Orderly data processing events occur in both the retina and tectum.

Of particular importance, experimentally, is the fact that the optic nerve of the goldfish is capable of extensive regeneration after parts have been destroyed. Attardi and Sperry (1963) used this system to study the specificity of nerve regeneration. Fibers in the optic trunk were scrambled, and then discrete segments of the retina were excised. Regenerating

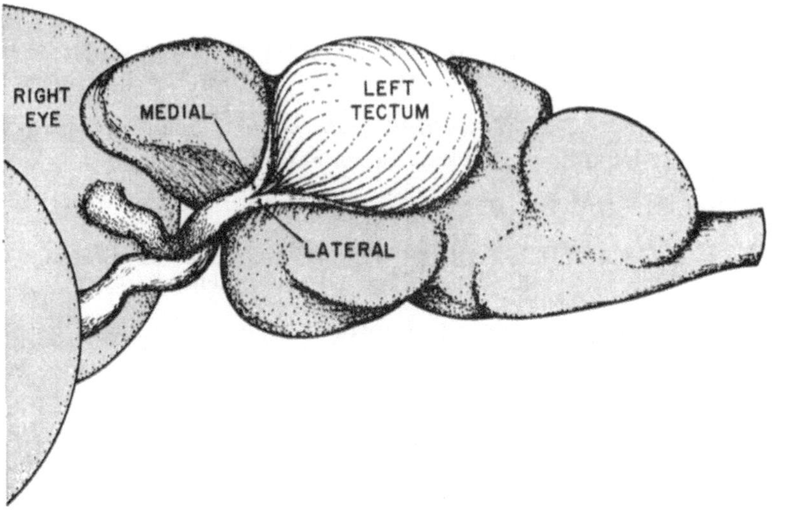

Fig. 1. Schematic drawing of the goldfish brain and optic system. About 500,000 fibers are bundled in the optic nerve. Each seems to make a precise and genetically controlled connection from a cell body in the retina to some point in the tectum. (After R. W. Sperry, 1965.)

fibers from the remaining areas followed defined routes, and made final conections only within specific regions of the tectum. The final pattern of connections made was such as to create an orderly topographic map of the retina within the tectum. The same relationships were maintained, even when one half of the nerve was excised. From these and other similar studies Sperry concluded that a highly sophisticated chemospecific recognition system was involved.

In addition we find it attractive to propose that the formation of this eye-brain system in the embryo occurs as a result of an accurately programmed commitment of cell lines to specific chemoreceptors (see below).

A further illustration of the high degree of order generated by these "chromosomal programming" and "surface display" mechanisms is seen in Figure 2, which is a photograph of a section taken through a deep layer of the tectum. This order seems exactly analogous to a complex three-dimensional printed circuit and could only result from an orderly molecular recognition process.

B. Surface Displays of Specific Molecules in Other Excitable Cells

Careful examination of time lapse photographs of regenerating nerve fibers in tissue culture has led to the conclusion that the growth cones or tactoid processes of these amoeboid cells are rather similar to those

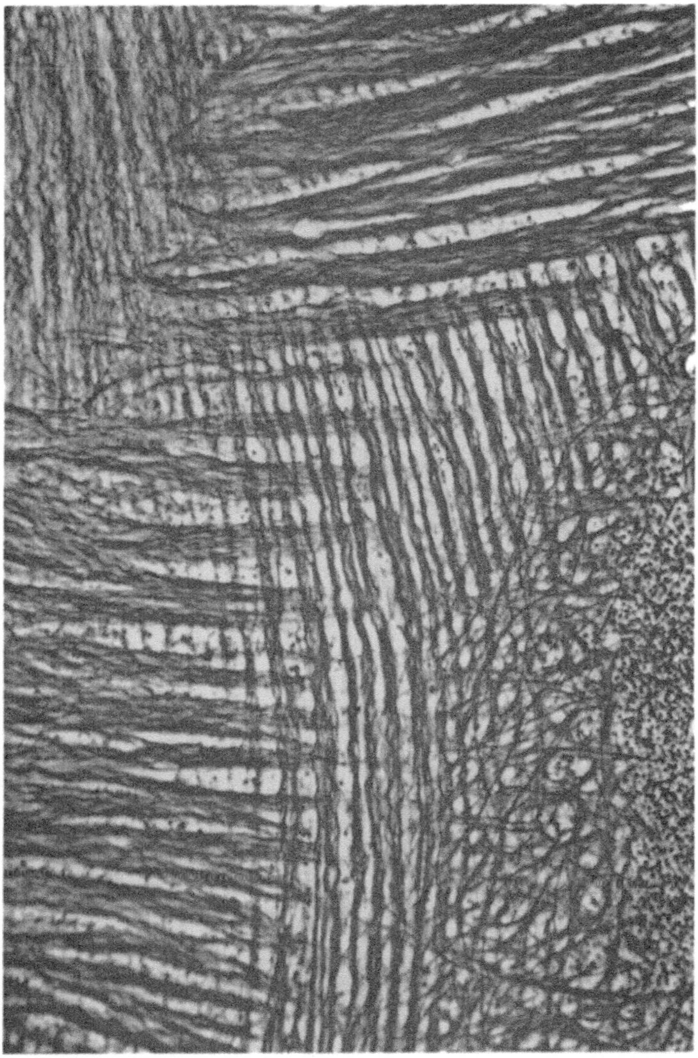

Fig. 2. Section through deep layer of optic tectum of goldfish. Note the precisely ordered arrays of three series of fibers, resembling a three-dimensional printed circuit. (Photograph by courtesy of Dr. E. Hibbard, California Institute of Technology.)

seen in many other types of motile cells, including lymphocytes. All of these amoeboid cells reveal sensing movements and specific chemotactic responses.

We propose that a similar and general chromosomal mechanism is utilized in various types of highly differentiated cells so as to commit them, essentially irreversibly, to the production of precoded specific protein molecules. These molecules then interact with others to form the sensing elements in excitable cell membranes.

Consider, for example, the heritable character of the sense of smell, whether in a moth or a hound. Males of both species have sensitive receptors for highly specific molecules released at great distances by the respective females. A slightly different molecule released by a closely related species of moth fails to trigger these *genetically controlled* receptor molecules.

One wonders whether migratory homing of such creatures as the Monarch Butterfly to the Monterey Peninsula might be genetically encoded in a similar way. Another example of inherited homing instincts concerns two closely related species of eels, one coming from New England, the other from northern Europe, and both spawning in the same locality in the Sargasso Sea. Fingerlings of the two species migrate separately to the home territories of the parents, even though they have never been there. It seems plausible to one of us (W. D.) that this behavior might be due in part to genetically determined chemoreceptors in the two species, recognizing different "chemoattractant" molecules in the sea. Many phenomena concerning instinctive behavior might be dependent to some extent on this mechanism.

An example of chromosomally programmed chemotaxis is seen in the amoeboid forms of primordial heart cells which must sense their individual paths as they migrate from the periphery by peripatetic routes and coalesce so as to form the primitive heart (Trinkaus, 1965).

Membranes of other highly differentiated cells must detect particular peptide hormones and react instantly in a variety of ways. Such is the case when capillary contraction is triggered instantly by the octapeptide hormone, angiotensin. Peptide hormones of similar size are released as neurosecretory products of very specific nerve cells. These neurohormones then react virtually immediately with their respective specific target cells in the pituitary so as to trigger the release of the desired pituitary hormone. These molecules, in turn, react with highly specific receptors on (or in) other cells throughout the body. Even the inflammatory response may well involve chemotactic movements wherein polypeptide hormones ("leucotaxin") and/or antigen molecules (see below) are sensed by cell surfaces.

A more concrete example of the concept of surface display of specific molecules is illustrated in Figure 3. This photograph is of an amoeboid blast cell in the erythroid series. There are specific receptor areas on the membrane of such cells which, at a precise stage in differentiation, recognize ferritin molecules. The protein coat which surrounds the iron in ferritin is, evidently, responsible for the binding of large numbers of these particles to specific surface patches. Micropinocytosis is triggered so as to draw the iron containing particles into the cell for use in heme synthesis. Transferrin (cf. London *et al.*, 1964) must also be recognized and taken up in order to permit the proper utilization of iron. It is a

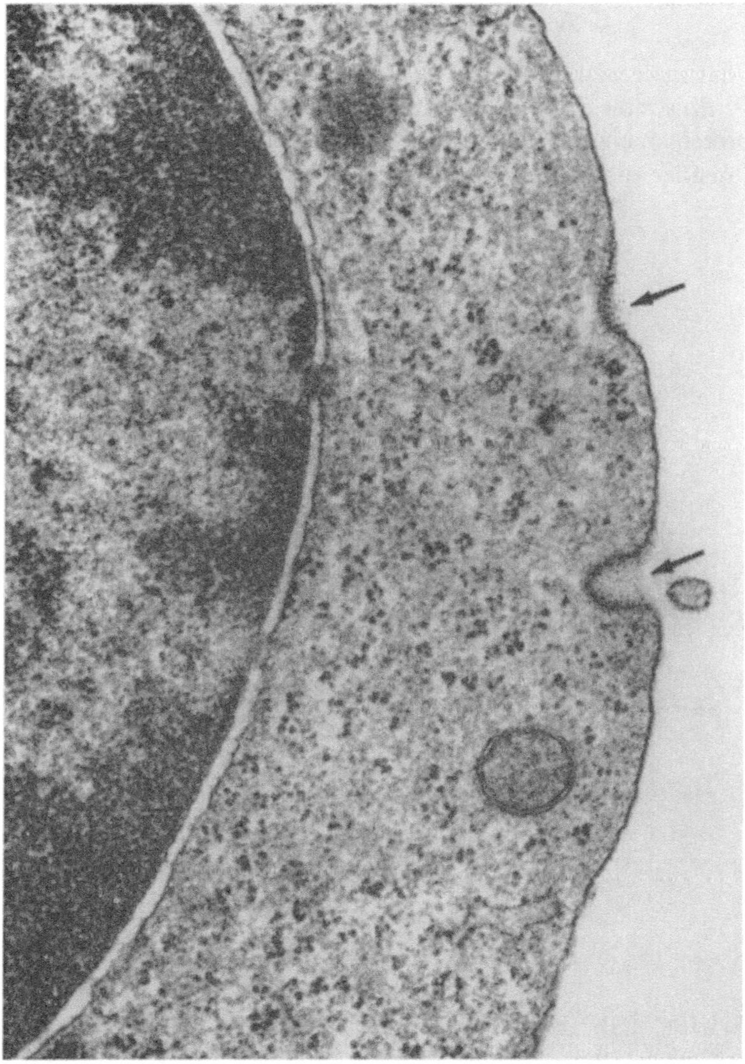

Fig. 3. Uptake of ferritin by erythroblast from guinea pig bone marrow. The adherence of ferritin only to the specialized patches of membrane (arrows) suggests that there is an array of specific receptor molecules ("surface display") able to find the ferritin and facilitate its uptake by micropinocytosis. (After Fawcett, 1965.)

fortunate circumstance that ferritin is electron opaque; thus this process can be visualized by electron microscopy. One can easily imagine an expanded version of this type of surface display such that one membrane might have a more elaborate array of specificities. Certain features of this concept have been proposed many times in the past (e.g., Sperry, 1963; Tyler, 1946, 1965), to explain various aspects of cell–cell interactions.

C. A Hypothetical Receptor Molecule

What molecular form might such chemoreceptor molecules take? Figure 4 illustrates a rather fanciful possibility. In this drawing two types of protein molecules interact so as to form a complex molecule. The two smaller subunits are assumed to result from the translation of per-

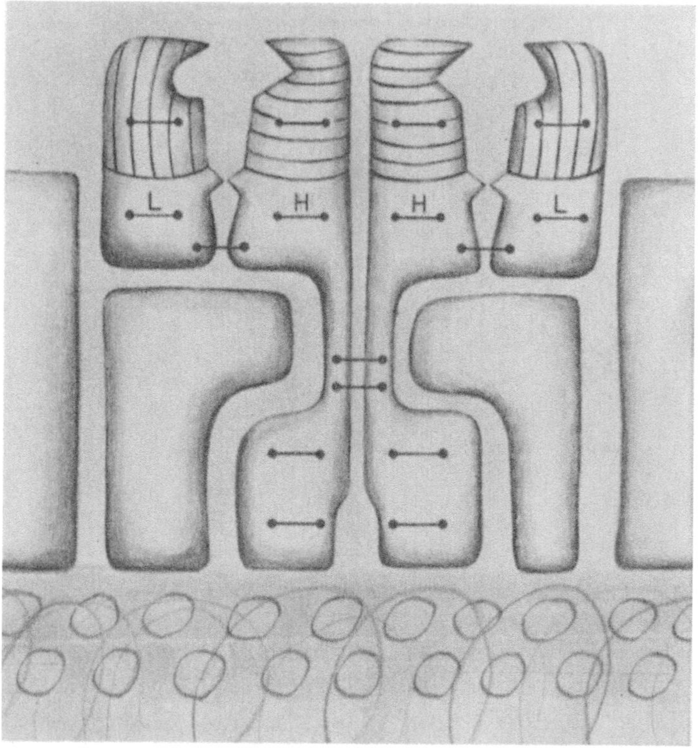

Fig. 4. Hypothetical receptor/transducer molecule. We envisage the molecule as being a large multi-chain protein, built into the surface membrane of a differentiated cell. The molecule is specific for sensing particular chemical stimuli in the environment—perhaps also a complementary structure on the surface of another cell. Only part of the molecule (cross-hatched areas protruding out from the membrane) confers this specificity, while the bulk of it occupies a well-defined site in the membrane. Upon interaction of the specificity end with its target molecule, a conformational change takes place, producing corresponding changes in properties of the membrane such as permeability, contractility, etc. These in turn lead to alterations of the cell's behavior such as the initiation of cell division, or establishment of permanent links with the target cell. The part of the molecule within the membrane would vary from one type of receptor to another, depending on the nature of the *response* desired for the stimulus; within each class of receptor, however, this would be common to all molecules. Thus *specificity* is determined by the part of the molecule which protrudes from the membrane, while the *response* is determined by the part which interacts with the membrane. The dots joined by lines represent disulfide crosslinkages.

haps two hundred triplets contained in the DNA of the cell which made this chain. The usual route of messenger RNA and ribosomal protein synthesis is presumed. Furthermore, it is assumed that the resulting sequence of about two hundred amino acids serves to direct the folding of the molecule to form a precise shape with a highly specific recognition site. This binding area is exactly analogous to the conformation of the portion of a lysozyme molecule which recognizes a specific sequence of polysaccharides. Lysozyme is chosen as an example since its crystal structure is known and its size is approximately that of one-half of the "light chain" molecule; that is, the region illustrated with cross-hatching and referred to as a "specificity region." The other half of our hypothetical light chain molecule serves to interact with a larger, but similar, chain of fifty thousand molecular weight (heavy chain). A different cross-hatching on the specificity region of a heavy chain is used to suggest that it recognizes entirely different properties of the molecule that is to be bound. A multi-chain structure of this type enables a much more versatile series of recognition sites to be made using a given number of genes. Concerted binding by both light and heavy chains might easily trigger a configurational change in the double molecule, provided that binding constants are adequate and other properties of the molecules to be bound (such as charge) are suitable. The common regions of the heavy chain molecules must, in addition to playing a role in configurational changes, serve to bind to a second heavy chain to form the 150,000 M.W. structure illustrated. The common regions must also interact in a variety of highly specific ways with further protein and lipoprotein molecules so as to form a complicated "building block" structure which can easily move into position in the surface membrane of a cell.

Appropriate triggering of the specific configurational change within the complex membrane structure might, for example, lead to altered ion fluxes and potentials in the membrane. Such a primary event might be imagined to trigger virtually all of the chemotactic and sensory phenomena which have been alluded to thus far.

D. Genetic and Chromosomal Controls of Receptor Synthesis

It is clear that specificity of the type referred to would of necessity require an orderly and exactly timed chromosomal programming process which commits a given cell line to a particular specificity; this specificity should be a stable, heritable property of that cell line. What sort of molecular events might occur at the level of DNA so as to permit the virtually irreversible commitment of a nerve cell or a heart cell to a particular specificity? A hypothetical mechanism is illustrated in Figure 5. This figure is intended to convey the impression that precise and orderly recombinational events can occur during differentiation in such a way as

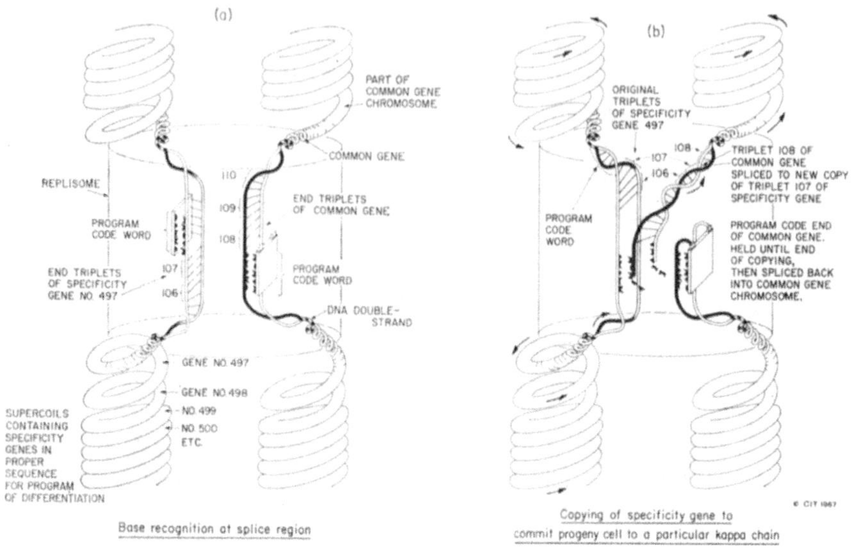

Fig. 5. Proposed "copy-splice" mechanism for production of membrane-bound receptor molecules. The process is depicted as a conservative copying, involving two different stretches of double-stranded DNA. The specificity strand (left in a and b) contains many individual specificity genes separated by "program code" sequences of DNA; the other strand (right in a and b) contains the unique common region gene, plus its particular program code DNA.

(a) Base recognition at splice region. The replisome is shown as containing two adapters containing anticodons which are able to recognize and pair with the program code words on the DNA strands, thus bringing the specificity and common genes into the correct alignment. These adapters might be considered somewhat analogous to ribosomal transfer RNA molecules. The specificity gene adapter is positioned at the start of a particular gene (No. 497), after having been involved in the copy-splicing of gene No. 496 in the previous round of cell division. It thus acts as a program reader, enabling the various specificity genes to be expressed in an orderly sequence. The recognition process is shown as being exclusively between genes and adapters; this is certainly not obligatory, and we do not wish to imply that there cannot be direct pairing between common and specificity strands in the splice region or elsewhere.

(b) Copying of specificity gene to commit progeny cell to a particular receptor. The strand containing the common gene has been severed immediately after triplet 108, and the distal ends have been temporarily attached to the replisome. For clarity, we have shown the remainder of this strand as being stationary during the copying process—as would be probable if the replisome were membrane-bound. The free ends of the common gene then act as growing points for the addition of nucleotides, and a copy of the specificity gene is built up onto it, working back towards the first triplet. Some of the free energy of reaction is used to drive the DNA strands through the replisome. As the strands emerge they take up the characteristic double-helix structure, producing coils and supercoils and turning the DNA in the direction shown. When the copy is complete, the ends of the common gene strand are joined ("spliced"), remaking a continuous stretch of chromosomal DNA. The net result of these processes is that the common gene chromosome contains a stretch of DNA coding for a specific, intact light sub-unit, and the program for receptor production has progressed one more step.

Though not illustrated, an asymmetric cell division is assumed. (Taken from Dreyer *et al.*, 1967.)

to incorporate a *copy* of information contained in a particular specificity region gene into a common region gene found in an entirely different location, perhaps in another chromosome. The net result would be a new stretch of coded DNA which is thereafter a stable, heritable sequence used to code for a particular light chain protein molecule in that cell line. The next cell (produced in an asymmetric cell division) might be committed to a different specificity by utilizing the succeeding stretch of chromosomal DNA in combination with the common region. Such a process might well utilize very large stretches of DNA so as to introduce a great variety of specificity regions onto the same common region gene as new differentiated cells are committed. This process would be rather analogous to manufacturing tumblers in locks in a systematic way. Just as the casing in a given style of lock remains constant and thus can always fit into a standard door, the common region of these molecules would remain constant (within a given class of molecules) so as to be inserted into a standard membrane subunit building block.

It should be noted that the critical and admittedly very unorthodox assumption made in this hypothesis is that *most highly differentiated somatic cell lines are generated, in part, by essentially irreversible alterations in chromosomal DNA. That is, coded information contained in two distinct germ line genes is combined to form a new sequence coding for one polypeptide chain.*

II. AN EXPERIMENTAL SYSTEM

To a chemical biologist the problem of isolating molecules of the type proposed from one particular type of committed nerve cell represents a most formidable task. If, however, the basic mechanism is a general one, we might look toward recently evolved organisms such as vertebrates for highly specialized variants of the chromosomal programming process. With luck a suitable experimental system might be found.

A. Cells of the Immune System

One highly specialized type of cellular differentiation is, of course, the process which leads to the immune response and antibody formation. In the immune system, some cells do in fact behave as though they contain chemoreceptors in their surface membranes. This certainly seems to be the case with thymus derived cells, active in the graft versus host response and in delayed hypersensitivity (see, for example, Cooper *et al.*, 1967). Other cell lines which appear to be derived from the mammalian equivalent of the bursa are capable of differentiating from very small lymphocytes (which are non-secreting, long-lived memory cells) into large secretory cells producing the circulating immunoglobulin molecules. One can obtain gram quantities of "gamma globulin" thus produced. However, this too remains a nightmare to the chemical biologist since this protein fraction contains thousands of different amino acid sequences.

B. Transplantable Plasma Cell Tumors in Inbred Strains of Mice

The plasma cell neoplasms of mice produce a variety of proteins related to the various classes of immunoglobulin molecules and their polypeptide chain subunits (see, for example, Potter *et al.*, 1964.) Any given plasma cell tumor can be propagated indefinitely by transplantation within the inbred line of mice in which it originated. Each tumor is committed to a specific program for protein secretion. Some tumors secrete a homogeneous IgG molecule while others may secrete an IgA molecule. It is also common to find tumor lines which secrete only light chain subunits from immunoglobulins (kappa chains or lambda chains). These are small enough to pass the kidney barrier and are therefore secreted in the urine. Gram quantities of these urinary "Bence Jones" proteins can be isolated from mice bearing specific tumor lines.

Figure 6 illustrates the various electrophoretic mobilities observed when the products of ten different tumor lines were compared. Each of these kappa chain protein molecules represents a stable, heritable character of the somatic cell line which secreted the protein in question.

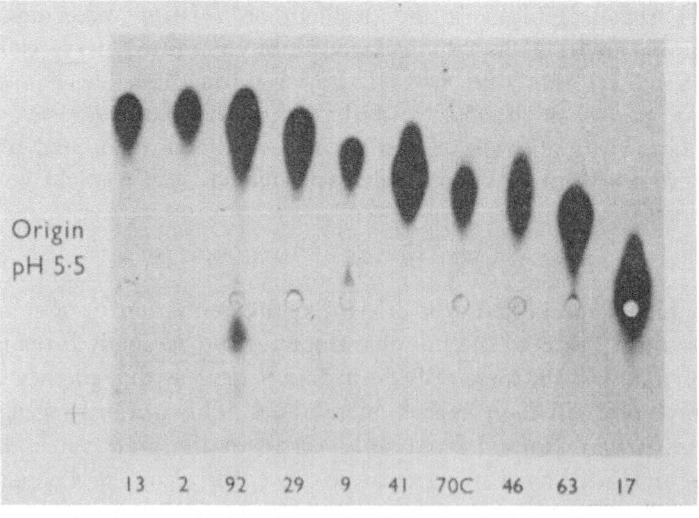

Fig. 6. Electrophoretic comparison of ten kappa chain type Bence-Jones proteins. Each of these proteins was secreted by a separate plasma cell tumor line, which had been serially transplanted in BALB/C mice. (After Potter *et al.*, 1964.)

The heritable nature of the commitment of these specific cell lines was shown many years ago when it was found that protein products, which had been collected from very early transplant generations, were identical with those produced after large numbers of successive transplantation and growth cycles in separate mice (Potter *et al.*, 1964). In other words, *the chromosomal events which have led to the commitment of these plasma cells to production of a specific protein product are, for practical purposes, irreversible.* Furthermore, there is no known case of a commitment of one of these tumors to produce more than a single light chain molecule with its particular precisely defined amino acid sequence. Structural studies on proteins have yet to detect evidence for identical kappa chain sequences in tumors of independent origin. This fact suggests that perhaps thousands of different distinct amino acid sequences may be found in kappa chains, even in inbred mice.

Similar studies on the stability of somatic cell lines have been carried out recently on plasma cell tumors derived from humans. In this species results are obtained by tissue culture techniques rather than by transplantation of tumor lines.

C. Structural Studies of Kappa and Lambda Chains from Mice and Men

The extraordinary structural variability observed in these molecules is necessitated by the requirement for conformational groupings appropriate for interaction with a multitude of antigenic substances. The molecular basis for this specificity derives from the amino acid sequence of the particular polypeptide chain. Knowledge of the exact amino acid sequence can be translated into knowledge of base sequence in the chromosomal DNA contained in the differentiated cell lines which coded for the protein chains. Thus, a study of the nature of amino acid variability in the specificity region of light chains provides an experimentally feasible approach to the question of genetic mechanisms responsible for antibody diversity. One can determine the amino acid sequences of a number of light chains and compare them with one another to look for meaningful patterns of variation which might reflect the underlying mechanism responsible for antibody diversity (see Dreyer and Bennett, 1965).

Two general approaches to such a study were taken. On the one hand we determined the entire amino acid sequence of three kappa type light chains; two were derived from mouse myeloma lines and one was obtained from a human patient (Gray *et al.*, 1967a, b; Dreyer *et al.*, 1967). On the other hand we determined amino-terminal sequences for fifteen different light chains (both lambda and kappa classes) in order to obtain a more complete picture of the nature of the variability of the specificity region (Hood *et al.*, 1967). The experimental procedures utilized in these studies have been described elsewhere and will not be discussed here. Amino terminal sequences from twenty different light chains are given in Figure 7, which includes results from several laboratories.

Figure 8 shows the amino acid sequences of four immunoglobulin kappa chains. Two were obtained from separate lines of transplantable cell tumors of mice. The other two sequences are of human kappa chains obtained from separate individuals suffering from multiple myeloma. Note that the sequence of one protein (Ag) was determined by Titani *et al.* (1966) and the other three by Gray *et al.* (1967a, b).

The chains illustrated in Figures 7 and 8 all relate to each other in a manner which resembles *in all respects* sets of proteins which bear an evolutionary relationship to one another. Just as when hemoglobin chains are compared with one another, many of the changes which occur in amino acids found at particular positions are very conservative in terms of amino acid chemistry. Radical changes are a minority, but positions are known which can accept one of several different amino acids. There is certainly no evidence to suggest that the changes are due to any particular type of hypermutational mechanism, and no need to postulate any unusual amount of crossing over between the genes involved. It is, of course, assumed that gene duplication and recombination do occur on a continuing basis, as a major part of the organism's dynamic response to a changing environment.

AMINO TERMINAL LIGHT CHAIN SEQUENCES

		5 · 10 · 15 · 20 · 25
SL	HBJ 11*	Glp-Ser-Val-Leu-Thr-Gln-Pro-Pro-Ser-[ ]-Ala-Ser-Gly-Thr-Pro-Gly-Gln (Arg)
	HBJ 8*	Glp-Ser-Ala-Leu-Ala-Gln-Pro-Ala-Ser-[ ]-Val-Ser-Gly-Ser-Pro-Gly-Gln-Ser-Ile-Thr
	HBJ 15*	Glp-Ser-Ala-Leu-Thr-Gln-Pro-Ala-Ser-[ ]-Val-Ser-Gly-Ser-Pro-Gly-Gln-Thr-Ile-Thr
	HBJ 2*	Glp-Ser-Ala-Leu-Thr-Gln-Pro-Pro-Ser-[ ]-Ala-Ser-Gly-Ser-Pro-Gly-Gln-Ser-Val-Thr-Ile-Ser-Cys-Thr-Gly-Thr
	HBJ 7*	Glp-Ser-Val-Leu-Thr-Gln-Pro-Pro-Ser-[ ]-Ala-Ser-Gly-Thr-Pro-Gly-Gln-Gly-Val-Thr-Ile-Ser-Cys-Ser-Gly-Ser
	Sh‡	[ ]-Ser-Glu-Leu-Thr-Gln-Asp-Pro-Ala-[ ]-Val-Ser-Val-Ala-Leu-Gly-Gln-Thr-Val-Arg-Ile-Thr-Cys-Gln (Gly-Asp
		5 · 10 · 15 · 20 · 25
SK_I	MBJ 41*	Asp-Ile-Gln-Met-Thr-Gln-Ser-Pro-Ser-Ser-Leu-Ser-Ala-Ser-Leu-Gly-Glu-Arg-Val-Ser-Leu-Thr-Cys-Arg-Ala-Ser
	Ag‡	Asp-Ile-Gln-Met-Thr-Gln-Pro-Ser-Ser-Ser-Leu-Ser-Ala-Ser-Val-Gly-Asp-Arg-Val-Thr-Ile-Thr-Cys-Gln-Ala-Ser
	Ker†	Asx (Ile, Glx) Met (Thr, Gln, Ser, Pro, Ser, Ser, Leu) Ser-Ala-Ser-Val-Gly-Asp-Arg-Ile-Thr-Ile-Thr-Cys-Gln-Ala-Ser
	BJ†	Asx (Val, Glx) Met-Thr-Gln (Ser, Pro, Ser, Ser, Leu)(Ser, Ala, Ser, Val, Gly, Asp) Arg-Val-Thr-Ile-Thr-Cys-Gln-Ala-Ser
	Roy**	Asx (Ile, Glx, Met, Thr, Glx, Ser, Pro, Ser, Ser) Leu-Ser (Ala, Ser, Val, Gly) Asx-Arg (Ile, Thr, Ile, Thr, Cys, Glx, Ala, Ser,
	HBJ 10*	Asp-Ile-Gln-Met-Thr-Gln-Ser-Pro-Ser-Ser-Leu-Ser-Ala-Ser (Val, Gly, Asx)(Arg)
	HBJ 1*	Asp-Ile-Leu-Met-Thr-Gln-Ser-Pro-Thr-Ser-Leu-Ser-Ala-Ser-Val-Gly-Asp (Arg)
	HBJ 4*	Asp-Ile-Gln-Met-Thr-Gln-Ser-Pro (Ser, Ser) Leu
		5 · 10 · 15 · 20 · 25
SK_II	MBJ 70*	Asp-Ile-Val-Leu-Thr-Gln-Ser-Pro-Ala-Ser-Leu-Ala-Val-Ser-Leu-Gly-Gln-Arg-Ala-Thr-Ile-Ser-Cys-Arg-Ala-Ser
	HBJ 3*	Asp-Ile-Val-Leu-Thr-Gln-Ser-Pro-Leu-Ser-Leu-Pro-Val-Thr-Pro-Gly-Glu-Pro-Ala-Ser-Ile-Ser-Cys-Arg-Ser-Ser
	HBJ 5*	Glu-Ile-Val-Leu-Thr-Gln-Ser-Pro-Asx-Thr-Leu-Ser-Leu-Ser-Pro-Gly-Glx (Arg)
	HS 4*	Glu-Ile-Val-Leu-Thr-Gln-Ser-Pro-Gly-Thr-Leu-Ser-Leu-Ser-Pro-Gly-Glu (Arg)
	MBJ 6*	Asp-Ile-Val-Val-Thr-Gln
	HBJ 12*	Glu-Ile-Val-Val-Thr-Gln

Fig. 7. Amino terminal sequences of myeloma light chains. Data have been taken from: * Hood *et al.* (1967), Gray *et al.* (1967a & b), ‡ Wikler *et al.* (1967), Titani *et al.* (1966); † Milstein (1966); ** Hilschmann and Craig (1965). A much more complete table of data is given by Cohn (1968, this volume).

COMPARISON OF MOUSE AND HUMAN K-CHAINS

```
            1         5         10         15         20
Mouse 41    Asp-Ilu-Gln-Met-Thr-Gln-Ser-Pro-Ser-Ser-Leu-Ser-Ala-Ser-Leu-Gly-Glu-Arg-Val-Ser-Leu-Thr-Cys-Arg-
Mouse 70    Asp-Ilu-Val-Leu-Thr-Gln-Ser-Pro-Ala-Ser-Leu-Ala-Val-Ser-Leu-Gly-Gln-Arg/Ala-Thr-Ilu-Ser-Cys-Arg/
Human Ag    Asp-Ilu-Gln-Met-Thr-Gln-Ser-Pro-Ser-Ser-Leu-Ser-Ala-Ser-Val-Gly-Asp-Arg-Val-Thr-Ilu-Thr-Cys-Gln-
Human Mil   Asp-Ilu-Val-Leu-Thr-Gln-Ser-Pro-Leu-Ser-Leu-Pro-Val-Thr-Pro-Gly-Glu-Pro-Ala-Ser-Ilu-Ser-Cys-Arg/

25  30  a b c d  31  35  40  45
Ala-Ser-Gln (Asx, Ilu, Gly, a-b-c-d)Ser-Leu-Ser-Asx-Trp-Leu-Glx-Glx (Gly, Pro, Asx, Thr, Glx) Ilu-Lys-Arg-Leu-Ilu-Tyr-
Ala-Ser-Glu-Ser-Val-Asx-Asx-Ser-Gly-Ilu-Ser-Phe-Met-Asn(Trp, Phe, Glx) Glx-Lys-Pro-Gly-Glx-Pro-Pro-Lys/Leu-Leu-Ilu-Tyr-
Ala-Ser-Gln (Asx, Ilu, Asx, ———————— Ser, Phe) Leu-Asn-Trp-Tyr-Gln-Gln-Gly-Pro-Lys-Lys-Ala-Pro-Lys-Ilu-Leu-Ilu-Tyr-
Ser-Ser-Gln-Asn-Leu-Leu/Glx-Ser-Asx-Gly (Asx) Tyr-Leu-Asp-Trp-Tyr/Leu-Glx-Lys-Pro-Gly-Glx-Ser-Pro-Glx-Leu-Leu-Ilu-Tyr/

50  55  60  65  70  75
Ala-Thr-Ser-Ser-Leu-Asx-Ser-Gly-Val-Pro-Lys-Arg-Phe-Ser-Gly-Ser-Arg-Ser-Gly-Ser-Asp/Tyr-Ser-Leu-Thr/Ilu-Ser-Ser-Leu/
Ala-Ala-Ser-Asn-Gln-Gly-Ser-Gly-Val-Pro-Ala-Arg/Phe-Ser-Gly-Ser-Gly-Ser-Gly-Thr-Asp-Phe-Ser-Leu-Asn-Ilu-His-Pro-Met-
Asp-Ala-Ser-Asn-Leu-Glu-Thr-Gly-Val-Pro-Ser-Arg-Phe-Ser-Gly-Ser-Gly-Phe-Gly-Thr-Asp-Phe-Thr-Phe-Thr-Ilu-Ser-Gly-Leu-
Leu-Gly-Ser-Asn-Arg/Ala-Ser-Gly-Val-Pro-Asn-Arg/Phe-Ser-Gly-Ser-Gly-Ser-Gly-Thr-Asx-Phe-Thr-Leu-Lys/Ilu-Ser-Arg/Val-

80  85  90  95  100  105  107
Glu-Ser-Glu-Asp-Phe-Val-Asp-Tyr/ ? Cys-Leu-Gln-Tyr-Ala-Ser-Ser-Pro-Trp-Thr-Phe-Gly-Gly-Gly-Thr-Lys-Leu-Glu-Ilu-Lys-
Glx-Glx-Asx-Asx-Thr-Ala-Met-Tyr-Phe-Cys-Glx-Glx-Ser-Lys/Glu-Val-Pro-Trp-Thr-Phe-Gly-Gly-Gly-Thr-Lys-Leu-Glu-Ilu-Lys-
Gln-Pro-Glu-Asp-Ilu-Ala-Thr-Tyr-Tyr-Cys-Gln-Gln-Tyr-Asp-Thr-Leu-Pro-Arg-Thr-Phe-Gly-Gln-Gly-Thr-Lys-Leu-Glu-Ilu-Lys-
Glx-Ala-Glx-Asx-Val-Gly-Val-Tyr-Tyr-Cys/Met-Gln-Ala-Leu-Gln-Thr-Pro-Leu-Thr-Phe-Gly-Gly-Gly-Thr-Asn-Val-Glu-Ilu-Lys-

108  110  115  120  125  130  135
Arg-Ala-Asp-Ala-Ala-Pro-Thr-Val-Ser (Ilu, Phe, Pro, Pro, Ser, Ser) Glu-Gln-Leu-Thr-Gly (Gly, Ser, Ala, Ser) Val-Val-Cys-Phe-Leu-
Arg (Ala, Asx, Ala, Ala, Pro, Thr, Val, Ser, Ilu, Phe, Pro, Pro, Ser, Ser, Glx, Glx, Leu, Thr, Gly, Gly, Ser, Ala, Ser, Val, Val) Cys-Phe(Leu,
Arg-Thr-Val-Ala-Ala-Pro-Ser-Val-Phe-Ilu-Phe-Pro-Pro-Ser-Asn-Glu-Gln-Leu-Lys-Ser-Gly-Thr-Ala-Ser-Val-Val-Cys-Leu-Leu-
Arg-Thr (Val, Ala, Ala, Pro, Ser, Val, Phe, Ilu, Phe, Pro, Pro, Ser, Asx, Glx, Glx, Leu) Lys/Ser (Gly, Thr, Ala, Ser, Val, Val) Cys/Leu-Leu-

140  145  150  155  160  165
Asn-Asn-Phe-Tyr-Pro-Lys-Asp-Ilu-Asn-Val-Lys-Trp-Lys-Ilu-Asp-Gly-Ser-Glu-Arg-Gln-Asn-Gly-Val-Leu-Asx-Ser-Trp-Thr-Asx-
Asx, Asx, Phe, Tyr, Pro) Lys/Asp-Ilu-Asn-Val-Lys/Trp-Lys-Ilu (Asp, Gly, Ser, Glu) Arg (Glx, Asx, Gly, Val, Leu, Asx, Ser, Trp, Thr, Asx,
Asn-Asn-Phe-Tyr-Pro-Arg-Glu-Ala-Lys-Val-Gln-Trp-Lys-Val-Asp-Asn-Ala-Leu-Gln-Ser-Gly-Asn-Ser-Gln-Glu-Ser-Val-Thr-Glu-
Asn-Asn-Phe-Tyr-Pro-Arg/Glu-Ala-Lys/Val-Gln-Trp-Lys/Val (Asx, Asx, Ala, Leu, Glx, Ser, Gly, Asx, Ser, Glx, Glx, Ser, Val, Thr, Glx,

170  175  180  185  190
Glx-Asp-Ser-Lys-Asp-Ser-Thr-Tyr-Ser-Met-Ser-Ser-Thr-Leu-Thr-Leu-Thr-Lys-Asp-Glu-Tyr-Glu-Arg-His-Asx-Ser-Tyr-Thr-Cys-
Glx, Asp, Ser) Lys/Asp (Ser, Thr, Tyr, Ser, Met, Ser, Ser, Thr, Leu, Thr, Leu, Thr) Lys/Asx (Glx, Tyr, Glx) Arg/His (Asx, Ser) Tyr-Thr-Cys-
Gln-Asp-Ser-Lys-Asp-Ser-Thr-Tyr-Ser-Leu-Ser-Ser-Thr-Leu-Thr-Leu-Ser-Lys-Ala-Asp-Tyr-Glu-Lys-His-Lys-Val-Tyr-Ala-Cys-
Glx, Asx, Ser) Lys/Asp (Ser, Thr, Tyr, Ser, Leu, Ser, Ser, Thr, Leu, Thr, Leu, Ser) Lys/Ala (Asx, Tyr, Glx) Lys/His-Lys/Val (Tyr, Ala) Cys/

195  200  205  210  214
Glx-Ala-Thr-His-Lys-Thr-Ser-Thr-Ser-Pro-Ilu-Val-Lys-Ser-Phe-Asn-Arg-Asn-Glu-Cys
Glx (Ala, Thr, His) Lys/Thr (Ser, Thr, Ser, Pro, Ilu, Val) Lys/Ser-Phe-Asn-Arg-Asn-Glu-Cys
Glu-Val-Thr-His-Gln-Gly-Leu-Ser-Ser-Pro-Val-Thr-Lys-Ser-Phe-Asn-Arg-Gly-Glu-Cys
Glx (Val, Thr, His, Glx, Gly, Leu, Ser, Ser, Pro, Val, Thr) Lys/Ser (Phe, Asn) Arg-Gly-Glu-Cys
```

Fig. 8. Amino acid sequences of kappa light chains from mice and humans. / indicates that overlaps between primary peptides have not been established independently. / indicates that arrangement of secondary fragments of a tryptic peptide (residues 25–51 of Mil) is not complete. Three proteins were investigated in our laboratory (mouse 41, mouse 70, and human Mil: Gray, Dreyer, and Hood, 1967a, b); the fourth was investigated by Putnam *et al.* (1966).

One remarkable fact that is immediately apparent on examination of Figure 8 is the presence of additional residues in both the mouse protein 70 and the human protein Mil. The number of additional residues in Mil is not yet established with certainty, but it may be the same as in MBJ 70, and the addition is in the same position in the molecule. This, perhaps more than any other single fact derived from the sequence data, is exceedingly difficult to reconcile with any of the many proposed hypermutability mechanisms (see for example, Lennox and Cohn, 1967). These size differences can be used, along with many other criteria (such as coordinated changes of amino acids at positions 3, 4, etc.) to divide the kappa chains into subclasses. Further division into sub-subclasses is becoming apparent as more data are acquired. In this regard the reader should examine the data of Hilschmann (1967) which have appeared since these figures were prepared.

It is particularly important to notice that the division into two major subclasses of kappa chains is so clear that it is apparent even after the tens of millions of years which have passed since the evolutionary divergence of the lines which gave rise to mice and men. These and many other considerations have led us to conclude that *one must assume that different germ-line genes code for different molecules.*

(1) *Observations on the Sequences Within the Common Regions*

In marked contrast to the results just discussed for the specificity regions, all BALB/c kappa chains studied to date by means of peptide map techniques (Potter *et al.,* 1963, 1965; Bennett *et al.,* 1965; Hood *et al.,* 1966a, b) may contain identical sequences from residues 108 to 214 (the common region). Many human proteins studied also contain a common region through positions 108 to 214 (Hilschmann and Craig, 1965; Titani *et al.,* 1965; Hilschmann, 1967). The human common region sequence is homologous with but not identical to the mouse common region sequence.

In the human population there is a specific variation which has been detected within the common region of kappa chains. This single amino acid substitution occurs at position 191 where either valine or leucine may be found depending on the individual. The genetics of this Inv "marker" have been studied (Ropartz *et al.,* 1962; Ceppellini, 1967), and it is clear that the common region sequence is coded by a single gene which follows ordinary Mendelian rules of segregation. This is in strong contrast to what has just been concluded about the specificity regions of these same proteins, which are coded by multiple genes.

In summary, the experimental evidence available at this time permits us to conclude with a high level of confidence that:

1. *There is no known single mutational or recombinational mechanism which can generate the observed specificity region sequences from a single germ line gene.*

2. *The pattern of amino acid variation is similar in all respects to that found in sets of evolutionarily related proteins.*

3. *The common regions of human kappa chains are encoded by a single germ-line gene which segregates as expected for diploid organisms.*

4. *The two ends of such molecules violate the classical one-gene one-polypeptide chain rule since the above data force us to conclude that the two ends of one molecule are under the control of separate genes.*

5. *The somatic cell lines which have been transplanted for so many years in the BALB/c mice appear to contain separate and independent genes coding for the entire sequence of 214 residues of the kappa chains produced by any given cell line.*

D. Evolutionary Origins of the Basic Chromosomal Mechanisms

The experimental facts which have been described above make it abundantly clear that the immune response of mammals represents an extremely complex phenomenon when viewed at the genetic, molecular and cellular levels. One assumes, quite naturally, that the genes and genetic mechanisms required to manufacture the sophisticated molecules found in the immune system required an extended period of time for evolution.

The phylogeny of the immune response has been studied extensively (see Good and Papermaster, 1964, for review), and it appears that the ability to synthesize circulating antibody is a relatively recent development in terms of geological time. There is a sharp division between those species which can mobilize an immune response and those which cannot. The thymus gland, circulating lymphocytes, and adaptive immunity appear to be restricted to the vertebrates. Nevertheless, when the most primitive of vertebrates are studied the molecules found appear fully as complex as those found in the mammals. Thus, the sharks have both heavy chains and light chains; the latter are of both the kappa and lambda types and show a high degree of heterogeneity (Suran and Papermaster, 1967; Hood *et al.,* 1967).

The evolutionary process is a slow one and it is not normal to find cataclysmic events leading to entirely new molecules or structures without precedent in simpler forms. It is, therefore, pertinent to ask what type of molecules might have been the ancestral precursors of antibodies, and what function they might have served in their ancestral cellular environment. Our knowledge of the immune response (which will be discussed later) leads us to feel quite certain that antibodies are found as components of surface membranes in small lymphocytes wherein they can act as chemoreceptors triggering a given cell to undergo developmental changes upon interaction with an antigen.

It is the purpose of the following discussion to elaborate the hypothesis that the immune system represents a late evolutionary off-shoot of a more basic type of membrane-bound chemo-receptor system involved as an essential part of the differentiation of multicellular organisms.

(1) *Protein Evolution*

An understanding of some of the ground rules of protein evolution is essential in order to evaluate intelligently the mass of experimental facts that are now available. For this reason the following passage will attempt to convey some of these general principles.

Because of the direct relationship between protein structure and gene structure we are now able to take a genetic code dictionary in hand and translate relatively freely from one to the other and vice versa. Each protein molecule may be considered to be a cluster of several hundred phenotypic characters (amino acid residues). Thus the genetic information derived from sequence studies is enormous and can be expressed in quantitative form. We can determine the degree of relatedness between any two proteins, and can construct a phylogenetic tree to show evolutionary relationships between proteins.

When new proteins evolve they often appear as modifications of existing molecules, usually after gene duplication has given rise to redundant copies of the relevant genetic information. Very often the divergent proteins retain considerable identity of sequence, and major structural and binding features as well. In this regard we may cite the retention of tertiary structure, heme- and oxygen-binding capacity of the various globins (see Eck and Dayhoff, 1967, for references); the extensive sequence homology and catalytic similarity of pancreatic enzymes (Walsh and Neurath, 1964; Hartley *et al.,* 1967); and the close relationship between lysozyme (an enzyme that hydrolyzes glycosidic bonds) and lactose synthetase (an enzyme that synthesizes glycosidic bonds) (Brew *et al.,* 1967). It is for this reason that we expect proteins with recently evolved functions to have been derived from proteins involved in older and more basic cellular processes.

The evolution of proteins occurs mainly by accumulation of random point mutations (substitution of one DNA base by another) in the structural genes, followed by competitive selection of those populations of organisms whose proteins enable them to reproduce most effectively in their particular environment. At the molecular level this hinges upon the ability of the protein to carry out its function most effectively within its own micro-environment. In some cases, where there are distinctive micro-environments within one organism, two forms of a protein may exist, each of these having been particularly selected for functioning in one environment. An example of such a situation is seen in the red blood cells of developing mammals, where a programmed development occurs so that cells making fetal hemoglobin become replaced by cells making the adult form. Other examples are seen in the various isozymes, and also in the classes and subclasses of immunoglobulins.

(2) *The Polypeptide Chain of Antibody Molecules: An Example of Typical Protein Evolution*

Proteins act as extremely complex mechanical, chemical, or electronic components of sophisticated machines. For this reason their function is intimately dependent on a three dimensional configuration of the peptide chain, which is in turn determined by the pattern of interaction of the twenty different amino acid side chains. The rules governing these

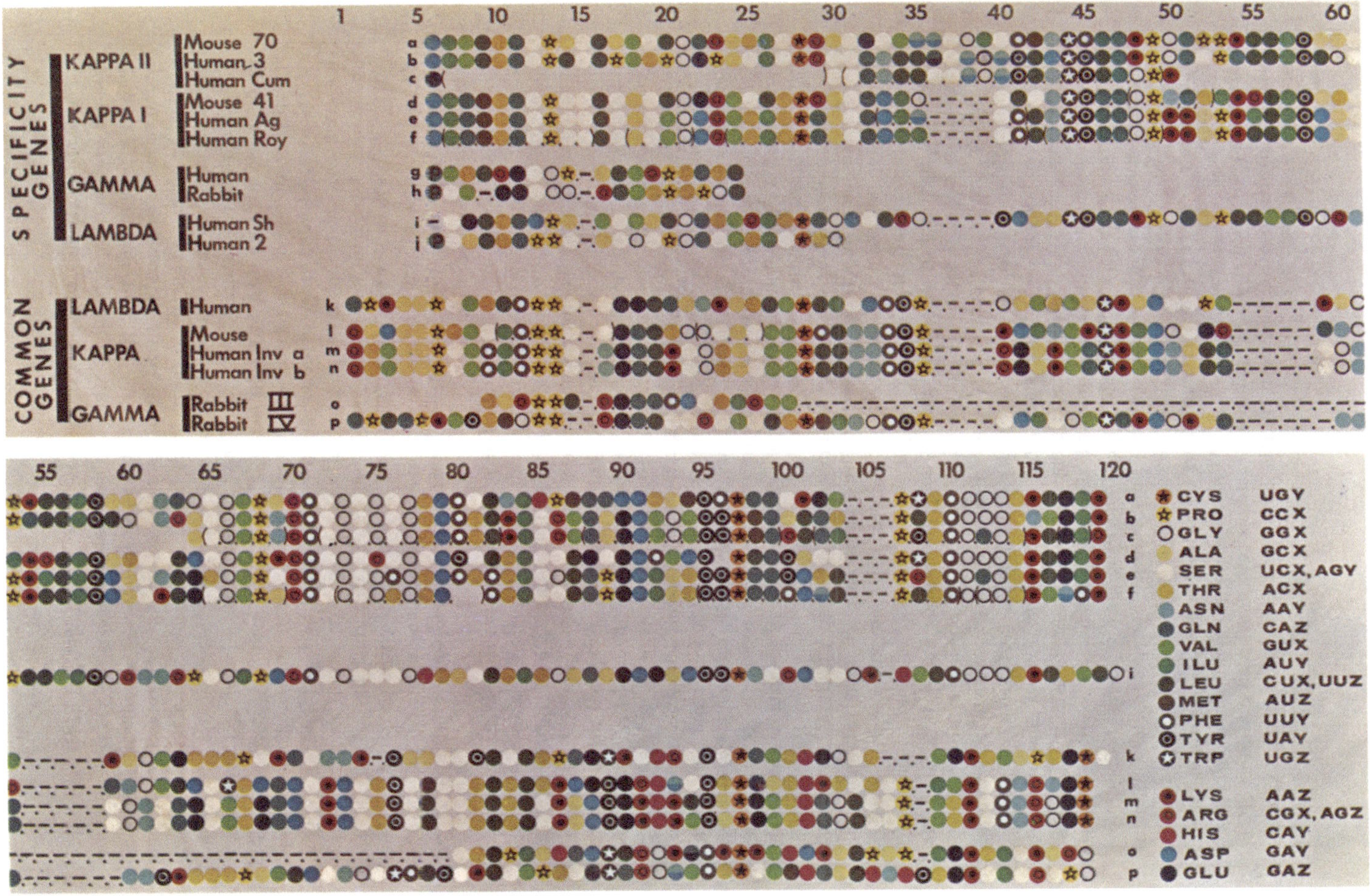
SPECIFICITY GENES
KAPPA II
Mouse 70
Human 3
Human Cum
KAPPA I
Mouse 41
Human Ag
Human Roy
GAMMA
Human
Rabbit
LAMBDA
Human Sh
Human 2
COMMON GENES
LAMBDA
Human
KAPPA
Mouse
Human Inv a
Human Inv b
GAMMA
Rabbit III
Rabbit IV
CYS UGY
PRO CCX
GLY GGX
ALA GCX
SER UCX, AGY
THR ACX
ASN AAY
GLN CAZ
VAL GUX
ILU AUY
LEU CUX, UUZ
MET AUZ
PHE UUY
TYR UAY
TRP UGZ
LYS AAZ
ARG CGX, AGZ
HIS CAY
ASP GAY
GLU GAZ

Fig. 9. Relationships between various regions of immunoglobulin chains. *Symbols:* Colored spots may be decoded as either triplets in a nucleic acid sequence or as amino acids in a protein sequence. Symbols are chosen so that chemical similarities between amino acids are suggested by similarity of color (see key on right of figure). Cysteine and proline, as mentioned in the text, are of particular significance, and have been given starred symbols. Glycine, with no side chain, is often found where polypeptide chains have to cross in close contact, and is coded by an empty circle. Alanine and serine, other small amino acids, are coded by light colors. Threonine has both a hydroxyl group and a methyl group and is thus related to both serine and valine. In immunoglobulins there is very frequent interchange of serine and threonine (e.g., positions 19, 25, 27, 81, etc.) suggesting that the hydrogen-bonding properties of the hydroxyl group are of particular importance. Asparagine and glutamine are other amino acids which have marked hydrogen-bonding character. The amino acids with predominantly hydrophobic side-chains are coded in various shades of green (aliphatic—valine, leucine, isoleucine, and methionine) and dark (aromatic—phenylalanine, tyrosine, and tryptophan; white marks have been added to assist identification).

Sources and arrangement of material: The peptide chains are regarded as made up of basic units of approximately 107 amino acids, coded by 107 triplets in nucleic acid chains. Thus, the rows of colored spots refer to both amino acid sequences and base sequences. It appears that these sequences evolved from common ancestors. The numbering system used is chosen so as to apply to both specificity genes and common genes, and is referred to as the "reference number system." Each light chain contains one specificity region and one common region. For example, protein Ag (a human kappa chain of the Inv b type) is shown as being encoded by the specificity gene Ag (line "e"; reference numbers 6–119, corresponding to amino acid residues 1–107) and the common gene kappa human Inv b (line "n"; reference numbers 1–118, corresponding to amino acid residues 108–214). Heavy chains may contain more than one of the basic common region units. Thus the partial carboxyl terminal sequence of rabbit gamma chain is described as common gamma rabbit III and IV, representing a continuous sequence from reference number 9, line o to reference number 118, line p (ref. no. 118, line o is followed immediately by ref. no. 1, line p). Note that we have left some redundancy between the top and bottom halves of the figure. Specificity genes and common genes are aligned at the cysteine residues, placing a gap between reference numbers 53 and 59. All other gaps (indicated by dashes) are inserted only to maximize homology within either specificity or common regions. Parentheses indicate areas within which the amino acid composition, but not the sequence, is known. Possible sequences of such areas were inserted by reference to the homologous sequences of the most closely related chains, and the periods indicate that the individual symbols were placed in this way. The same procedures have been used to assign amide groups in positions of uncertainty; where no basis for assignment exists we have inserted composite symbols (e.g., reference number 102 in the specificity gene Roy indicates that there is no good basis for distinguishing aspartic acid from asparagine). The figure is thus intended to convey general relationships between proteins, and should not be viewed as a direct reference source for sequence data. The protein sequences used are taken from Gray *et al.* (1967a, b; lines a, b, d, l, n); Piggot and Press (1967; line g); Wilkinson *et al.* (1966; line h); Wikler *et al.* (1967; lines i, k); Hood *et al.* (1966; line j); Milstein *et al.* (1967; line k); Milstein (1966; lines k, m, n); Hill *et al.* (1966; lines o, p); Hilschmann and Craig (1965; lines c, f, m, n); and Titani *et al.* (1966; lines e, m, n). In some cases we have used different alignments from those used by the original authors.

interactions and the folding of protein molecules are only beginning to be understood, but some of the concepts can be illustrated by means of the colored plate (Fig. 9). This plate was prepared in May of 1967 and thus lacks much new information. Nevertheless, the mass of data is so great that we have attempted to take advantage of the human's excellent ability to process color information so as to make it possible for the reader to evaluate more easily the relationships to be discussed. We have tried to convey some impression of chemical relationship between amino acid side chains by the choice of related colors.

The color code is discussed in detail in the figure legend, but a few residues deserve special comment as being of particular significance in determining or stabilizing the structure. The functional role of cysteine residues (orange with black star) in these proteins is to form disulfide bonds crosslinking critical parts of the molecule and stabilizing structures. It is for this reason that we have used the cysteine residues as the prime reference for aligning the specificity and common genes in Figure 9.

Another example of a particularly important function is seen in the case of proline residues (yellow with open star) which are known to introduce bends into polypeptide chains. Quite obviously, these amino acid side chains could not easily be changed without some disruption in the general structure, so they tend to be highly conserved throughout evolution.

The polypeptide chains illustrated are arranged in approximate order according to the degree to which they are related. A uniform numbering system is used to facilitate comparison of various chains. Since this reference system is chosen so as to relate to primordial genes coding for proteins of approximately one half the size of the immunoglobulin light chains, the same numbers are used twice within light chain molecules and three or more times within heavy chain sequences. For example, reference position 44 relates to both amino acid residue 35 (Trp) and residue 146 (Val) of kappa chains (see Fig. 9).

The relationships among the polypeptide sequences (and therefore the genes which coded for them) can be visualized qualitatively by considering the patterns of gaps and colors. Notice that the colored spots can be decoded as either amino acids or as nucleotide triplets in nucleic acids as indicated in the lower right hand corner of the colored figure. Examine, for example, the lambda chain labelled human Sh. This protein lacks the usual cyclic or "blocked" form of glutamic acid usually found in lambda chains. Thus, the first amino acid residue in the chain is serine, which appears at reference position 7. The specificity region of this chain which is labelled (i) is shown as extending on to reference position 121. The same covalently linked polypeptide chain then continues on as a common region sequence listed as common gene: lambda:

human. The reference position here is row (k) number 1. The second half of the lambda chain sequence is then completed at the end of row (k) with a serine residue at reference position 119.

Cysteine residues are found at positions 28 and 97 within each segment of chain according to the number-reference system. Whether found in the specificity regions (assembled in the upper rows within the colored figure), or within the common regions (collected together in the lower rows), these residues form a disulfide bridge within the portion in which they are found. In other words the first two cysteines in the kappa chain crosslink with each other in the folded protein molecule, as do the third and fourth cysteines. The proline residues at positions 12 and 13 have also been strongly conserved; in many other instances conservation is detected within either the specificity regions or the common regions but not throughout the entire array of the chains.

A second type of evolutionary selective pressure is exerted in such a manner that amino acids tend to be replaced by others with similar chemical functions. Thus an isoleucine may be replaced by another amino acid with a large aliphatic side chain (shades of green) without rendering the molecule so ineffectual that the organism containing that gene would be selected against. A change to a basic residue (red) in the same position might be disastrous thus placing the creature containing that gene at a selective disadvantage, however slight. The information given in the legend should aid in further and more detailed analysis of the figure.

The dark dashes in the figure are extremely important in examining these protein comparisons. Each of these dashes indicates the absence of three nucleotides in the DNA which codes for these chains relative to other chains. The addition or deletion of four triplets (twelve base pairs) at precise positions in the DNA molecule is virtually certain to represent a major event in the evolution of a polypeptide chain, since the probability of such an occurrence is very much less than that of a single base change converting a leucine to an isoleucine, for example. As we have discussed before, these patterns of deletions must have been present in genes prior to the divergence of mice and men if we are to explain the evolutionary origin of classes of kappa chains containing these gaps (compare mice and men in the figure).

Careful examination of the relationships that exist between the specificity and the common portions derived from the single chain reveals that any homology is rather tenuous. Nevertheless, it seems possible that true homology may exist in the cysteines, prolines and a few other residues, and that the gaps such as the ones indicated at position 15, and also starting at position 36, may be meaningful. In contrast, one can easily see close homologies when comparing the various common genes, whether they are derived from mice, from men, or from rabbits. Furthermore, when the human lambda specificity region is compared with the kappa I class of specificity regions from mice and men, a close homology is detected. Relationships are also clear between the lambda specificity gene and the kappa class II proteins. However, the gap starting at position 36 is absent in the kappa II proteins, as discussed before. When one considers the gap that starts at position 104, it is clear that the kappa chains are more closely related to each other than to the lambda chain. This is also easily apparent from the patterns of colored spots seen in this area. Examples of selection for the function of amino acid side chains are found at position 80 in the specificity region and at position 89 in the common region where interchanges have been tolerated only between the large aromatic amino acids (tyrosine, phenylalanine, and tryptophan). Furthermore, all specificity regions have very similar sequences between reference numbers 81 and 100; all common regions also have similar sequences, but similarities between the specificity and common regions are much less marked.

One way in which these relationships may be expressed quantitatively is in terms of the percentage identity of amino acid sequence. The degree of absolute identity (ignoring even the most conservative changes) is approximately 12% between specificity and common regions, 40% between specificity lambda and specificity kappa, and 60% between specificity kappa (I) and specificity kappa (II).

We suggest that the reader use this colored chart as a do-it-yourself theory testing kit. It should, for example, be carefully consulted and thoughtfully analyzed while reading the paper contributed by Dr. Cohn (p. 671). The reader must ask, for example, what sort of somatic mutational mechanism might be found in both mouse and man so as to give, in each, the patterns of gaps and colors which are characteristic of kappa class I and kappa class II polypeptide chains? What selective forces in the soma might lead to these patterns of amino acid sequences and functional proteins?

(3) *The Time-Scale of the Evolutionary Process*

We have concluded that these proteins have arisen by the normal process of chemical evolution, and we have utilized the information to draw out the tentative evolutionary tree presented in Figure 10. This is a relatively complex subject which will be discussed in another manuscript dealing specifically with more quantitative aspects of the evolution of these molecules (Gray, 1967).

Qualitatively speaking, it seems almost certain that the chains evolve from a primordial gene pool coding for molecules of other kinds, perhaps within membrane surfaces. At the first branch point, we assume that the copy-splice mechanism evolved so as to permit systems of chromosomal programming to carry out carefully controlled recombinational

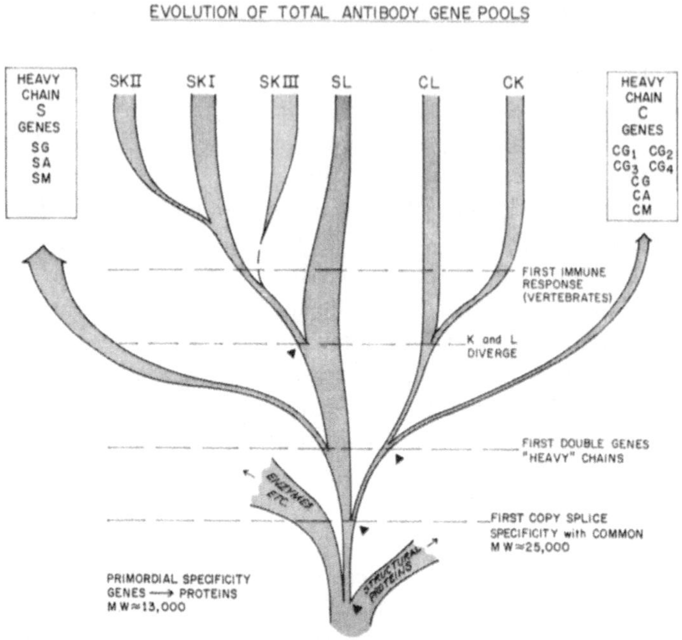

Fig. 10. Evolution of immunoglobulin gene pools. Relative order of branch points has been deduced from consideration of the extent of divergence of the various regions of peptide chains. These are fairly reliable as far back as the branch point marked "First double genes: Heavy chains." The data suggest that gene-doubling, to produce large common regions in gamma chains, occurred after divergence of the primordial CG gene from the primordial CL gene. The previous branch point is based on the assumption that the observed relationship between S genes and C genes is due to divergent rather than convergent evolution. It is likely that, in either case, S and C genes were independent before the gene doubling which occurred in the CG line, and that a mechanism must have existed for joining the two. Of particular note is the very late appearance of the immune response, when several major classes of peptide chains had already evolved extensively. Other evolutionary schemes have been proposed (Singer and Doolittle, 1966; Hill *et al.*, 1966; Ceppelini, 1967). Figure taken from Dreyer *et al.* (1967).

events during cellular differentiation. Such copying of genetic information and splicing into new regions within a chromosome would serve to commit a given differentiated cell line to the production of chains of approximately 25,000 molecular weight. These would be the protein molecules which represent the early ancestors of antibody chains. The primordial differentiated cells would be fully committed to the production of a particular type of molecule, and thus a particular specificity. Presumably the commitment would be quite as permanent as the commitment of the plasma cell tumors to a particular protein product.

Subsequent to the evolution of the first copy-splice mechanism in chromosomal programming, we assume that the molecules were approximately the size of present-day light chains. The next step might well have involved the development of ability to dimerize, to form functional molecules of 50,000 molecular weight. Another branch point must have occurred when larger common region genes appeared which were to become the precursors of today's heavy chains. More elaborate heavy chain regions then developed by contiguous duplication of the common gene units. The present-day heavy chains show evidence of this gene duplication phenomenon, as seen in the gamma chain sequences illustrated in the colored figure. The kappa and lambda chain divergence occurred later. Finally, at the time of the first immune response (the appearance of vertebrates), one finds the huge, multi-chain type of protein molecules characteristic of contemporary immune systems, including those of man.

Reasoning such as this has led us to feel virtually certain that molecules will be found in invertebrates which display clear evolutionary relationship to the immunoglobulin chains depicted here.

III. FACT OR FANTASY?

It should be obvious to the reader that the fictitious replisome illustrated in Figure 5 was invented in an attempt to resolve the paradox seen in structures of antibody molecules. The triplet numbering system refers to the kappa chain molecules, and the particular cell which is about to be produced by asymmetric cell division was intended to be a committed lymphocyte.

At first sight, such a mechanism seems far-fetched. Let us, however, consider this figure in somewhat greater detail. The process as illustrated is a "conservative copying" in that the individual strands of DNA wind up with their original partners. Closely related events involving base pairing in recombinational and copying mechanisms have turned up repeatedly in molecular genetics. Examples are the synthesis of RNA on a DNA template (another conservative system) and the replication of bacterial chromosomes (semi-conservative).

The replication of the circular chromosome of *E. coli* (Cairns, 1963) appears to be very closely analogous to the "copy-splice" process we envisage for light chain production. In order to reproduce itself semiconservatively this piece of double-stranded DNA must be broken at a precisely determined position, copies must be built up on each of the two strands, and ends must be spliced so as to circularize the copies. A second, somewhat different, but related copy-splice mechanism is seen in the phenomenon of ϕX bacterial virus replication (Sinsheimer *et al.*, 1967). In this case a single-stranded DNA molecule contained in the virus particle is replicated in such a way as to produce a double-stranded "replicative form" of the viral DNA. This two-stranded ring can be reproduced so that many copies are generated. These rings are then utilized in the synthesis of single-stranded DNA rings which are encapsulated in protein to form the virus particle. Further examples are seen in the replication of lambda viruses (Campbell, 1962) and in the incorporation of lambda DNA into *E. coli* chromosomes (Smith and Levine, 1967).

Even more strikingly similar processes must occur during reversion from the lysogenic state. Lambda viral DNA is then synthesized under the control of a stretch of prophage DNA contained in a very specific short region of the ring which constitutes the *E. coli* chromosome. In this case it is clear that the equivalent of "program code words" must be recognized in order to permit the precise programming of lambda virus production. It is likely that program code word recognition is sometimes accomplished by means of protein specificity and sometimes by nucleic acid specificity as indicated in Figure 5. The general principle is the same.

We thus assume that a number of distinctly different but closely related types of "replisomes" must exist in order to carry out the various exact functions required in different situations.

We must also anticipate a very great expansion of the DNA genetic code word dictionary in the future. The code words ("break strand," "start," "stop," "splice," "count cell division," "time," etc.) used in the various processes are surely not confined to the 64 primitive three-letter words used for protein synthesis. It seems likely that a whole new set of DNA code word sequences (and possibly of anti-codon sequences) will be uncovered in future years since it is virtually impossible to explain the known facts relating to application of nucleic acids without the existence of a far more sophisticated language than has been decoded to date.

The extrapolation of such thoughts and mechanisms to complex chromosomes may seem fanciful to some, but many of the known properties of chromosomes are readily thought of in terms of similar processes. Examples include the phenomenon of DNA puffing (Beerman, 1967), which involves the precisely programmed copying and splicing of DNA at exact positions within chromosomes of diptera; the beautifully programmed events known to occur in lampbrush loops; the formation of highly sophisticated polytene chromosomes; a variety of DNA ring-making activities well documented in chromosomes (Callan, 1966, for example); carefully controlled cutting and removal of DNA from chromosomes as is seen in chromatin diminution (Beerman, 1966); the superbly regulated processes of messenger RNA synthesis; as well as a large number of other phenomena which are known to occur in vertebrate and invertebrate chromosomes.

The various puffing, ring-making, and other forms of cutting, copying and/or splicing processes that occur in chromosomes must be accurately programmed and timed during somatic differentiation. The remarkable genetic elements which have been studied for many years in maize (see, for example, McClintock, 1965) represent excellent candidates for programming and timing units (Ac and Ds gene control systems respectively). This exceedingly complex genetic behavior can be interpreted relatively simply by invoking specific chromosomal programming and copy-splice mechanisms.

IV. FROM NUCLEIC ACIDS TO ANTIBODY MOLECULES

Suppose, for the moment, that the germ-line theory of the origin of genes coding for specificity is approximately correct, and that a specialized type of recombinational event in somatic cells does indeed provide the correct basis for the resolution of the genetic paradox discussed earlier. Can one then develop a theory of the immune response which can explain most of the known facts? It should be noted that theories are rarely, if ever, able to explain all of the experimental results reported at a meeting such as this. We have admittedly been selective in accepting ideas expressed by a great many of those in the field. Our goal has been to arrive at a composite picture that appears compatible with the facts which have been obtained by chemical studies. Many of the ideas expressed are not new but it is hoped that their selection and juxtaposition in this particular way will stimulate experimental testing of several crucial points.

We assume that a single mammalian germ cell carries two alleles of each common region gene, and a very large number of light chain specificity region genes (10^4 such genes would occupy less than 0.1% of the human genome). Ordinary genetic techniques would fail to detect alleles in specificity genes since the loss of one functional light chain out of the large number would be difficult to detect. There are regions of chromosomes referred to as interstitial heterochromatin in which one does not normally find genetic markers of the conventional sort. We might assume, therefore, that it is in these heterochromatin regions that our specificity genes are found.

A. The First Event

The maturation of the immune system is considered to be a precisely controlled process which starts with the fertilization of a single cell. During embryogenesis bone marrow is seeded by cells derived from the embryonic liver. Such cells are pluripotential (Till *et al.*, 1964) and may give rise to separate lines which eventually produce either erythrocytes or lymphocytes. In other words, the remarkable genetic events that result in the commitment of lymphocytes to production of specific light chain molecules occur during the relatively short period of time after the bone marrow is seeded and the time of appearance of the first immune response. Bone marrow cells from non-irradiated animals can repopulate the thymus and other lymphatic organs of animals which have been lethally radiated by X-rays, and thus provide a functional immune system after a period of time which permits only a relatively small number of cell divisions. In our opinion this time-scale is too small to provide the number of cell generations which would be required for somatic mutation and selection mechanisms.

B. The Thymus

In the thymus (normal or repopulated) there is a rapid production of cells (10^9 thymocytes/day/gram of thymic tissue) which is essentially independent of the age and antigenic status of the animal. Antigen-free, germ-free animals have very few lymphocytes in the peripheral nodes and lymphatic system (see, for example, Bauer *et al.*, 1963) but the rapid production of thymocytes continues. We propose that it is in the thymus (and in organs with similar functions such as the Bursa of Fabricius or its mammalian equivalent) that the programmed copy-splice differentiation occurs on a continuing basis, giving rise to a complete spectrum of committed cells.

C. The First Exposure to Antigen

In the absence of antigen the committed thymocytes are reabsorbed *in situ* and do not necessarily migrate into the periphery. When an antigen is present in proper physical form both to reach the thymocytes and to interact in the correct fashion, there is a rapid population of the peripheral lymphatics as well as the appearance of an immune response. It has been shown that only a small percentage of cells leaves the thymus even under these circumstances (Weissman, 1967). These observations seem compatible with the assumption that thymocytes are committed according to a chromosomal program and not by the directive influence resulting from exposure to antigen. The stretches of DNA that are now present in various committed cells might make messenger RNA and then protein molecules which are to become receptor molecules in the surface membranes of thymocytes and lymphocytes. It is proposed that these receptors might resemble those diagrammed in Figure 4 and are the antibody molecules themselves. These are combined with other macromolecules and lipids so as to perform the necessary receptor function. Antigen, when on the correct carrier and in the right concentration range, acts as a trigger for these amoeboid cells.

D. Immunological Memory

A cell, once excited, migrates from the thymus and populates the periphery as a medium-sized lymphocyte. Such cells then continue to divide and produce the highly specific antibody molecules to which they are committed. Disappearance of the antigen, for example by neutralization by antibody, permits these dividing cells to revert to small lymphocytes. Such long-lived lymphocytes constitute the pool of receptive target cells (Ellis *et al.,* 1967) which are able to respond directly to subsequent challenge with antigen.

E. Production of Circulating Antibodies

In a similar fashion a few cells out of the enormous population might be selected from the mammalian equivalent of the Bursa of Fabricius so as to form the nucleus of a clone which gives rise to a germinal center in a lymph node. Cells of this lineage may be the ones which are triggered to undergo further differentiation to form large secretory plasma cells no longer capable of cell division. Such cells would produce large amounts of circulating free antibody molecules.

F. The "Secondary Response"

The kinetic aspects of the immune response might be explained in several ways. Important in this regard would be the competition of cell receptors for limited quantities of antigen. Clearly those cells which possess receptors (antibody molecules) which bind with the highest affinity to antigen in a given form would be triggered most often. Thus the population of antibody molecules produced would shift with time toward those with ever greater binding constants. According to these speculations a "secondary" response curve which shows the same slope (rate of increase of cell population) as the primary response is easily explained. We simply assume that the same general type of response occurs, but that the number of target cells available depends to a great extent on the nature of the previous exposure to antigenic stimulation.

G. Tolerance

Antigen dosage and chemical form are clearly of critical importance (see, for example, Nossal *et al.,* 1967; Ada *et al.,* 1967). An excellent example of this requirement is seen in the work of the Australian group just quoted. When the 25,000 molecular weight protein, flagellin, is chemically cleaved into large peptide fragments, which are then purified, an interesting phenomenon results. Exceedingly small quantities of the peptide fragment (microgram amounts) induce "low zone tolerance" when injected into an animal. After a time lapse of twenty-four hours or so, such an animal is not able to respond to doses of the parent antigen which would normally stimulate the immune response to flagella. This phenomenon can be envisioned as due to a strong binding of the peptide fragment to trigger molecules in such a way that a critical configurational change cannot occur in these membrane bound "chemo-receptors." Metabolic changes leading to cell death might easily be imagined to follow such an event. Alternatively, antigen in certain physical states and quantities seems capable of causing differentiation of medium-sized lymphocytes into dead-end cell lines (Sterzl, 1967).

V. CONCLUDING COMMENT

An important goal in human development is to acquire the ability to distinguish the world of fact from the world of fantasy, and yet recognize the values of both. To the former belongs the hard core of amino acid sequence data. We feel that many of the conclusions drawn directly from the sequence data also belong there. A particular example is the conclusion that the two ends of a light-chain molecule are under the control of different genes. The mechanism by which these two genes become expressed as one is clearly an integral part of the differentiation process in the immune system. Our suggested mechanism should be viewed as fanciful, but not arbitrary; it might approximate the true mechanism.

References

Ada, G. L., C. R. Parrish, G. J. V. Nossal, and A. Abbot (1967). The tissue localization, immunogenic, and tolerance-inducing properties of antigens and antigenic fragments. Cold Spring Harbor Symp., *32*:381–394.

Attardi, D., and R. W. Sperry (1963). Preferential selection of central pathways by regenerating optic fibers. Exptl. Neurol., *7*:46.

Bauer, Heinz, R. E. Horowitz, S. M. Levenson, and Hans Popper (1963). The response of the lymphatic tissue to the microbial flora: Studies on germ-free mice. Am. Jour. of Pathol., *42*(4), 471–483.

Beerman, S. (1966). A quantitative study of chromatin diminution in embryonic mitoses of *Cyclops Furcifer*. Genetics, *54*:567–576.

Beerman, W. (1967). Gene action at the level of the chromosome. In: Heritage from Mendel. R. A. Brink (ed.). The University of Wisconsin Press, 179–201.

Bennett, J. Claude, L. Hood, W. J. Dreyer, and M. Potter (1965). Evidence for amino acid sequence differences among proteins resembling the L-chain subunits of immunoglobulins. J. Mol. Biol., *12*:81–87.

Brew, K., T. C. Vanaman, and R. L. Hill (1967). Comparison of the amino acid sequence of bovine α-lactalbumin and hens egg white lysozyme. J. Biol. Chem., *242*:3747–3749.

Cairns, J. (1963). The chromosome of *Escherichia coli.* Cold Spring Harbor Symp., *28*:43.

Callan, H. G. (1966). Chromosomes and nucleoli of the axolotl, *Ambystoma mexicanum.* J. Cell Sci., *1*:85–108.

Campbell, A. M. (1962). Episomes. Advan. Genet., *11*:101.

Ceppellini, R. (1966). Genetica Delle Immunoglobuline. Opening lecture at XII Annual Meeting of Associazione Genetica Italiana—Parma.

Cooper, M. D., Ann E. Gabrielsen, and R. A. Good (1967). Role of the thymus and other central lymphoid tissues in immunological disease. Ann. Rev. of Med., *8*:113–138.

Dreyer, W. J., and J. C. Bennett (1965). The molecular basis of antibody formation: a paradox. Proc. Nat. Acad. Sci., *54*:864–869.

Dreyer, W. J., W. R. Gray, and L. Hood (1967). The genetic, molecular, and cellular basis of antibody formation: some facts and a unifying hypothesis. Cold Spring Harbor Symp., *32*:353–368.

Eck, R., and M. O. Dayhoff (1966). Atlas of Protein Sequence and Structure, p. 146.

Ellis, S. T., J. L. Gowans, and J. C. Howard (1967). Cellular events during the induction of antibody formation. Cold Spring Harbor Symp., *32*:395–406.

Fahey, J. L., and M. Potter (1959). Bence Jones proteinuria associated with a transplantable mouse plasma cell neoplasm. Nature, *184*:654–655.

Good, R. A., and B. W. Papermaster (1964). Ontogeny and phylogeny of adaptive immunity. Advances in Immunol., *4*:1–15

Gray, W. R., W. J. Dreyer, and L. Hood (1967a). Mechanism of antibody synthesis: Size difference between mouse kappa chains. Science, *155*:465–467.

Gray, W. R., W. J. Dreyer, and L. Hood (1967b). Correlated variations in immunoglobulins: Two major subclasses of kappa chains, in press.

Gray, W. R. (1967). Evolutionary relationships in immunoglobulins. Manuscript in preparation.

Hartley, B. S., and D. L. Kaufmann (1966). Amino acid sequence of bovine chymotrypsinogen A. Biochem. J., *101*:229.

Hill, R. L., R. Delaney, R. E. Fellows, and M. E. Lebovitz (1966). The evolutionary origins of the immunoglobulins. Proc. Nat. Acad. Sci., *56*:1762.

Hilschmann, N. (1967). Hoppe-Seyler Phys. Chem., *348*:1077–1080.

Hilschmann, N., and L. C. Craig (1965). Amino acid sequence studies with Bence Jones proteins. Proc. Nat. Acad. Sci., *53*:1403.

Hood, L., W. R. Gray, and W. J. Dreyer (1966). On the mechanism of antibody synthesis: A species comparison of L-chains. Proc. Nat. Acad. Sci., *55*(4), 826–832.

Hood, L., W. R. Gray, and W. J. Dreyer (1966). On the evolution of antibody light chains. J. Mol. Biol., *22*:179–182.

Hood, L., W. R. Gray, W. R. Sanders, and W. J. Dreyer (1967). Light chain evolution. Cold Spring Harbor Symp., *32*:133–146.

Lennox, E. S., and M. Cohn (1967). Immunoglobulins. Ann. Rev. of Biochem., *36*(1), 365–406.

London, I. M., G. P. Bruns, and D. Karibian (1964). The regulation of hemoglobin synthesis and the pathogenesis of some hypochromic anemias. Medicine, *43*(6), 789–802.

McClintock, B. (1965). The control of gene action in maize. Brookhaven Symposia in Biol., No. 18.

Milstein, C., J. B. Clegg, and J. M. Jarvis (1967). C-terminal half of immunoglobulin λ chains. Nature, *214*:270.

Milstein, C. (1966). Chemical structure of light chains. Roy. Soc. Discussion, *166*:138–149.

Nossal, G. J. V., K. Shortman, A. P. Miller, G. F. Mitchell, and S. Haskill (1967). The target cell in the induction of immunity and tolerance. Cold Spring Harbor Symp., *32*:369–380.

Piggot, P. J., and E. M. Press (1967). Cyanogen bromide cleavage and partial sequence of the heavy chain of a pathological immunoglobulin G. Biochem. J., *104*:616–626.

Potter, M., W. J. Dreyer, E. L. Kuff, and K. R. McIntire (1964). Heritable variation in Bence Jones protein structure in an inbred strain of mice. J. Mol. Biol., *8*:814–822.

Ropartz, C., L. Rivat, and P. Y. Rousseau (1962). Deux nouveaux facteurs des systemes hereditaires des gamma-globulins le Gm(e) et l'Inv(l). Proc. 9th Congress of the Internal. Soc. of Blood Transfusion, Mexico, p. 445.

Sinsheimer, R. L., C. A. Hutchison, III, and B. H. Lindqvist (1967). Bacteriophage φX174: Viral functions. In: The Molecular Biology of Viruses, J. S. Colter (ed.), Academic Press, New York.

Smith, H. O., and M. Levine (1967). A phage P22 gene controlling integration of prophage. Virology, *31*(2), 207–216.

Sperry, R. W. (1963). Chemoaffinity in the Growth of Nerve Fiber Patterns and Connections. Proc. Nat. Acad. Sci., *50*(4), 703–710.

——— (1965). Embryogenesis of behavioral nerve nets. In: Organogenesis. R. L. DeHaan and H. Ursprung (eds.), Holt, Rinehart & Winston, New York, pp. 161–186.

Sterzl, J. (1967). Studies on the regulation of antibody synthesis. Cold Spring Harbor Symp., *32*:493–506.

Suran, A. A., and B. W. Papermaster (1967). N-terminal sequence of heavy and light chains of leopard shark immunoglobulins: evolutionary implications. In press.

Szilard, L. (1960). The molecular basis of antibody formation. Proc. Nat. Acad. Sci., *46*:293.

Till, J. E., E. A. McCullock, and L. Siminovitch (1964). A stochastic model of stem cell proliferation based on the growth of spleen colony forming cells. Proc. Nat. Acad. Sci., *51*:29.

Titani, K., E. Whitley, and F. W. Putnam (1966). Immunoglobulin structure: Variation in the sequence of Bence-Jones proteins. Science, *152*:1513.

Titani, K., E. Whitley, L. Avogardo, and F. W. Putnam (1965). Immunoglobulins structure: partial amino acid sequence of a Bence-Jones protein. Science, *149*:1090.

Titani, K., M. Wikler, and F. W. Putnam (1967). Evolution of immunoglobulins: structural homology of kappa and lambda Bence-Jones proteins. Science, *155*:828.

Trinkaus, J. P. (1965). Mechanisms of morphogenetic movements. In: Organogenesis. R. L. DeHaan and R. Ursprung (eds.), Holt, Rinehart & Winston, New York.

Tyler, A. (1946). An auto-antibody concept of cell structure, growth and differentiation. Growth, *10*(suppl.), 7–19.

——— (1965). The biology and chemistry of fertilization. Amer. Naturalist, *99*: 309–334.

Walsh, K., and H. Neurath (1964). Trypsinogen and chymotrypsinogen as homologous proteins. Proc. Nat. Acad. Sci., *52*:884–889.

Weissman, I. L. (1967). Thymus cell migration. J. Exptl. Med., *126*:291.

Wikler, M., K. Titani, T. Shinoda, and F. W. Putnam (1967). The complete amino acid sequence of a lambda type Bence-Jones protein. J. Biol. Chem., *242*:1668.

Wilkinson, J., E. M. Press, and R. R. Porter (1966). The N-terminal sequence of the heavy chain of rabbit immunoglobulin IgG. Biochem. J., *100*:303.

STUDIES ON THE MULTIPLE ISOACCEPTING TRANSFER RIBONUCLEIC ACIDS IN MOUSE PLASMA CELL TUMORS *

WEN-KUANG YANG † AND G. DAVID NOVELLI

Biology Division, NCI-AEC Carcinogenesis Program
Oak Ridge National Laboratory, Oak Ridge, Tennessee

I. INTRODUCTION

It is well recognized that immunoglobulins consist of heterogeneous populations of antibody proteins which are synthesized following a complicated series of cellular events induced by antigen stimulation. Chemical studies of purified antibodies suggested that antibodies of different specificities may differ from one another in their primary protein structure (Koshland and Englberger, 1963). The mechanism by which this tremendous variety of immunoglobulins are synthesized has thus been a very interesting biological problem. Biochemical knowledge concerning transcription and translation of the genetic messages for the synthesis of these proteins is, however, extremely deficient. This is due mainly to the difficulty in obtaining for study a sufficient quantity of lymphoid tissue with a given homogeneity of antibody production.

Recent genetic and functional studies of mouse plasma cell tumors have demonstrated that these transplantable tumors may represent different clones of protein-producing cells and synthesize individually homogeneous populations of myeloma globulins which are chemically similar, if not identical, to the normal immunoglobulins (Potter, 1962; Cohn,

* Research jointly sponsored by the National Cancer Institute, and the U. S. Atomic Energy Commission under contract with the Union Carbide Corporation.

The authors express thanks to Dr. M. Potter of the National Cancer Institute, National Institutes of Health, for giving us the cell lines MOPC 31B and 31C, to Drs. Hooper and Albright for giving us the cell lines MPC 47 and MPC 62, to Dr. A. N. Best for providing *E. coli* tRNA adenylyl transferase and to Dr. C. J. Wust for his constant interest in this project.

† This investigation was conducted during the tenure of a Damon Runyon Fellowship for Cancer Research.

1967). Knowledge of the chemical structure of immunoglobulins has been advanced greatly by several groups of investigators working on the myeloma proteins (Titani *et al.*, 1965; Potter *et al.*, 1965; Doolittle and Singer, 1965; Hood *et al.*, 1965; Milstein, 1966; and Apella, 1967). In view of this, we have selected plasma cell tumors as models for the study of the mechanism of immunoglobulin biosynthesis. A few cell lines of mouse plasma cell tumors have been examined for amino acid activation, and the chromatographic profile has been determined for certain of the transfer ribonucleic acids (tRNAs). The results will be summarized in the present report.

II. AMINO ACID ACTIVATION

A. General Features

It is well known that before an amino acid can be incorporated into protein it must first be activated, that is, the amino acid must be esterified to its specific tRNA by its specific aminoacyl-tRNA synthetase using energy from adenosine triphosphate (ATP). The existence of such a mechanism of amino acid activation in the plasma cell tumors was investigated. A standard procedure was used to extract tRNA from the tumor tissue; the procedure includes a sub-cellular fractionation, extraction with phenol and 1 *M* NaCl and column chromatography on DEAE-cellulose. The aminoacyl-tRNA synthetases were prepared by sub-cellular fractionation, gel filtration through Sephadex G100 and column chromatography on DEAE cellulose. Reaction of these preparations with ^{14}C-amino acids and ATP showed that plasma cell tumors have various tRNAs and the synthetases for amino acid activation (Table 1). A high content of tRNA (60–70 mg per 100 grams of wet tissue weight) and very active synthetases, comparable to those of liver, were observed in these plasma cell tumors and were found mainly in the post-microsomal supernatant of the tissue homogenate. High activity of amino acid activation in these tissues may well reflect the synthesis of myeloma proteins which constitute 30–40% of the newly formed proteins in the tissue.

A few more properties of amino acid activation in the plasma cell tumor were characterized: (1) Under optimal reaction conditions and at a concentration of the enzyme in excess, the cross-reaction involving activation of leucine and tryptophan was almost 100% between the plasma cell tumor and the BALB/c mouse liver; that between the tissues and *Escherichia coli* was very low (Table 2). These findings showed that the tRNA and synthetases of plasma cell tumor and liver were closely related. As will be shown, however, tRNA of the two tissues may differ in their pattern of the isoaccepting species for certain amino acids. (2) The optimal ratio of Mg^{++}/ATP seemed to be the same for the tumor and the liver (1 for leucine, 2 for glutamate and 3–5 for most of the other amino acids). Alteration of the ratio of Mg^{++}/ATP, however, had a marked effect on the maximal charging of tRNA in the liver system and only a slight effect in the tumor system. (3) The aminoacyl-tRNA synthetases from these tumor tissues were included by Sephadex G-200 but mainly excluded by Sephadex G-100, suggesting that they are of about the same molecular weight as has been commonly reported for

Table 1

AMINO ACID ACCEPTING ACTIVITIES OF TWO tRNA PREPARATIONS FROM MOPC 31B AND MOPC 31C TUMORS USING HOMOLOGOUS SYNTHETASES

Amino Acid	MOPC 31B	MOPC 31C	31B/31C
Ala	59	70	0.84
Asn	41	60	0.68
Asp	51	61	0.84
Arg	35	41	0.85
Gly	55	71	0.77
His	30	20	1.50
Ile	34	40	0.85
Leu	42	43	0.98
Lys	52	66	0.79
Met	39	42	0.93
Phe	21	24	0.88
Pro	18	24	0.75
Ser	80	90	0.89
Thr	51	57	0.89
Try	28	30	0.93
Tyr	16	18	0.89
Val	63	58	1.10

The reaction mixture contained per ml: Tris-HCl, pH 7.5, 100 μmoles; KCl, 10 μmoles; β-mercaptoethanol, 1 μmole; ATP, 4 μmoles; Mg-acetate, 4–20 μmoles; 1–2 μc of ^{14}C-amino acid; 1 $O.D._{260}$ unit of tRNA and 0.2 mg of mixed synthetase protein. Mg/ATP ratio in the reaction mixture was 1 for leu, 2.5 for arg, asn, gly, his, met, pro, ser, thr, tyr, and val and 5 for the remaining amino acids.

0.1 ml of the reaction mixture was spotted onto a filter paper disc at various incubation time intervals at 30°C. The radioactivity was assayed according to the method of Mans and Novelli (Arch. Biochem. Biophys. *94,* 48, 1961). The values given were calculated from the plateau segment of each individual kinetic curve and expressed as μμmoles/$O.D._{260}$.

Table 2

CROSS-REACTIONS OF THE AMINO ACID ACTIVATION FOR TRYPTOPHAN AND LEUCINE IN MOPC 31C MOUSE PLASMA CELL TUMOR, BALB/c MOUSE LIVER AND *E. coli* B *

	tRNA from		
	MOPC 31C	Liver	*E. coli*
Tryptophan Activation with †			
MOPC 31C Synthetase	980	1041	177
Liver Synthetase	968	995	—
E. coli Synthetase	473	398	1245
Leucine Activation with †			
MOPC 31C Synthetase	1854	2210	208
Liver Synthetase	1839	2194	—
E. coli Synthetase	357	288	2844

* The incubation mixtures were the same as described in Table 1, except that in reaction using *E. coli* synthetases 60 μmoles of Mg-acetate were employed.

† Activation activity for tryptophan is expressed as counts per minute per one $O.D._{260}$ unit of tRNA, and that for leucine, counts per minute per 0.1 $O.D._{260}$ unit.

synthetases of other systems. Also, they are eluted from DEAE-cellulose at the same range of salt concentration as is required to elute the same enzymes from other tissues. These experimental results demonstrated that the plasma cell tumor has a mechanism of amino acid activation for protein synthesis and that the general features of this mechanism are similar to those for other mammalian tissues.

B. MULTIPLE ISOACCEPTING tRNAs

The fact that specific tRNAs are required for amino acid activation during protein synthesis has been established (Hoagland *et al.,* 1957). For each amino acid there are multiple synonym codons (Marshall *et al.,* 1967) and possibly physically separable isoaccepting tRNAs recognizing different synonym codons (Weisblum *et al.,* 1962; Bennett *et al.,* 1963; Nathenson *et al.,* 1965; Kellog *et al.,* 1966). It was of interest therefore to see if such multiplicity of isoaccepting tRNAs exists in plasma cell tumors. A newly developed column chromatographic system for separating tRNAs, a reversed-phase Freon column, was used (Weiss and Kelmers, 1967). Figure 1 shows the complete profile of various isoaccepting tRNAs of a mouse plasma cell tumor line, MOPC 31C, as detected by post-chromatography aminoacylation. It can be seen that this chromatographic

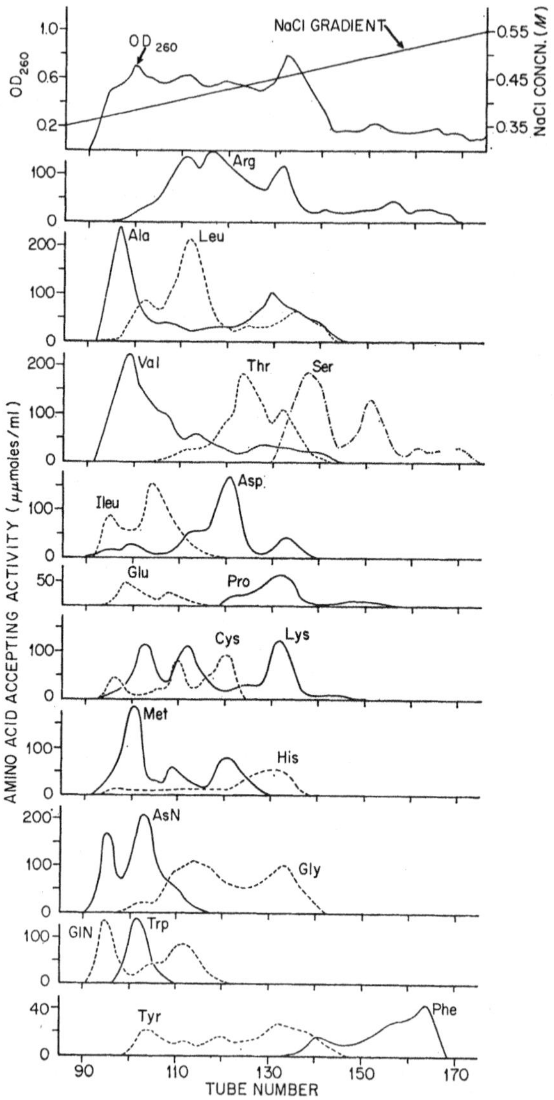

Fig. 1. The complete elution profiles of all 20 amino acid tRNAs from mouse plasma cell tumor, MOPC-31C, on the reversed phase Freon column chromatography. 207 O.D.$_{260}$ units were applied on a column of 1 × 240 cm and eluted with 2 liters of 0.25 to 0.65 *M* NaCl gradient containing 0.01 *M* Na acetate (pH 4.5), 0.01 *M* $MgCl_2$, 0.001 *M* EDTA. 10 ml fractions were collected and concentrated 10× for the amino acid-accepting assay. The reaction mixture contained per ml: Tris-HCl, pH 7.5, 100 μmoles; KCl, 10 μmoles; β-mercaptoethanol, 1 μmole; ATP, 4 μmoles; Mg acetate, 4–20 μmoles; 1–2 μc of ^{14}C amino acid; 0.1–0.2 ml (0.5–1.4 O.D.$_{260}$ units) of the 10× concentrated tRNA fraction; and 0.2 mg of mixed synthetase protein. The incubation was performed at 22°C for 45 minutes. 0.1 ml of the reaction mixture was pipetted onto filter paper discs which were subsequently washed in cold trichloroacetic acid solution (10%, 1×; 5%, 2×), alcohol-ether solution (1×), ether (2×), dried and the radioactivity was determined by using a liquid scintillation spectrometer.

system was efficient for separating mammalian tRNAs and served well for analytical purposes. Except for tryptophan, multiple peaks were observed for each of the other 19 amino acids, each having its own characteristic elution profile. The multiplicity of isoaccepting tRNAs observed for an amino acid in this tumor cell line seems to be related to the number of its known synonym codons (Marshall *et al.*, 1967). For arginine, leucine, and serine, which have six codons, several peaks were detected; for tryptophan having only one codon, only a single peak was observed. There were, however, exceptions in the cases of lysine, aspartic acid, tyrosine and methionine; for these, the number of observed peaks were more than the number of the corresponding codons for the amino acid. Another interesting feature of the profile in Figure 1 is that these plasma cell tumor tRNAs seemed to be more heterogeneous with regard to most of the amino acid accepting activity when compared with the profile of tRNAs obtained from *E. coli* under similar chromatographic conditions (Weiss and Kelmers, 1967). Recent data that we obtained showed that some of the aminoacyl tRNAs of the plasma cell tumor and those of *E. coli* migrated differently on the Freon column (Yang and Waters, 1967, unpublished observations). This suggests that, although the genetic code may not be changed throughout evolution, there may be phylogenetically evolved alterations in the chemical structure of the isoaccepting tRNAs for an amino acid.

The possibility that the observed multiplicity of isoaccepting tRNAs in MOPC 31C might be due to artifacts was considered and examined experimentally. Various tests showed that the multiplicity was not due to aggregation, to partial degradation of the tRNAs by ribonuclease, or to differential removal of the pCpCpA terminus of some molecules of tRNA. (For details see Yang and Novelli, 1968.)

C. Comparisons of Isoaccepting tRNAs Among Various Tissues

Recently, it was realized that modifications in the system of amino acid activation, due either to aminoacyl-tRNA synthetases or to isoaccepting tRNAs, may play a role in some cellular regulatory mechanisms (see review by Novelli, 1967). Since the recognition of the genetic message depends upon the tRNA molecule which carries the amino acid and not upon the amino acid to be incorporated, there has been speculation that changes in tRNA may result in alterations in the translation of the genetic message, thus producing proteins with altered amino acid residues. In this regard, Potter *et al.* (1965) postulated the possible important role played by tRNAs and aminoacyl-tRNA synthetases in the functional differences of plasma cells. Accordingly, a comparative study of various tRNAs was initiated: (1) those of a cell line (MOPC 31B) showing an unbalanced production of L and H chains and an excretion of Bence

Jones protein, and those of a cell line (MOPC 31C) producing complete myeloma immunoglobulin; and (2) tRNAs of myeloma IgG producers (MOPC 31C and MPC 47) and a myeloma IgA producer (MPC 62).

1. *MOPC 31B and 31C*

On the basis of relative accepting activity, the tRNAs prepared from 31B and 31C plasma cell tumors showed some quantitative differences in their ability to accept certain amino acids in the presence of the homologous synthetases (Table 1). Representative amino acids, histidine, lysine, leucine, aspartate and methionine, were selected for studying the precharged aminoacyl-tRNAs on the reversed phase Freon column. The two tumors were found to have nearly identical patterns for these aminoacyl-tRNAs except in methionyl tRNAs, which showed small quantitative differences in the three isoaccepting species detected by chromatography. Further chromatographic studies for the other aminoacyl-tRNAs are now in progress.

2. *Myeloma IgA and IgG Producers*

Aminoacylation reactions suggested that there were more differences between the IgA and the IgG producers. Co-chromatography of H^3-aminoacyl-tRNA of one tumor and ^{14}C-aminoacyl-tRNA of the other tumor was carried out for serine, glycine, tyrosine, threonine, methionine and leucine. Minor differences were found in most cases. Of particular interest was the result with the seryl-tRNAs of MPC 62 and MPC 47. As demonstrated in Figure 2, a prominent peak (seryl-tRNA^{IV}) was detected in the tRNAs of MPC 62; it was nearly absent from MPC 47. With improvement in the chromatographic conditions, seryl-tRNAs could be resolved into 4 main peaks which were designated as seryl-tRNA^{I}, seryl-tRNA^{II}, seryl-tRNA^{III} and seryl-tRNA^{IV}. The MPC 62 tumor, the IgA myeloma protein producer, was found to have a very minor seryl-tRNA^{I} and a prominent seryl-tRNA^{IV} whereas the MPC 47 tumor, the IgG producer, was shown to have a prominent seryl-tRNA^{I} and a nearly absent seryl-tRNA^{IV}.

3. *Authenticity of the Observations on Seryl-tRNAs*

Before these interesting observations on the seryl-tRNAs were accepted as real, the possibility that they might be artifacts were considered. Experiments were devised to answer the following questions: (a) Is the difference due to the different labeled (^{14}C and ^{3}H) amino acids used? (b) Since the tRNAs were prepared from the post-mitochondrial supernatant, is it possible that the difference observed is caused by compartmentalization of seryl-tRNAs within the subcellular components of the two tumors? We isolated tRNAs from the 34,000 $\times$ g pellets of the two tissues and found them to be active in accepting serine, although the

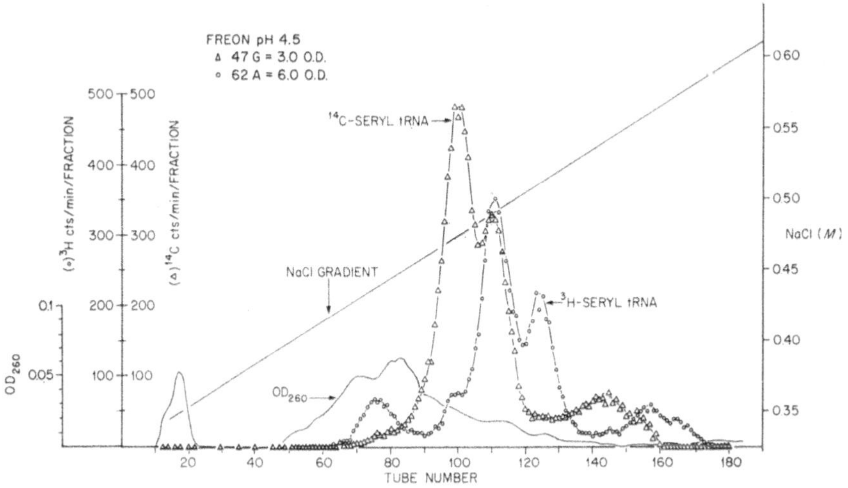

Fig. 2. Co-chromatography of ^{14}C-seryl-tRNAs from the 34,000 × g supernatant of MPC 47 tumor (IgG producer) and ^{3}H-seryl tRNAs from the 34,000 × g supernatant of MPC 62 tumor (IgA producer) on the reversed phase Freon column. To prepare the labeled seryl-tRNAs, the 20 ml reaction mixture contained Mg acetate, 12 μmoles/ml; ^{14}C-serine (116 mc/mM), 2 μc/ml, or ^{3}H-serine (330 mc/mM), 15 μc/ml; 1.0 μmoles/ml of each of the other 19 ^{12}C-amino acids; tRNA, 1.0 O.D.$_{260}$ unit/ml and other materials as listed in the legend of Fig. 1. After incubation at 30°C for 45 minutes, the mixture was poured onto a DEAE-cellulose column (0.8 × 4 cm) which had been pre-equilibrated and was subsequently washed with 0.25 *M* NaCl buffered solution containing 0.01 *M* $MgCl_2$, 0.001 *M* EDTA and 0.01 *M* Na-acetate (pH 4.5). The labeled seryl-tRNAs were eluted in a small volume of 0.70 *M* NaCl buffered solution. In this co-chromatography, 3 O.D.$_{260}$ units containing ^{14}C-seryl-tRNAs and 6 O.D.$_{260}$ units containing ^{3}H-seryl-tRNAs were used. Each of the 10 ml chromatographic fractions was chilled and 4 O.D.$_{260}$ units of calf-thymus DNA and 2 ml of cold 50% trichloroacetic acid solution added. The precipitates were collected on Millipore filters and washed with 75% ethanol. The Millipore filter was dried and radioactivity was determined as before.

specific activities were low (due to contamination with other RNA). The labeled serines were reversed for the charging of these pellet tRNA preparations and they were co-chromatographed on the reversed phase Freon column (Figure 3). It was found that a small peak in the front portion was due to a contaminant in the tritiated serine, but differences observed in tRNA preparations from the 34,000 × g supernatant of the two tissues were persistent in these pellet preparations. (c) Could this phenomenon be due to seryl-tRNA synthetases rather than to the tRNAs in the two tissues? Thus, tRNA from MPC 62 was charged with ^{14}C-serine by MPC 47 synthetase and tRNA from MPC 47 was charged with ^{3}H-serine by MPC 62 synthetase and then co-chromatographed. From this study it was evident that the differences were not due to the synthetases (Figure 4). (d) Could it be that most of serine-tRNA^{I} in MPC 62

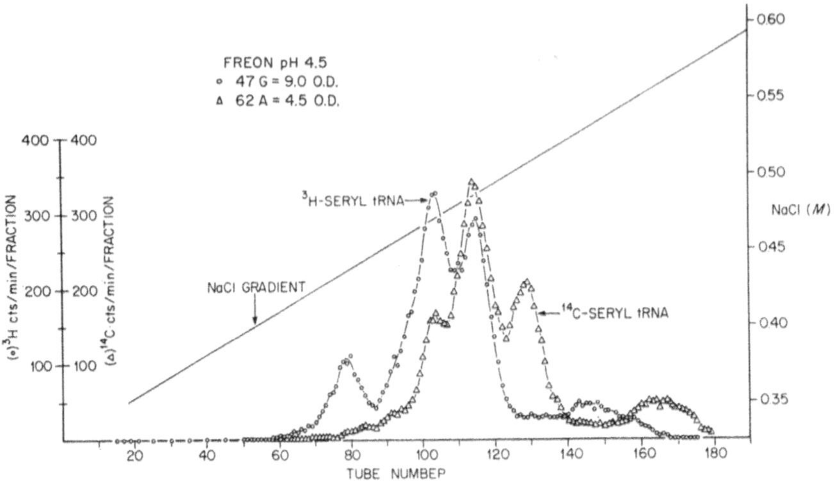

Fig. 3. Co-chromatography of ^{3}H-seryl-tRNAs from 34,000 × g pellet of MPC 47 tumor (4.5 O.D.$_{260}$) and ^{14}C-seryl-tRNAs from 34,000 × g pellet of MPC 62 tumor (9.0 O.D.$_{260}$) on the reversed phase Freon column. The methods of preparing labeled seryl-tRNAs and assaying the chromatographic fractions were the same as described in the legend of Fig. 2.

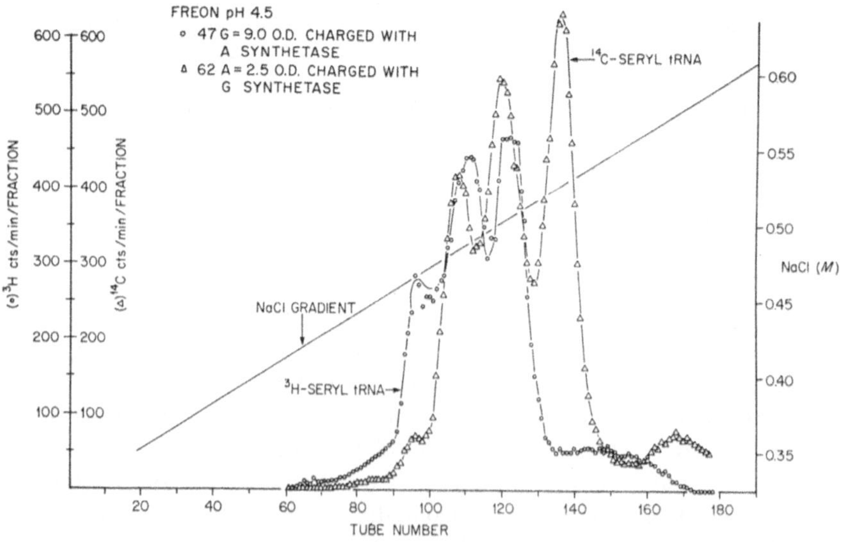

Fig. 4. Co-chromatography of ^{3}H-seryl-tRNAs (9.0 O.D.$_{260}$) and ^{14}C-seryl-tRNAs (2.5 O.D.$_{260}$) on the reversed phase Freon column. tRNA from 34,000 × g supernatant of MPC 47 was charged with ^{3}H-serine by using synthetase preparation from MPC 62; and tRNA from 34,000 × g supernatant of MPC 62 was charged with ^{14}C-serine by using synthetase preparation from MPC 47. In the NaCl gradient solutions, 0.003 *M* of β-mercaptoethanol was included. Other experimental conditions were the same as described in Fig. 2 legend.

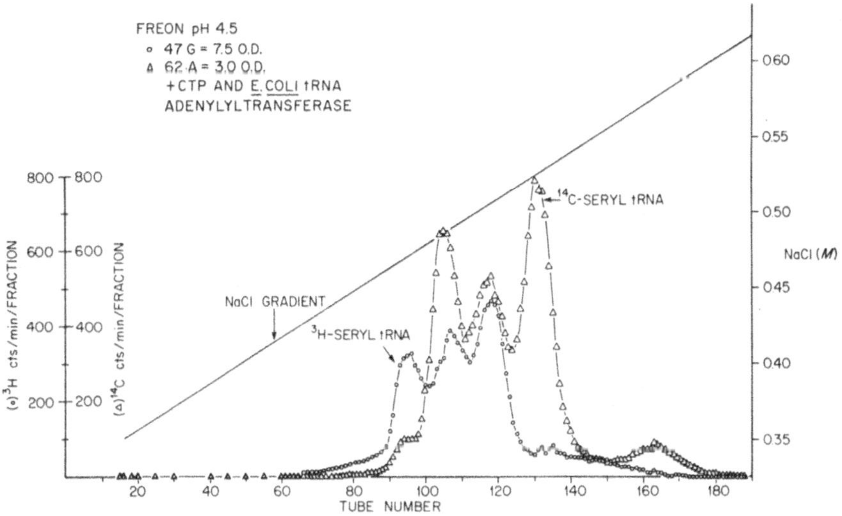

Fig. 5. Co-chromatography of ^{3}H-seryl-tRNAs from MPC 47 tumor (7.5 O.D.$_{260}$) and ^{14}C-seryl-tRNAs (3.0 O.D.$_{260}$) MPC 62. The labeled seryl-tRNAs were prepared by using homologous synthetases and also in the presence of *E. coli* tRNA-adenylyl transferase and cytidine triphosphate. Other experimental conditions were the same as those in Fig. 4 legend.

and most of serine-tRNA^{IV} of MPC 47 were lacking their pCpCpA terminus, thus causing them not to be detected by *in vitro* charging with labeled serine? Charging reactions were incubated in the presence of cytidine triphosphate and *E. coli* tRNA adenylyl transferase which is capable of repairing any tRNA with missing pCpCpA terminus. Despite the fact that a slight increase of labeled serine charging was observed under these conditions, the co-chromatographic patterns of the seryl-tRNAs from the two tissues were unaltered (Figure 5). (e) Could there be some other factors in the two tissues which might cause such changes during the isolation process of the tRNAs? Fresh tissue of MPC 47 and MPC 62 was obtained, and divided into half. Half of each were combined. The combined mixture and the two individual halves were subjected to isolation procedures for tRNA in parallel. The tRNA from the combined mixture was charged with ^{3}H-serine and co-chromatographed separately with ^{14}C-seryl-tRNAs from the two individual halves of the tumor tissues. As shown in Figure 6, an intermediate pattern of seryl-tRNAs was observed in the combined tissue. Analyses by means of a Dupont 310 curve resolver clearly demonstrated that the curve of seryl-tRNAs from the combined tissue was a pure summation of those of seryl-tRNAs from the two individual halves of MPC 47 and MPC 62 tissues. Authenticity of the observation on the differences of seryl-tRNAs be-

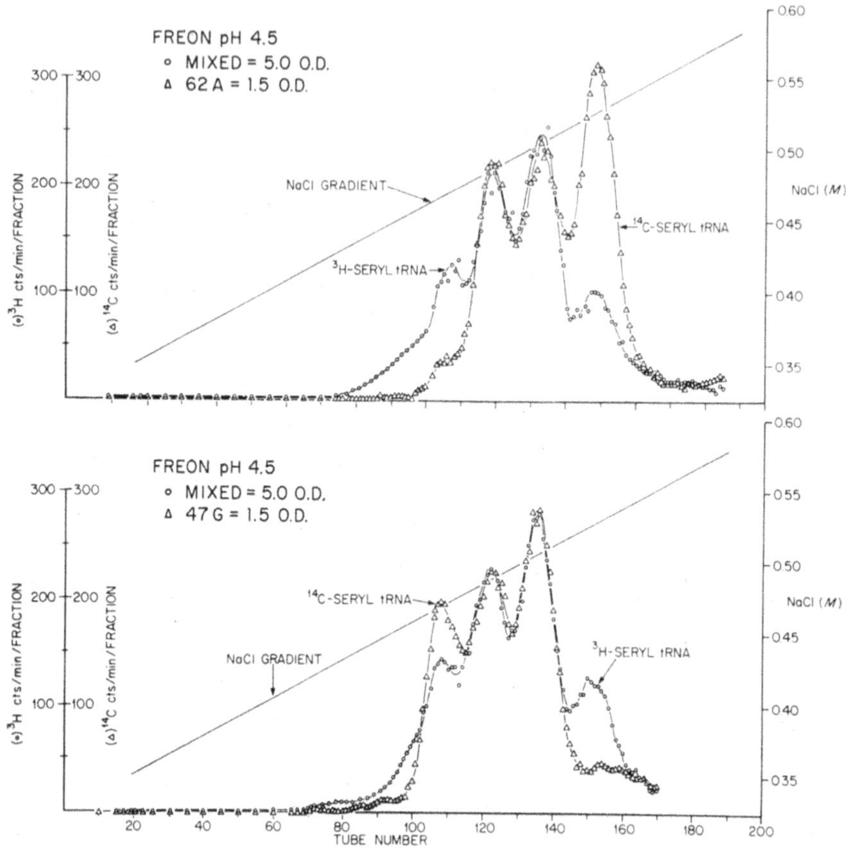

Fig. 6. Two sets of co-chromatographs of ^{3}H-seryl-tRNAs prepared by using tRNA (5.0 O.D.$_{260}$) from the mixed tumor tissue, (MPC 47:MPC 62 = 2:1) and ^{14}C-seryl-tRNAs prepared either by using tRNA (1.5 O.D.$_{260}$) from MPC 62 (upper graph) or by using tRNA (1.5 O.D.$_{260}$) from MPC 47 (lower graph). For other experimental conditions, see legends of Fig. 2 and Fig. 4.

tween an IgA producer, MPC 62 tumor, and an IgG producer, MPC 47 tumor, was thus well established.

4. *Examination of Seryl-tRNAs in Other Tissues*

The interesting observations on the seryl-tRNAs described above led us to examine the seryl-tRNAs in other tissues. The labeled seryl-tRNAs from another myeloma IgG producer (MOPC 31C), BALB/c mouse liver, and mouse L-cells grown in synthetic culture medium were co-chromatographed separately on the reversed phase Freon column, with seryl-tRNAs from either MPC 62 or from MPC 47 tumor tissue (Figs. 7–9).

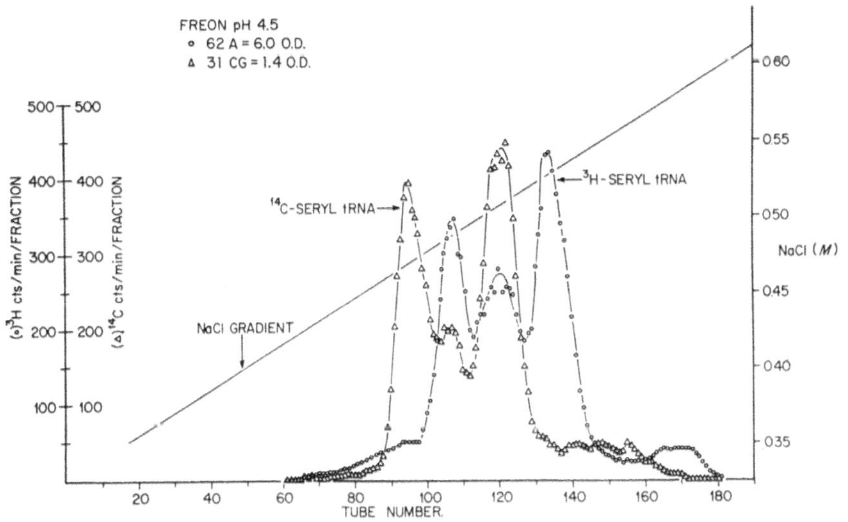

Fig. 7. Co-chromatography of ^{14}C-seryl-tRNAs (1.4 O.D.$_{260}$) from 34,000 × g supernatant of MOPC 31C tumor (IgG producer) and ^{3}H-seryl-tRNAs (6.0 O.D.$_{260}$) from 34,000 × g supernatant of MPC 62 (IgA producer). The ^{3}H-seryl-tRNA from MPC 62 tumor was the same preparation employed in the experiment of Fig. 2. For other experimental conditions, see legends of Fig. 2 and Fig. 4.

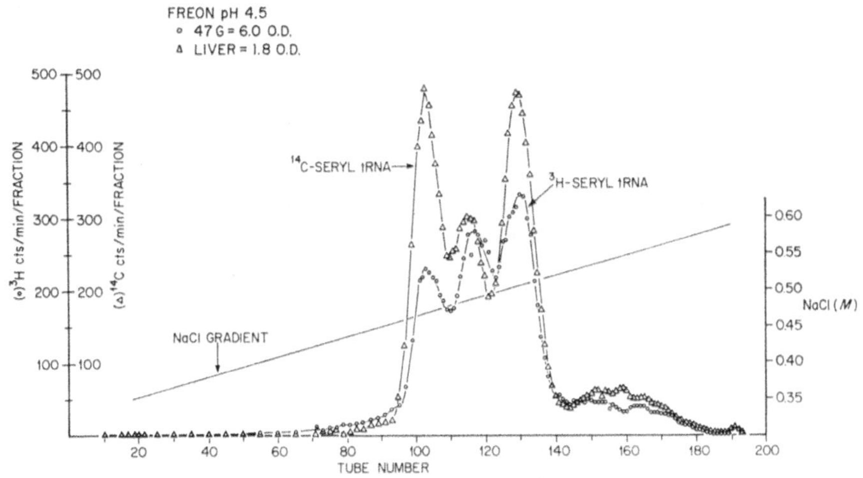

Fig. 8. Co-chromatography of ^{14}C-seryl-tRNAs from BALB/c mouse liver (1.8 O.D.$_{260}$) and ^{3}H-seryl-tRNAs from MPC 47 tumor (6.0 O.D.$_{260}$). For other experimental conditions, see legends of Fig. 2 and Fig. 4.

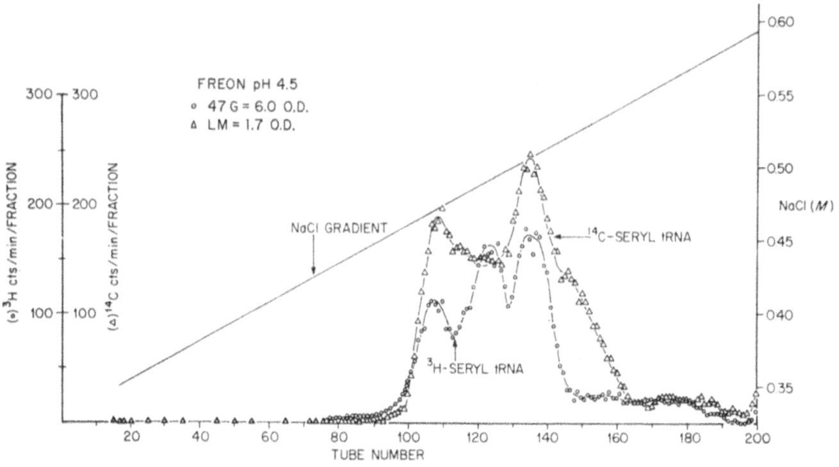

Fig. 9. Co-chromatography of ^{14}C-seryl-tRNAs from mouse L-cells grown in synthetic culture medium (1.7 $O.D._{260}$) and ^{3}H-seryl-tRNAs from MPC 47 tumor. The ^{3}H-seryl-tRNAs from MPC 47 tumor was the same preparation employed in the experiment of Fig. 3. For other experimental conditions, see legends of Fig. 2 and Fig. 4.

Both MOPC 31C and BALB/c mouse liver have no observable seryl-tRNA^{IV}, but have a more prominent seryl-tRNA^{I} than MPC 47. A small seryl-tRNA^{IV}, detected as a shoulder of seryl-tRNA^{III}, was present in the mouse L-cells. The relative quantity of the seryl-tRNAs also varied among these different tissues (Table 3).

Table 3

Percentages of Isoaccepting Seryl-tRNAs of Various Tissues as Separated on the Reversed Phase Freon Column and Determined by the Dupont 310 Curve Resolver

Tissues	Seryl-tRNAs *			
	I	II	III	IV
MPC 62 (IgA Tumor)	4	26	24	35
MPC 47 (IgG Tumor)	27	27	33	2
MOPC 31C (IgG Tumor)	31	17	43	0
BALB/c Mouse Liver	29	22	31	16

* The values were the average of 2–4 chromatographic separations and expressed as percentages of total seryl-tRNAs. Percentages of other minor components are not listed in the table.

III. COMMENTS

From the present investigation, it has been established that a common mechanism of amino acid activation, the first step of protein biosynthesis, exists in the plasma cell tumors which produce myeloma immunoglobulins in the BALB/c mouse. Modifications occurring either in the multiple isoaccepting tRNAs or in the aminoacyl-tRNA synthetases could distinguish this kind of tissue from other mammalian tissues. It has also been established that plasma cell tumors with different functions of protein synthesis may have different patterns of certain of the multiple isoaccepting tRNAs. It was found that an IgA producer, MPC 62, shows a prominent seryl-$tRNA^{IV}$, whereas two IgG producers, MPC 47 and MOPC 31C, show a marked seryl-$tRNA^{I}$ and a nearly absent seryl-$tRNA^{IV}$; these tumors thus should provide a good model for studying the role of tRNAs in the translation process of gene expression. It is interesting to ask whether the modification of isoaccepting tRNAs is a means devised by the living cell in response to the necessity of expressing the existing genetic message or whether it is a means for modifying the reading of the genetic code and altering the protein primary structure. We are currently pursuing this question. However, the answer to this question may not be able to shed light on the hypothesis of Potter *et al.* (1965), which has been expanded to include an explanation of the heterogeneity of immunoglobulins and the specificity of antibodies (Mach, 1967; Campbell, 1967). Since the immunoglobulins are not the only protein produced by these tumors, it is not certain whether these modifications of seryl-tRNAs in the different tumor tissues are directly related to the synthesis of different immunoglobulins or are related to the synthesis of other cellular proteins. If a very efficient cell-free system for the synthesis of immunoglobulin can be developed (an attempt has been made by Williamson, p. 590), the question regarding the role of isoaccepting tRNAs in the synthesis of immunoglobulin and whether it is regulatory for the "permissive translation" of the genetic message for immunoglobulins, may be answered.

The finding by Koshland (1966) that antibodies of different specificities from the same rabbit differed significantly in their content of serine residues in the active fragment of the protein strongly suggests the importance of this amino acid for antibody activity. If the present findings with respect to the seryl-tRNAs in the myeloma tissues were also applicable to antibody-synthesizing lymphoid tissue under various experimental manipulations, further implications for the role of tRNA in specific antibody synthesis could be made.

IV. SUMMARY

Amino acid activation, the first step in protein synthesis, has been studied in a few cell lines of mouse plasma cell tumors. Suitable methods have been developed for the preparation of transfer-ribonucleic acid (tRNA) and aminoacyl-tRNA synthetases from these tissues. A complete study of isoaccepting tRNAs specific for each of the 20 amino acids has been performed in MOPC 31C tumor by employing reversed phase Freon column chromatography. The chromatographic profile of specific tRNAs accepting an amino acid showed a unique pattern and the multiplicity of these isoaccepting tRNAs seemed to be related to the number of synonym codons for the amino acid. The isoaccepting tRNAs of mouse plasma cell tumors are in general similar to those of the BALB/c mouse liver, but are remarkably different from and more complicated than those of *E. coli*. This study demonstrated that the mouse plasma cell tumor has the same mechanism of amino acid activation for protein synthesis as do other tissues. Careful comparisons were performed on tRNAs of MOPC 31B and 31C, which produce the same immunoglobulin G molecule but different amounts of L and H chains, and of an IgA producer (MPC 62) and of IgG producers (MPC 47 and MOPC 31C). No major difference was found between MOPC 31B and 31C for the several amino acid-specific tRNAs examined. MPC 62 and MPC 47 were shown to have very different patterns of the four seryl-tRNA peaks demonstrable on the Freon reversed phase column. Seryl-$tRNA^{I}$ was a minor compound in MPC 62 but a major one in MPC 47, MOPC 31C and BALB/c mouse liver; seryl-$tRNA^{IV}$ was the major peak in MPC 62 but could not be detected in the latter three tissues. The possibility that this observation might be due to differences in the aminoacyl-tRNA synthetase, differential removal of the pCpCpA terminus or artifacts caused by the isolation procedures, has been excluded experimentally.

References

Apella, E. (1967). Amino acid sequence of immunoglobulin light chain. Cold Spring Harbor Symp. Quant. Biol. "Antibodies," *32*:37–44.

Bennett, T. P., J. Goldstein, and F. Lipmann (1963). Coding properties of *E. coli* leucyl-sRNA's charged with homologous or yeast activating enzymes. Proc. Natl. Acad. Sci., USA, *49*:850–857.

Campbell, J. (1967). Permissive translation of nucleic acid: A mechanism for antibody synthesis. J. Theoret. Biol., *16*:321–331.

Cohn, M. (1967). Natural history of the myeloma. Cold Spring Harbor Symp. Quant. Biol. "Antibodies," *32*:211–222.

Doolittle, R. F., and S. J. Singer (1965). Tryptic peptides from the active sites of antibody molecules. Proc. Natl. Acad. Sci. USA, *54*:1773–1779.

Hoagland, M. B., P. C. Zamecnik, and M. L. Stephenson (1957). Intermediate reaction in protein biosynthesis. Biochim. Biophys. Acta, *24*:215–216.

Hood, L. E., W. R. Grey, and W. J. Dreyer (1966). On the mechanism of antibody synthesis: A species comparison of L-chain. Proc. Natl. Acad. Sci. USA, *55*:826–832.

Kellog, D. A., B. P. Doctor, J. E. Loebel, and M. W. Nirenberg (1966). RNA codons and protein synthesis, IX. Synonym codon recognition by multiple species of valine-, alanine-, and methionine-sRNA. Proc. Natl. Acad. Sci. USA, *55*:912–919.

Koshland, M. E., and F. M. Englberger (1963). Differences in the amino acid composition of two purified antibodies from the same rabbit. Proc. Natl. Acad. Sci. USA, *50*:61–68.

Koshland, M. E. (1966). Primary structure of immunoglobulins and its relationship to antibody specificity. J. Cell. Physiol., *67* (Suppl. 1):33–50.

Mach, B. (1967). Studies on the cell-free biosynthesis of immunoglobulins. Cold Spring Harbor Symp. Quant. Biol. "Antibodies," *32*:269–276.

Marshall, R. E., C. T. Caskey, and M. Nirenberg (1967). Fine structure of RNA codewords recognized by bacterial, amphibian, and mammalian transfer RNA. Science, *155*:820–825.

Milstein, C. (1966). Variations in amino acid sequence near the disulfide bridges of Bence-Jones protein. Nature, *209*:370–373.

Nathenson, S. G., F. C. Dohan, H. H. Richards, and G. L. Cantoni (1965). Partition chromatography of yeast and *Escherichia coli* soluble ribonucleic acid. Relation of coding properties to fractionation. Biochemistry, *4*:2412–2418.

Novelli, G. D. (1967). Amino acid activation for protein synthesis. Annual Rev. Biochem., *36*(II):449–484.

Potter, M. (1962). Plasma cell neoplasia in a single host: A mosaic of different protein producing cell types. J. Exp. Med., *115*:339–356.

Potter, M., E. Apella, and S. Geisser (1965). Variations in the heavy polypeptide chain structure of gamma myeloma immunoglobulins from an inbred strain of mice and a hypothesis as to their origin. J. Mol. Biol., *14*:360–372.

Titani, K., E. Whitley, Jr., L. Avagardo, and F. W. Putnam (1965). Immunoglobulin structure: partial amino acid sequence of a Bence Jones protein. Science, *149*:1090–1092.

Weisblum, B., S. Benzer, and R. W. Holley (1962). A physical basis for degeneracy in the amino acid code. Proc. Natl. Acad. Sci., USA, *48*:1449–1454.

Weiss, J. F., and A. D. Kelmers (1967). A new chromatographic system for increased resolution of transfer ribonucleic acids. Biochemistry, *6*:2507–2513.

Yang, W. K., and G. D. Novelli (1968). Multiple isoaccepting transfer RNAs in a mouse plasma cell tumor. Proc. Natl. Acad. Sci. U. S., *59*:208–215.

ISOLATION AND CHARACTERIZATION OF RNA PRODUCED IN RESPONSE TO ANTIGEN *

PETER MCNAMARA AND KATHERINE D. PIH

Institute of Microbiology, Rutgers, The State University
New Brunswick, New Jersey

We have begun a program to isolate from lymphocytes of mice messenger RNAs specific for the synthesis of antibodies. To do this, specific messenger RNA must be separated from the bulk RNA and must be distinguished from the other messenger RNAs present. The technique of double labeling and methylated albumin-kieselguhr (MAK) column chromatography (Kano-Sueoka and Spiegelman, 1962) have enabled us to approach these problems and we have succeeded in identifying RNAs from mouse spleens elicited specifically by immunization with either of two antigens, namely sheep red blood cells or chicken red blood cells. These RNAs, we believe, may represent messenger RNAs for the synthesis of specific antibodies.

Figure 1 shows the general design of the experiments. Two groups of mice, each immunized twice against either sheep red blood cells (sRBC) or chicken red blood cells (cRBC) were labeled with, respectively, C^{14} uridine and H^3 uridine at a time when there was active antibody synthesis. Two hours after labeling, the mice were killed, their spleens removed and pooled, and the RNA from them was extracted with phenol-sodium dedocyl sulfate, and was fractionated on a MAK column. Such columns separate RNAs not only on the basis of size but also by nucleotide composition.

A typical fractionation pattern is shown in Figure 2.

Figure 3 shows the ratios of the cpm of C^{14} to the cpm of H^3 in RNAs eluted from MAK columns by salt solutions of increasing concentration.

* Supported by a grant from the Ortho Pharmaceutical Corporation. We thank Dr. Werner Braun for the valuable help and encouragement he gave us during the course of this work.

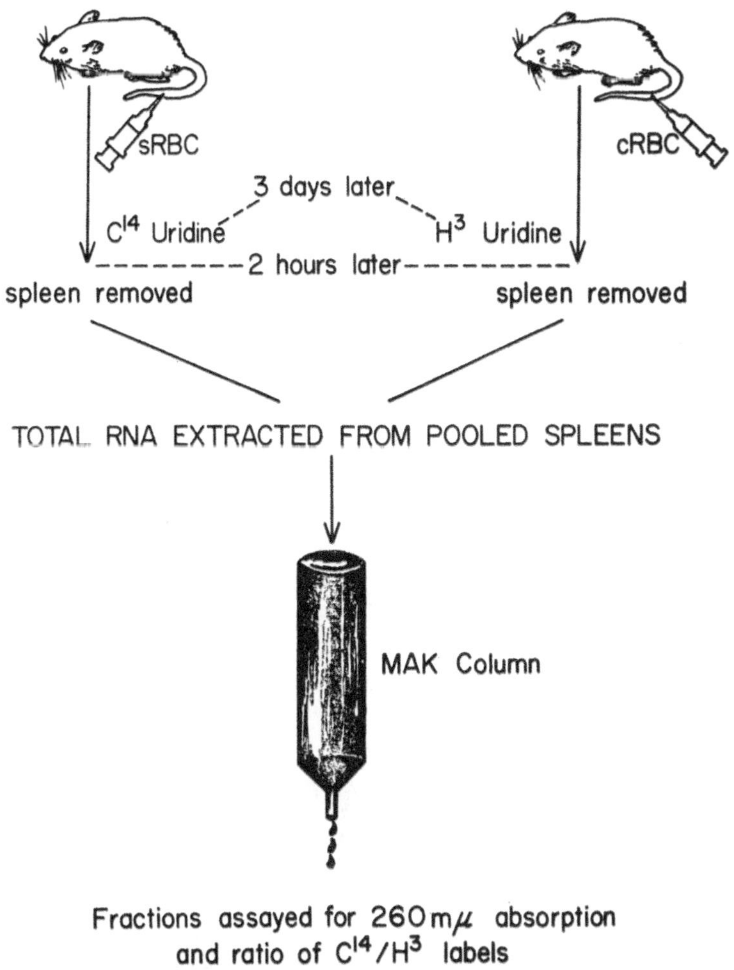

Fig. 1. General design of the experiment.

A control graph is at the bottom of the figure. For the controls, two groups of mice were both immunized against sheep red blood cells. Two hours before sacrifice one group was injected with C^{14} uridine, the other with H^3 uridine. The ratios of C^{14} to H^3 in the fractions eluted from a column charged with a mixture of RNAs from both groups do not vary significantly, thus showing that the RNAs from the two groups are identical. However, when the RNA from one group of mice immunized with sRBC and labeled with C^{14} uridine was mixed with the RNA from a second group of mice immunized with cRBC and labeled with H^3 uridine, the ratio of C^{14} to H^3 in this RNA mixture showed considerable variation after the MAK column fractionation of the RNA. If it is assumed that the labeling of the bulk RNA and of the messenger RNAs, not associated with the synthesis of the antibodies under consideration, is identical in both groups of mice, then these ratio variations may be regarded as reflecting an enrichment for RNAs specific for the synthesis of antibodies against sheep and chicken red blood cells, recoverable from the fractions deviating from the normal ratio. The variations in many cases encountered in such fractions are nearly 50% of the control ratio. It is unlikely that such large variations could be due to experimental artifact.

We intend to use larger columns in the future with more precise salt gradients in order to isolate more of the specific RNAs in purer form. The availability of larger quantities of the specific RNA should help us to determine (1) if antigen is associated with the pertinent RNA fraction, and (2) if the RNA in such fractions has biological activity. In regard

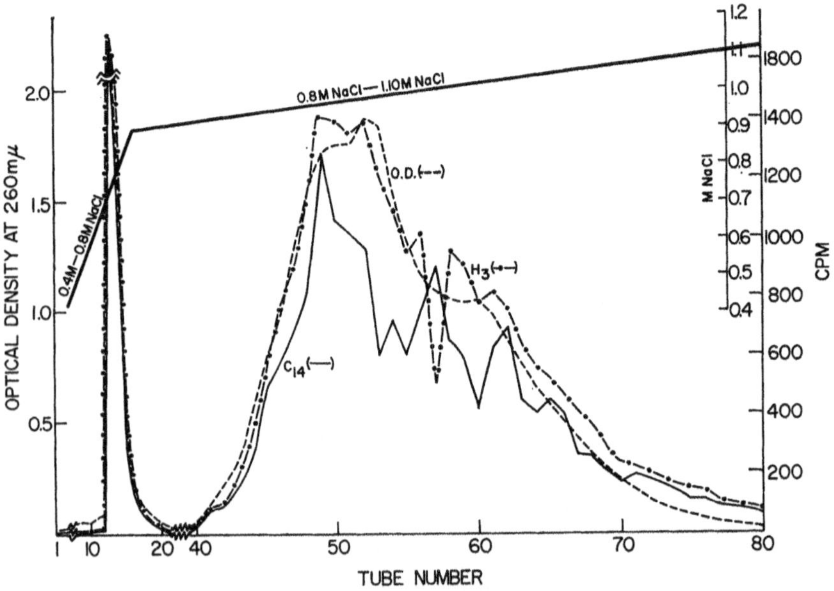

Fig. 2. MAK column chromatography of a mixture of RNAs from mice immunized with different antigens. C^{14} RNA was from mice immunized with sRBC. H^3 RNA was from mice immunized with cRBC. The column was prepared according to the method of Monier *et al.* (1962) and was run at 37 °C (Kubinski *et al.,* 1962). The salt gradient was in a buffer containing .05M $NaPO_4$, pH 6.7 and 10 γ/ml. PVS. 4 ml. fractions were collected and the absorbance at 260 mμ was determined. Approximately 0.5 mg. carrier RNA was added to each tube and 1 ml. of 50% cold TCA was added to each tube. The tubes were kept at 4 °C in an ice bath for five minutes. The precipitate was collected on a membrane filter and washed once with 10% TCA. The filters were dried and then placed in vials containing 10 ml. of a scintillation fluid containing .022% POPOP and 0.5% PPO in toluene. The radioactivity on the filters was measured in a Packard tricarb scintillation counter.

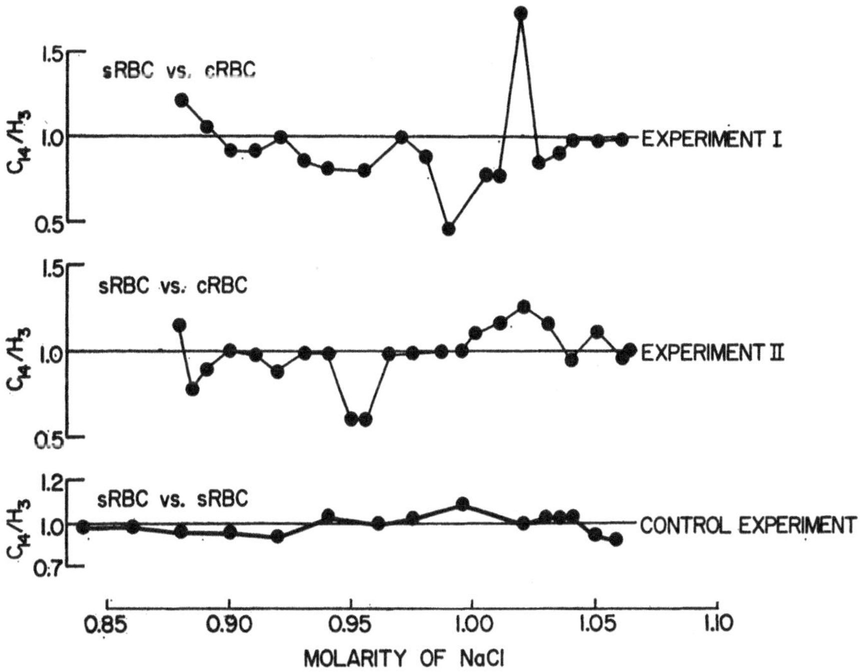

Fig. 3. The ratios of C^{14} to H^3 in mixtures of RNAs from two groups of mice immunized with different antigens are shown in the two experiments at the top of the figure. Mice immunized with sRBC were labelled with uridine-2-C^{14}. Mice immunized with cRBC were labelled with uridine-6-H^3. In the control at the bottom of the figure two groups of mice were both immunized with sRBC. The first group was labelled with uridine-2-C^{14} and the second group with uridine-6-H^3.

to the latter question, preliminary experiments have shown that when an RNA fraction enriched for C^{14}, and thus presumably for RNA specific for the synthesis of antibodies against sheep red blood cells, is injected into non-immune mice, spleen cells forming antibodies to sRBC but not to cRBC can be isolated 48 hours later. These tests will require further substantiation but, nevertheless, we have been able to demonstrate that specific RNAs produced in response to particular antigens can be distinguished from non-specific RNAs.

References

Kano-Sueoka, T., and S. Spiegelman (1962). Evidence for a non-random reading of the genome. Proc. Natl. Acad. Sci. U. S., *48*:1942–1949.

Kubinski, H., G. Koch, and O. Dress (1962). Fractionation of ribonucleic acid from poliovirus-infected cells by column chromatography. Biochim. Biophys. Acta, *61*:332–339.

Monier, R., S. Naono, D. Hayes, and F. Gros (1962). Studies on the heterogeneity of messenger RNA from *E. coli*. J. Mol. Biol., *5*:311–324.

DISCUSSION

Chairman: G. EDELMAN

EDELMAN: I would like to start this discussion by asking Dr. Williamson if he has any remarks about the possibility of the existence of an LH intermediate.

WILLIAMSON: As Dr. Uhr has reported, there are plasma cell tumors which make LH dimers and there is no tumor known that makes just H chains. The actual order of assembly of the chains is probably not too important to the overall concept of the synthesis of antibody. It is only of interest to those of us concerned with protein biosynthesis as a model for the assembly of other proteins. The intermediate H dimer that we find is not just a random aggregate of two H chains but is a specific dimer of two H chains covalently bound together. As such it represents the formation of a disulfide bond which may vary from one molecular species to another within the same animal species. We have evidence that the lability to reduction of the disulfide bond in immunoglobulin G varies from one molecular species to another.

EDELMAN: I had a question about that. Correct me if I am wrong, but I know of no cell-free experiments in which a dimer of H chains linked by disulfide bonds was formed. How have you shown this?

WILLIAMSON: Well, you are probably aware, Dr. Edelman, of Dr. Nisonoff's work on the mild reduction of rabbit gamma globulin. We were able to repeat his observations in tests with rabbit gamma globulin. However, with mouse protein we find exactly the opposite, namely relative stability of the disulfide bonds to mild reduction. We can reduce away the two LH disulfide bonds almost completely and leave the HH bond intact. This enabled us to prepare the HH marker which we used along with an LH marker from another mouse tumor in our identification procedures.

EDELMAN: Very good. Does it stay as a dimer?

WILLIAMSON: In neutral solutions we have as yet not too much evidence about the dimer's properties. Our identifications were performed in either SDS, urea or acid urea in which there is no aggregation of free H chains or of H dimers.

EDELMAN: Now let us turn to the question of diversity at the genetic level. This diversity may be explained in at least three different ways. Dr. Dreyer assumes a single gene for each variable region and he was responsible for the suggestion that the constant portion of the L chain may be represented by a single gene and that this single gene somehow fuses with each of the genes controlling the variable portion. One can also assume that the variations are the result of random somatic mutations and finally one could assume that recombination of somatic mutations may play a role. Perhaps we can make some attempt to relate the question of informational RNA to the question of diversity.

DREYER: One question that is important is how can one retain genes once they have been acquired. Some people have worried about the validity of the multi-gene hypothesis because they assume that during one's life-time one would use only a small number of available genes, for example the kappa chain genes. I am not convinced that this is true. I envision using the same kappa chain again and again for different antibodies, for example chain 99 might be used with heavy chain number 402, or it might also be used with heavy chain number 900. The binding site might be in part the same and in part different. Therefore, I think it is quite possible that there are constant selective pressures from antigens to which both mouse and man have been exposed through evolution and thus have retained sequences in much the same way as cytochrome has been retained in evolution from geese up to man.

EDELMAN: I think this brings us to a question that I can direct to Dr. E. Cohen. The question is: can the DNA-RNA hybridization technique help us to answer the question of the percentage of the genome involved or is that really beyond the technique at the present moment?

E. COHEN: We are working towards something like this and the main problem is that when we talk about an RNA preparation we really mean a mixture of molecular species of RNA of different specific activities. If one tries to relate the micrograms or the proportion of the input material which actually hybridizes at saturation, one obtains a figure which has very little meaning. But with such reservations in mind, we nevertheless did some calculations and came up with a figure that said that .08% of the genome was turned on in response to an exposure of cells to sheep red blood cells. This is, of course, extraordinarily large, but Russel Little of Washington University independently came up with a figure of .1%. Now let us say we are off by a factor of a thousand.

This is still an extraordinarily large number. We hope to be able to fractionate the rapidly labelled material prior to hybridization and thus to come up with a more meaningful figure.

EDELMAN: I suppose the real experiment would be to take the RNA from polyribosomes. Is that correct?

E. COHEN: That would be, of course, most desirable but since at this point we require many micrograms to saturate one filter it is technically not feasible.

MAURER: I would like to address my question to Dr. Cohen relative to the data he presented with our material. I was wondering whether results of hybridization experiments could reflect the results that Dr. Pinchuck obtained *in vivo*. If you use non-immunogenic partially deaminated polylysine-DNP, and if you use immunogenic phosphorylated albumin-polylysine-DNP, would you be able to detect any differences in hybridization?

E. COHEN: We have tried an analogous experiment using the Dresser type antigen, i.e., centrifuged BSA compared to aggregated BSA, and unfortunately we detected new species of RNA in both cases.

THOR: We have been investigating, at the University of Illinois Medical Center, some *in vitro* correlates of delayed hypersensitivity with our principal interest being in man and in a possible mechanism of getting at transplantation and immune diseases. We recently examined 72-hour old mononuclear cell cultures, isolated from human lymph nodes, by the migration inhibition technique in a capillary tube as used by David, Bloom and others. We have been able to transfer this correlate of delayed hypersensitivity using an RNA extract from sensitized lymph node cells and we have been able to show that this correlate is specific in the human system for three different antigens, mainly PPD, histoplasma and coccidioidin. The interesting thing about these 72 hours mononuclear cell cultures is that we can see both a lymphoblast type of cell and a macrophage type of cell. Interesting also is that some of the morphological characteristics of the lymphoblasts are rather confusing in that they engulf micro-size iron particles. These cells do not conform in all respects to current concepts of morphology of macrophages. Dr. Alexander suggested earlier that macrophages may not always be involved in the processing of all antigens. As an example he cited the ability of lymphocytes to mediate graft versus host reactions. Perhaps the lymphoblast can also process a type of antigen. It is very conceivable then that the lymphoblast type of cell or the macrophage in the small lymphocyte cultures, which are dynamic, changing populations of cells, may not meet all of the criteria that we are trying to apply for macrophages nor can they meet all of the criteria for lymphocytes. I wonder if Dr. Mannick would have a comment on this, mainly because his RNA preparations were made from lymph nodes which contain both macrophage and lymphocyte type cells.

MANNICK: I do think that it is certainly possible that the macrophage or some similar cell may participate in the processing of whatever antigen is responsible for the transplantation immune response. There are almost no data on this subject. The only data that may be relevant are some unpublished observations by Dr. Julius Gordon of McGill University. He has used the reaction of small lymphocytes in tissue culture against transplantation antigens as a test, and he has treated populations of small lymphocytes so as to remove as many macrophages as possible. Assaying one way reactivity, that is whether cells from one individual are capable of responding to the antigens of cells of another individual but not vice-versa, he found that macrophages are necessary for the typical proliferative response of the lymphocytes against foreign antigens. I know of very little other work bearing on this question. Dr. Alexander has expressed doubts about the role of the macrophage in such a response, and he quoted Gowan's work for showing that graft versus host reactivity can be transferred by very pure populations of small lymphocytes. Well, Gordon has other information with regard to this problem; he finds that it does not matter where the macrophage comes from, the macrophages can even come from the individual against whom the response is produced. So perhaps graft versus host reactivity takes place on the basis of foreign lymphocytes and host macrophages.

E. COHEN: Appropos somatic mutation versus germ line changes, I would like to point out that experiments on the proportion of cells in the peritoneal cavity of mice have shown that the number of cells that respond to specific antigen by forming antibody is far too large to be explained by any somatic theory. Bussard found up to 10% of the mouse peritoneal cell population could respond to sheep cells *in vitro*.

COHN: The difference between the germ line theory and any somatic mutation theory resides only in the number of genes in the germ line. It is not a problem of selection. It is not a problem of hanky-panky. Now I would like to make this clear because repeated changes in the germ line presuppose at least as many assumptions as are necessary in the somatic recombination model. The reason for considering other models is precisely the large number of suppositions that one has to make in order to rationalize the germ-line model. It is the same kind of hanky-panky. For example, what would prevent a repeated series of genes, say a hundred of them, from recombining and mixing. You would have to assume the existence of recombinationless areas or else all of these genes would be mixed up rapidly. What stabilizes the genetic markers in the variable region at position 9 of the rabbit immunoglobulin, where the arginine goes to threonine? You would have to make very special assumptions not only about recombinations but also about mutations in that area, so that arginine does not go to threonine in one of these genes by chance. You have to make assumptions about how the system evolved from the precursor of the rabbit to the rabbit, and how it continued. Kabat has made the argument very strongly in comparing the mouse and human sequences that although they are very similar, there are a number of positions that are species-specific. Thus one has again to repeat the mutation or the events in evolution. How can one minimize the number of assumptions that have to be made? I have pointed out that the model proposed by Smithies requires that you throw out 40% of the known sequences, and the assumptions that have to be made in any germ-line model are of such a magnitude to make these models unlikely (see pp. 671–715).

DREYER: Now if we restrict ourselves to kappa chains where we have a lot of structural information, are there really a number of assumptions that one must make in order to accept the germ line theory?

COHN: I will give you the Doolittle-Kabat argument. Kabat made the point that in comparing variable sequences in mouse and humans there are six positions that are characteristic of the mouse and characteristic of the human. It is possible by these six positions to tell whether the globulin is of human or mouse origin. Now let us ask the question, if I repeat these genes in the germ line, how do I go from one species to the other? To do this, you have to make a certain number of assumptions about repeated mutations and that is the fundamental contradiction.

TEXT OF SPEECH PRESENTED AT THE SYMPOSIUM BANQUET

THE MOLECULAR BIOLOGY OF EXPECTATION

MELVIN COHN *

The Salk Institute for Biological Studies
San Diego, California

I. INTRODUCTION

I have listened to but never given an after-dinner talk. I can remember, however, being grateful when the speaker was witty and as far as possible, not controversial. This combination plus the food and drink means that nobody feels threatened and the audience is relaxed and receptive. In this mood, I intend to speculate about our world which took only six days to fashion. While this is a *tour de force,* it is even more of a miracle that after 6000 years of civilization we can still live in it. This is a tribute to the extraordinary variety of learned and adaptive responses which we are capable of performing. In fact, man, throughout history, has so marvelled at his own prowess to learn and adapt that he created God in his own image. There is no question that animals behave as though there were a point to their actions; they are machines with the ability to show adaptive behavior. The vitalists, therefore, asked the right questions but gave the wrong answers.

Tonight I would like to deal with a straightforward question of vitalism: "What is the mechanism by which an individual can react in an adaptive way to an unexpected stimulus? The three systems responding to unexpected stimuli are the immune, detoxifying and learning mechanisms.

The unexpected stimulus has two characteristics. First, a priori, it does not appear to have been a selective force in evolution. Secondly, an unexpected stimulus need not act over long enough periods of time or on

* My own work mentioned here was supported by the National Institutes of Health Grant No. A105875 and the American Cancer Society. The Salk Institute has received a generous contribution from the National Foundation part of which was used to establish my laboratory.

a large enough population to be selective on the germ-line. For example, each of you can make antibody to crocodile albumin, detoxify mescaline, or learn to speak Lilliputian.

In the past, I could have given a Lamarckian or a Darwinian answer to the question "How does the individual respond to the unexpected?" I need not dwell on this problem today although immunologists are, next to neuropsychiatrists, the last stronghold of Lamarckism. Even as late as 1967, antibody synthesis was explained in that olympian journal, Nature, as the antigen-directed sequencing of nucleic acid (Mekler, 1967). This is like planting toads to grow toadstools.

It is important that I emphasize the distinction I will be making between the evolution of genes coding for proteins with combining properties, like the structural proteins (e.g., in membranes or ribosomes) and those coding for antibodies, or for the normal enzymes of metabolism and detoxification enzymes, or for instinctual and learned or abstractive behavior. No doubt, genes coding for structural proteins, normal enzymes of metabolism and instinctual behavior are all carried in the germ-line. The problem of distinguishing germ-line from somatically derived genes only arises when antibodies, detoxification enzymes and learned behavior are considered. The need to consider somatic hypotheses is most obvious for antibodies, less so for learned behavior and much less so for the detoxification mechanism. I wish to develop in detail the factors involved in considering germ-line or somatic hypotheses to explain response to the unexpected.

If the capacity to make every constituent of an animal were to be determined only by genes carried in the germ-line, then the response to any rare stimulus could result only from the expression of germ-line genes which are ordinarily silent. There is no doubt that the switching on of silent genes by external stimuli accounts for many responses, adaptive enzyme synthesis being an example. There is a storage of potentially useful genes during evolution and these genes account for properties of the animal that range from its large variety of proteins to complex instinctual behavior. What surprises us is that higher animals have systems for immunity, detoxification and learning that respond in a specific way to an enormous variety of stimuli which hardly could have been selective forces during evolution. Not only that, but on repeated stimulation, the response improves and is "learned," often increasing the chances of survival of the individual. There is selection for "fitting" between the stimulus and the response.

The simple fact that an individual can show an enormous variety of responses is not sufficient reason to doubt that all of the information for them is stored in the germ-line. The combining of gene products gives a tremendous amplification to responsiveness as illustrated by antibody specificity resulting from interaction of its subunits, the light and heavy (Edelman and Benacerraf, 1962) polypeptide chains.

There are two reasons to doubt that all responsiveness is determined in the germ-line. The first reason is that a completely programmed system of specialized cells has an inherent limitation on the number of germ-line genes which can be stored for adaptation to the unexpected. Since most responses are not revealed during the lifetime of an individual nor, for that matter, during many successive generations, the corresponding genes are virtually never expressed. The chances are very low that during your lifetime, you will reveal your antibodies to hydroxypyrene or your enzymes to detoxify chloroxylenol, or your ability to learn Scribbledehobble. Silence in a gene means that selection cannot act and it drifts due to mutation and is lost. A steady state level of genes is maintained by gain of genes through selection when they are expressed and by loss of genes through mutational drift when they are silent. The answer to whether this steady state number will be high enough to account for responsiveness to the unexpected is what distinguishes the two classes of theories, germ-line and somatic. The second reason is that complete programming does not permit us to understand how an organism can anticipate truly *de novo* stimuli and respond meaningfully to them. The population, of course, meets the unexpected by variation and selection but what about the individual?

I would imagine that it was for these reasons that many biologists, first led by Burnet (1959) and Lederberg (1959), proposed theories that transposed variation and selection to the soma by assuming that certain germ-line genes are varied and distributed in cells such that each of them expresses only one gene product. The stimulus acts to select cells with gene products "complementary" to the stimulus. Whether or not an individual animal can respond to an unexpected stimulus depends upon the size of the varied cell population from which the stimulus can select.

The two hypotheses, germ-line and somatic, are analogous in that variation and selection of individuals in a population is paralleled by variation and selection of cells in an individual. During germ-line evolution, a potentially useful family of genes generated during evolution is stored in the germ-line under regulatory control and during somatic evolution, cells are stored in the soma, each one carrying a unique somatically generated potentially useful gene. Germ-line evolution is the way *populations* adapt to *de novo* stimuli and somatic evolution is the way *individuals* adapt. Since both hypotheses are compatible with a clonal distribution of properties amongst the somatic cells, this fact does not distinguish them (Cohn, 1967b).

My doubts that germ-line hypotheses are sufficient might be answered by the classical geneticist (Dreyer *et al.*, 1967) as follows: first of all, it is a play on words to speak of *de novo* stimuli. Any stimulus to which an animal responds today could be analogous to one which was of selective value sometime in the past. Secondly, the gene in question does not have to be put under regulatory wraps during eons of evolution and then chance the hazards of mutational drift. The gene might be expressed primarily in functions we have not yet revealed and only secondarily be used for immunity. You simply do not know whether a given gene is silent.

The classical geneticist has a strong argument. After all, we do carry an enormous load of anachronistic but essential genes simply because decisions made early in evolution limit the solutions which can be envisaged later on. We do go through useless embryonic stages of gills, tails, webbed feet and fur because we are caught up like the "Daughters of the American Revolution" in the need to express our fossil genes. Evolution did not arrive at man by the hindsight of theology but rather solved problems as they arose by trial and error. As a result we express fossil genes which a rational approach to ontogeny would have eliminated. The optimist among you would hope that the Creator left them around for our amusement knowing that protein sequencing was difficult with fossils and that the intermediate animal evolutionary stages are extinct. He might be allowing us a last chance to trace and resurrect our ancestral immunoglobulins by looking in embryos.

I can see no formal arguments to answer the hypothesis that apparently silent genes are actually being expressed in an essential but unknown function. However, the argument is not entirely satisfying because selection on germ-line genes for an unknown function could not optimize them for the one we are studying, the immune system, for example, which can respond to any determinant I can construct and which therefore I presume to be a *de novo* stimulus in the world. It is unlikely that a species could long survive playing double bluff, keeping a huge library of genes selected for one purpose and being incidentally essential for another. Therefore, it is reasonable to consider the "neoradical" idea of somatic evolution.

Somatic evolution is not autonomous since it depends on the evolution of germ-line genes to give it guide-lines. Whatever the mechanism of variation, the germ-line gene determines the range and mean of the antibody response. However, in this sense, neither would the germ-line evolution of antibody specificity be autonomous. It would depend upon pre-existing genes. The characteristics of somatic evolution are that a limited number of germ-line genes are varied and expressed on a one cell-one specificity basis. The response to a *de novo* stimulus consists in selecting for the complementary cells and storing them in a somatic arsenal awaiting any future encounter with that stimulus.

A summary of the present situation then would be as follows:

Classical Cartesian geneticists, like Dreyer and Bennett (1965), would say, "Under the sun, each individual can do nothing new; it has all been invented. Nature never blunders; when she makes a fool, she means it."

Romantic Hegelian geneticists, like Edelman (1967), Smithies (1967), and Brenner (1966), would say, "New things, by each individual, under the sun are invented. Alas! In order for evolution to be intelligent, some individuals must be stupid."

So tempting is this hypothesis of somatic evolution that we would like to search the molecular biology of systems which cope with the unexpected and produce formal disproof of the germ-line hypothesis. I will disappoint you because we still lack the formal evidence necessary to decide between these two theories. However, I think that somatic models are much simpler and for this reason I prefer them. In order to evaluate the state of our present understanding I will consider the immune system first, in considerable detail, and will finally propose what I feel is a simple theory accounting for the range of its responses. Then I will take a cursory look at the detoxification and learning mechanisms.

II. THE IMMUNE SYSTEM

A. The Paradoxes of the Origin of the Specificity: "Krapp's Last Tape"

I will be dealing almost exclusively with amino acid sequence data and their implications. I should begin by stressing that all of our analyses are on serum immunoglobulins or myeloma proteins which are a product of two processes. One is the generation of diversity and the other is the selection by antigen of those cells which express a complementary product. Therefore, somatic evolution so mimics germ-line evolution that the amino acid sequences cannot distinguish the two, unless some very unusual relationships emerge. In any case, the arguments are indirect. In spite of the fact that we cannot analyze the products of unselected cells generated during somatic or germ-line evolution, we should be able to say what the process of diversification is. Under a germ-line model, any given family of sequences appears to be conserved because evolution proceeds via the random process of mutation followed by selection. The same could be true of somatic selective processes which act on mutationally diversified somatic genes. However, in addition, under somatic models, there is the possibility (in fact, a very popular one) that selected germ-line genes are varied by an orderly process of recombination or translation and antigenic selection does not distort this primary process. In any case, selection on one basic pattern of variation or on another has very different consequences and we should be able to read through this superimposed filter with as much success as has been encountered in analyzing the evolution of the hemoglobins, cytochromes or lysozymes.

The data are most complete for the variability of kappa light chains of immunoglobulins. I will begin by considering hypotheses in which the gene coding for the constant sequence (c) is in one copy in the haploid germ-line. The arguments for this are convincing (Dreyer and Bennett, 1965; Cohn, 1967; Lennox and Cohn, 1967). How many variable sequence genes (v) must we suppose and how are they associated in the germ-line? The answer to this question is the starting point for arriving at a mechanism of variability.

The repeating of all of the v-genes in the germ-line eliminates the "hanky-panky," to quote Dreyer's comment at the present conference, of a mechanism for generating diversity. This is true, but the germ-line model requires so many other special assumptions that it is no less complicated than somatic models. I will recall these without amplification as they are discussed elsewhere (Cohn, 1966; Lennox and Cohn, 1967).

1. The Lennox-Cohn (1967) argument: each one of the 100 or more repeated v-regions must have an identical sequence which permits specific association with the single gene for the constant region of kappa. Mutational drift should affect this sequence which would be expected to code for a portion of the kappa sequence at the amino-terminal end (position 108+) of the constant region. In other words, the same problem of mutation drift which led all of us to assume one gene-copy in the haploid germ-line for the constant sequence is, at best, only partially resolved if v and c are separated.

2. The Kabat (1967) argument: the variable regions of mouse and man are almost indistinguishable whereas the constant regions are quite distinct. Therefore, selection on the variable regions was more restrictive than that maintaining constant regions. If v were in many copies, e.g., 100–1000 in the germ-line, this would be improbable.

3. The Doolittle (1966) argument: there are indications, however, that in the variable region of kappa chains, some species-specific positions exist. As an example, for all kappa chains, position 72 is Ser in mouse and Thr in human (Dreyer *et al.*, 1967), position 1 is Ala or Ile in rabbit (Doolittle, 1966) and Asp or Glu in mouse and human, and position 5 is Thr in human or mouse and Glu in rabbit (Doolittle, 1966). Possibly also the rabbit kappa chain has an extra disulfide bridge in its variable region (Crumpton and Wilkinson, 1963). The evolution from a primordial, antibody-producing ancestor to a rabbit, mouse or human would require precise repeated events so that each of the hundred ancestral v-genes would give one hundred rabbit, mouse or human genes, each with their repeated identical species-specific positions. This too is improbable for germ-line evolution.

4. The Lennox-Cohn (1967) argument: the Doolittle (1966) analysis applies also to the allelic genetic markers A1, A2, A3 in the variable region of the heavy chain of rabbits. For example, position 9 in the $H\gamma$-chain is Arg for allele A1 or A2 and is replaced by an as yet unidentified amino acid for allele A3 (Porter, 1967a, b). In order to evolve from the rabbit ancestor with 100 v_H-genes to a rabbit with its v_H-alleles, tandem repeated genes with identical markers must evolve. This requires that polymorphism preceded duplication which has occurred independently three times. This seems to me to be very unlikely.

5. The Suran-Papermaster argument (Suran and Papermaster, 1967): the leopard shark has the following amino-terminal sequences for pooled subunits:

	1	2	3	4	5	6
Heavy Chain	Glu	Ile	Val	Leu	Thr	Gln
Light Chain	Asp Glu	Ile	Val	Leu Val Gly	Thr	Gln
Rabbits have:						
Heavy Chain	PCA	Ser Gln	Val Leu	Glu	Ser	Gly
Light Chain	Ala Ile	Val	Val Leu	Val	Gln Ala	Gln Ala

	1	2	3	4	5	6
Humans have:						
Heavy Chain (one myeloma) Daw	PCA	Val	Thr	Leu	Arg	Glu
Light Chain	Asp Glu	Ile	Gln Val	Met	Thr Leu	Gln

The sequences from the leopard shark show that light and heavy chain must have had a common evolutionary origin. Since the leopard shark makes antibodies it must have variable light and heavy chain sequences which by assumption are all carried in the germ-line as tandem sequences. The light and heavy chain variable sequences diverged in evolution to rabbits and men such that their patterns are different. It is unlikely that repeated sequences could have undergone a process of selection to become entirely distinct classes.

6. The additional-trouble argument: once evolved, special mechanisms must be envisaged to prevent recombination and mutation, for the alleles would become linked and the species-specific amino acids lost.

Therefore, even if all immunoglobulin-genes were to be carried in the germ-line, "hanky panky" is necessary to explain the evolution and genetics of the system.

Can we escape these dilemmas of germ-line models by proposing an evolutionary scheme which is reasonable? Whether the diversity is generated in the soma or germ-line, the evolution of the v_κ, v_λ, v_H, c_κ, c_λ and c_H must be coordinated because the interaction between them yielding a functional immunoglobulin is what is selected for. Since there are two subunits, the favored mutations in each are such that the evolution of them proceeds by alternating reciprocal improvements in their interactions; these may take place (1) in the variable region sequence of light and heavy subunits, (2) in the pairing regions for a specific translocation linking v_X to c_X, (3) in the complementing region in c_{light} and c_{heavy} responsible for light-heavy chain interaction, and (4) in the recognition site for secretion (Cohn, 1967a). A most likely attempt to avoid some of the above listed dilemmas of germ-line theories would be to assume that only *one* of the repeated v genes underwent speciation or divergence to the next evolutionary basal pattern and then duplicated. However, there being many v-genes but only one c-gene per class, the chosen v_L-gene, for example, could diverge but had to keep its pairing region unchanged. The sequences of the constant regions of mouse and man makes this event unlikely. Furthermore, the diverging v_L-gene had to fit with the one v_H-gene which was simultaneously undergoing divergence to the next species. This concordant divergence of the two unique genes for v_L and v_H had to go on while all of the other v-genes were functioning in antibody synthesis. Therefore, some selection must be envisaged for the one gene undergoing divergence to the next stage without its being effectively expressed as antibody.

The second problem is that gene polymorphism yielding the rabbit A1, A2, A3 markers had to precede gene duplication which had to occur three times, once for each marker. If another interpretation of the rabbit genetics were possible this complexity could be avoided. One could postulate, as I discussed elsewhere (Cohn, 1967c), that the genetic markers of the rabbit are in the constant cistron which codes for a sequence 1 to 9 and 221 to end. The variable gene could be translocated and inserted between these two positions. However, in order to do this one must in addition assume that the "Todd phenomenon" (Todd, 1963), i.e., all heavy chain classes carry the same genetic markers, is in doubt and that the genetic markers are associated only with positions 9 and 220.

This detailed argument points out that even if we try to make the evolution of repeated germ-line genes plausible by assuming that one v-gene in each class diverges first for speciation, then duplicates in the germ-line before undergoing diversification for antibody specificity, the picture is not simple. The evolution of one germ-line gene in a somatic model solves the evolutionary complexity because the duplication of the germ-line gene occurs in the soma by duplicating cells each expressing one immunoglobulin. Furthermore, the paradox of the germ-line model, due to one diverging gene being selected upon in the face of a functioning antibody-producing mechanism, is avoided. For me these paradoxes kill germ-line theories but I am prepared to admit that a good experiment would be satisfying.

The fact that the evolutionary dilemma is resolved by having only one or two germ-line v-genes for each constant-class now raises the question of how somatic diversification is generated. In order to discuss this I will ignore, for the moment, the constant region and the relationship between v and c. Let us start with the assumption that there is only one v-gene per constant-class carried in each individual's germ-line. As a result of this assumption, the mechanism of variation is virtually limited to mutation. Since the sequences known for human and mouse light chains make it clear that in the population at least there are $\geq$ three classes of v_κ for example, the simplest hypothesis would be that they are allelic and I will return to this point later. A somatic mutational hypothesis introduces a new factor. Random mutation would lead to less than 1 in 10^4 cells with a functional immunoglobulin (Lennox and Cohn, 1967). The waste of individuals which occurred during the eons of germ-line evolution is now transposed to the soma as a waste of cells. I should point out that a major source of somatic cell waste, even if all genes were in the germ-line, would be the generating of specificities (LH combination) for which no corresponding antigen was present. This is why most repeated germ-line genes would be silent. The mutational waste due to nonfunctional subunits would be superimposed on this. It is because of the waste-factor that it is worth examining the two or more gene-model in which waste is minimized by the orderly process of recombination.

If we postulate two genes which undergo recombination (Smithies, 1967), understanding either the evolution or the genetics suffers no difficulties. We need only admit that one of the two recombining v-genes carries the rabbit type (A1, A2, A3) allelic markers and that the species-specific positions are identical in both v-genes. Since two genes would allow for strong germ-line selection, the evolution is no problem. The variants on one theme would produce one spectrum of combining sites and the variants on another a different spectrum, both spectra of specificity being essential to the animal (Cohn, 1967b).

However, when we consider three or more tandem v-genes which undergo mutation or recombine (Edelman and Gally, 1967) to generate variability, we begin to feel the noose of the germ-line paradoxes which I have discussed. The only way to rationalize even a small number of v-genes is to see them as classes with special functions upon which selection acts. I will turn to this question now.

Hood, Gray, Sanders and Dreyer, in an elegant investigation of light chain evolution (Hood *et al.*, 1967), were the first to notice that there were at least three major classes of kappa chain which they called κI,

Position	0	1	2	3	4	5	6	7	8	9	10	11	12	13	14	15	16	17	18	19	20	21	22	23	24	25	26	27
KAPPA I	–	Asp	Ile	Gln	Met	Thr	Gln	Ser	Pro	Ser	Ser	Leu	Ser	Ala	Ser	Val	Gly	Asp	Arg	Val	Thr	Ile	Thr	Cys	Gln	Ala	Ser	Gln
AG	—	—	—	—	—	—	—	—	—	—	—	—	—	—	—	—	—	—	—	—	—	—	—	—	—	—	—	—
ROY	—	—	—	—	—	—	—	—	—	—	—	—	—	—	—	—	—	—	—	—	—	—	—	—	—	—	—	—
EU	—	—	—	—	—	—	—	—	—	—	Thr	—	—	—	—	—	—	—	—	—	—	—	—	—	Arg	—	—	—
BJ	—	—	Val	—	—	—	—	—	—	—	—	—	—	—	—	—	—	—	—	—	—	—	—	—	—	—	—	—
KER	—	—	—	—	—	—	—	—	—	—	—	—	—	—	—	—	—	—	—	Ile	—	—	—	—	—	—	—	—
DAY	–	.	.	.	.	.	.	.	.	.	.	.	.	.	.	.	.	.	.	—	—	—	—	—	—	—	—	—
CRA	—	—	—	—	—	—	—	—	—	—	—	—	—	—	—	Leu	Arg	—	—	—	—	—	—					
PAP	—	—	—	—	—	—	—	—	—	—	—	—	—	Val	—	—	—	—	—	—	—	—	Ala					
LUX	—	—	—	—	Leu	—	—	—	—	—	Phe	—	—	—	—	—	—	—	—	—	—	—	—					
MON	—	—	—	—	—	—	—	—	—	—	Thr	—	—	—	—	—	—	—	—	—	—	—	—					
CON	—	—	—	—	—	—	—	—	—	—	—	—	—	—	—	—	—	—	—	—	—	—	—					
TRA	—	—	—	—	—	—	—	—	—	—	—	—	—	—	—	—	—	—	—	—	—	—	—					
HBJ10	—	—	—	—	—	—	—	—	—	—	—	—	—	—	—	—	—	—	—									
HBJ1	—	—	—	Leu	—	—	—	—	—	Thr	—	—	—	—	—	—	—	—	—									
HBJ4	—	—	—	—	—	—	—	—	—	—	—	—																
HBJ6	—	—	—	—	—	—	—																					
MOUSE 41	—	—	—	—	—	—	—	—	—	—	—	—	—	—	—	Leu	—	—	—	—	Ser	Leu	—	—	Arg	—	—	—
KAPPA II	–	Asp	Ile	Val	Leu	Thr	Gln	Ser	Pro	Leu	Ser	Leu	Pro	Val	Thr	Pro	Gly	Glu	Pro	Ala	Ser	Ile	Ser	Cys	Arg	Ser	Ser	Gln
HBJ3 (Mil)	—	—	—	—	—	—	—	—	—	—	—	—	—	—	—	—	—	—	—	—	—	—	—	—	—	—	—	—
MAN	—	—	.	.	.	.	.	.	.	.	—	—	—	—	—	—	—	—	—	—	—	—	—	—	—	.	.	.
CUM	Glu	—	—	—	Met	—	—	Thr	—	—	—	—	—	—	—	—	—	—	—	—	—	—	—	—	—	—	—	—
MOUSE 70	—	—	—	—	—	—	—	—	—	Ala	—	—	Ala	—	Ser	Leu	—	Arg	—	Thr	—	—	—	—	—	Ala	—	Glu
MOUSE 6	—	—	—	—	Val	—	—																					
MOUSE 157	—	—	—	—																								
KAPPA III	–	Glu	Ile	Val	Leu	Thr	Gln	Ser	Pro	Gly	Thr	Leu	Ser	Leu	Ser	Pro	Gly	Glu	Arg	Ala	Thr	Leu	Ser	Cys	Arg	Ala	Ser	Gln
RAD	—	—	—	—	—	—	—	—	—	—	—	—	—	—	—	—	—	Asp	—	—	—	—	—	—	—	—	—	—
Fr4	—	—	—	—	—	—	—	—	—	—	—	—	—	—	—	—	—	—	—	—	—	—	—	—	—	.	.	.
NIG	—	Lys	—	—	—	—	—	—	—	—	—	—	—	—	—	—	—	—	—	—	—	—	—					
CAS	—	—	—	—	—	—	—	—	—	—	—	—	—	—	—	—	—	Asp	—	—	—	—	—					
HBJ5	—	—	—	—	—	—	—	—	—	Asx	—	—	—	—	—	—	—	—	—									
HS4	—	—	—	—	—	—	—	—	—	—	—	—	—	—	—	—	—	—	—									
GRA	—	—	Met	—	Met	—	—	—	—	Ala	—	—	—	Met	—	—	—	—	—	—	—	—	—					
WIN	—	Asp	—	—	—	—	—	—	—	Ala	—	—	—	—	—	—	—	—	—	—	—	—	—					
SMI	—	—	—	—	—	—	—	—	—	Ala	—	—	—	—	—	—	—	—	—	—	—	—	—					
HBJ12	—	—	—	—	Val	—	—																					
MOUSE 46	—	Asp	—	—	—	—	—	—	—	Ala	Ser	—	—	Val	Thr	—	—	Asp	(Ser,	Val,	Ser)	—	—	—	—	—	—	—
MOUSE 9	—	PCA	—	—	—																							
HUMAN POOL	–	Asp Glu	Ile	Gln Val	Met Leu	Thr	Gln	Ser	Pro	Ser Gly Ala Val Leu Pro	Ser Thr	Leu Val	Ser	Ala Val Leu	Ser Thr Val	Val Pro Leu	Gly	Asp Glu	Arg									
LAMBDA I	–	PCA	Ser	Ala	Leu	Thr	Gln	Pro	Pro	Ser	–	Gly	Ser	Gly	Ser	Pro	Gly	Gln	Ser	Val	Thr	Ile	Ser	Cys	Thr	Gly	Thr	Ser
BO	—	—	—	—	—	—	—	—	—	—	—	—	—	—	—	—	—	—	—	—	—	—	—	—	—	—	—	—
HBJ2	—	—	—	—	—	—	—	—	—	—	—	Ala	—	—	—	—	—	—	—	—	—							
HBJ8	—	—	—	—	—	Ala	—	—	Ala	—	—	Val	—	—	—	—	—	—	—	Ile	—							
LAMBDA II	–	PCA	Ser	Val	Leu	Thr	Gln	Pro	Pro	Ser	–	Gly	Ser	Gly	Thr	Pro	Gly	Gln	Arg	Val	Thr	Ile	Ser	Cys	Ser	Gly	Gly	Ser
HA	—	—	—	—	—	—	—	—	—	—	—	—	—	—	—	—	—	—	—	—	—	—	—	—	—	—	—	—
HBJ7	—	—	—	—	—	—	—	—	—	—	—	Ala	—	—	—	—	—	—	Gly	—	—							
HBJ98	—	—	—	—	—	—	—	—	—	—	—	Val	—	Ala	Ala	Asx	—	—	Ala	—	—	Ser	Ile	—				
HBJ11	—	—	—	—	—																							
LAMBDA III	–	–	Ser	Glu	Leu	Thr	Gln	Asp	Pro	Ala	–	Val	Ser	Val	Ala	Leu	Gly	Gln	Thr	Val	Arg	Ile	Thr	Cys	Gln	Gly	Asp	Ser
SH	—	—	—	—	—	—	—	—	—	—	—	—	—	—	—	—	—	—	—	—	—	—	—	—	—	—	—	—

Table 1. Comparison of amino acid sequences in human and mouse light chains. The human sequences CRA, PAP, LUX, MON, CON, TRA, NIG, CAS, GRA, WIN, SMI, and the HUMAN POOL were taken from Niall and Edman (1967). AG, BO, HA and SH are from Putnam (1967). ROY is from Hilschmann (1967) and CUM is from Hilschmann (1967a). BJ, KER, DAY, RAD, MAN and Fr4 are from Milstein (1967). HBJ3 (Mil), MOUSE 41 and MOUSE 70 are from Dreyer *et al.* (1967). HBJ10,

27a	27b	27c	27d	27e	27f	28	29	30	31	32	33	34	35	91	92	93	94	95	96	104	105	106	107	108	109	110	110a	Inv type
–	–	–	–	–	–	Asp	Ile	Ser	Ile	Phe	Leu	Asn	Trp	Tyr	Asp	Asn	Leu	Pro	Arg	Leu	Glu	Ile	Lys	Arg	Thr	Val	–	
–	–	–	–	–	–	___	Asx	___	___	___	___	___	___	___	Glu	Thr	___	___	___	___	___	___	___	___	___	___	–	Inv (3)
–	–	–	–	–	–	___	___	___	___	___	___	___	___	Phe	___	___	___	___	Leu	Val	Asp	Phe	___	___	___	___	–	Inv (1, 2)
–	–	–	–	–	–	Ser	___	Asx	Thr	Trp	___	Ala	___	.	.	.	.	.	.	Val	___	Val	___	Gly	Thr	Val	–	Inv (3)
–	–	–	–	–	–	___	___	Asn	Lys	Tyr	.	.	.	___	Glu	___	___	___	Tyr	___	___	___	___	.	.	.	–	Inv (1, 2)
–	–	–	–	–	–	___	___	Lys	Asn	___	.	.	.	___	___	Asp	___	___	Pro	Val	Asp	Leu	___	.	.	.	–	Inv (1, 2)
–	–	–	–	–	–	___	___	___	Asx	___	.	.	.	.	.	.	.	.	.	___	Asp	___	___	.	.	.	–	Inv (3)
–	–	–	–	–	–	___	___	Gly	Ser	Leu	Ser	___	___	___	Ala	Ser	Ser	___	Trp	___	___	___	___	___	Ala	Asx	–	
–	–	Asn	Leu	Leu	Glx	Ser	Asx	Gly	Asp	Tyr	Leu	Asp	Trp	Ser	Leu	Glu	Thr	Pro	Leu	Leu	Glu	Ile	Lys	Arg	Thr	Val	–	
–	–	___	___	___	___	___	___	___	___	___	___	___	___	Ala	___	___	___	___	___	Val	___	___	___	.	.	.	–	Not Inv (1, 2)
.	.	.	.	.	.	.	.	.	.	.	.	.	.	.	.	.	.	.	.	Val	___	___	___	.	.	.	–	Inv (3)
Ser	Leu	Leu	Asp	Ser	Gly	Asp	Gly	Asn	Thr	___	___	Asn	___	Arg	___	___	Ile	___	Tyr	___	___	___	Arg	___	___	___	–	Inv (3)
–	–	Ser	Val	Asx	Asx	___	Gly	Ile	Ser	Phe	Met	Asn	___	___	Lys	___	Val	___	Trp	___	___	___	___	___	Ala	Asx	–	
–	–	–	–	–	–	Ser	Ser	Asn	Ser	Tyr	Leu	Asp	Trp	Tyr	Glu	Thr	Ser	Pro	Thr	Leu	Glu	Ile	Lys	Arg	Thr	Val	–	
–	–	–	–	–	Val	___	___	___	___	___	___	.	.	___	___	___	___	___	___	___	Asp	___	___	.	.	.	–	Inv (3)
.	.	.	.	.	.	.	.	.	.	.	.	.	.	.	.	.	.	.	.	___	___	___	___	.	.	.	–	Inv (1, 2)
–	–	–	–	–	–	His	Ile	Ser	Asn	Asn	___	___	___															
–	–	–	Ser	Asp	Val	Gly	Asx	Asx	Lys	Tyr	Val	Ser	Trp	Tyr	Val	Asx	Asx	Asx	Asx	Leu	Thr	Val	Leu	Arg	Glx	Pro	Lys	
–	–	–	Ser	Asn	Gly	Thr	Gly	Asn	Asn	Tyr	Val	Tyr	Trp	Trp	Asp	Tyr	Arg	Leu	Ser	Leu	Thr	Val	Leu	Arg	Glx	Pro	Lys	
–	–	–	–	–	–	Leu	Arg	Gly	Tyr	Asx	Ala	Ala	Trp	Arg	Asp	Ser	Ser	Gly	Lys	Leu	Thr	Val	Leu	Gly	Glx	Pro	Lys	

HBJ4, HBJ1, HBJ5, HS4, HBJ12, HBJ2, HBJ8, HBJ7, and HBJ11 are from Hood *et al.* (1967). MOUSE 46 is an unpublished sequence of Melchers and Lennox (Salk Institute for Biological Studies). HBJ98 is from Baglioni (1967). HBJ6 and MOUSE 6 are from Hood *et al.* (1966). MOUSE 157 and MOUSE 9 are from Perham *et al.* (1966). EU is from Edelman (1967). *Inv* type taken from Milstein (1967), except for HBJ3 (Mil), which is a personal communication from W. R. Gray (California Institute of Technology).

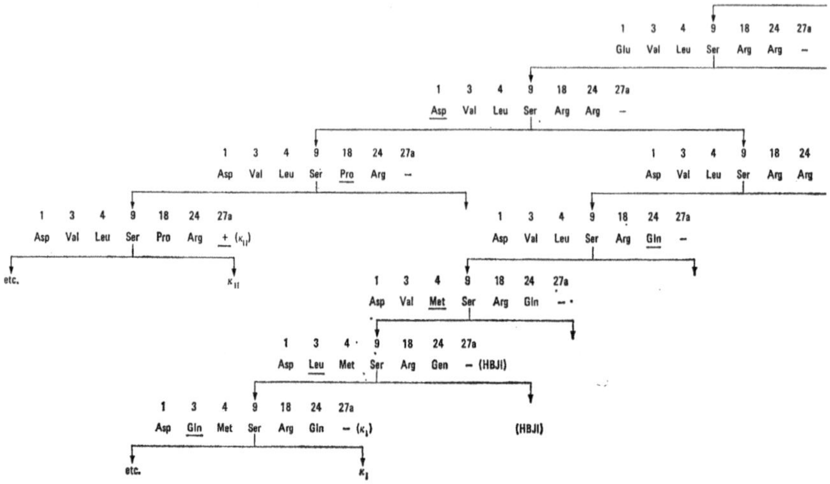

κII, and κIII. However, since they were arguing that all genes for the v-region are carried in the germ-line, a possible classification into three or three hundred categories was implicit and the degree of homology (defined as classes or sub-classes) is simply interpreted as a map guiding us through their pathway of evolution. In general, this observation was confirmed by Niall and Edman (1967) using data only from humans, as well as by Milstein (1967). Niall's and Edman's sequences fall clearly into two classes whereas the total available sequence data fall into at least three, as Milstein so clearly demonstrates. However, there is a certain amount of arbitrariness in stopping at three. Nevertheless, as a first approximation I have taken Milstein's classification and extended it slightly (Table 1).

How real are the classes? A glance at the data tells you that κI and κIII are major categories and evidently each cluster of sequences within the class is closely related. Niall and Edman (1967) did not find the κII class but their data on pooled human sera are not incompatible with a κII class in low concentration in such sera. Milstein's κII category is certainly distinct, above all, because it has been found in inbred mice (Mouse 70). κIV is a guess based on the frequency of Ala in position 9 seen in GRA, WIN and SMI (Table 1).

Dreyer and his colleagues might view the data in Table 1 as follows: Since they postulate "one V-sequence one germ-line" gene, the grouping into classes reveals nothing about the genetics of the system but is a question of convenience in visualizing a probable pathway of germ-line evolution by single mutational events.

All sequences should be arrangeable into a sequential series of hierarchical evolutionary events in the germ-line (Table 2). Each step in-

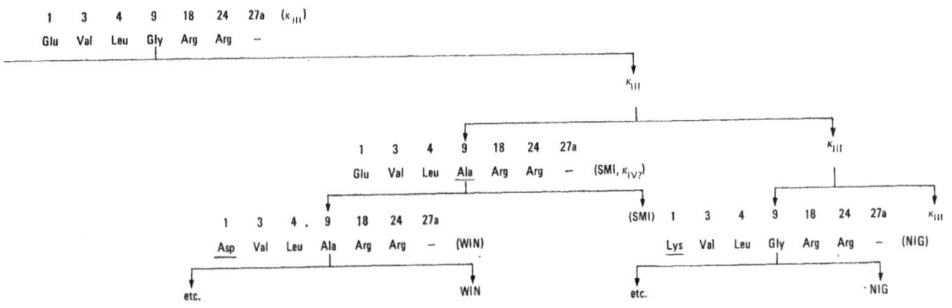

Table 2. A possible germ-line evolution by mutation of kappa genes to the various v_κ-classes ($v_{\kappa I}$, $v_{\kappa II}$, $v_{\kappa III}$) with examples of possible somatic mutation within the v_κ-classes. Underlined amino acid means replacement. + means insertion between positions 27 and 28 (see Table 1).

volves a gene duplication and a subsequent mutational event as I have illustrated for a small number of the known sequences. These germ-line events can be ordered differently and varied in many ways but the principle would be the same. A reciprocal reassortment of sequences as would be generated by recombination, e.g., ABCDEF × 123456 → ABC456 + 123DEF would look entirely different than random mutation if one tried to fit them on a tree generated by one-step events like those illustrated in Table 2.

Brenner, Milstein, and others who argue somatic mutation would say that germ-line evolution generated three or four genes (kappa I, II, III, IV) and these are now varied by somatic events. Clearly these data do not distinguish germ-line from somatic hypotheses. However, a somatic theory which deals with one or two germ-line v-genes or with three or four has different consequences as regards their relationship to the c-genes. Therefore, if one wishes to consider a simple somatic theory, it is critical to decide whether the genes for the three kappa classes are carried by all individuals or whether they are alleles.

Are the v_κ-classes alleles? It is truly unfortunate that Niall and Edman did not analyze individuals instead of a pool. However, Milstein (47) points to a study on individuals showing that of 12 analyzed, all possess light chains with Asp and Glu amino-termini. If one ignores the rare Glu N-terminal in κII and the rare Asp N-terminal in κIII this result seems to imply that either κI or κII and κIII are made by all individuals. Therefore, it can be argued that κI or κII and κIII are not alleles and that at least κI and κIII are carried by all individuals since these are the major classes. In humans, the κII class might be allelic to κI or κIII but since inbred mice of the BALB/c strain seem to show all three classes, e.g., mouse 41, 70, and 46, at least in this case they do not appear to be alleles but genes, possibly tandem, carried by all individuals. I am, for the moment, putting aside but will return to the possibility that inbred mice can be polymorphic at the v_κ-locus.

Let us, for arguments sake, accept three tandem genes in the germ-line and ask how they can be varied. The recombination mechanism is obviously a priori the most attractive (Edelman and Gally, 1967; Smithies, 1967) so that I will first show why it is untenable.

The three genes suspected of recombining would be κI, κII and κIII. Since κII is of different length than κI or κIII by an insertion precisely between positions 27 and 28, it is unlikely to enter into recombination with the others. Obviously the elegance of the recombination model is lost if unequal length pairings lead to waste. This might be circumvented by supposing recombinations under conditions where the insertions buckle out

but such a situation means that recombination in the buckled-out part should not occur or at least very infrequently. There are three variations on the numbers of amino acids in the insertion region of CUM (6 amino acids), HBJ 3 (4 amino acids), RAD (1 amino acid) (Table 1). This fact is in contradiction with a simple model and whether such variation implies classes of v will become obvious when inbred mouse sequences in this region are determined on several chains.

A possible recombination between $v_{\kappa I}$ and $v_{\kappa III}$ still remains. Consider six positions (Table 1) in which the amino acid replacements are not compatible with recombination.

Position	1	2	4	9	10	13
			↓			↓
κI	Asp GA^{U}_{C}	Ile $AU\overset{U}{\underset{A}{C}}$	Met AUG	Ser UCX / AG^{U}_{C}	Ser UCX / AG^{U}_{C}	Ala GCX
			↓			↓
κIII	Glu GA^{A}_{G}	Ile $AU\overset{U}{\underset{A}{C}}$	Leu UU^{A}_{G} / CUX	Gly GGX	Thr ACX	Leu UU^{A}_{G} / CUX
			↓			↓
Replacement	Lys AA^{A}_{G}	Met AUG	Val GUX	Asp GA^{A}_{G} / AA^{U}_{C}	Phe UU^{U}_{C}	Met AUG

The amino acid replacements in positions 4 and 13 are incompatible with a recombination because three different bases are involved in the first letter indicated by an arrow. However, since there is a possibility that UUG might code for methionine in animal cells (Nirenberg *et al.*,

1966), this contradiction could disappear. For positions 1, 2, 9 and 10, one could imagine a recombination between two codons yielding the observed amino acid replacements. However, considering position 1 as an example, >14/15 κI have Asp, whereas 9/11 κIII have Glu. It is reasonable to assume then that the codons for Asp and Glu are in the germline. The Lys replacement in κIII Nig then is not derivable by recombination between the codons for Asp and Glu. A similar argument holds for positions 2, 9 and 10.

This rules out a recombination model either of the Hilschmann (1967) or Whitehouse (1967) or Edelman-Gally (1967) or Smithies type (1967). The Edelman-Gally model could be modified to account for this contradiction by increasing (as they did) the number of recombining genes to ⩾ six and the Smithies model could be modified by postulating three recombining pairs of genes. The increase to six, however, makes it virtually a germ-line hypothesis as far as the previously discussed difficulties are concerned and the postulating of three recombining pairs is eliminated by the sequences:

κI × κI is eliminated by considering replacements at position 96

	Arg ${}^{C}_{A}$GX
	Leu ${}^{U}_{C}$UX
	Tyr UA${}^{U}_{C}$
	Pro CCX

κII × κII at positions 91		and	96
	Tyr UA${}^{U}_{C}$		Trp UGG
	Ala GCX		Leu ${}^{U}_{C}$UX
	Arg ${}^{C}_{A}$GX		Tyr UA${}^{U}_{C}$
κIII × κIII at position 4		and	9
	Leu ${}^{U}_{C}$UX		Gly GGX
	Met AUG		Asx AA${}^{U}_{C}$ GA${}^{U}_{C}$
	Val GUG		Ala GCX

As I mentioned before, the Leu-Met-Val contradiction would be resolved if animal cells use UUG as a codon for Met (Nirenberg *et al.*, 1966). The contradiction at position 9 could be resolved by assuming another class κIV for chains with Ala at position 9 and which now give four pairs of recombining genes, but this escape not even ardent proponents of somatic recombination would take.

Smithies could now answer as follows: the contradictions are due to two kinds of additional variation superimposed on recombination, assuming, of course, that the *Escherichia coli* code is universal.

1. Mutations could have arisen in the myelomas subsequent to the initial event of diversification. This probability is increased since the myelomas chosen for analysis were defective in that they secreted light chain only. We know that such mutations to defective synthesis do occur and are selected for (Cohn, 1967a). Therefore, many Bence Jones protein secreting tumors could produce defective light chains unable to complement with heavy chain. This argument could account for the incompatibility of the replacement data with recombinational diversification but not for the variable length of the insertion which is always precisely between positions 27 and 28 in a postulated κII $\times$ κII recombination.

2. The insertions could be due to polymorphism in the germ-line and the incompatible replacements could be the sum of subsequent mutations in the myeloma, ambiguity in translation and polymorphism. Since the major data are from humans, all this is possible, in fact satisfying. However, Gray, Dreyer and Hood (Gray *et al.*, 1967) showed that an insertion between positions 27 and 28 is also found in myelomas derived from mice of the inbred strain BALB/c. Of course, there being only three known sequences in mice (Table 1) one can either assume crossing-over with the insertion region not participating, since the contradiction of variation in the length of the insertion region has not yet been found in mice, or three pairs of recombining genes in the mouse population. Whether or not they are tandem or allelic depends upon how much faith one has that inbred mice are not polymorphic at the v-locus. Clearly as the contradictions mount this possibility merits testing.

One last argument against the recombination model for diversification was alluded to previously.

If the classification into κI, κII and κIII is accepted then recombination models are virtually ruled out because of the lack of symmetrical reassortment of sequences, for example:

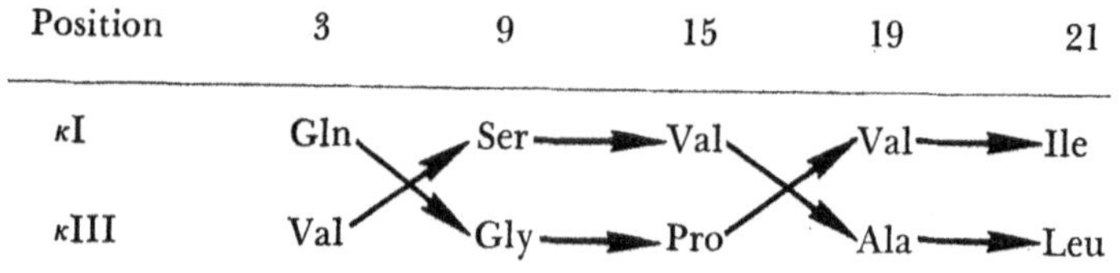

No recombinatory sequences of the following kinds are found:

1	Gln	Gly	Pro	Val	Ile
	Val	Ser	Val	Ala	Leu
2	Gln	Ser	Pro	Ala	Leu
	Val	Gly	Val	Val	Ile
3	Gln	Ser	Val	Ala	Leu
	Val	Gly	Pro	Val	Ile

The shape of the variations are the branched trees of mutation not the knotted braids of recombination. Therefore, I would conclude that the sequences in Table 1 were not derived by recombination between any two of the postulated germ-line genes for the classes.

The ruling out of recombination as a generator of diversity now makes mutation a good candidate. Even before the sequences began to be a graduate student parlor game, there was evidence for a mutational mechanism derived from the characteristic banding pattern of light chains seen on electrophoretic analysis (Cohen, 1965, 1966). Each band differs from its neighbor by one charge and the pattern shows a major central band of protein with progressively decreasing amounts in each band separated from it either in the negative or positive direction. The distribution of amounts in each band (equivalent to rates of synthesis of each band) is roughly Gaussian, peaking at the central band. Such a finding is easily interpretable as due to a mutational pattern affecting charged amino acids starting from the sequences involved in the central band and each band from it involving one, two, three, etc., sequential mutations. The bands at the extremes may be very minor because many mutational steps from the mean are involved. Of course, what happens to charged amino acids could also be happening to the others but may not be resolvable in the electrophoretic system. When the sequences appeared in sufficient number to make the case for mutation clear, the recognition of classes, as Dreyer and his colleagues stress, only reinforced the conclusion since replacements involving more than one base change within a class are infrequent. This fact, extended by new sequences and recognition of variable classes, even for lambda (Table 1) and probably for heavy chains, has not simplified the assumptions (Lennox and Cohn, 1967) necessary for meaningful translation level models (Campbell, 1967; Mach *et al.*, 1967; Potter *et al.*, 1965) which do not predict that the amino acid replacements correspond to one base changes in the DNA.

At this point along with Lederberg (1959), Dreyer and Bennett (1965), Brenner and Milstein (1960) and Lennox and Cohn (1967), I think we could agree that variation is due to mutation and concentrate on the arrangements of genes in the germ-line and on possible somatic mechanisms of generating a diversified population of antigen-sensitive cells.

The attractiveness of any somatic theory will depend upon how plausible a model of variation by mutation one can propose. In order to do this I must first return to the relationship of the v-gene to the c-gene in the germ-line which thus far has only been alluded to.

The human kappa chain has two groups of genetic determinants, one in the v-region, $v_{\kappa I}$, $v_{\kappa II}$, $v_{\kappa III}$, and the other in the c_κ-regions, *Inv(1, 2)* and *Inv(3)*. There are two ways that these determinants can be associated in the germ-line, as a vc-gene or as a v-gene which at the DNA, RNA, or protein level is associated with the c-gene or its product.

Let us dwell a moment on this point comparing somatic and germ-line views of the association of genes, v and c. A vc-germ-line gene is ruled out if it is to be repeated several times in the germ-line and under a simple model, more than once (Lennox and Cohn, 1967). The separation of v from c entails some step linking them at the DNA, RNA, or protein level. Recent data (Fleischman, 1967; Knopf *et al.*, 1967; Lennox *et al.*, 1967) argue against the joining of polypeptide chains leaving linking at the RNA level which is unlikely, or a DNA-translocation. This latter possibility, proposed by Dreyer and Bennett (1965) because they favor a germ-line model, would be mediated by a constant pairing region in both v and c but distinct for kappa, lambda and heavy chain classes. It is obvious that this region must cover a very long stretch not only for accurate pairing (Lennox and Cohn, 1967) but as indicated by the sequences. If the pairing region involved only the two amino acids of kappa positions:

Positions	109 110
	. . . Thr.Val . . .
	ACX GUX
and lambda positions	. . . Pro.Lys . . .
	CCX AA_G^A

it would be too small. A longer stretch in either direction, on the amino terminal side, is variable and on the carboxyl terminal side, identical for both kappa and lambda, . . . Ala.Ala.Pro.Ser.Val. . . . Therefore, these stretches cannot be the total pairing region. Pairing must extend over a very long sequence and must be sensitive to small differences, since between positions 108 and 168 (61 amino acids), 26/61 are identical, 24/61 are 1-base replacements and 11/61 are 2-base replacements. The reason for counting between 108–168, and not further, is the necessity to introduce a deletion at 169 to maximize the homology (Putnam, 1967) between kappa and lambda. These data make it unlikely that there is a simple translocation mechanism in which the beginning of the c-gene is the pairing region (Lennox and Cohn, 1967). One can make more complicated models in which the pairing region is not translated (Dreyer *et al.*, 1967) or is not needed (Hilschmann, 1967). Certain data which I will analyze now argue that it may be necessary to do so, since they imply that a translocation (or its equivalent) does in fact occur.

These data are:

1. $v_{\kappa I}$ and $v_{\kappa III}$ seem to be "tandem" genes, not alleles (Milstein, 1967).
2. Inv(1, 2) and Inv(3) appear to be alleles (Ropartz, 1964; Steinberg, 1962).
3. The $v_{\kappa I}$ and $v_{\kappa III}$ genes are each associated with both polymorphic forms of c_κ, Inv(1, 2) and Inv(3) (see Table 1).
4. v_κ of mouse and man are nearly indistinguishable whereas the c_κ is very different in the two species (Kabat, 1967a, b).
5. The $v_{\kappa I}$, $v_{\kappa II}$, and $v_{\kappa III}$ classes appear to be present in the inbred BALB/c mouse (Table 1).

6. All classes of rabbit heavy chain seem to carry the same allelic genetic markers A1, A2, A3 in the amino-terminal half of the molecule (the Todd-phenomenon, 1963).

The association of each v_κ-class with either allele of c_κ is formally equivalent to that found for the rabbit heavy chains where allelic markers in the v-region are associated with the non-allelic genes of the classes.

Taken at face value then these data say that the germ-line has three tandem v_κ-genes and one c_κ-gene which are linked together to result in a $v_\kappa c_\kappa$ cistron by translocation. Since v_λ also seems to have classes (Table 1) and c_λ is polymorphic (Appella and Ein, 1967; Ein and Fahey, 1967; Quattrocchi, 1967) a similar assumption would be made for lambda cistrons. It might be simplest in the case of v_H to postulate one gene which is translocated to the various c_H-genes for the classes of γ, α and μ (Todd, 1963; Lennox and Cohn, 1967).

Is there any way to save a *minimum model* of one vc gene per class in the germ-line?

It would be possible to question the conclusion, hinted at by Milstein (1967), that the $v_{\kappa I}$ and $v_{\kappa III}$ are not alleles if it is based solely on the finding of amino-terminal Glu and Asp in the light chains of twelve individuals. These two amino acids are not sufficient to define a class since we can see already that both κII and κIII have Asp and Glu members (Table 1). However, it is very difficult to explain the finding that inbred mice possess at least $v_{\kappa I}$ and $v_{\kappa III}$ and probably also $v_{\kappa II}$, except by admitting that they are non-allelic. I had assumed previously (Cohn, 1967b) that we are misled by inbreeding and that at certain loci the mice are polymorphic because "phenotypic-haploidy" in mice would be lethal. This postulate might be disproven, and if so, a rather new look at inbred laboratory animals from Drosophila to mice will be necessitated. Nevertheless, I must admit that this assumption is not very probable.

Even were we to ignore the above arguments, and assume $v_{\kappa I}$, $v_{\kappa II}$, and $v_{\kappa III}$ to be alleles, how could the associations of $v_{\kappa I}$ and $v_{\kappa III}$ with both alleles of c_κ come about other than by translocation? We might assume, for example, recurrent identical mutations which were highly selected for. Suppose that the primordial gene was $v_\kappa c_\kappa$ which mutated to give the alleles $v_{\kappa I} c_\kappa$ and $v_{\kappa III} c_\kappa$ which comprise one class of polymorphisms. Each v-class of $v_\kappa c_\kappa$ then diverged into *Inv(1, 2)* and *Inv(3)* to give, for example, $v_{\kappa I} c_{\kappa Inv(1,2)}$ or $v_{\kappa I} c_{\kappa Inv(3)}$. If this repeated event sounds improbable, consider the translocation model. V-diverged to give $v_{\kappa I}$ and $v_{\kappa III}$. In this respect, both models are identical. C_κ diverged to give $c_{\kappa Inv(1,2)}$ or $c_{\kappa Inv(3)}$. Obviously, if a rare mutation occurs to yield a polymorphic gene, one would never see this event unless there were strong selection for the variant in order to bring it up to the steady state level in the population at which the Hardy-Weinberg law applies. It is a minor extrapolation in the face of this necessitated strong selection pressure to assume that the mutation and its spread in the population happened twice, once for $v_{\kappa I}c_{\kappa}$ and once for $v_{\kappa III}c_{\kappa}$. However, Dr. William B. Gray pointed out to me that this argument explains the last step of the evolution but ignores the preceding ones. The $v_{\kappa I}$ and $v_{\kappa III}$ genes of man and mouse are almost identical whereas the c_{κ} genes are very different. If throughout evolution there were a vc germ-line gene, then as one example, the following improbable pathway of evolution can be proposed:

$$\begin{array}{ccccc} v_p c_p & \longrightarrow & v_{\kappa I} c_p & \longrightarrow & v_{\kappa III} c_p \\ \text{primordial (p)} & & \downarrow & & \downarrow \\ & & v_{\kappa I} c_{mouse} & & v_{\kappa III} c_{mouse} \\ & & \downarrow & & \downarrow \\ & & v_{\kappa I} c_{man} & & v_{\kappa III} c_{man} \end{array}$$

In this example, the evolution of the c_p gene to c_{mouse} and c_{man} must occur twice, independently. The argument is that there is no pathway which can be drawn from a primordial vc gene to the vc genes of mouse and man that do not require repeated unlikely events due to the parallel identical evolution of either the v or c gene halves. Gray, therefore, argues strongly for separate v and c genes in the germ-line.

An identical situation exists for the heavy chains where the "Todd phenomenon" (Todd, 1963), analyzed previously (Lennox and Cohn, 1967), implies a translocation of v_H carrying the alleles A1, A2, A3 to c_H of the various classes $C\gamma$, $C\alpha$, $C\mu$, etc. However, this conclusion will be formally established only when it has been shown by sequence studies that the allelic markers A1, A2, and A3, associated with positions 9 and 220 on the rabbit $H\gamma$ chains (Porter, 1967a, b) are also associated with positions 9 and 220 on $H\alpha$ and $H\mu$.

In summary, then, the data are all pointing in the same direction. Separate v and c genes seem to be carried in the germ-line. This conclusion, necessitated by germ-line models, is not a priori required by somatic models and therefore comes as a surprise which could well bear on the question of how diversity is selected for. I will assume *contre-coeur* for the remainder of the discussion that there are three v_{kappa} genes, three v_{lambda} genes, and one or several v_H genes carried by all individuals. A linking of v and c or their products would be necessitated. This result predicted by germ-line theories must eventually be rationalized by any somatic theory. However, the mechanism of somatic diversification which I will discuss could cope either with the more elaborate translocation (Dreyer *et al.*, 1967; Hilschmann, 1967) or the *minimum model.*

There are two ways to arrive at a somatic cell population varied by mutation; either you take the road of Brenner and Milstein and postulate a mechanism which differentially hypermutates v and not c or you allow an average uniform mutation rate throughout the gene vc and postulate selection of a unique kind. The latter, of course, is what germ-line theories propose happened during evolution but arriving at a plausible selection mechanism for somatic models is more difficult because of the time scale and the absence of somatic sexual mechanisms.

What is most bothersome about a mechanism of differential hypermutation such as the one presented by Brenner & Milstein (or for that matter any so far discussed or published (Hechter and Halkerston, 1964)) is its unlikelihood (Cohn, 1967b). This, however, is a question of definition. Nevertheless a hypermutation mechanism generates antigen-sensitive cells of many specificities so that, without any special assumptions, antigen can be postulated to select amongst them. The waste of such a mechanism does not allow a disproof of it but it is interesting to estimate as I will do now.

Consider the κI class up to position 27 and assume that 23 to 27 is invariant (see Table 1). In 10 sequences there are 9 replacements or an average of 1 replacement per 25 amino acids and from positions 80 to 107 there are 11 replacements in four sequences or 1 replacement per 10 amino acids. Whether this suggests a gradient of mutation remains to be seen. On an average in any one class, I would estimate that one v-region chosen at random from one class in a population differs from the other and from the germ-line by five amino acids (1 in 20 being replaced). I will ignore the small factor involved in mutations that do not lead to an amino acid replacement, i.e., $\sim 2/3$ mutations lead to amino acid replacements. If a germ-line κI chain has 50 out of 100 constant positions and the probability of any amino acid being replaced is 1 in 20 (0.05), then the probability of making a functional light chain is $(0.95)^{50} \sim 0.01$. If the probability of making a functional heavy chain is of the same order then $(0.01)^2 = 1$ in 10^4 cells would produce a functional immunoglobulin. This result is not very different than that arrived at from earlier data without considering the kappa classes (Lennox and Cohn, 1967). The number of functional cells as a maximum value could be as high as 1 in 10^3 and as a minimum as low as 1 in 10^5. Clearly, this seems to be a large waste. However, as one reduces the mutational frequency from that postulated above, 5 per cistron per division, to that normally encountered in known organisms, 10^{-4} per cistron per division, then clearly the variability would be insufficient and the waste would be much larger since cells would either express, as the general class, germ-line genes that were unchanged or, in very rare cases, somatic variants with only one mutation corresponding to an amino acid replacement in only one of the two immunoglobulin subunits. Therefore, if we wish to avoid fancy-foot-work-hypermutation-mechanisms then we must substitute a simple and probable selection-mechanism. Let us see what that looks like.

B. New Somatic Models of Diversification—"The End Game"

The alternate class of hypotheses, which I prefer, postulates no special mechanism for generating diversity and this in itself is refreshing. The v-gene, like every other one in an animal, undergoes spontaneous mutation throughout its length at the rate of 10^{-4}/cistron/generation. If there is a steady state population of N cells turning over once each day then there will be $10^{-4}N\ln 2 \sim 10^{-4}N$ mutants, M, generated each day in each v-class. This number M in each individual will increase linearly with time, t; $M \sim 10^{-4}Nt$. The problem now is to envision a selective mechanism which would amplify the diversity. There are two general classes of selective mechanisms, the first class utilizes two independent steps of antigenic selection and the second exploits the interaction between the subunits, light and heavy, as a pre-selection step to antigenic selection.

1. *Antigen-selection Models*

The assumption of models of this category is that those cells which have at least one amino acid replacement in the variable region of a light or heavy chain subunit will be selected for by antigen. In other words, those cells which express unchanged germ-line v-genes have a negligible chance of meeting a corresponding antigenic determinant and it is for this property that germ-line v-genes were selected (see Cohn, 1967b, page 637, Evolution of immunoglobulin and tolerance). Cells with one mutation have a better chance and if they react with antigen are selected for because they differentiate and do not turn over as does the unselected population. This new selected population now undergoes more divisions leading to a repetition of the above and therefore accumulates selected for mutations in the v region, *not the c region* which remains either unselected or selected against if the mutation affects complementation with a heavy chain. This latter point is critical to understanding the argument. In order to "know" that a mutation had occurred, it must be expressed as antibody, so that this model if it were spelled out would have to deal in detail with the pathway of differentiation. I will not do this in the present discussion for reasons that will be evident later. Instead, I would like to look at some details of the mutational and selection patterns.

Assume that every step, including which class is expressed, is random, and that the population N dividing once per day generates $10^{-4}N$ mutants per day in each v-class, heavy and light. The following will be the distribution of light chain mutants: One half will be in critical codons of v and not functional; only one sixth of the light chain mutants will be revealed because a given cell expresses only one v class of the total of six (three v_κ and three v_λ); and one-half will be lost due to allelic functional light chain subunit possessing one mutation in the v_L-region and an unchanged heavy chain subunit would be

$$10^{-4}\text{ N per v-gene per generation} \times \tfrac{1}{2}\begin{bmatrix}\text{mutation in}\\ \text{a forbidden}\\ \text{codon}\end{bmatrix}$$

$$\times \tfrac{1}{6}\begin{bmatrix}\text{expression of one v-}\\ \text{kappa or v-lambda by}\\ \text{each cell}\end{bmatrix}$$

$$\times \tfrac{1}{2}\,[\text{allelic exclusion}] \sim 10^{-6}\text{N}\begin{bmatrix}\text{cells with functionally}\\ \text{mutated light chain subunits}\\ \text{in each v-class per generation}\end{bmatrix}.$$

For heavy chain subunits the calculation is roughly the same except that the expression factor is one since we are assuming one v_H shared by all c_H. Therefore, the fraction of cells with functional heavy chain subunits and one mutation in the v_H region will be $\sim 10^{-5}$N.

If I can reasonably assume that any antibody-producing animal has a population N of 10^7 cells with a generation time of one day in its steady state population then the distribution of varied cells generated in each v_{light} class will be $\geqq 10$ and in the v_H class will be $\geqq 100$. Of course each singly mutated light or heavy subunit will be associated with an unchanged partner in the first round of diversification before antigenic selection. These mutant cells accumulate at a linear rate per day in the absence of selection, so that the population would accumulate in 10 days $\geqq 100$ light chain and $\geqq 1000$ heavy chain variants. With this as a starting point, it is not too difficult to imagine a series of successive selections which would generate cells of wide diversity. However, I will not spell this out here because other considerations seem to point up serious weaknesses in a simple antigen selection model. These are:

1. It appears improbable that any one-step mutation in either subunit will convey an enormous advantage in recognizing an antigen. This means that some large waste factor must be added because most mutants will never "meet" antigen. We need a theory and more data concerning the relationship of primary to tertiary structure in order to evaluate this objection and I will leave it at this.

2. It is well established that the light and heavy chains coming from a given cell complement preferentially with each other. The evidence for this optimization of fit is based on the "Mannik phenomenon" (Mannik, 1967) where the light and heavy subunit of any given myeloma have a

much higher affinity for each other than light or heavy subunits taken at random. This was interpreted (Lennox and Cohn, 1967) as due to the contribution of the variable region to the interchain binding energy and was assumed to be the result of selection by antigen because good fitting antibody implied good fitting subunits. The antigen selection theory necessitates this assumption to explain the results on preferential complementation. However, Dr. Jon Singer has given me strong arguments why this cannot be so and insists that the contribution of the variable region to interchain binding must be selected for prior to or independently of that for a combining site. His argument is quite simple. The antibody combining site is a cavity in a specific region of the molecule and therefore, if anything, the site would decrease the surface over which close interaction between light and heavy chains must occur to give preferential complementary fits. Selection to improve subunit interaction should involve "touching" surfaces. Selection for a cavity, the combining site, would not necessarily imply optimization of subunit interaction. If Singer's conclusion were correct, antigen selection mechanisms as sole guides through the jungle of diversity would be improbable. I might stress as an aside that the Singer-argument is most critical for germ-line models in which genes are activated randomly and the "Mannik phenomenon" could only be the result of antigenic selection.

3. If one were to make a guess, applying to whatever selection model is preferred, as to which rapidly renewed population is providing the diversified mutants, it would be the cells of the thymus, bursa or their equivalent. It is a mystery why such an enormous proportion (99%) of the cells are born and die in the mammalian thymus without ever leaving it (Metcalf, 1966). A small fraction (~1%) leave this organ per day and seed the lymph nodes (Metcalf, 1966). Assigning a diversifying role to this population would necessitate that antigenic selective interactions occur in this organ to permit diversified cells to leave it (Dreyer *et al.*, 1967) and from our present understanding this appears unlikely.

It is in order to come to grips with these contradictions that two-step selection models, of the kind I will mention now, are appealing.

2. *Two-step Subunit Selection Models*

I had long ago learned from pleasant experience that when the "chips were down," it was wise to discuss the situation with Dr. Leslie Orgel and Dr. Edwin Lennox. Both of them suggested this class of model and proposed two variants of it.

The assumption of models of this category is that the expression of germ-line genes yields subunits which do not interact effectively with each other. However, mutations which increase the light-heavy interaction are sensed by the cell in such a way that diversity is selected for. The Orgel-model is that secreted antibody subunits control the division of cells producing that antibody, but have no effect on cells which have undergone a mutation and produce a variant antibody. The Lennox-model is that (1) the LH complex internally regulates division and (2) the subunits are mutual repressors of the synthesis of each other. The initial diversifying population produces an L^0 and H^0 which do not interact and mutations to L^1 or H^1 which improves the LH fit tends to turn off the division of the cell.

In order to block the replication of the cell, the concentration of a secreted or internal variant subunit must reach a certain threshold which depends upon the affinity between L and H. The system *selects for diversity* and for good fit because those cells not turned off by this feedback divide and turn over until they produce a mutated subunit which effectively inhibits their growth and stores them as long-lived antigen-sensitive cells. The immunoglobulin, with its combining site accessible, might be stored in the membrane where interaction with antigen triggers division and an antibody response. These models differing in detail are identical in principle.

A good deal of the interaction between light and heavy subunits resides in the constant sequences. The subunit selection model depends upon optimization of detection of variable region interactions. Therefore, a very good rationalization for the separation of v and c genes is possible because it is V-region complementation that the selective mechanism searches for and this interaction would be best seen in the absence of a C-sequence. However, such a proposal implies that the v-gene is expressed as such and that the polypeptide products of the v and c cistrons are joined. The evidence is against this (Fleischman, 1967; Knopf *et al.*, 1967; Lennox *et al.*, 1967). Another explanation must be sought.

The Orgel-model is analogous to those used to explain how tissues and organs sense their size and regulate it (reviewed in *Control of Cellular Growth in Adult Organisms*, 1967). The Lennox-model involves a regulatory circuitry of the kind I have discussed elsewhere (Cohn, 1967d). This class of models is elegant because they *select for diversity* by selecting for good fit between subunits independent of antigen and therefore they do not face the above objections to antigen-selection models. The rate at which mutants are generated in the primary population is the same whatever the selection model and has been dealt with already. The development of subunit-selection models requires that a structural basis for the selection between subunits be established as well as a detailed description of the pathway of differentiation.

I wish to stress that under both models the antigen-sensitive cell results from a pre-selection for diversity by either antigen or subunits prior to the second step of induction of antibody synthesis by antigen. The selection models deal only with how spontaneous mutation is exploited to give a variety of cells responsive to induction. The later stages of antigen induction will be referred to subsequently.

Although I have thought through both selection models, I have not described them in detail because I am not certain which is incorrect. The antigen selection model seems simpler because all of the elements are known and fewer assumptions are needed. Visualizing a mechanism for selecting mutations to improved fit between subunits should present no more difficulty than that encountered in understanding the selective forces in hemoglobin evolution which I will come to later on. The major difficulty with subunit selection models is that the light and heavy subunits already interact rather well due to the constant sequences. A mechanism for sensing the improved fitting mutations in the V-region is needed. For example, a high affinity interaction in the V-regions could result in a conformational change in the C-region which is detected by the cell. A similar mechanism might also explain how antigen might select for tight fit in the V-region (Mannik phenomenon, 1967). Since I have great faith in simplicity, I would not be surprised (a) if antigen were to interact with the cells in the thymus possibly permitting them to leave (Dreyer *et al.*, 1967), and (b), if antigenic selection for high affinity antibody actually concomitantly selected for high affinity subunit interaction as originally proposed (Lennox and Cohn, 1967). If this were so, the previously discussed objections to antigen selection models would disappear. One rather easy way out of all dilemmas is to have both antigen and interaction of subunits act simultaneously on the initially diversifying population to select for variety but I will avoid this hedge at present. I will leave the choice between these alternatives to the future since my main point now is that *spontaneous somatic mutation* followed by *regulatory selection* is the source of the highly diversified antigen-sensitive population of cells.

These models whether they suppose antigen or subunit selection require an enormous waste of cells in the initial N^0 population and this waste is minimized in the final steady state population when the animal has an almost maximally diversified antigen-sensitive-population. This latter situation is identical to that with which germ-line models deal in that a part of the cell waste necessitated by somatic models is eliminated because it occurred as a waste of individuals during the eons of evolution.

It is reasonable to ask why this selection might occur in the immune cell system (or others able to respond to the unexpected) and not in every other cell population in the body which is presumably also undergoing in every gene, mutations at the rate of 10^{-4}/cistron/generation. The secret is not in the generator of diversity but in the selection. The antibody forming cell expresses its product in a unique way such that selection by regulatory mechanisms can act on it. An erythrocyte or epithelial cell does not exhibit its specialized product so that selection is possible and it is turned over like most differentiated cells (Cohn, 1967d). By this formulation, then, the problem has shifted from the generation of diversity per se to the mechanism of exploitation of the results of spontaneous mutation.

There are the germs of a serious contradiction with cell selection models in the literature (Appella *et al.,* 1967) thus raising the question, can one distinguish somatic hypermutation from selection models. The lambda chains of two independently arising mouse myelomas have been reported to be identical (Appella *et al.,* 1967). If this result implies that lambda chains in the immunoglobulins of the mouse are invariant, then a strong argument against selection hypotheses is provided. Spontaneous mutation varies all genes. It therefore occurs in the v_λ-gene of the mouse as it does in humans where the lambda chain is known to be varied. Since lambda is associated with immunoglobulin in the serum of the mouse, the v_λ-gene is expressed and by selection models should be variable. It is unlikely that a spontaneously varied gene which is expressed would be associated with antibody activity in man and not in mouse. If the mouse lambda chain were to be invariant, then the operation of selection on a population of cells varied by spontaneous mutation is unlikely as a means of variability generation. We would be forced to a hypermutation model and conclude that the generator of diversity does not act on the v_λ-gene because of its sequence or position in the chromosome.

Another type of experiment might bear on the hypermutation versus the selection model. The N^0 population with which the animal begins is at best able to recognize one antigenic determinant. As mutations are introduced to give the N^1, N^2, N^3, . . . N^n populations, the range of recognizable determinants increases. Quite possibly the N^1 population responds to a limited class of determinants. This means that there might be a sequence to the time of appearance of responsiveness to the individual members of a chosen set of antigens. Certain antibodies corresponding to the N^1 class would appear before those of N^2, N^3, N^4, . . . N^n class. The hypermutation model (or for that matter any model thus far proposed, germ-line or somatic) would have the maximum range of antibody specificity appear the moment antibody activity was detectable for any one antigen. The antigen selection model implies a progression in potentiality to respond whereas the subunit selection hypothesis can have it either way. The presently available evidence (Silverstein, 1963) that the ability to respond to certain antigens is precisely programmed needs discussion in terms of the selection model. The precise and reproducible order in time with which foetuses mature to respond to different antigens has been used as an argument against somatic mutational models (Silverstein, 1963). It does seem unlikely, at first glance, that this "abrupt appearance of competence" to respond should reside in a "mechanism involving statistical chance" (Silverstein, 1963). However, three points should be made. First, the difficulty in explaining the stepwise attainment of immunological competence is the same under any mechanism of diversification, even in germ-line models where any given gene would be activated randomly. Secondly, one can always assume that the antigen to which a foetus responds is closely related to one which it experienced previously, or that any one of a large number of amino acid replacements in the germ-line sequence yields an antibody directed against a given test antigen whereas two or more replacements are necessary for another test antigen. Thirdly, the phenomenon of ordered appearance of responsive ability could be irrelevant because it is due to antigen-processing ability, not to the mechanism of generating diversity. It is difficult, but not impossible, at present to rule out the role of the processing system in determining the order of the response. If amino acid sequences were determined on antibodies of foetuses which first show a response to one and not the other determinant, the range of variability would be much more limited than that seen later. In fact the sequences corresponding to the germ-line genes should be easily obtainable since most of the subunits would have no substitutions.

At this point let us return to the question of how interactions of subunits are selected for in evolution, somatic or germ-line, and the difficulties of the concepts.

C. An Aside Concerning the General Problem of Evolution

1. *"That-missing-something"*

There has been a recent symposium on the "Mathematical Challenges to the Neo-Darwinian Interpretation of Evolution" (Moorhead and Kaplan, eds., 1967) where sharp feelings were expressed that the theory of germ-line evolution lacks something. It seemed obvious to the heretics led by Eden, Ulam, Weisskopf and Schutzenberger that evolutionary theory is missing a critical step because all of the possibilities resulting from single base changes in the DNA could not have been tested by spontaneous mutation and selection during the evolutionary time we are dealing with. Yet very complicated systems of immunity, detoxification and learning have evolved. There must have been short cuts. In order to have intricate functioning systems many genes are needed. Clearly the more genes you have the more you can do but the lesser becomes your potential to evolve. The fewer the genes the easier it is to evolve. The missing link in the above argument is that wonderful property of a somatic mechanism, namely, that evolution acts on a germ-line gene to perfect it so that its expression as a series of variants around the mean gives maximum universality as well as greater specificity. A somatic mechanism for varying few germ-line genes optimizes the adaptive potential to cope with the unexpected. This point is *not* simply a transposition of the problem of germ-line evolution to the soma. There is an enormous synergistic amplification due to the interaction of germ-line and somatic selection on variants in each genome, germ-line and soma.

2. *How Does Selection Act in the Immune System in Order to Optimize a Somatically Varied Germ-line Gene?*

Although I will discuss the v and c genes separately, it should be understood that they could be a single cistron. The separation is a matter of emphasis because I am not certain that the question is closed.

a. The v-Genes. I have already stressed our difficulty in understanding how selection can act on a repeated series of genes most of which are silent. The greater the number of germ-line genes, the less the chance that a given antigenic determinant on a complex molecule of many determinants can be selective in maintaining its corresponding germ-line genes. If it should ever be shown that the genetically determined specific non-responders (Lennox, 1966) result from v-gene mutations, a germ-line hypothesis could not account for them because the response to any known antigen is always heterogeneous. A somatic hypothesis does account for non-responders. Take the extreme case of a mutation in a critical codon in the v-region, e.g., cysteine 23 or 88. Somatic mutation would correct this very rarely. If not corrected, the kappa light chain would be unable to complement with a heavy chain to make an antibody. Therefore there would be no cells which express the kappa class with all of the variety of antibody specificities and functions that it normally carries. This dysgammaglobulinemia of a v_κ-class is sufficiently detrimental to that individual to select against it. If the mutation occurs in a codon for an amino acid which is not essential to the structure of the immunoglobulin there is an effect on the range of specificities which can be generated, because single step mutations from one codon could give a different series of amino acid replacements than from another. Since non-response to one group of antigenic determinants and response to another would be the phenotypic expression of such mutations, the germ-line gene is optimized for the most expected stimuli as well as for rare but very important ones, e.g., a toxin, a pathogen, etc., because somatic variants can deal also with unexpected stimuli at the extremes of the distribution of specificities. Each germ-line gene then for a variable region, v_κ, v_λ, v_H, etc., can drift apart to optimize variety until cell waste becomes selective or the frequency of cells expressing critical antibodies is too low. I will return to the interaction between the v_L and v_H at the end of the discussion of the c-genes.

b. The c-Genes. The selection of the c-genes has somewhat different ground rules (Lennox and Cohn, 1967). The constant sequence is selected in the germ-line by virtue of its ability to complement with the other subunits and by its unique function. The latter is obvious; the former needs some comment, since there are analogies between selected complementation of subunits in the germ-line evolution and the selected complementation of v_L and v_H or any other subunits in somatic evolution. Many oligomeric proteins are constructed of subunits which are identical, e.g., β-galactosidase or alkaline phosphatase of *Escherichia coli,* or evolutionarily related because they were derived from the same primordial genes, e.g., hemoglobin, immunoglobulins. There is a strong selection for oligomeric proteins necessary for structures, e.g., membranes and for regulatory and transport functions, e.g., feedback-regulated enzymes, hemoglobin. Why is the selection often on a pair of identical duplicated genes and not upon what a priori seems more reasonable, unrelated genes? I will illustrate with hemoglobin. The first event in evolution is a functional structure, however primitive. After all, in the development of a complex structure, as an arm for example, the intermediate stages like limb buds must have some selective advantage or they would be lost. In the case of hemoglobin it was the monomeric globin able to react with heme to produce an oxygen binding heme protein exemplified by myoglobin today. Therefore, this already improbable event determines the structure upon which selection can act to yield the more efficient transporter, hemoglobin, made up of subunits and able to undergo conformational transitions with different affinities for oxygen. If the primordial gene for a protein capable of reacting with heme duplicates, then the genes diverge and a protein comprised of non-identical but evolutionarily related subunits emerges. If the gene does not duplicate but remains unique, then the selection for an oligomer is *per force* for one made of identical subunits. The protein that functions today is then the result of a series of reciprocal mutations for fitting. The rate of selection on such a series of events is probably autocatalytic since the first mutation would give a marginal selective advantage but each successive amino acid replacement would greatly increase the efficacy of the molecule and be more easily refereed by selection. There are many factors in the evolution of this alternating variation improving the fit between subunits since the more efficient transport of oxygen means more CO_2 is released by cells and the selection continues for a transporter which in an acid environment releases its oxygen more readily and in an alkaline one binds oxygen more tightly. In other words, the environment is changed by the mutational event and hits back (to use a Popperism). In the case of the c-genes of immunoglobulins, non-identical evolutionarily related subunits were selected upon, possibly because the generator of diversity or the regulatory mechanism for expression was that unique evolutionary event fixing the kind of selection that could follow. The identical principles which have guided germ-line evolution to select oligomers also guides somatic evolution and this point will be particularly important in discussing the learning and abstractive process.

In the immune system the serum antibodies are optimized in the fit between their v_{light} and v_{heavy} subunits. I do not yet know which structural considerations would lead to such a result if selection were uniquely due to antigen. If selection is by subunit interaction, of course, this problem is resolved.

Clearly, however, v and c are not independent units but are selected together since certain v-sequences must be incompatible with an immunoglobulin structure. The more different vc classes in the germ-line for light and heavy subunits, the broader the range of specificities and functions the animal can express. This number can increase until the silence of the gene becomes a counter-selective force.

D. The Expression of the Genes

The extrapolation of the somatic mechanism described here to the detoxification and learning process requires that I say a brief word concerning the expression of the gene. This subject was masterfully discussed at the Cold Spring Harbor Symposium this year by Jerne (1967) in his summing up address called "Waiting for the End." Clearly it would take a Mitchison (1967) to equal it and this too was done this year at the Gatlinburg meeting of the International Society of Cell Biology. Having been at both meetings I will "abashedly" plagiarize. The three key points are:

1. *One Cell–One Immunoglobulin*

If a large variety of specificities is somatically generated, there must be a mechanism for selecting the one that is relevant. This has been done as it was for germ-line evolution with little that is new. The single cell expresses only one immunoglobulin. The stimulus complementary to the specificities it carries selects for the cell expressing it.

2. *Antibody Recognizes Antigen; How Not to Ignore the Unexpected*

Whereas the dictum, one cell-one immunoglobulin, is an experimental fact, the assumption that it is antibody itself which recognizes antigen simply has no competing hypotheses. I will not argue the experimental basis of this. It has been done by Jerne, Mitchison and others (Jerne, 1967; Mitchison, 1967). I wish to use this occasion to point out that it could not have been otherwise in any system geared to cope with the unexpected.

Clearly, an animal ignores what it does not expect. This concept has been elegantly analyzed by the philosopher Karl Popper. The only way to interpret your environment is in terms of an expectation which is equivalent to making a theory. What you do not expect because you do not have a theory is likely to be overlooked. It is this theory which makes for those prepared souls which chance favors. I might point out that the greatest generalization of modern biology, A=T, G=C, was an established fact stumbled on and ignored until it became part of a general solution to the problem of heredity.

So in a profound sense a given antibody is a theory made by the animal about what is in its environment. Like any given theory, most often it is wrong, i.e., the antibody fits no determinant stimulus in the environment and is eliminated. When it does fit, the antibody like the theory is amplified by selection and increases in its approximation to the truth, which by analogy is equivalent to the selection for improved fit to the antigenic determinant. The concept is a completely general one which I will refer you to in the writings of Karl Popper, particularly his charming essay "Of Clouds and Clocks" (1966).

3. *Memory*

The secret of germ-line evolution is gene duplication and divergence by mutation and exchange. A "successful" gene is put away in the germ-line under the wraps of control and revealed when appropriate stimuli are applied. This is germ-line memory. Similarly, the secret of somatic evolution, as applied to antibodies, is the selection, from a pool of duplicated divergent genes, of those whose products have proven "usefulness." These genes are then stored in a long-lived memory cell. The test of usefulness is the primary response to antigen and the secondary response tells us what was catalogued in the library. Therefore, germ-line and somatic evolution are identical in this respect, in one case the unit is the individual and in the other it is the cell. The mechanism of this process is one major subject of this symposium.

Now, I would like to ask if an attempt to analyze the detoxification and learning mechanisms in terms of somatic evolution has any point. I did this previously (Cohn, 1967b, c) and would like to amplify a bit here. Clearly, a somatic system for diversification, whatever its mechanism, could be so effective that I wonder if some generalization would not be of value. In the beginning, I rather casually put the learning process and the detoxification mechanism in the same class as the immune system by assuming that the adaptive variety of all three systems was due to a large number of proteins coded by somatically generated genes. I did this because I was struck by the superficial resemblance of the three systems.

III. THE DETOXIFICATION SYSTEM

The liver responds to an enormous number of drugs by producing increased amounts of enzymes which act on them to produce so-called detoxification reactions. Examples of such variety are systems which oxidize, hydrolyze, acetylate, glucuronidate steroids, aromatics, dyestuffs, terpenes, camphors, heterocyclics, etc. (Williams, 1959). Of course, there is no evidence that, like the antibody systems, the enzyme systems referred to here vary their combining sites against a constant protein background. Since structures capable of enzymatic activities are very exigent compared to antibodies with combining activity only, there is less probability that useful reactions will be catalyzed by proteins whose sequences are varied during somatic evolution by even such controlled mechanisms as recombination. Nevertheless, useful and functioning enzymes were generated by gene duplication and divergence during germ-line evolution. In principle, somatic evolution is analogous, so that before discarding the idea, classes of liver enzymes with similar detoxifying activities should be examined to see what structures and sequences they share. I can imagine a situation analogous to antibodies where a constant portion determines a restricted class of reaction, e.g., glucuronidation, and variation of a basic pattern gives a wide range of specificities. Varied combinations of subunits could change the nature of the enzymatic reactions. Enzymes of this type might be expected to show a wide distribution of affinities for a given substrate.

An example might be of interest. The liver responds to many polycyclic hydrocarbons and carcinogens by hydroxylating them. The response to a given drug, phenobarbital or 3-methyl-cholanthrene for example, adaptively and specifically increases the rate of hydroxylation of that drug by the liver (Takeshita and Tanaka, 1966). A known cytochrome, P-450, is increased in amount whenever any one of the wide variety of hydroxylation substrates is used as inducer. This unique cytochrome involved in known detoxification reactions might be revealing a recognition protein of the type implied by analogy with immune systems. P-450 might be a "subunit" of a somatically varied polypeptide responsible for recognition of the inducer-substrate or it might be simply a substrate of this enzymatically active molecule selected for by the drug and concomitantly induced (Estabrook *et al.,* 1966). The wide variety of specifities with which these inducible microsomal oxidases interact could result from having a protein with a constant P-450 recognition region and a variable site responsible for combination with the inducer-substrate. This system is a good candidate for analysis as a "responder to the unexpected."

In order for selection to act, I would anticipate one such detoxifying enzyme to be expressed per cell. Using a one step model, interaction of the substrate with the enzyme in or on the cell would select for that given class of cells exactly as postulated for antibody-producing cells. A two step model implies subunit interaction already discussed for antibodies. There is, for example, a clonal distribution of steroid reductases in the liver (Bakemeier, 1961). However, the distribution of enzymatic activities among single parenchymal cells has not been adequately studied due to technical limitations to the analyses of single cells. The use of hepatomas in a way already so successful with myelomas has not been exploited. One might predict that when a library of hepatomas derived from parenchymal cells are studied, the clonal distribution of detoxifying enzymes will emerge, possibly along with classes of enzymes of varied specificity associated with a constant protein background analogous to antibodies. Since the highly restrictive requirements for specific catalytic activity make the extrapolation of somatic evolution to enzymes appear unlikely, the finding of such a phenomenon will have enormous implications.

IV. THE LEARNING SYSTEM

The parallel between immune processes and those of learning and memory has repeatedly been pointed out (Anker, 1960; Hechter and Halkerstron, 1964; Silverstein, 1963; Smith, 1962; Szilard, 1960, 1964). The extension of the idea of somatic evolution from the immune to the learning and memory processes can be formulated only in general terms at this moment; yet there are some fascinating consequences.

There is good evidence in the case of instinctual or reflex actions that there must be predetermined neuronal connections (Attardi and Sperry, 1963; Sperry, 1951). It is an obvious guess that the specificity of the neuronal connections depends upon surface membrane interactions of a complementary kind involving much the same mechanism as that used for all cell-cell interactions seen by studying specific aggregation phenomena (Gaze, 1967). Specific recognition sites in membranes undoubtedly reside in proteins. In order to sidestep the argument of tapes versus circuits I will simply assume that learned and "abstractive" behavior uses circuitry involving membrane recognitions of the same basic kind as is employed for instinctual circuitry. At the molecular level this distinction between tapes and circuits is not too important to us because in any case we must postulate a molecule or combination of molecules unique to a given cell responsible for the tape of circuit function. The question is whether, like instinctual or reflex circuits, the learned pathway is also determined by germ-line genes or are somatically derived genes coding for membrane proteins responsible for recognition. There would not be any reason to raise this heretical view for the learning system were it not for an analogy with the immune system where the problem has crystallized.

I will assume by analogy with the antibody forming cell that each neurone involved in making a circuit expresses one membrane protein involved in recognition of its partner. I am aware that this hypothesis is more extreme than one in which the topology of fit between cells were to be determined by the combinations of several proteins because the assumption, one cell-one protein, necessitates that a large number of genes $>10^4$ be used to code for circuitry of a mammalian brain. This number can be reduced by use of proteins made of complementing subunits, and further reduced by use of combinations of proteins which comprise a unit configuration. However, the analysis will be sharper if I assume one cell-one recognition protein.

I should re-emphasize that I am making a distinction between instinctive or reflex and learned or "abstractive" behavior and that I will be assuming that all genes are carried in the germ-line for instinctive processes whereas a few somatically varied genes are responsible for both learned behavior and that unique human quality, abstraction. We do not understand how the information is stored, nor how the neuronal connections are made, nor for that matter their ontogeny. Therefore it is even difficult to find direct evidence that the complete control of the instinctive process is by the germ-line genes, not to speak of learned processes.

Nevertheless, the fact that instinctive or reflex action is not modifiable except by mutation is one argument that it is entirely determined by germ-line genes which were selected by the good adaptation of the given behavior pattern to the environment. Learned behavior, like antibody synthesis, appears to be only in a limited and trivial sense controlled by germ-line genes. Any behavior pattern you learn can be re-learned as a directly opposite pattern. I am always impressed by the rapidity with which we can learn to ride trick bicycles that have a front wheel which turns to the left when the handle-bar is turned to the right. The instinctual behavior is eventually subordinated to the learned one not by selection on the population but on the individual. I doubt that germ-line genes can specify any learned behavior pattern as well as its anti-pattern because I suspect that only one or the other would have been of selective value in evolution. However, these comments are but prejudices of somatic evolutionists only discussable after dinner.

One extreme position might be that all neuronal circuits are coded for in the germ-line and the entire circuitry of the brain is laid down following a blueprint during differentiation and development. Learning would be limited by the number of pathways hereditarily available. This assumption could be examined by calculating whether a large enough series of connections could be established by germ-line gene-products. Unfortunately, this calculation is too marginal (as in the case of the immune system) to enable one to rule out even the extreme form of the germ-line hypothesis because of the combinatorial factors in variety derived from complementation of gene-products. Nevertheless when this calculation is made ignoring complementation of gene-products (Bremermann, 1967), the conclusion is that the number of neuron connections in the human brain exceeds the total number of genes by four orders of magnitude. Another suggestive argument against this germ-line theory (the same as I used previously) is that most of the relevant genes would be silent during an individual's lifetime, e.g., the genes determining the circuitry for learning of Boolean algebra or pig Latin or how to select TV dinners or to repair lasers. These genes would drift making the learning of most things impossible. Furthermore, once again, this form of the germ-line hypothesis only enables expectation for *de novo* stimuli of the kind which overlap with a previous stimulus.

Another point which comes out of an exciting essay by Natapoff (1967) is that the "human qualities" of the brain evolved in a very short time of the order of 30,000 generations. He argues from this that there could not have been many mutational combinations tried in order to get from *Australopithecus* to *Homo*. What I am pointing out is that the optimization of a somatically varied germ-line gene takes a minimum of tries for a maximum of variety. In fact the arguments which led Natapoff to his view of the evolution of the higher brain were exactly the same ones which led to the view that the immune system evolved a somatic mechanism as discussed here and elsewhere (Cohn, 1967b; Lennox and Cohn, 1967).

Once a one neuron-one protein hypothesis is assumed, the way in which circuitry is made need not be of the extreme form I postulated previously. Learning circuits could be formed by neurones which interact to produce short-lived networks. If a learning stimulus coming via the senses or via consciousness is encountered, its effect is to select those cells arranged in a "complementary" short-lived circuit and establish a long-lived stable network of neuronal associations. One could visualize stabilization of the synaptic connections via differentiation of the cells which were stimulated to become long-lived neurones with irreversibly fixed synapses. The same terminology which is used to describe a short-lived virgin cell being stimulated to become an experienced long-lived memory cell (Cohn, 1967b) may be applicable here. Just as the antigen-sensitive cell is the unit of selection for the immune process, so the idea-sensitive packet of connected cells is the unit in the learning process. If the brain of an individual evolves as learning experience accumulates this process might be pictured as the branching off from an instinctual innate circuit, of new circuits comprised of short-lived connections which become long-lived on learning stimulation. These learning circuits are interconnectedly layered one on top of the other as more and more learning events are experienced. The cells from which the new circuits are formed could derive their unique proteins from germ-line or somatically varied genes.

One distinction between germ-line and somatic mechanisms is revealed by the way germ-line mutations affect it and in the waste of cells. The germ-line theory would postulate hundreds of genes being expressed randomly on a one neuron-one cistron basis. These neurons make connections with other neurons if their surfaces are complementary (like antigen-antibody reactions) but these connections are short-lived and the cells turn over if no learning stimulus stabilizes the given neuronal connections. The waste of cells in this process is due to the revealing of specificities for which no corresponding learning stimulus has occurred. The somatically derived genes are revealed identically but if the variation is based upon mutation then an additional factor of waste is introduced by having membrane proteins incapable of functioning in recognition. However, exploitation of somatic genes enormously amplifies the diversity so that the increased waste is made up for by greater variation. Since there is one germ-line gene per class of membrane recognition proteins, selection is strong to maintain it such that variety is maximized and waste minimized, remembering of course that these two parameters are mutually dependent, more variety, more waste. The trick is to separate them as far as possible. There are no available physiological or biochemical data which distinguish these two theories for the learning process.

The unit of interaction is between the two cells whose membrane proteins are complementary. Since the probability of a somatically and randomly varied protein, interacting with another randomly varied protein, seems even more remote here than in the immune system, there must be a restraint on the system. I have tried to analyze such a restraint when I discussed the evolution of the c-genes of the immune system. A series of reciprocal fittings derived from successive one step germ-line mutations in one or two cistrons has resulted in oligomeric proteins made up of subunits derived from a single ancestral gene. The same process is visualized in the selection for interaction in the brain. Once a fit is established, however poor, it can be autocatalytically improved by subsequent selections of the short-lived circuits. Thus, I envisage a pathway of selection exactly as that postulated in my model for the immune system. Different classes of interacting membrane proteins could be expressed locally in a given pocket of short-lived interacting cells. This is analogous to the IgA class of cells being expressed uniquely on secretory surfaces. Thus special functions can be assigned to individual pockets in which many circuits can be established by abstraction and learning stimuli. Any short-lived neuronal circuit which is stabilized differentiates to a long-lived one and then makes further connections with the central integrative system of long-term memory and learning.

If mutants with specific learning defects were to be observed, a very strong argument for a somatic, not germ-line, model would be available. Clearly, if there were 100–1000 repeated genes, a mutation in one would be unlikely to be observed since a given stimulus provokes a heterogeneous response. However, if one germ-line gene were mutated in a critical codon inactivating, for example, a class of proteins, this error would be so rarely corrected by somatic reverse mutations or modulations that it would be seen as a phenotype with a defect in specific class of learning. This point was made more convincingly in the immune system (Cohn, 1967b; Lennox and Cohn, 1967).

Natapoff (1967) takes as the key step in the evolution of the human brain the ability to "generate at will errors of a suitably restricted kind." These errors must be tested by "desirability" and if chosen, preserved from further change. He points out that "random processes play a role in our model, ultimately, because they provide a simple mechanism that might be rapidly evolved." This, of course, is what I have been postulating that the immune system does via a somatic mechanism. The question of how this system operates in the brain is left open but if I were to sense his position, Natapoff would be closer to Brenner and Milstein in seeking a mechanism for error-generation by hypermutation rather than strong selection on a normal spontaneous rate as I have proposed.

In either case a germ-line mutation preventing the somatic system from generating diversity and/or its expression in the case of the immune system leads to complete agammaglobulinemia but in the human brain it might lead to an individual whose brain lacks a unique class of proteins. It is not possible to say whether the dysgammaglobulinemia of the immune system is analogous to the loss of ability to abstract in certain areas, specifically, e.g., mathematics, poetry, music and painting or to a complete loss of learning ability.

Are there any relevant data and is there an experimental handle?

First of all I would expect the brain to contain a population of rapidly turning over neurones which on stimulus become long-lived cells of the kind which we associate with the brain. A search through the literature revealed the suggestive work of Altman and his coworkers (Altman and Das, 1965, 1966a, b; Altman *et al.*, 1967) who found such a short-lived class which are called microneurons because their axons make connections over a limited distance, defining some kind of anatomical pocket. Altman *et al.* (1967) could show that a neuroblast generates by division these microneurons which accumulate paralleling the improved responses of an animal. These experiments show that postnatal neurogenesis occurs and is affected by learning but do not say whether the microneuron itself is directly involved. However, this is the kind of cell upon which learning stimuli can select and has the characteristics of the one I am postulating in this discussion. The short-lived neurone might show short term memory (analogous to the IgM response in the immune system). It is this short-term expectation or theory making network which enables the animal to perceive a *de novo* stimulus. Short-term memory becomes long-term memory when the learning stimulus successfully finds its way through a short-lived circuit and stabilizes the cells involved, by their conversion to neurones with fixed long-lived connections. The critical point is that the learning stimulus selects out of a pool of cells those whose somatically derived genes correspond to the learning event.

Since dividing cells would be the source of the new neuronal connections of learning, these cells should be exquisitely sensitive to agents such as X-rays or cytosine arabinoside whereas non-dividing long-lived neurones should be resistant. Treatment of trained animals with such agents should block *de novo* learning but have no effect on already established behavior.

The second property I would expect is that the brain contains groups of closely related but distinguishable proteins like the immunoglobulins. In fact, I might guess that the same mechanism generating variety for the immune system might be plugged into the neuronal system, be associated with another constant region and produce variations on another basic theme. The search in the brain for amino acid sequences varying in an analogous way to those found in the variable regions associated with the immune system might be successful. Each neurone-tumor might produce unique membrane proteins analogous to the secreted immunoglobulin of myelomas. If microchemistry were possible, one unique protein per neurone might be revealed by direct assay.

I have already made the error of extrapolating too far. The only point to my hypothesis is to suggest as a starting point for the analysis of the learning process at the molecular and genetic level, one that has been so successful in the immune system.

There is a closing remark applicable to all three systems, immune, detoxification and learning. If one responds to the unexpected by a "random" process, it is essential to have a recognition mechanism for distinguishing responses to "self" or "lethal" responses from those which are "not self" and of "adaptive" value. In the immune system the recognition is antibody itself and the decision to make antibody (induction) or destroy the cell (tolerance) is the property of a superimposed regulatory mechanism which has in fact occupied us during this symposium. This secondary regulatory mechanism must play the same role whether variability is a germ-line or somatic event. It is not a part of the generator of diversity itself except in a minor way (see Lennox and Cohn, 1967). So with the detoxifying mechanism, cells generating enzymes that destroy essential metabolites, hormones and organelles are to be ridded also. The nervous system requires a way to eliminate noise or confusion generated by random connections that interfere with reflex or instinctual behavior. All of these systems must have a superimposed regulatory mechanism analogous to that responsible for tolerance in immunity. The one point essential to the somatic as well as germ-line hypothesis is that the recognition of the inductive or tolerance stimulus be due to the same gene product which will either be amplified by induction or eliminated by tolerance.

V. "WAITING FOR GODOT"

Biology is the understanding of how living things make choices. These are made by specific proteins which interact with given factors in the environment like "locks and keys." An animal can respond to a stimulus only because it possesses these choice-molecules. We have only a Darwinian-mechanism for generating and preserving these exquisitely selective proteins. Once of successful adaptive value, the information to make a given protein is stored in the germ-line as a gene responsive to stimuli because of superimposed regulatory mechanisms. The surprise comes only when we consider how vast and unexpected are the stimuli to which an individual can respond.

In this talk I have attempted to deal with this problem by reviewing what we know about the way in which animals respond to stimuli which do not appear to have direct selective influences in evolution. These responses are of the kind seen in the immune, nervous and detoxification systems. The range of potential responses (expectation) would be relatively limited if only germ-line genes were expressed as such. Somatic variation of these genes extends the range of response and therefore of adaptation some several orders of magnitude. Somatic evolution based on variation of chosen germ-line genes and their expression in different short-lived cells is the raw material upon which antigenic, chemical and learning stimuli act to select functional patterns of corresponding cells. These experienced cells are now long-lived and can respond on second encounter with the same stimulus. The process of somatic evolution is analogous to that of germ-line evolution, the unit of selection being cells rather than individuals. What is critical to realize is that the response to the unexpected is the property of a population. If an individual responds, it is his cell population which is being selected upon. This means that the phenomenon is a statistical one.

The extrapolation of the concept of somatic evolution to the detoxification, learning and abstraction processes is obviously important. The assumption that these processes derive their variety from a family of somatically diversified germ-line genes leads to experiments designed to reveal classes of proteins with a large number of combining sites, associated with a constant background, cells which express them on a one cell-one combining site basis, and a pathway of differentiation, signaled by interaction with the combining site which converts a short-lived into a long-lived cell responsible for memory.

I have discussed a relatively simple mechanism by which diversity can be achieved. Instead of trying to find a new mechanism for specific hypermutation, I showed that spontaneous somatic mutation exploited by a regulatory mechanism which permitted *selection for diversity* was adequate to account for the immune response. It is only this point I wish to insist upon, leaving detailed mechanisms for future discussion. This implies that it is not a special "Generator of Diversity" that prepares us for the unexpected but a regulator of the growth and expression of germ-line genes which determines whether the spontaneous variation of them is amplified and used adaptively or suppressed and eliminated. This proposal is so simple and obvious that I do not think that it is of any use "Waiting for Godot."

References

Altman, J., and G. D. Das (1965). Post-natal origin of microneurones in the rat brain. Nature, *207*:953–956.

——— (1966a). Autoradiographic and histological studies of postnatal neurogenesis. I. A longitudinal investigation of the kinetics, migration and transformation of cells incorporating tritiated thymidine in neonate rats, with special reference to postnatal neurogenesis in some brain regions. J. Comp. Neur., *126*:337–389.

——— (1966b). Autoradiographic and histological studies of postnatal neurogenesis. II. A longitudinal investigation of the kinetics, migration and transformation of cells incorporating tritiated thymidine in infant rats, with special reference to postnatal neurogenesis in some brain regions. J. Comp. Neur., *128*:431–474.

Altman, J., W. J. Anderson, and K. A. Wright (1967). Selective destruction of precursors of microneurons of the cerebellar cortex with fractionated low-dose X-rays. Exp. Neur., *17*:481–497.

Anker, H. S. (1960). A possible biochemical mechanism for memory. Nature, *188*:938.

Appella, E., K. R. McIntire, and R. N. Perham (1967). Lambda Bence Jones proteins of the mouse: chemical and immunological characterization. J. Mol. Biol., *27*:391–394.

Appella, E., and D. Ein (1967). Two types of lambda polypeptide chains in human immunoglobulins based on an amino acid substitution at position 190. Proc. Natl. Acad. Sci. U. S., *57*:1449–1454.

Attardi, D. G., and R. W. Sperry (1963). Preferential selection of central pathways by regenerating optic fibers. Exp. Neur., *7*:46–64.

Baglioni, C. (1967). Homologies in the position of cysteine residues of K and L type chains of human immunoglobulins. Biochem. Biophys. Res. Commun., *26*:82–89.

Bakemeier, R. F. (1961). A possible cellular explanation of the multiplicity of steroid reductases. Cold Spring Harbor Symp. Quant. Biol., *26*:379–387.

Bremermann, H. (1967). Quantitative aspects of goal-seeking self-organizing systems. Progress in Theoretical Biology, *1*:59–77 (F. M. Snell, ed.), New York: Academic Press.

Brenner, S., and C. Milstein (1966). Origin of antibody variation. Nature, *211:* 242–243.

Burnet, M. (1959). Clonal selection theory of acquired immunity. Nashville, Tennessee: Vanderbilt Univ. Press.

Campbell, J. (1967). Permissive translation of nucleic acid: a mechanism for antibody synthesis. J. Theoret. Biol., *16:*321–331.

Cohen, S. (1965). The *in vitro* synthesis of immunoglobulin light chains. Immunopathology, IV Intl. Symp. (P. Grabar and P. A. Miescher, eds.), 243–254, New York: Grune and Stratton, Inc.

——— (1966). General structure and heterogeneity of immunoglobulins. Proc. Roy. Soc. (London) Ser. B, *166:*114–123.

Cohn, M. (1967a). Natural history of the myeloma. Cold Spring Harbor Symp. Quant. Biol., *32:*211–221.

——— (1967b). Antibody synthesis: the take home lesson. Gamma Globulins Nobel Symp. III (J. Killander, ed.), 615–640, New York: Interscience Pub.

——— (1967c). Reflections on a discussion with Karl Popper: the molecular biology of expectation. Bulletin of the All India Institute of Medical Sciences, *1:*8–16.

——— (1967d). What can *Escherichia coli* and the plasmacytoma contribute to understanding differentiation and immunology? Symp. Intl. Soc. for Cell Biology, in press.

Crumpton, M. J., and J. M. Wilkinson (1963). Amino acid compositions of human and rabbit γ-globulins and of the fragments produced by reduction. Biochem. J., *88:*228–234.

Doolittle, R. F. (1966). The amino-terminal amino acid sequences of rabbit immunoglobulin light chains. Proc. Nat. Acad. Sci. U. S., *55:*1195–1201.

Dreyer, W. J., and J. C. Bennett (1965). The molecular basis of antibody formation: a paradox. Proc. Nat. Acad. Sci. U. S., *54:*864–869.

Dreyer, W. J., W. R. Gray, and L. Hood (1967). The genetic, molecular and cellular basis of antibody formation: some facts and a unifying hypothesis. Cold Spring Harbor Symp. Quant. Biol., *32:*353–367.

Edelman, G. M., and B. Benacerraf (1962). On structural and functional relations between antibodies and proteins of the gamma-system. Proc. Nat. Acad. Sci. U. S., *48:*1035–1042.

Edelman, G. M., and J. A. Gally (1967). Somatic recombination of duplicated genes: an hypothesis of the origin of antibody diversity. Proc. Nat. Acad. Sci. U. S., *57:*353–358.

Edelman, G. M. (1967). Studies on the primary and tertiary structure of γG immunoglobulin. Gamma Globulins Nobel Symp. III (J. Killander, ed.), 89–108, New York: Interscience Pub.

Ein, D., and J. L. Fahey (1967). Two types of lambda polypeptide chains in human immunoglobulins. Science, *156:*947–948.

Estabrook, R. W., J. B. Schenkman, W. Cammer, H. Remmer, D. Y. Cooper, S. Narasimhulu, and O. Rosenthal (1966). Cytochrome P-450 and mixed function oxidations. Biological and Chem. Aspects of Oxygenases. Proc. of United States-Japan Symp. on Oxygenases. (K. Bloch and O. Hayaishi, eds.), 153–178. Tokyo: Maruzen Co., Ltd.

Fleischman, J. (1967). Synthesis of the γG heavy chain in rabbit lymph node cells. Biochemistry, *6*:1311–1320.

Gaze, R. M. (1967). Growth and differentiation. Ann. Rev. Physiol., *29*:59–86.

Gray, W. R., W. J. Dreyer, and L. Hood (1967). Mechanism of antibody synthesis: size differences between mouse kappa chains. Science, *155*:465–467.

Hechter, O., and D. K. Halkerston (1964). On the nature of macromolecular coding in neuronal memory. Perspectives Biol. & Med., *7*:183–198.

Hilschmann, N. (1967). The structure of the immunoglobulins (L-chains) and the antibody problem. Gamma Globulin Nobel Symp. III (J. Killander, ed.), 33–44, New York: Interscience Pub.

——— (1967). Die vollständige Aminosäuresequenz des Bence-Jones-Proteins Cum. (κ-Typ). Z. Physiol. Chem., *348*:1718–1722.

Hood, L. E., W. R. Gray, and W. J. Dreyer (1966). On the mechanism of antibody synthesis: a species comparison of L-chains. Proc. Nat. Acad. Sci. U. S., *55*:826–832.

Hood, L., W. R. Gray, B. G. Sanders, and W. J. Dreyer (1967). Light chain evolution. Cold Spring Harbor Symp. Quant. Biol., *32*:133–145.

Jerne, N. K. (1967). Waiting for the end. Cold Spring Harbor Symp. Quant. Biol., *32*:591–603.

Kabat, E. A. (1967a). The paucity of species-specific amino acid residues in the variable regions of human and mouse Bence-Jones proteins and its evolutionary and genetic implications. Proc. Nat. Acad. Sci. U. S., *57*:1345–1349.

——— (1967b). A comparison of invariant residues in the variable and constant regions of human K, human L, and mouse K Bence-Jones proteins. Proc. Nat. Acad. Sci. U. S., *58*:229–233.

Knopf, P., R. M. E. Parkhouse, and E. S. Lennox (1967). Biosynthetic units of an immunoglobulin heavy chain. Proc. Nat. Acad. Sci. U. S., *58*:2288–2296.

Lederberg, J. (1959). Genes and antibodies. Science, *129*:1649–1653.

Lennox, E. S. (1966). The genetics of the immune response. Proc. Roy. Soc. (London) Ser. B, *166*:222–231.

Lennox, E. S., and M. Cohn (1967). Immunoglobulins. Ann. Rev. Biochem., *36*: 365–406.

Lennox, E. S., P. M. Knopf, A. J. Munro, and R. M. E. Parkhouse (1967). A search for biosynthetic subunits of light and heavy chains of immunoglobulins. Cold Spring Harbor Symp. Quant. Biol., *32*:249–254.

Mach, B., H. Koblet, and D. Gross (1967). Studies on the cell free biosynthesis of immunoglobulins. Cold Spring Harbor Symp. Quant. Biol., *32*:269–275.

Mannik, M. (1967). Variability in the specific interactions of H and L chains of γG-Globulins. Biochemistry, *6*:134–142.

Mekler, L. B. (1967). Mechanism of biological memory. Nature, *215*:481–484.

Metcalf, D. (1966). The Thymus. Recent Results in Cancer Research. New York: Springer-Verlag.

Milstein, C. (1967). Linked groups of residues in immunoglobulin κ chains. Nature, *216*:330–332.

Mitchison, N. A. (1967). Recognition of antigen. Symp. Intl. Soc. for Cell Biology, in press.

Mathematical Challenges to the Neo-Darwinian Interpretation of Evolution

(1967), P. S. Moorhead and M. M. Kaplan, eds., Wistar Inst. Symp. Monograph. Philadelphia, Pa.: Wistar Inst. Press.

Natapoff, A. (1967). The consideration of evolutionary conservatism toward a theory of the human brain. Persp. Biol. Med., *10*:445–461.

Niall, H. D., and P. Edman (1967). Two structurally distinct classes of kappa-chains in human immunoglobulins. Nature, *216*:262–263.

Nirenberg, M., T. Caskey, R. Marshall, R. Brimacombe, D. Kellogg, B. Doctor, D. Hatfield, J. Levin, F. Rottman, S. Pestka, M. Wilcox, and F. Anderson (1966). The RNA code and protein synthesis. Cold Spring Harbor Symp. Quant. Biol., *31*:11–24.

Perham, R., E. Appella, and M. Potter (1966). Light chains of mouse myeloma proteins: partial amino acid sequence. Science, *154*:391–393.

Popper, K. R. (1966). Of clouds and clocks: an approach to the problem of rationality and the freedom of man. St. Louis, Mo.: Washington Univ. Press.

Porter, R. R. (1967a). The structure of the heavy chain of immunoglobulin and its relevance to the nature of the antibody-combining site. Biochem. J., *105*:417–426.

——— (1967b). The chemical structure of the heavy chain of immunoglobulins. Gamma Globulins Nobel Symp. III (J. Killander, ed.), 81–88, New York: Interscience Pub.

Potter, M., E. Appella, and S. Geisser (1965). Variations in the heavy polypeptide chain structure of gamma myeloma immunoglobulins from an inbred strain of mice and a hypothesis as to their origin. J. Mol. Biol., *14*:361–372.

Putnam, F. W. (1967). Structural evolution of kappa and lambda light chains. Gamma Globulins Nobel Symp. III (J. Killander, ed.), 45–70, New York: Interscience Pub.

Quattrocchi, R., D. Cioli, C. Baglioni (1967). Distribution of an arginine-lysine interchange in the invariable half of human L-type Bence-Jones proteins. Nature, *216*:56–57.

Ropartz, C. (1964). Les Systèmes Héréditaires des Gamma Globulines Humaines. Ann. Biol. Clin., *22*:445–473.

Silverstein, A. M. (1963). Immunologic and psychic memory. Neurosciences Res. Prog. Bull. *1* (1), 1–29.

——— (1967). Immunological differentiation in ontogeny. Symp. Intl. Soc. for Cell Biology, in press.

Smith, C. E. (1962). Is memory a matter of enzyme induction? Science, *138:* 889–890.

Smithies, O. (1967). Antibody variability. Science, *157*:267–273.

Sperry, R. W. (1951). Mechanisms of neural maturations, *in* S. S. Stevens (ed.), Handbook of Experimental Psychology, 236–280. New York: John Wiley & Sons.

Steinberg, A. G. (1962). Progress in the study of genetically determined human gamma globulin types (The Gm and Inv groups). Progress Med. Genetics, *2*:1–33.

Suran, A. A., and B. W. Papermaster (1967). N-terminal sequences of heavy and light chains of leopard shark immunoglobulins: evolutionary implications. Proc. Nat. Acad. Sci. U. S., *58*:1619–1623.

Szilard, L. (1960). The control of the formation of specific proteins in bacteria and in animal cells. Proc. Nat. Acad. Sci. U. S., *46*:277–292.

——— (1964). On memory and recall. Proc. Nat. Acad. Sci. U. S., *51*:1092–1099.

Takeshita, M., and S. Tanaka (1966). The induction of liver microsomal hydroxylating enzyme by phenobarbital and polycyclic hydrocarbons. Biological and Chem. Aspects of Oxygenases. Proc. of United States-Japan Symp. on Oxygenases. (K. Bloch and O. Hayaishi, eds.), 145–150. Tokyo: Maruzen Co., Ltd.

Control of cellular growth in adult organisms (1967). (H. Teir and T. Tytömaa, eds.), New York: Academic Press.

Todd, C. W. (1963). Allotypy in Rabbit 19S protein. Biochem. Biophys. Res. Commun., *11*:170–175.

Whitehouse, H. L. K. (1967). Crossover model of antibody variability. Nature, *215*:371–374.

Williams, R. T. (1959). Detoxification Mechanisms, 2nd ed. New York: John Wiley & Sons.

INDEX

GPSR Compliance

The European Union's (EU) General Product Safety Regulation (GPSR) is a set of rules that requires consumer products to be safe and our obligations to ensure this.

If you have any concerns about our products, you can contact us on ProductSafety@springernature.com

In case Publisher is established outside the EU, the EU authorized representative is:

Springer Nature Customer Service Center GmbH
Europaplatz 3
69115 Heidelberg, Germany

Batch number: 10761593

Printed by Printforce, the Netherlands